CHILTON'S™

POWERTRAIN CODE & OXYGEN SENSOR
SERVICE MANUAL

COVERS ALL 1990-99 U. S. AND CANADIAN VEHICLES
DOMESTIC AND IMPORTED

C.E.O. WITHDRAWN
Rick Van Dalen

President
Dean F. Morgantini, S.A.E.

Vice President–Finance
Barry L. Beck

Vice President–Sales
Glenn D. Potere

Executive Editor
Kevin M. G. Maher, A.S.E.

Manager—Consumer Automotive
Richard Schwartz, A.S.E.

Manager—Marine/Recreation
James R. Marotta, A.S.E.

Electronic Fulfillment Manager
Will Kessler, A.S.E., S.A.E.

Production Specialists
Brian Hollingsworth, Melinda Possinger

Project Managers
Thomas A. Mellon, A.S.E., S.A.E., Richard J. Rivele, Christine Sheeky,
S.A.E., Todd W. Stidham, A.S.E., Ron Webb

Schematics Editor
Christopher G. Ritchie

Editors
Tim Crain, A.S.E., Scott A. Freeman, Eric Michael Mihalyi, A.S.E., S.A.E., S.T.S., Richard T. Smith,
Joseph D'Orazio, A.S.E., Paul D'Santo, A.S.E.

CHILTON™ Automotive Books
PUBLISHED BY W. G. NICHOLS, INC.

Manufactured in USA, © 1999 W. G. Nichols, 1020 Andrew Drive, West Chester, PA 19380
ISBN 0-8019-9127-7
Library of Congress Catalog Card No. 98-74943
2345678901 9876543210

Table of Contents

SECTION 1—POWERTRAIN CODES AND MAINTENANCE LIGHT RESETTING

Table of Contents

Table of Contents

Table of Contents

Table of Contents

Table of Contents

Table of Contents

Table of Contents

Table of Contents

Table of Contents

HOW TO USE THIS MANUAL

Locating Information

The Table of Contents, located at the front of the book, lists each section in this manual.

To find where a particular model specific section is located in the book, you need only look in the Table of Contents. Once you have found the proper section, you may wish to find where specific procedures are located in that section. Either use those listed in the Table of Contents or turn to the Index at the front of each of the two sections. Each section Index lists all of the procedures within that section and the applicable page numbers.

Safety Notice

Proper service and repair procedures are vital to the safe, reliable operation of all motor vehicles, as well as the personal safety of those performing the repairs. This manual outlines procedures for servicing and repairing vehicles using safe effective methods. The procedures contain many NOTES, WARNINGS and CAUTIONS which should be followed along with standard safety procedures to eliminate the possibility of personal injury or improper service which could damage the vehicle or compromise its safety.

It is important to note that repair procedures and techniques, tools and parts for servicing vehicles, as well as the skill and experience of the individual performing the work vary widely. It is not possible to anticipate all of the conceivable ways or conditions under which vehicles may be serviced, or to provide cautions as to all of the possible hazards that may result. Standard and accepted safety precautions and equipment should be used when handling toxic or flammable fluids, and safety goggles or other protection should be used during cutting, grinding, chiseling, prying, or any other process that can cause material removal or projectiles.

Some procedures require the use of tools specially designed for a specific purpose. Before substituting another tool or procedure, you must be completely satisfied that neither your personal safety, nor the performance of the vehicle will be endangered.

Although information in this manual is based on industry sources and is as complete as possible at the time of publication, the possibility exists that some vehicle manufacturers made later changes which could not be included here. Information on very late models may not be available in some circumstances. While striving for total accuracy, Nichols Publishing cannot assume responsibility for any errors, changes, or omissions that may occur in the compilation of this data.

Part Numbers

Part numbers listed in this book are not recommendations by Nichols Publishing for any product by brand name. They are references that can be used with interchange manuals and aftermarket supplier catalogs to locate each brand supplier's discrete part number.

Special Tools

Special tools are recommended by the vehicle manufacturer to perform their specific job. Use has been kept to a minimum, but where absolutely necessary, they are referred to in the text by the part number of the tool manufacturer. These tools may be purchased, under the appropriate part number, from your local dealer or regional distributor, or an equivalent tool can be purchased locally from a tool supplier or parts outlet. Before substituting any tool for the one recommended, read the previous Safety Notice.

Acknowledgments

This publication contains material that is reproduced and dsitributed under a license from Ford Motor Company. No further reproduction or distribution of the Ford Motor Company material is allowed without the expressed written permission from Ford Motor Company.

Portions of the material contained herein have been reprinted with permission of General Motors Corporation, Service Technology Group.

Nichols Publishing would like to express thanks to all of the fine companies who participate in the production of our books. Hand tools supplied by Craftsman are used during all phases of our vehicle teardown and photography. Many of the fine specialty tools used in our procedures were provided courtesy of Lisle Corporation. Lincoln Automotive Products (1 Lincoln Way, St. Louis, MO 63120) has provided their industrial shop equipment, including jacks (engine, transmission and floor), engine stands, fluid and lubrication tools, as well as shop presses. Rotary Lifts, the largest automobile lift manufacturer in the world, offering the biggest variety of surface and in-ground lifts available (Rotary 1-800-640-5438), has fulfilled our shop's lift needs. Much of our shop's electronic testing equipment was supplied by Universal Enterprises Inc. (UEI).

Copyright Notice

POWERTRAIN CODES AND MAINTENANCE LIGHT RESETTING

1

GENERAL INFORMATION

Introduction

OBD I CODES

For years, vehicles have been capable of storing diagnostic trouble codes. Codes prior to the 1996 OBD II legislation have been proprietary to the vehicle manufacturer. In some cases, the codes are specific to the individual make and model.

Furthermore, some manufacturers have developed specialized devices to read their codes. This complicates code reading and clearing.

OBD II CODES

Federal law required all vehicle manufacturers to meet On Board Diagnostics, Second Generation or OBD II standards by 1996. In order to meet this standard, the automobile's on-board computer must monitor and perform diagnostic tests on vehicle emissions to ensure that the vehicle is operating at an acceptable (legal) emission level. The maximum allowable emission level is set by the Federal Test Procedure (FTP).

Some 1995 and all 1996–99 vehicles are OBD II compliant. All OBD II vehicles have the same 16 pin diagnostic connector or DLC. This eliminates the need to have a manufacturer specific connector to plug a scan tool into your vehicle.

➡**Many 1995 vehicles have a 16 pin OBD II connector, however, this does not mean that the vehicle is OBD II compliant.**

To comply with the Federal On-Board Diagnostic 2nd Generation (OBD II) regulations, vehicle control computers are equipped with software designed to allow it to monitor vehicle emission control systems and components. Once the ignition is turned **ON** or the engine is started, and certain test conditions are met, the PCM runs a series of monitors to test the emission control systems and components. Test conditions include different inputs such as time since startup, run-time, engine speed and temperature, transaxle gear position, and the engine open or closed loop status. Once the monitor is started, the control module attempts to run it to completion. If a particular monitor fails a test, a code is set and operating conditions at that time are recorded in memory. If the same component or system fails twice in succession, the Malfunction Indicator Lamp (MIL) is activated.

Monitors are divided into two types: Main Monitors and the Comprehensive Component Monitors.
- Catalyst Monitor
- EGR Monitor
- EVAP Monitor
- Fuel System Monitor
- Misfire Monitor
- Oxygen Sensor Monitor
- Oxygen Sensor Heater Monitor

Certain monitors, in particular the fuel system and misfire monitors, have limitations that are different from any of the other monitors. The first time either of these monitors fail, the MIL is activated, and engine conditions at the time of the fault are recorded. In order

for the control module to turn off an MIL related to these two monitors, it must determine that no faults are present with engine operating conditions similar to when it detected the fault. To qualify, the engine must be operated within a specified speed range, engine load range and temperature range.

A warm-up cycle is considered to be vehicle operation after the engine has been turned off for a period of time, with the ECT input rising a specified amount and reaching normal operating temperature. When the MIL is turned off because a fault is no longer present, most OBD II codes will be erased after a minimum of 40 warm-up cycles. Misfire and fuel system codes require a minimum of 80 warm-up cycles before they clear.

OBD II Systems use a standardized test connector, called the Data Link Connector (DLC). It is usually located beneath the left side of the instrument panel. The DLC is located out of the line of sight of vehicle passengers, but is easily viewable from a kneeling position outside the vehicle. The connector is rectangular in design and contains up to 16 terminals. It has keying features to allow for easy connection. Both the DLC and Scan Tool connectors have latching features that ensure the scan tool will remain properly connected.

Some common uses of the Scan Tool are to identify and clear Diagnostic Trouble Codes (DTC's) and to read control module freeze frame.

The Malfunction Indicator Lamp (MIL) looks similar to the "Check Engine" lamp. However, on OBD II Systems, it is controlled under a strict set of guidelines that dictate when the MIL is illuminated. If any of the control module monitors detects a fault that could impact vehicle emissions, a fault code is set. A One-Trip Monitor requires that a test fail once, a Two-Trip Monitor requires a test fail twice in succession, and a Three-Trip Monitor requires that a test fail three times in succession to activate the MIL.

The MIL is mounted in the instrument panel and has two functions: To act as a bulb check at key On and to inform the driver that an emissions fault has occurred.

Once the engine is started, if no faults are detected, the control module should extinguish the MIL after a few seconds. If the MIL remains On or flashes with the engine running a driveability symptom is present.

Trouble Code Description

In the past, trouble code numbers varied between manufacturers, years, makes and models. OBD II requires that all vehicle manufacturers use a common Diagnostic Trouble Code (DTC) numbering system. Since the generic listing was not specific enough, most manufacturers came up with their own DTC listings which are called manufacturer specific codes. Both generic and manufacturer specific codes are 5 digits. The numbers can be decoded as follows:

The first digit is a letter which identifies the function of the device or circuit which has the fault. This digit can be either:
- P—Powertrain
- B—Body
- C—Chassis
- U—Network or data link code

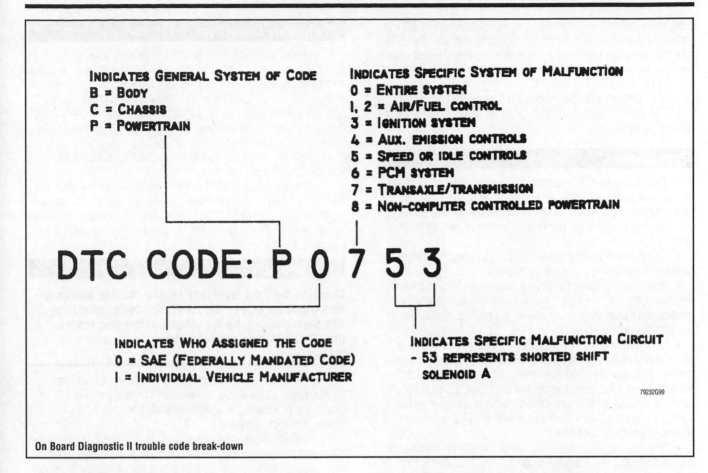

INDICATES GENERAL SYSTEM OF CODE
B = BODY
C = CHASSIS
P = POWERTRAIN

INDICATES SPECIFIC SYSTEM OF MALFUNCTION
0 = ENTIRE SYSTEM
1, 2 = AIR/FUEL CONTROL
3 = IGNITION SYSTEM
4 = AUX. EMISSION CONTROLS
5 = SPEED OR IDLE CONTROLS
6 = PCM SYSTEM
7 = TRANSAXLE/TRANSMISSION
8 = NON-COMPUTER CONTROLLED POWERTRAIN

DTC CODE: P 0 7 5 3

INDICATES WHO ASSIGNED THE CODE
0 = SAE (FEDERALLY MANDATED CODE)
1 = INDIVIDUAL VEHICLE MANUFACTURER

INDICATES SPECIFIC MALFUNCTION CIRCUIT
- 53 REPRESENTS SHORTED SHIFT
SOLENOID A

79232G99

On Board Diagnostic II trouble code break-down

The second digit is either a 0 or 1 and indicates whether the code is generic or manufacturer specific.
- 0—Generic (Federally-mandated codes)
- 1—Manufacturer Specific

The third digit represents the specific vehicle circuit or system that has the fault. Listed below are the number identifiers for the powertrain system.
- 1—Fuel and Air Metering
- 2—Fuel and Air Metering (Injector Circuit Malfunctions Only)
- 3—Ignition System or Misfire
- 4—Auxiliary Emission Control
- 5—Vehicle Speed Control and Idle Control System
- 6—Computer and Auxiliary Outputs
- 7—Transmission
- 8—Transmission

The last two digits indicate the specific trouble code.

On OBD II vehicles there are two different types of DTC's: Stored and Pending. For a DTC to become Stored, certain malfunction conditions must occur. The condition(s) required to Store codes are different for every DTC and vary by vehicle manufacturer.

In order for some DTC's to become Stored, a malfunction condition has to happen more than once. If the malfunction conditions are required to occur more than once, the potential malfunction is called a Pending DTC. The DTC remains pending until the malfunction condition occurs the required number of times to make the code stored. If the malfunction condition does not occur again after a set time the pending DTC will be cleared.

MAINTENANCE INDICATOR LIGHTS (MIL)

Maintenance lights are used to indicate to the operator of the vehicle that some type of routine maintenance should be performed. Unlike a Check Engine light that will be displayed when there is a fault with the engine management system, the maintenance light will be displayed when an engine or transmission oil change is recommended according to driving conditions. Also, the light will be displayed to indicate when the emission control system needs to be serviced.

Vehicle Diagnostic Hardware

VEHICLE COMPUTERS

The actual hardware (computers, wires, sensors, etc.) used in the vehicles' control systems have changed over the years as well as the software used (DTC retrieval, clearing, etc.). However, despite the large number of manufacturers and vehicles designed between the mid-1980's and today, there are generally only three control computer configurations used. Back when the first computer-controlled vehicles were introduced, they used a simple feedback system consisting of one computer (usually known as the Engine Control Module or ECM) used exclusively for the engine systems (air/fuel and ignition primarily). As the systems grew and became more complicated, a transmission/transaxle control computer (often called the Transmission Control Module or TCM) was added to the vehicle. This computer communicated with the ECM. Together, based on predetermined parameters, they decided what was best for the entire powertrain (engine & transmission). Eventually, the ECM and the TCM were formed into one computer, most commonly referred to as the Powertrain Control Module (PCM). The PCM configuration is used on most, if not all, vehicles manufactured today.

Precautions

The electronic control systems used by the vehicles covered in this section are very delicate and complicated. Please, to save yourself aggravation, money and time, be sure to adhere to the following points when working on your vehicle's control system:

• Unless otherwise instructed, always disconnect the battery cables when servicing the electronic system.

• When disconnecting the battery, always be sure to detach the negative battery cable FIRST, then the positive cable. This simple practice will almost completely prevent the chance of arcing or shorting the system.

• Never pierce or cut the insulation off of a wire for testing purposes. Many of the control systems wires are designed to handle a precise amount of electrical resistance, and the computer expects to see a certain predetermined amount of resistance. If you pierce or cut the insulation off of a wire, corrosion can build up in the wiring, leading to decreased control system efficiency, DTC storing or possibly even component damage.

• Whenever handling (that includes just touching) the system's control computer, use a ground strap to prevent accidental static electricity charges from damaging the computer's electronic chips.

• Never subject any control computer to excessive jolts (such as dropping).

• If welding on the vehicle (don't worry, nothing in this section requires welding), always disconnect the computer from the vehicle's wiring harness.

• Always be careful when working around a running engine. Loose clothing or long hair can easily become tangled in moving components. Hot exhaust system parts can cause painful burns.

• Always wear eye protection when servicing a vehicle.

• On vehicles equipped with an air bag system (make sure you check for this!), it is necessary to properly disarm the system before working on or around any air bag components, since the air bag system can often deploy even if the negative battery cable has been disconnected. This includes the steering wheel, front impact sensors, system wiring, and the instrument panel. If you do not know the proper disarming procedure, refer to the Chilton's Total Car Care manual for your specific vehicle.

• Never detach a wiring harness connector when the ignition switch is turned **ON**.

• Always operate the vehicle in a well-ventilated area. Also, never smoke around the battery or fuel system.

• Always keep a dry chemical (class B) fire extinguisher at hand.

Preliminary Inspection

Prior to trying to determine what components are the cause of DTC's, it is always a good idea to do a preliminary inspection. Often a loose or corroded wire or terminal can cause the control computer to store "ghost" DTC's. A repair can take a much longer amount of time if a general inspection is skipped before servicing or repairing components and circuits indicated by stored DTC's. Perform a general inspection, as follows:

1. Retrieve and write down the stored DTC's as described through-out this section, then clear them.

2. Check the engine and transmission/transaxle fluid level and fill it if necessary. While the engine is running, to heat the transmission/transaxle up, watch for fluid leaks. Any leaks must be repaired immediately, otherwise engine and/or transmission/transaxle damage can occur.

✳✳ CAUTION

Checking the fluid level may require that the transmission/transaxle be hot. So take this into consideration when working around the vehicle during this inspection. Painful burns can result from the exhaust system.

3. Check the battery voltage with a voltmeter; it should be approximately 12.6 volts. If it lower than 11 volts, you should charge it with a battery charger. If the battery will not hold a full charge, it may be defective.

4. Inspect every connector you can get at (especially the transmission/transaxle connectors) to ensure that they are tightly connected. Separate the connector halves and ensure that the terminals inside are corrosion-free and not loose. Re-engage all of the connector halves making sure they are properly attached. If any of the connectors are faulty, fix or replace them.

5. If possible, trace the wires to ensure that they are not cut, burned or similarly damaged. If any such damage is found, the wire must be repaired.

6. Check all vacuum hoses to ensure they are properly attached. Replace any vacuum line that is cracked or crumbling; this can cause a vacuum leak, which can adversely affect the engine and/or transmission/transaxle.

7. Inspect all engine and transmission-related linkage to ensure that it actuates smoothly and properly. Maladjusted linkage can cause early or late shifts, excessively hard shifts and other problems.

8. Check all of the transmission/transaxle fluid lines for kinks, cracks, or leaks. Fix any defects.

9. Raise and safely support the vehicle. Inspect the undervehicle wires, connectors, lines, linkage and fluid lines as previously described.

10. Once you have satisfactorily inspected the engine and transmission/transaxle and fixed any problems found, test drive the vehicle (for ½ hour if possible) and recheck for DTC's.

OBD I POWERTRAIN DIAGNOSTIC TROUBLE CODES

Acura

ENGINE CODES

General Information

Programmed Fuel Injection (PGM-FI) System is a fully electronic microprocessor based engine management system. The Electronic Control Unit (ECU) is given responsibility for control of injector timing and duration, intake air control, ignition timing, cold start enrichment, fuel pump control, fuel cut-off, NC compressor operation, alternator control as well as EGR function and canister purge cycles.

The ECU receives electric signals from many sensors and sources on and around the engine. The signals are processed against pre-programmed values; correct output signals from the ECU are determined by these calculations. The ECU contains additional memories, back-up and fail-safe functions as well as self diagnostic capabilities.

Service Precautions

- Do not operate the fuel pump when the fuel lines are empty.
- Do not operate the fuel pump when removed from the fuel tank.
- Do not reuse fuel hose clamps.
- The washer(s) below any fuel system bolt (banjo fittings, service bolt, fuel filter, etc.) must be replaced whenever the bolt is loosened. Do not reuse the washers; a high-pressure fuel leak may result.
- Make sure all ECU harness connectors are fastened securely. A poor connection can cause an extremely high voltage surge and result in damage to integrated circuits.
- Keep all ECU parts and harnesses dry during service. Protect the ECU and all solid-state components from rough handling or extremes of temperature.
- Use extreme care when working around the ECU or other components; the airbag or SRS wiring may be in the vicinity. On these vehicles, the SRS wiring and connectors are yellow; do not cut or test these circuits.

- Before attempting to remove any parts, turn the ignition switch **OFF** and disconnect the battery ground cable.
- Always use a 12 volt battery as a power source for the engine, never a booster or high-voltage charging unit.
- Do not disconnect the battery cables with the engine running.
- Do not disconnect any wiring connector with the engine running or the ignition **ON** unless specifically instructed to do so.
- Do not apply battery power directly to injectors.
- Whenever possible, use a flashlight instead of a drop light.
- Keep all open flame and smoking material out of the area.
- Use a shop cloth or similar to catch fuel when opening a fuel system. Consider the fuel-soaked rag to be a flammable solid and dispose of it in the proper manner.
- Relieve fuel system pressure before servicing any fuel system component.
- Always use eye or full-face protection when working around fuel lines, fittings or components.
- Always keep a dry chemical (class B-C) fire extinguisher near the area.

Reading Codes

1986–90 LEGEND AND 1986–91 INTEGRA

When a fault is noted, the ECU stores an identifying code and illuminates the CHECK ENGINE light. The code will remain in memory until cleared; the dashboard warning lamp may not illuminate during the next ignition cycle if the fault is no longer present. Not all faults noted by the ECU will trigger the dashboard warning lamp although the fault code will be set in memory. For this reason, troubleshooting should be based on the presence of stored codes, not the illumination of the warning lamp while the car is operating.

Stored codes are displayed by a flashing LED on the ECU. When the CHECK ENGINE warning lamp has been on or reported on, lift or remove the carpet from the right front passenger footwell. The ECU is below a protective cover; the LED may be viewed through a small window without removing the ECU cover. Turn the ignition switch **ON**; the LED will display any stored codes by rhythmic flashing. Note that 1986–90 Legends have two LEDs on the controller; one is red and one is amber. The red one will flash the fault codes;

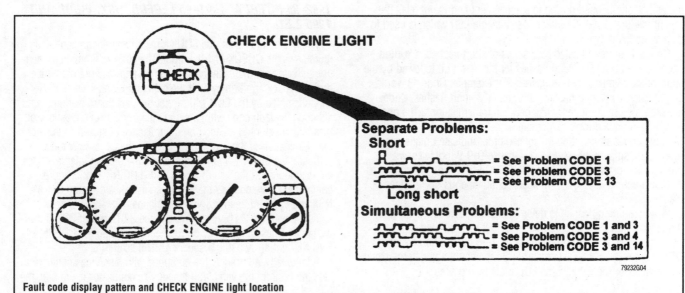

Fault code display pattern and CHECK ENGINE light location

79232G04

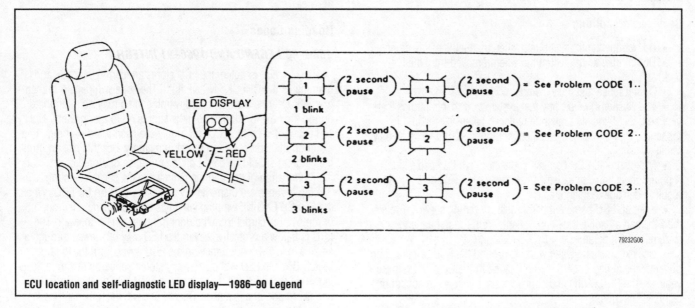

ECU location and self-diagnostic LED display—1986–91 Integra

ECU location and self-diagnostic LED display—1986–90 Legend

the amber one is used during idle adjustment and is not related to this procedure. 1986–91 Integra use a single LED which is used to display codes only.

Codes 1–9 are indicated by a series of short flashes; two-digit codes use a number of long flashes for the first digit followed by the appropriate number of short flashes. For example, Code 43 would be indicated by 4 long flashes followed by 3 short flashes. Codes are separated by a longer pause between transmissions. The position of the codes during output can be helpful in diagnostic work. Multiple codes transmitted in isolated order indicate unique occurrences; a display of showing 1-1-1-pause-9-9-9 indicates two problems or problems occurring at different times. An alternating display, such as 1-9-1-9-1, indicates simultaneous occurrences of the faults.

When counting flashes to determine codes, a code not valid for the vehicle may be found. In this case, first recount the flashes to confirm an accurate count. If necessary, turn the ignition switch **OFF**, then recycle the system and begin the count again. If the Code is not valid for the vehicle, the ECU must be replaced.

1992–95 INTEGRA, 1991–95 LEGEND, NSX, VIGOR AND 1995 2.5TL

When a fault is noted, the ECU stores an identifying code and illuminates the CHECK ENGINE light. The code will remain in memory until cleared; the dashboard warning lamp may not illuminate during the next ignition cycle if the fault is no longer present. Not all faults noted by the ECU will trigger the dashboard warning lamp although the fault code will be set in memory. For this reason, troubleshooting should be based on the presence of stored codes, not the illumination of the warning lamp while the car is operating.

Beginning in 1991 on the Legend and in 1992 on Integra, codes are read thorough the use of the CHECK ENGINE light or more commonly know today as the Malfunction Indicator Lamp (MIL). NSX and Vigor are read in the same manner. The 1995 Legend equipped with the 2.7L V6 engine and the 2.5TL utilize OBD II trouble codes. These codes may be read either through the CHECK ENGINE light or by using a special OBD II scan tool.

Additionally, all models are equipped with a service connector in side the cabin of the vehicle. If the service connector is jumped, the

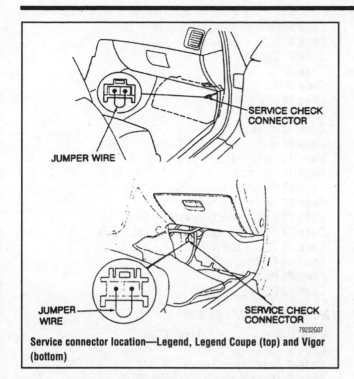

Service connector location—Legend, Legend Coupe (top) and Vigor (bottom)

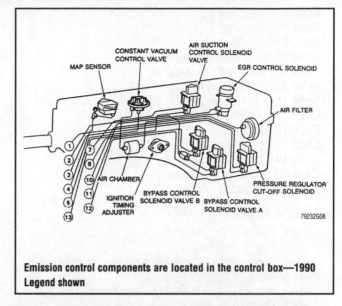

Emission control components are located in the control box—1990 Legend shown

CHECK ENGINE lamp will display the stored codes in the same fashion. The 2-pin service connector is located under the extreme right dashboard on Integra, Legend, 2.5TL and NSX; on Vigor models, it is found behind the right side of the center console well under the dashboard.

Codes 1–9 are indicated by a series of short flashes; two-digit codes use a number of long flashes for the first digit followed by the appropriate number of short flashes. For example, Code 43 would be indicated by 4 long flashes followed by 3 short flashes. Codes are separated by a longer pause between transmissions. The position of the codes during output can be helpful in diagnostic work. Multiple codes transmitted in isolated order indicate unique occurrences; a display of showing 1-1-1-pause-9-9-9 indicates two problems or problems occurring at different times. An alternating display, such as 1-9-1-9-1, indicates simultaneous occurrences of the faults.

When counting flashes to determine codes, a code not valid for the vehicle may be found. In this case, first recount the flashes to confirm an accurate count. If necessary, turn the ignition switch **OFF**, then recycle the system and begin the count again. If the Code is not valid for the vehicle, the ECU must be replaced.

➡**On vehicles with automatic transaxles, the S, D or D4 lamp may flash with the CHECK ENGINE lamp if certain codes are stored. For Legend and NSX, this may occur with Codes 6, 7 or 17. On Vigor and Integra it may occur with codes 6, 7 or 13. In addition, the TCS lamp on NSX may flash with codes 3, 5, 6,13,15,16,17, 35 or 36. In all cases, proceed with the diagnosis based on the engine code shown. After repairs, recheck the lamp. If the additional warning lamp is still lit, proceed with diagnosis for that system.**

Clearing Codes

1986–95 VEHICLES

Stored codes are removed from memory by removing power to the ECU. Disconnecting the power may also clear the memories used for other solid-state equipment such as the clock and radio. For this reason, always make note of the radio presets before clearing the system. Additionally, some radios contain anti-theft programming; obtain the owner's code number before clearing the codes. While disconnecting the battery will clear the memory, this is not the recommended procedure. The memory should be cleared after the ignition is switched **OFF** by removing the appropriate fuse for at least 10 seconds. The correct fuses and their locations are:

- 1990 and earlier Legend—ALTERNATOR SENSE, in the underhood fuse and relay panel.
- 1991 and later Legend—ACG, in the dashboard fuse panel. Removing this fuse also cancels the memory for the power seats.
- Integra, U.S. vehicles—BACK UP, in the underhood fuse and relay panel.
- Integra, Canadian vehicles—HAZARD/BACK UP in the underhood fuse and relay panel.
- Vigor—BACK UP, located in the underhood fuse and relay panel. Removing this fuse will cancel memories for the clock and radio.
- 1991–94 NSX—CLOCK, located in the main fuse and relay panel in the front luggage compartment, right side. Removing this fuse will cancel memories for the clock and radio.
- 1995 NSX—CLOCK, located in the main fuse and relay panel in the front luggage compartment, right side. Removing this fuse will cancel memories for the clock and radio. Codes may also be cleared using the OBD II scan tool or an equivalent tool, using the tool manufacturer's directions.
- 1995 2.5TL—BACK UP, located in the underhood fuse and relay panel. Removing this fuse will cancel memories for the clock and radio. Codes may also be cleared using the OBD II scan tool or an equivalent tool, using the tool manufacturer's directions.

Diagnostic Trouble Codes

1986–93 1.6L ENGINE

Code 0 Electronic Control Unit
Code 1 Oxygen Content
Code 2 Replace Engine Control Unit with known-good unit
Code 3 Manifold Absolute Pressure
Code 4 Replace Engine Control Unit with known-good unit
Code 5 Manifold Absolute Pressure
Code 6 Coolant Temperature

Code 7 Throttle Angle
Code 8 Crank Angle (TDC)
Code 9 Crank Angle (Cyl.)
Code 10 Intake Air Temperature
Code 12 EGR system (if equipped)
Code 13 Atmospheric Pressure
Code 14 Electronic Air Control
Code 15 Ignition output signal
Code 16 Fuel injector
Code 17 Vehicle speed sensor
Code 19 Lock up solenoid valve
Code 20 Electric load (to 1989)
Code 43 Fuel supply system

1986–87 2.5L C25A1 ENGINE

Code 0 Electronic Control Unit
Code 1 Oxygen Content
Code 2 Replace Engine Control Unit with known-good unit
Code 3 Manifold Absolute Pressure
Code 4 Replace Engine Control Unit with known-good unit
Code 5 Manifold Absolute Pressure
Code 6 Coolant Temperature
Code 7 Throttle Angle
Code 8 Crank Angle (TDC)
Code 9 Crank Angle (Cyl.)
Code 10 Intake Air Temperature
Code 12 EGR system (if equipped)
Code 13 Atmospheric Pressure
Code 14 Electronic Air Control
Code 15 Ignition output signal
Code 16 Fuel injector
Code 17 Vehicle speed sensor
Code 20 Electric load (to 1989)
Code 43 Fuel supply system

1987–90 2.7L C27A1 ENGINE

Code 0 Electronic Control Unit (ECU)
Code 1 Front Oxygen Content
Code 2 Rear Oxygen Content
Code 3 Manifold Absolute Pressure (MAP)
Code 4 Crank angle
Code 5 Manifold Absolute Pressure (MAP)
Code 6 Coolant Temperature
Code 7 Throttle angle
Code 8 Crank Angle-Top Dead Center
Code 9 Crank Angle-Number 1 Cylinder
Code 10 Intake Air temperature
Code 12 Exhaust Gas Recirculation system
Code 13 Atmospheric pressure
Code 14 Electronic Idle Control
Code 15 Ignition output signal
Code 17 Vehicle speed pulser
Code 18 Ignition timing adjustment
Code 30 A/T FI Signal A (if equipped)
Code 31 A/T FI Signal B (if equipped)

1991–93 1.8L B18A1, 1992–93 1.7L B17A1, 1994–95 1.8L B18B1 AND 1994–95 1.8L B18C1 ENGINES

Code 0 Electronic Control Unit
Code 1 Oxygen Content
Code 3 Manifold Absolute Pressure
Code 4 Crank Angle sensor

Code 5 Manifold Absolute Pressure
Code 6 Coolant Temperature
Code 7 Throttle Angle
Code 8 TDC Position
Code 9 No. 1 Cylinder Position
Code 10 Intake Air Temperature
Code 12 EGR system
Code 13 Atmospheric Pressure
Code 14 Electronic Air Control
Code 15 Ignition output signal
Code 16 Fuel injector
Code 17 Vehicle speed sensor
Code 20 Electric Load Detector
Code 21 VTEC Solenoid Valve (1.8L GS-R)
Code 22 VTEC Oil Pressure Switch (1.8L GS-R)
Code 30 TCM Signal A
Code 31 TCM Signal B
Code 41 H025 Heater
Code 43 Fuel supply system

1991–95 3.2L C32A1 ENGINE

Code 0 Electronic Control Unit (ECU)
Code 1 Left Oxygen Sensor
Code 2 Right Oxygen Sensor
Code 3 Manifold Absolute Pressure (MAP)
Code 4 Crank angle 1
Code 5 Manifold Absolute Pressure (MAP)
Code 6 Coolant Temperature
Code 7 Throttle angle
Code 9 Crank Angle-Number 1 Cylinder
Code 10 Intake Air temperature
Code 12 Exhaust Gas Recirculation (EGR) system
Code 13 Atmospheric pressure
Code 14 Electronic Air Control (EACV)
Code 15 Ignition output signal
Code 17 Vehicle speed pulser
Code 18 Ignition timing adjustment
Code 23 Left Knock Sensor
Code 30 A/T FI Signal A
Code 35 Traction Control System Circuit
Code 36 Traction Control System Circuit
Code 41 Left Oxygen Sensor Heater
Code 42 Right Oxygen Sensor Heater
Code 43 Left Fuel Supply System
Code 44 Right Fuel Supply System
Code 45 Left Fuel Supply Metering
Code 46 Right Fuel Supply Metering
Code 53 Right Knock Sensor
Code 54 Crank angle 2
Code 59 No. 1 Cylinder Position 2 (Cylinder Sensor)

1991–95 3.0L C30A1 ENGINE

Code 0 CU
Code 1 Front Oxygen Sensor
Code 2 Rear Oxygen Sensor
Code 3 Manifold Absolute Pressure (MAP)
Code 4 Crank angle A
Code 5 Manifold Absolute Pressure (MAP)
Code 6 Coolant Temperature
Code 7 Throttle angle
Code 9 Crank Angle-Number 1 Cylinder/Position A
Code 10 Intake Air temperature

Code 12 Exhaust Gas Recirculation (EGR) system
Code 13 Atmospheric pressure
Code 14 Electronic Air Control (EACV)
Code 15 Ignition output signal
Code 16 Fuel Injector
Code 17 Vehicle speed pulser
Code 18 Ignition timing adjustment
Code 22 VTEC System; front, bank 2
Code 23 Front Knock Sensor
Code 30 A/T FI Signal A
Code 31 NT FI Signal B
Code 35 TC STB signal
Code 36 TCFC signal
Code 37 Accelerator Position; Sensors 1, 2 or 1 and 2 circuits
Code 40 Throttle Position or Throttle Valve Control Motor Circuits 1 or 2
Code 41 Front Oxygen Sensor Heater; (circuit malfunction; bank 2 sensor 1)
Code 42 Rear Primary Heated Oxygen Sensor Heater (circuit malfunction)
Code 43 Front Fuel Supply System
Code 44 Rear Fuel Supply System
Code 45 Front Fuel Supply Metering; front bank 2
Code 46 Rear Fuel Supply Metering; rear bank 1
Code 47 Fuel Pump
Code 51 Rear Spool Solenoid Valve
Code 52 VTEC System; rear, bank 1
Code 53 Rear Knock Sensor
Code 54 Crank Angle B
Code 59 No. 1 Cylinder Position B (Cylinder Sensor)
Code 61 Front Heated Oxygen Sensor (slow response; bank 2 sensor 1)
Code 62 Rear Primary Heated Oxygen Sensor (slow response; bank 1 sensor 1)
Code 63 Front Secondary Oxygen Sensor (slow response or circuit voltage high or low)
Code 65 Front Secondary Heated Oxygen Sensor (circuit malfunction; bank 2 sensor 2)
Code 64 Rear Secondary Oxygen Sensor (slow response or circuit voltage high or low)
Code 66 Rear Secondary Heated Oxygen Sensor (circuit malfunction; bank 1 sensor 2)
Code 67 Front Catalytic Converter System
Code 68 Rear Catalytic Converter System
Code 80 Exhaust Gas Recirculation (EGR) system
Code 86 Coolant temperature
Code 70 Automatic Transaxle; the D indicator light and MIL may come on simultaneously.
Code 71 Misfire detected; cylinder No. 1 or random misfire
Code 72 Misfire detected; cylinder No. 2 or random misfire
Code 73 Misfire detected; cylinder No. 3 or random misfire
Code 74 Misfire detected; cylinder No. 4 or random misfire
Code 75 Misfire detected; cylinder No. 5 or random misfire
Code 76 Misfire detected; cylinder No. 6 or random misfire
Code 79 Spark Plug Voltage Detection; circuit malfunction; (Front Bank (Bank 2) or (Rear Bank (Bank 1)
Code 79 Spark Plug Voltage Detection; circuit malfunction; (Front Bank (Bank 2) or (Rear Bank (Bank 1)
Code 79 Spark Plug Voltage Detection Module; reset circuit malfunction; (Front Bank (Bank 2)) or (Rear Bank (Bank 1)
Code 92 Evaporative Emission Control System

1992–94 2.5L (G25A1) ENGINE

Code 0 Electronic Control Unit
Code 1 HO_25 circuit
Code 3 Manifold Absolute Pressure
Code 4 Crank Angle Sensor
Code 5 Manifold Absolute Pressure
Code 6 Coolant Temperature
Code 7 Throttle Angle
Code 8 TDC and or Crankshaft Position sensors
Code 9 No. 1 Cylinder Position
Code 10 Intake Air Temperature
Code 12 EGR system
Code 13 Atmospheric Pressure
Code 14 Electronic Air Control
Code 15 Ignition output signal
Code 16 Fuel injector
Code 17 Vehicle speed sensor
Code 18 Ignition Timing Adjuster
Code 20 Electric Load Detector
Code 30 NT FI Signal
Code 31 NT FI Signal
Code 41 HO_25 Heater
Code 43 Fuel supply system
Code 45 Fuel Supply Metering
Code 50 Mass Air Flow (MAF) circuit—2.5TL
Code 53 Rear Knock Sensor
Code 54 Crankshaft Speed Fluctuation sensor—2.5TL
Code 61 HO_25 sensor heater—2.5TL
Code 65 Secondary HO_25 sensor—2.5TL
Code 67 Catalytic Converter System—2.5TL
Code 70 Automatic transaxle or NT FI Data line—2.5TL
Code 71 Misfire detected; cylinder No. 1 or random misfire
Code 72 Misfire detected; cylinder No. 2 or random misfire
Code 73 Misfire detected; cylinder No. 3 or random misfire
Code 74 Misfire detected; cylinder No. 4 or random misfire
Code 75 Misfire detected; cylinder No. 5 or random misfire
Code 76 Random misfire detected—2.5TL
Code 80 EGR system—2.5TL
Code 86 Coolant Temperature circuit—2.5TL
Code 92 Evaporative Emission Control System—2.5TL

TRANSMISSION CODES

Reading and Retrieving Codes

EARLY MODELS

➡**Early models refers to the following vehicles: 1990–93 Integra and 1988–90 Legend models.**

All of these models display stored DTC's through a Light Emitting Diode (LED) located on the transmission/transaxle control computer, which constantly flashes out the DTC's whenever the ignition key is **ON**. When the computer stores a transmission-related DTC, the S3 light in the instrument cluster illuminates all the time, regardless of the gear selected, to let the driver know there is a stored DTC.

These vehicles use one and two-digit DTC's, which are displayed through the LED on the transmission/transaxle computer in a series of flashes. For example, if the LED flashes five times, pauses, then flashes six times and pauses, this indicates Codes 5 and 6 have been stored by the transmission/transaxle computer. For two-digit codes, the tens digits are displayed as long flashes and the ones digits are displayed as shorter flashes. Therefore, if the LED dis-

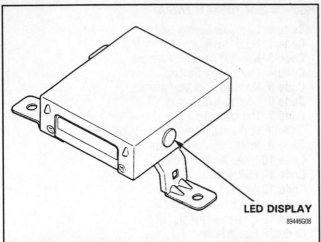

LED DISPLAY

89446G08

The LED is usually located on the side of the transaxle control computer, and, if a fault is detected, flashes whenever the ignition key is ON—early Acura models

plays one long flash and four short flashes followed by a pause, this indicates a Code 14. The system displays the codes from the lowest to the highest, and when it reaches the last code it returns to the lowest and cycles through the codes again.

To retrieve the codes, perform the following:

1. Perform the preliminary inspection, located earlier in this section. This is very important, since a loose or disconnected wire, or corroded connector terminals can cause a whole slew of unrelated DTC's to be stored by the computer; you will waste a lot of time performing a diagnostic "goose chase."

2. Grab some paper and a pencil or pen to write down the DTC's when they are flashed out.

3. Locate the transmission/transaxle control computer. The computer is generally located in the following positions:

- 1987–90 Legend Sedan—under the driver's seat
- 1988–90 Legend Coupe—under the passenger's seat
- 1990–93 Integra—up under the dashboard to the left side of the instrument cluster

4. Turn the ignition switch **ON** and observe the computer's LED. Note all of the DTC's.

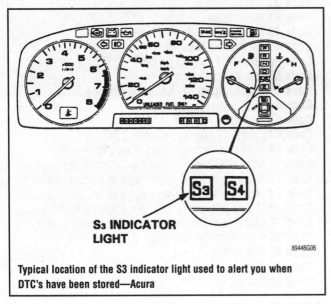

S3 INDICATOR LIGHT

89446G06

Typical location of the S3 indicator light used to alert you when DTC's have been stored—Acura

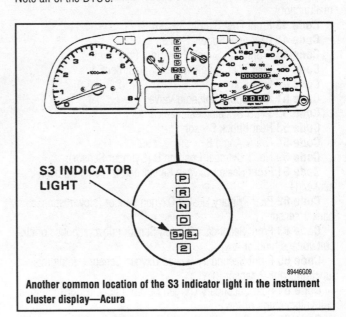

S3 INDICATOR LIGHT

89446G09

Another common location of the S3 indicator light in the instrument cluster display—Acura

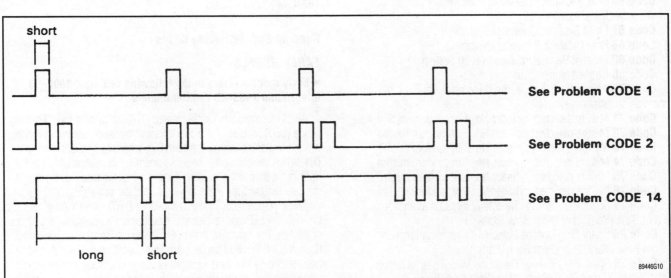

short

See Problem CODE 1

See Problem CODE 2

See Problem CODE 14

long   short

89446G10

The computer displays the DTC's as a logical series of long and short flashes, as shown—early Acura models

5. Once all of the DTC's have been noted, turn the ignition switch **OFF**.

LATE MODELS

➥**Late models refers to the following vehicles: 1994–96 Integra, 1991–94 Legend, 1995–97 2.5TL, 1996–97 3.2TL, 1992–94 Vigor, 1996–97 3.5RL models.**

Unlike the other Acura models described earlier in this section, these models display the transmission-related DTC's through either the S or D4 light on the instrument cluster. These models use 2-digit trouble codes, which are displayed in a series of flashes. The tens digits are displayed as long flashes and the ones digits are displayed as shorter flashes. Therefore, if the S or D4 light displays one long flash and four short flashes followed by a pause, this indicates a Code 14.

➥**Some of the early models of these vehicles have the added benefit, in which you can read the DTC's through the S/D4 light and also from the LED on the computer. The LED**

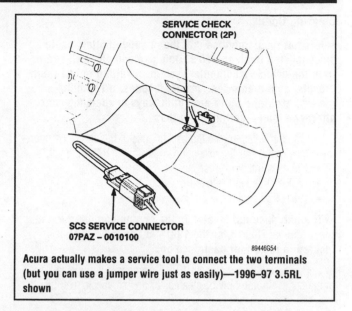

SERVICE CHECK CONNECTOR (2P)

SCS SERVICE CONNECTOR 07PAZ – 0010100

89446G54

Acura actually makes a service tool to connect the two terminals (but you can use a jumper wire just as easily)—1996–97 3.5RL shown

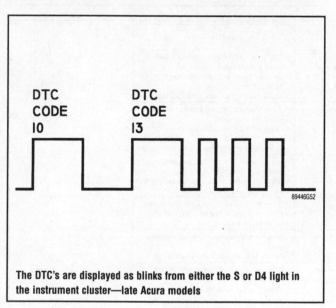

DTC CODE 10

DTC CODE 13

89446G52

The DTC's are displayed as blinks from either the S or D4 light in the instrument cluster—late Acura models

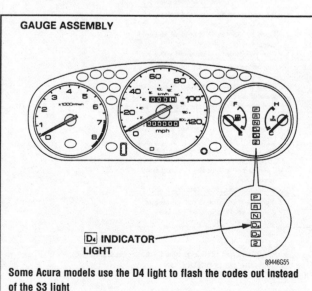

GAUGE ASSEMBLY

D4 INDICATOR LIGHT

89446G55

Some Acura models use the D4 light to flash the codes out instead of the S3 light

located on the computer was slowly phased out during these model years.

To retrieve the DTC's, perform the following:

1. Perform the preliminary inspection, located earlier in this section. This is very important, since a loose or disconnected wire, or corroded connector terminals can cause a whole slew of unrelated DTC's to be stored by the computer; you will waste a lot of time performing a diagnostic "goose chase."

2. Grab some paper and a pencil or pen to write down the DTC's when they are flashed out.

✷✷ WARNING

The Diagnostic Link Connector (DLC) is usually situated next to the service check connector; do not jump the terminals of the DLC by mistake. This could damage the vehicle's computer.

3. Locate the service check connector. The service check connector is either a single connector with two wires running to it or it is two separate one-wire connectors next to each other. The service check connector(s) are not attached to anything. It can be located according to the model of the vehicle at hand, as follows:
   • 1994–97 Integra—behind the passenger's side lower right kickpanel
   • 1991–94 Legend—beneath the passenger's side instrument panel
   • 1995–97 2.5TL—beneath the passenger's side instrument panel
   • 1996–97 3.2TL—beneath the passenger's side instrument panel
   • 1992–94 Vigor—beneath the passenger's side instrument panel, against the firewall
   • 1996–97 3.5RL—beneath the passenger's side instrument panel, against the firewall

4. Using a jumper wire or bent paper clip, connect the two terminals of the service check connector.

5. Turn the ignition switch **ON** and observe the S or D4 light in the instrument cluster. Note all of the DTC's.

6. Once all of the DTC's have been noted, turn the ignition switch **OFF** and remove the jumper wire/paper clip.

Clearing Codes

➡️It is not recommended that the negative battery cable be disconnected to clear DTC's. Doing so will clear all settings from the vehicle's computer system, resulting in lost radio presets, seat memories, anti-theft codes, driveability parameters, etc., and there are better ways of spending your afternoon than resetting all of these.

For most models, locate the backup fuse in the engine compartment fuse block. For the exceptions, refer to the following list:

- 1991–93 Integra—fuse No. 34
- 1988–90 Legend—fuse No. 22
- 1991–95 Legend—fuse No. 15

➡️The only fuse not located in the engine compartment fuse block is fuse No. 15 for the 1991–95 Legend, which is located in the under dash fuse box.

Remove the fuse for at least 10–15 seconds, then reinstall it. All of the codes should now be cleared. To double check, road test the

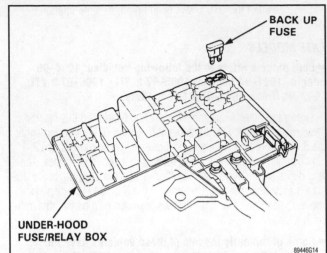

To clear the DTC's from most Acura models' memories, remove the backup fuse from the underhood fuse/relay box

Number of LED display blinks	S₃ indicator light	Symptom	Probable Cause
1	Blinks	• Lock-up clutch does not engage. • Lock-up-clutch does not disengage. • Unstable idle speed.	• Disconnected lock-up control solenoid valve A connector • Open or short in lock-up control solenoid valve A wire • Faulty lock-up control solenoid valve A
2	Blinks	• Lock-up clutch does not engage.	• Disconnected lock-up control solenoid valve B connector • Open or short in lock-up control solenoid valve B wire • Faulty lock-up control solenoid valve B
3	Blinks or OFF	• Lock-up clutch does not engage.	• Disconnected throttle angle sensor connector • Open or short in throttle angle sensor wire • Faulty throttle angle sensor
4	Blinks	• Lock-up clutch does not engage.	• Disconnected speed pulser connector • Open or short in speed pulser wire • Faulty speed pulser
5	Blinks	• Fails to shift other than 2nd ↔ 4th gear. • Lock-up clutch does not engage.	• Short in shift position console switch wire • Faulty shift position console switch
6	OFF	• Fails to shift other than 2nd ↔ 4th gear. • Lock-up clutch does not engage. • Lock-up clutch engages and disengages alternately.	• Disconnected shift position console switch connector • Open in shift position console switch wire • Faulty shift position console switch.
7	Blinks	• Fails to shift other than 1st ↔ 4th, 2nd ↔ 4th, or 2nd ↔ 3rd gears. • Fails to shift (stuck in 4th gear).	• Disconnected shift control solenoid valve A connector • Open or short in shift control solenoid valve A wire • Faulty shift control solenoid valve A
8	Blinks	• Fails to shift (stuck in 1st gear or 4th gear).	• Disconnected shift control solenoid valve B connector • Open or short in shift control solenoid valve B wire • Faulty shift control solenoid valve B
9	Blinks	• Lock-up clutch does not engage.	• Disconnected A/T speed pulser • Open or short in A/T speed pulser wire • Faulty A/T speed pulser
10	Blinks	• Lock-up clutch does not engage.	• Disconnected coolant temperature sensor connector • Open or short in coolant temperature sensor wire • Faulty coolant temperature sensor
11	OFF	• Lock-up clutch does not engage.	• Disconnected ignition coil connector • Open or short in ignition coil wire • Faulty ignition coil

NOTE:
- If a customer describes the symptoms for codes 3, 6 or 11, yet the LED is not blinking, it will be necessary to recreate the symptom by test driving, and then checking the LED with the ignition STILL ON.
- If the LED display blinks 12 or more times, the control unit is faulty.

Acura DTC's—1990–93 Integra and 1988–90 Legend models

Number of D4 indicator light blinks while Service Check Connector is jumped.	D4 indicator light	Possible Cause	Symptom
1	Blinks	•Disconnected lock-up control solenoid valve A connector •Short or open in lock-up control solenoid valve A wire •Faulty lock-up control solenoid valve A	•Lock-up clutch does not engage. •Lock-up clutch does not disengage. •Unstable idle speed.
2	Blinks	•Disconnected lock-up clontrol solenoid valve B connector •Short or open in lock-up control solenoid valve B wire •Faulty lock-up control solenoid valve B	•Lock-up clutch not engage.
3	Blinks or OFF	•Disconnected throttle position (TP) sensor connector •Short or open in TP sensor wire •Faulty TP sensor	•Lock-up clutch does not engage.
4	Blinks	•Disconnected vehicle speed sensor (VSS) connector •Short or open in VSS wire •Faulty VSS	•Lock-up clutch does not engage.
5	Blinks	•Short in A/T gear position switch wire •Faulty A/T gear position switch	•Fails to shift other than 2nd ↔ 4th gears. •Lock-up clutch does not engage.
6	OFF	•Disconnected A/T gear position switch connector •Open in A/T gear position switch wire •Faulty A/T gear position switch	•Fails to shift other than 2nd ↔ 4th gears. •Lock-up clutch does not engage. •Lock-up clutch engages and disengages alternately.
7	Blinks	•Disconnected shift control solenoid valve A connector •Short or open in shift control solenoid valve A wire •Faulty shift control solenoid vavle A	•Fails to shift (between 1st ↔ 4th, 2nd ↔ 4th or 2nd ↔ 3rd gears only). •Fails to shift (stuck in 4th gear)
8	Blinks	•Disconnected shift control solenoid valve B connector •Short or open in shift control solenoid valve B wire •Faulty shift control solenoid valve B	•Fails to shift (stuck in 1st or 4th gears).
9	Blinks	•Disconnected countershaft speed sensor connector •Short or open in the countershaft speed sensor wire •Faulty countershaft speed sensor	•Lock-up clutch does not engage.

89446G02

Acura DTC's, 1 of 2—1994–95 Integra, 1991–94 Legend, 1995 2.5TL, and 1992–94 Vigor models

Number of [D₄] indicator light blinks while Service Check Connector is jumped.	[D₄] indicator light	Possible Cause	Symptom
10	Blinks	• Disconnected engine coolant temperature (ECT) sensor connector • Short or open in the ECT sensor wire • Faulty ECT sensor	• Lock-up clutch does not engage.
11	OFF	• Disconnected ignition coil connector • Short or open in ignition coil wire • Faulty ignition coil	• Lock-up clutch does not engage.
14	OFF	• Short or open in FAS (BRN/WHT) wire between the D16 terminal and ECM • Trouble in ECM	• Transmission jerks hard when shifting.
15	OFF	• Disconnected mainshaft speed sensor connector • Short or open in mainshaft speed sensor wire • Faulty mainshaft speed sensor	• Transmission jerks hard when shifting.

89446G03

Acura DTC's, 2 of 2—1994–95 Integra, 1991–94 Legend, 1995 2.5TL, and 1992–94 Vigor models

Diagnostic Trouble Code (DTC)*	D4 Indicator Light	Symptom	Possible Cause
P1753 (1)	Blinks	• Lock-up clutch does not engage. • Lock-up clutch does not disengage. • Unstable idle speed.	• Disconnected lock-up control solenoid valve A connector • Short or open in lock-up control solenoid valve A wire • Faulty lock-up control solenoid valve A
P1758 (2)	Blinks	• Lock-up clutch does not engage.	• Disconnected lock-up control solenoid valve B connector • Short or open in lock-up control solenoid valve B wire • Faulty lock-up control solenoid valve B
P1791 (4)	Blinks	• Lock-up clutch does not engage.	• Disconnected vehicle speed sensor (VSS) connector • Short or open in VSS wire • Faulty VSS
P1705 (5)	Blinks	• Fails to shift other than 2nd – 4th gears. • Lock-up clutch does not engage.	• Short in A/T gear position switch wire • Faulty A/T gear position switch
P1706 (6)	OFF	• Fails to shift other than 2nd – 4th gears. • Lock-up clutch does not engage. • Lock-up clutch engages and disengages alternately.	• Disconnected A/T gear position switch connector • Open in A/T gear position switch wire • Faulty A/T gear position switch
P0753 (7)	Blinks	• Fails to shift (between 1st – 4th, 2nd – 4th or 2nd – 3rd gear only). • Fails to shift (stuck in 4th gear).	• Disconnected shift control solenoid valve A connector • Short or open in shift control solenoid valve A wire • Faulty shift control solenoid valve A
P0758 (8)	Blinks	• Fails to shift (stuck in 1st or 4th gears).	• Disconnected shift control solenoid valve B connector • Short or open in shift control solenoid valve B wire • Faulty shift control solenoid valve B
P0720 (9)	Blinks	• Lock-up clutch does not engage.	• Disconnected countershaft speed sensor connector • Short or open in countershaft speed sensor wire • Faulty countershaft speed sensor
P0715 (15)	OFF	• Transmission jerks hard when shifting.	• Disconnected mainshaft speed sensor connector • Short or open in mainshaft speed sensor wire • Faulty mainshaft speed sensor
P1768 (16)	Blinks	• Transmission jerks hard when shifting. • Lock-up clutch does not engage.	• Disconnected linear solenoid connector • Short or open in linear solenoid wire • Faulty linear solenoid

(DTC)*: The DTCs in parentheses are the number of the D4 indicator light blinks when the service check connector is connected with the special tool (SCS service connector).

89446G04

Diagnostic Trouble Code (DTC)*	D4 Indicator Light	Symptom	Possible Cause
P0740 (40)	OFF	• Lock-up clutch does not engage. • Lock-up clutch does not disengage. • Unstable idle speed.	• Faulty lock-up control system
P0730 (41)	OFF	• Fails to shift (between 1st – 2nd, 2nd – 4th or 2nd – 3rd gears only). • Fails to shift (stuck in 1st or 4th gears).	• Faulty shift control system

(DTC)*: The DTCs in parentheses are the number of the D4 indicator light blinks when the service check connector is connected with the special tool (SCS service connector).

89446G05

Acura DTC's, 2 of 2—1996–97 Integra, 1996–97 2.5TL, 1996–97 3.2TL and 1996–97 3.5RL models

vehicle and perform the code retrieval procedure again. If there are still codes present, either the codes were not cleared (Are the codes identical to those flashed out previously?) or the underlying problem is still there (Are only some of the codes the same as previously?)

Audi

ENGINE CODES

General Information

MOTRONIC AND MULTI-POINT INJECTION (MPI) SYSTEMS

The Motronic and MPI fuel injection systems are similar and share most components and modes of operation. Audi uses Motronic to describe the fuel injection system on the V8 Quattro, S4, 200 Quattro and the 200 Quattro Wagon. Audi uses MPI to describe the systems used on the 90 Quattro, Coupe Quattro and the 2.8L V6 equipped 100 series vehicles.

The Motronic and MPI fuel injection systems are self-learning adaptive systems. They continuously learn using a sophisticated feedback system that readjusts various control settings. These new values are then stored in the ECU memory. The adaptive capability allows the systems to compensate for changes in the engine's operating conditions, such as intake leaks, altitude changes or any other system malfunction. If the battery or ECU is disconnected, the vehicle must be driven so ECU can relearn its operating conditions.

Operation of the fuel injection system is based on the information received by the various sensors. This keeps the system constantly updated on engine speed, coolant temperature, throttle position and the intake air volume.

On the V8, the power supply to ECU is at terminal 18, through a 5 amp fuse (S27) in the main fuse/relay panel. Power from fuse S27 energizes the power supply relay in the ECU when engine speed reaches 25 rpm. The main fuse/relay panel is located behind the side kick panel cover on the passenger's side.

A Hall effect signal from the right distributor helps the ECU establish a reference point to start the fuel injection process. After the engine is running, the reference sender and speed sensor provide the necessary information to the ECU for ignition and fuel injection.

The ECU has a self-diagnostic feature. Any faults detected by the sensors are sent to the ECU and are recorded in the ECU memory. Fault codes can be displayed using LED tester US 1115 and a jumper wire.

CONTINUOUS INJECTION SYSTEM-ELECTRONIC (CIS-E)

The CIS-E system incorporates 2 control units. An Ignition Control Unit (ICU) or Knock Sensor Control Unit (KSCU, on SOQOS only) and a Fuel Injection Control Unit (FICU).

The CIS-E system also has self-diagnosis and troubleshooting capabilities. Input and output signals from various sensors, switches and signaling devices are constantly monitored for faults. These faults are stored in the control unit memory. Faults can be displayed by a flashing 4 digit code sequence from an LED light located on the instrument panel.

CIS-MOTRONIC FUEL INJECTION SYSTEM

The CIS Motronic system used on Audi 80 and 90 models use a single Electronic Control Unit (ECU), located behind the NC evaporator assembly. The ECU controls the fuel delivery, ignition system and operation of the emission control components. The CIS Motronic system also incorporates self-diagnostic capabilities. The CIS-Motronic system consists of the following components:

- Ignition coil with power stage
- Differential pressure regulator
- Cold start valve
- Idle stabilizer valve
- Ignition distributor with Hall sender
- Knock sensor
- Coolant temperature sensor
- Idle/Full throttle switches
- Air sensor potentiometer
- Oxygen sensor
- Carbon canister frequency valve
- Carbon canister ON/OFF valve
- CIS Motronic control unit

The ECU receives signals from various sensors, switches and signaling components which are constantly monitored for faults. These faults are stored in the ECU memory. Faults can be displayed by using a suitable test light connected between the battery positive terminal and the test lead, located next to the fuel distributor in the engine compartment. Characteristics of the CIS-Motronic system are as follows:

- Fuel injection control
- Oxygen sensor regulation with adaptive learning capability
- MAP type ignition control with individual cylinder knock regulation
- Idle speed control
- Fuel tank ventilation control
- Permanent fault memory for self-diagnosis

Service Precautions

- Do not disconnect the battery or power to the control module before reading the fault codes. On the Motronic SMPI and Audi SMPI systems, fault code memory is erased when power is interrupted.
- Make sure the ignition switch is **OFF** before disconnecting any wiring.
- Before removing or installing a control module, disconnect the negative battery cable. The unit receives power through the main connector at all times and will be permanently damaged if improperly powered up or down.
- Keep all parts and harnesses dry during service. Protect the control module and all solid-state components from rough handling or extremes of temperature.
- Do not apply voltage to engine control module to simulate output signals.

- When coil wire, terminal 4, is disconnected from distributor, always ground using a jumper wire.
- Do not try to start the engine with the fuel injectors removed.
- In emergency starting situations, use a fast charge for cranking up to 15 seconds only and not more than 16.5 volts; allow at least 1 minute between attempts.

Generating Codes

Before attempting to read trouble codes, the vehicle must be driven for at least 5 minutes to set codes in the computer's memory. This procedure is referred to as 'generating codes'.

CIS SYSTEMS

1. The engine must be running to generate fault codes. Read all the Steps in this procedure before starting.

2. On 1985–88 models, turn the ignition switch **ON** without starting the engine to make sure the engine warning light works, if equipped. If it does not light with the ignition **ON**, engine not running, but does light when attempting to retrieve fault codes, either the wiring between the control units is faulty or the ignition control unit is faulty.

3. On 1989–92 California models only, turn the ignition switch **ON** without starting the engine to make sure the engine warning light On Board Diagnostics (OBD) works. If it does not light with the ignition **ON**, engine not running, then turn ignition **OFF** and bridge terminals of diagnostic connectors using adapter (cable 357 971 51 4E) or an equivalent tool. Now turn ignition **ON**, but do not start the engine. The engine warning light OBD will light up, if it does not

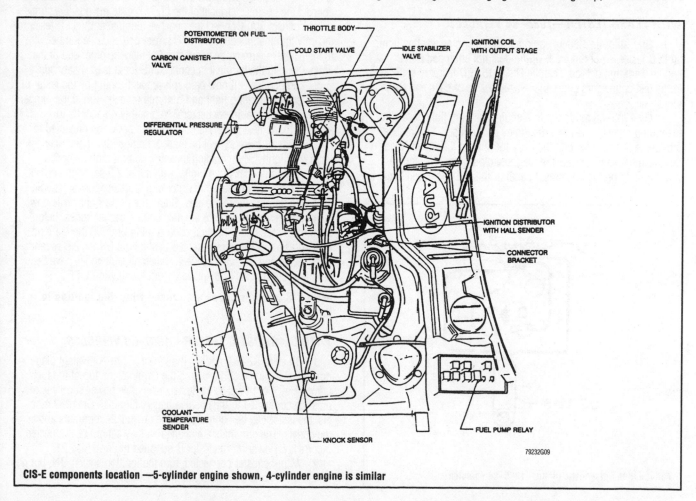

CIS-E components location —5-cylinder engine shown, 4-cylinder engine is similar

79232G09

light the wiring between the control modules is faulty or the ignition control module is faulty. Consult a wiring diagram and complete the necessary repairs before continuing.

4. Fuel pump relay and fuses 13, 19, 24 and 28 on 1985–88 models or 13, 19, 21, 24, 27 and 28 on 1989–92 models, must be good and all ground connections in the engine compartment must be good. Also, make sure the air conditioning is OFF.

5. To generate fault codes, the vehicle must be driven with air conditioner OFF, for at least 5 minutes at normal operating temperature. The engine must be kept above 3000 RPM for the majority of the test drive with at least one full throttle application. On turbocharged engines, full boost should be reached during the full throttle acceleration. After the test drive, allow the engine to idle for at least 2 minutes before retrieving codes.

6. Do not turn the ignition switch **OFF** or the temporary memory will be erased. If the engine stalls, do not restart it. The codes will still be in memory.

7. If the engine will not run, operate the starter for at least 6 seconds and leave the ignition switch **ON**.

Reading Codes

➡ **Vehicles that do not have a CHECK ENGINE light or Malfunction Indicator Lamp (MIL) codes can only be accessed by the use of special equipment. US 1115 LED tester, VAG 1551 diagnostic tester or equivalent special testers can only be used to retrieve diagnostic codes from these vehicles. When using special diagnostic equipment, always observe the tool manufacturer's instructions.**

WITH FLASH TESTER—1985–94 VEHICLES

1. On California models equipped with a engine warning light, an LED tester is not required. On models not equipped with a engine warning light, connect the US 1115 LED tester or equivalent to the test connectors under the left-hand side of dash and above the pedals.

2. On 1985–88 models with a engine warning light, locate the fuel pump relay on the main fuse/relay panel. Insert a spare fuse into the terminals on top of the relay for at least 4 seconds, then remove the fuse to activate the diagnostic program. The engine warning light on the instrument panel or the LED tester will begin to

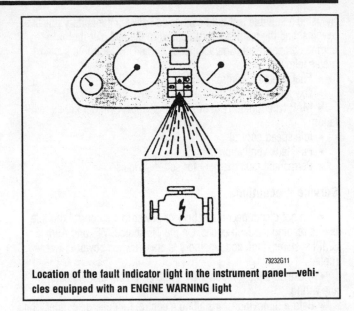

Location of the fault indicator light in the instrument panel—vehicles equipped with an ENGINE WARNING light

flash the first code. It will continue to flash this code until the fuse is installed.

3. On all models, codes should be retrieved with the engine running at idle. If engine will not start, operate the starter for approximately 6 seconds and leave the ignition switch **ON**. On CIS systems the engine should be left running after the generating codes procedure.

4. On 1989–94 models, so equipped, locate the test connectors above the pedals and connect the tester. To connect the LED tester, connect the positive terminal of the LED tester to the positive terminal in connector A. Connect the negative terminal of the LED tester to the only terminal in connector B. Connect one end of a jumper wire to the negative terminal in connector A, touch the other end of the jumper wire to the terminal in connector B for at least 4 seconds.

5. Fault codes will now be displayed as flashing by the tester or by the engine warning light on California models. Touch the jumper wire to the terminal in connector B for another 4 seconds to advance to the next code. Do not leave jumper wire connected for ten seconds or memory will be erased. Engine idle speed may increase slightly when reading injection control module codes.

6. All flash codes are 4 digits, with about 2.5 seconds between digits. Codes are displayed in order of importance, usually beginning with ignition system codes. Count the flashes and write down the code, then proceed onto the next code. Read all codes, before starting any repairs. If the first code is 4444 or 0000 there are no faults present. 0000 is represented by the light ON for 2.5 seconds with 2.5 second intervals between. When all fault codes have been reported (0000 or 4444 displayed), turn the ignition OFF.

➡ **On CIS systems codes are erased when the ignition is turned OFF.**

WITH DIAGNOSTIC TESTER—1987–89 VEHICLES

1. On all 1987–88 models, the VAG 1551 or equivalent diagnostic tester can be connected to the terminals on top of the fuel pump relay. For power to the tester, a separate power supply wire must be connected to the positive battery terminal. On 1989 models, the tester can be connected to the diagnostic terminals above the pedals. The terminals are shaped so they cannot be connected incorrectly. Power for the tester is supplied through fuse 21.

2. With the tester connected, turn the ignition switch **ON**, but do not start the engine or the codes will be erased.

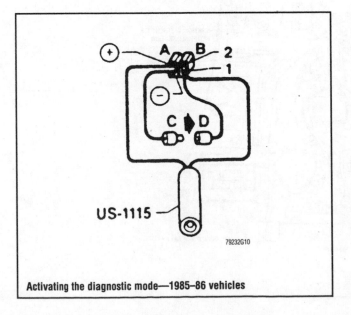

Activating the diagnostic mode—1985–86 vehicles

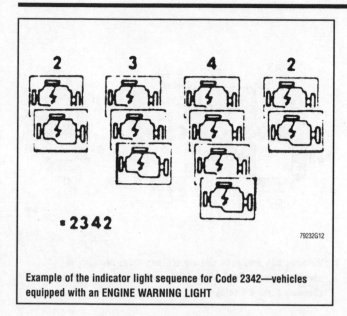

Example of the indicator light sequence for Code 2342—vehicles equipped with an ENGINE WARNING LIGHT

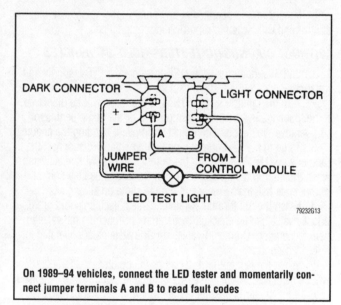

On 1989–94 vehicles, connect the LED tester and momentarily connect jumper terminals A and B to read fault codes

3. Select menu option 2, Blink Code Output. Press and release, then press and hold the run (arrow) key until the program starts, then release the key. An asterisk (*) will appear and flash the codes, which the tester will count and report on the screen as numbers. If Code 4444 is displayed, no faults are found in memory.

4. Press the run key to advance to the next code. Read all the fault codes before starting repairs.

5. All engine system fault codes will be displayed first. If there are other control units on the vehicle, press the run key again to access those codes.

6. When the End of the Report Code 0000 is displayed and there are no other control units on the vehicle, pressing the run key again may return to the main menu or it may erase the codes. To stop the program without erasing the codes, turn the ignition switch **OFF** and press the clear button once.

WITH DIAGNOSTIC TESTER—1990–91 VEHICLES

1. Connect the VAG 1551 or equivalent diagnostic tester to the diagnostic terminals above the pedals or in the passenger

side foot well. The terminals are shaped so they cannot be connected incorrectly. Power for the tester is supplied through fuse 21 or 27.

2. On all models, codes should be retrieved with the engine running at idle. If the engine will not start, operate the starter for at least 5 seconds and leave the ignition switch **ON**. On CIS systems, the engine should be left running after the procedure for generating codes.

3. Operate the tester to select menu option 2, Blink Code Output or Fault Memory Recall. An asterisk (*) will appear and flash the codes, which the tester will count and report on the screen as numbers. If Code 4444 is displayed, no faults are found in memory.

4. If the engine is not running, some codes will be displayed. These codes can be ignored if the engine has been intentionally stalled but should be investigated if the engine will not start.

5. Press the run key to advance to the next code. Read all the fault codes before starting repairs.

6. When the End of the Report Code 0000 is displayed and there are no other control units on the vehicle, pressing the run key again may return to the main menu or it may erase the codes. To stop the program without erasing the codes, turn the ignition switch **OFF** and press the clear button once.

WITH DIAGNOSTIC TESTER—1992–94 VEHICLES WITH 2.2L AND 2.8L ENGINES

Diagnostic trouble codes (DTC's) may be accessed using the VAG 1551 scan tool.

1. Turn the ignition switch to the **OFF** position.

2. Connect the VAG1551/1 diagnostic lead to the Data Link Connector (DLC) in the underhood relay box.

➡ **Observe the connector shape when connecting diagnostic leads.**

3. Connect the black lead of the VAG1551/1 diagnostic lead to the DLC 1, and the white lead to the DLC 4.

4. Connect the VAG15S1/1 diagnostic lead to the VAG 1551 scan tool. The scan tool should read—VAG self diagnosis—1 Rapid Data Transmission or 2 Flash Code Output.

5. Additional operating instructions may be accessed by pressing the help key on the VAG 1551 scan tool. press the arrow key to continue fault tracing.

WITH DIAGNOSTIC TESTER—1992–94 VEHICLES WITH 4.2L ENGINE

Diagnostic Trouble Codes (DTC's) may be accessed using the VAG 1551 scan tool.

1. Turn the ignition switch to the **OFF** position.

2. Connect the VAG1551/1 diagnostic lead to the Data Link Connectors (DLC's) under the passenger side footwell carpet.

➡ **Observe the connector shape when connecting diagnostic leads.**

3. Connect the black lead of the VAG1551/1 diagnostic lead to the DLC 1, and the white lead to the DLC 2 and the blue lead to the DLC 4.

4. Connect the VAG1551/1 diagnostic lead to the VAG 1551 scan tool. The scan tool should read—VAG self diagnosis—Rapid Data Transmission or 2 Flash Code Output.

5. Additional operating instructions may be accessed by pressing the help key on the VAG 1551 scan tool. Press the arrow key to continue fault tracing.

WITH ON BOARD DIAGNOSTIC (OBD) DISPLAY—1992–94 VEHICLES

The air conditioning system On-Board Diagnostic (OBD) can be accessed without the need of a scan tool. The air conditioning control head contains a 61 channel OBD display.

1. To start the display, turn the ignition **ON** or start the engine.
2. Press and hold down RECIRCULATION button 1 and press and hold down upper AIR DISTRIBUTION button 2.
3. Release both buttons and "O1c" will be displayed, O1c indicates channel 1, O2c indicates channel 2, etc.
4. To change to a different channel, press the temperature + button to go to the next higher channel or the temperature—button to go to the next lower channel.
5. To call up information about a particular channel, select the desired channel and press RECIRCULATION button 1.

➡**Diagnostic channel 1 O1c contains the DTC's. There are also graphics channels 1 and 2 in diagnostic channel 52, to aid in diagnosis.**

6. When using channel 52, graphics channels 1 and 2, a segment of an 88.8 display will appear. This appears when there is a compressor off situation. Each segment has an alpha numeric denomination, which can be used to diagnose a particular air conditioning compressor off problem.
7. To leave the memory display, press AUTO button or switch the ignition **OFF**.
8. The VAG 1551 is needed to erase codes from memory. See clearing codes using diagnostic tester section.

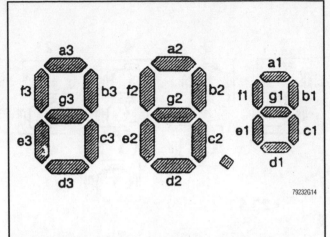

Vehicles with automatic climate control, codes can also be accessed through the On Board Diagnostic display through graphics channels 1 and 2 display—1992–94 vehicles

Generating Output Signals

WITHOUT DIAGNOSTIC TESTER—1985–86 VEHICLES

1. Insert the fuse in the opening on top of the fuel pump relay for 4 seconds.
2. Remove the fuse from the fuel pump relay.
3. The fault code will be displayed by observing the indicator light in the instrument cluster and counting the flashes.
4. To display the next code repeat Steps 1 and 2.
5. Each code will repeat until the fuse is inserted into the fuel pump relay.
6. The diagnosis procedure will be canceled if the engine speed is raised above 2000 rpm or the ignition switch is turned **OFF**.

WITHOUT DIAGNOSTIC TESTER—1987–89 VEHICLES

1. With the ignition switch **OFF**, insert the spare fuse in the top of the fuel pump relay or connect the jumper wire to the test connectors.
2. Turn the ignition switch **ON**. The first code will be displayed. If the first code is for the fuel pump relay, or the pump begins to run, remove the fuse or jumper wire quickly to prevent flooding the engine.
3. To go to the next output signal, momentarily remove the fuse from the fuel pump relay or disconnect the jumper wire. The next item on the output code list will be activated when the full throttle switch is closed. Be sure to use the correct output code list, the sequence is not the same on all engines.
4. When the full throttle switch is closed, each solenoid or frequency valve can be checked by listening or touching the valve to detect operation. Cold start valves are operated for only 10 seconds.
5. When the last output test has been completed, Code 0000 will be displayed. At this time, the fault code memory can be erased or the test can be repeated by turning the ignition switch **OFF** and **ON** again.

6. If the starter is operated at any point in the output test, the control unit will switch to reporting input signal codes.

WITHOUT DIAGNOSTIC TESTER—1990–94 VEHICLES

1. With the ignition switch **OFF**, connect the LED test light and the jumper wire to the test connectors.
2. Turn the ignition switch **ON**. The first code will be displayed. If the first code is for the fuel pump relay, or the pump begins to run, remove the jumper wire quickly to prevent flooding the engine.
3. To go to the next output signal, momentarily disconnect the jumper wire. The next item on the output code list will be activated when the full throttle switch is closed. Be sure to use the correct output code list, the sequence is not the same on all engines.
4. When the full throttle switch is closed, each solenoid or frequency valve can be checked by listening or touching the valve to detect operation. Cold start valves on CIS systems are operated for only 10 seconds.
5. When the last output test has been completed, Code 0000 will be displayed. At this time, the fault code memory can be erased.
6. If the starter is operated at any point in the output test, the control module will switch to reporting input signal codes.

WITH DIAGNOSTIC TESTER—1987–91 VEHICLES

1. Follow the procedure for retrieving fault codes. After all codes have been reported, turn the ignition switch **OFF** and press the clear button.
2. Select the Blink Code Output program on the tester. Press and hold the run key until the Continuous Short Circuit message appears on the screen, then turn the ignition switch **ON** without starting the engine.
3. Press and release the run key. The first output code should appear on the display. If the first output signal is for the fuel pump relay, remove the fuse for the fuel pump quickly after that test to avoid flooding the engine.
4. The code for each item in the output signal test will appear in the order listed. Press the run key to change to the next item on the list.
5. Except for the cold start valve, the output signal will be activated, as long as the code appears on the screen. As each item is activated, touch each valve to physically check that it is vibrating or humming.

6. When the test is completed, turn the ignition switch **OFF** to stop the test. The test can be repeated by turning the ignition switch **ON**.

WITH DIAGNOSTIC TESTER—1992–94 VEHICLES

1. Go to Step 3, if not using Rapid Data Transfer. Turn the ignition **ON**, but do not start the car.

➡**The engine must not be running when output checks are being performed. The output check mode will not work if the car is running.**

2. After selecting mode 1, Rapid Data Transfer, and the Select Function display appears, enter 01 for engine electronics. Now the control module coding and engine identification numbers will appear. After the coding is deciphered and the information matches your engine, press the Run key to continue. If the Fault In Communication display appears, one of the four displays indicating an open/short wire will appear or possibly the control module may be defective. Pushing the Help button will give you a list of possible causes for this problem. This problem must be corrected before continuing.

3. When the Select Function message appears, select 03 for Output Check Diagnosis. Each output check being tested will be displayed on the VAG 1551. Press the Run key to advance the next output check. The output check displayed on the screen will be performed until the next output check is selected.

➡**When testing the fuel pump, do not run the test too long or the engine could become flooded. When the output checks are finished the Select Function Menu will appear. The output tests can be run again by selecting function 03. Turn the ignition OFF for approximately 20 seconds, before selecting Output Check Diagnosis again.**

4. Follow the procedure for retrieving fault codes. After all codes have been reported, turn the ignition switch **OFF** and press the clear C button.

5. Select the Blink Code Output program on the tester, then turn the ignition **ON**, but do not start the engine. Output checks can only be performed with the engine NOT RUNNING. The tests will be stopped if the engine is started or a speed impulse is recognized.

6. Press and release the Run key. The first output code should appear on the display. If the first output signal is for the fuel pump relay, remove the fuse for the fuel pump quickly after that test to avoid flooding the engine.

➡**During output checks diagnosis, the carbon canister solenoid valve, idle stabilizer valve and cold start valve are checked audibly or by touch. Avoid background noise while audibly checking these components.**

7. The code for each item in the output signal test will appear in the order listed for your particular engine. Press the Run key to change to the next item on the list.

8. Except for the cold start valve, the output signal will be activated, as long as the code appears on the screen. As each item is activated, touch each valve to physically check that it is vibrating or humming.

9. When the test is completed, turn the ignition switch **OFF** to stop the test. The test can be repeated by turning the ignition switch **ON**.

Clearing Codes

WITHOUT DIAGNOSTIC TESTER—1985–88 VEHICLES

1. On California vehicles, the output signals must be tested before codes can be erased. Leave the ignition switch **ON**.

2. After the last output signal code is displayed, install the fuse for at least 4 seconds and remove it again. The engine warning light should come **ON** for 2.5 seconds, then go OFF for 2.5 seconds, displaying Code 0000.

3. Install the fuse again for at least 10 seconds, then remove it. If the engine warning light stays ON, all codes have been erased.

4. On Federal vehicles, after activating the fault code memory, the codes will automatically be erased when the ignition switch is turned **OFF** or when the engine is started.

WITHOUT DIAGNOSTIC TESTER—1989–94 VEHICLES

1. The output signals must be tested before codes can be erased. Leave the ignition switch **ON**.

2. After the last output signal code is displayed, connect the test connector jumper wire for at least 4 seconds and remove it again. The engine warning light or flash tester light should come ON for 2.5 seconds, then go OFF for 2.5 seconds, displaying Code 0000.

3. Connect the jumper wire again for at least 10 seconds, then remove it. If the engine warning light or tester light stays ON, all codes have been erased.

WITH DIAGNOSTIC TESTER—1985–88 VEHICLES

When all control unit memory codes have been retrieved and Code 0000 is displayed, press and hold the run key with the ignition switch **ON** to clear all codes.

WITH DIAGNOSTIC TESTER—1989–91 VEHICLES WITHOUT THE CIS-E III SYSTEM

When all control module memory codes have been retrieved and code 0000 or End of output is displayed, press the run key, now press 05 and the codes will be erased using mode 1 (Rapid Data Transfer). Mode 1 will display a message saying: Fault memory is erased! after erasing the codes. In mode 2 (Blink Code Output), press and hold the run key with the ignition **ON** and all codes will be cleared.

➡**This procedure erases all control module codes, make sure you have checked all control modules for DTC's before performing the erasing codes procedure.**

WITH DIAGNOSTIC TESTER—1992–94 VEHICLES WITHOUT THE CIS-E III SYSTEM

Diagnostic Trouble Codes (DTC's) may be erased after they are retrieved using the VAG 1551 scan tool.

1. Turn the ignition switch to the **OFF** position.

2. Connect the VAG 1551 scan tool as outlined in Generating Output Signals.

3. Press 01 on the VAG scan tool to select VAG address 'Engine Electronics'.

4. Press the arrow button until the display reads 'Select Function XX'.

5. Press the Q on the VAG scan tool and view all DTC's.

6. Press the 05 on the VAG scan tool to select 02—Cancel Fault Code Memory. Press Q to erase all DTC's.

7. Road test the vehicle and reactivate DTC memory to ensure all faults have been eliminated.

WITH DIAGNOSTIC TESTER—1989–92 VEHICLES WITH THE CIS-E III SYSTEM

After activating the fault code memory, the codes will automatically be erased when the ignition switch is turned **OFF** or when the engine is started.

➡**CIS-E III California models have permanent memory and will retain the fault codes after the ignition is turned off. This memory can be erased by following the erasure method using the VAG 1551 diagnostic tester. Control module part numbers can be used to identify the control modules and determine whether the CIS-E III system has permanent or temporary memory.**

Diagnostic Trouble Codes

1985–89 VEHICLES

2.0L (3A), 2.1L (MC), 2.2L (MC), and 2.3L (NG, NF) Engines:

Code 1111 Ignition control unit or fuel injection control unit
Code 1231 Transmission speed sensor
Code 2111 Engine speed sensor
Code 2112 Ignition reference sensor
Code 2113 Hall sensor
Code 2121 Idle switch
Code 2122 Engine speed/Hall sensor
Code 2123 Full throttle switch
Code 2132 No data being transmitted from fuel injection control unit to ignition control unit
Code 2141 Knock control 1, knock sensor 1 for cylinder 2; knock control 2, knock sensor 2 for cylinder 4
Code 2142 Knock sensor 1 on cylinder 2; knock sensor 2 on cylinder 4
Code 2143 Knock control 1, knock sensor 1 for cylinder 2; knock control 2, knock sensor 2 for cylinder 4
Code 2144 Knock sensor 1 on cylinder 2; knock sensor 2 on cylinder 4
Code 2212 Throttle valve potentiometer (position sensor)
Code 2221 Vacuum hose to pressure sensor in control unit
Code 2222 Pressure sensor in control unit
Code 2223 Altitude sensor
Code 2232 Air sensor potentiometer (position sensor)
Code 2233 Reference (supply) voltage
Code 2234 MPI control unit supply voltage
Code 2242 CO potentiometer
Code 2312 Engine coolant temperature (ECT) sensor
Code 2322 Intake air temperature (IAT) sensor
Code 2341 Oxygen sensor control unit is at its limit
Code 2342 Oxygen sensor (does not control)
Code 4431 Idle stabilizer valve
Code 4444 No faults stored in memory
Code 0000 End of diagnosis

1990–94 VEHICLES

2.0L (3A), 2.2L (MC, 3B, AAN), 2.3L (NG, NF, 7A), 2.8L (AAH), 3.6L (PT) and 4.2L (ABH) Engines:

Code 00000 or 0000 No faults in memory (1992–93 all engines except 2.8L and 1992 2.3L)
Code 00000 or 0000 End of diagnosis (1990–91 all engines; 1992–94 2.8L and 1992 2.3L)
Code 00000 or 4444 No faults in memory (1990–91 all engines; 1992–94 2.8L and 1992 2.3L)

Code 00281 or 1231 Vehicle speed sender signal is missing
Code 00513 or 2111 Engine speed (RPM) sensor has no change in signal
Code 00513 or 2231 Air mass sensor has open/short in circuit
Code 00514 or 2112 Crankshaft position (CKP) sensor has no change in signal
Code 00515 or 2113 Hall sender has fault in basic setting or open/short in circuit
Code 2114 Hall sender is not on reference point or out of adjustment
Code 00516 or 2121 Idle switch (closed throttle position switch 4.2L ABH) has open/short in circuit
Code 00517 or 2123 Full throttle switch
Code 00518 or 2212 Throttle position (TP) sensor has open/short in circuit
Code 00519 or 2222 Manifold vacuum sensor signal is out of range
Code 00520 or 2232 Air mass sensor signal is missing/signal out of limit
Code 00521 or 2242 CO potentiometer position sensor (2.3L 7A 1990–91)
Code 00522 or 2312 Engine coolant temperature (ECT) sensor signal is out of range
Code 00523 or 2322 Intake air temperature (IAT) sensor has open/short in circuit
Code 00524 or 2142 Knock sensor (KS) 1 has no change in signal, possible open/short between KS and ECM
Code 00525 or 2342 Oxygen sensor signal is out of range
Code 00528 or 2223 Pressure sensor (altitude sensor 1990–91) has open/short in circuit
Code 00529 or 2122 Engine RPM signal missing (2.3L NG,NF)
Code 00531 or 2233 Air mass sensor reference voltage signal missing; voltage high (2.3L 7A 1990–91)
Code 00532 or 2234 Supply voltage signal is too high or low
Code 00533 or 2231 Idle speed regulation, the idle speed is too low or too high
Code 00535 or 2141 Knock sensor regulation has exceeded its maximum control limit (1992 2.3L NG)
Code 00536 or 2141 First & second knock regulation, the maximum control limits have been exceeded (4.2L ABH engine)
Code 00536 or 2143 Second knock control has exceeded its control limits (1992–93 2.8L AAH)
Code 00537 or 2341 Oxygen sensor signal is out of range
Code 2343 Air/fuel mixture rich (2.0L engine)
Code 2344 Air/fuel mixture lean (2.0L engine)
Code 00538 or 2241 Second knock control (1991 2.2L 3B)
Code 00540 or 2144 Knock sensor 2 has no change in signal, possible open/short between KS and ECM
Code 00543 or 2214 Engine speed signal is too high, the RPM exceeds maximum limit
Code 00544 or 2224 Wastegate frequency valve has exceeded maximum boost pressure (Manifold dump valve 1990–91 2.2L MC)
Code 00545 or 2314 Engine/Transmission electrical connection has ground between ECM and TCM
Code 00546 or 2132 Fuel injection/ignition control data link (2.3L NG, NF)
Code 00553 or 2324 Mass air flow (MAF) sensor signal
Code 00554 or 2331 Oxygen control for cylinders (4-6) exceeded control limits (1992–93 2.8L AAH)

Code 00555 or 2332 Oxygen sensor 2 (G108) signal is missing (1992–93 2.8L AAH)

Code 00560 or 2411 EGR system not working properly

Code 00560 or 2441 EGR system has false readings (1992–93 2.8L AAH, California)

Code 00561 or 2413 Fuel mixture too rich

Code 00575 or 222 Manifold pressure signal missing (1990–91 2.2L MC)

Code 00577 or 2141 Knock regulation cylinder 1 has exceeded control limit

Code 00578 or 2141 Knock exceeded control limit

Code 00579 or 2141 Knock exceeded control limit

Code 00580 or 2143 Knock exceeded control limit

Code 00581 or 2143 Knock regulation cylinder 5 has exceeded control limit

Code 00824 or 3424 Engine warning light is defective

Code 4312 EGR frequency valve (2.3L 7A 1990–91)

Code 4331 Carbon canister solenoid valve 2

Code 01242 or 4332 Ignition final control circuit problem (1992–93 2.8L AAH)

Code 01247 or 4343 Carbon canister solenoid has short/open in circuit

Code 01249 or 4411 Fuel Injector cylinder 1 (& 5 on 3.6/4.2L) open/short injector circuit, fuse 23 open (fuse 13 on 2.8L)

Code 01250 or 4412 Fuel Injector cylinder 2 (& 7 on 3.6/4.2L) open/short injector circuit, fuse 23 open (fuse 13 on 2.8L)

Code 01251 or 4413 Fuel Injector cylinder 3 (& 6 on 3.6/4.2L) open/short injector circuit, fuse 23 open (fuse 13 on 2.8L)

Code 01252 or 4414 Fuel Injector cylinder 4 (& 8 on 3.6/4.2L) open/short injector circuit, fuse 23 open (fuse 13 on 2.8L)

Code 01253 or 4415 Fuel Injector cylinder 5 (code applies to 2.8L) has open/short in injector circuit, fuse 13 is open, ECM

Code 01253 or 4416 Fuel Injector cylinder 6 (code applies to 2.8L) has open/short in injector circuit, fuse 13 is open, ECM

Code 01253 or 4421 Fuel Injector cylinder 5 (code does not apply to 6 & 8 cylinder engines) has open/short in injector circuit, fuse 23 is open

Code 01254 or 4422 Fuel Injector cylinder 6 has open/short in circuit (1992–93 2.8L AAH)

Code 01257 or 4431 Idle air control (IAC) has open/short in circuit, fuse 2 is open

Code 01262 or 4442 Boost pressure limiting valve has open/short in circuit, thermofuse S75 for EVAP frequency valve is blown

Code 01265 or 4312 EGR valve has open/short in circuit (1992–93 2.8L AAH)

Code 65535 or 1111 Control module is defective

Code 65535 or 2324 Engine control module (ECM) is defective (1992–93 4.2L ABH) Ignore this code if displayed as an intermittent

TRANSMISSION CODES

Reading Codes

It is necessary to use Audi's VAG 1551 scan tool/tester to retrieve/read transmission related Diagnostic Trouble Codes (DTC's). Follow Audi's instructions to properly connect the tester/scan tool, then to retrieve the transmission codes.

Clearing Codes

It is necessary to use Audi's VAG 1551 scan tool/tester to clear transmission related Diagnostic Trouble Codes (DTC's) from the Transmission Control Module's (TCM's) memory. Follow Audi's instructions to properly connect the tester/scan tool, then to retrieve the transmission codes.

Diagnostic Trouble Codes

Code 0000 or 4444 No fault recognized

Code 00258 or 1113 Solenoid valve 1, or open circuit or short to ground

Code 00260 or 1121 Solenoid valve 2, or open circuit or short to ground

Code 00262 or 1123 Solenoid valve 3, or open circuit or short to ground

Code 00263 or 1124 Mechanical or hydraulic transmission fault

Code 00264 or 1131 Solenoid valve 4, or open circuit or short to ground

Code 00266 or 1133 Solenoid valve 5, or open circuit or short to ground

Code 00268 or 1141 Solenoid valve 6, or open circuit or short to ground

Code 00270 or 1143 Solenoid valve 7, or open circuit or short to ground

Code 00281 or 1231 Vehicle Speed Sensor (VSS) and/or circuit

Code 00293 or 1314 Multi-Function switch and/or circuit

Code 00296 or 1323 Kickdown switch and/or circuit

Code 00299 or 1332 Trans-Program switch and/or circuit

Code 00300 or 1333 Trans oil temperature sender and/or circuit

Code 00518 or 2212 Throttle valve potentiometer and/or circuit

Code 00526 or 2131 Brake light switch and/or circuit

Code 00529 or 2122 Engine speed signal missing

Code 00532 or 2234 Supply voltage, voltage for all values too low

Code 00545 or 2314 Engine/transmission electrical connection

Code 00638 Engine/transmission electrical connection

Code 00641 Trans oil temperature and/or circuit

Code 00652 Gear monitoring implausible signal—electric/hydraulic fault

Code 00660 Kickdown switch/Throttle valve potentiometer and/or circuits

Code 01236 or 4314 Selector lever lock solenoid and/or circuit

Code 65535 or 1111 Control unit—electrical influences from outside sources, bad ground connection and/or defective Transmission Control Module (TCM)

Chrysler Corporation

ENGINE CODES

General Information

The Chrysler fuel injection systems combine electronic spark advance and fuel control. At the center of these systems is a digital, pre-programmed computer, known as an Powertrain Control Module (PCM). The PCM can also be referred to as the Single Module Engine Controller (SMEC) or as the Single Board Engine Controller (SBEC). The PCM regulates ignition timing, air-fuel ratio, emission control devices, cooling fan, charging system idle speed and speed control. It has the ability to update and revise its commands to meet changing operating conditions.

Various sensors provide the input necessary for PCM to correctly regulate fuel flow at the injectors. These include the Manifold Absolute Pressure (MAP), Throttle Position Sensor (TPS), oxygen sensor, coolant temperature sensor, charge temperature sensor, and vehicle speed sensors.

In addition to the sensors, various switches are used to provide important information to the PCM. These include the neutral safety switch, air conditioning clutch switch, brake switch and speed control switch. These signals cause the PCM to change either the fuel flow at the injectors or the ignition timing or both.

The PCM is designed to test it's own input and output circuits, If a fault is found in a major system, this information is stored in the PCM for eventual display to the technician. Information on this fault can be displayed to the technician by means of the instrument panel CHECK ENGINE light or by connecting a diagnostic read-out tester and reading a numbered display code, which directly relates to a general fault. Some inputs and outputs are checked continuously and others are checked under certain conditions. If the problem is repaired or no longer exists, the PCM cancels the fault code after approximately 50 key **ON/OFF** cycles.

When a fault code is detected, it appears as either a flash of the CHECK ENGINE light on the instrument panel or by watching the Diagnostic Readout Box version II (ORB-II). This indicates that an abnormal signal in the system has been recognized by the PCM. Fault codes do indicate the presence of a failure but they don't identify the failed component directly.

FAULT CODES

Fault codes are 2 digit numbers that tell the technician which circuit is bad. Fault codes do indicate the presence of a failure but they don't identify the failed component directly. Therefore a fault code a result and not always the reason for the problem.

INDICATOR CODES

Indicator codes are 2 digit numbers that tell the technician if particular sequences or conditions have occurred. Such a condition where the indicator code will be displayed is at the beginning or the end of a diagnostic test. Indicator codes will not generate a CHECK ENGINE light or engine running test code.

ACTUATOR TEST MODE (ATM) CODES

Starting in 1985, ATM test codes are 2 digit numbers that identify the various circuits used by the technician during the diagnosis procedure. In 1989 the PCM and test equipment changed design. The actuator test functions where expanded, but access to these functions may have changed, dependent on vehicle or test equipment being used.

ENGINE RUNNING TEST CODES

Engine running test codes where introduced on fuel injected vehicles. These are 2 digit numbers. The codes are used to access sensor readouts while the engine is running and place the engine in particular operating conditions for diagnosis. Feedback carburetor system does not offer engine running sensor test mode.

CHECK ENGINE LIGHT

This is possibly the most critical step of diagnosis. A detailed examination of connectors, wiring and vacuum hoses can often lead to a repair without further diagnosis. A careful inspector will check the undersides of hoses as well as the integrity of hard-to-reach hoses blocked by the air cleaner or other component. Wiring should be checked carefully for any sign of strain, burning, crimping, or terminals pulled-out from a connector. Checking connectors at components or in harnesses is required; usually, pushing them together will reveal a loose fit.

The CHECK ENGINE or Maintenance Indicator Lamp (MIL) light has 2 modes of operation: diagnostic mode and switch test mode.

If a ORB-II diagnostic tester is not available, the PCM can show the technician fault codes by flashing the CHECK ENGINE light on the instrument panel in the diagnostic mode. In the switch test mode, after all codes are displayed, switch function can be confirmed. The light will turn on and off when a switch is turned ON and OFF.

Even though the light can be used as a diagnostic tool, it cannot do the following: Once the light starts to display fault codes, it cannot be stopped. If the technician loses count, he must start the test procedure again. The light cannot display all of the codes or any blank displays.

The light cannot tell the technician if the oxygen feed-back system is lean or rich and if the idle motor and detonation systems are operational. The light cannot perform the actuation test mode, sensor test mode or engine running test mode.

➡**Be advised that the CHECK ENGINE light can only perform a limited amount of functions and is not to be used as a substitute for a diagnostic tester. All diagnostic procedure described herein are intended for use with a Diagnostic Readout Box II (DRB-II) or equivalent tool.**

LIMP-IN MODE

The limp-in mode is the attempt by the PCM to compensate for the failure of certain components by substituting information from other sources. If the PCM senses incorrect data or no data at all from the MAP sensor, throttle position sensor or coolant temperature sensor, the system is placed into limp-in mode and the CHECK ENGINE light on the instrument panel is activated. This mode will keep the vehicle drive able until the customer can get it to a service facility.

TEST MODES

There are 5 modes of testing required for the proper diagnosis of the system. They are as follows:

Diagnostic Test Mode This mode is used to access the fault codes from the PCM's memory.

Circuit Actuation Test Mode (ATM Test) This mode is used to turn a certain circuit on and off in order to test it. ATM test codes are used in this mode.

Switch Test Mode This mode is used to determine if specific switch inputs are being received by the PCM.

Sensor Test Mode This mode looks at the output signals of certain sensors as they are received by the PCM when the engine is not running. Sensor access codes are read in this mode. Also this mode is used to clear the PCM memory of stored codes.

Engine Running Test Mode This mode looks at sensor output signals as seen by the PCM when the engine is running. Also this mode is used to determine some specific running conditions necessary for diagnosis.

Reading Codes

OBTAINING TROUBLE CODES

Entering the Jeep or Eagle self-diagnostic system requires the use of a special adapter that connects with the Diagnostic Readout Box II (DRB-II). These systems require the adapter because all of the system diagnosis is done Off-Board instead of On-Board like most vehicles. The adapter, which is a computer module itself, measures signals at the diagnostic connector and converts the signals into a form which the ORBII can use to perform tests. On vehicles other than Jeep and Eagle the following procedures will obtain stored Diagnostic Trouble Codes (DTC).

Using The CHECK ENGINE Lamp:

Codes display on vehicles built before 1989 are displayed in numerical order, after 1989 codes are displayed in order of occurrence.

1. Connect the readout box to the diagnostic connector located in the engine compartment near PCM.

2. Start the engine, if possible, cycle the transmission selector and the A/C switch if applicable. Shut off the engine.

3. Turn the ignition switch **ON - OFF, ON - OFF, ON - OFF, ON** within 5 seconds.

4. Observe the CHECK ENGINE light on the instrument panel.

5. Just after the last **ON** cycle, the dash warning (MIL) lamp will begin flashing the stored codes.

6. The codes are transmitted as two digit flashes.

7. Example would be Code 21 will be displayed as a FLASH FLASH pause FLASH.

8. Be ready to write down the codes as they appear; the only way to repeat the codes is to start over at the beginning.

Using a Scan Tool:

The scan tool is the preferred choice for fault recovery and system diagnosis. Some hints on using the ORB-II include:

• To use the HELP screen, press and hold F3 at any time.

• To restart the DRB-II at any time, hold the MODE button and press ATM at the same time.

• Pressing the up or down arrows will move forward or backward one item within a menu.

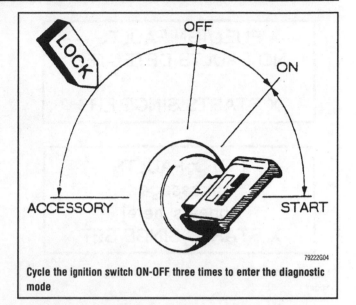

Cycle the ignition switch ON-OFF three times to enter the diagnostic mode

• To select an item, either press the number of the item or move the cursor arrow to the selection, then press ENTER.

• To return to the previous display (screen), press ATM.

• Some test screens display multiple items. To view only one, move the cursor arrow to the desired item, then press ENTER.

To read stored faults with the DRB-II:

1. With the ignition switch **OFF**, connect the tool to the diagnostic connector near the engine controller under the hood. On some 1988 and earlier models, cycling the ignition key **ON-OFF** three times may be necessary to enter the diagnostics. On 1989 and newer models, simply turn the ignition switch **ON** to access the read fault code data.

2. Start the engine if possible. Cycle the transmission from Park to a forward gear, then back to Park. Cycle the air conditioning ON and OFF. Turn the ignition switch **OFF**.

3. Turn the ignition switch ON but do not start the engine. The DRB-II will begin its power-up sequence; do not touch any keys on the scan tool during this sequence.

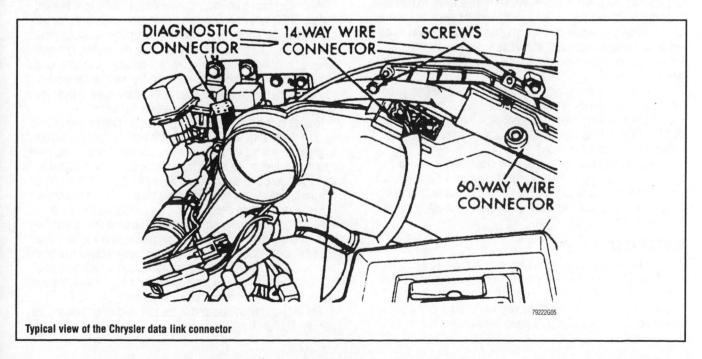

Typical view of the Chrysler data link connector

```
----- FUEL/IGN FAULTS -----
NO FAULTS DETECTED

X STARTS SINCE ERS
```

```
1 OF X FAULTS
[message
appears here]
X STARTS SINCE SET
```

79222G06

Example of the DRB-II display screen while reading the trouble codes

4. Reading faults must be selected from the FUEL/IGN MENU. To reach this menu on the DRB-II:

a. When the initial menu is displayed after the power-up sequence, use the down arrow to display choice 4) SELECT SYSTEM and select this choice.

b. Once on the—SELECT SYSTEM—screen, choose 1) ENGINE. This will enter the engine diagnostics section of the program.

c. The screen will momentarily display the engine family and SBEC identification numbers. After a few seconds the screen displays the choices 1) With A/C and 2) Without A/C. Select and enter the correct choice for the vehicle.

d. When the—ENGINE SYSTEM—screen appears, select 1) FUEL/IGNITION from the menu.

e. On the next screen, select 2) READ FAULTS.

5. If any faults are stored, the display will show how many are stored (1 of 4 faults, etc.) and issue a text description of the problem, such as COOLANT SENSOR VOLTAGE TOO LOW. The last line of the display shows the number of engine starts since the code was set. If the number displayed is 0 starts, this indicates a hard or current fault. Faults are displayed in reverse order of occurrence; the first fault shown is the most current and the last fault shown is the oldest.

6. Press the down arrow to read each fault after the first. Record the screen data carefully for easy reference.

7. If no faults are stored in the controller, the display will state NO FAULTS DETECTED and show the number of starts since the system memory was last erased.

8. After all faults have been read and recorded, press ATM.

9. Refer to the appropriate diagnostic chart for a diagnostic path. Remember that the fault message identifies a circuit problem, not a component. Use of the charts is required to sequentially test a circuit and identify the fault.

SWITCH TEST

The PCM only recognizes 2 switch input states—HI and LOW. For this reason the PCM cannot tell the difference between a selected switch position and an open circuit, short circuit or an open switch. However, if one of the switches is toggled, the controller does have the ability to respond to the change of state in the switch. If the change is displayed, it can be assumed that entire switch circuit to the PCM is operational.

1988 and Earlier Models:

After all codes have been shown and has indicated Code 55 end of message, actuate the following component switches. The digital display must change its numbers between 00 and 88 and the CHECK ENGINE light will blink when the following switches are activated and released:

- Brake pedal
- Gear shift selector
- A/C switch
- Electric defogger switch (1984)

1989 and Newer Models:

To enter the switch test mode, activate read input states or equivalent function on the readout box for the following switch tests:

- Z1 Voltage Sense
- Speed Control Set
- Speed Control ON/OFF
- Speed Control Resume
- A/C Switch Sense
- Brake Switch
- Park/neutral Switch

Scan Tool Functions

✷✷ CAUTION

Always apply the parking brake and block the wheels before performing any diagnostic procedures with the engine running. Failure to do so may result in personal injury and/or property damage.

After stored faults have been read and recorded, the scan tool may be used to investigate states and functions of various components. This ability compliments but does not replace the use of diagnostic charts. The ORB-II functions are useful in identifying circuits which are or are not operating correctly as well as checking component function or signal.

When diagnosing an emissions-related problem, keep in mind that the SBEC system only enters closed loop mode under certain conditions. The single most important criteria for entry into closed loop operation is that the engine be at normal operating temperature; i.e., fully warmed up. The engine is considered to be at normal operating temperature if any of the following are true: the electric cooling fan cycles on at least once or the upper radiator hose is hot to the touch or the heater is able to deliver hot air.

In open loop operation, the signal from the oxygen sensor is ignored by the engine controller and the fuel injection is controlled by pre-programmed values within the computer. Once closed loop operation is begun, the signal from the oxygen sensor is used by the engine controller to constantly adjust the fuel injection to maintain the proper air/fuel ratio. The system will switch in and out of closed loop operation depending on sensor signals and driver input. In most cases, the system will be in closed loop operation during normal driving, acceleration or deceleration and idle. Wide open throttle will cause the system to momentarily switch to open loop operation. Additionally, some engine control systems will momentarily switch to open loop under hard acceleration or deceleration until the MAP sensor signal stabilizes.

The ORB-II may be operated in the following diagnostic modes from the FUEL/IGN MENU screen.

SENSORS

This function displays current data being transmitted from the fuel and ignition sensors to the engine controller. Examples of sensor data available include MAP voltage, throttle position sensor voltage and percentage, RPM, coolant temperature, voltage sensor, total spark advance, and vehicle speed. Many other sensors may be monitored depending on engine/transmission combinations.

Data for each sensor is displayed in the appropriate units, such as volts, mph, in. Hg, degrees F, etc.

1988 and Earlier Models:

1. Put the system into the diagnostic test mode and wait for Code 55 to appear on the display screen.

2. Press the ATM button on the diagnostic tool to activate the display. If a specific sensor read test is desired, hold the ATM button down until the desired test code appears.

3. Slide the READ/HOLD switch to the HOLD position to display the corresponding sensor output.

Sensor Read Test Display codes:

Code 01 Battery temperature sensor; display voltage divided by 10 equals sensor temperature

Code 02 Oxygen sensor voltage; display number divided by 10 equals sensor voltage

Code 03 Charge temperature sensor voltage; display number divided by 10 equals sensor voltage

Code 04 Engine coolant temperature sensor; display number multiplied by 10 equals degrees of engine coolant sensor

Code 05 Throttle position sensor voltage; display number divided by 10 equals sensor voltage or temperature

Code 06 Peak knock sensor voltage; display number is sensor voltage

Code 07 Battery voltage; display number is battery voltage

Code 08 Map sensor voltage; display number divided by 10 equals sensor voltage

Code 09 Speed control switches:
- Display is blank—Cruise OFF
- Display shows 00—Cruise ON
- Display shows 10- Cruise SET
- Display shows 01—Cruise RESUME

Code 10 Fault code erase routine; display will flash 0's for 4 seconds

The State Display programs allow the operator to view the present conditions in the SBEC system. These choices are displayed on the FUEL IGN STATE screen and offer the choices of MODULE INFO, SENSORS, INPUTS/OUTPUTS or MONITORS. Viewing system data through these windows can be helpful in observing the effects of repairs or to compare the problem vehicle to a known-good vehicle.

1989 and Newer Models:

To enter the sensor test mode, activate read sensor voltage or read sensor values or equivalent functions on the readout box for the following sensor displays:

Read Sensor Voltage
- Battery temperature sensor
- Oxygen sensor input
- Throttle body temperature sensor
- Coolant temperature sensor
- Throttle position
- Minimum throttle
- Battery voltage
- MAP sensor voltage

Read Sensor Values
- Throttle body temperature
- Coolant temperature
- MAP gauge reading
- AIS motor position
- Added adaptive fuel
- Adaptive fuel factor
- Barometric pressure
- Engine speed
- Module spark advance
- Vehicle speed
- Oxygen sensor state

ENGINE RUNNING TEST MODE

1988 and Earlier Models:

The Engine Running Test Mode monitors the sensors on the vehicle which check operating conditions while the engine is running. The engine running test mode can be performed with the engine idling in NEUTRAL and with parking brake set or under actual driving conditions. With the diagnostic readout box READ/HOLD switch in the READ position, the engine running test mode is initiated after the engine is started.

Select a test code by switching the READ/HOLD switch to the READ position and pressing the actuator button until the desired code appears. Release actuator button and switch the READ/HOLD switch to the HOLD position. The logic module will monitor that system test and results will be displayed.

Only fuel injected engines offer this function. The Feedback carburetor system does not offer engine running sensor test mode.

Engine Running Test Display Codes:

Code 61 Battery temperature sensor; display number divided by 10 equals voltage

Code 62 Oxygen sensor; display number divided by 10 equals voltage

Code 63 Fuel injector temperature sensor; display number divided by 10 equals voltage

Code 64 Engine coolant temperature sensor; display number multiplied by 10 equals degrees F

Code 65 Throttle position sensor; display number divided by 10 equals voltage

Code 67 Battery voltage sensor; display is voltage

Code 68 Manifold vacuum sensor; display is in. Hg code 69—Minimum throttle position sensor; display number divided by 10 equals voltage

Code 70 Minimum airflow idle speed sensor; display number multiplied by 10 equals rpm (see minimum air flow check procedure)

Code 71 Vehicle speed sensor; display is mph Code 72—Engine speed sensor; display number multiplied by 10 equals rpm Fuel/Ignition Input/Output:

The engine controller recognizes only two states of electrical signals, voltage high or low. In some cases this corresponds to a switch or circuit being on or off; in other circuits a voltage signal may change from low voltage to higher voltage as a sensor opens. The controller cannot recognize the difference between a selected switch position and an open or shorted circuit.

In this test mode, the change in the circuit may be viewed as the switch is operated. For example, if the BRAKE SWITCH state is selected, the display should change from Low to High as the brake pedal is pressed. If a change in a circuit is displayed as the switch is used, it may be reasonably assumed that the entire switch circuit into the engine controller is operating correctly.

Depending on the engine/transmission in the vehicle, some of the switch states which may be checked include the air conditioning switch, brake switch, park/neutral switch, fuel flow signal, air conditioning clutch relay, radiator fan relay, CHECK ENGINE lamp, overdrive solenoid(s), lock-up solenoid and the speed control vent or vacuum solenoids. The scan tool will recognize the correct choices for each vehicle and only offer the appropriate systems on the screen.

Monitors:

On vehicles built before 1991, this display is called ENGINE PARAMETERS. 1991 and newer vehicles name the screen MONITORS.

This display allows close observation of groups of related signals. For example, if RPM is chosen, the screen will display data for many of the factors affecting the rpm such as throttle position sensor, advance, air conditioning status, park/neutral status, AIS status and coolant temperature.

One of the screens within this test is NO START. When this display is selected, the screen shows the initial data sent to the engine controller during cranking. Using this screen to identify missing or unusual signals can shorten diagnostic time.

ACTUATOR TESTS

The purpose of the circuit actuation mode test is to check for proper operation of the output circuits that the PCM cannot internally recognize. The PCM can attempt to activate these outputs and allow the technician to affirm proper operation. Most of the tests performed in this mode issue an audible click or visual indication of component operation (click of relay contacts, injector spray, etc.). Except for intermittent conditions, if a component functions properly when it is tested, it can be assumed that the component, attendant wiring and driving circuit are functioning properly.

1988 and Earlier Models:

The Actuator Test Mode 10 Code number was introduced in 1985. In 1983–84 ATM function only provided 3 ignition sparks, 2 AIS motor cycles and 1 injector pulse.

1. Put the system into the diagnostic test mode and wait for Code 55 to appear on the display screen.

2. Press ATM button on the tool to activate the display. If a specific ATM test is desired, hold the ATM button down until the desired test code appears.

3. The computer will continue to turn the selected circuit on and off for as long as 5 minutes or until the ATM button is pressed again or the ignition switch is turned to the OFF position.

4. If the ATM button is not pressed again, the computer will continue to cycle the selected circuit for 5 minutes and then shut the system off. Turning the ignition to the OFF position will also turn the test mode off.

Actuator Test Display Codes:

Code 01 Spark activation—once every 2 seconds

Code 02 Injector activation—once every 2 seconds

Code 03 AIS activation—one step open, one step closed every 4 seconds

Code 04 Radiator fan relay—once every 2 seconds

Code 05 A/C WOT cutout relay—once every 2 seconds

Code 06 ASD relay activation—once every 2 seconds

Code 07 Purge solenoid activation—one toggle every 2 seconds (The A/C fan will run continuously and the A/C switch must be in the ON position to allow for actuation)

Code 08 Speed control activation—speed control vent and vacuum every 2 seconds (Speed control switch must be in the ON position to allow for activation)

Code 09 Alternator control field activation—one toggle every 2 seconds

Code 10 Shift indicator activation—one toggle every 2 seconds

Code 11 EGR diagnosis solenoid activation—one toggle every 2 seconds

1989 and Newer Models:

This family of tests is chosen from the FUEL/IGN MENU screen. The actuator tests allow the operation of the output circuits not recognized by the engine controller to be checked by energizing them on command. Testing in this fashion is necessary because the controller does not recognize the function of all the external components. If an output to a relay is triggered, and the relay is heard to click, it may be reasonably assumed that both the output circuit and the relay are operating properly. In this mode, most of the tests cause a response that may be seen or heard, although close attention may be necessary to notice the change.

Once selected, the ACTUATOR TEST screen offers a choice of items to be activated. Depending on engine and fuel system, some of the choices include:

- Stop all tests
- Engine rpm
- Ignition coil
- Fuel injector
- Fuel system
- Solenoid/relay
- AIS motor

The engine speed may be set to a desired level through the ENGINE RPM screen. Once a system is chosen, related screens will appear allowing detailed selection of which relay, injector or component is to be operated.

EXITING DIAGNOSTIC TEST

By turning the ignition switch to the OFF position, the test mode system is exited. With a Diagnostic Readout Box attached to the system and the ATM control button not pressed, the computer will continue to cycle the selected circuits for 5 minutes and then automatically shut the system down.

Clearing Codes

Stored faults should only be cleared by use of the ORB-II or similar scan tool. Disconnecting the battery will clear codes but is not recommended as doing so will also clear all other memories on the vehicle and may affect drive ability. Disconnecting the PCM connector will also clear codes, but on newer models it may store a power loss code and will affect driveability until the vehicle is driven and the PCM can relearn it's drive ability memory.

The—ERASE—screen will appear when ATM is pressed at the end of the stored faults. Select the desired action from ERASE or DON'T ERASE. If ERASE is chosen, the display asks ARE YOU SURE? Pressing ENTER erases stored faults and displays the message FAULTS ERASED. After the faults are erased, press ATM to return the FUEL/IGN MENU.

Chrysler Domestic Built Vehicle Codes

FUEL INJECTION SYSTEM

Code 88 Display used for start of test

Code 11 Camshaft signal or Ignition signal—no reference signal detected during engine cranking

Code 12 Memory to controller has been cleared within 50-100 engine starts

Code 13 MAP sensor pneumatic signal—no variation in MAP sensor signal is detected or no difference is recognized between the engine MAP reading and the stored barometric pressure reading

Code 14 MAP voltage too high or too low

Code 15 Vehicle speed sensor signal—no distance sensor signal detected during road load conditions

Code 16 Knock sensor circuit—Open or short has been detected in the knock sensor circuit

Code 16 Battery input sensor—battery voltage sensor input below 4 volts with engine running

Code 17 Low engine temperature—engine coolant temperature remains below normal operating temperature during vehicle travel; possible thermostat problem

Code 21 Oxygen sensor signal—neither rich or lean condition is detected from the oxygen sensor input

Code 22 Coolant voltage low—coolant temperature sensor input below the minimum acceptable voltage/Coolant voltage high—coolant temperature sensor input above the maximum acceptable voltage

Code 23 Air Charge or Throttle Body temperature voltage HIGH/LOW—charge air temperature sensor input is above or below the acceptable voltage limits

Code 24 Throttle Position sensor voltage high or low. Code 25—Automatic Idle Speed (AIS) motor driver circuit—short or open detected in 1 or more of the AIS control circuits

Code 26 Injectors No. 1, 2, or 3 peak current not reached, high resistance in circuit

Code 27 Injector control circuit—bank output driver stage does not respond properly to the control signal

Code 27 Injectors No. 1, 2, or 3 control circuit and peak current not reached

Code 31 Purge solenoid circuit—open or short detected in the purge solenoid circuit

Code 32 Exhaust Gas Recirculation (EGR) solenoid circuit—open or short detected in the EGR solenoid circuit EGR system failure—required change in fuel/air ratio not detected during diagnostic test

Code 32 Surge valve solenoid—open or short in turbocharger surge valve circuit—some 1993 vehicles

Code 33 Air conditioner clutch relay circuit—open or short detected in the air conditioner clutch relay circuit. If vehicle doesn't have air conditioning ignore this code

Code 34 Speed control servo solenoids or MUX speed control circuit HIGH/LOW—open or short detected in the vacuum or vent solenoid circuits or speed control switch input above or below allowable voltage

Code 35 Radiator fan control relay circuit—open or short detected in the radiator fan relay circuit

Code 35 Idle switch shorted—switch input shorted to ground—some 1993 vehicles

Code 36 Wastegate solenoid—open or short detected in the turbocharger wastegate control solenoid circuit

Code 37 Part Throttle Unlock (PTU) circuit for torque converter clutch—open or short detected in the torque converter part throttle unlock solenoid circuit

Code 37 Baro Reed Solenoid—solenoid does not turn off when it should

Code 37 Shift indicator circuit (manual transaxle)

Code 37 Transaxle temperature out of range—some 1993 models

Code 41 Charging system circuit—output driver stage for generator field does not respond properly to the voltage regulator control signal

Code 42 Fuel pump or no Auto shut-down (ASD) relay voltage sense at controller

Code 43 Ignition control circuit—peak primary circuit current not respond properly with maximum dwell time

Code 43 Ignition coil #1, 2, or 3 primary circuits—peak primary was not achieved within the maximum allowable dwell time

Code 44 Battery temperature voltage—problem exists in the PCM battery temperature circuit or there is an open or short in the engine coolant temperature circuit

Code 44 Fused J2 circuit in not present in the logic board; used on the single engine module controller system

Code 45 Turbo boost limit exceeded—MAP sensor detects overboost

Code 44 Overdrive solenoid circuit—open or short in overdrive solenoid circuit

Code 46 Battery voltage too high—battery voltage sense input above target charging voltage during engine operation

Code 47 Battery voltage too low—battery voltage sense input below target charging voltage

Code 51 Air/fuel at limit—oxygen sensor signal input indicates LEAN air/fuel ratio condition during engine operation

Code 52 Air/fuel at limit—oxygen sensor signal input indicates RICH air/fuel ratio condition during engine operation

Code 52 Logic module fault—1984 vehicles. Code 53—Internal controller failure—internal engine controller fault condition detected during self test

Code 54 Camshaft or (distributor sync.) reference circuit—No camshaft position sensor signal detected during engine rotation

Code 55 End of message

Code 61 Baro read solenoid—open or short detected in the baro read solenoid circuit

Code 62 EMR mileage not stored—unsuccessful attempt to update EMR mileage in the controller EEPROM

Code 63 EEPROM write denied—unsuccessful attempt to write to an EEPROM location by the controller

Code 64 Flex fuel sensor—Flex fuel sensor signal out of range—(new in 1993)- CNG Temperature voltage out of range—CN gas pressure out of range

Code 65 Manifold tuning valve—an open or short has been detected in the manifold tuning valve solenoid circuit (3.3L and 3.5L LH-Platform)

Code 66 No CCO messages or no BODY CCO messages or no EATX CCO messages—messages from the CCO bus or the BODY CCD or the EATX CCO were not received by the PCM

Code 76 Ballast bypass relay—open or short in fuel pump relay circuit

Code 77 Speed control relay—an open or short has been detected in the speed control relay

Code 88 Display used for start of test

Code Error Fault code error—Unrecognized fault 10 received by ORBII

➡ **This list is for reference and does not mean that a component is defective. The code identifies the circuit and component that require further testing.**

FEEDBACK CARBURETOR SYSTEM

Code 88 Display used for start of test—must appear or other codes aren't valid

Code 11 Carburetor oxygen solenoid

Code 12 Transmission unlock relay—3.7L and 5.2L

Code 13 Air switching solenoid—3.7L and 5.2L—or Vacuum operated secondary control solenoid—2.2L

Code 14 Battery feed to computer disconnected with 20–40 engine starts

Code 16 Ignore

Code 17 Electronic throttle control solenoid

Code 18 EGR or Purge control solenoid

Code 21 Distributor pick-up signal

Code 22 Oxygen feedback stays rich or lean too long -3.7L and 5.2L—or Oxygen feedback is LEAN too long—2.2L

Code 23 Oxygen feedback is RICH too long—2.2L

Code 24 Vacuum transducer signal problem

Code 25 Charge temperature switch signal—3.7L and 5.2L engine—or Radiator fan temperature switch signal—2.2L engine

Code 26 Charge temperature sensor signal—3.7L and 5.2L engine—or Engine temperature sensor signal—2.2L

Code 28 Speed sensor circuit (if equipped)

Code 31 Battery feed to computer

Code 32 Computer can't enter diagnostics

Code 33 Computer can't enter diagnostics

Code 55 End of message

Code 88 Display used for start of test

Code 00 Diagnostic readout box is powered up and waiting for codes

➡ This list is for reference and does not mean that a component is defective. The code identifies the circuit and component that require further testing.

Chrysler Import Built Vehicle Codes

1984–1988 COLT, VISTA, SUMMIT AND D50

Code 1 Oxygen sensor

Code 2 Crank angle sensor—or Ignition signal

Code 3 Air flow sensor

Code 4 Barometric pressure sensor

Code 5 Throttle Position Sensor (TPS)

Code 6 Motor Position Sensor (MPS)—or Idle Speed Control (ISC) position sensor

Code 7 Engine Coolant Temperature Sensor

Code 8 No. 1 cylinder TDC Sensor—or Vehicle speed sensor

➡ Some 1988 Multi-Point injected vehicles use 1989 2-digit codes.

1989–93 COLT, SUMMIT, VISTA, LASER, TALON, STEALTH AND D50

Code 11 Oxygen sensor

Code 12 Air flow sensor

Code 13 Intake Air Temperature Sensor

Code 14 Throttle Position Sensor (TPS)

Code 15 SC Motor Position Sensor (MPS)

Code 21 Engine Coolant Temperature Sensor

Code 22 Crank angle sensor

Code 23 No. 1 cylinder TOC (Camshaft position) Sensor

Code 24 Vehicle speed sensor

Code 25 Barometric pressure sensor

Code 31 Knock (KS) sensor

Code 32 Manifold pressure sensor

Code 36 Ignition timing adjustment signal

Code 39 Oxygen sensor (rear—turbocharged)

Code 41 Injector

Code 42 Fuel pump

Code 43 EGR-California

Code 44 Ignition Coil—power transistor unit (No. 1 and No. 4 cylinders) on 3.0L Stealth

Code 52 Ignition Coil—power transistor unit (No. 2 and No. 5 cylinders) on 3.0L Stealth

Code 53 Ignition coil, power transistor unit (No. 3 and No. 6 cylinders)

Code 55 IAC valve position sensor

Code 59 Heated oxygen sensor

Code 61 Transaxle control unit cable (automatic transmission)

Code 62 Warm up control valve position sensor (non-turbo)

Jeep and Eagle Built Vehicle Codes

1988–90 2.5L, 3.0L AND 4.0L ENGINES

Code 1000 Ignition line low

Code 1001 Ignition line high

Code 1002 Oxygen heater line

Code 1004 Battery voltage low

Code 1005 Sensor ground line out of limits

Code 1010 Diagnostic enable line low

Code 1011 Diagnostic enable line high

Code 1012 MAP line low

Code 1013 MAP line high

Code 1014 Fuel pump line low

Code 1015 Fuel pump line high

Code 1016 Charge air temperature sensor low

Code 1017 Charge air temperature sensor high

Code 1018 No serial data from the ECU

Code 1021 Engine failed to start due to mechanical, fuel, or ignition problem

Code 1022 Start line low

Code 1024 ECU does not see start signal

Code 1025 Wide open throttle circuit low

Code 1027 ECU sees wide open throttle

Code 1028 ECU does not see wide open throttle

Code 1031 ECU sees closed throttle

Code 1032 ECU does not see closed throttle

Code 1033 Idle speed increase line low

Code 1034 Idle speed increase line high

Code 1035 Idle speed decrease line low

Code 1036 Idle speed decrease line high

Code 1037 Throttle position sensor reads low

Code 1038 Park/Neutral line high

Code 1040 Latched B+ line low

Code 1041 Latched B+ line high

Code 1042 No Latched B+ 1/2 volt drop

Code 1047 Wrong ECU

Code 1048 Manual vehicle equipped with automatic ECU

Code 1949 Automatic vehicle equipped with manual ECU

Code 1050 Idle RPM less than 500

Code 1051 Idle RPM greater than 2000

Code 1052 MAP sensor out of limits

Code 1053 Change in MAP reading out of limits

Code 1054 Coolant temperature sensor line low

Code 1055 Coolant temperature sensor line high

Code 1056 Inactive coolant temperature sensor

Code 1057 Knock circuit shorted

Code 1058 Knock value out of limits

Code 1059 A/C request line low

Code 1060 A/C request line high
Code 1061 A/C select line low
Code 1062 A/C select line high
Code 1063 A/C clutch line low
Code 1064 A/C clutch line high
Code 1065 Oxygen reads rich
Code 1066 Oxygen reads lean
Code 1067 Latch relay line low
Code 1068 Latch relay line high
Code 1070 A/C cutout line low
Code 1071 A/C cutout line high
Code 1073 ECU does not see speed sensor signal
Code 1200 ECU defective
Code 1202 Injector shorted to ground
Code 1209 Injector open
Code 1218 No voltage at ECU from power latch relay
Code 1220 No voltage at ECU from EGR solenoid
Code 1221 No injector voltage
Code 1222 MAP not grounded
Code 1223 No ECU tests run

➡ **Prior to 1988 vehicles used an Off-Board Diagnostic system which required special diagnostic equipment to read codes. After 1991 Jeep and Eagle vehicles used the Chrysler Domestic Built Engine Control system. The code list for Chrysler Built Domestic Fuel injection System also covers 1991 and newer Jeep and Eagle vehicles.**

TRANSMISSION CODES

Reading and Retrieving Codes

Chrysler-built transmission/transaxle diagnostic systems use two-digit DTC's, as with most pre-OBD II vehicles, to inform to the mechanic or technician the problem circuit or component.

Unfortunately, it is necessary to use a diagnostic scan tool to retrieve DTC's from almost all of the Chrysler automatic transmis-sions and transaxles, with the exception of the 42RH and 46RH assemblies. The other transmissions/transaxles use a separate computer (TCM) to control the electronic transmission/transaxle , and require a scan tool with a model-specific cartridge to retrieve and read the DTC's. However, the 42RH and 46RH units are con-trolled by the Powertrain Control Module, and, therefore, their DTC's can be read along with the engine-related DTC's. Chrysler models which utilize either the 42RH or 46RH use a light, located in the instrument cluster and usually referred to as the Malfunction Indicator Lamp (MIL), CHECK ENGINE light or POWER LOSS light (on earlier models), to display the DTC's as a series of flashes or blinks.

The two-digit codes are flashed out in a logical series of blinks. For example, when first starting to retrieve the DTC's the MIL will flash eight times, pause briefly, then flash eight more times followed by a longer pause. This indicates a Code 88. Therefore, a flash pat-tern of two flashes-pause-three flashes-long pause would indicate a Code 23.

➡ **Code 88 should always be the first DTC flashed out by the MIL, since it indicates that the diagnostic system is ready to start showing the DTC's. When the system is finished flash-ing all of the DTC's, it will indicate a Code 55.**

The DTC's are displayed in numerical sequence starting with the lowest code number and moving to the highest.

➡ **When reading the codes be sure not to miss any of them, otherwise the entire sequence must be repeated. This can be tiresome if a large number of codes was stored, and you missed the last one.**

Unlike most other manufacturers, the Chrysler diagnostic com-puter system is very easy to actuate, simply turn the ignition key **ON**, **OFF**, **ON**, **OFF**, **ON** within 2 seconds. The DTC's will start to flash after turning the ignition key to the **ON** position for the third time. Code 55 indicates that all of the DTC's have been flashed out, and that the DTC cycle is done.

Clearing Codes

➡ **It is not recommended that the negative battery cable be disconnected to clear DTC's. Doing so will clear all settings from the vehicle's computer system, resulting in lost radio presets, seat memories, anti-theft codes, driveability para-meters, etc., and there are better ways of spending your afternoon than resetting all of these.**

There are two methods for clearing the DTC's from Chrysler models: the first is by using a scan tool, the second is slightly more time consuming. To erase the DTC's without a scan tool, you must turn the ignition switch **ON** and **OFF** 51 times! Yes, you read correctly, 51 times! Since the system is designed to erase codes if they have not been evident during the last 50 ignition key cycles, you must either cycle the ignition 51 times at one sitting, or wait until you have started and driven the vehicle 51 times or a period of time. (For diagnostic purposes, it is best to clear the codes right away so that you can ensure the problems have been properly fixed.) This is one reason (not to mention the added bonus of ease of diagnosing the systems) that a scan tool would be more advan-tageous.

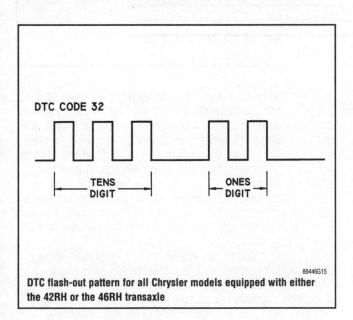

DTC CODE 32

TENS DIGIT | ONES DIGIT

89446G15

DTC flash-out pattern for all Chrysler models equipped with either the 42RH or the 46RH transaxle

DIAGNOSTIC TROUBLE CODES—CHRYSLER 42RH AND 46RH TRANSMISSIONS

Code	Component/Circuit Fault
11	No distributor reference signal
12	Number of ignition key cycles since the last fault was erased
13	MAP sensor or circuit
14	MAP sensor or circuit
15	Vehicle Speed Sensor (VSS) or circuit
16	Battery Input Sense
17	Engine temperature too low
21	Oxygen sensor or circuit
22	Engine Coolant Temperature (ECT) sensor or circuit
23	Throttle body temperature sensor or circuit
24	TP sensor or circuit
25	AIS motor circuits
26	Fuel injectors or circuits
27	Fuel injectors or circuits
31	Purge solenoid or circuit
32	EGR system or solenoid circuit
33	A/C clutch relay or circuit
34	S/C servo solenoids or circuit
35	Idle switch
36	Air switch solenoid or circuit
37	PTU solenoid circuit
41	Alternator Field circuit
42	ASD relay or circuit
43	Ignition control circuit
44	FJ2 voltage sense
45	Overdrive solenoid or circuit
46	Battery voltage too high
47	Low charging output
51	Lean air/fuel mixture
52	Rich air/fuel mixture
53	Internal self-test
55	Not used
62	EMR miles not stored
63	EEPROM write denied

89446G11

Chrysler 42RH and 46RH transaxle DTC's

Ford Motor Company

ENGINE CODES

Introduction To Ford Self-Diagnostics

The engine control systems are used in conjunction with either a throttle body (CFI) injection or multi-point (EFI and SEFI) injection fuel delivery system or feedback carburetor systems depending on the year, model and powertrain. Although the individual system components vary slightly, the electronic control system operation is basically the same. The major difference is the number and type of output devices being controlled by the ECA.

Automotive manufacturers have developed on-board computers to control engines, transmissions and many other components. These on-board computers with dozens of sensors and actuators have become almost impossible to test without the help of electronic test equipment.

One of these electronic test devices has become the on-board computer itself. The Powertrain Control Modules (PCM), sometimes called the Electronic Control Assembly (ECA), used on toadies vehicles has a built in self testing system. This self test ability is called self-diagnosis. The self-diagnosis system will test many or all of the sensors and controlled devices for proper function. When a malfunction is detected this system will store a fault code in memory that's related to that specific circuit. You can access the computer to obtain fault codes recorded in memory by using an analog voltmeter or special diagnostic scan tool. This will help narrow down what area to begin testing.

Fault code meanings can vary from year to year even on the same model. It is extremely important after retrieving a fault code to verify its meaning with a proper manual. Servicing a fault code incorrectly will not only lead to the wrong conclusion but could also cause damage if tested or serviced incorrectly. There is a list of general code descriptions provide later in this manual.

WHAT SYSTEM IS ON THE VEHICLE?

There are 3 electronic fuel control systems used by Ford Motor Company. These systems all operate using similar components and on-board computers. Self-Diagnostic on these systems will vary, but, the basic fuel control operation is the same. Ford uses the following systems:

• **EEC-IV and EEC-V** engine control system: used on most domestic built Ford vehicles since 1984.

• **Non-NAAO EEC** engine control system: used on import built Ford vehicles, referred to as Non-NAAO cars.

• **MCU** feedback carburetor system: used on most Ford vehicles before 1984 and some later model vehicles equipped with a V8 engine and feedback carburetor.

Most Ford vehicles made after 1983 use the 4th generation Electronic Engine Control system, commonly designated EEC-IV.

If you own a vehicle with a 2.0L, 2.2L, or 2.5L engine, then the fuel control system is referred to as NON-NAAO (Not North American Automotive Operations produced vehicles) system. The fuel system used on these vehicles is called Electronic Engine Control (EEC). This Non-NAAO EEC system components and operation are basically the same as the EEC-IV system. The self-diagnostic function on the EEC system differs from the EEC-VI system and is covered under NON-NAAO vehicle.

Most 1984–94 Ford domestic built vehicles employ the 4th generation Electronic Engine Control system, commonly called EEC-IV, to manage fuel, ignition and emissions on vehicle engines. In 1994 the EEC-V system was introduced on some models. The diagnostic system on EEC-V provides 3 digit codes in place of 2 digit codes, and it is capable of monitoring more inputs and outputs.

If your vehicle was made before 1984, or has a feedback carburetor equipped V8 engine, then it probably uses the Microprocessor Control Unit (MCU). The MCU system was used on most 1981-83 carburetor equipped vehicles, and 1984 and newer V8 engines with feedback carburetors. The MCU system uses a large six sided connector, identical to the one used with EEC-IV systems. The MCU system does NOT use the small single wire connector, like the EEC-IV system. The MCU system is covered in greater detail later in this manual.

EEC-IV And EEC-V Diagnostic Systems

Most 1984–94 Ford domestic built vehicles employ the 4th generation Electronic Engine Control system, commonly designated EEC-IV, to manage fuel, ignition and emissions on vehicle engines. In 1994 the EEC-V system was introduced on some models. The diagnostic system on EEC-V provides 3 digit codes in place of 2 digit codes and monitors more components.

ENGINE CONTROL SYSTEM

The Powertrain Control Modules (PCM), usually referred to as the Electronic Control Assembly (ECA) by Ford, is given responsibility for the operation of the emission control devices, cooling fans, ignition and advance and in some cases, automatic transmission functions. Because the EEC-IV oversees both the ignition timing and the fuel injector operation, a precise air/fuel ratio will be maintained under all operating conditions. The ECA is a microprocessor or small computer which receives electrical inputs from several sensors, switches and relays on and around the engine.

Based on combinations of these inputs, the ECA controls outputs to various devices concerned with engine operation and emissions. The engine control assembly relies on the signals to form a correct picture of current vehicle operation. If any of the input signals is incorrect, the ECA reacts to what ever picture is painted for it. For example, if the coolant temperature sensor is inaccurate and reads too low, the ECA may see a picture of the engine never warming up. Consequently, the engine settings will be maintained as if the engine were cold. Because so many inputs can affect one output, correct diagnostic procedures are essential on these systems.

One part of the ECA is devoted to monitoring both input and output functions within the system. This ability forms the core of the self-diagnostic system. If a problem is detected within a circuit, the controller will recognize the fault, assign it an identification code, and store the code in a memory section. Depending on the year and model, the fault code(s) may be represented by two or three digit numbers. The stored code(s) may be retrieved during diagnosis.

When the term Powertrain Control Module (PCM) is used in this manual it will refer to the engine control computer regardless that it may also be called an Electronic Control Assembly (ECA).

While the EEC-IV system is capable of recognizing many internal faults, certain faults will not be recognized. Because the computer system sees only electrical signals, it cannot sense or react to mechanical or vacuum faults affecting engine operation. Some of these faults may affect another component which will set a code. For example, the ECA monitors the output signal to the fuel injectors, but cannot detect a partially clogged injector. As long as the output driver responds correctly, the computer will read the system as functioning correctly. However, the improper flow of fuel may result in a lean mixture. This would, in turn, be detected by the oxygen sensor and noticed as a constantly lean signal by the ECA. Once the signal falls outside the pre-programmed limits, the engine control assembly would notice the fault and set an identification code.

Additionally, the EEC-IV system employs adaptive fuel logic. This process is used to compensate for normal wear and variability within the fuel system. Once the engine enters steady-state operation, the engine control assembly watches the oxygen sensor signal for a bias or tendency to run slightly rich or lean. If such a bias is detected, the adaptive logic corrects the fuel delivery to bring the air/fuel mixture towards a centered or 14.7:1 ratio. This compensating shift is stored in a non-volatile memory which is retained by battery power even with the ignition switched off. The correction factor is then available the next time the vehicle is operated.

➡**If the battery is disconnected for longer than 5 minutes, the adaptive fuel factor will be lost. After repair it will be necessary to drive the car at least 10 miles to allow the processor to relearn the correct factors. The driving period should include steady-throttle open road driving if possible. During the drive, the vehicle may exhibit driveability symptoms not noticed before. These symptoms should clear as the ECA computes the correction factor. The ECA will also store Code 19 indicating loss of power to the controller.**

Failure Mode Effects Management (FMEM)

The engine controller assembly contains back-up programs which allow the engine to operate if a sensor signal is lost. If a sensor input is seen to be out of range—either high or low—the FMEM program is used. The processor substitutes a fixed value for the missing sensor signal. The engine will continue to operate, although performance and driveability may be noticeably reduced. This function of the controller is sometimes referred to as the limp-in or fail-safe mode. If the missing sensor signal is restored, the FMEM system immediately returns the system to normal operation. The dashboard warning lamp will be lit when FMEM is in effect.

Hardware Limited Operation Strategy (HLOS)

This mode is only used if the fault is too extreme for the FMEM circuit to handle. In this mode, the processor has ceased all computation and control; the entire system is run on fixed values. The vehicle may be operated but performance and driveability will be greatly reduced. The fixed or default settings provide minimal calibration, allowing the vehicle to be carefully driven in for service. The dashboard warning lamp will be lit when HLOS is engaged. Codes cannot be read while the system is operating in this mode.

DASHBOARD WARNING LAMP

The CHECK ENGINE or SERVICE ENGINE SOON dashboard warning lamp is referred to as the Malfunction Indicator Lamp (MIL). The lamp is connected to the engine control assembly and will alert the driver to certain malfunctions within the EEC-IV system. When the lamp is lit, the ECA has detected a fault and stored an identity code in memory. The engine control system will usually enter either FMEM or HLOS mode and driveability will be impaired.

The light will stay on as long as the fault causing it is present. Should the fault self-correct, the MIL will extinguish but the stored code will remain in memory.

Under normal operating conditions, the MIL should light briefly when the ignition key is turned ON. As soon as the ECA receives a signal that the engine is cranking, the lamp will be extinguished. The dash warning lamp should remain out during the entire operating cycle.

➡**On Continental, the CHECK ENGINE message is displayed on the message center. When a fault is detected, the message is accompanied by a 1 second tone every 5 seconds. The tone stops after 1 minute. When the Continental system enters HLOS, the additional message CHECK DCL is displayed. DCL refers to the Data Communications Link running between the engine controller and the message center.**

EEC-IV and EEC-V Scan Tool Functions

Although stored codes may be read by using a analog voltmeter, the use of hand-held scan tools such as Ford's Self-Test Automatic Readout (STAR) tester or the second generation SUPER STAR II tester or their equivalent is recommended. There are many manufacturers of these tools; the purchaser must be certain that the tool is proper for the intended use.

Both the STAR and SUPER STAR testers are designed to communicate directly with the EEC-IV system and interpret the electrical signals. The SUPER STAR tester may be used to read either 2 or 3 digit codes; the original STAR tester will not read the 3 digit codes used on many 1990 and newer vehicles.

The scan tool allows any stored faults to be read from the engine controller memory. Use of the scan tool provides additional data during troubleshooting but does not eliminate the use of the charts. The scan tool makes collecting information easier; the data must be correctly interpreted by an operator familiar with the system.

ELECTRICAL TOOLS

The most commonly required electrical diagnostic tool is the Digital Multimeter, allowing voltage, resistance and amperage to be read by one instrument. Many of the diagnostic charts require the use of a volt or ohmmeter during diagnosis.

The multimeter must be a high impedance unit, with 10 megohms of impedance in the voltmeter. This type of meter will not place an additional load on the circuit it is testing; this is extremely important in low voltage circuits. The multimeter must

be of high quality in all respects. It should be handled carefully and protected from impact or damage. Replace the batteries frequently in the unit.

Additionally, an analog (needle type) voltmeter may be used to read stored fault codes if the STAR tester is not available. The codes are transmitted as visible needle sweeps on the face of the instrument. Almost all diagnostic procedures will require the use of the Breakout Box, a device which connects into the EEC-IV harness and provides testing ports for the 60 wires in the harness. Direct testing of the harness connectors at the terminals or by back-probing is not recommended; damage to the wiring and terminals is almost certain to occur.

Other necessary tools include a quality tachometer with inductive (clip-on) pickup, a fuel pressure gauge with system adapters and a vacuum gauge with an auxiliary source of vacuum.

EEC-IV and EEC-V Self-Diagnostics

Diagnosis of a driveability problem requires attention to detail and following the diagnostic procedures in the correct order. Resist the temptation to begin extensive testing before completing the preliminary diagnostic steps. The preliminary or visual inspection must be completed in detail before diagnosis begins. In many cases this will shorten diagnostic time and often cure the problem without electronic testing.

VISUAL INSPECTION

This is possibly the most critical step of diagnosis. A detailed examination of all connectors, wiring and vacuum hoses can often lead to a repair without further diagnosis. Performance of this step relies on the skill of the technician performing it; a careful inspector will check the undersides of hoses as well as the integrity of hard-to-reach hoses blocked by the air cleaner or other components. Wiring should be checked carefully for any sign of strain , burning, crimping or terminal pull-out from a connector.

Checking connectors at components or in harnesses is required; usually, pushing them together will reveal a loose fit. Pay particular attention to ground circuits, making sure they are not loose or corroded. Remember to inspect connectors and hose fittings at components not mounted on the engine, such as the evaporative canister or relays mounted on the fender aprons. Any component or wiring in the vicinity of a fluid leak or spillage should be given extra attention during inspection.

Additionally, inspect maintenance items such as belt condition and tension, battery charge and condition and the radiator cap carefully. Any of these very simple items may affect the system enough to set a fault.

DIAGNOSTIC CONNECTOR LOCATION

The Diagnostic Link Connectors (DLC) are located a 6 basic locations:
- Near the bulkhead (right or left side of vehicle)
- Near the wheel well (right or left side of vehicle)
- Near the front corner of the engine compartment (right or left side of vehicle)

EEC-IV and EEC-V Reading Codes

The EEC-IV system may be interrogated for stored codes using the Quick Test procedures. These tests will reveal faults immediately present during the test as well as any intermittent codes set within the previous 80 warm up cycles. If a code was set before a problem self-corrected (such as a momentarily loose connector), the code

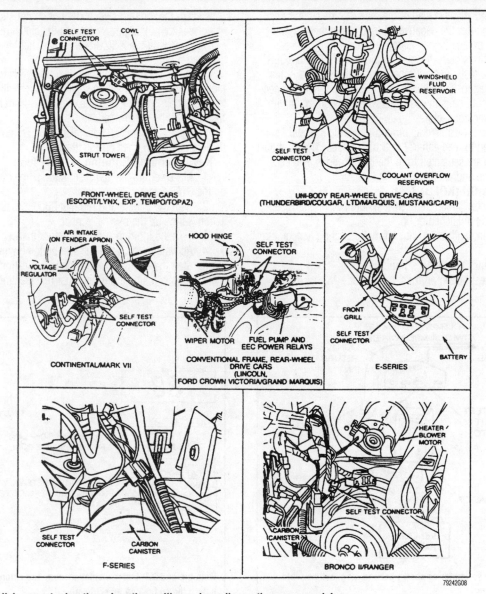

Typical diagnostic link connector locations. Locations will vary depending on the year or model

will be erased if the problem does not reoccur within 80 warm-up cycles.

The Quick Test procedure is divided into 2 sections, Key On Engine Off (KOEO) and Key On Engine Running (KOER). These 2 procedures must be performed correctly if the system is to run the internal self-checks and provide accurate fault codes. Codes will be output and displayed as numbers on the hand scan tool, i.e. 23. Code 23 would be displayed as 2 needle sweeps and pause and 3 more needle sweeps. For codes being read on an analog voltmeter, the needle sweeps indicate the code digits in the same manner as the lamp flashes on other systems.

In all cases, the codes 11 or 111 are used to indicate PASS during testing. Note that the PASS code may appear, followed by other stored codes. These are codes from the continuous memory and may indicate intermittent faults, even though the system does not presently contain the fault. The PASS designation only indicates the system passes all internal tests at the moment.

Once the Quick Test has been performed and all fault codes recorded, refer to the code charts. The charts direct the use of spe-

cific pinpoint tests for the appropriate circuit and will allow complete circuit testing.

✲✲ CAUTION

To prevent injury and/or property damage, always block the drive wheels, firmly apply the parking brake, place the transmission in Park or Neutral and turn all electrical loads off before performing the Quick Test procedures.

READING CODES WITH ANALOG VOLTMETER

➡**There are inexpensive tools available at auto parts stores that make reading and clear Ford engine codes very easy. Reading the voltmeter needle sweeps is sometimes difficult. Always check the code more than once to make certain it was read correctly.**

In the absence of a scan tool, an analog voltmeter may be used to retrieve stored fault codes. Set the meter range to read DC 0–15 volts. Connect the positive (+) lead of the meter to the battery positive terminal and connect the negative (-) lead of the meter to the self-test output pin of the diagnostic connector.

Follow the directions given for performing the KOEO and KOER tests. To activate the tests, use a jumper wire to connect the signal return pin on the diagnostic connector to the self-test input connector. The self-test input line is the separate wire and connector with or near the diagnostic connector.

The codes will be transmitted as groups of needle sweeps. This method may be used to read either 2 or 3 digit codes. The Continuous Memory codes are separated from the KOEO codes by 6 seconds, a single sweep and another 6 second delay.

Key On Engine Off (KOEO) Test:
1. Connect the scan tool to the self-test connectors. Make certain the test button is unlatched or up.
2. Start the engine and run it until normal operating temperature is reached.
3. Turn the engine OFF for 10 seconds.
4. Activate the test button on the STAR tester.

5. Turn the ignition switch **ON** but do not start the engine. For vehicles with 4.9L engines, depress the clutch during the entire test. For vehicles with the 7.3L diesel engine, hold the accelerator to the floor during the test.
6. The KOEO codes will be transmitted. Six to nine seconds after the last KOEO code, a single separator pulse will be transmitted. Six to nine seconds after this pulse, the codes from the Continuous Memory will be transmitted.
7. Record all service codes displayed. Do not depress the throttle on gasoline engines during the test.

Key On Engine Running (KOER) Test:
1. Make certain the self-test button is released or de-activated on the STAR tester.
2. Start the engine and run it at 2000 rpm for two minutes. This action warms up the oxygen sensor.
3. Turn the ignition switch OFF for 10 seconds.
4. Activate or latch the self-test button on the scan tool.
5. Start the engine. The engine identification code will be transmitted. This is a single digit number representing ½ the number of

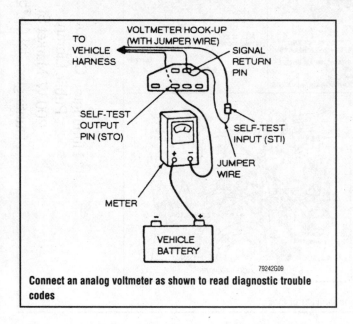

Connect an analog voltmeter as shown to read diagnostic trouble codes

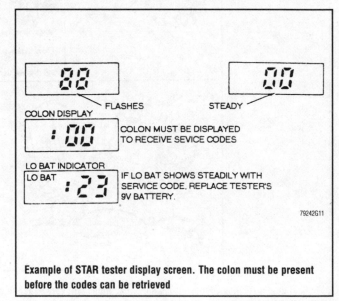

Example of STAR tester display screen. The colon must be present before the codes can be retrieved

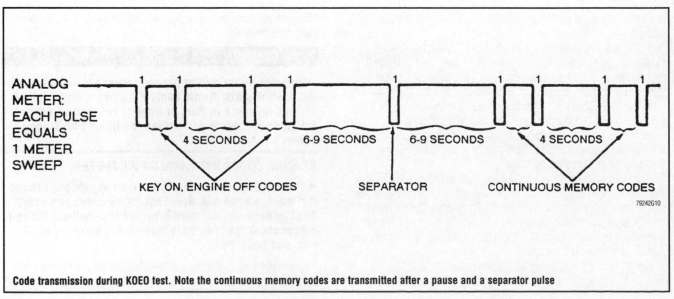

Code transmission during KOEO test. Note the continuous memory codes are transmitted after a pause and a separator pulse

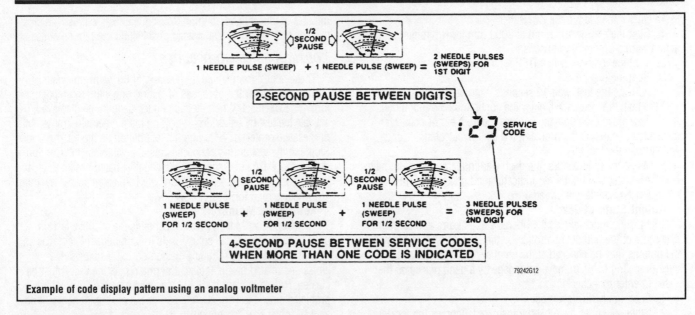

Example of code display pattern using an analog voltmeter

cylinders in a gasoline engine. On the STAR tester, this number may appear with a zero, i.e., 20 = 2. For 7.3L diesel engines, the 10 code is 5. The code is used to confirm that the correct processor is installed and that the self-test has begun.

6. If the vehicle is equipped with a Brake On/Off (BOO) switch, the brake pedal must be depressed and released after the 10 code is transmitted.

7. If the vehicle is equipped with a Power Steering Pressure Switch (PSPS), the steering wheel must be turned at least '/2 turn and released within 2 seconds after the engine 10 code is transmitted.

8. If the vehicle is equipped with the E400 transmission, the Overdrive Cancel Switch (OCS) must be cycled after the engine 10 code is transmitted.

9. Certain Ford vehicles will display a Dynamic Response code 6–20 seconds after the engine 10 code. This will appear as one pulse on a meter or as a 10 on the STAR tester. When this code appears, briefly take the engine to wide open throttle. This allows the system to test the throttle position, MAF and MAP sensors.

10. All relevant codes will be displayed and should be recorded. Remember that the codes refer only to faults present during this test cycle. Codes stored in Continuous Memory are not displayed in this test mode.

11. Do not depress the throttle during testing unless a dynamic response code is displayed.

Testing With Continental Message Center:
The stored fault codes may be displayed on the electronic message screen in Continentals so equipped. To perform the KOEO test, press all 3 buttons on the electronic instrument cluster (GAUGE SELECT, ENGLISH/METRIC, SPEED ALARM or SELECT, RESET and SYSTEM CHECK) simultaneously. Turn the ignition switch **ON** and release the buttons; stored codes will be displayed on the screen.

To perform the KOER test:

1. Hold in all 3 buttons, start the engine and release the buttons.

2. Press the SELECT or GAUGE SELECT button 3 times. The message DEALER 4 should appear at the bottom of the message panel.

3. Initiate the test by using a jumper wire to connect the signal return pin on the diagnostic connector to the self-test input connector. The self-test input line is the separate wire and connector with or near the diagnostic connector.

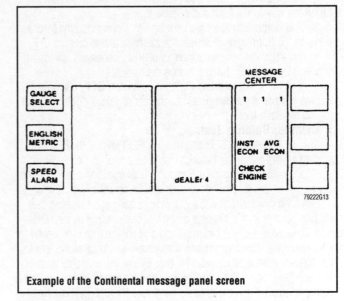

Example of the Continental message panel screen

4. The stored codes will be output to the vehicle display.

5. To exit the test, turn the ignition switch OFF and disconnect the jumper wire.

ADVANCED TEST MODES

Continuous Monitor or Wiggle Test Mode:
Once entered, this mode allows the technician to attempt to recreate intermittent faults by wiggling or tapping components, wiring or connectors. The test may be performed during either KOEO or KOER procedures. The test requires the use of either an analog voltmeter or a hand scan tool.

To enter the continuous monitor mode during KOEO testing, turn the ignition switch ON. Activate the test, wait 10 seconds, then deactivate and reactivate the test; the system will enter the continuous monitor mode. Tap, move or wiggle the harness, component or connector suspected of causing the problem; if a fault is detected, the code will store in the memory. When the fault occurs, the dash warning lamp will illuminate, the STAR tester will light a red indicator (and possibly beep) and the analog meter needle will sweep once.

To enter this mode in the KOER test:

1. Start the engine and run it at 2000 rpm for two minutes. This action warms up the oxygen sensor.

2. Turn the ignition switch **OFF** for 10 seconds.

3. Start the engine.

4. Activate the test, wait 10 seconds, then deactivate and reactivate the test; the system will enter the continuous monitor mode.

5. Tap, move or wiggle the harness, component or connector suspected of causing the problem; if a fault is detected, the code will store in the memory.

6. When the fault occurs, the dash warning lamp will illuminate, the STAR tester will light a red indicator (and possibly beep) and the analog meter needle will sweep once.

Output State Check:

This testing mode allows the operator to energize and de-energize most of the outputs controlled by the EEC-IV system. Many of the outputs may be checked at the component by listening for a click or feeling the item move or engage by a hand placed on the case. To enter this check:

1. Enter the KOEO test mode.

2. When all codes have been transmitted, depress the accelerator all the way to the floor and release it.

3. The output actuators are now all ON. Depressing the throttle pedal to the floor again switches the all the actuator outputs OFF.

4. This test may be performed as often as necessary, switching between ON and OFF by depressing the throttle.

5. Exit the test by turning the ignition switch **OFF**, disconnecting the jumper at the diagnostic connector or releasing the test button on the scan tool.

Cylinder Balance Test:

This test is only for SEFI engines. On SEFI engine the EEC-IV system allows a cylinder balance test to be performed on engines equipped with the Sequential Electronic Fuel Injection system. Cylinder balance testing identifies a weak or non-contributing cylinder.

Enter the cylinder balance test by depressing and releasing the throttle pedal within 2 minutes of the last code output in the KOER test. The idle speed will become fixed and engine mm is recorded for later reference. The engine control assembly will shut off the fuel to the highest numbered cylinder (4, 6 or 8), allow the engine to stabilize and then record the rpm. The injector is turned back on and the next one shut off and the process continues through cylinder No. 1.

The controller selects the highest rpm drop from all the cylinders tested, multiplies it by a percentage and arrives at an rpm drop value for all cylinders. For example, if the greatest drop for any cylinder was 150 rpm, the processor applies a multiple of 65% and arrives at 98 mm. The processor then checks the recorded rpm drops, checking that each was at least 98 rpm. If all cylinders meet the criteria, the test is complete and the ECA outputs Code 90 indicating PASS.

If one cylinder did not drop at least this amount, then the cylinder number is output instead of the 90 code. The cylinder number will be followed by a zero, so 30 indicates cylinder No. 3 did not meet the minimum rpm drop.

The test may be repeated a second time by depressing and releasing the throttle pedal within 2 minutes of the last code output. For the second test, the controller uses a lower percentage (and thus a lower rpm) to determine the minimum acceptable rpm drop. Again, either Code 90 or the number of the weak cylinder will be output.

Performing a third test causes the ECA to select an even lower percentage and rpm drop. If a cylinder is shown as weak in the third test, it should be considered non-contributing. The tests may be repeated as often as needed if the throttle is depressed within two

minutes of the last code output. Subsequent tests will use the percentage from the third test instead of selecting even lower values.

CONTINUOUS MEMORY CODES

These codes are retained in memory for 80 warm-up cycles. To clear the codes for the purposes of testing or confirming repair, perform the KOEO test. When the fault codes begin to be displayed, deactivate the test by either disconnecting the jumper wire (meter, MIL or message center) or releasing the test button on the hand scanner. Stopping the test during code transmission will erase the Continuous Memory. Do not disconnect the negative battery cable to clear these codes; the Keep Alive memory will be cleared and a new code, 19, will be stored for loss of ECA power.

Keep Alive Memory

The Keep Alive Memory (KAM) contains the adaptive factors used by the processor to compensate for component tolerances and wear. It should not be routinely cleared during diagnosis. If an emissions related part is replaced during repair, the KAM must be cleared. Failure to clear the KAM may cause severe driveability problems since the correction factor for the old component will be applied to the new component.

To clear the Keep Alive Memory, disconnect the negative battery cable for at least 5 minutes. After the memory is cleared and the battery reconnected, the vehicle must be driven at least 10 miles so that the processor may relearn the needed correction factors. The distance to be driven depends on the engine and vehicle, but all drives should include steady-throttle cruise on open roads. Certain driveability problems may be noted during the drive because the adaptive factors are not yet functioning.

To prevent the replacement of good components, remember that the EEC-IV system has no control over the following items:

- Fuel quantity and quality
- Damaged or faulty ignition components
- Internal engine condition—rings, valves, timing belt, etc.
- Starter and battery circuit
- Dual Hall sensor
- TFI or DIS module
- Distributor condition or function
- Camshaft sensor
- Crankshaft sensor
- Ignition or DIS coil
- Engine governor module

Any of these systems can cause erratic engine behavior easily mistaken for an EEC-IV problem.

Non-NAAO Diagnostic System

The 2.0L, 2.2L and 2.5L engines are referred to by Ford Motor Company as NON-NAAO, indicating the vehicles and/or their engines originate outside North American Automotive Operations.

Although these vehicles share many similarities in their engine control systems, differences must also be considered. While the fault codes are almost standardized (i.e., Code 14 indicates the barometric pressure sensor), not all engines use the same components so a code may be unique to a particular engine or family. These procedures encompass both turbocharged and non-turbocharged engines.

Beside the engine diagnostic function, these procedures will also display codes related to the 4-speed Electronically-controlled Automatic Transaxle (4EAT) used in these vehicles. Note that the 4EAT codes are displayed by these procedures even though retrieving the engine fault codes may require the North American procedures described at the beginning of this section.

ENGINE CONTROL SYSTEM

These vehicles employ the Electronic Engine Control system, commonly designated EEC, to manage fuel, ignition and emissions on vehicle engines. This system is not EEC-IV, but does share some similarities.

The engine control assembly (ECA) is given responsibility for the operation of the emission control devices, cooling fans, ignition and advance and in some cases, automatic transmission functions. Because the EEC oversees both the ignition timing and the fuel injector operation, a precise air/fuel ratio will be maintained under all operating conditions. The ECA is a microprocessor or small computer which receives electrical in-puts from several sensors, switches and relays on and around the engine.

Based on combinations of these inputs, the ECA controls outputs to various devices concerned with engine operation and emissions. The engine control assembly relies on the signals to form a correct picture of current vehicle operation. If any of the input signals is incorrect, the ECA reacts to what ever picture is painted for it. For example, if the coolant temperature sensor is inaccurate and reads too low, the ECA may see a picture of the engine never warming up. Consequently, the engine settings will be maintained as if the engine were cold. Because so many inputs can affect one output, correct diagnostic procedures are essential on these systems.

One part of the ECA is devoted to monitoring both input and output functions within the system. This ability forms the core of the self-diagnostic system. If a problem is detected within a circuit, the controller will recognize the fault, assign it an identification code, and store the code in a memory section. Most NON-NAAO vehicles use two-digit codes for both engine and 4EAT transaxle faults. The stored code(s) may be retrieved during diagnosis.

➡**When the term Powertrain Control Module (PCM) is used in this manual it will refer to the engine control computer regardless that it may also be called an Electronic Control Assembly (ECA).**

While the EEC system is capable of recognizing many internal faults, certain faults will not be recognized. Because the computer system sees only electrical signals, it cannot sense or react to mechanical or vacuum faults affecting engine operation. Some of these faults may affect another component which will set a code. For example, the ECA monitors the output signal to the fuel injectors, but cannot detect a partially clogged injector. As long as the output driver responds correctly, the computer will read the system as functioning correctly. However, the improper flow of fuel may result in a lean mixture. This would, in turn, be detected by the oxygen sensor and noticed as a constantly lean signal by the ECA. Once the signal falls outside the pre-programmed limits, the engine control assembly would notice the fault and set an identification code.

DASHBOARD WARNING LAMP

The CHECK ENGINE dashboard warning lamp is referred to as the Malfunction Indicator Lamp (MIL). The lamp is connected to the engine control assembly and will alert the driver to certain malfunctions within the EEC system. When the lamp is lit, the ECA has detected a fault and stored an identity code in memory.

The light will stay on as long as the fault causing it is present. Should the fault self-correct, the MIL will extinguish but the stored code will remain in memory.

Under normal operating conditions, the MIL should light briefly when the ignition key is turned ON. As soon as the ECA receives a signal that the engine is running, the lamp will be extinguished. The dash warning lamp should remain out during the entire operating cycle.

Vehicles with a 4EAT transaxle also provide a manual shift light, indicating when the transmission is in manual shift mode.

Non-NAAO Scan Tool Functions

Although stored codes may be read by using an analog voltmeter by counting the needle sweeps, the use of hand-held scan tools such as Ford's second generation SUPER STAR II tester or equivalent is recommended. There are many manufacturers of these tools; the purchaser must be certain that the tool is proper for the intended use.

➡**The engine and 4EAT fault codes on NON-NAAO vehicles may only be read with the SUPER STAR II or its equivalent. The regular STAR tester or voltmeter may be capable not retrieve the stored codes.**

The SUPER STAR II tester is designed to communicate directly with the EEC system and interpret the electrical signals. The scan tool allows any stored faults to be read from the engine controller memory. Use of the scan tool provides additional data during troubleshooting but does not eliminate the use of the charts. The scan tool makes collecting information easier; the data must be correctly interpreted by an operator familiar with the system.

An adapter cable will be required to connect the scan tool to the vehicle; the adapter(s) may differ depending on the vehicle being tested.

ELECTRICAL TOOLS

The most commonly required electrical diagnostic tool is the Digital Multimeter, allowing voltage, resistance and amperage to be read by one instrument. Many of the diagnostic charts require the use of a voltmeter or ohmmeter during diagnosis.

The multimeter must be a high impedance unit, with 10 megohms of impedance in the voltmeter. This type of meter will not place an additional load on the circuit it is testing; this is extremely important in low voltage circuits. The multimeter must be of high quality in all respects. It should be handled carefully and protected from impact or damage. Replace the batteries frequently in the unit.

Additionally, an analog (needle type) voltmeter may be used to read stored fault codes if the SUPER STAR II tester is not available. The codes are transmitted as visible needle sweeps on the face of the instrument.

Almost all diagnostic procedures will require the use of the Breakout Box, a device which connects into the EEC harness and provides testing ports for the 60 wires in the harness. Direct testing of the harness connectors at the terminals or by backprobing is not recommended; damage to the wiring and terminals is almost certain to occur.

Other necessary tools include a quality tachometer with inductive (clip-on) pickup, a fuel pressure gauge with system adapters and a vacuum gauge with an auxiliary source of vacuum.

Non-NAAO Self-Diagnostics

Diagnosis of a driveability problem requires attention to detail and following the diagnostic procedures in the correct order. Resist the temptation to begin extensive testing before completing the preliminary diagnostic steps. The preliminary or visual inspection must be completed in detail before diagnosis begins. In many cases this will shorten diagnostic time and often cure the problem without electronic testing.

Keep in mind that all the things that previously went wrong with vehicles, before the age of electronics, can still go wrong and are still the cause of the majority of the driveability problems. The best diagnosis starts with a list of symptoms and possible causes, followed by careful checking of those causes in the most likely order. Eliminate all the possible mechanical causes before considering electrical faults.

VISUAL INSPECTION

This is possibly the most critical step of diagnosis. A detailed examination of all connectors, wiring and vacuum hoses can often lead to a repair without further diagnosis. Performance of this step relies on the skill of the technician performing it; a careful inspector will check the undersides of hoses as well as the integrity of hard-to-reach hoses blocked by the air cleaner or other components. Wiring should be checked carefully for any sign of strain , burning, crimping or terminal pull-out from a connector.

Checking connectors at components or in harnesses is required; usually, pushing them together will reveal a loose fit. Pay particular attention to ground circuits, making sure they are not loose or corroded. Remember to inspect connectors and hose fittings at components not mounted on the engine, such as the evaporative canister or relays mounted on the fender aprons. Any component or wiring in the vicinity of a fluid leak or spillage should be given extra attention during inspection.

Additionally, inspect maintenance items such as belt condition and tension, battery charge and condition and the radiator cap carefully. Any of these very simple items may affect the system enough to set a fault.

Non-NAAO READING Codes

The EEC system may be interrogated for stored codes using the Quick Test procedures. If a code was set before a problem self-corrected (such as a momentarily loose connector), the code will remain in memory until cleared.

The Quick Test procedure is divided into 3 sections, Key On Engine Off (KOEO), Key On Engine Running (KOER) and the Switch Monitor test. These 3 procedures must be performed correctly if the system is to run the internal self-checks and provide accurate fault codes. Codes will be output and displayed as numbers on the hand scan tool, i.e. 23. If the codes are being read by an analog voltmeter, the codes will be displayed as groups of needle sweeps separated by pauses.

Code 23 would be shown as two sweeps, a pause and three more sweeps. A longer pause will occur between codes. Unlike the EEC-IV system, the EEC system does not broadcast a PASS designator or code. If no fault codes are stored, the display screen of the hand

scanner will remain blank. Additionally, the EEC system does not operate switches or sensors during KOEO or KOER testing.

Once the Quick Test has been performed and all fault codes recorded, refer to the service code charts. The charts direct the use of specific pinpoint tests for the appropriate circuit and will allow complete circuit testing.

The EEC diagnostic connector is located at the left rear corner of the engine compartment on most vehicles. When connecting the test equipment and adapters, note that the Self-Test Input (STI) connector is separate from the main diagnostic connector on all NON-NAAO engines except for the 1.8L engine. The Self-Test Output (STO) connector is contained within the main diagnostic connector.

✳✳ CAUTION

To prevent injury and/or property damage, always block the drive wheels, firmly apply the parking brake, place the transmission in Park or Neutral and turn all electrical loads off before performing the Quick Test procedures.

READING CODES WITH ANALOG VOLTMETER

In the absence of a scan tool, an analog voltmeter may be used to retrieve stored fault codes. Set the meter range to read DC 0–20 volts. Connect the + lead of the meter to the STO pin in the diagnostic connector and connect the—lead of the meter to the negative battery terminal or a good engine ground.

Follow the directions given for performing the KOEO and KOER tests. To activate the tests, use a jumper wire to connect the STI connector to ground. The codes will be transmitted as groups of needle sweeps.

Key On Engine Off (KOEO) Test:

1. Make certain the scan tool is OFF; connect it to the self-test connectors. Switch the scan tool to the MECS position. Except on 1.8L engines, make certain the adapter ground cable is connected to the negative battery terminal. On the 1.8L engine, make certain the switch on the adapter is set to EEC or ECA if engine codes are to be retrieved. The other switch position will retrieve codes from the 4EAT.

2. Make certain the scan tool test button is ON or latched down.

3. For all engine or 4EAT codes except 1.8L and 1 .9L engines, turn the ignition switch **ON** but do not start the engine, then turn the scan tool ON. On 1.8L and 1 .9L engines, turn the scan tool ON first, then turn the ignition switch ON.

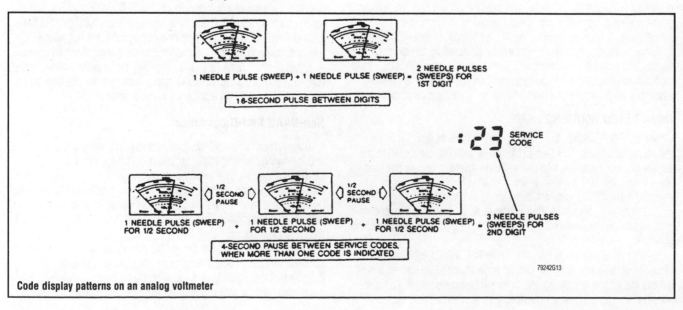

1 NEEDLE PULSE (SWEEP) + 1 NEEDLE PULSE (SWEEP) = 2 NEEDLE PULSES (SWEEPS) FOR 1ST DIGIT

1 6-SECOND PULSE BETWEEN DIGITS

:23 SERVICE CODE

1 NEEDLE PULSE (SWEEP) FOR 1/2 SECOND + 1/2 SECOND PAUSE + 1 NEEDLE PULSE (SWEEP) FOR 1/2 SECOND + 1/2 SECOND PAUSE + 1 NEEDLE PULSE (SWEEP) FOR 1/2 SECOND = 3 NEEDLE PULSES (SWEEPS) FOR 2ND DIGIT

4-SECOND PAUSE BETWEEN SERVICE CODES, WHEN MORE THAN ONE CODE IS INDICATED

79242G13

Code display patterns on an analog voltmeter

4. Once energized, the tester should display 888 and beep for 2 seconds. Release the test button; 00 should appear, signifying the tool is ready to read codes.

5. Re-engage the test button.

6. The KOEO codes will be transmitted.

7. Record all service codes displayed.

8. After all codes are received, release the test button to review all the codes retained in tester memory.

9. Make sure all codes displayed are recorded. Clear the ECA memory and perform the KOEO test again. This will isolate hard faults from intermittent ones. Any hard faults will cause the code(s) to be repeated in the 2nd test. An intermittent which is not now present will not set a new code.

10. Record all codes from the 2nd test. After repairs are made on hard fault items, the intermittent ones must be recreated by tapping suspect sensors, wiggling wires or connectors or reproducing circumstances on a test drive.

➡**For both KOEO and KOER tests, the message STO LO always displayed on the screen indicates that the system** cannot initiate the Self-Test. The message STI LO displayed with an otherwise blank screen indicates Pass or No Codes Stored.

Key On Engine Running (KOER) Test:

1. Make certain the self-test button is released or de-activated on the SUPER STAR II tester and that the tester is properly connected.

2. Start the engine and run it at 2000 rpm for 2 minutes. This action warms up the oxygen sensor.

3. Turn the ignition switch OFF.

4. Turn the ignition switch **ON** for 10 seconds but do not start the engine.

5. Start the engine and run it at idle.

6. Activate or latch the self-test button on the scan tool.

7. All relevant codes will be displayed and should be recorded.

Switch Monitor Tests:

This test mode allows the operator to check the input signal from individual switches to the ECA. All switches to be tested must be OFF at the time the test begins; if one switch is on, it will affect the

Switch	1.3L	1.8L	2.2L	2.2L Turbo	SUPER STAR II Tester LED or Analog VOM Indications
Clutch engage Switch/ Neutral Gear Switch (CES/NGS) (MTX only)	X	X	X	X	LED on or 12V in gear and clutch pedal released
Manual Lever Position Switch (MLP) (ATX Only)	X	X	X		LED on or 12V in P or N
Idle Switch (IDL)	X	X	X	X	LED on or 12V with accelerator pedal depressed
Brake On-Off Switch (BOO)	X	X MTX	X	X	LED on or 12V with brake pedal depressed
Headlamps Switch (HLDT)	X	X	X	X	LED on or 12V with headlamp switch on
Blower Motor Switch (BLMT)	X	X	X	X	LED on or 12V with blower switch at 2nd or above position
A/C Switch (ACS)	X	X			LED on or 12V with A/C switch on and blower on
Defrost Switch (DEF)	X	X	X	X	LED on or 12V with defrost switch on
Coolant Temperature Switch (CTS)	X	X		X	LED on or 12V with cooling fan on
Wide Open Throttle Switch (WOT)	X	X MTX			LED off or 0V with accelerator pedal fully depressed

79242G14

Switch tests for 1990 and early Ford Non-NAAO vehicles

Switch	1.3L	1.6L	1.8L	2.2L	2.2L Turbo	SUPER STAR II Tester LED or Analog VOM Indications
Clutch engage Switch/ Neutral Gear Switch (CES/NGS) (MTX only)	X	X	X	X	X	LED on or less than 1.5V in gear and clutch pedal released
Manual Lever Position Switch (MLP) (ATX Only)	X	X	X	X	X	LED on or less than 1.5V in P or N
Idle Switch (IDL)	X	X	X	X	X	LED on or less than 1.5V with accelerator pedal depressed
Brake On-Off Switch (BOO)	X	X	X MTX	X	X	LED on or less than 1.5V with brake pedal depressed (not fully)
Headlamps Switch (HLDT)	X	X	X	X	X	LED on or less than 1.5V with headlamp switch on
Blower Motor Switch (BLMT)	X	X	X	X	X	LED on or less than 1.5V with blower switch at 2nd or above position
A/C Switch (ACS)	X	X	X	X	X	LED on or less than 1.5V with A/C switch on and blower on
Defrost Switch (DEF)	X	X	X	X	X	LED on or less than 1.5V with defrost switch on
Coolant Temperature Switch (CTS)	X	X	X	X	X	LED on or less than 1.5V with cooling fan on
Wide Open Throttle Switch (WOT)	X		X			LED off or 0V with accelerator pedal fully depressed
Knock Control (KC)					X	LED on or less than 1.5V while tapping on engine

79242G15

Switch tests for 1991 and newer Ford Non-NAAO vehicles

testing of another. The test must begin with the engine cool. The tests may be performed with either the SUPER STAR II tester or an analog voltmeter. When using the scan tool, the small LED on the adapter cable will light to show that the ECA has received the switch signal. If the voltmeter is used, the voltage will change when the switch is engaged or disengaged.

1. The engine must be off and cooled. Place the transmission in Park or Neutral.

2. Turn all accessories OFF.

3. If using the SUPER STAR II, connect it properly. If using an analog voltmeter, use a jumper to ground the STI terminal. Connect the positive (+) voltmeter lead to the SML terminal of the diagnostic connector and connect the negative (-) lead to a good engine ground.

4. Turn the ignition switch **ON.** Engage the center button on the SUPER STAR II. Most switches can be exercised without starting the engine.

5. Operate each switch according to the test chart and note the response either on the LED or the volt scale. Remember that an

improper response means the ECA did not see the switch operation; check circuitry and connectors before assuming the switch is faulty.

6. Turn the ignition switch **OFF** when testing is complete.

CLEARING CODES

Codes stored within the memory must be erased when repairs are completed. Additionally, erasing codes during diagnosis can separate hard faults from intermittent ones.

To erase stored codes, disconnect the negative battery cable, then depress the brake pedal for at least 10 seconds. Reconnect the battery cable and recheck the system for any remaining or newly-set codes.

MCU Carbureted Diagnostic System

The Microprocessor Control Unit (MCU) system was used on most 1981–83 carburetor equipped vehicles, and 1984 and newer V8 engines with feedback carburetors. The MCU system uses a large six sided connector, identical to the one used with EEC-IV systems. The MCU system does NOT use the small single wire connector, like the EEC-IV system.

This system has limited ability to diagnose a malfunction within itself. Through the use of trouble codes, the system will indicate where to test. When an analog voltmeter or special tester is connected to the diagnostic link connector and the system is triggered, the self-test simulates a variety of engine operating conditions and evaluates all the responses received from the various MCU components, so any abnormal operating conditions can be detected.

MCU Carbureted Self-Diagnostics

Diagnosis of a driveability problem requires attention to detail and following the diagnostic procedures in the correct order. Resist the temptation to begin extensive testing before completing the preliminary diagnostic steps. The preliminary or visual inspection must be completed in detail before diagnosis begins. In many cases this will shorten diagnostic time and often cure the problem without electronic testing.

VISUAL INSPECTION

This is possibly the most critical step of diagnosis. A detailed examination of all connectors, wiring and vacuum hoses can often lead to a repair without further diagnosis. Performance of this step relies on the skill of the technician performing it; a careful inspector will check the undersides of hoses as well as the integrity of hard-to-reach hoses blocked by the air cleaner or other components. Wiring should be checked carefully for any sign of strain , burning, crimping or terminal pull-out from a connector.

Checking connectors at components or in harnesses is required; usually, pushing them together will reveal a loose fit. Pay particular attention to ground circuits, making sure they are not loose or corroded. Remember to inspect connectors and hose fittings at components not mounted on the engine, such as the evaporative canister or relays mounted on the fender aprons. Any component or wiring in the vicinity of a fluid leak or spillage should be given extra attention during inspection.

Additionally, inspect maintenance items such as belt condition and tension, battery charge and condition and the radiator cap carefully. Any of these very simple items may affect the system enough to set a fault.

MCU Carbureted Reading Codes

PREPARATION FOR READING CODES

1. Turn OFF all electrical equipment and accessories in vehicle.
2. Follow all safety precautions during testing.
3. Make sure all fluids are at proper levels.
4. Perform 'Visual Inspection' as detailed in EEC-IV system testing, earlier in this section.
5. Start the engine and let it idle, until the engine reaches normal operating temperature. This is when the upper radiator hose is Hot and engine RPM has dropped to its normal warm idle speed.
6. Turn ignition switch OFF.

✳✳ CAUTION

Always operate the vehicle in a well ventilated area. Exhaust gases are very poisonous.

Inline 4- and 6-Cylinder Engines:

On Inline 4- and 6-cylinder engines with canister control valves, remove the hose that goes to the carbon canister (this simulates a clean carbon canister). Do NOT plug this hose for the remainder of

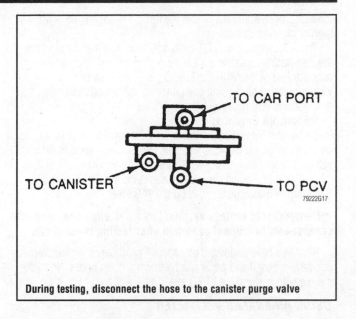

During testing, disconnect the hose to the canister purge valve

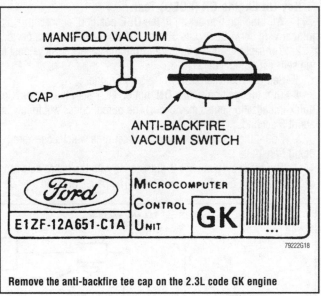

Remove the anti-backfire tee cap on the 2.3L code GK engine

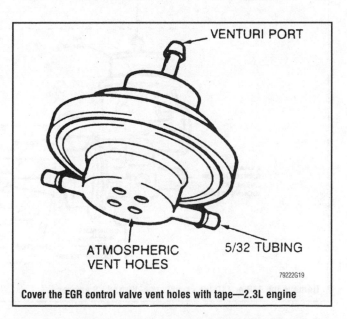

Cover the EGR control valve vent holes with tape—2.3L engine

the test procedure. Make certain the throttle linkage is off of the high choke cam setting.

The 2.3L engines with GK code, you must remove the cap from the anti-backfire vacuum switch tee during testing. The switch is near the rear of the MCU module. On 2.3L engines with an EGR vacuum load control (wide open throttle) valve, you must cover the atmospheric vent holes with a piece of tape.

V6 and V8 Engines:

On V6 and V8 engines, remove the PCV valve from breather cap on valve cover. On the 4.2L and 5.8L engines with vacuum delay valves, uncap the restrictor on the thermactor vacuum control line. On the V6 4.2L engine the vacuum cap is on the TAD line on the 5.8L engine the vacuum cap is on the TAB line.

➡ **Remember to replace vacuum lines, tee caps and return all components to original condition after testing is complete.**

After you have performed any special procedures for your vehicle, have a pencil and paper nearby to write down codes. Now you are ready to perform the KOEO test.

USING AN ANALOG VOLTMETER

Key On Engine Off (KOEO) Test:

1. With the ignition switch in the **OFF** position, connect a jumper wire between circuits 60 and 201 on the self-test connector.
2. Connect the analog voltmeter from the battery positive post to the self-test output connector.
3. Self the voltmeter scale to 0-15 volt range.
4. Turn the ignition switch **ON**, but do NOT start the engine. One quick initialization pulse may occur. The output codes will follow in about 5 seconds.
5. Count the voltmeter sweeps, to determine which codes are being transmitted.
6. The MCU system uses 2-digit codes with the pause between each digit being about 2 seconds long. The pause between the two

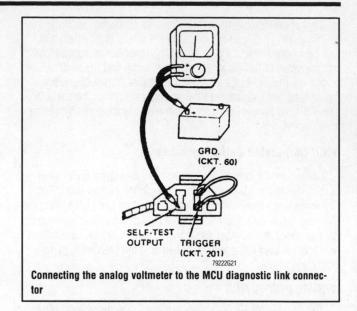

Connecting the analog voltmeter to the MCU diagnostic link connector

different codes is about 4 seconds long. The code group is sent twice. This allows you to check the accuracy of the codes as you record them.

7. Once this test has been performed and all fault codes recorded, you can refer to the 'Code Descriptions' in this manual for the meaning of the fault code(s). For more detailed code retrieval information and to repair faults refer to your specific vehicle service manual.

Key On Engine Running (KOER) Test:

1. Turn OFF all electrical equipment and accessories in vehicle.
2. Follow all safety precautions during testing.
3. Make sure all fluids are at the proper levels.
4. Perform 'Visual Inspection' as detailed earlier in this manual.

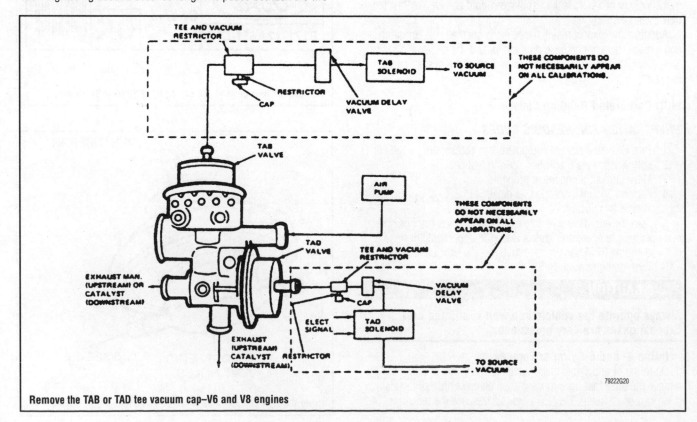

Remove the TAB or TAD tee vacuum cap–V6 and V8 engines

The following steps involve servicing the engine with the engine running. Observe safety precautions.

- Apply the parking brake.
- Put the shift lever in P (automatic transmission) or NEUTRAL (manual transmission).
- Block the drive wheels.
- Always operate vehicle in well ventilated area. Exhaust gases are very poisonous.
- Stay clear of hot and moving engine parts.

5. The engine should be at normal operating temperature for this test. If not, start the engine and let it idle, until the engine reaches normal operating temperature. This is when the upper radiator hose is Hot and engine RPM has dropped to its normal warm idle speed and repeat Key ON Engine OFF test again.

6. If engine is warm, after codes have been retrieved, Start the engine.

Always operate the vehicle in a well ventilated area. Exhaust gases are very poisonous.

7. To extract the fault codes:

Inline 4- and 6-cylinder engines: Start the engine and raise the idle to 3000 RPM within 20 seconds of starting vehicle. Hold at 3000 RPM until codes are sent. When codes are sent release throttle and let engine return to idle speed.

V6 and V8 engines: Start the engine are raise to 2000 RPM for 2 minutes and turn OFF ignition. Immediately restart engine and allow to idle. Some engines equipped with throttle kicker will increase idle during the testing, this is normal.

8. If your vehicle is equipped with a knock sensor perform the following test, if not skip to Step 10. Simulate a spark knock by placing a 4 inch socket extension (or similar tool) on the manifold near the base of knock sensor. Tap on the end of extension lightly with a 2-6 oz. hammer for approximately 15 seconds. Do NOT hit on the knock sensor itself. Count the voltmeter sweeps to determine which codes are being sent.

9. The first series of sweeps should be the engine ID code, ignore any sweeps that last any longer than 1 second. The engine ID code will be ½ the number of cylinders. For example, a 4 cylinder would appear as 2 sweeps and a 6 cylinder as 3 sweeps and an 8 cylinder as 4 sweeps.

10. If no sweeps occur repeat KOER test procedures, starting with Step 1. If the meter still does not sweep, you have a problem which must be repaired before proceeding. Refer to your specific vehicle service manual.

11. Count the sweeps on the meter to find out which codes are being sent. All codes are 2-digits long and will appear the same way as in KOEO Self-Test. Ignore any sweeps lasting more than 1 second. Write codes down on a piece of paper, codes will be sent twice so you can check your list for accuracy. Write codes down in the order they appear. Turn the ignition switch OFF when codes are finished and remove jumper wire.

Diagnostic Trouble Codes (DTC's)

MCU SYSTEM

The code definitions listed are general for Ford Vehicles using the Microprocessor Control Unit (MCU) engine control system.

Most Ford vehicles up to 1983 and feedback carburetor equipped V8 engines into the 1990's use the MCU engine control system. For a specific code definition or component test procedure consult service manual for your vehicle. A diagnostic code does not mean the component is defective. For example a Code 44 is an oxygen sensor code (rich oxygen sensor signal). This code may set if a carburetor is flooding or has a very restricted air cleaner. Replacing the oxygen sensor would not fix the problem.

➡**When the term Powertrain Control Module (PCM) is used in this manual it will refer to the engine control computer regardless that it may be a Powertrain Control Module (PCM) or Electronic Control Module (ECM) or Electronic Control Assembly (ECA).**

Code 11 System Pass—Except High Altitude—or Altitude (ALT) circuit is open—High Altitude

Code 12 RPM out of specification (throttle kicker system)

Code 25 Knock Sensor (KS) signal is not detected during Key On Engine Running (KOER) Self-Test

Code 33 Key On Engine Running (KOER) Self-Test not initiated

Code 41 Oxygen sensor voltage signal always Lean (low value)—does not switch

Code 42 Oxygen sensor voltage signal always Rich (high value)—does not switch

Code 44 Oxygen sensor signal indicates Rich—excessive fuel, restricted air intake—or Inoperative Thermactor System

Code 45 Thermactor Air flow is always upstream (going into exhaust manifold)

Code 46 Thermactor Air System unable to bypass air (vent to atmosphere)

Code 51 Low or Mid Temperature vacuum switch circuit is open when engine is hot on Inline 4 and 6 cylinder engines—or HI or HI/LOW vacuum switch circuit is always open on V6 or V8 engines

Code 52 Idle Tracking Switch (ITS) voltage does not change from closed to open throttle (Closed throttle checked during KOEO condition. Open throttle checked during KOER conditions) on 4 cylinder car—or Idle/Decel Vacuum switch circuit always open -on 4 cylinder truck—or Wide Open Throttle vacuum switch circuit always open—on Inline 6 cylinder engine

Code 53 Wide Open Throttle vacuum switch circuit always open on 4 cylinder engine—or Crowd vacuum switch circuit is always open—on Inline 6 cylinder engine—or Dual temperature switch circuit is always open -on V6 and V8 engines

Code 54 Mid temperature switch circuit is always open

Code 55 Road load vacuum switch circuit is always open—on 4 cylinder engine—or Mid vacuum switch circuit is always open—on V6 and V8 engines

Code 56 Closed throttle vacuum switch circuit is always open

Code 61 Hi/Low Vacuum switch circuit is always closed

Code 62 Idle Tracking Switch (ITS) circuit is closed at idle—or Idle/Decel vacuum switch circuit is always closed—on 4 cylinder car—or Wide Open Throttle vacuum switch circuit always closed -on 4 cylinder truck—or System Pass—High Altitude; Altitude (ALT) circuit is open except High Altitude on V6 and V8 engines

Code 63 Wide Open Throttle (WOT) vacuum switch circuit is always closed—on 4 cylinder engine—or Crowd vacuum switch circuit is always closed -on 6 cylinder engine

Code 65 System pass—on 4 cylinder engine (High Altitude)—or Altitude (ALT) circuit is open—4 cylinder engine (except High Altitude)—or Mid vacuum circuit is always closed—V6 and V8 engines

Code 66 Closed Throttle Vacuum switch circuit is always closed

➡This list is for reference and does not mean a specific component is defective. NOTE: High Altitude refers to vehicles with computer adjusted for operation at high elevations as in mountain regions.

EEC-IV SYSTEM

The code definitions listed general 2-digit codes for Ford Vehicles using the Ford EEC-IV engine control system. In 1991 Ford started introducing vehicles that use 3-digit codes. The code definitions for both the 2 and 3-digit codes are found in this section. For a specific code definition or component test procedure consult your 'Chilton Total Car Care' manual for your vehicle. A diagnostic code does not mean the component is defective. For example a Code 29 is a vehicle speed sensor code. This does not mean the sensor is defective, but to check the sensor and related components. A defective speedometer cable or transmission problem will also set this code.

➡'When the term Powertrain Control Module (PCM) is used in this manual it will refer to the engine control computer regardless that it may be a PCM or Electronic Control Module (ECM) or Electronic Control Assembly (ECA).

2-Digit DTC's—1981–94 Passenger Cars and 1984–94 Light Trucks:

Code 11 System Pass

Code 12 (R) Idle control fault—RPM Unable To Reach Upper Limit Self-Test

Code 13 (C) DC Motor Did Follow Dashpot

Code 13 (O) DC Motor Did Not Move

Code 13 (R) Idle control fault—Cannot control RPM during Self-Test low RPM check

Code 14(C)—Engine RPM signal fault—Profile Ignition Pickup (PIP) circuit failure or RPM sensor.

Code 15 (C) EEC Processor, power to Keep Alive Memory (KAM) interrupted or test failed

Code 15 (O) Power Interrupted To Processor or EEC Processor ROM Test failure

Code 16 (O,R) RPM too low to perform Exhaust Gas Oxygen (EGO) sensor test or fuel control error.

Code 1 (O)7 CFI Fuel Control System fault—Rich/Lean condition indicated; 3.8L V-6/5.0LV-8 (1984).

Code 17 (R) RPM Below Self-Test Limit, Set Too Low. Code 18 (C)—Ignition diagnostic monitor (1DM) circuit failure, loss of RPM signal or SPOUT circuit grounded

Code 18 (O)—Ignition Diagnostic Monitor (1DM) circuit

Code 18 (R) SPOUT or SAW circuit open

Code 19 (C) Cylinder Identification (CID) Sensor Input failure

Code 19 (O) Failure in EEC Processor internal voltage. Code 19 (R)—Erratic RPM During EGR Test or RPM Too Low During ISC Off Test

Code 21 Engine Coolant Temperature (ECT) out of Self-Test range

Code 22 (O, R) Manifold Absolute Pressure (MAP)/Barometric Pressure (BP/BARO) Sensor circuit out of Self-Test range

Code 23 Throttle Position (TP) Sensor out of Self-Test range

Code 24 (O, R) Air Charge (ACT) or Intake Air (IAT) Temperature out of Self-Test range

Code 25 (R) Knock not sensed during dynamic response test

Code 26 (O, R) Transmission Fluid Temp (TFT) out of Self-Test range

Code 26 (O, R) Vane Air (VAF) or Mass Air (MAF) sensor out of self-test range

Code 28 (C) Loss Of Primary Tach, Right Side.

Code 29 (C) Insufficient input from Vehicle Speed Sensor (VSS) or Programmable Speedometer/Odometer Module (PSOM)

Code 31 EGR valve position sensor circuit below minimum voltage

Code 32 EGR Valve Position (EVP) sensor circuit voltage below closed limit

Code 33 (C) Throttle Position (TP) sensor noisy/harsh on line

Code 33 (R, C) EGR valve position sensor circuit, EGR valve opening not detected

Code 34 EGR valve circuit out of self-test range or valve not closing

Code 35 EGR valve circuit above maximum voltage -except 2.3L HSC with Feedback Carburetor System—or—Throttle Kicker on 2.3L HSC with Feedback Carburetor System.

Code 38 (C) Idle Track Switch Circuit Open.

Code 39 (C) AXOD Torque Converter or Bypass Clutch Not Applying Properly

Code 41 (R,C) Oxygen Sensor circuit indicates system always lean

Code 42 (R,C) Oxygen Sensor circuit indicates system always rich, right side if 2 sensors used

Code 43 (C) Oxygen Sensor Out Of Test Range—on 1992 and earlier vehicles—or—Throttle Position Sensor failure—on 1993 and newer vehicles

Code 43 (R) Exhaust Gas Oxygen (EGO) sensor cool down has occurred during testing—2.3L HSC and 2.8L FBC truck

Code 44 (R) Air injection control system failure (right side cylinders, if a split system)

Code 45 (C) Coil 1 primary circuit failure

Code 45 (R) Air injection control system air flow misdirected

Code 46 (C) Coil Primary Circuit failure

Code 46 (R) Thermactor air not bypassed during Self-Test

Code 47 (C) 4x4 switch is closed—on Truck.

Code 47 (R) Airflow low at idle—on fuel injected engines—or—4 x 4 switch is closed—on Truck—or—Fuel control system/Exhaust Gas Oxygen (EGO) Sensor fault—on 2.3L HSC and 2.8L FBC truck

Code 48 (C) Coil Primary Circuit failure; Except 2.3L Truck—or—Loss Of Secondary Tach, Left Side—with 2.3L Truck engine

Code 48 Airflow high at base idle

Code 49 (C) El electronic Transmission Shift Error—on Truck and 1992 and later cars—or—SPOUT Signal Defaulted To 10 Degrees BTDC or SPOUT Open—Up to 1991 passenger cars

Code 51 (O, C) Engine Coolant Temperature (ECT) circuit open or out of range during self-test

Code 52 (O) Power Steering Pressure Switch (PSPS) circuit open

Code 52 (R) Power Steering Pressure Switch (PSPS) circuit did not change states

Code 53 (O, C) Throttle Position (TP) circuit above maximum voltage

Code 54 (O, C) Air Charge (ACT) or Intake Air (IAT) Temperature circuit open

Code 55 (R) Key Power Input To Processor—open circuit

Code 56 (O, C) Mass Air (MAF) or Vane Air (VAF) Flow circuit above maximum voltage—Port fuel injected engines—or—Transmission oil temperature (TOT) circuit open—on vehicles with automatic transaxle

Code 57 (C) AXOD Circuit failure—on vehicles with automatic overdrive transaxle—or—Octane Adjust Circuit failure—on some 1992 and newer cars

Code 58 (R) Idle Tracking Switch circuit fault.

Code 59 (C) Automatic Transmission Shift Error—on 1991 and newer—or—AXOD 4/3 or Neutral Pressure Switch Failed Open—on 3.0L EFI and 3.8L AXOD—vehicles with automatic overdrive transaxle

Code 59 (O) AXOD 4/3 Pressure Switch Failed Closed -on 3.8L engine AXOD—vehicles with automatic transaxle—or—Idle Adjust Service Pin In Use—on 2.9L EFI engine—or—Low Speed Fuel Pump Circuit failure—on 3.0L SHO engine

Code 61 (O, C) Engine Coolant Temperature (ECT) circuit grounded

Code 62 (C) Converter clutch error

Code 62 (O) Electronic Transmission Shift Error.

Code 63 (O, C) Throttle Position (TP) circuit below minimum voltage

Code 64 (O, C) Air Charge (ACT) or Intake Air (IAT) Temperature circuit grounded

Code 65 (C) Fuel System Failed To Enter Closed Loop Mode or key power

Code 65 (O) Key Power Check—Possible Charging System overvoltage condition

Code 65 (R) Overdrive Cancel Switch (OCS) circuit did not switch

Code 66 (C) Mass Air (MAF) or Vane Air (VAF) Flow circuit below minimum voltage—engine with Port fuel injection—or—Transmission Oil Temperature (TOT) circuit grounded—vehicles with automatic transaxle

Code 67 (O, C) Manual Lever Position (MLP) sensor out of range and A/C ON

Code 67 (O, C) Neutral/Drive Switch (NDS) circuit open/A/C on during Self-Test

Code 67 (O, R) Neutral Drive Circuit Failed or A/C Input High—or—Clutch Switch Circuit failed—on vehicles with manual transaxle—or—Manual Lever Position Sensor out of range—on vehicles with automatic transaxle

Code 68 (C) Transmission Fluid Temp (TFT) transmission over temp (over heated)

Code 68 (O) Idle Tracking Switch circuit—on 2.8L FBC truck only—or—Air temperature sensor—except FBC truck.

Code 68 (R, C) Air Temperature Sensor Circuit failure -on 1.9L EFI engine—or—Idle Tracking Switch Circuit failure—on CFI engine—or—Transmission Temperature Circuit

Code 69 (O, C) Transmission Shift Error

Code 70 (C) Data Communications Link Circuit failure

Code 71(C) Software Re-Initialization Detected—on 1 .9L EFI and 2.3L Turbo—or—Idle Tracking Switch failure—on CFI engine—or—Message Center Control Circuit failure—on vehicles with Message Center Control Center—or—Power Interrupt Detected—except vehicles with 3.8L AXOD (automatic overdrive transaxle)

Code 72 (R) Insufficient Manifold Absolute Pressure (MAP) change during Dynamic Response Test

Code 73 (R) Insufficient Throttle Position (TP) change during Dynamic Response Test

Code 74 (R, C) Brake On/Off (BOO) circuit open/not actuated during Self-Test

Code 75 (R) Brake On/Off (BOO) circuit closed/EEC processor input open

Code 76 (R) Insufficient Airflow Output Change During Test

Code 77 (R) Brief Wide Open Throttle (WOT) not sensed during Self-Test/operator error (Dynamic Response/Cylinder Balance Tests)

Code 78 (C) Power Interrupt Detected

Code 79 (O) A/C on/Defrost on during Self-Test

Code 81(C) MAP Sensor Has Not Changing Normally

Code 81(O) Air Management Circuit failure

Code 82 (O) Supercharger Bypass Circuit failure, 3.8L SC engine—or—Air Management Circuit failure, Except 3.8L SC engine—or—EGR Solenoid Circuit failure, 2.3L OHC engine

Code 83 OIC—Low speed fuel pump relay circuit failure

Code 83 (O) High Speed Electro Drive Fan Circuit failure, Except 2.3L OHC and 3.0L SHO engine—or—Low Speed Fuel Pump Relay Circuit failure, 3.0L SHO engine

Code 84 (O) EGR Vacuum Regulator (EVR) circuit failure

Code 84 (R) EGR Solenoid Circuit failure

Code 85 (C) Adaptive Lean Limit Reached

Code 85 (O) Canister Purge (CANP) circuit failure

Code 86 (C) Adaptive Rich Limit Reached

Code 86 (O) Shift Solenoid (SS) circuit failure—or—Wide Open Throttle (WOT) A/C Cutoff Solenoid circuit—on Carbureted engine

Code 87 Fuel Pump circuit fault

Code 88 (C) Loss Of Dual Plug Input control

Code 88 (O) Electro Drive Fan Circuit failure—fuel injected engine—or—Throttle Kicker, feedback carburetor system

Code 89 (O) Transmission solenoid circuit failure.

Code 89 (O) Clutch Converter Override (CCO) circuit failure—or—Exhaust Heat Control (EHC) Solenoid circuit—3.8L CFI engine

Code 91(C) No Heated Exhaust Gas Oxygen (HEGO) sensor switching detected—left HEGO

Code 91(O) Shift Solenoid 1 (SS1) circuit failure.

Code 91(R) Heated Exhaust Gas Oxygen (H EGO) sensor circuit indicates system lean—left HEGO

Code 92 (O) Shift Solenoid Circuit failure

Code 92 (R) Oxygen Sensor Circuit failure

Code 93 (O) Throttle Position Sensor (TPS) input low at maximum DC motor extension—OR—Shift solenoid circuit failure

Code 93 (O) Coast Clutch Solenoid (CCS) circuit failure

Code 94 (O) Torque Converter Clutch (TCC) solenoid circuit failure

Code 94 (O) Converter Clutch Control (CCC) Solenoid circuit failure

Code 94 (R) Thermactor Air System inoperative, left side

Code 95 (O, C) Fuel Pump secondary circuit failure/Fuel Pump circuit open—EEC processor to motor ground

Code 96 (O, C) Fuel Pump secondary circuit failure/Fuel Pump circuit open—battery to EEC processor

Code 97 (O) Overdrive Cancel Indicator Light (OCIL) circuit failure

Code 98 (R) Electronic control assembly failure

Code 98 (O) Electronic Pressure Control (EPC) Driver open in EEC processor

Code 98 (R) Hard fault is present—FMEM mode

Code 99 (0,C) Electronic Pressure Control (EPC) circuit failure

Code 92 (O) Shift Solenoid 2 (SS2) circuit failure

Code 92 (R) Heated Exhaust Gas Oxygen (HEGO) sensor circuit indicates system rich—left HEGO

Code 93 (O) Throttle Position Sensor Input Low At Max DC Motor Extension, CFI engine—or—Shift Solenoid Circuit failure—Except CFI engine

Code 94 (O) Converter Clutch Solenoid Circuit failure

Code 94 (R) Thermactor Air System Inoperative

Code 95 (O, C) Fuel Pump Circuit failure, ECA To ground

Code 96 (O, C) Fuel Pump Circuit failure

Code 97 (O) Transmission Indicator Circuit failure

Code 98 (O) Electronic Pressure Control Circuit failure

Code 98 (R) Electronic Control Assembly failure

Code 99 (O, C) Electronic Pressure Control Circuit or Transmission Shift failure

Code 99 (R) EEC System Has Not Learned To Control Idle: Ignore Codes 12 & 13

No Code–Unable to Run Self Test or Output Codes, or list does not apply to vehicle tested, refer to service manual.

➡ **This list is to be used as a reference for testing and does not mean a specific component Is defective.**

(O)—Key On, Engine Off
(R)—Engine running
(C)—Continuous Memory

3-Digit DTC's—1991–95 Vehicles:

Code 111 System pass

Code 112 Intake Air Temperature (IAT) Sensor circuit below minimum voltage

Code 113 Intake Air Temperature (IAT) Sensor circuit above maximum voltage

Code 114 Intake Air Temperature (IAT) higher or lower than expected

Code 116 Engine Coolant Temperature (ECT) higher or lower than expected

Code 117 Engine Coolant Temperature (ECT) Sensor circuit below minimum voltage

Code 118 Engine Coolant Temperature (ECT) Sensor circuit above maximum voltage

Code 121 Closed throttle voltage higher or lower than expected

Code 121 Indicates Throttle Position voltage inconsistent with Mass Air Flow (MAF) Sensor

Code 122 Throttle Position (TP) Sensor circuit below minimum voltage

Code 123 Throttle Position (TP) Sensor circuit above maximum voltage

Code 124 Throttle Position (TP) Sensor circuit voltage higher than expected

Code 125 Throttle Position (TP) Sensor circuit voltage lower than expected

Code 126 Manifold Absolute Pressure/Barometric Pressure (MAP/BARO) Sensor higher or lower than expected

Code 128 Manifold Absolute Pressure (MAP) Sensor vacuum hose damaged/disconnected

Code 129 Insufficient Manifold Absolute Pressure (MAP)/Mass Air Flow (MAF) change during Dynamic Response Test-KOER

Code 136 Lack of Heated Oxygen Sensor (HO2S-2) switches during KOER, indicates lean—Bank # 2

Code 137 Lack of Heated Oxygen Sensor (HO2S-2) switches during KOER, indicates rich—Bank # 2

Code 138 Cold Start Injector (CSI) flow insufficient—KOER

Code 139 No Heated Oxygen Sensor (HO2S-2) switches detected—Bank # 2

Code 141 Fuel system indicates lean

Code 144 No Heated Oxygen Sensor (HO2S-1) switches detected—Bank # 1

Code 157 Mass Air Flow (MAF) Sensor circuit below minimum voltage

Code 158 Mass Air Flow (MAF) Sensor circuit above maximum voltage

Code 159 Mass Air Flow (MAF) higher or lower than expected

Code 167 Insufficient Throttle Position (TP) change during Dynamic Response Test—KOER

Code 171 Fuel system at adaptive limits, Heated Oxygen Sensor (HO2S-I) unable to switch—Bank # 1

Code 172 Lack of Heated Oxygen Sensor (HO2S-1) switches, indicates lean—Bank # 1

Code 173 Lack of Heated Oxygen Sensor (HO2S-1) switches, indicates rich—Bank # 1

Code 174 Heated Oxygen Sensor (HO2S) switching time is slow—Right side—1992 vehicles only

Code 175 Fuel system at adaptive limits, Heated Oxygen Sensor (HO2S-2) unable to switch—Bank # 2

Code 176 Lack of Heated Oxygen Sensor (HO2S-2) switches, indicates lean—Bank # 2

Code 177 Lack of Heated Oxygen Sensor (HO2S-2) switches, indicates rich—Bank # 2

Code 178 Heated Oxygen Sensor (HO2S) switching time is slow—Left side—1992 vehicles only

Code 179 Fuel system at lean adaptive limit at part throttle, system rich—Bank # 1

Code 181 Fuel system at rich adaptive limit at part throttle, system lean—Bank # 1

Code 182 Fuel system at lean adaptive limit at idle, system rich—Right side—1992 vehicles only

Code 183 Fuel system at rich adaptive limit at idle, system lean—Right side—1992 vehicles only

Code 184 Mass Air Flow (MAF) higher than expected

Code 185 Mass Air Flow (MAF) lower than expected

Code 186 Injector pulse width higher or Mass Air Flow (MAF) lower than expected (without BARO Sensor)

Code 187 Injector pulse width lower than expected (with BARO Sensor)

Code 187 Injector pulse width lower or Mass Air Flow (MAF) higher than expected (without BARO Sensor)

Code 188 Fuel system at lean adaptive limit at part throttle, system rich—Bank # 2

Code 189 Fuel system at rich adaptive limit at part throttle, system lean—Bank # 2

Code 191 Adaptive fuel lean limit is reached at idle—Left side—1992 vehicles only

Code 192 Adaptive fuel rich limit is reached at idle—Left side—1992 vehicles only

Code 193 Flexible Fuel (FF) Sensor circuit failure

Code 211 Profile Ignition Pickup (PIP) circuit failure

Code 212 Loss of Ignition Diagnostic monitor (1DM) input to Powertrain Control Module (PCM)/SPOUT circuit grounded

Code 213 SPOUT circuit open

Code 214 Cylinder Identification (CID) circuit failure

Code 215 Powertrain Control Module (PCM) detected Coil 1 Primary circuit failure (EI)

Code 216 Powertrain Control Module (PCM) detected Coil 2 Primary circuit failure (EI)

Code 217 Powertrain Control Module (PCM) detected Coil 3 Primary circuit failure (EI)

Code 218 Loss of Ignition Diagnostic Monitor (1DM) signal left side (dual plug EI)

Code 219 Spark Timing defaulted to 10 degrees—SPOUT circuit open (EI)

Code 221 Spark Timing error (EI)

Code 222 Loss of Ignition Diagnostic Monitor (1DM) signal—right side (dual plug EI)

Code 223 Loss of Dual Plug Inhibit (DPI) control (Dual Plug EI)

Code 224 Powertrain Control Module (PCM) detected Coil 1, 2, 3 or 4 Primary circuit failure (Dual Plug EI)

Code 225 Knock not sensed during Dynamic Response Test—KOER

Code 226 Ignition Diagnostic Monitor (1DM) signal not received (EI)

Code 232 Powertrain Control Module (PCM) detected Coil 1, 2, 3 or 4 Primary circuit failure (EI)

Code 238 Powertrain Control Module (PCM) detected Coil 4 Primary circuit failure (EI)

Code 241 Ignition Control Module (1CM) to Powertrain Control Module (PCM) Ignition Diagnostic Monitor (1DM) Pulse Width Transmission error (EI)

Code 244 Cylinder Identification (CID) circuit fault present when Cylinder Balance Test requested

Code 311 Secondary Air Injection (AIR) system inoperative during KOER Bank # 1 with dual HO₂S

Code 312 Secondary Air Injection (AIR) misdirected during KOER

Code 313 Secondary Air Injection (AIR) not bypassed during KOER

Code 314 Secondary Air Injection (AIR) system inoperative during KOER—Bank # 2 with dual HO₂S

Code 326 EGR (PFE/DPFE) circuit voltage lower than expected

Code 327 EGR (EVP/PFE/DPFE) circuit below minimum voltage

Code 328 EGR (EVP) closed valve voltage lower than expected

Code 332 Insufficient EGR flow detected/EGR Valve opening not detected (EVP/PFE/DPFE)

Code 334 EGR (EVP) closed valve voltage higher than expected

Code 335 EGR (PFE/DPFE) Sensor voltage higher or lower than expected during KOEO

Code 336 Exhaust pressure high/EGR (PFE/DPFE) circuit voltage higher than expected

Code 337 EGR (EVP/PFE/DPFE) circuit above maximum voltage

Code 338 Engine Coolant Temperature (ECT) lower than expected (thermostat test)

Code 339 Engine Coolant Temperature (ECT) higher than expected (thermostat test)

Code 341 Octane Adjust service pin open

Code 411 Cannot control RPM during KOER low rpm check

Code 412 Cannot control RPM during KOER high rpm check

Code 415 Idle Air Control (IAC) system at maximum adaptive lower limit

Code 416 Idle Air Control (IAC) system at upper adaptive learning limit

Code 452 Insufficient input from Vehicle Speed Sensor (VSS) to PCM

Code 453 Servo leaking down (KOER IVSC test)

Code 454 Servo leaking up (KOER IVSC test)

Code 455 Insufficient RPM increase (KOER IVSC test)

Code 456 Insufficient RPM decrease (KOER IVSC test)

Code 457 Speed Control Command Switch(s) circuit not functioning (KOEO IVSC test)

Code 458 Speed Control Command Switch(s) stuck/circuit grounded (KOEO IVSC test)

Code 459 Speed Control ground circuit open (KOEO IVSC test)

Code 511 Powertrain Control Module (PCM) Read Only Memory (ROM) test failure (KOEO)

Code 512 Powertrain Control Module (PCM) Keep Alive Memory (KAM) test failure

Code 513 Powertrain Control Module (PCM) internal voltage failure (KOEO)

Code 519 Power Steering Pressure (PSP) Switch circuit open—KOEO

Code 521 Power Steering Pressure (PSP) Switch circuit did not change states—KOER

Code 522 Vehicle not in park or neutral during KOEO/Park/Neutral Position (PNP) Switch circuit open

Code 524 Low speed Fuel Pump circuit open—battery to PCM

Code 525 Indicates vehicle in gear/A/C on

Code 526 Neutral Pressure Switch (NPS) circuit closed; A/C on—1992 vehicles only

Code 527 Park/Neutral Position (PNP) Switch open—A/C on, KOEO

Code 528 Clutch Pedal Position (CPP) switch circuit failure

Code 529 Data Communications Link (DCL) or PCM circuit failure

Code 532 Cluster Control Assembly (CCA) circuit failure

Code 533 Data Communications Link (DCL) or Electronic Instrument Cluster (EIC) circuit failure

Code 536 Brake On/Off (BOO) circuit failure/not actuated during KOER

Code 538 Insufficient RPM change during KOER Dynamic Response Test

Code 538 Invalid Cylinder Balance Test due to throttle movement during test—SFI only

Code 538 Invalid Cylinder Balance test due to Cylinder Identification (CID) circuit failure

Code 539 A/C on/Defrost on during Self-Test

Code 542 Fuel Pump secondary circuit failure

Code 543 Fuel Pump secondary circuit failure

Code 551 Idle Air Control (IAC) circuit failure—KOEO

Code 552 Secondary Air Injection Bypass (AIRB) circuit failure—KOEO

Code 553 Secondary Air Injection Diverter (AIRD) circuit failure—KOEO

Code 554 Fuel Pressure Regulator Control (FPRC) circuit failure

Code 556 Fuel Pump Relay primary circuit failure

Code 557 Low speed Fuel Pump primary circuit failure

Code 558 EGR Vacuum Regulator (EVR) circuit failure -KOEO

Code 559 Air Conditioning On (ACON) Relay circuit failure-KOEO

Code 563 High Fan Control (HFC) circuit failure—KOEO

Code 564 Fan Control (FC) circuit failure—KOEO

Code 565 Canister Purge (CANP) circuit failure—KOEO

Code 566 3–4 Shift Solenoid circuit failure, A4LD transmission—KOEO

Code 567 Speed Control Vent (SCVNT) circuit failure -KOEO IVSC test

Code 568 Speed Control Vacuum (SCVAC) circuit failure—KOEO IVSC test

Code 569 Auxiliary Canister Purge (CANP2) circuit failure-KOEO

Code 571 EGRA solenoid circuit failure KOEO

Code 572 EGRV solenoid circuit failure KOEO

Code 578 A/C Pressure Sensor circuit shorted (VCRM) mode

Code 579 Insufficient A/C pressure change (VCRM) mode

Code 581 Power to fan circuit over current (VCRM) mode

Code 582 Fan circuit open (VCRM) mode

Code 583 Power to Fuel Pump over current (VCRM) mode

Code 584 Power ground circuit open (Pin 1) (VCRM) mode

Code 585 Power to A/C Clutch over current (VCRM) mode

Code 586 A/C Clutch circuit open (VCRM) mode

Code 587 Variable Control Relay Module (VCRM) communication failure

Code 593 Heated Oxygen Sensor Heater (HO₂S HTR)

Code 617 1–2 Shift error

Code 618 2–3 Shift error

Code 619 3–4 Shift error

Code 621 Shift Solenoid 1 (SS1) circuit failure—KOEO

Code 622 Shift Solenoid 2 (SS2) circuit failure—KOEO

Code 623 Transmission Control Indicator Lamp (TCIL) circuit failure

Code 624 Electronic Pressure Control (EPC) circuit failure

Code 625 Electronic Pressure Control (EPC) driver open in PCM

Code 626 Coast Clutch Solenoid (CCS) circuit failure—KOEO

Code 627 Torque Converter Clutch (TCC) solenoid circuit failure

Code 628 Excessive Converter Clutch slippage

Code 629 Torque Converter Clutch (TCC) solenoid circuit failure

Code 631 Transmission Control Indicator Lamp (TCIL) circuit failure—KOEO

Code 632 Transmission Control Switch (TCS) circuit did not change states during KOER

Code 633 4 x 4L Switch closed during KOEO

Code 634 Manual Lever Position (MLP) voltage higher or lower than expected/ error in Transmission Select Switch (TSS) circuit(s)

Code 636 Transmission Oil Temperature (TOT) higher or lower than expected

Code 637 Transmission Oil Temperature (TOT) Sensor circuit above maximum voltage/circuit open

Code 638 Transmission Oil Temperature (TOT) Sensor circuit below minimum voltage/circuit shorted

Code 639 Insufficient input from Transmission Speed Sensor (TSS)

Code 641 Shift Solenoid 3 (SS3) circuit failure

Code 643 Torque Converter Clutch (TCC) circuit failure

Code 645 Incorrect gear ratio obtained for first gear

Code 646 Incorrect gear ratio obtained for second gear

Code 647 Incorrect gear ratio obtained for third gear

Code 648 Incorrect gear ratio obtained for fourth gear

Code 649 Electronic Pressure Control (EPC) higher or lower than expected

Code 651 Electronic Pressure Control (EPC) circuit failure

Code 652 Torque Converter Clutch (TCC) Solenoid circuit failure

Code 654 Manual Lever Position (MLP) Sensor not indicating park during KOEO

Code 655 Manual Lever Position (MLP) Sensor indicating not in neutral during Self-Test

Code 656 Torque Converter Clutch (TCC) continuous slip error

Code 657 Transmission Over Temperature condition occurred

Code 659 High vehicle speed in park indicated

Code 667 Transmission Range sensor circuit voltage below minimum voltage

Code 668 Transmission Range sensor circuit voltage above maximum voltage

Code 675 Transmission Range sensor circuit voltage out of range

Code 691 4x4 Low switch open or short circuit

Code 692 Transmission state does not match calculated ratio

Code 998 Hard fault present—FMEM Mode

➡**If specific cylinder banks or sides are referred to in any of the above codes, but the vehicle code is being obtained from has a 4 cylinder engine, or only one Oxygen Sensor, disregard the bank side reference, but the code definition and components it pertains to is always the same.**

EEC-V SYSTEM

1994 Passenger Cars and Light Trucks:

DTC P0102 Mass Air Flow (MAF) Sensor circuit low input

DTC P0103 Mass Air Flow (MAF) Sensor circuit high input

DTC P0112 Intake Air Temperature (IAT) Sensor circuit low input

DTC P0113 Intake Air Temperature (IAT) Sensor high input

DTC P0117 Engine Coolant Temperature (ECT) low input

DTC P0118 Engine Coolant Temperature (ECT) Sensor circuit high input

DTC P0122 Throttle Position (TP) Sensor circuit low input

DTC P0123 Throttle Position (TP) Sensor high input

DTC P0125 Insufficient coolant temperature to enter closed loop fuel control

DTC P0132 Upstream Heated Oxygen Sensor (HO$_2$S 11) circuit high voltage (Bank #1)

DTC P0135 Heated Oxygen Sensor Heater (HTR 11) circuit malfunction

DTC P0138 Downstream Heated Oxygen Sensor (HO$_2$S 12) circuit high voltage (Bank #1)

DTC P0140 Heated Oxygen Sensor (HO$_2$S 12) circuit no activity detected (Bank #1)

DTC P0141 Heated Oxygen Sensor Heater (HTR 12) circuit malfunction

DTC P0152 Upstream Heated Oxygen Sensor (HO$_2$S 21) circuit high voltage (Bank #2)

DTC P0155 Heated Oxygen Sensor Heater (HTR 21) circuit malfunction

DTC P0158 Downstream Heated Oxygen Sensor (HO$_2$S 22) circuit high voltage (Bank #2)

DTC P0160 Heated Oxygen Sensor (HO$_2$S 12) circuit no activity detected (Bank #2)

DTC P0161 Heated Oxygen Sensor Heater (HTR 22) circuit malfunction

DTC P0171 System (adaptive fuel) too lean (Bank #1)

DTC P0172 System (adaptive fuel) too lean (Bank #1)

DTC P0174 System (adaptive fuel) too lean (Bank #1)

DTC P0175 System (adaptive fuel) too lean (Bank #1)

DTC P0300 Random misfire detected

DTC P0301 Cylinder #1 misfire detected

DTC P0302 Cylinder #2 misfire detected

DTC P0303 Cylinder #3 misfire detected

DTC P0304 Cylinder #4 misfire detected

DTC P0305 Cylinder #5 misfire detected

DTC P0306 Cylinder #6 misfire detected

DTC P0307 Cylinder #7 misfire detected

DTC P0308 Cylinder #8 misfire detected

DTC P0320 Ignition engine speed (Profile Ignition Pickup) input circuit malfunction

DTC P0340 Camshaft Position (CMP) sensor circuit malfunction (CID)

DTC P0402 Exhaust Gas Recirculation (EGR) excess flow detected (valve open at idle)

DTC P0420 Catalyst system efficiency below threshold (Bank #1)

DTC P0430 Catalyst system efficiency below threshold (Bank #2)

DTC P0443 Evaporative emission control system Canister Purge (CANP) Control Valve circuit malfunction

DTC P0500 Vehicle Speed Sensor (VSS) malfunction

DTC P0505 Idle Air Control (IAC) system malfunction

DTC P0605 Powertrain Control Module (PCM)—Read Only Memory (ROM) test error

DTC P0703 Brake On/Off (BOO) switch input malfunction
DTC P0707 Manual Lever Position (MLP) sensor circuit low input
DTC P0708 Manual Lever Position (MLP) sensor circuit high input
DTC P0720 Output Shaft Speed (OSS) sensor circuit malfunction
DTC P0741 Torque Converter Clutch (TCC) system incorrect mechanical performance
DTC P0743 Torque Converter Clutch (TCC) system electrical failure
DTC P0750 Shift Solenoid #1(SS1) circuit malfunction
DTC P0751 Shift Solenoid #1(SS1) performance
DTC P0755 Shift Solenoid #2 (SS2) circuit malfunction
DTC P0756 Shift Solenoid #2 (SS2) performance
DTC P1000 OBD II Monitor Testing not complete
DTC P1100 Mass Air Flow (MAF) sensor intermittent
DTC P1101 Mass Air Flow (MAF) sensor out of Self-Test range
DTC P1112 Intake Air Temperature (IAT) sensor intermittent
DTC P1116 Engine Coolant Temperature (ECT) sensor out of Self-Test range
DTC P1117 Engine Coolant Temperature (ECT) sensor intermittent
DTC P1120 Throttle Position (TP) sensor out of range low
DTC P1121 Throttle Position (TP) sensor inconsistent with MAF sensor
DTC P1124 Throttle Position (TP) sensor out of Self-Test range
DTC P1125 Throttle Position (TP) sensor circuit intermittent
DTC P1130 Lack of HO2S 11 switch, adaptive fuel at limit
DTC P1131 Lack of HO2S 11 switch, sensor indicates lean (Bank #1)
DTC P1132 Lack of HO2S 11 switch, sensor indicates rich (Bank #1)
DTC P1137 Lack of HO2S 12 switch, sensor indicates lean (Bank #1)
DTC P1138 Lack of HO2S 12 switch, sensor indicates rich (Bank #1)
DTC P1150 Lack of HO2S 21 switch, adaptive fuel at limit
DTC P1151 Lack of HO2S 21 switch, sensor indicates lean (Bank #2)
DTC P1152 Lack of HO2S 21 switch, sensor indicates rich (Bank #2)
DTC P1157 Lack of HO2S 22 switch, sensor indicates lean (Bank #2)
DTC P1158 Lack of HO2S 22 switch, sensor indicates rich (Bank #2)
DTC P1351 Ignition Diagnostic Monitor (1DM) circuit input malfunction
DTC P1352 Ignition coil A primary circuit malfunction
DTC P1353 Ignition coil B primary circuit malfunction
DTC P1354 Ignition coil C primary circuit malfunction
DTC P1355 Ignition coil D primary circuit malfunction
DTC P1364 Ignition coil primary circuit malfunction
DTC P1390 Octane Adjust (OCT ADJ) out of Self-Test range
DTC P1400 Differential Pressure Feedback Electronic (DPFE) sensor circuit low voltage detected
DTC P1401 Differential Pressure Feedback Electronic (DPFE) sensor circuit high voltage detected
DTC P1403 Differential Pressure Feedback Electronic (DPFE) sensor hoses reversed
DTC P1405 Differential Pressure Feedback Electronic (DPFE) sensor upstream hose off or plugged

DTC P1406 Differential Pressure Feedback Electronic (DPFE) sensor downstream hose off or plugged
DTC P1407 Exhaust Gas Recirculation (EGR) no flow detected (valve stuck closed or inoperative)
DTC P1408 Exhaust Gas Recirculation (EGR) flow out of Self-Test range
DTC P1473 Fan Secondary High with fan(s) off
DTC P1474 Low Fan Control primary circuit malfunction
DTC P1479 High Fan Control primary circuit malfunction
DTC P1480 Fan Secondary low with low fan on
DTC P1481 Fan Secondary low with high fan on
DTC P1500 Vehicle Speed Sensor (VSS) circuit intermittent
DTC P1505 Idle Air Control (IAC) system at adaptive clip
DTC P1605 Powertrain Control Module (PCM)—Keep Alive Memory (KAM) test error
DTC P1703 Brake On/Off (BOO) switch out of Self-Test range
DTC P1705 Manual Lever Position (MLP) sensor out of Self-Test range
DTC P1711 Transmission Fluid Temperature (TFT) sensor out of Self-Test range
DTC P1742 Torque Converter Clutch (TCC) solenoid mechanically failed (turns MIL on)
DTC P1743 Torque Converter Clutch (TCC) solenoid mechanically failed (turns TCIL on)
DTC P1744 Torque Converter Clutch (TCC) system mechanically stuck in off position
DTC P1746 Electronic Pressure Control (EPC) solenoid circuit low input (open circuit)
DTC P1747 Electronic Pressure Control (EPC) solenoid circuit high input (short circuit)
DTC P1751 Shift Solenoid #1(SS1) performance
DTC P1756 Shift Solenoid #2 (SS2) performance
DTC P1780 Transmission Control Switch (TCS) circuit out of Self-Test range
1995 Light Trucks:
DTC P0102 Mass Air Flow (MAF) Sensor circuit low input
DTC P0103 Mass Air Flow (MAF) Sensor circuit high input
DTC P0112 Intake Air Temperature (IAT) Sensor circuit low input
DTC P0113 Intake Air Temperature (IAT) Sensor high input
DTC P0117 Engine Coolant Temperature (ECT) low input
DTC P0118 Engine Coolant Temperature (ECT) Sensor circuit high input
DTC P0121 In range operating Throttle Position (TP) sensor circuit failure
DTC P0122 Throttle Position (TP) Sensor circuit low input
DTC P0123 Throttle Position (TP) Sensor high input
DTC P0125 Insufficient coolant temperature to enter closed loop fuel control
DTC P0126 Insufficient coolant temperature for stable operation
DTC P0131 Upstream Heated Oxygen Sensor (HO2S 11) circuit out of range low voltage (bank #1)
DTC P0132 Upstream Heated Oxygen Sensor (HO2S 11) circuit high voltage (Bank #1)
DTC P0133 Upstream Heated Oxygen Sensor (HO2S 11) circuit slow response (Bank #1)
DTC P0135 Heated Oxygen Sensor Heater (HTR 11) circuit malfunction
DTC P0136 Downstream Heated Oxygen Sensor (HO2S 12) circuit malfunction (Bank #1
DTC P0138 Downstream Heated Oxygen Sensor (HO2S 12) circuit high voltage (Bank #1)

DTC P0140 Heated Oxygen Sensor (HO2S 12) circuit no activity detected (Bank #1)

DTC P0141 Heated Oxygen Sensor Heater (HTR 12) circuit malfunction

DTC P0151 Upstream Heated Oxygen Sensor (HO2S 21) circuit out of range low voltage (Bank #2)

DTC P0152 Upstream Heated Oxygen Sensor (HO2S 21) circuit high voltage (Bank #2)

DTC P0153 Upstream Heated Oxygen Sensor (HO2S 21) circuit slow response (Bank #2)

DTC P0155 Heated Oxygen Sensor Heater (HTR 21) circuit malfunction

DTC P0156 Downstream Heated Oxygen Sensor (HO2S 22) circuit malfunction (Bank #2)

DTC P0158 Downstream Heated Oxygen Sensor (HO2S 22) circuit high voltage (Bank #2)

DTC P0160 Heated Oxygen Sensor (HO2S 12) circuit no activity detected (Bank #2)

DTC P0161 Heated Oxygen Sensor Heater (HTR 22) circuit malfunction

DTC P0171 System (adaptive fuel) too lean (Bank #1)

DTC P0172 System (adaptive fuel) too rich (Bank #1)

DTC P0174 System (adaptive fuel) too lean (Bank #2)

DTC P0175 System (adaptive fuel) too rich (Bank #2)

DTC P0222 Throttle Position Sensor B (TP-B) circuit low input

DTC P0223 Throttle Position Sensor B (TP-B) circuit high input

DTC P0230 Fuel pump primary circuit malfunction

DTC P0231 Fuel pump secondary circuit low

DTC P0232 Fuel pump secondary circuit high

DTC P0300 Random misfire detected

DTC P0301 Cylinder #1 misfire detected

DTC P0302 Cylinder #2 misfire detected

DTC P0303 Cylinder #3 misfire detected

DTC P0304 Cylinder #4 misfire detected

DTC P0305 Cylinder #5 misfire detected

DTC P0306 Cylinder #6 misfire detected

DTC P0307 Cylinder #7 misfire detected

DTC P0308 Cylinder #8 misfire detected

DTC P0320 Ignition engine speed (Profile Ignition Pickup) input circuit malfunction

DTC P0340 Camshaft Position (CMP) sensor circuit malfunction (CID)

DTC P0350 Ignition Coil primary circuit malfunction

DTC P0351 Ignition Coil A primary circuit malfunction

DTC P0352 Ignition Coil B primary circuit malfunction

DTC P0353 Ignition Coil C primary circuit malfunction

DTC P0354 Ignition Coil D primary circuit malfunction

DTC P0400 Exhaust Gas Recirculation (EGR) flow malfunction

DTC P0401 Exhaust Gas Recirculation (EGR) flow insufficient detected

DTC P0402 Exhaust Gas Recirculation (EGR) excess flow detected (valve open at idle)

DTC P0411 Secondary Air Injection system incorrect flow detected

DTC P0412 Secondary Air Injection system control valve malfunction

DTC P0420 Catalyst system efficiency below threshold (Bank #1)

DTC P0430 Catalyst system efficiency below threshold (Bank #2)

DTC P0443 Evaporative emission control system Canister Purge (CANP) Control Valve circuit malfunction

DTC P0500 Vehicle Speed Sensor (VSS) malfunction

DTC P0505 Idle Air Control (IAC) system malfunction

DTC P0603 Powertrain Control Module (PCM)—Keep Alive Memory (KAM) test error

DTC P0605 Powertrain Control Module (PCM)—Read Only Memory (ROM) test error

DTC P0704 Clutch Pedal Position (CPP) switch input circuit malfunction

DTC P0703 Brake On/Off (BOO) switch input malfunction

DTC P0707 Manual Lever Position (MLP) sensor circuit low input

DTC P0708 Manual Lever Position (MLP) sensor circuit high input

DTC P0712 Transmission Fluid Temperature (TFT) sensor circuit low input

DTC P0713 Transmission Fluid Temperature (TFT) sensor circuit high input

DTC P0715 Turbine Shaft Speed (TSS) sensor circuit malfunction

DTC P0720 Output Shaft Speed (OSS) sensor circuit malfunction

DTC P0731 Incorrect ratio for first gear

DTC P0732 Incorrect ratio for second gear

DTC P0733 Incorrect ratio for third gear

DTC P0734 Incorrect ratio for fourth gear

DTC P0736 Reverse incorrect gear

DTC P0741 Torque Converter Clutch (TCC) system incorrect mechanical performance

DTC P0746 Electronic Pressure Control (EPC) solenoid performance

DTC P0743 Torque Converter Clutch (TCC) system electrical failure

DTC P0750 Shift Solenoid #1(SS1) circuit malfunction

DTC P0751 Shift Solenoid #1(SS1) performance

DTC P0755 Shift Solenoid #2 (SS2) circuit malfunction

DTC P0756 Shift Solenoid #2 (SS2) performance

DTC P0760 Shift Solenoid #3 (SS3) circuit malfunction

DTC P0761 Shift Solenoid #3 (SS3) performance

DTC P0781 1 to 2 shift error

DTC P0782 2 to 3 shift error

DTC P0783 3 to 4 shift error

DTC P0784 4 to 5 shift error

DTC P1000 OBD II Monitor Testing not complete

DTC U1039 OBD II Monitor not complete

DTC U1051 Brake switch signal missing or incorrect

DTC P1100 Mass Air Flow (MAF) sensor intermittent

DTC P1101 Mass Air Flow (MAF) sensor out of Self-Test range

DTC P1112 Intake Air Temperature (IAT) sensor intermittent

DTC P1116 Engine Coolant Temperature (ECT) sensor out of Self-Test range

DTC P1117 Engine Coolant Temperature (ECT) sensor intermittent

DTC P1120 Throttle Position (TP) sensor out of range low

DTC P1121 Throttle Position (TP) sensor inconsistent with MAF sensor

DTC P1124 Throttle Position (TP) sensor out of Self-Test range

DTC P1125 Throttle Position (TP) sensor circuit intermittent

DTC P1130 Lack of HO2S 11 switch, adaptive fuel at limit

DTC P1131 Lack of HO2S 11 switch, sensor indicates lean (Bank #1)

DTC P1132 Lack of HO2S 11 switch, sensor indicates rich (Bank #1)

DTC U1135 Ignition switch signal missing or incorrect
DTC P1137 Lack of HO$_2$S 12 switch, sensor indicates lean (Bank #1)
DTC P1138 Lack of HO$_2$S 12 switch, sensor indicates rich (Bank #1)
DTC P1150 Lack of HO$_2$S 21 switch, adaptive fuel at limit
DTC P1151 Lack of HO$_2$S 21 switch, sensor indicates lean (Bank #2)
DTC P1152 Lack of HO$_2$S 21 switch, sensor indicates rich (Bank #2)
DTC P1157 Lack of HO$_2$S 22 switch, sensor indicates lean (Bank #2)
DTC P1158 Lack of HO$_2$S 22 switch, sensor indicates rich (Bank #2)
DTC P1220 Series Throttle Control malfunction
DTC P1224 Throttle Position Sensor (TP-B) out of Self-test range
DTC P1233 Fuel Pump driver Module off-line
DTC P1234 Fuel Pump driver Module off-line
DTC P1235 Fuel Pump control out of range
DTC P1236 Fuel Pump control out of range
DTC P1237 Fuel Pump secondary circuit malfunction
DTC P1238 Fuel Pump secondary circuit malfunction
DTC P1260 THEFT detected—engine disabled
DTC P1270 Engine RPM or vehicle speed limiter reached
DTC P1351 Ignition Diagnostic Monitor (1DM) circuit input malfunction
DTC P1352 Ignition coil A primary circuit malfunction
DTC P1353 Ignition coil B primary circuit malfunction
DTC P1354 Ignition coil C primary circuit malfunction
DTC P1355 Ignition coil D primary circuit malfunction
DTC P1358 Ignition Diagnostic Monitor (1DM) signal out of Self-Test range
DTC P1359 Spark output circuit malfunction
DTC P1364 Ignition coil primary circuit malfunction
DTC P1390 Octane Adjust (OCT ADJ) out of Self-Test range
DTC P1400 Differential Pressure Feedback Electronic (DPFE) sensor circuit low voltage detected
DTC P1401 Differential Pressure Feedback Electronic (DPFE) sensor circuit high voltage detected
DTC P1403 Differential Pressure Feedback Electronic (DPFE) sensor hoses reversed
DTC P1405 Differential Pressure Feedback Electronic (DPFE) sensor upstream hose off or plugged
DTC P1406 Differential Pressure Feedback Electronic (DPFE) sensor downstream hose off or plugged
DTC P1407 Exhaust Gas Recirculation (EGR) no flow detected (valve stuck closed or inoperative)
DTC P1408 Exhaust Gas Recirculation (EGR) flow out of Self-Test range
DTC P1409 Electronic Vacuum Regulator (EVR) control circuit malfunction
DTC P1414 Secondary Air Injection system monitor circuit high voltage
DTC P1443 Evaporative emission control system—vacuum system purge control solenoid or purge control valve malfunction
DTC P1444 4- Purge Flow Sensor (PFS) circuit low input
DTC P1445 Purge Flow Sensor (PFS) circuit high input
DTC U1451 Lack of response from Passive Anti-Theft system (PATS) module—engine disabled
DTC P1460 Wide Open Throttle Air Conditioning Cut-off (WAC) circuit malfunction

DTC P1461 Air Conditioning Pressure (ACP) sensor circuit low input
DTC P1462 Air Conditioning Pressure (ACP) sensor circuit high input
DTC P1463 Air Conditioning Pressure (ACP) sensor insufficient pressure change
DTC P1469 Low air conditioning cycling period
DTC P1473 Fan Secondary High with fan(s) off
DTC P1474 Low Fan Control primary circuit malfunction
DTC P1479 High Fan Control primary circuit malfunction
DTC P1480 Fan Secondary low with low fan on
DTC P1481 Fan Secondary low with high fan on
DTC P1500 Vehicle Speed Sensor (VSS) circuit intermittent
DTC P1505 Idle Air Control (IAC) system at adaptive clip
DTC P1506 Idle Air control (IAC) over speed error
DTC P1518 Intake Manifold Runner Control (IMRC) malfunction (stuck open)
DTC P1519 Intake Manifold Runner Control (IMRC) malfunction (stuck closed)
DTC P1520 Intake Manifold Runner Control (IMRC) circuit malfunction
DTC P1507 Idle Air control (IAC) under speed error
DTC P1605 Powertrain Control Module (PCM)—Keep Alive Memory (KAM) test error
DTC P1650 Power steering Pressure (PSP) switch out of Self-Test range
DTC P1651 Power steering Pressure (PSP) switch input malfunction
DTC P1701 Reverse engagement error
DTC P1703 Brake On/Off (BOO) switch out of Self-Test range
DTC P1705 Manual Lever Position (MLP) sensor out of Self-Test range
DTC P1709 Park or Neutral Position (PNP) switch out of Self-test range
DTC P1729 4X4 Low switch error
DTC P1711 Transmission Fluid Temperature (TFT) sensor out of Self-Test range
DTC P1741 Torque Converter Clutch (TOC) control error
DTC P1742 Torque Converter Clutch (TCC) solenoid mechanically failed (turns MIL on)
DTC P1743 Torque Converter Clutch (TCC) solenoid mechanically failed (turns TOIL on)
DTC P1744 Torque Converter Clutch (TCC) system mechanically stuck in off position
DTC P1748 Electronic Pressure Control (EPC) solenoid circuit low input (open circuit)
DTC P1747 Electronic Pressure Control (EPC) solenoid circuit high input (short circuit)
DTC P1749 Electric Pressure Control (EPC) solenoid failed low
DTC P1751 Shift Solenoid #1(SS1) performance
DTC P1756 Shift Solenoid #2 (SS2) performance
DTC P1780 Transmission Control Switch (TCS) circuit out of Self-Test range

TRANSMISSION CODES

Reading and Retrieving Codes

EXCEPT PROBE AND 1991–92 ESCORT MODELS

Compared with the other domestic manufacturers, the Ford diagnostic computer system is either the most comprehensive or the most difficult to work with (depending on your outlook-glass half full

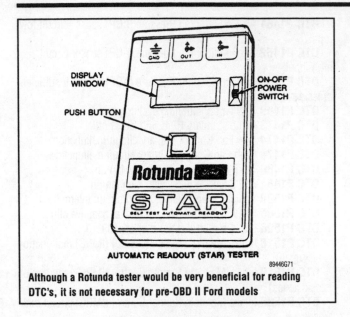

AUTOMATIC READOUT (STAR) TESTER

89446G71

Although a Rotunda tester would be very beneficial for reading DTC's, it is not necessary for pre-OBD II Ford models

or half empty?). The Ford system can use either 2-digit or 3-digit codes, depending on the year and/or model of the vehicle at hand. And, as with Chrysler and Jeep vehicles, it is most beneficial if you have access to a scan tool. But, if you do not have a scan tool, don't worry the DTC's can still be read by more conventional methods.

Some Ford models use a light, located in the instrument cluster and usually referred to as the Malfunction Indicator Lamp (MIL), CHECK ENGINE light or a similarly-named light, to display the DTC's as a series of flashes or blinks. Unfortunately, not all Ford vehicles are equipped with such a light. In these cases, you will need either an analog voltmeter (with an arm-not the digital read-out type), a test probe light, or any other device capable of reading voltage sweeps.

➡Although a scan tool is not necessary to retrieve and read the DTC's, it can be very helpful, especially on vehicles which use 3-digit trouble codes. Reading Code 586 as a series of flashes (flash, flash, flash, flash, flash, pause, flash, flash, flash, flash, flash, flash, flash, flash, pause, flash, flash, flash, flash, flash, flash-you get the idea) can be irritating. Be sure you have enough paper and a pen/pencil that works!

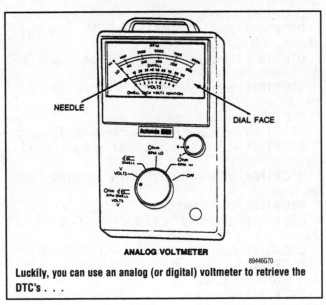

ANALOG VOLTMETER

89446G70

Luckily, you can use an analog (or digital) voltmeter to retrieve the DTC's . . .

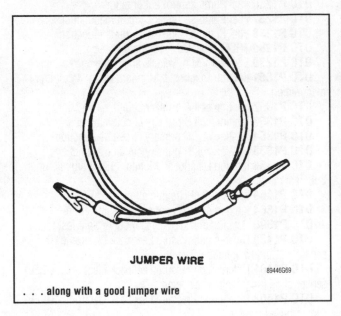

JUMPER WIRE

89446G69

. . . along with a good jumper wire

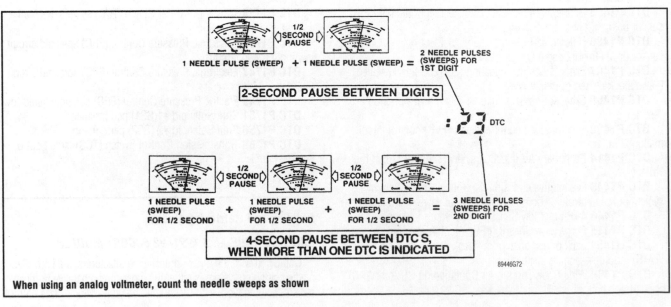

When using an analog voltmeter, count the needle sweeps as shown

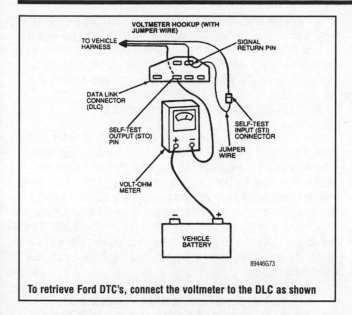

To retrieve Ford DTC's, connect the voltmeter to the DLC as shown

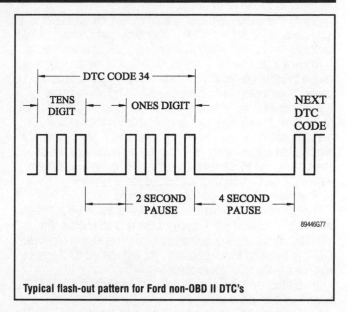

Typical flash-out pattern for Ford non-OBD II DTC's

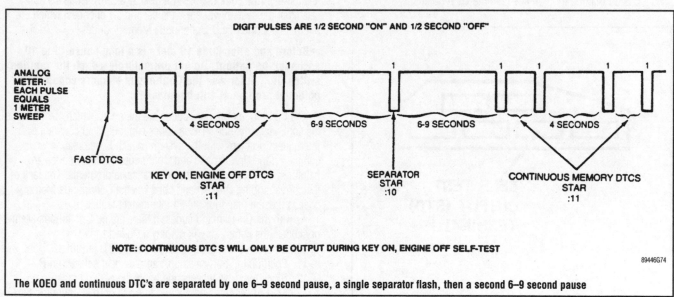

The KOEO and continuous DTC's are separated by one 6–9 second pause, a single separator flash, then a second 6–9 second pause

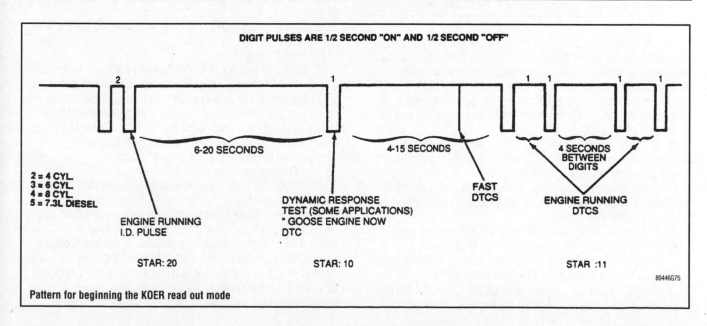

Pattern for beginning the KOER read out mode

The MIL or analog voltmeter (analog voltmeter will be used as an example of any voltage measuring device you plan on using to read the codes for vehicles which do not utilize a MIL) will indicate the DTC's in a logical series of flashes, or sweeps. Individual digits (say, a 5) are displayed as a series of flashes with a ½ second pause between each flash. Therefore, the digit 5 would be five flashes separated by ½ second pauses. Individual digits belonging to the same 2- or 3-digit number (say the 2 and the 3 of digit 23) are separated by two second pauses. Therefore, Code 23 would be indicated as follows: flash, ½ second pause, flash, 2 second pause, flash, ½ second pause, flash, ½ second pause, flash. In between each individual trouble code a 4 second pause is used.

To retrieve the codes, perform the following:

1. Perform the preliminary inspection, located earlier in this section. This is very important, since a loose or disconnected wire, or corroded connector terminals can cause a whole slew of unrelated DTC's to be stored by the computer; you will waste a lot of time performing a diagnostic "goose chase."

2. Grab some paper and a pencil or pen to write down the DTC's when they are flashed out.

3. Ensure that the ignition key is in the **OFF** position.

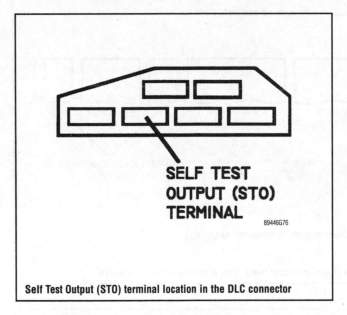

Self Test Output (STO) terminal location in the DLC connector

4. Locate the Self-Test Connector (STC), which is comprised of two separate connectors: the Self-Test Output (STO) connector and the Self-Test Input (STI) connector. The STC is usually located underhood near one of the strut towers or near the firewall. If you cannot locate the STC either ask someone at a local dealership for help with its location, or refer to the Ford model-specific Chilton's Total Car Care manual.

5. Once the STC is located, identify the two connectors. The smaller of the two, with only one terminal, is the STI connector. The larger connector is the STO.

6. If the vehicle at hand is not equipped with a MIL, the analog voltmeter must be connected to the computer system. To attach the voltmeter, perform the following:

　a. Using a jumper wire, attach the STI terminal to a good engine ground or to the negative terminal of the battery.

　b. Attach the negative voltmeter lead to the STO terminal and the positive lead to the battery positive terminal.

7. For models equipped with a MIL, simply jumper the STI terminal to a good engine ground or to the negative battery cable.

8. Turn the ignition switch **ON**, but do not start the engine. This condition is known as Key On-Engine Off (KOEO). After the ignition is turned **ON**, depending on the instrument used to read the codes, either the MIL will start flashing the DTC's as a series of blinks or the voltmeter needle will start moving between 0 and 12 volts in a series sweeps.

9. Count and note the number and sequence of the flashes/sweeps.

The DTC system will progress through two sets of codes at this time. First, all of the KOEO On-Demand codes will be displayed. These codes are not normally stored in the computer memory. Rather, the computer performs a quick self-diagnostic check once the ignition key is turned **ON** and displays any applicable codes for faults that are currently just detected. If there are no current trouble faults in the computer system, only Code 11 (Code 111 or vehicles which use 3-digit codes) will be displayed. Code 11 or Code 111 indicates that there are no "hard," or current malfunction, faults.

After the system displays all of the ON-Demand codes or Code 11 (111), it will display a Code 10, which appears as a single flash/sweep (the computer cannot display a zero). Code 10 indicates that the system will enter the second set of codes, which are referred to as the KOEO Continuous Memory Codes.

➡**Before and after Code 10 there is a long pause (5 to 10 seconds); be patient. Do not prematurely switch the ignition switch OFF, otherwise the Continuous Memory codes (and potential problems) will be missed.**

The Continuous Memory codes are the codes which are stored in the computer's memory. These codes can indicate a constant or intermittent problem with the system. Which is one reason why comparing the On-Demand and Continuous Memory codes are helpful in diagnosing transmission/transaxle problems. The lack of the corresponding On-Demand code when a Continuous Memory code is present may indicate an intermittent fault.

As with the On-Demand codes, if there are no Continuous Memory codes, the computer will display a Code 11 (111).

10. Disconnect the ground wire from the STI terminal.

11. Position the transmission/transaxle gear selector in **P**.

12. Block the drive wheels and apply the parking brake.

13. Start the engine and allow it to reach normal operating temperature.

14. Turn the engine **OFF**, then reconnect the ground wire to the STI terminal.

15. Start the engine.

16. Watch for a series of flashes/sweeps, which are the engine ID code. 4-cylinder engines will display a two flash/sweep code, a 6-cylinder engine will display a three flash/sweep code, and an 8-cylinder engine will display a four flash/sweep code-the code is always half the total number of cylinders of the engine. As soon as the engine ID code is displayed, perform the following:

　a. Turn the steering wheel at least ½ turn.

　b. Depress the brake pedal.

　c. Cycle the transmission/transaxle control switch ON and OFF, if equipped.

17. On some models, after completing the previous three sub-steps, there should be a long pause (possibly as long as 20-30 seconds), then a single flash/sweep signal. When this signal occurs, depress the accelerator to the floor and release it quickly. This is a dynamic response test; the computer system uses it in its diagnostic self-test. After depressing the accelerator, there may be a small series of slight voltage changes, which is simply the computer

sending its information to the scan tool, in case one is being used. In our case, just ignore these small signals.

On models which do not display the dynamic response test single flash/sweep, the system will go into the KOER On-Demand code display without you having to depress the accelerator pedal.

18. Six seconds after the small voltage signals are detected, the system will start to display the KOER On-Demand. As with the KOEO ON-Demand codes, these codes are not stored in the computer; they are on-the-spot codes and will be erased when the DTC retrieval cycle is finished. Therefore, be sure to pay attention as they are displayed to avoid having to perform this entire procedure again. As with the other two DTC sets, a Code 11 (111) will be displayed if no DTC's are found.

19. After noting all of the displayed DTC's, turn the ignition switch **OFF**, remove the STI terminal ground wire, and the voltmeter, test light, etc. (if used).

The Ford computer system is also capable of numerous self-diagnostic tests to aid in trouble-shooting system faults. Some of the tests include the Wiggle Test and the Output State Test. If you are interested in these more specific test subroutines, refer to the Chilton's Total Car Care manual specific to your vehicle or the Ford factory manual.

1991–92 ESCORT MODELS

Early Escort models equipped with a 1.8L or 1.9L engine were provided with the F4EAT transaxle, which were manufactured by Mazda. Although there is no outward way of knowing when or if the vehicle has detected a DTC, there is a way to retrieve codes. The 1991–92 Escort models use two-digit DTC's, not the more complex three-digit ones available in some Ford vehicles. An analog voltmeter or similar voltage measuring device is necessary to read the DTC's.

The two-digit codes are flashed out in a logical series of voltmeter arm sweeps. The tens digits are displayed as long sweeps (in duration of time, not distance the arm moves), and the ones digits are displayed as short sweeps. For example, when first starting to retrieve the DTC's the voltmeter sweeps one long time, pause briefly, then sweep two short times followed by a longer pause. This indicates a Code 12 (long sweep-pause-two short sweeps-long pause). Therefore, a pattern of five long sweeps-pause-five short sweeps-long pause would indicate a Code 55.

Each DTC is displayed three times, then the next code is displayed out (if more than one code has been stored). The codes are displayed in numerical sequence starting with the lowest code number and moving to the highest.

When reading the codes be sure not to miss any of them, otherwise the entire sequence must be repeated. Although the codes will continue to be read until the ignition key is turned **OFF**, this can be tiresome if a large number of codes was stored, and you missed the last one.

To retrieve the codes, perform the following:

1. Perform the preliminary inspection, located earlier in this section. This is very important, since a loose or disconnected wire, or corroded connector terminals can cause a whole slew of unrelated DTC's to be stored by the computer; you will waste a lot of time performing a diagnostic "goose chase."

2. Grab some paper and a pencil or pen to write down the DTC's when they are flashed out.

3. Ensure that the ignition key is in the **OFF** position.

4. Locate the diagnostic connector (STC). The diagnostic connector is usually located underhood near one of the strut towers or near the firewall. If you cannot locate the diagnostic connector either

ask someone at a local dealership for help with its location, or refer to the Ford model-specific Chilton's Total Car Care manual.

5. Attach a negative voltmeter lead to the STO terminal and the positive lead to the battery positive terminal.

6. Turn the ignition switch **ON**, but do not start the engine. This condition is known as Key On-Engine Off (KOEO). After the ignition is turned **ON**, the voltmeter needle will start moving between 0 and 12 volts in a series sweeps.

7. Count and note the number, sequence and duration of the sweeps.

PROBE MODELS

As with the 1991–92 Escort models, the Probe vehicles also utilize Mazda-built transaxles. The computer diagnostic system also uses 2-digit trouble codes, which are displayed through flashing of the MANUAL SHIFT light or the CHECK ENGINE light.

The Probe DTC's are displayed in the same format (with the exception that they are flashes of a light rather than sweeps of an analog voltmeter needle), in that the tens digits are long flashes and the ones digits are short flashes. For example, when first starting to retrieve the DTC's the MANUAL SHIFT or CHECK ENGINE light flashes one long time, pauses briefly, then flash two short times followed by a longer pause. This indicates a Code 12 (long flash-pause-two short flash-long pause). Therefore, a pattern of five long flashes-pause-five short flashes-long pause would indicate a Code 55.

The codes are displayed from the lowest number on through to the highest number, and three times in a row. Therefore the first code is flashed three times, then the second code is flashed three times, etc., until the highest code has also been flashed three times. After the highest code number is displayed, the system begins with the lowest again. The system will continue to cycle through all of the codes in this manner; so if you missed a code or two, you can wait until it starts the cycle over again.

To retrieve the codes for 1989–92 models, perform the following:

1. Perform the preliminary inspection, located earlier in this section. This is very important, since a loose or disconnected wire, or corroded connector terminals can cause a whole slew of unrelated DTC's to be stored by the computer; you will waste a lot of time performing a diagnostic "goose chase."

2. Grab some paper and a pencil or pen to write down the DTC's when they are flashed out.

3. Locate the Diagnostic Request (DR) terminal, which is usually located next to or near the transmission/transaxle control computer. If the DR terminal cannot be found, either call your local Ford dealer or refer to the applicable Chilton's Total Car Care manual for your vehicle.

4. Turn the ignition key **OFF**, if not already done.

5. Using a jumper wire, attach the DR terminal to a good chassis or engine ground.

6. Without starting the engine, turn the ignition switch **ON**.

7. The MANUAL SHIFT light should start to flash out any stored DTC's.

To retrieve the codes for 1993 and newer non-OBD II models, perform the following:

1. Perform the preliminary inspection, located earlier in this section. This is very important, since a loose or disconnected wire, or corroded connector terminals can cause a whole slew of unrelated DTC's to be stored by the computer; you will waste a lot of time performing a diagnostic "goose chase."

2. Grab some paper and a pencil or pen to write down the DTC's when they are flashed out.

3. Locate the Diagnostic Link Connector (DLC), which is usually located near the left-hand side of the engine compartment and the firewall. If the DLC cannot be found, either call your local Ford dealer or refer to the applicable Chilton's Total Car Care manual for your vehicle.

4. Turn the ignition key **OFF**, if not already done.

5. Using a jumper wire, attach the TAT terminal to a good chassis or engine ground.

➡**Some vehicles may not use the TAT terminal in the DLC. If this is the case with the vehicle at hand, ground the TEN terminal of the DLC. If this is done, however, watch the CHECK ENGINE light for the DTC's, not the MANUAL SHIFT light .**

6. Without starting the engine, turn the ignition switch **ON**.

7. The MANUAL SHIFT light should start to flash out any stored DTC's.

Clearing Codes

EXCEPT PROBE AND 1991–92 ESCORT MODELS

To clear the DTC's from the vehicle's memory, simply disconnect the ground jumper wire from the STI terminal while the retrieval cycle is running. The computer will erase all applicable codes.

PROBE AND 1991–92 ESCORT MODELS

To clear the DTC's from the vehicle's memory, disconnect the negative battery cable for at least one minute. Reconnect the cable and all of the DTC's should be purged from the computer's memory.

DIAGNOSTIC TROUBLE CODES—FORD MOTOR CO.		
2-DIGIT KOEO CODES		
Codes	**Ford Pinpoint Test Designation**	**System/Component/Circuit Fault**
11	—	All Systems Functioning Normally
19	—	Faulty processor
21	DA	Intake Air Temperature (IAT)/Engine Coolant Temperature (ECT) Sensors
22	DF	Manifold Absolute Pressure (MAP)/Barometric Pressure (BARO) Sensor
23	DH	Throttle Position (TP) Sensor (Gasoline Engines)
24	DA	Intake Air Temperature (IAT)/Engine Coolant Temperature (ECT) Sensors
26	DC	Mass Air Flow (MAF) Sensor—Cars
26	TE	Transmission Fluid Temperature (TFT) Sensor—Trucks
31	DN	EGR Valve Position (EVP) Sensor/EGR Vacuum Regulator (EVR) Solenoid
32	DN	EGR Valve Position (EVP) Sensor/EGR Vacuum Regulator (EVR) Solenoid
34	DN	EGR Valve Position (EVP) Sensor/EGR Vacuum Regulator (EVR) Solenoid
35	DN	EGR Valve Position (EVP) Sensor/EGR Vacuum Regulator (EVR) Solenoid
47	TB	4x4 Low/Transmission Control Switch TCIL-TCS-TCSM
51	DA	Intake Air Temperature (IAT)/Engine Coolant Temperature (ECT) Sensors
52	FF	Power Steering Pressure (PSP) Switch
53	DH	Throttle Position (TP) Sensor (Gasoline Engines)
53	DQ	Throttle Position (TP) Sensor (7.3L Diesel)
54	DA	Intake Air Temperature (IAT)/Engine Coolant Temperature (ECT) Sensors
56	DC	Mass Air Flow (MAF) Sensor—Cars
56	TE	Transmission Fluid Temperature (TFT) Sensor—Trucks
57	KP	Octane Adjust (OCT ADJ)
61	DA	Intake Air Temperature (IAT)/Engine Coolant Temperature (ECT) Sensors
63	DH	Throttle Position (TP) Sensor—Gasoline Engines
63	DQ	Throttle Position (TP) Sensor—7.3L Diesel Engines
64	DA	Intake Air Temperature (IAT)/Engine Coolant Temperature (ECT) Sensors
66	TE	Transmission Fluid Temperature (TFT) Sensor—Trucks
67	KM	WOT A/C Cutout (WAC) A/C Demand
67	TA	Park/Neutral Position (PNP)/Clutch Pedal Position (CPP) Switches
79	KM	WOT A/C Cutout (WAC) A/C Demand
81	KC	Secondary Air Injection (AIRB)/(AIRD) Solenoids
82	KC	Secondary Air Injection (AIRB)/(AIRD) Solenoids
84	DN	EGR Valve Position (EVP) Sensor/EGR Vacuum Regulator (EVR) Solenoid
85	KD	Canister Purge (CANP) Solenoid
86	TC	Transmission Solenoids
87	J	Fuel Pump or Circuit
89	TC	Transmission Solenoids
91	TC	Transmission Solenoids
92	TC	Transmission Solenoids
93	TC	Transmission Solenoids
94	TC	Transmission Solenoids
98	TC	Transmission Solenoids
99	TC	Transmission Solenoids
All Other Codes Not Listed	QA	No Diagnostic Trouble Codes (DTC)/DTC Not Listed

89446G60

Ford Diagnostic Trouble Codes—2-digit KOEO codes

DIAGNOSTIC TROUBLE CODES—FORD MOTOR CO.
2-DIGIT KOER CODES

Codes	Pinpoint Test Designation	System/Component/Circuit Fault
11	—	All Systems Functioning Normally
12	KE	Idle Air Control (IAC) Solenoid
13	KE	Idle Air Control (IAC) Solenoid
16	KE	Idle Air Control (IAC) Solenoid
18	PA	Spark Timing Check—Distributor Ignition (DI)
19	—	Faulty processor
21	DA	Intake Air Temperature (IAT)/Engine Coolant Temperature (ECT) Sensors
22	DF	Manifold Absolute Pressure (MAP)/Barometric Pressure (BARO) Sensor
23	DH	Throttle Position (TP) Sensor (Gasoline Engines)
24	DA	Intake Air Temperature (IAT)/Engine Coolant Temperature (ECT) Sensors
25	DG	Knock Sensor (KS)
26	DC	Mass Air Flow (MAF) Sensor—Cars
26	TE	Transmission Fluid Temperature (TFT) Sensor—Trucks
31	DN	EGR Valve Position (EVP) Sensor/EGR Vacuum Regulator (EVR) Solenoid
32	DN	EGR Valve Position (EVP) Sensor/EGR Vacuum Regulator (EVR) Solenoid
33	DN	EGR Valve Position (EVP) Sensor/EGR Vacuum Regulator (EVR) Solenoid
34	DN	EGR Valve Position (EVP) Sensor/EGR Vacuum Regulator (EVR) Solenoid
35	DN	EGR Valve Position (EVP) Sensor/EGR Vacuum Regulator (EVR) Solenoid
41	H	Fuel Control System
42	HA	Adaptive Fuel System
44	KC	Secondary Air Injection (AIRB)/(AIRD) Solenoids
45	KC	Secondary Air Injection (AIRB)/(AIRD) Solenoids
46	KC	Secondary Air Injection (AIRB)/(AIRD) Solenoids
52	FF	Power Steering Pressure (PSP) Switch
65	TB	4x4 Low/Transmission Control Switch TCIL-TCS-TCSM
72	DF	Manifold Absolute Pressure (MAP)/Barometric Pressure (BARO) Sensor
73	DH	Throttle Position (TP) Sensor (Gasoline Engines)
74	FD	Brake On/Off (BOO) switch
75	FD	Brake On/Off (BOO) switch
77	M	Dynamic Response Test
91	H	Fuel Control System
92	H	Fuel Control System
94	KC	Secondary Air Injection (AIRB)/(AIRD) Solenoids
All Other Codes Not Listed	QA	No Diagnostic Trouble Codes (DTC)/DTC Not Listed

KOER - Key On, Engine Running

89446G61

Ford Diagnostic Trouble Codes—2-digit KOER codes

DIAGNOSTIC TROUBLE CODES—FORD MOTOR CO.
2-DIGIT CONTINUOUS CODES

Codes	Pinpoint Test Designation	System/Component/Circuit Fault
11	—	All Systems Functioning Normally
14	NA	Ignition Diagnostic Monitor (IDM)/Distributor Ignition(DI)
15	QB	Continuous Memory Diagnostic Trouble Code (DTC)
18	NA	Ignition Diagnostic Monitor (IDM)/Distributor Ignition(DI)
22	DF	Manifold Absolute Pressure (MAP)/Barometric Pressure (BARO) Sensor
29	DP	Vehicle Speed Sensor (VSS)—Cars
29	DS	Programmable Speedometer/Odometer Module (PSOM)—Trucks
31	DN	EGR Valve Position (EVP) Sensor/EGR Vacuum Regulator (EVR) Solenoid
32	DN	EGR Valve Position (EVP) Sensor/EGR Vacuum Regulator (EVR) Solenoid
33	DN	EGR Valve Position (EVP) Sensor/EGR Vacuum Regulator (EVR) Solenoid
34	DN	EGR Valve Position (EVP) Sensor/EGR Vacuum Regulator (EVR) Solenoid
35	DN	EGR Valve Position (EVP) Sensor/EGR Vacuum Regulator (EVR) Solenoid
41	H	Fuel Control System
49	TG	Electronic Transmission/Continuous Memory Diagnostic Trouble Codes (DTC)
51	DA	Intake Air Temperature (IAT)/Engine Coolant Temperature (ECT) Sensors
53	DH	Throttle Position (TP) Sensor (Gasoline Engines)
53	DQ	Throttle Position (TP) Sensor (7.3L Diesel)
54	DA	Intake Air Temperature (IAT)/Engine Coolant Temperature (ECT) Sensors
56	DC	Mass Air Flow (MAF) Sensor—Cars
56	TG	Electronic Transmission/Continuous Memory Diagnostic Trouble Codes (DTC)
59	TG	Electronic Transmission/Continuous Memory Diagnostic Trouble Codes (DTC)
61	DA	Intake Air Temperature (IAT)/Engine Coolant Temperature (ECT) Sensors
62	TG	Electronic Transmission/Continuous Memory Diagnostic Trouble Codes (DTC)
63	DH	Throttle Position (TP) Sensor (Gasoline Engines)
63	DQ	Throttle Position (TP) Sensor (7.3L Diesel)
64	DA	Intake Air Temperature (IAT)/Engine Coolant Temperature (ECT) Sensors
66	DC	Mass Air Flow (MAF) Sensor—Cars
66	TG	Electronic Transmission/Continuous Memory Diagnostic Trouble Codes (DTC)
67	TA	Park/Neutral Position (PNP)/Clutch Pedal Position (CPP) Switches
67	TG	Electronic Transmission/Continuous Memory Diagnostic Trouble Codes (DTC)
69	TG	Electronic Transmission/Continuous Memory Diagnostic Trouble Codes (DTC)
81	DF	Manifold Absolute Pressure (MAP)/Barometric Pressure (BARO) Sensor
87	J	Fuel Pump or Circuit
91	H	Fuel Control System
95	J	Fuel Pump or Circuit
96	J	Fuel Pump or Circuit
99	TG	Electronic Transmission/Continuous Memory Diagnostic Trouble Codes (DTC)

89446G62

Ford Diagnostic Trouble Codes—2-digit continuous codes

DIAGNOSTIC TROUBLE CODES—FORD MOTOR CO.
3-DIGIT KOEO CODES

Codes	Pinpoint Test Designation	System/Component/Circuit Fault
111	—	All Systems Functioning Normally
112	DA	Intake Air Temperature (IAT)/Engine Coolant Temperature (ECT) Sensors
113	DA	Intake Air Temperature (IAT)/Engine Coolant Temperature (ECT) Sensors
114	DA	Intake Air Temperature (IAT)/Engine Coolant Temperature (ECT) Sensors
116	DA	Intake Air Temperature (IAT)/Engine Coolant Temperature (ECT) Sensors
117	DA	Intake Air Temperature (IAT)/Engine Coolant Temperature (ECT) Sensors
118	DA	Intake Air Temperature (IAT)/Engine Coolant Temperature (ECT) Sensors
121	DH	Throttle Position (TP) Sensor (Gasoline Engines)
122	DH	Throttle Position (TP) Sensor (Gasoline Engines)
123	DH	Throttle Position (TP) Sensor (Gasoline Engines)
126	DF	Manifold Absolute Pressure (MAP)/Barometric Pressure (BARO) Sensor
158	DC	Mass Air Flow (MAF) Sensor—Cars
159	DC	Mass Air Flow (MAF) Sensor—Cars
226	NC	Ignition Diagnostic Monitor (IDM)/Distributor Ignition(DI)—EDIS Ignition System
327	DL	Pressure Feedback EGR (PFE)/Differential PFE (DPFE) Sensor/EGR Vacuum Regulator (EVR) Solenoid—except Mustang, 5.0L Thunderbird/Cougar & Trucks
327	DN	EGR Valve Position (EVP) Sensor/EGR Vacuum Regulator (EVR) Solenoid—Mustang, 5.0L Thunderbird/Cougar & Trucks
328	DN	EGR Valve Position (EVP) Sensor/EGR Vacuum Regulator (EVR) Solenoid
334	DN	EGR Valve Position (EVP) Sensor/EGR Vacuum Regulator (EVR) Solenoid
335	DL	Pressure Feedback EGR (PFE)/Differential PFE (DPFE) Sensor/EGR Vacuum Regulator (EVR) Solenoid
336	DL	Pressure Feedback EGR (PFE)/Differential PFE (DPFE) Sensor/EGR Vacuum Regulator (EVR) Solenoid
337	DL	Pressure Feedback EGR (PFE)/Differential PFE (DPFE) Sensor/EGR Vacuum Regulator (EVR) Solenoid—except Mustang, 5.0L Thunderbird/Cougar & Trucks
337	DN	EGR Valve Position (EVP) Sensor/EGR Vacuum Regulator (EVR) Solenoid—Mustang, 5.0L Thunderbird/Cougar & Trucks
341	KP	Octane Adjust (OCT ADJ)
511	—	Faulty processor
513	—	Faulty processor
519	FF	Power Steering Pressure (PSP) Switch
524	X	Constant Control Relay Module (CCRM)
525	TA	Park/Neutral Position (PNP)/Clutch Pedal Position (CPP) Switches
539	KM	WOT A/C Cutout (WAC) A/C Demand
551	KT	Intake Manifold Runner Control (IMRC) System
552	KC	Secondary Air Injection (AIRB)/(AIRD) Solenoids
553	KC	Secondary Air Injection (AIRB)/(AIRD) Solenoids
554	KN	Fuel Pressure Regulator Control (FPRC) Solenoid
556	J	Fuel Pump or Circuit—except Tempo/Topaz, 2.3L Mustang, Probe, Taurus/Sable, Continental
556	X	Constant Control Relay Module (CCRM)—Tempo/Topaz, 2.3L Mustang, Probe, Taurus/Sable, Continental
557	X	Constant Control Relay Module (CCRM)
558	DL	Pressure Feedback EGR (PFE)/Differential PFE (DPFE) Sensor/EGR Vacuum Regulator (EVR) Solenoid—except Mustang, 5.0L Thunderbird/Cougar & Trucks
558	DN	EGR Valve Position (EVP) Sensor/EGR Vacuum Regulator (EVR) Solenoid—Mustang, 5.0L Thunderbird/Cougar & Trucks
559	KM	WOT A/C Cutout (WAC) A/C Demand

Ford Diagnostic Trouble Codes—3-digit KOEO codes

89446G67

DIAGNOSTIC TROUBLE CODES—FORD MOTOR CO.
3-DIGIT KOEO CODES

Codes	Pinpoint Test Designation	System/Component/Circuit Fault
563	KF	Low Fan Control (LFC)/High Fan Control (HFC)—Excort/Tracer
563	X	Constant Control Relay Module (CCRM)—except Excort/Tracer
564	KF	Low Fan Control (LFC)/High Fan Control (HFC)—Excort/Tracer
564	X	Constant Control Relay Module (CCRM)—except Excort/Tracer
565	KD	Canister Purge (CANP) Solenoid
566	TC	Transmission Solenoids
569	KD	Canister Purge (CANP) Solenoid
621	TC	Transmission Solenoids
622	TC	Transmission Solenoids
624	TC	Transmission Solenoids
625	TC	Transmission Solenoids
626	TC	Transmission Solenoids
627	TC	Transmission Solenoids
629	TC	Transmission Solenoids
631	TB	4x4 Low/Transmission Control Switch TCIL-TCS-TCSM
633	TB	4x4 Low/Transmission Control Switch TCIL-TCS-TCSM
634	TD	Manual Lever Position (MLP) Sensor
636	TE	Transmission Fluid Temperature (TFT) Sensor—Trucks
637	TE	Transmission Fluid Temperature (TFT) Sensor
638	TE	Transmission Fluid Temperature (TFT) Sensor
641	TC	Transmission Solenoids
643	TC	Transmission Solenoids
652	TC	Transmission Solenoids
998	TC	Transmission Solenoids
Codes Not Listed	QA	No Diagnostic Trouble Codes (DTC)/DTC Not Listed

KOEO - Key (Ignition) On, Engine Off

89446G68

Ford Diagnostic Trouble Codes—3-digit KOEO codes continued

DIAGNOSTIC TROUBLE CODES—FORD MOTOR CO.
3-DIGIT KOER CODES

Codes	Pinpoint Test Designation	System/Component/Circuit Fault
111	—	All Systems Functioning Normally
114	DA	Intake Air Temperature (IAT)/Engine Coolant Temperature (ECT) Sensors
116	DA	Intake Air Temperature (IAT)/Engine Coolant Temperature (ECT) Sensors
121	DH	Throttle Position (TP) Sensor (Gasoline Engines)
126	DF	Manifold Absolute Pressure (MAP)/Barometric Pressure (BARO) Sensor
129	DC	Mass Air Flow (MAF) Sensor—Cars, Ranger, Aerostar and Explorer
129	DF	Manifold Absolute Pressure (MAP)/Barometric Pressure (BARO) Sensor—Trucks except Ranger, Aerostar and Explorer
136	H	Fuel Control System
137	H	Fuel Control System
159	DC	Mass Air Flow (MAF) Sensor—Cars
167	DH	Throttle Position (TP) Sensor (Gasoline Engines)
172	H	Fuel Control System
173	H	Fuel Control System
213	PA	Spark Timing Check—Distributor Ignition (DI)
213	PB	Spark Timing Check—Electronic Ignition (Low Data Rate)
213	PC	Spark Timing Check—Electronic Ignition (High Date Rate)
225	DG	Knock Sensor (KS)
311	KC	Secondary Air Injection (AIRB)/(AIRD) Solenoids
312	KC	Secondary Air Injection (AIRB)/(AIRD) Solenoids
313	KC	Secondary Air Injection (AIRB)/(AIRD) Solenoids
314	KC	Secondary Air Injection (AIRB)/(AIRD) Solenoids
326	DL	Pressure Feedback EGR (PFE)/Differential PFE (DPFE) Sensor/EGR Vacuum Regulator (EVR) Solenoid
327	DL	Pressure Feedback EGR (PFE)/Differential PFE (DPFE) Sensor/EGR Vacuum Regulator (EVR) Solenoid—except Mustang, 5.0L Thunderbird/Cougar & Trucks
327	DN	EGR Valve Position (EVP) Sensor/EGR Vacuum Regulator (EVR) Solenoid—Mustang, 5.0L Thunderbird/Cougar & Trucks
328	DN	EGR Valve Position (EVP) Sensor/EGR Vacuum Regulator (EVR) Solenoid
332	DL	Pressure Feedback EGR (PFE)/Differential PFE (DPFE) Sensor/EGR Vacuum Regulator (EVR) Solenoid—except Mustang, 5.0L Thunderbird/Cougar & Trucks
332	DN	EGR Valve Position (EVP) Sensor/EGR Vacuum Regulator (EVR) Solenoid—Mustang, 5.0L Thunderbird/Cougar & Trucks
334	DN	EGR Valve Position (EVP) Sensor/EGR Vacuum Regulator (EVR) Solenoid
336	DL	Pressure Feedback EGR (PFE)/Differential PFE (DPFE) Sensor/EGR Vacuum Regulator (EVR) Solenoid
337	DL	Pressure Feedback EGR (PFE)/Differential PFE (DPFE) Sensor/EGR Vacuum Regulator (EVR) Solenoid—except Mustang, 5.0L Thunderbird/Cougar & Trucks
337	DN	EGR Valve Position (EVP) Sensor/EGR Vacuum Regulator (EVR) Solenoid—Mustang, 5.0L Thunderbird/Cougar & Trucks
411	KE	Idle Air Control (IAC) Solenoid
412	KE	Idle Air Control (IAC) Solenoid
511	—	Faulty processor
513	—	Faulty processor
521	FF	Power Steering Pressure (PSP) Switch
536	FD	Brake On/Off (BOO) switch
538	M	Dynamic Response Test
632	TB	4x4 Low/Transmission Control Switch TCIL-TCS-TCSM
636	TE	Transmission Fluid Temperature (TFT) Sensor—Trucks
639	TF	Transmission Shaft Speed Sensor (TSS)/Output Shaft Speed (OSS) Sensor

89446G63

Ford Diagnostic Trouble Codes—3-digit KOER codes

DIAGNOSTIC TROUBLE CODES—FORD MOTOR CO.
3-DIGIT CONTINUOUS CODES

Codes	Pinpoint Test Designation	System/Component/Circuit Fault
111	—	All Systems Functioning Normally
112	DA	Intake Air Temperature (IAT)/Engine Coolant Temperature (ECT) Sensors
113	DA	Intake Air Temperature (IAT)/Engine Coolant Temperature (ECT) Sensors
117	DA	Intake Air Temperature (IAT)/Engine Coolant Temperature (ECT) Sensors
118	DA	Intake Air Temperature (IAT)/Engine Coolant Temperature (ECT) Sensors
121	G	In Range MAF/TP/Fuel Injector Pulse Width Test
122	DH	Throttle Position (TP) Sensor (Gasoline Engines)
123	DH	Throttle Position (TP) Sensor (Gasoline Engines)
124	G	In Range MAF/TP/Fuel Injector Pulse Width Test
125	G	In Range MAF/TP/Fuel Injector Pulse Width Test
126	DF	Manifold Absolute Pressure (MAP)/Barometric Pressure (BARO) Sensor
128	DF	Manifold Absolute Pressure (MAP)/Barometric Pressure (BARO) Sensor
139	H	Fuel Control System
144	H	Fuel Control System
157	DC	Mass Air Flow (MAF) Sensor—Cars, Ranger, Aerostar and Explorer
158	DC	Mass Air Flow (MAF) Sensor—Cars
171	H	Fuel Control System
172	H	Fuel Control System
173	H	Fuel Control System
174	H	Fuel Control System
175	H	Fuel Control System
176	H	Fuel Control System
177	H	Fuel Control System
178	H	Fuel Control System
179	HA	Adaptive Fuel System
181	HA	Adaptive Fuel System
182	HA	Adaptive Fuel System
183	HA	Adaptive Fuel System
184	G	In Range MAF/TP/Fuel Injector Pulse Width Test
185	G	In Range MAF/TP/Fuel Injector Pulse Width Test
186	G	In Range MAF/TP/Fuel Injector Pulse Width Test
187	G	In Range MAF/TP/Fuel Injector Pulse Width Test
188	HA	Adaptive Fuel System
189	HA	Adaptive Fuel System
191	HA	Adaptive Fuel System
192	HA	Adaptive Fuel System
194	H	Fuel Control System
195	H	Fuel Control System
211	NA	Ignition Diagnostic Monitor (IDM)/Distributor Ignition(DI)—TFI Ignition System
211	NB	Ignition Diagnostic Monitor (IDM)/Distributor Ignition(DI)—DIS Ignition System
211	NC	Ignition Diagnostic Monitor (IDM)/Distributor Ignition(DI)—EDIS Ignition System
212	NA	Ignition Diagnostic Monitor (IDM)/Distributor Ignition(DI)—TFI Ignition System
214	DR	Cylinder Identification (CID) Circuits
219	PB	Spark Timing Check—Electronic Ignition (Low Data Rate)
327	DN	EGR Valve Position (EVP) Sensor/EGR Vacuum Regulator (EVR) Solenoid—Mustang, 5.0L Thunderbird/Cougar & Trucks
327	DN	EGR Valve Position (EVP) Sensor/EGR Vacuum Regulator (EVR) Solenoid—except Mustang, 5.0L Thunderbird/Cougar & Trucks

Ford Diagnostic Trouble Codes—3-digit KOER codes continued

89446G64

DIAGNOSTIC TROUBLE CODES—FORD MOTOR CO.
3-DIGIT CONTINUOUS CODES

Codes	Pinpoint Test Designation	System/Component/Circuit Fault
328	DN	EGR Valve Position (EVP) Sensor/EGR Vacuum Regulator (EVR) Solenoid
332	DL	Pressure Feedback EGR (PFE)/Differential PFE (DPFE) Sensor/EGR Vacuum Regulator (EVR) Solenoid—except Mustang, 5.0L Thunderbird/Cougar & Trucks
332	DN	EGR Valve Position (EVP) Sensor/EGR Vacuum Regulator (EVR) Solenoid—Mustang, 5.0L Thunderbird/Cougar & Trucks
334	DN	EGR Valve Position (EVP) Sensor/EGR Vacuum Regulator (EVR) Solenoid
336	DL	Pressure Feedback EGR (PFE)/Differential PFE (DPFE) Sensor/EGR Vacuum Regulator (EVR) Solenoid
337	DL	Pressure Feedback EGR (PFE)/Differential PFE (DPFE) Sensor/EGR Vacuum Regulator (EVR) Solenoid—except Mustang, 5.0L Thunderbird/Cougar & Trucks
337	DN	EGR Valve Position (EVP) Sensor/EGR Vacuum Regulator (EVR) Solenoid—Mustang, 5.0L Thunderbird/Cougar & Trucks
338	DA	Intake Air Temperature (IAT)/Engine Coolant Temperature (ECT) Sensors
339	DA	Intake Air Temperature (IAT)/Engine Coolant Temperature (ECT) Sensors
452	DP	Vehicle Speed Sensor (VSS)—Cars
452	DS	Programmable Speedometer/Odometer Module (PSOM)—Trucks
511	—	Faulty processor
512	QB	Continuous Memory Diagnostic Trouble Code (DTC)
513	—	Faulty processor
524	X	Constant Control Relay Module (CCRM)
525	TA	Park/Neutral Position (PNP)/Clutch Pedal Position (CPP) Switches
528	TA	Park/Neutral Position (PNP)/Clutch Pedal Position (CPP) Switches
529	ML	Self-Test Output (STO)/Malfunction Indicator Lamp (MIL)
533	ML	Self-Test Output (STO)/Malfunction Indicator Lamp (MIL)
536	FD	Brake On/Off (BOO) switch
556	J	Fuel Pump or Circuit—except Tempo/Topaz, 2.3L Mustang, Probe, Taurus/Sable, Continental
556	X	Constant Control Relay Module (CCRM)—Tempo/Topaz, 2.3L Mustang, Probe, Taurus/Sable, Continental
557	X	Constant Control Relay Module (CCRM)
617	TG	Electronic Transmission/Continuous Memory Diagnostic Trouble Codes (DTC)
618	TG	Electronic Transmission/Continuous Memory Diagnostic Trouble Codes (DTC)
619	TG	Electronic Transmission/Continuous Memory Diagnostic Trouble Codes (DTC)
621	TC	Transmission Solenoids
622	TC	Transmission Solenoids
624	TG	Electronic Transmission/Continuous Memory Diagnostic Trouble Codes (DTC)
625	TG	Electronic Transmission/Continuous Memory Diagnostic Trouble Codes (DTC)
628	TG	Electronic Transmission/Continuous Memory Diagnostic Trouble Codes (DTC)
634	TD	Manual Lever Position (MLP) Sensor—Escort/Tracer, E-Series/F-Series Trucks, Bronco
634	TG	Electronic Transmission/Continuous Memory Diagnostic Trouble Codes (DTC)—except Escort/Tracer, E-Series/F-Series Trucks, Bronco
637	TE	Transmission Fluid Temperature (TFT) Sensor—Escort/Tracer
637	TG	Electronic Transmission/Continuous Memory Diagnostic Trouble Codes (DTC)—except Escort/Tracer
638	TE	Transmission Fluid Temperature (TFT) Sensor—Escort/Tracer
638	TG	Electronic Transmission/Continuous Memory Diagnostic Trouble Codes (DTC)—except Escort/Tracer
639	TF	Transmission Shaft Speed Sensor (TSS)/Output Shaft Speed (OSS) Sensor
641	TC	Transmission Solenoids—Escort/Tracer

89446G65

Ford Diagnostic Trouble Codes—3-digit continuous codes

DIAGNOSTIC TROUBLE CODES—FORD MOTOR CO.
3-DIGIT CONTINUOUS CODES

Codes	Pinpoint Test Designation	System/Component/Circuit Fault
643	TC	Transmission Solenoids
645	TG	Electronic Transmission/Continuous Memory Diagnostic Trouble Codes (DTC)
646	TG	Electronic Transmission/Continuous Memory Diagnostic Trouble Codes (DTC)
647	TG	Electronic Transmission/Continuous Memory Diagnostic Trouble Codes (DTC)
648	TG	Electronic Transmission/Continuous Memory Diagnostic Trouble Codes (DTC)
649	TG	Electronic Transmission/Continuous Memory Diagnostic Trouble Codes (DTC)
651	TG	Electronic Transmission/Continuous Memory Diagnostic Trouble Codes (DTC)
654	TD	Manual Lever Position (MLP) Sensor
656	TG	Electronic Transmission/Continuous Memory Diagnostic Trouble Codes (DTC)

Ford Diagnostic Trouble Codes—3-digit continuous codes continued

89446G66

General Motors Corporation

ENGINE CODES

Self-Diagnostics

Automotive manufacturers have developed on-board computers to control engines, transmissions and many other components. These on-board computers with dozens of sensors and actuators have become almost impossible to test without the help of electronic test equipment.

One of these electronic test devices has become the on-board computer itself. The Powertrain Control Modules (PCM), sometimes called the Electronic Control Module (ECM), used on toadies vehicles has a built in self testing system. This self test ability is called self-diagnosis. The self-diagnosis system will test many or all of the sensors and controlled devices for proper function. When a malfunction is detected this system will store a code in memory that's related to that specific circuit. The computer can later be accessed to obtain fault codes recorded in memory using the procedures for Reading Codes. This helps narrow down what area to begin testing.

Fault code meanings can vary from year to year even on the same model. It is extremely important after retrieving a fault code to verify its meaning with a proper manual. Servicing a code incorrectly will not only lead to the wrong conclusion but could also cause damage if tested or serviced incorrectly.

Since the control module is programmed to recognize the presence and value of electrical inputs, it will also note the lack of a signal or a radical change in values. It will, for example, react to the loss of signal from the vehicle speed sensor or note that engine coolant temperature has risen beyond acceptable (programmed) limits. Once a fault is recognized, a numeric code is assigned and held in memory. The dashboard warning lamp—CHECK ENGINE or SERVICE ENGINE SOON—will illuminate to advise the operator that the system has detected a fault.

More than one code may be stored. Although not every engine uses every code and the same code may carry different meanings relative to each engine or engine family. For example, on the 3.3L (VIN N), Code 46 indicates a fault found in the power steering pressure switch circuit. The same code on the 5.7L (VIN F) engine indicates a fault in the VATS anti-theft system. The list of codes and descriptions can be found in the 'Code Descriptions' section of the manual.

In the event of an PCM failure, the system will default to a pre-programmed set of values. These are compromise values which allow the engine to operate, although possibly at reduced efficiency. This is also known as the default, limp-in or back-up mode. Driveability is almost always affected when the PCM enters this mode.

SERVICE PRECAUTIONS

• Protect the on-board solid-state components from rough handling or extremes of temperature.

• Always turn the ignition OFF when connecting or disconnecting battery cables, jumper cables, or a battery charger. Failure to do this can result in PCM or other electronic component damage.

• Remove the PCM before any arc welding is performed to the vehicle

• Electronic components are very susceptible to damage caused by electrostatic discharge (static electricity). To prevent electronic component damage, do not touch the control module connector pins or soldered components on the control module circuit board.

VISUAL INSPECTION

This is possibly the most critical step of diagnosis. A detailed examination of all connectors, wiring and vacuum hoses can often lead to a repair without further diagnosis. Also, take into consideration if the vehicle has been serviced recently? Sometimes things get reconnected in the wrong place, or not at all. A careful inspector will check the undersides of hoses as well as the integrity of hard-to-reach hoses blocked by the air cleaner or other components. Correct routing for vacuum hoses can be obtained from your specific vehicle service manual or Vehicle Emission Control Information (VECI) label in the engine compartment of the vehicle. Wiring should be checked carefully for any sign of strain, burning, crimping or terminals pulled-out from a connector.

Checking connectors at components or in harnesses is required; usually, pushing them together will reveal a loose fit. Also, check electrical connectors for corroded, bent, damaged, improperly seated pins, and bad wire crimps to terminals. Pay particular attention to ground circuits, making sure they are not loose or corroded. Remember to inspect connectors and hose fittings at components not mounted on the engine, such as the evaporative canister or relays mounted on the fender aprons. Any component or wiring in the vicinity of a fluid leak or spillage should be given extra attention during inspection.

➡There are many problems with connectors on electronic engine control systems. Due to the low voltage signals that these systems use any dirt, corrosion or damage will affect their operation. Note that some connectors use a special grease on the contacts to prevent corrosion. Do not wipe this grease off, it is a special type for this purpose. You can obtain this grease from your vehicle dealer.

Additionally, inspect maintenance items such as belt condition and tension, battery charge and condition and the radiator cap carefully. Any of these very simple items may affect the system enough to set a fault.

DASHBOARD WARNING LAMP

The primary function of the dash warning lamp is to advise the operator that a fault has been detected, and, in most cases, a code stored. Under normal conditions, the dash warning lamp will illuminate when the ignition is turned ON. Once the engine is started and running, the PCM will perform a system check and extinguish the warning lamp if no fault is found.

Additionally, the dash warning lamp can be used to retrieve stored codes after the system is placed in the Diagnostic Mode. Codes are transmitted as a series of flashes with short or long pauses. When the system is placed in the Field Service Mode (available on fuel injected model), the dash lamp will indicate open loop or closed loop function.

The Check Engine or Service Engine Soon (MIL) lamp is used for reading trouble codes

INTERMITTENT PROBLEMS

If a fault occurs intermittently, such as a loose connector pin breaking contact as the vehicle hits a bump, the PCM will note the fault as it occurs and energize the dash warning lamp. If the problem self-corrects, as with the terminal pin again making contact, the dash lamp will extinguish after 10 seconds but a code will remain stored in the PCM memory. When an unexpected code appears during an intermittent failure that self-corrected; the codes are still useful in diagnosis and should not be discounted.

DIAGNOSTIC CONNECTOR LOCATION

The Assembly Line Communication Link (ALCL) or Assembly Line Diagnostic Link (ALDL) is a Diagnostic Link Connector (DLC) located in the passenger compartment. It has terminals which are used in the assembly plant to check that the engine is operating properly before it leaves the plant.

This DLC is where you connect you jumper the terminals to place the engine control computer into self-diagnostic mode. The standard term DLC is sometimes referred to as the ALCL or the ALDL in different manuals. Either way it is referred to, they all still perform the same function.

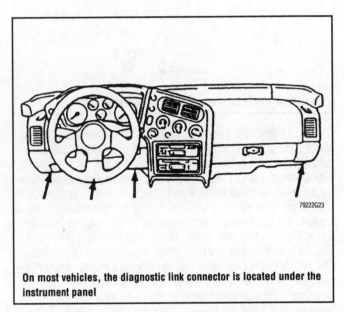

On most vehicles, the diagnostic link connector is located under the instrument panel

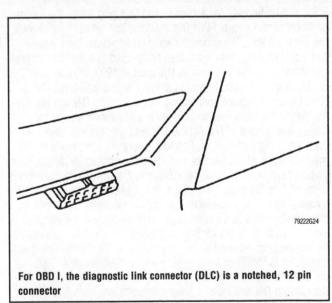

For OBD I, the diagnostic link connector (DLC) is a notched, 12 pin connector

Reading & Clearing Codes

READING CODES (EXCEPT CADILLAC)

Since the inception of electronic engine management systems on General Motors vehicles, there has been a variety of connectors provided to the technician for retrieving Diagnostic Trouble Codes

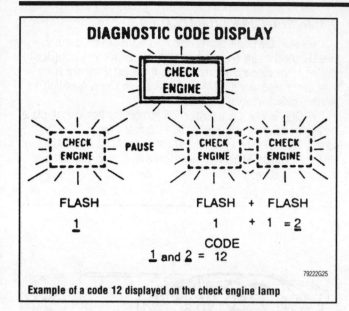

DIAGNOSTIC CODE DISPLAY

CHECK ENGINE

CHECK ENGINE — PAUSE — CHECK ENGINE — CHECK ENGINE

FLASH
<u>1</u>

FLASH + FLASH
1 + 1 = <u>2</u>

CODE
<u>1</u> and <u>2</u> = 12

79222G25

Example of a code 12 displayed on the check engine lamp

(DTC)s. Additionally, there have been a number of different names given to these connectors over the years; Assembly Line Communication Link (ALCL), Assembly Line Diagnostic Link (ALDL), Data Link Connector (DLC). Actually when the system was initially introduced to the 49 states in 1979, early 1980, there was no connector used at all. On these early vehicles there was a green spade terminal taped to the ECM harness and connected to the diagnostic enable line at the computer. When this terminal was grounded with the key ON, the system would flash any stored diagnostic trouble codes. The introduction of the ALOL was found to be a much more convenient way of retrieving fault codes. This connector was located underneath the instrument panel on most GM vehicles, however on some models it will not be found there. On early Corvettes the ALOL is located underneath the ashtray, it can be found in the glove compartment of some early FWD Oldsmobiles, and between the seats in the Pontiac Fiero. The connector was first introduced as a square connector with four terminals, then progressed to a flat five terminal connector, and finally to what is still used in 1993, a 12 terminal double row connector. To access stored Diagnostic Trouble Codes (DTC) from the square connector, turn the ignition **ON** and identify the diagnostic enable terminal (usually a white wire with a black tracer) and ground it. The flat five terminal connector is identified from left to right as A, B, C, D, and E. There is a space between terminal D and E which permits a spade to be inserted for the purposes of diagnostics when the ignition key is ON. On this connector terminal D is the diagnostic enable line, and E is a ground. The 12 terminal double row connector has been continually expanded through the years as vehicles acquired more on-board electronic systems such as Anti-lock Brakes. Despite this the terminals used for engine code retrieval have remained the same. The 12 terminal connector is identified from right-to-left on the top row A-F, and on the bottom row from left-to-right, G-L. To access engine codes turn the ignition **ON** and insert a jumper between terminals A and B. Terminal A is a ground, and terminal B is the diagnostic request line. Stored trouble codes can be read through the flashing of the Check Engine Light or on later vehicles the Service Engine Soon lamp. Trouble codes are identified by the timed flash of the indicator light. When diagnostics are first entered the light will flash once, pause; then two quick flashes.

This reads as DTC 12 which indicates that the diagnostic system is working. This code will flash indefinitely if there are no stored trouble codes. If codes are stored in memory, Code 12 will flash three times before the next code appears. Codes are displayed in the next highest numerical sequence. For example, Code 13 would be displayed next if it was stored in memory and would read as follow: flash, pause, flash, flash, flash, long pause, repeat twice. This sequence will continue until all codes have been displayed, and then start all over again with Code 12.

CLEARING CODES (EXCEPT CADILLAC)

Except Riviera, Toronado and Trofeo:
To clear any Diagnostic Trouble Codes (DTC's) from the PCM memory, either to determine if the malfunction will occur again or because repair has been completed, power feed must be disconnected for at least 30 seconds. Depending on how the vehicle is equipped, the system power feed can be disconnected at the positive battery terminal pigtail, the inline fuse holder that originates at the positive connection at the battery, or the ECM/PCM fuse in the fuse block. The negative battery terminal may be disconnected but other on-board memory data such as preset radio tuning will also be lost. To prevent system damage, the ignition switch must be in the OFF position when disconnecting or reconnecting power.

When using a Diagnostic Computer such as Tech 1, or equivalent scan tool to read the diagnostic trouble codes, clearing the codes is done in the same manner. On some systems, OTC's may be cleared through the Tech 1, or equivalent scan tool.

On Riviera, Toronado and Trofeo, clearing codes is part of the dashboard display menu or diagnostic routine. Because of the amount of electronic equipment on these vehicles, clearing codes by disconnecting the battery is not recommended.

Riviera, Toronado and Trofeo (Non-CRT/DID Vehicles)—Using The On-Board Diagnostic Display System:
First turn the ignition to the **ON** position. On Riviera depress the OFF and TEMP buttons on the ECCP at the same time and hold until all display segments light. This is known as the Segment Check. On Toronado and Trofeo follow the same procedure, however, depress the OFF and WARMER buttons on the ECOP instead. After diagnostics is entered, any OTC's stored in computer memory will be displayed. Codes may be stored for the PCM, BCM, PC or SIR systems. Following the display of OTC's, the first available system for testing will be displayed. For example, 'EC?' would be displayed on Riviera for EOM testing, while on Toronado and Trofeo the message 'ECM?' will appear. The message is more clear on these vehicles due to increased character space in the IPO display area.

1. Depress the 'FAN UP' button on the ECCP until the message 'DATA EC?' appears on the display for Riviera, or 'ECM DATA?' is displayed on Toronado and Trofeo.

2. Depress the 'FAN DOWN' button on the ECCP until the message 'CLR E CODE' appears on the display for Riviera, or 'ECM CLEAR CODES?' is displayed on Toronado and Trofeo.

3. Depressing the 'FAN UP' button on the ECCP will result in the message 'E CODE CLR' or 'E NOT CLR' on Riviera, 'EOM CODES CLEAR' or 'ECM CODES NOT CLEAR' on Toronado and Trofeo. This message will appear for 3 seconds. After 3 seconds the display will automatically return to the next available test type for the selected system. It is a good idea to either cycle the ignition once or test drive the vehicle to ensure the code(s) do not reset.

Toronado and Trofeo (CRT/DID Equipped)—Using The On-Board Diagnostic Display System:
First turn the ignition switch to the **ON** position. Depress the 'OFF' hard key and 'WARM' soft key on the CRT/DID at the same time and hold until all display segments light. This is the 'Segment Check.' During diagnostic operation, all information will be displayed on the

Driver Information Center (DIC) located in the Instrument Panel Cluster (IPC). Because of the limited space available single letter identifiers are often used for each of the major computer systems. These are: E for ECM, B for 6CM, I for IPC and R for SIR. After diagnostics is entered, any OTC's stored in computer memory will be displayed. Codes may be stored for the PCM, BCM, PC or SIR systems. Following the display of OTC's, the first available system for testing will be displayed. This will be displayed as 'ECM?'.

1. Depress the 'YES' soft key until the display reads 'ECM DATA?'.

2. Depress the 'NO' soft key until the display reads 'ECM CLEAR CODES?'.

3. Depressing the 'YES' soft key will result in either the message 'ECM CODES CLEAR' or 'ECM CODES NOT CLEAR' being displayed, indicating whether or not the codes were successfully cleared. This message will appear for 3 seconds. After 3 seconds the display will automatically return to the next available test type for the selected system. It is a good idea to either cycle the ignition once or test drive the vehicle to ensure the code(s) do not reset.

READING AND CLEARING CADILLAC ENGINE CODES

➡ **The Cadillac Cimmaron used the 12 terminal DLC and codes can be accessed in the conventional manner as all other General Motors vehicles. The rear wheel drive Cadillac equipped with either the 4.1L V6, 5.0L V8, or the 5.7L V8 all can also be accessed in the conventional manner using the DLC.**

1980–1983 Digital Fuel Injection:

1. Turn the ignition switch **ON**.

2. Depress the OFF and WARMER buttons on the Electronic Climate Control (ECO) panel simultaneously and hold until .. is displayed.

3. The numerals 88 should then appear indicating that all display segments are functional. Diagnosis should not be attempted unless the entire 88 is displayed or misdiagnosis will result.

4. If trouble codes are present they will appear on the digital ECO panel as follows:

   a. The lowest numbered code will be displayed for approximately two seconds.

   b. Progressively higher numbered codes, if present, will be displayed consecutively for two second intervals until the highest code has been displayed.

   c. 88 is again displayed.

   d. Parts A, B, and C will be repeated a second time.

   e. After the trouble codes have been displayed Code 70 will then appear. 70 indicates that the ECM is prepared for the switch test procedure.

5. If no trouble codes are stored in memory, 88 will appear for a longer time, and then the ECM will display Code 70.

To clear the codes, while still in the diagnostic mode, press the OFF and HIGH buttons simultaneously until 00 appears. Trouble codes are now removed from the system memory.

➡ **The fuel data panel will go blank when the system is displaying in the diagnostic mode.**

1984 Digital Fuel Injection:

1. Turn the ignition switch **ON**.

2. Depress the OFF and WARMER buttons on the Electronic Climate Control (ECC) panel simultaneously and hold until .. is displayed.

3. —1.8.8 should then appear indicating that all display segments are functional. Diagnosis should not be attempted unless the entire -1.8.8 is displayed or misdiagnosis will result.

4. If trouble codes are present they will appear on the digital ECO panel as follows:

   a. The lowest numbered code will be displayed for approximately two seconds.

   b. Progressively higher numbered codes, if present, will be displayed consecutively for two second intervals until the highest code has been displayed.

   c. 88 is again displayed.

   d. Parts A, B, and C will be repeated a second time.

   e. After the trouble codes have been displayed Code 70 will then appear. 70 indicates that the ECM is prepared for the switch test procedure.

5. If no trouble codes are stored in memory, —1.8.8 will appear for a longer time, and then the ECM will display Code 70.

To clear the codes, while still in the diagnostic mode, press the OFF and HIGH buttons simultaneously until .0.0 appears. Trouble codes are now removed from the system memory.

➡ **The fuel data panel will go blank when the system is displaying in the diagnostic mode.**

1985–1986 Digital Fuel Injection:

1. Turn the ignition switch **ON**.

2. Depress the OFF and WARMER buttons on the Climate Control Panel (CCP) simultaneously and hold until −188 is displayed.

3. −188 should then appear indicating that all display segments are functional. Diagnosis should not be attempted unless the entire −188 is displayed or misdiagnosis will result.

4. If trouble codes are present they will appear on the Fuel Data Center (FDC) panel as follows:

   a. Display of trouble codes will begin with an 8.8.8 on the FDC panel for approximately one second. ...E will then be displayed which indicates beginning of the ECM stored trouble codes. The initial pass of ECM codes includes all the detected malfunctions whether or not they are currently present. If no ECM codes are stored the ...E display will be bypassed.

   b. Following the display of ...E the lowest numbered ECM code will be displayed for approximately two seconds. All ECM codes will be prefixed with an E (i.e. E12, E13, etc.).

   c. Progressively higher numbered codes, if present, will be displayed consecutively for two second intervals until the highest code has been displayed.

   d. .E.E is again displayed which indicates the start of the second pass of ECM trouble codes. On the second pass only current faults (hard codes) will be displayed. Codes displayed on the first pass are history failures or (soft codes). If all displayed codes were history codes the .E.E will be bypassed.

   e. When all ECM codes have been displayed, BCM codes will appear with the prefix F in the same manner as the ECM did.

   f. After the display of all codes, or if no codes were stored, Code .7.0 will appear indicating the start of the switch tests.

To clear the codes, while still in the diagnostic mode, press the OFF and HIGH buttons simultaneously until E.0.0 appears. Trouble codes are now removed from the ECM memory.

1987–1993 Deville and Fleetwood:

1. Turn the ignition switch **ON**.

2. Depress the OFF and WARMER buttons on the Climate Control Panel (CCP) simultaneously and hold until -188 is displayed.

3. -188 should then appear indicating that all display segments are functional. Diagnosis should not be attempted unless the entire -188 is displayed or misdiagnosis will result.

4. If trouble codes are present they will appear on the Fuel Data Center (FDC) panel as follows:

   a. Display of trouble codes will begin with an 8.8.8 on the FDC panel for approximately one second. ...E will then be displayed which indicates beginning of the ECM stored trouble codes. The initial pass of ECM codes includes all the detected malfunctions whether or not they are currently present. If no ECM codes are stored the ...E display will be bypassed.

b. Following the display of ...E the lowest numbered ECM code will be displayed for approximately two seconds. All ECM codes will be prefixed with an E (i.e. E12, E13, etc.).

c. Progressively higher numbered codes, if present, will be displayed consecutively for two second intervals until the highest code has been displayed.

d. .E.E is again displayed which indicates the start of the second pass of ECM trouble codes. On the second pass only current faults (hard codes) will be displayed. Codes displayed on the first pass are history failures or (soft codes). If all displayed codes were history codes the .E.E will be bypassed.

e. When all ECM codes have been displayed, BCM codes will appear with the prefix F in the same manner as the ECM did.

5. After the display of all codes, or if no codes were stored, Code .7.0 will appear indicating the start of the switch tests.

To clear the codes, while still in the diagnostic mode, press the OFF and HIGH buttons simultaneously until E.0.0 appears. Trouble codes are now removed from the ECM memory.

1987–1993 Allante, Eldorado and Seville:

1. Turn the ignition switch **ON**.

2. Depress the OFF and WARMER buttons on the Climate Control Panel (CCP) simultaneously and hold until the segment check appears on the Instrument Panel Cluster (IPC) and the Climate Control Driver Information Center (CODIC).

3. Diagnosis should not be attempted unless all of the segments of the vacuum fluorescent display are working as this could lead to misdiagnosis. On the PC however the turn signal indicators do not light during this check.

4. After the service mode is entered, any trouble codes stored in the computer memory will be displayed, starting with ECM codes prefixed with an E.

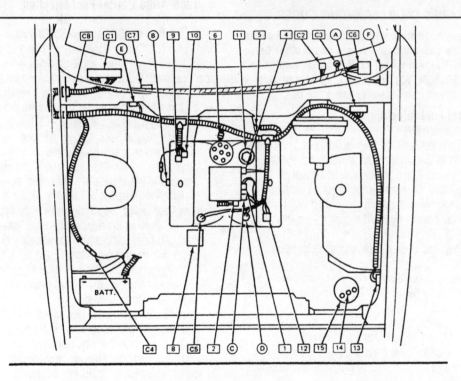

"F" SERIES
2.8L (173 CID) V6   RPO: LC1/LL1 H.O.   V.I.N. CODE: 1/L

COMPUTER SYSTEM		TRANSMISSION CONVERTER CLUTCH CONTROL SYSTEM		SEM S/SWITCHES	
C1	Electronic Control Module (ECM)	5	Trans. Conv. Clutch Connector	A	Differential Pressure Sensor
C2	ALCL Connector	**IGNITION SYSTEM**		B	Exhaust Oxygen Sensor
C3	"CHECK ENGINE" Light	6	Electronic Spark Timing Connector	C	Throttle Position Sensor
C4	System Power	**AIR INJECTION SYSTEM**		D	Coolant Sensor
C5	System Ground	8	Air Injection Pump	E	Barometric Pressure Sensor
C6	Fuse Panel	9	Air Control Solenoid Valve (Divert)	F	Vehicle Speed Sensor
C7	Lamp Driver	10	Air Switching Solenoid Valve		
C8	Computer Control Harness	**EXHAUST GAS RECIRCULATION CONTROL SYSTEM**			
AIR/FUEL SYSTEM		11	Exhaust Gas Recirculation Valve		
1	Mixture Control	12	Exhaust Gas Recirculation Solenoid Valve		
2	Idle Speed Solenoid	**FUEL VAPOR CONTROL SYSTEM**			
4	Heated Grid EFE	13	Canister Purge Solenoid Valve		
		14	From Fuel Tank		
		15	Vapor Canister		

79222G26

Example of component locator on a vehicle with electronic engine controls—2.8L Firebird shown

5. If no trouble codes are present, the message NO ECM CODES will be displayed. Some later systems will display NO X CODES present, with X representing the system selected such as ECM, BOM, SIR, etc.

6. To clear the codes, while still in the service mode, and the ECM diagnostic code display has been completed press the HI button on the CCP.

7. This action should cause the display to read 'ECM DATA?'.

8. Press the LO button on the COP until the display reads 'ECM CLEAR CODES?'.

9. Press the HI button on the CCP and the display should read 'ECM CODES CLEAR'.

10. After approximately 3 seconds, all stored ECM codes will be erased.

➡The Cadillac Cimmaron used the 12 terminal DLC and codes can be accessed in the conventional manner. The rear wheel drive Cadillacs equipped with either the 4.1 L V6, 5.0L V8, or the 5.7L V8 all can be accessed in the conventional manner using the DLC.

Diagnostic Trouble Codes

EXCEPT FRONT WHEEL DRIVE CADILLAC

Code 12 No engine RPM reference pulses—System Normal

Code 13 Oxygen Sensor (O$_2$S) circuit open—left side on 2 sensor system

Code 14 Engine Coolant Temperature (ECT) sensor -possible circuit high or shorted sensor

Code 15 Engine Coolant Temperature (ECT) sensor -circuit low or open circuit

Code 16 Direct ignition system (DIS), fault line circuit or Distributor ignition system (low resolution pulse) or Missing 2x reference circuit or OPTI-Spark ignition timing system (low resolution pulse) or System voltage out of range

Code 17 Camshaft Position Sensor (OPS) or spark reference circuit error

Code 18 Crank/Cam error

Code 19 Crankshaft Position Sensor (CPS) circuit

Code 21 Throttle Position (TP) sensor circuit—signal voltage out of range, probably high

Code 22 Throttle Position (TP) sensor circuit—signal voltage low

Code 23 Intake Air Temperature (IAT or MAT) sensor circuit temperature out of range, low or Open or grounded M/C solenoid Feedback Carburetor system

Code 24 Vehicle Speed Sensor (VSS) circuit

Code 25 Intake Air Temperature (IAT or MAT) sensor circuit temperature out of range, high

Code 26 Quad-Driver Module #1 circuit or Transaxle gear switch circuit

Code 27 Quad-Driver Module circuit or Transaxle gear switch, probably 2nd gear switch circuit

Code 28 Quad-Driver Module (QDM) #2 circuit or Transaxle gear switch, probably 3rd gear switch circuit

Code 29 Transaxle gear switch, probably 4th gear switch circuit

Code 31 Camshaft sensor circuit fault or Park/Neutral Position (PNP) switch circuit or Wastegate circuit signal

Code 32 Exhaust Gas Recirculation (EGR) circuit fault or Barometric Pressure Sensor circuit low Feedback Carburetor system

Code 33 Manifold Absolute Pressure (MAP) sensor—signal voltage out of range, high or Mass Air Flow (MAE) sensor—signal voltage out of range, probably high

Code 34—Manifold Absolute Pressure (MAP) sensor—circuit out of range voltage, low or Mass Air Flow (MAF) sensor circuit (gm/sec low)

Code 35—Idle Air Control (IAC) or idle speed error or Idle Speed Control (ISO) circuit throttle switch shorted Feedback Carburetor system

Code 36 Ignition system circuit error or Transaxle shift problem—4T60E Transaxle

Code 38 Brake input circuit fault—Torque converter clutch signal

Code 39 Clutch input circuit fault—Torque converter clutch signal

Code 41 Cam sensor or cylinder select circuit fault ignition control (IC) reference pulse system fault or Electronic Spark Timing (EST) circuit open or shorted

Code 42 Electronic Spark Timing (EST) circuit grounded or Ignition Control (IC) circuit grounded or faulty bypass line

Code 43 Knock Sensor (KS) or Electronic Spark Control (ESC) circuit fault

Code 44 Oxygen Sensor (O$_2$S), left side on 2 sensor system lean exhaust indicated

Code 45 Oxygen Sensor (O$_2$S), left side on 2 sensor system rich exhaust indicated

Code 46 Personal Automotive Security System (PASSKey II) circuit or Power Steering Pressure Switch (PSPS) circuit

Code 47 PCM-BCM data circuit

Code 48 Misfire diagnosis

Code 51 Calibration error, faulty MEM-CAL, ECM or EEPROM failure

Code 52 Engine oil temperature sensor circuit, low temperature indicated or Fuel Calpac missing or Over voltage condition or EGR Circuit fault

Code 53 Battery voltage error or EGR problem or Personal Automotive Security System (PASS-Key) circuit

Code 54 EGR #2 problem or Fuel pump circuit (low voltage) or Shorted mixture control solenoid circuit Feedback Carburetor system

Code 55 A/D Converter error, PCM error or not grounded, EGR #3 problem, Fuel lean monitor, Grounded voltage reference, faulty oxygen sensor or fuel lean Feedback Carburetor system

Code 56 Quad-Driver Module (QDM) #2 circuit or Secondary air inlet valve actuator vacuum sensor circuit signal high 5.7L (VIN J)

Code 57 Boost control problem

Code 58 Vehicle Anti-theft System fuel enable circuit

Code 61 A/C system performance or Cruise vent solenoid circuit fault or Oxygen Sensor (O$_2$S) degraded signal or Secondary port throttle valve system fault 5.7L (VIN J) or Transaxle gear switch signal

Code 62 Cruise vacuum solenoid circuit fault or Engine oil temperature sensor, high temperature indicated or Transaxle gear switch signal circuit fault

Code 63 Oxygen Sensor (O$_2$S), right side circuit open or Cruise system problem (speed error) or Manifold Absolute Pressure (MAP) sensor circuit out of range

Code 64 Oxygen Sensor (O$_2$S), right side—lean exhaust indicated

Code 65 Oxygen Sensor (O$_2$S), right side—rich exhaust indicated or Cruise servo position circuit or Fuel injector circuit low current

Code 66 A/C pressure sensor circuit fault, probably low pressure or Engine power switch, voltage high or low or PCM fault 5.7L (VIN J)

Code 67 A/C pressure sensor circuit, sensor or A/C clutch circuit failure or Cruise switch circuit fault

Code 68 A/C compressor relay (shorted circuit) or Cruise system fault

Code 69 A/C clutch circuit or head pressure high

Code 70 A/C refrigerant pressure sensor circuit (high pressure)

Code 71 A/C evaporator temperature sensor circuit (low temperature)

Code 72 Gear selector switch circuit

Code 73 A/C evaporator temperature sensor circuit (high temperature)

Code 75 Digital EGR #1 solenoid error

Code 76 Digital EGR #2 solenoid error

Code 77 Digital EGR #3 solenoid error

Code 79 Vehicle Speed Sensor (VSS) circuit signal high

Code 80 Vehicle Speed Sensor (VSS) circuit signal low

Code 81 Brake input circuit fault—Torque converter clutch signal

Code 82 Ignition Control (IC) 3X signal error

Code 85 PROM error

Code 86 Analog/Digital ECM error

Code 87 EEPROM error

Code 99 Power management

➡This list is for reference and does not mean a specific component is defective.

FRONT WHEEL DRIVE CADILLAC

Cadillac Codes may start with an 'E', 'EO', 'P' or 'PO' dependent on model or type of code display. This prefix has been left off the following code description list.

Code 12 No spark reference from ignition control module or distributor

Code 13 Oxygen sensor No.1 not ready

Code 14 Engine Coolant Temperature (ECT) sensor circuit shorted

Code 15 Engine Coolant Temperature (ECT) sensor circuit open

Code 16 System voltage out of range

Code 17 Oxygen sensor No.2 not ready

Code 19 Fuel pump circuit shorted

Code 20 Fuel pump circuit open

Code 21 Throttle Position Sensor (TPS) circuit shorted

Code 22 Throttle Position Sensor (TPS) circuit open

Code 23 Electronic Spark Timing (EST) circuit fault or Ignition Control (IC) circuit problem

Code 24 Vehicle Speed Sensor (VSS) circuit problem

Code 26 Throttle Position (TP) switch circuit shorted

Code 27 Throttle Position (TP) switch circuit open

Code 28 Transaxle pressure switch problem

Code 29 Transaxle shift 'B' solenoid problem

Code 30 Idle Speed Control (ISO) RPM out of range

Code 31 Manifold Absolute Pressure (MAP) sensor circuit shorted

Code 32 Manifold Absolute Pressure (MAP) sensor circuit open

Code 33 Extended travel brake switch input circuit problem

Code 34 Manifold Absolute Pressure (MAP) sensor signal too high

Code 35 Ignition ground voltage out of range

Code 36 EGR valve pintle position out of range

Code 37 Intake Air Temperature (IAT) Manifold Air Temperature (MAT) circuit shorted

Code 38 Intake Air Temperature (IAT) sensor, Manifold Air Temperature (MAT) circuit open

Code 39 Torque Converter Clutch (TCC) engagement problem

Code 40 Power Steering Pressure Switch (PSPS) open

Code 41 Cam sensor circuit fault

Code 42 Oxygen sensor No.1 LEAN exhaust signal

Code 43 Oxygen sensor No.1 RICH exhaust signal

Code 44 Oxygen sensor No.2 LEAN exhaust signal

Code 45 Oxygen sensor No.2 RICH exhaust signal

Code 46 Bank-to-bank fueling difference

Code 47 ECM Body Control Module (BCM) or IPC/PCM data fault

Code 48 EGR control system fault

Code 50 2nd gear pressure circuit fault

Code 51 MEM-CAL error or PROM checksum mismatch

Code 52 ECM memory reset indicator or PCM keep alive memory reset

Code 53 Spark reference signal interrupt from Ignition Control (IC) module

Code 55 Closed throttle angle out of range or Throttle Position Sensor (TPS) maladjusted

Code 56 Transaxle input speed sensor circuit problem

Code 57 Shorted transaxle temperature sensor circuit

Code 58 Personal Automotive Security System (PASS) control fault

Code 59 Open transaxle temperature sensor circuit

Code 60 Cruise transaxle not in drive

Code 61 Cruise vent solenoid circuit fault

Code 62 Cruise vacuum solenoid circuit fault

Code 63 Cruise vehicle speed and set speed difference

Code 64 Cruise vehicle acceleration too high

Code 65 Cruise servo position sensor failure

Code 66 Cruise engine RPM too high

Code 67 Cruise set/coast or resume/accel input shorted

Code 68 Cruise Control Command (CCC) fault or servo position out of range

Code 69 Traction control active in cruise

Code 70 Intermittent Throttle Position (TP) sensor signal

Code 71 Intermittent Manifold Absolute Pressure (MAP) sensor signal

Code 73 Intermittent Engine Coolant Temperature (ECT) sensor signal

Code 74 Intermittent Intake Air Temperature (IAT) sensor signal

Code 75 Vehicle Speed Sensor (VSS) signal intermittent

Code 76 Transaxle pressure control solenoid circuit malfunction

Code 80 Fuel system rich or TP Sensor/idle learn not complete

Code 81 Cam to 4X reference correlation problem

Code 83 24X Reference signal high

Code 85 Idle throttle angle too high, Throttle body service required

Code 86 Undefined gear ratio

Code 88 Torque Converter Clutch (TOO) not disengaging

Code 89 Long shift and maximum adapt

Code 90 Viscous Converter Clutch (VOC) brake switch input fault

Code 91 Park/neutral switch fault

Code 92 Heated windshield fault

Code 93 Traction control system PWM link failure

Code 94 Transaxle shift 'A' solenoid problem

Code 95 Engine stall detected

Code 96 Torque converter overstress

Code 97 Park/neutral to drive/reverse at high throttle angle

Code 98 High RPM P/N to D/R shift under Idle Speed Control (ISO)

Code 99 Cruise control servo not applied in cruise
Code P102 Shorted Brake Booster Vacuum (BBV) sensor
Code P103 Open Brake Booster Vacuum (BBV) sensor
Code P105 Brake Booster Vacuum (BBV) too low
Code P106 Stop lamp switch input circuit problem
Code P107 PCM/BCM data link problem
Code P108 PROM checksum mismatch
Code P109 POM keep alive memory reset
Code P110 Generator L-terminal circuit problem
Code P112 Total EEPROM failure
Code P117 Shift 'A'/Shift 'B' circuit output open or shorted
Code P131 Active Knock Sensor (KS) failure
Code P132 Knock Sensor (KS) circuit failure
Code P137 Loss of ABS/TCS data

TRANSMISSION CODES

Reading and Retrieving Codes

EXCEPT GEO

General Motors (GM) transmission/transaxle diagnostic systems use two-digit DTC's, as with most pre-OBD II vehicles, to inform to the mechanic or technician the problem circuit or component.

Fortunately, it is not necessary to use a diagnostic scan tool to retrieve DTC's from a pre-OBD II GM vehicle. General Motors models use a light, located in the instrument cluster and usually referred to as the Malfunction Indicator Lamp (MIL), CHECK ENGINE light or SERVICE ENGINE SOON light, to display the DTC's as a series of flashes or blinks.

The two-digit codes are flashed out in a logical series of blinks. For example, when first starting to retrieve the DTC's the MIL will flash once, pause briefly, then flash two more times followed by a longer pause. This indicates a Code 12 (one flash-pause-two flashes-long pause). Therefore, a flash pattern of two flashes-pause-three flashes-long pause would indicate a Code 23.

➡**Code 12 should always be the first DTC flashed out by the MIL, since it indicates that the diagnostic system is functioning properly.**

Each DTC is flashed out by the MIL three times, then the next code is flashed out (if more than one code has been stored). The codes are displayed in numerical sequence starting with the lowest code number and moving to the highest. Therefore, if Codes 21, 14, and 16 were stored by the vehicle computer, the MIL would display them in the following order: 12 (always the first code—it indicates that the ignition key is **ON**, but the engine is not running), 12, 12, 14, 14, 14, 16, 16, 16, 21, 21, 21. Remember that each code is flashed three times.

When reading the codes be sure not to miss any of them, otherwise the entire sequence must be repeated. This can be tiresome if a large number of codes was stored, and you missed the last one.

To retrieve the codes, perform the following:

1. Perform the preliminary inspection, located earlier in this section. This is very important, since a loose or disconnected wire, or corroded connector terminals can cause a whole slew of unrelated DTC's to be stored by the computer; you will waste a lot of time performing a diagnostic "goose chase."

2. Grab some paper and a pencil or pen to write down the DTC's when they are flashed out.

3. Locate the Diagnostic Link Connector (DLC) or Assembly Line Diagnostic Link (ALDL). This is the connector that provides a direct link with the vehicle's computer system. It is usually located

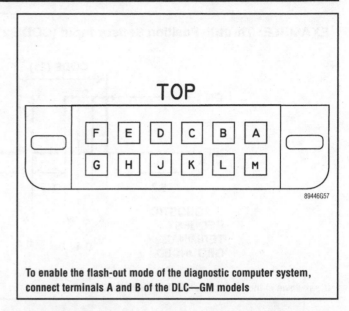

To enable the flash-out mode of the diagnostic computer system, connect terminals A and B of the DLC—GM models

underneath the instrument panel or in the glove box, but occasionally can be located in the engine compartment, near the firewall or relay box. If the DLC/ALDL cannot be found, either call your local GM dealer or refer to the applicable Chilton Total Car Care manual for your vehicle.

4. Using a jumper wire or paper clip (you have to bend the paper clip into a large U shape), connect DLC terminals **A** and **B**, which grounds the test terminal and switches the computer system into the diagnostic display mode.

5. Turn the ignition switch **ON**, but DO NOT start the engine.

➡**Remember that when the ignition is ON, but the engine is not running, a Code 12 is normal.**

6. At this point, write down the codes flashed out by the MIL.
7. Turn the ignition switch **OFF**.
8. Fix any problems indicated by the DTC's.

GEO STORM

The Geo Storm uses two-digit DTC's, which can be read by having the ECONOMY light flash the codes out as a series of flashes or blinks.

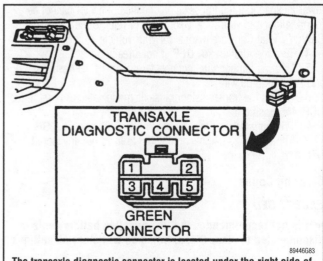

The transaxle diagnostic connector is located under the right side of the instrument panel, as shown—Geo models

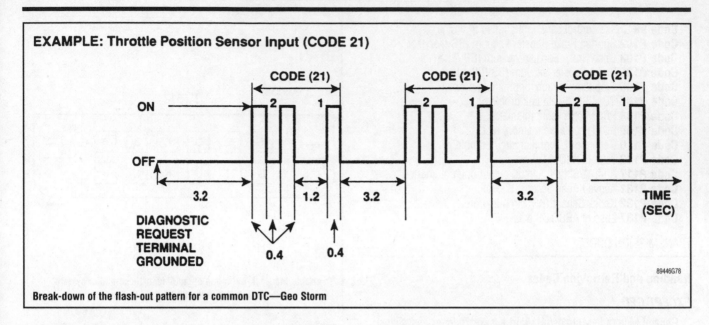

EXAMPLE: Throttle Position Sensor Input (CODE 21)

Break-down of the flash-out pattern for a common DTC—Geo Storm

The two-digit codes are flashed out in a logical series of blinks. For example, when first starting to retrieve the DTC's the MIL will flash once, pause briefly, then flash three more times followed by a longer pause. This indicates a Code 13 (one flash-pause-three flashes-long pause). Therefore, a flash pattern of two flashes-pause-four flashes-long pause would indicate a Code 24.

Each DTC is displayed from the lowest number on through to the highest number. After the highest code number is displayed, the system begins with the lowest again. The system will continue to cycle through all of the codes in this manner; so if you missed a code or two, you can wait until it starts the cycle over again.

To retrieve the codes, perform the following:

1. Perform the preliminary inspection, located earlier in this section. This is very important, since a loose or disconnected wire, or corroded connector terminals can cause a whole slew of unrelated DTC's to be stored by the computer; you will waste a lot of time performing a diagnostic "goose chase."

2. Grab some paper and a pencil or pen to write down the DTC's when they are flashed out.

3. Locate the transaxle diagnostic connector, which is a 5-terminal green connector and is usually located behind the right-hand lower kickpanel. If the transaxle diagnostic connector cannot be found, either call your local Geo dealer or refer to the applicable Chilton's Total Car Care manual for your vehicle.

4. Turn the ignition key **OFF**, if not already done.

5. Using a jumper wire, connect terminals 1 (black/white wire) and 2 (black/blue wire).

6. Ensure the transmission/transaxle mode selector is set to the NORMAL position.

7. Without starting the engine, turn the ignition switch **ON**.

8. The ECONOMY light should start to flash out any stored DTC's.

Clearing Codes

EXCEPT GEO

➡It is not recommended that the negative battery cable be disconnected to clear DTC's. Doing so will clear all settings from the vehicle's computer system, resulting in lost radio presets, seat memories, anti-theft codes, driveability parameters, etc., and there are better ways of spending your afternoon than resetting all of these.

1. Locate the vehicle's fuse box. The fuse box is usually located under the driver's side of the instrument panel or in the engine compartment.

2. If applicable, remove the fuse box cover.

3. Look for a PCM, ECM or similarly marked fuse (computer, battery +, etc.). This fuse is used for the battery voltage circuit that supplies the PCM with power even when the ignition is switched **OFF**.

4. Remove this fuse from the fuse box, and wait at least 10 seconds.

5. Reinstall the fuse, and install the fuse box cover (if equipped).

The transmission/transaxle DTC's should now be cleared. To double check, perform the code retrieval procedure again. Only Code 12 should be displayed, if not either the codes were not cleared (Are the codes identical to those flashed out previously?) or the underlying problem is still there (Are only some of the codes the same as previously?)

GEO STORM

➡It is not recommended that the negative battery cable be disconnected to clear DTC's. Doing so will clear all settings from the vehicle's computer system, resulting in lost radio presets, seat memories, anti-theft codes, driveability parameters, etc., and there are better ways of spending your afternoon than resetting all of these.

To clear the DTC's from the vehicle's memory, locate the 4AT fuse in the fuse block. Remove the fuse for approximately 60 seconds, then reinstall it. All of the codes should now be cleared. To double check, perform the code retrieval procedure again. If there are still codes present, either the codes were not cleared (Are the codes identical to those flashed out previously?) or the underlying problem is still there (Are only some of the codes the same as previously?)

DIAGNOSTIC TROUBLE CODES—GENERAL MOTORS

Code	Component/Circuit Fault
12	No distributor or speed reference signal
13	Oxygen (O_2) sensor circuit (open circuit)
14	Coolant Temperature Sensor (CTS) circuit (high temperature indicated)
15	Coolant Temperature Sensor (CTS) Circuit (low temperature indicated)
16	System voltage high/low
17	Spark reference circuit
18	Crank signal circuit
19	Fuel pump or circuit
20	Fuel pump or circuit
21	Throttle Position Sensor (TPS) circuit (signal voltage high)
22	Throttle Position Sensor (TPS) circuit (signal voltage low)
23	Intake Air Temperature (IAT) sensor circuit (low temperature indicated)
24	Vehicle Speed Sensor (VSS) circuit
25	Intake Air Temperature (IAT) sensor or circuit (high temperature indicated)
26	Quad-Driver (QDM) fault (1 of 3)
27	2nd gear switch or circuit; 4th gear switch or circuit
28	3rd gear switch or circuit; 4th gear switch or circuit
29	4th gear switch or circuit; canister purge solenoid or circuit; 4th gear switch or circuit
30	ISC or circuit
31	PRNDL switch circuit
32	Barometric pressure sensor or circuit; EGR system; MAP sensor or circuit
33	MAP sensor or circuit; MAF sensor or circuit
34	Mass Air Flow (MAF) sensor or circuit
35	ISC or circuit; IAC or circuit; Barometric pressure sensor or circuit
36	Transaxle shift control problem
37	MAT sensor or circuit; Brake switch stuck on
38	TCC brake input circuit
39	TCC or VCC engagement fault
40	Power Steering Pressure (PSP) circuit
41	Camshaft sensor or circuit (Type 1 Ignition)
41	Camshaft sensor or circuit (Type 2 Ignition)
42	Electronic Spark Timing (EST) circuit
43	Electronic Spark Control (ESC) circuit
44	Oxygen (O_2) sensor or circuit (lean exhaust indicated)
45	Oxygen (O_2) sensor or circuit (rich exhaust indicated)
46	PSP switch or circuit; Anti-theft system; Left-to-right bank fuel delivery difference
47	ECM/PCM/BCM fault
48	Misfire detected
49	Air management system
50	Second gear pressure circuit
51	MEM-CAL error (faulty or incorrect MEM-CAL)
52	Mem-Cal CalPal ECM fault; ECM/PCM memory reset indicator; EOT sensor or circuit
53	System voltage too high; distributor signal interrupt; EGR system; Anti-theft system; Air switch solenoid or circuit; Reference voltage overload (Diesel)
54	M/C solenoid or circuit; Fuel pump or circuit
55	ECM fault; Air switch valve or circuit; TPS or circuit
56	4th gear switch or circuit; 3-4 shift solenoid or circuit; vacuum sensor or circuit
57	PCM/BCM fault; 4th gear switch or circuit; 3-4 shift solenoid or circuit
58	Personal Automotive Security System (PASS-Key) fuel enable circuit
59	Transmission temperature circuit
60	Cruise control—transmission not in drive

89446G86

General Motors Diagnostic Trouble Codes, 1 of 2—except Geo Storm

DIAGNOSTIC TROUBLE CODES—GENERAL MOTORS

Code	Component/Circuit Fault
61	Cruise vent solenoid circuit
62	Cruise vacuum solenoid circuit
63	Cruise control—vehicle speed and set speed difference; MAP sensor or circuit; EGR flow check fault; Right oxygen sensor or circuit
64	Cruise control—vehicle acceleration too high; EGR flow check fault; Right oxygen sensor or circuit
65	Cruise servo position sensor or circuit
66	Cruise control—engine speed too high; A/C pressure sensor or circuit; Power switch or circuit; 3-2 control solenoid or circuit
67	Cruise control switches or circuits
68	Cruise control system fault
69	A/C head pressure switch or circuit
70	Intermittent TPS sensor or circuit fault
71	Intermittent MAP sensor or circuit fault
72	Vehicle speed control circuit
73	Force motor current error
74	Intermittent MAT sensor or circuit fault
75	Intermittent VSS sensor or circuit fault
79	Transmission fluid too hot
80	Fuel system overly rich
81	Shift solenoid B circuit
82	Shift solenoid A or circuit
83	TCC solenoid or circuit
85	Throttle angle incorrect range
86	Low ratio error—solenoid B closed
87	High ratio error—solenoid B open
90	VCC or circuit
91	PRNDL switch or circuit
92	Heated windshield request problem
96	Torque converter overstress
97	P/N to D/R at high throttle angle
98	High RPM from P/N to D/R in ISC range
99	Cruise control servo application fault

89446G56

General Motors Diagnostic Trouble Codes, 2 of 2—except Geo Storm

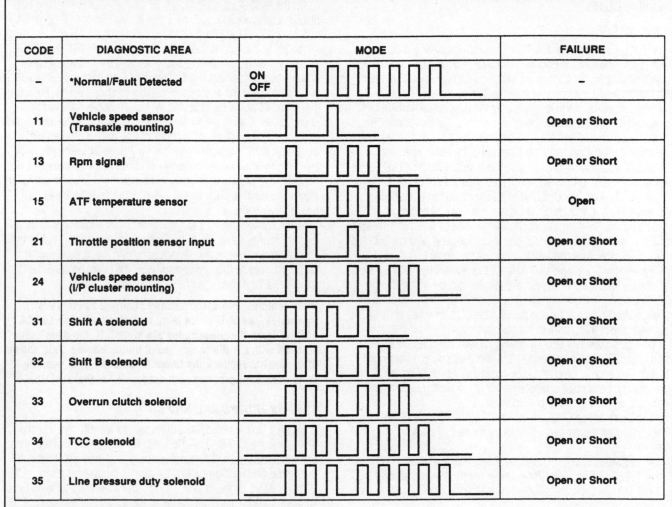

CODE	DIAGNOSTIC AREA	MODE	FAILURE
–	*Normal/Fault Detected	ON / OFF	–
11	Vehicle speed sensor (Transaxle mounting)		Open or Short
13	Rpm signal		Open or Short
15	ATF temperature sensor		Open
21	Throttle position sensor input		Open or Short
24	Vehicle speed sensor (I/P cluster mounting)		Open or Short
31	Shift A solenoid		Open or Short
32	Shift B solenoid		Open or Short
33	Overrun clutch solenoid		Open or Short
34	TCC solenoid		Open or Short
35	Line pressure duty solenoid		Open or Short

*Normal with diagnostic request terminal grounded.
Fault detected without diagnostic request terminal grounded.

89446G79

Geo Storm Diagnostic Trouble Codes

Honda

ENGINE CODES

General Information

Honda utilizes 2 types of fuel systems. The first is the feedback carburetor system of which there are 2 types; a 2 barrel down draft-fixed venturi type, and 2 side draft carburetors variable venturi type. The feedback carburetor was in use up to 1991 in Honda vehicles.

The second type fuel system is Programmed Fuel Injection (PGM-FI) system. This system began in 1985 and was available in the Accord and Civic. As of 1992 all Hondas are fuel injected.

Service Precautions

• Make sure all ECM harness connectors are fastened securely. A poor connection can cause an extremely high voltage surge and result in damage to integrated circuits.

• Keep all ECM parts and harnesses dry during service. Protect the ECM and all solid-state components from rough handling or extremes of temperature.

• Use extreme care when working around the ECM or other components. The airbag or SRS wiring may be in the vicinity. On these vehicles, the SRS wiring and connectors are yellow. Do not cut or test these circuits.

• Before attempting to remove any parts, turn the ignition switch **OFF** and disconnect the battery ground cable.

• Always use a 12 volt battery as a power source for the engine, never a booster or high-voltage charging unit.

• Do not disconnect the battery cables with the engine running.

• Do not disconnect any wiring connector with the engine running or the ignition **ON** unless specifically instructed.

• Do not apply battery power directly to injectors.

• Whenever possible, use a flashlight instead of a drop light.

• Relieve fuel system pressure before servicing any fuel system component.

• Always use eye or full-face protection when working around fuel lines, fittings or components.

Reading Codes

1985–89 CARS

When a fault is noted, the ECU stores an identifying code and illuminates the CHECK ENGINE light. The code will remain in memory until cleared; the dashboard warning lamp may not illuminate during the next ignition cycle if the fault is no longer present. Not all faults noted by the ECU will trigger the dashboard warning lamp although the fault code will be set in memory. For this reason, troubleshooting should be based on the presence of stored codes, not the illumination of the warning lamp while the car is operating.

Stored codes are displayed by either a single flashing LED (Light Emitting Diode) light, or an illuminated light pattern of 4 LED lights on the ECU. When the CHECK ENGINE warning lamp has been on or reported on, check the ECU LED for presence of codes.

The location of the malfunction is determined by observing the LED display. Earlier Hondas used 2 types of LED displays: a single LED and a 4 LED display. After 1987 all models use the single LED display.

Systems with a single LED indicate the malfunction with a series of flashes. The number of flashes indicates a code which identifies the location of the component or system malfunction. The code will flash, followed by a 2 second pause, repeat, followed by another 2 second pause, then move to the next code.

On systems with 4 LED's a display pattern identifies the malfunction. The LED's are numbered 1, 2, 4 and 8 as counted from right-to-left. The code is determined by observing which LED's are lit on the display. Each code is displayed once, followed by a 2 second pause, then the next code is displayed.

The LED's are part of the Electronic Control Module (ECM). Depending on the vehicles, the ECU is located in the following places:

- 1985–89 Accord—Under the driver side front seat
- 1985–87 Civic and CRX—Under the passenger side seat
- 1988–89 Civic and CRX—Under the passenger side foot-well, below the dashboard
- 1987 Prelude—Behind driver side rear seat trim panel
- 1988–89 Prelude—Under the passenger side footwell, below the dash. (The LED may be viewed through a small window without removing the ECU cover).

Turn the ignition switch **ON**; the LED will display any stored codes.

On the 1985 Accord and 1985–87 Civic/CRX having the 4 LED display, codes are indicated by a specific pattern of LED lights illuminated on the ECU.

On 1986–89 Accord, 1988–89 Civic, and Prelude having the single LED display, codes 1–9 are indicated by a series of short flashes; two-digit codes use a number of long flashes for the first digit followed by the appropriate number of short flashes. For example, Code 43 would be indicated by 4 long flashes followed by 3 short flashes. Codes are separated by a longer pause between transmissions. The position of the codes during output can be helpful in diagnostic work. Multiple codes transmitted in isolated order indicate unique occurrences; a display of showing 1-1-1 pause 9-9-9 indicates two problems or problems occurring at different times. An alternating display, such as 1-9-1-9-1, indicates simultaneous occurrences of the faults.

When counting flashes to determine codes, a code not valid for the vehicle may be found. In this case, first recount the flashes to confirm an accurate count. If necessary, turn the ignition switch **OFF**, then recycle the system and begin the count again. If the Code is not valid for the vehicle, the ECU must be replaced.

➡**On vehicles with electronically controlled automatic transaxles, the 5, D or D4 lamp may flash with the CHECK ENGINE lamp if certain codes are stored. If this does occur, proceed with the diagnosis based on the engine code shown. After repairs, recheck the lamp. If the additional warning lamp is still lit, proceed with diagnosis for that system.**

1985–89 TRUCKS AND SUV'S

When a fault is noted, the ECU stores an identifying code and illuminates the CHECK ENGINE light. The code will remain in memory until cleared; the dashboard warning lamp may not illuminate during the next ignition cycle if the fault is no longer present. Not all faults noted by the ECU will trigger the dashboard warning lamp although the fault code will be set in memory. For this reason, troubleshooting should be based on the presence of stored codes, not the illumination of the warning lamp while the car is operating.

Stored codes are displayed by either a single flashing LED (Light Emitting Diode) light, or an illuminated light pattern of 4 LED lights on the ECU. When the CHECK ENGINE warning lamp has been on or reported on, check the ECU LED for presence of codes.

The location of the malfunction is determined by observing the LED display. Earlier Hondas used 2 types of LED displays: a single LED and a 4 LED display. After 1987 all models use the single LED display.

Systems with a single LED indicate the malfunction with a series of flashes. The number of flashes indicates a code which identifies the location of the component or system malfunction. The code will flash, followed by a 2 second pause, repeat, followed by another 2 second pause, then move to the next code.

On systems with 4 LED's a display pattern identifies the malfunction. The LED's are numbered 1, 2, 4 and 8 as counted from right-to-left. The code is determined by observing which LED's are lit on the display. Each code is displayed once, followed by a 2 second pause, then the next code is displayed.

The LED's are part of the Electronic Control Module (ECM).

Turn the ignition switch ON; the LED will display any stored codes.

When counting flashes to determine codes, a code not valid for the vehicle may be found. In this case, first recount the flashes to confirm an accurate count. If necessary, turn the ignition switch OFF, then recycle the system and begin the count again. If the Code is not valid for the vehicle, the ECU must be replaced.

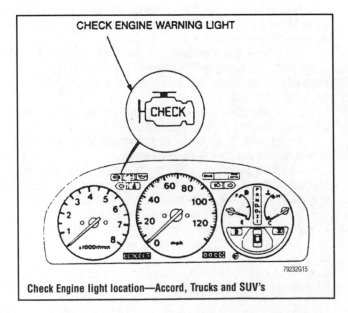

CHECK ENGINE WARNING LIGHT

79232G15

Check Engine light location—Accord, Trucks and SUV's

➡On vehicles with electronically controlled automatic transaxles, the 5, D or D4 lamp may flash with the CHECK ENGINE lamp if certain codes are stored. If this does occur, proceed with the diagnosis based on the engine code shown. After repairs, recheck the lamp. If the additional warning lamp is still lit, proceed with diagnosis for that system.

1990–95 VEHICLES

When a fault is noted, the ECM stores an identifying code and illuminates the CHECK ENGINE light. The code will remain in memory until cleared. The dashboard warning lamp may not illuminate during the next ignition cycle if the fault is no longer present. Not all faults noted by the ECM will trigger the dashboard warning lamp although the fault code will be set in memory. For this reason, troubleshooting should be based on the presence of stored codes, not the illumination of the warning lamp while the car is operating.

In 1990, the Accord and Prelude were equipped with a 2-pin service connector in addition to the LED. if the service connector is jumpered, with the ignition key in the **ON** position, the CHECK ENGINE lamp will display the stored codes in a series of flashes. The 2-pin service connector is located under the passenger side of dash on the Accord and behind the center console on the Prelude.

As of 1992, the LED on the ECU was eliminated and all vehicles obtain codes by jumping the 2-pin connector when the ignition switch is **ON**. The CHECK ENGINE light will then flash codes present in the ECU memory.

Diagnostic Codes 1–9 are indicated by a series of short flashes; two-digit codes use a number of long flashes for the first digit followed by the appropriate number of short flashes. For example, Code 43 would be indicated by 4 long flashes followed by 3 short flashes. Codes are separated by a longer pause between transmissions. The position of the codes during output can be helpful in diagnostic work. Multiple codes transmitted in isolated order indicate unique occurrences; a display of showing 1-1-1 pause 9-9-9 indicates two problems or problems occurring at different times. An alternating display, such as 1-9-1-9-1, indicates simultaneous occurrences of the faults.

When counting flashes to determine codes, a code not valid for the vehicle may be found. In this case, first recount the flashes to confirm an accurate count. If necessary, turn the ignition switch **OFF**, then recycle the system and begin the count again. If the code is not valid for the vehicle, the ECM must be replaced.

➡On vehicles with electronically controlled automatic transaxles, the D4 lamp may flash with the CHECK ENGINE lamp if certain codes are stored. If this does occur, proceed with the diagnosis based on the engine code shown. After repairs, recheck the lamp. If the additional warning lamp is still lit, proceed with diagnosis for that system.

Clearing Codes

1985–87 VEHICLES

The memory for the PGM-FI CHECK ENGINE lamp on the dashboard will be erased when the ignition switch is turned **OFF**; however, the memory for the LED display will not be canceled. Thus, the CHECK ENGINE lamp will not come on when the ignition switch is again turned **ON** unless the trouble is once more detected. Troubleshooting should be done according to the LED display even if the CHECK ENGINE lamp is off.

After making repairs, disconnect the battery negative cable from the battery negative terminal for at least 10 seconds and reset the

ECU memory. After reconnecting the cable, check that the LED display is turned off.

Turn the ignition switch **ON**. The PGM-FI CHECK ENGINE lamp should come on for about 2 seconds. If the CHECK ENGINE lamp won't come on, check for:—Blown CHECK ENGINE lamp bulb—Blown fuse (causing faulty back up light, seat belt alarm, clock, memory function of the car radio) -Open circuit in Yellow wire—Open circuit in wiring and control unit.

After the PGM-FI CHECK ENGINE lamp and self-diagnosis indicators have been turned on, turn the ignition switch **OFF**. If the LED display fails to come on when the ignition switch is turned **ON** again, check for:—Blown fuses, especially No. 10 fuse—Open circuit in wire between ECU fuse.

Replace the ECU only after making sure that all couplers and connectors are connected securely.

1988–90 VEHICLES

The memory for the PGM-CARB and PGM-FI CHECK ENGINE lamp on the dashboard will be erased when the ignition switch is turned **OFF**; however, the memory for the LED display will not be

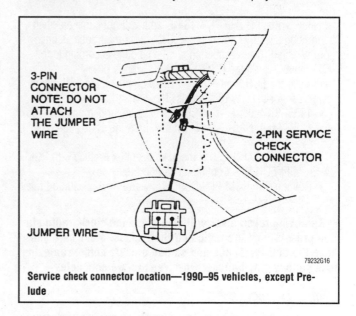

Service check connector location—1990–95 vehicles, except Prelude

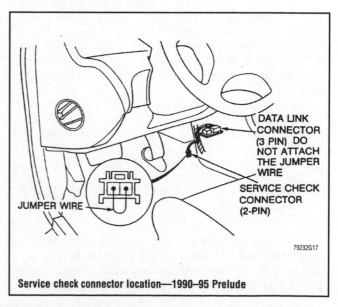

Service check connector location—1990–95 Prelude

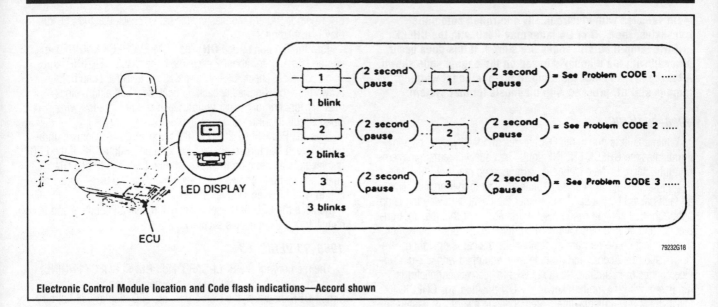

Electronic Control Module location and Code flash indications—Accord shown

canceled. Thus, the CHECK ENGINE lamp will not come on when the ignition switch is again turned **ON** unless the trouble is once more detected. Troubleshooting should be done according to the LED display even if the CHECK ENGINE lamp is off.

To clear the ECU trouble code memory, remove the ECU memory power fuse for at least 10 seconds. The correct fuse to remove is:
- 1988–89 Accord—CLOCK fuse from the underhood relay box
- 1988–90 Accord—BACK UP fuse from the underhood relay box
- Civic—HAZARD fuse at the main fuse box Prelude with PGM-CARB—EFI-ECU fuse from the underhood relay box
- Prelude with PGM-FI—CLOCK fuse from the underhood relay box

➡**Removing these fuses will also erase the clock, radio station presets, and the radio anti-theft codes. Make sure you have the anti-theft code and station presets before removing fuse so they may be reset when repairs are complete.**

1990–95 VEHICLES

Stored codes are removed from memory by removing power to the ECU. Disconnecting the power may also clear the memories used for other solid-state equipment such as the clock and radio. For this reason, always make note of the radio presets before clearing the system. Additionally, some radios contain anti-theft programming; obtain the owner's code number before clearing the codes.

While disconnecting the battery will clear the memory, this is not the recommended procedure. The memory should be cleared after the ignition is switched **OFF** by removing the appropriate fuse for at least 10 seconds. The correct fuses and their locations are:
- Accord—BACK UP fuse from the underhood relay box
- Civic—HAZARD fuse at the main fuse box
- Civic del Sol—BACK UP fuse from the underhood relay box
- Prelude with PGM-FI—CLOCK fuse from the underhood relay box

➡**Removing these fuses will also erase the clock, radio station presets, and the radio anti-theft codes. Make sure you have the anti-theft code and station presets before removing fuse so they may be reset when repairs are complete.**

Diagnostic Trouble Codes

1985–87 4-LED SYSTEM

Accord:

➡**This list includes vehicles with fuel injection engines: 1 .5L (EW3), 1 .5L (DI 5A3), 1.8L (E53) Engine. Code definition as shown in illustration for the 4-LED type ECM.**

Civic and CRX:
This list includes the following carbureted engines: 1 .5L (EW1 and D15A2) engines. Code definition as shown in illustration for the 4-LED type ECM.

1986–93 SINGLE LED SYSTEM

1988–91 Prelude:
This list includes all of the following carbureted engines: 2.0L (B20A5, B20A3) Engines
Code 1 Oxygen count
Code 2 Vehicle speed pulser
Code 3 Manifold Absolute Pressure (MAP)
Code 4 Vacuum switch signal
Code 5 Manifold solute Pressure (MAP)
Code 6 Coolant temperature
Code 7 Coolant switch signal (MT) or Shift position switch signal (AT) (1990–91 engines only)
Code 8 Ignition coil signal
Code 9 No. 1 cylinder position sensor
Code 10 Intake air temperature sensor (IAT sensor)
Code 12 Exhaust Gas Recirculation (EGR) System (except Del Sol and Civic & CRX 1.6L D16A6)
Code 13 Barometric pressure sensor (BARO sensor)
Code 14 Idle air control (IAC valve) except 1987- A20A3 engine or—1986 BS, BT and 1987 A20A3 Engines, Code 14 or high is possible faulty Electronic Control Module (ECM)
Code 14 Electronic Air Control
Code 15 Ignition output signal
Code 16 Fuel Injector
Code 17 Vehicle Speed sensor (VSS)
Code 19 NT lock-up control solenoid valve NB (DI 5B1, D15B2, D15B6, D15B7, D15B8, D15Z1, D16A6, D16Z6)
Code 20 Electric load detector (ELD)

ECM TROUBLE CODES

Code	Explanation
○ ○ ○ ○ (Dash Warning Light ON only)	Loose or poorly connected power line to Electronic Control Unit (ECU). Short circuit in combination meter or warning light wire. Faulty ECU
○ ○ ○ ● (1)	Disconnected oxygen sensor coupler. Spark plug misfire. Short or open circuit in oxygen sensor circuit. Faulty oxygen sensor
○ ○ ● ○ (2)	Faulty Electronic Control Unit (ECU)
○ ○ ● ● (2 1)	Disconnected Manifold Absolute Pressure (MAP) sensor coupler. Short or open circuit in MAP sensor wire. Faulty MAP sensor
○ ● ○ ○ (4)	Faulty Electronic Control Unit (ECU)
○ ● ○ ● (4 1)	Disconnected Manifold Absolute Pressure (MAP) sensor piping
○ ● ● ○ (4 2)	Disconnected coolant temperature sensor coupler. Open circuit in coolant temperature sensor wire. Faulty coolant temperature sensor (thermostat housing)
○ ● ● ● (4 2 1)	Disconnected throttle angle sensor coupler. Open or short circuit in throttle angle sensor wire. Faulty throttle angle sensor
● ○ ○ ○ (8)	Short or open circuit in crank angle sensor wire. Crank angle sensor wire interfering with high tension wire. Crank angle sensor at fault
● ○ ○ ● (8 1)	Short or open circuit in crank angle sensor wire. Crank angle sensor wire interfering with high tension wire. Crank angle sensor at fault
● ○ ● ○ (8 2)	Disconnected intake air temperature sensor. Open circuit in intake air temperature sensor wire. Faulty intake air temperature sensor
● ○ ● ● (8 2 1)	Disconnected idle mixture adjuster sensor coupler. Shorted or disconnected idle mixture adjuster sensor wire. Faulty idle mixture adjuster sensor
● ● ○ ○ (8 4)	Disconnected Exhaust Gas Recirculation (EGR) control system coupler. Shorted or disconnected EGR control wire. Faulty EGR control system
● ● ○ ● (8 4 1)	Disconnected atmospheric pressure sensor coupler. Shorted or disconnected atmospheric pressure sensor wire. Faulty atmospheric pressure sensor
● ● ● ○ (8 4 2)	Faulty Electronic Control Unit (ECU)
● ● ● ● (8 4 2 1)	Faulty Electronic Control Unit (ECU)

79232G19

4-LED code definition—1985 Accord 18L (E53), 1985–86 Civic Si 1.5L (EW3), 1987 Civic CRX Si 1.5L (D15A3) fuel injected engines

ENGINE CODES

Code (8 4 2 1)	Explanation
○ ○ ○ ●	Short circuit in frequency solenoid valve B (brown/black) wire
○ ○ ● ○	Short circuit in frequency solenoid valve A (green/white) wire
○ ○ ● ●	Disconnected Manifold Absolute Pressure (MAP) sensor connector. Short or open circuit in MAP sensor (black/wnite, green/white, yellow/white) wires. Faulty MAP sensor
○ ● ○ ○	Short circuit in ignition timing control unit (red/white, blue/white) wires. Faulty ignition timing control unit
○ ● ● ○	Disconnected Coolant Temperature Sensor (CTS) A connector. Short or open circuit in coolant temperature sensor A (light blue) wire. Faulty coolant temperature sensor A
● ◐ ○ ○	Short or open circuit in ignition coil (blue) wire
● ● ○ ○	Disconnected Exhaust Gas Recirculation (EGR) lift sensor connector. Short or open circuit in EGR lift sensor (yellow, green/white, yellow/white) wires. Faulty EGR lift sensor
● ● ○ ●	Disconnected atmospheric pressure sensor connector. Short or open circuit in atmospheric pressure sensor (green/black, green/white, yellow/white) wires. Faulty atmospheric pressure sensor
15 +	If the LED display pattern differs from those listed above, the Electronic Control Unit (ECU) is faulty

79232G20

4-LED code definition—1985–87 Civic/CRX with feedback carburetor—1.5 (EW1 and D15A2) engines

Code 21 V-TEC control solenoid (D15Z1, D16Z6, H22A1)
Code 22 V-TEC pressure switch (D15Z1, D16Z6, H22A1)
Code 23 Knock sensor (H22A1-DOHC—VTEC)
Code 30 NT FI Signal A (F22A1, F22A4, F22A6)
Code 31 A/T FI Signal B (F22A1, F22A4, F22A6)
Code 41 Heated Oxygen Sensor Heater (F22A1, F22A4)
Code 43 Fuel supply system (except D15B1, D15B2, D15B6, B20A5, B21A, D16A6)
Code 48 Heated oxygen sensor (D15Z1 engine only, except Calif. emission)

1985–95 VEHICLES

Accord, Civic, Del Sol, Odyssey and Prelude:
This list includes all of the following fuel injected engines:
1.5L (D15B1, D15B2, D15B6, D15B7, D15B8, D15Z1), 1.6L (B16A3, D16A6, D16Z6), 2.0L (A20A3, BS, BT, B20A5) 2.1L (B21A) 2.2L (F22A1, F22A4, F22A6, F22B1, F22B2, H22A1), 2.3L (H23A1) and 2.7L (C27A4) Engines
Code 0 Electronic Control Module (ECM)
Code 1 Heated oxygen sensor (or Oxygen content)—or—Oxygen content A (A20A3, B20A5)
Code 2 Oxygen content B (A20A3, B20A5)—or—Electronic Control Module (ECM) (BS, BT—1986 only) and (A20A3—1987 only)

Code 3 Manifold Absolute Pressure (MAP)
Code 4 Crankshaft position sensor or—Faulty ECU (BS, BT—1986 only) and (A20A3 -1987 only, B20A5, B21A1)
Code 5 Manifold Absolute Pressure (MAP)
Code 6 Engine coolant temperature (ECT)
Code 7 Throttle position sensor (TP sensor)
Code 8 Top dead center sensor (TDC sensor)
Code 9 No. 1 cylinder position sensor
Code 10 Intake air temperature sensor (IAT sensor)
Code 11 Electronic Control Module (ECM) (BS, BT -1986 only) and (A20A3—1987 only)
Code 12 Exhaust Gas Recirculation (EGR) System (except Del Sol and Civic & CRX 1 .6L DI 6A6)
Code 13 Barometric pressure sensor (BARO sensor)
Code 14 Idle air control (IAC valve) except 1987 -A20A3 engine. or—1986 BS, BT and 1987 A20A3 Engines, Code 14 or high is possible faulty Electronic Control Module (ECM)
Code 15 Ignition output signal
Code 16 Fuel Injector
Code 17 Vehicle Speed sensor (VSS)
Code 19 NT lock-up control solenoid valve NB (D15B1, D15B2, D15B6, D15B7, D15B8, D15Z1, D16A6, D16Z6)
Code 20 Electric load detector (ELD)
Code 21 V-TEC control solenoid (D15Z1, D16Z6, H22A1)

Code 22 V-TEC pressure switch (D15Z1, D16Z6, H22A1)

Code 23 Knock sensor (H22A1-DOHC—VTEC)

Code 30 NT FI Signal A (F22A1, F22A4, F22A6)

Code 31 NT FI Signal B (F22A1, F22A4, F22A6)

Code 41 Heated Oxygen Sensor Heater (F22A1, F22A4)

Code 43 Fuel supply system (except D1SB1, D15B2, D15B6, B20A5, B21A, D16A6)

Code 45 Fuel supply metering

Code 48 Heated oxygen sensor (D15Z1 engine only, except Calif. emission)

Code 61 Front Heated Oxygen Sensor

Code 63 Rear Heated Oxygen Sensor

Code 65 Rear Heated Oxygen Sensor Heater

Code 67 Catalytic Converter System

Code 70 Automatic Transaxle or A/F FI Data line

Code 71 Misfire detected; cylinder No. 1 or random misfire

Code 72 Misfire detected; cylinder No. 2 or random misfire

Code 73 Misfire detected; cylinder No. 3 or random misfire

Code 74 Misfire detected; cylinder No. 4 or random misfire

Code 75 Misfire detected; cylinder No. 5 or random misfire

Code 76 Misfire detected; cylinder No. 6 or random misfire

Code 80 Exhaust Gas Recirculation (EGR) system

Code 86 Coolant temperature

Code 92 Evaporative Emission Control System

TRANSMISSION CODES

Reading and Retrieving Codes

EARLY MODELS

➡Early models refers to the following vehicles: 1989–91 Civic 4WD and 1988–91 Prelude models.

All of these models display stored DTC's through a Light Emitting Diode (LED) located on the transmission/transaxle control computer, which constantly flashes out the DTC's whenever the ignition key is **ON**. When the computer stores a transmission-related DTC, the S3 light in the instrument cluster illuminates all the time, regardless of the gear selected, to let the driver know there is a stored DTC.

These vehicles use one and two-digit DTC's, which are displayed

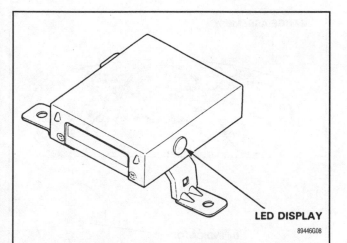

LED DISPLAY

89446G08

The LED is usually located on the side of the transaxle control computer, and, if a fault is detected, flashes whenever the ignition key is ON—early Honda models

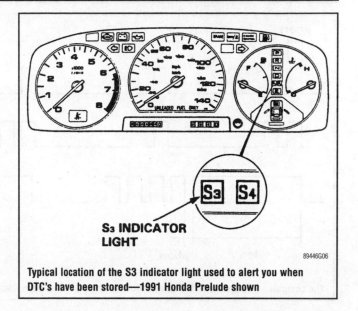

S₃ INDICATOR LIGHT

89446G06

Typical location of the S3 indicator light used to alert you when DTC's have been stored—1991 Honda Prelude shown

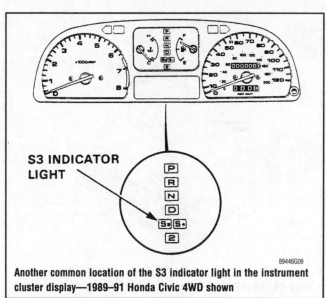

S3 INDICATOR LIGHT

89446G09

Another common location of the S3 indicator light in the instrument cluster display—1989–91 Honda Civic 4WD shown

through the LED on the transmission/transaxle computer in a series of flashes. For example, if the LED flashes five times, pauses, then flashes six times and pauses, this indicates Codes 5 and 6 have been stored by the transmission/transaxle computer. For two-digit codes, the tens digits are displayed as long flashes and the ones digits are displayed as shorter flashes. Therefore, if the LED displays one long flash and four short flashes followed by a pause, this indicates a Code 14. The system displays the codes from the lowest to the highest, and when it reaches the last code it returns to the lowest and cycles through the codes again.

To retrieve the codes, perform the following:

1. Perform the preliminary inspection, located earlier in this section. This is very important, since a loose or disconnected wire, or corroded connector terminals can cause a whole slew of unrelated DTC's to be stored by the computer; you will waste a lot of time performing a diagnostic "goose chase."

2. Grab some paper and a pencil or pen to write down the DTC's when they are flashed out.

3. Locate the transmission/transaxle control computer. The computer is generally located in the following positions:

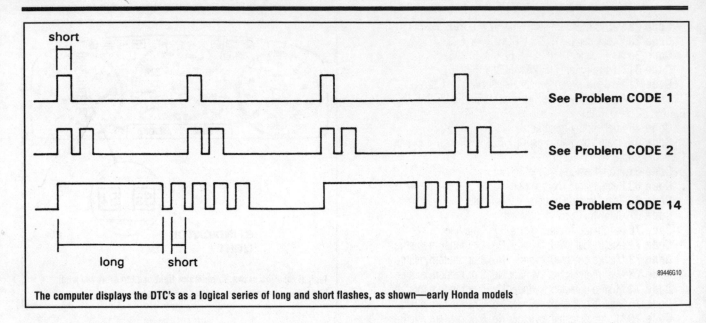

See Problem CODE 1

See Problem CODE 2

See Problem CODE 14

The computer displays the DTC's as a logical series of long and short flashes, as shown—early Honda models

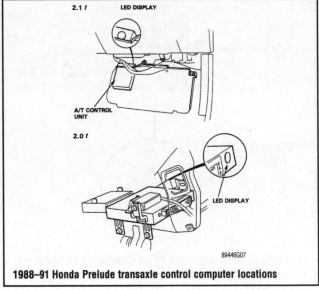

1988–91 Honda Prelude transaxle control computer locations

- 1988–90 Prelude—under the carpet in the passenger's side front footwell
- 1989–91 Civic 4WD—under the driver's seat
- 1991 Prelude with the 2.0L engine—behind the center console, below the radio (it may be visible without removing the center console)
- 1991 Prelude with the 2.1L engine—under the carpet in the passenger's side front footwell

4. Turn the ignition switch **ON** and observe the computer's LED. Note all of the DTC's.

5. Once all of the DTC's have been noted, turn the ignition switch **OFF**.

LATE MODELS

→**Late models refers to the following vehicles: 1990–96 Accord, 1996–97 Civic, 1992–96 Prelude models.**

Unlike the other Honda models described earlier in this section, these models display the transmission-related DTC's through either the S or D4 light on the instrument cluster. These models use 2-digit trouble codes, which are displayed in a series of flashes. The

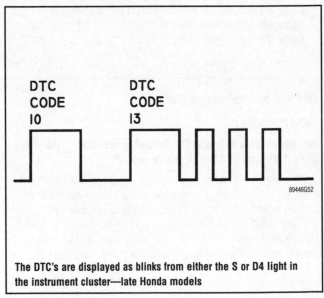

The DTC's are displayed as blinks from either the S or D4 light in the instrument cluster—late Honda models

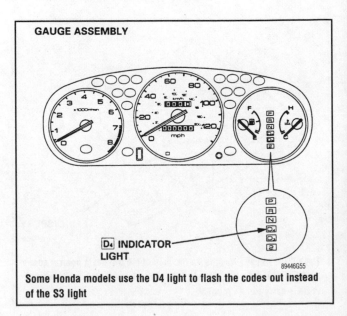

Some Honda models use the D4 light to flash the codes out instead of the S3 light

tens digits are displayed as long flashes and the ones digits are displayed as shorter flashes. Therefore, if the S or D4 light displays one long flash and four short flashes followed by a pause, this indicates a Code 14.

➡Some of the early models of these vehicles have the added benefit, in which you can read the DTC's through the S/D4 light and also from the LED on the computer. The LED located on the computer was slowly phased out during these model years.

To retrieve the DTC's, perform the following:

1. Perform the preliminary inspection, located earlier in this section. This is very important, since a loose or disconnected wire, or corroded connector terminals can cause a whole slew of unrelated DTC's to be stored by the computer; you will waste a lot of time performing a diagnostic "goose chase."

2. Grab some paper and a pencil or pen to write down the DTC's when they are flashed out.

❊❊ WARNING

The Diagnostic Link Connector (DLC) is usually situated next to the service check connector; do not jump the terminals of the DLC by mistake. This could damage the vehicle's computer.

3. Locate the service check connector. The service check connector is either a single connector with two wires running to it or it is two separate one-wire connectors next to each other. The service check connector(s) are not attached to anything. It can be located according to the model of the vehicle at hand, as follows:

• 1990–96 Accord—beneath the far right-hand side of the instrument panel

• 1996–97 Civic—beneath the passenger's side instrument panel

• 1992–96 Prelude—behind the left-hand side of the center console

4. Using a jumper wire or bent paper clip, connect the two terminals of the service check connector.

5. Turn the ignition switch **ON** and observe the S or D4 light in the instrument cluster. Note all of the DTC's.

6. Once all of the DTC's have been noted, turn the ignition switch **OFF** and remove the jumper wire/paper clip.

Clearing Codes

➡It is not recommended that the negative battery cable be disconnected to clear DTC's. Doing so will clear all settings from the vehicle's computer system, resulting in lost radio presets, seat memories, anti-theft codes, driveability parameters, etc., and there are better ways of spending your afternoon than resetting all of these.

For most models, locate the backup fuse in the engine compartment fuse block. For the exceptions, refer to the following list:

• 1989–91 Civic 4WD—fuse No. 34
• 1989–91 Prelude—fuse No. 35
• 1992–96 Prelude—clock/radio fuse

Remove the fuse for at least 10–15 seconds, then reinstall it. All of the codes should now be cleared. To double check, road test the vehicle and perform the code retrieval procedure again. If there are still codes present, either the codes were not cleared (Are the codes identical to those flashed out previously?) or the underlying problem is still there (Are only some of the codes the same as previously?)

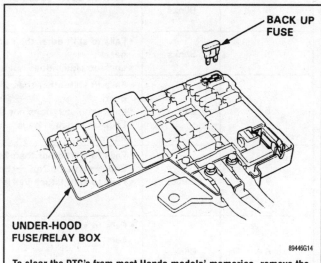

To clear the DTC's from most Honda models' memories, remove the backup fuse from the under hood fuse/relay box

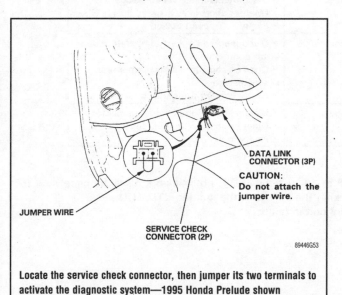

Locate the service check connector, then jumper its two terminals to activate the diagnostic system—1995 Honda Prelude shown

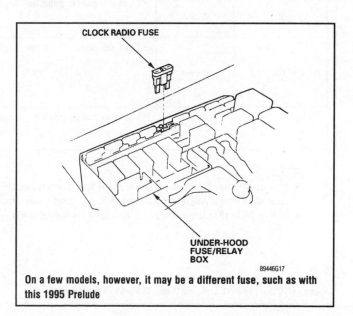

On a few models, however, it may be a different fuse, such as with this 1995 Prelude

Number of LED display blinks	S3 indicator light	Symptom	Probable Cause
1	Blinks	• Lock-up clutch does not engage. • Lock-up-clutch does not disengage. • **Unstable idle speed.**	• Disconnected lock-up control solenoid valve A connector • Open or short in lock-up control solenoid valve A wire • **Faulty lock-up control solenoid valve A**
2	Blinks	• Lock-up clutch does not engage.	• Disconnected lock-up control solenoid valve B connector • Open or short in lock-up control solenoid valve B wire • Faulty lock-up control solenoid valve B
3	Blinks or OFF	• Lock-up clutch does not engage.	• Disconnected throttle angle sensor connector • Open or short in throttle angle sensor wire • Faulty throttle angle sensor
4	Blinks	• Lock-up clutch does not engage.	• Disconnected speed pulser connector • Open or short in speed pulser wire • Faulty speed pulser
5	Blinks	• Fails to shift other than 2nd ←→ 4th gear. • Lock-up clutch does not engage.	• Short in shift position console switch wire • Faulty shift position console switch
6	OFF	• Fails to shift other than 2nd ←→ 4th gear. • Lock-up clutch does not engage. • Lock-up clutch engages and disengages alternately.	• Disconnected shift position console switch connector • Open in shift position console switch wire • Faulty shift position console switch.
7	Blinks	• Fails to shift other than 1st ←→ 4th, 2nd ←→ 4th, or 2nd ←→3rd gears. • Fails to shift (stuck in 4th gear).	• Disconnected shift control solenoid valve A connector • Open or short in shift control solenoid valve A wire • Faulty shift control solenoid valve A
8	Blinks	• Fails to shift (stuck in 1st gear or 4th gear).	• Disconnected shift control solenoid valve B connector • Open or short in shift control solenoid valve B wire • Faulty shift control solenoid valve B
9	Blinks	• Lock-up clutch does not engage.	• Disconnected A/T speed pulser • Open or short in A/T speed pulser wire • Faulty A/T speed pulser
10	Blinks	• Lock-up clutch does not engage.	• Disconnected coolant temperature sensor connector • Open or short in coolant temperature sensor wire • Faulty coolant temperature sensor
11	OFF	• Lock-up clutch does not engage.	• Disconnected ignition coil connector • Open or short in ignition coil wire • Faulty ignition coil

NOTE:
● If a customer describes the symptoms for codes 3, 6 or 11, yet the LED is not blinking, it will be necessary to recreate the symptom by test driving, and then checking the LED with the ignition STILL ON.
● If the LED display blinks 12 or more times, the control unit is faulty.

89446G01

Number of D4 indicator light blinks while Service Check Connector is jumped.	D4 indicator light	Possible Cause	Symptom
1	Blinks	• Disconnected lock-up control solenoid valve A connector • Short or open in lock-up control solenoid valve A wire • Faulty lock-up control solenoid valve A	• Lock-up clutch does not engage. • Lock-up clutch does not disengage. • Unstable idle speed.
2	Blinks	• Disconnected lock-up clontrol solenoid valve B connector • Short or open in lock-up control solenoid valve B wire • Faulty lock-up control solenoid valve B	• Lock-up clutch not engage.
3	Blinks or OFF	• Disconnected throttle position (TP) sensor connector • Short or open in TP sensor wire • Faulty TP sensor	• Lock-up clutch does not engage.
4	Blinks	• Disconnected vehicle speed sensor (VSS) connector • Short or open in VSS wire • Faulty VSS	• Lock-up clutch does not engage.
5	Blinks	• Short in A/T gear position switch wire • Faulty A/T gear position switch	• Fails to shift other than 2nd ↔ 4th gears. • Lock-up clutch does not engage.
6	OFF	• Disconnected A/T gear position switch connector • Open in A/T gear position switch wire • Faulty A/T gear position switch	• Fails to shift other than 2nd ↔ 4th gears. • Lock-up clutch does not engage. • Lock-up clutch engages and disengages alternately.
7	Blinks	• Disconnected shift control solenoid valve A connector • Short or open in shift control solenoid valve A wire • Faulty shift control solenoid vavle A	• Fails to shift (between 1st ↔ 4th, 2nd ↔ 4th or 2nd ↔ 3rd gears only). • Fails to shift (stuck in 4th gear)
8	Blinks	• Disconnected shift control solenoid valve B connector • Short or open in shift control solenoid valve B wire • Faulty shift control solenoid valve B	• Fails to shift (stuck in 1st or 4th gears).
9	Blinks	• Disconnected countershaft speed sensor connector • Short or open in the countershaft speed sensor wire • Faulty countershaft speed sensor	• Lock-up clutch does not engage.

Honda DTC's, 1 of 2—1990–95 Accord, 1992–95 Prelude models

89446G02

Number of [D4] indicator light blinks while Service Check Connector is jumped.	[D4] indicator light	Possible Cause	Symptom
10	Blinks	•Disconnected engine coolant temperature (ECT) sensor connector •Short or open in the ECT sensor wire •Faulty ECT sensor	•Lock-up clutch does not engage.
11	OFF	•Disconnected ignition coil connector •Short or open in ignition coil wire •Faulty ignition coil	•Lock-up clutch does not engage.
14	OFF	•Short or open in FAS (BRN/WHT) wire between the D16 terminal and ECM •Trouble in ECM	•Transmission jerks hard when shifting.
15	OFF	•Disconnected mainshaft speed sensor connector •Short or open in mainshaft speed sensor wire •Faulty mainshaft speed sensor	•Transmission jerks hard when shifting.

89446G03

Honda DTC's, 2 of 2—1990–95 Accord, 1992–95 Prelude models

Diagnostic Trouble Code (DTC)*	D4 Indicator Light	Symptom	Possible Cause
P1753 (1)	Blinks	• Lock-up clutch does not engage. • Lock-up clutch does not disengage. • Unstable idle speed.	• Disconnected lock-up control solenoid valve A connector • Short or open in lock-up control solenoid valve A wire • Faulty lock-up control solenoid valve A
P1758 (2)	Blinks	• Lock-up clutch does not engage.	• Disconnected lock-up control solenoid valve B connector • Short or open in lock-up control solenoid valve B wire • Faulty lock-up control solenoid valve B
P1791 (4)	Blinks	• Lock-up clutch does not engage.	• Disconnected vehicle speed sensor (VSS) connector • Short or open in VSS wire • Faulty VSS
P1705 (5)	Blinks	• Fails to shift other than 2nd – 4th gears. • Lock-up clutch does not engage.	• Short in A/T gear position switch wire • Faulty A/T gear position switch
P1706 (6)	OFF	• Fails to shift other than 2nd – 4th gears. • Lock-up clutch does not engage. • Lock-up clutch engages and disengages alternately.	• Disconnected A/T gear position switch connector • Open in A/T gear position switch wire • Faulty A/T gear position switch
P0753 (7)	Blinks	• Fails to shift (between 1st – 4th, 2nd – 4th or 2nd – 3rd gear only). • Fails to shift (stuck in 4th gear).	• Disconnected shift control solenoid valve A connector • Short or open in shift control solenoid valve A wire • Faulty shift control solenoid valve A
P0758 (8)	Blinks	• Fails to shift (stuck in 1st or 4th gears).	• Disconnected shift control solenoid valve B connector • Short or open in shift control solenoid valve B wire • Faulty shift control solenoid valve B
P0720 (9)	Blinks	• Lock-up clutch does not engage.	• Disconnected countershaft speed sensor connector • Short or open in countershaft speed sensor wire • Faulty countershaft speed sensor
P0715 (15)	OFF	• Transmission jerks hard when shifting.	• Disconnected mainshaft speed sensor connector • Short or open in mainshaft speed sensor wire • Faulty mainshaft speed sensor
P1768 (16)	Blinks	• Transmission jerks hard when shifting. • Lock-up clutch does not engage.	• Disconnected linear solenoid connector • Short or open in linear solenoid wire • Faulty linear solenoid

(DTC)*: The DTCs in parentheses are the number of the D4 indicator light blinks when the service check connector is connected with the special tool (SCS service connector).

89446G04

Honda DTC's, 1 of 2—1996–97 Accord, 1996–97 Civic, 1996–97 Prelude models

Diagnostic Trouble Code (DTC)*	D₄ Indicator Light	Symptom	Possible Cause
P0740 (40)	OFF	• Lock-up clutch does not engage. • Lock-up clutch does not disengage. • Unstable idle speed.	• Faulty lock-up control system
P0730 (41)	OFF	• Fails to shift (between 1st – 2nd, 2nd – 4th or 2nd – 3rd gears only). • Fails to shift (stuck in 1st or 4th gears).	• Faulty shift control system

(DTC)*: The DTCs in parentheses are the number of the D₄ indicator light blinks when the service check connector is connected with the special tool (SCS service connector).

89446G05

Honda DTC's, 2 of 2—1996–97 Accord, 1996–97 Civic, 1996–97 Prelude models

Hyundai

ENGINE CODES

General Information

Hyundai utilizes 2 fuel system types. The first is a feedback carburetor and the second is fuel injection. The feedback carburetor is used on 1986–93 Excel.

Multi-point Fuel Injection (MFI) was introduced in 1989 on the Sonata. In 1990 Excel was available with fuel injection also. Scoupe, Elantra and Accent came only with MFI.

Service Precautions

• Keep the ECU parts and harnesses dry during service. Protect the ECU and all solid-state components from rough handling or temperature extremes.
• Use extreme care when working around the ECU or other components.
• Disconnect the negative battery cable before attempting to disconnect or remove any electronic parts.
• Disconnect the negative battery cable and ECU connector before performing arc welding on the vehicle.
• Disconnect and remove the ECU from the vehicle before subjecting the vehicle to the temperatures experienced in a heated paint booth.

Reading Codes

➡**Hyundai utilized a Feedback carburetor system in the Excel. However, the self-diagnostic system is not used. Self-diagnosis pertains to fuel injected vehicles only. 1993–94 Scoupe and 1995 Accent have the ability to read diagnostic codes through the Malfunction Indicator Lamp (MIL) and therefore special equipment is not required to retrieve codes. All other vehicles however, do not process this ability, therefore either a Multi-Use Tester or and Analog voltmeter must be used to retrieve codes. When using special diagnostic equipment, always observe the tool manufacturer's instructions.**

1989–95 VEHICLES—USING THE MULTI-USE TESTER

1. Turn the ignition switch to the **OFF** position.
2. Connect the multi-use tester to the diagnosis connector in the fuse box.
3. Connect the power-source terminal of the multi-use tester to the cigar lighter.
4. Turn the ignition switch to the **ON** position.
5. Follow the manufacturer's instructions to retrieve the trouble codes. The codes will be displayed in numerical order.

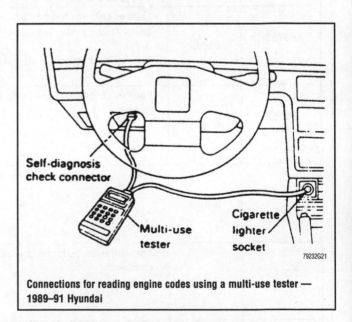

Connections for reading engine codes using a multi-use tester — 1989–91 Hyundai

1989–95 VEHICLES—USING ANALOG VOLTMETER

1. Connect the voltmeter to the self-diagnosis connector.
2. Turn the ignition switch to the **ON** position.
3. Observe the voltmeter to read the trouble codes. The code is determined by noting the duration of the voltmeter sweeps. A sweep of long duration indicates the multiple of ten digit, while a sweep of short duration indicates the single digit. For example, the code number 12 is indicated by 1 sweep of long duration followed by 2

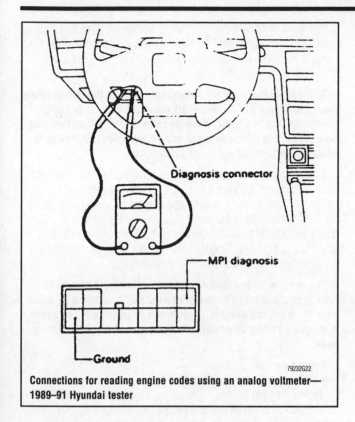

Connections for reading engine codes using an analog voltmeter—1989–91 Hyundai tester

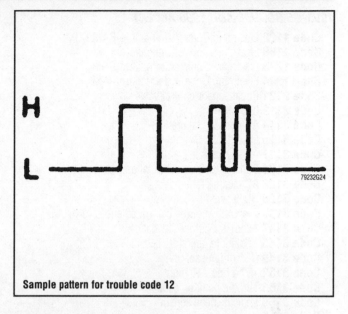

Sample pattern for trouble code 12

sweeps of short duration and so on. The trouble codes will be displayed in numerical order.

1989–95 VEHICLES—USING ENGINE (MIL) LAMP

The 1993–94 Scoupe and 1995 Accent has the ability to read codes by using the Maintenance Indicator Lamp (MIL).

1. Turn the ignition switch **ON** but do not start the vehicle.
2. Ground the L-wire (PIN 10) in the diagnostic terminal for 2½ seconds.
3. The first code to flash should be (4444) which will flash until the L-wire is disconnected.
4. Ground the L-wire again for 2½ seconds and record flash codes until the end of output code (3333) is flashed.

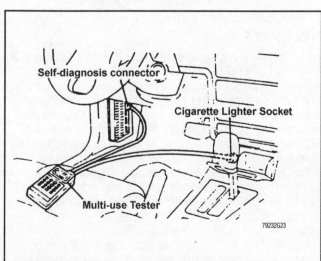

Connections for reading engine codes using a multi-use tester—1992–95 Hyundai

Clearing Codes

1989–95 VEHICLES

1. Turn the ignition switch to the **OFF** position.
2. Disconnect the negative battery cable for at least 15 seconds.
3. Disconnect the multi-use tester or analog voltmeter.
4. Reconnect the negative battery cable.

Diagnostic Trouble Codes

1989–94 EXCEL, 1992–95 ELANTRA, 1991–92 SCOUPE, 1989–95 SONATA

Code 11 Oxygen sensor—check harness and connector, fuel pressure, fuel injectors, oxygen sensor, check for intake air leaks

Code 12 Air flow sensor—check harness, connector and air flow sensor

Code 13 Air temperature sensor—check harness, connector and air temperature sensor

Code 14 Throttle position sensor—check harness and connector, throttle position sensor, idle position switch

Code 15 Motor position sensor—check harness and connector, motor position sensor

Code 21 Engine coolant temperature sensor—check harness and connector, engine coolant temperature sensor

Code 22 Crank angle sensor—check harness, connector and distributor assembly

Code 23 No. 1 cylinder top dead center sensor—check harness, connector and distributor assembly

Code 24 Vehicle speed sensor. (reed switch)—check harness and connector, and vehicle speed sensor

Code 25 Barometric pressure sensor—check harness, connector and barometric pressure sensor

Code 41 Injector—check harness and connector, injector coil resistance

Code 42 Fuel pump—check harness and connector, control relay

Code 43 EGR—check harness and connector, EGR temperature sensor, EGR valve, EGR control solenoid valve, EGR valve control vacuum (California)

Code 44 Ignition coil fault, faulty power transistor

Code 59 Oxygen (HO$_2$5) sensor fault

1993–95 SCOUPE AND 1995 ACCENT

Code 1122 Electronic Control Unit failure-RAM/ROM
Code 1169 Electronic Control Unit failure
Code 1233 Electronic Control Unit failure-ROM
Code 1234 Electronic Control Unit failure-RAM
Code 2121 Boost sensor control valve
Code 3112 Injector No.1
Code 3114 AC opening failure
Code 3116 Injector No. 3
Code 3117 Airflow sensor
Code 3121 Boost pressure sensor failure
Code 3122 AC closing failure
Code 3128 HO2S sensor
Code 3135 Evaporative purge control solenoid valve
Code 3137 Alternator output/low battery
Code 3145 Coolant temperature sensor
Code 3149 NC compressor
Code 3152 Turbo boost to high
Code 3153 Throttle position sensor
Code 3159 Vehicle speed sensor
Code 3211 Knock sensor
Code 3222 Phase sensor
Code 3224 ECM-knock evolution sensor
Code 3232 Crankshaft position sensor
Code 3233 ECM-knock evolution sensor
Code 3234 Injector No. 2
Code 3235 Injector No. 4
Code 3241 ECM-injector or purge control valve
Code 3242 ECM-IAC motor or NC relay
Code 3243 Electronic Control Unit failure
Code 4133 Electronic Control Unit failure
Code 4151 Air/fuel control
Code 4152 Air/fuel adaptive failure
Code 4153 Air/fuel adaptive (multiple) failure
Code 4154 Air/fuel adaptive (additive) failure
Code 4155 ECM-a/c relay, IAC motor, injector or PCV
Code 4156 Boost sensor control deviation failure

TRANSMISSION CODES

Reading and Retrieving Codes

GENERAL INFORMATION

➡**Hyundai vehicles utilize Mitsubishi manufactured transaxles and their electronic control systems.**

Most Mitsubishi/Hyundai computerized transaxle control systems have no way of indicating a problem in the system. If the computer detects a problem, it begins pulsing a code at the diagnostic link connector. The only way you'll know there's a code is to look for it; nothing lights up to tell you there's a problem.

The only exception to that is 1996 and newer Hyundai models (OBD II), which do have a Malfunction Indicator Lamp (MIL) in the instrument cluster display. The MIL illuminates when the diagnostic system detects a problem in the transaxle or control system.

Hyundai vehicles use Diagnostic Link Connectors (DLC) in a possible three different configurations, four different sets of diagnostic codes (types 1 through 4), and three methods for displaying DTC's. Mitsubishi transaxles appear in Mitsubishi (of course), Hyundai and Chrysler Import vehicles.

The Diagnostic Link Connector (DLC) must be located to retrieve any DTC's. Luckily, unlike some of the other import models covered

in this manual, there are only two locations for the DLC. The locations are as follows: 1989–95 Hyundai models, next to the fuse box inside the left side kick panel; all OBD II vehicles, under the left-hand side of the instrument panel.

➡**Remember that only OBD II models can let the driver know that codes have been stored; all other models do not alert the driver that any faults have been detected. Therefore, the only way to ascertain whether codes have been stored is to actually perform the retrieval procedure.**

Over the years, the transaxle computer control system used with Hyundai vehicles has been designed to display the transaxle DTC's in three different formats. All of these formats consist of changes to the on- and off-time of the signal. Also, four different sets of DTC's have been utilized to inform the person servicing the transaxle of any detected problems. The Hyundai models and their corresponding code types are presented in the following list:

➡**The precise date when models switched from one code type to another wasn't always consistent, and using this list as a Bible for the breaking points between code types could have you chasing down the wrong component for the wrong code.**

Type 1 codes:
- 1989 Sonata 4-cylinder models

Type 2 codes:
- 1989–93 Hyundai models, except for 1989 Sonata 4-cylinder models

Type 3 codes:
- 1994–95 Hyundai models
- 1996 and newer Hyundai Sonata models

Type 4 codes:
- 1996 and newer Hyundai models, except for Sonata

Since the precise date when models switched from one code type to another listed in the preceding list is not set in stone, record any codes in memory, clear the codes, then create a fault in the transaxle computer system. For example, if the shift solenoid code was not stored by the transaxle computer, disconnect the shift solenoid transmission/transaxle connector and restart the engine, after clearing the previous codes. The diagnostic system should store the appropriate code for a shift solenoid/circuit fault, which will allow you to ascertain precisely what set of codes the system being worked on uses.

TYPE 1 CODES

The first format only appeared in 1985–88 models, with the addition of the 4-cylinder 1989 Sonata. The Type 1 system was designed with what is known as the "Type 1" computer system. The Type 1 DTC's are made up of a series of signal pulses that varied in on-time and off-time. There is no regular pattern to these signals, except for a 4 second start and end signal, to indicate when each DTC starts and finishes. To identify the code, you must compare the signal to the accompanying code chart.

The Type 1 Mitsubishi/Hyundai transaxle diagnostic system functions so that once the system detects a malfunction, it pulses the DTC continuously through the DLC. There's no wire to ground, no special procedure to perform; just connect a DVOM to the respective DLC terminals and read the DTC's.

Because of the lack of a regular pattern, Type 1 codes can be difficult to read and interpret. One way to make these early codes easier to read is to use an oscilloscope or Digital Multi-Meter (DMM) with a graphic display instead of a voltmeter. By capturing the signal

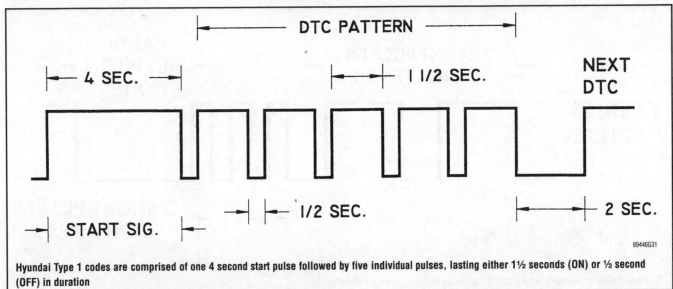

Hyundai Type 1 codes are comprised of one 4 second start pulse followed by five individual pulses, lasting either 1½ seconds (ON) or ½ second (OFF) in duration

as a single waveform, you can compare the waveform to the accompanying code chart.

The Type 1 code system utilized volatile code memory, which means that whenever the ignition key is turned **OFF**, the codes are cleared. Also, because of the simplistic design of the system, any code will force the vehicle to run in a failsafe mode.

➡**The Type 1 code system can only display one DTC at a time. Therefore, you must repair the problem for the first DTC before the second DTC can be read. So do not be surprised if you must fix several problems before there are no DTC's left.**

To retrieve the codes, perform the following:

1. Perform the preliminary inspection, located earlier in this section. This is very important, since a loose or disconnected wire, or corroded connector terminals can cause a whole slew of unrelated DTC's to be stored by the computer; you will waste a lot of time performing a diagnostic "goose chase."

2. Grab some paper and a pencil or pen to write down the DTC's when they are flashed out.

3. Locate the Diagnostic Link Connector (DLC), next to the fuse box inside the left side kick panel for 1989 Sonata 4-cylinder models.

4. Start the engine and drive the vehicle until the transaxle goes into the failsafe mode.

5. Park the vehicle, but do not turn the ignition **OFF**. Allow it to idle.

➡**It is important to allow the vehicle to idle while reading the DTC's, otherwise a Code 12—lack of an ignition signal—will be detected by the computer, leading to misdirected troubleshooting and wasted time.**

6. Attach a Digital Volt-Ohmmeter (DVOM), a Digital Multi-Meter (DMM), or an oscilloscope to the test terminals on the Diagnostic Link Connector (DLC). For 1989 Sonata 4-cylinder models, attach the negative lead to terminal 6 of the DLC and the positive lead to terminal 12.

7. Observe the DVOM/DMM/oscilloscope and note the wave pattern. Compare the pattern to the accompanying chart to determine which code it corresponds to.

➡**Be sure of which system the vehicle at hand uses, since Type 1 Code 12 looks the same as Types 2 and 3 Code 15.**

8. After all of the DTC(s) have been retrieved, fix the applicable problems, clear the codes, drive the vehicle, and perform the retrieval procedure again to ensure that all of the codes are gone.

TYPE 2 AND 3 CODES

➡**Reading and retrieving Type 2 and Type 3 codes is identical; the only difference between Type 2 and 3 codes is what each of the codes stands for.**

Unlike the Type 1 diagnostic system, the system used for Type 2 and 3 codes was designed with the capability of storing up to 10 DTC's. Therefore, you will not have to fix each problem one at a time. However, unlike other import manufacturers, the codes are not displayed from lowest number to highest number. Rather, they are flashed out in the order in which they detected. This means that if the same fault was detected numerous times, the same code will be flashed out the same number of times as detected. Also, since only 10 codes can be stored at one time, if the computer detects an eleventh code, the first code stored will be deleted to make room for the latest code. This makes it important to check for codes whenever you suspect a problem with the transaxle.

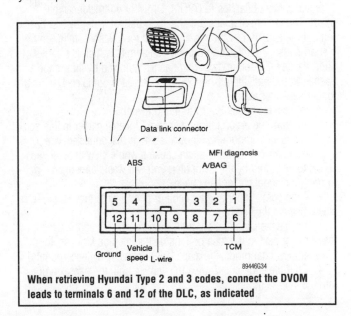

When retrieving Hyundai Type 2 and 3 codes, connect the DVOM leads to terminals 6 and 12 of the DLC, as indicated

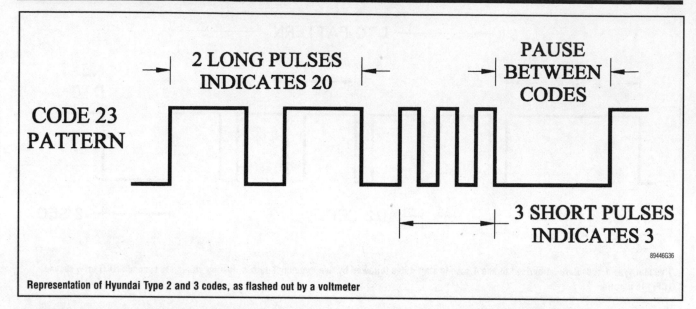

CODE 23 PATTERN

2 LONG PULSES INDICATES 20

PAUSE BETWEEN CODES

3 SHORT PULSES INDICATES 3

89446G36

Representation of Hyundai Type 2 and 3 codes, as flashed out by a voltmeter

➡Remember that only Type 4 (OBD II) models can let the driver know that codes have been stored; all other models do not alert the driver that any faults have been detected. Therefore, the only way to ascertain whether codes have been stored is to actually perform the retrieval procedure.

With the introduction of the Type 2 code system, Mitsubishi also developed Failsafe Codes. Failsafe codes are DTC's which, if detected, will switch the vehicle's management system into a failsafe mode (thus the name). The failsafe mode sets the control computer to run (albeit not as efficiently) on predetermined parameters, so that erroneous information from the vehicle's sensors will be ignored until the problem at hand can be repaired. Other manufacturers call the failsafe mode the Limp In mode.

➡Remember that since many codes are redundant, the system may store a non-failsafe code and a failsafe code for the same malfunction. In which case, the system will go into failsafe mode.

Mitsubishi also implemented an easier way for reading the DTC's in Type 2 and 3 codes. The DTC's are 2-digit codes represented by a logical series of flashes (a DVOM, DMM or voltmeter can be used). The system displays the codes on the voltmeter by first flashing out the tens digit, which is a series of long flashes, followed by a short pause, then flashing out the ones digits, which is a series of short flashes. Therefore, two long flashes followed by three short flashes would indicate a Code 23. Between each complete DTC is a longer separator code.

To retrieve the codes, perform the following:
1. Perform the preliminary inspection, located earlier in this section. This is very important, since a loose or disconnected wire, or corroded connector terminals can cause a whole slew of unrelated DTC's to be stored by the computer; you will waste a lot of time performing a diagnostic "goose chase."
2. Grab some paper and a pencil or pen to write down the DTC's when they are flashed out.
3. Locate the Diagnostic Link Connector (DLC), either in the fuse box, or next to the fuse box inside the left side kick panel for 1989–95 Hyundai models (except 1989 Sonata 4-cylinder models).
4. Start the engine and drive the vehicle until the transaxle goes into the failsafe mode.

5. Park the vehicle, but do not turn the ignition **OFF**. Allow it to idle.
6. Attach a voltmeter (analog or digital) to the test terminals on the Diagnostic Link Connector (DLC). The negative lead should be attached to terminal 6 and the positive lead to terminal 12.
7. Observe the voltmeter and count the flashes (or arm sweeps if using an analog voltmeter); note the applicable codes.
8. After all of the DTC(s) have been retrieved, fix the applicable problems, clear the codes, drive the vehicle, and perform the retrieval procedure again to ensure that all of the codes are gone.

TYPE 4 CODES (OBD-II CODES)

The Type 4 codes are covered later in this section as OBD II codes.

Clearing Codes

TYPE 1 CODES

It is not necessary to clear Type 1 codes, because they are stored in volatile memory and automatically erase whenever the ignition key is turned **OFF**.

TYPE 2, 3 AND 4 CODES

➡Type 4 codes are covered later in this section as OBD II codes. It is not recommended that the negative battery cable be disconnected to clear DTC's. Doing so will clear all settings from the vehicle's computer system, resulting in lost radio presets, seat memories, anti-theft codes, driveability parameters, etc., and there are better ways of spending your afternoon than resetting all of these.

1. Turn the ignition switch **OFF**.
2. Disconnect the wiring harness connector from the control system computer.
3. Wait at least one minute, then reconnect the computer. The codes should now be cleared.

If there are still codes present, either the codes were not properly cleared (Are the codes identical to those flashed out previously?) or the underlying problem is still there (Are only some of the codes the same as previously?)

DIAGNOSTIC TROUBLE CODES—MITSUBISHI/HYUNDAI TYPE 1 CODES

Code Pattern	Codes				Component/Circuit Fault
	1985	1986	1987	1988	
	1	1	1	1	Control computer fault
	2	2	2	2	First gear signal detected at high speed
	3	3	3	3	Vehicle speed detected by pulse generator B lower than actual vehicle speed
	4	4	4	4	Shift control solenoid A or circuit
	5	5	5	5	Shift control solenoid B or circuit
	6	—	6	6	Kickdown servo switch or circuit
	7	6	7	7	Slipping detected
	8	7	8	8	Pressure control solenoid or circuit
	9	—	—	—	Engine speed detected over 6,500 rpm
	10	—	—	—	Kickdown drum speed detected over 6,500 rpm
	11	8	9	—	Damper clutch control system
	12	9	10	9	No ignition signal

89446G18

Hyundai DTC's—Type 1 Codes

Fault code	Fault code (for voltmeter)	Cause	Remedy
21	5V ----- ⎍⎍⎍⎍ 0V -----	Abnormal increase of TPS output	o Check the throttle position sensor connector. o Check the throttle position sensor itself.
22	⎍⎍⎍⎍	Abnormal decrease of TPS output	o Adjust the throttle position sensor. o Check the accelerator switch (No.28: output or not).
23	⎍⎍⎍⎍⎍	Incorrect adjustment of the throttle-position sensor system	o Check the throttle position sensor output circuit harness.
24	⎍⎍⎍⎍⎍	Damaged or disconnected wiring of the oil temperature sensor system	o Check the oil temperature sensor circuit harness. o Check the oil temperature sensor connector. o Check the oil temperature sensor itself.
25	⎍⎍⎍⎍⎍⎍	Damaged or disconnected wiring of the kickdown servo switch system, or improper contact	o Check the kickdown servo switch output circuit harness. o Check the kickdown servo switch connector.
26	⎍⎍⎍⎍⎍⎍	Short circuit of the kickdown servo switch system	o Check the kickdown servo switch itself.
27	⎍⎍⎍⎍⎍⎍⎍	Damaged or disconnected wiring of the ignition pulse pick-up cable system	o Check the ignition pulse signal line.
28	⎍⎍⎍⎍⎍⎍⎍⎍	Short circuit of the accelerator switch system or improper adjustment	o Check the accelerator switch output circuit harness. o Check the accelerator switch connector. o Check the accelerator switch itself. o Adjust the accelerator switch.
31	⎍⎍⎍⎍	Malfunction of the microprocessor	o Replace the control unit.

89446G19

Hyundai DTC's, 1 of 4—Type 2 Codes

Fault code	Fault code (for voltmeter)	Cause	Remedy
32		First gear command during high-speed driving	o Replace the control unit
33		Damaged or disconnected wiring of the pulse generator B system	o Check the pulse generator B output circuit harness. o Check pulse generator B itself. o Check the vehicle speed reed switch (for chattering).
41		Damaged or disconnected wiring of the shift control solenoid valve A system	o Check the solenoid valve connector. o Check shift control solenoid valve A itself.
42		Short circuit of the shift-control solenoid valve A system	o Check the shift control solenoid valve A drive circuit harness
43		Damaged or disconnected wiring of the shift control solenoid valve B system	o Check the solenoid valve connector. o Check shift control solenoid valve B itself.
44		Short circuit of the shift control solenoid valve B system	o Check the shift control solenoid valve B drive circuit harness.
45		Damaged or disconnected wiring of the pressure control solenoid valve system	o Check the solenoid valve connector. o Check the pressure control solenoid valve itself.
46		Short circuit of the pressure control solenoid valve system	o Check the pressure control solenoid valve drive circuit harness.
47		Damaged or disconnected wiring of the damper clutch control solenoid valve system	o Check the solenoid valve connector. o Check the damper clutch control solenoid valve itself.
48		Short circuit of the damper clutch control solenoid valve system	o Check the damper clutch control solenoid valve drive circuit harness.

89446G20

FAIL-SAFE ITEM

Code No.	Output code Output pattern (for voltmeter)	Description	Fail-safe	Note (relation to fault code)
11	5V 0V	Malfunction of the microprocessor	3rd gear hold	When code No. 31 is generated a 4th time.
12		First gear command during high speed driving	3rd gear (D) or 2nd gear (2, L) hold	When code No. 32 is generated a 4th time.
13		Damaged or disconnected wiring of the pulse generator B system	3rd gear (D) or 2nd gear (2, L) hold	When code No. 33 is generated a 4th time.
14		Damaged or disconnected wiring, or short circuit, of shift control solenoid valve A	3rd gear hold	When code No. 41 or 42 is generated a 4th time.
15		Damaged or disconnected wiring, or short circuit, of shift control solenoid valve B	3rd gear hold	When code No. 43 or 44 is generated a 4th time.
16		Damaged or disconnected wiring, or short circuit, of the pressure control solenoid valve	3rd gear (D) or 2nd gear (2, L) hold	When code No. 45 or 46 is generated a 4th time.
17		Shift steps non-synchronous	3rd gear (D) or 2nd gear (2, L) hold	When code No. 51, 52 53 or 54 is generated a 4th time.

89446G21

Fault code	Fault code (for voltmeter)	Cause	Remedy
49		Malfunction of the damper clutch system	o Check the damper clutch control solenoid valve drive circuit harness. o Check the damper clutch hydraulic pressure system. o Check the damper clutch control solenoid valve itself. o Replace the control unit.
51		First gear non-synchronous	o Check the pulse generator output circuit harness. o Check the pulse generator connector. o Check pulse generator A and pulse generator B themselves. o Kickdown brake slippage.
52		Second gear non-synchronous	o Check the pulse generator A output circuit harness. o Check the pulse generator A connector. o Check pulse generator A itself. o Kickdown brake slippage.
53		Third gear non-synchronous	o Check the pulse generator A output circuit harness. o Check the pulse generator connector. o Check pulse generator A and pulse generator B themselves. o Front clutch slippage. o Rear clutch slippage.
54		Fourth gear non-synchronous	o Check the pulse generator A output circuit harness. o Check the pulse generator A connector. o Check pulse generator A itself. o Kickdown brake slippage.

89446G22

Hyundai DTC's, 4 of 4—Type 2 Codes

code	Diagnostic trouble code (for voltmeter)	Cause	Remedy
11		o Throttle position sensor malfunction	o Check the throttle position sensor connector.
12		o Open throttle position sensor circuit	o Check the throttle position sensor itself.
13		o Shorted throttle position sensor circuit	o Check the idle switch
14			o Check the throttle position sensor wiring harness
			o Check the wiring between ECM and throttle position sensor
15		Oil temperature sensor damaged or disconnected wiring	o Oil temperature sensor connector inspection
			o Oil temperature sensor inspection
			o Oil temperature sensor wiring harness inspection
21		Open kickdown servo switch circuit.	o Check the kickdown servo switch connector.
			o Check the kickdown servo switch.
			o Check the kickdown serro switch wiring harness
22		Shorted kickdown servo switch circuit.	
23		Open ignition pulse pickup cable circuit	o Check the ignition pulse signal line.
			o Check the wiring between ECM and ignition system
24		Open-circuited or improperly adjusted idle switch	o Check the idle switch connector
			o Check the idle switch itself
			o Adjust the idle switch.
			o Check the idle switch wiring harness

89446G23

Hyundai DTC's, 4 of 4—Type 2 Codes

code	Diagnostic trouble code (for voltmeter)	Cause	Remedy
31		Pulse generator A damaged or disconnectorted wiring	o Check the pulse generator A and pulse generator B.
32		Pluse generator B damaged or disconnected wiring	o Check the vehicle speed reed switch (for chattering) o Check the pulse generator A and B wiring harness
41		Open shift control solenoid valve A circuit	o Check the solenoid valve connector o Check the shift control solenoid valve A.
42		Shorted shift control solenoid valve A circuit	o Check the shift control solenoid valve A wiring harness
43		Open shift control solenoid valve B circuit	o Check the solenoid valve connector. o Check the shift control solenoid valve B wiring harness
44		Shorted shift control solenoid valve B circuit	o Check the solenoid valve connector.
45		Open pressure control solenoid valve circuit	o Check the pressure control solenoid valve. o Check the pressure control solenoid valve wiring harness
46		Shorted pressure control solenoid valve circuit	
47		Open circuit in damper clutch control control solenoid valve Short circuit in damper clutch control solenoid valve Defect in the damper clutch system	o Inspection of solenoid valve connector. o Individual inspection of damper clutch control solenoid valve
48			o Check the damper clutch control solenoid valve wiring harness o Check the TCM o Inspection of damper clutch hydraulic system
49			
51		Shifting to first gear does not match the engine speed.	o Check the pulse generator A and pulse generator B connector o Check the pulse generator A and pulse generator B o Check the one way clutch or rear clutch o Check the pulse generator wiring harness

89446G24

code	Diagnostic trouble code (for voltmeter)	Cause	Remedy
52		Shifting to second gear does not match the engine speed.	o　Check the pulse generator A and pulse generator B connector o　Check the pulse generator A and pulse generator B. o　Check the one way clutch or rear clutch o　Check the pulse generator wiring harness
53		Shifting to third gear does not match the engine speed	o　Check the rear clutch or control system o　Check the pulse generator A and pulse generator B connector o　Check the pulse generator A and pulse generator B. o　Check the front clutch slippage or control system o　Check the rear clutch slippage or control system o　Check the pulse generator wiring harness
54		Shifting to fourth gear does not match the engine speed.	o　Check the pulse generator A and B connector o　Check the pulse generator A and B o　Kickdown brake slippage. o　Check the end clutch or control system o　Check the pulse generator wiring harness
-		Normal	-
		Defective transaxle control module (TCM)	o　TCM power supply inspection o　TCM earth inspection o　TCM replacement

89446G25

FAIL-SAFE ITEM

Output code		Description	Fail-safe	Note (relation to diagnostic trouble code)
Code No.	Output pattern (for voltmeter)			
81		Open-circuited or damaged pulse generator A	Locked in third (D) or second (2,L)	When code No. 31 is generated fourth time
82		Open-circuited or damaged pulse generator B	Locked in third (D) or second (2, L)	When code No. 32 is generated fourth time
83		Open-circuited, shorted or damaged shift control solenoid valve A	Lock in third	When code No. 41 or 42 is generated fourth time
84		Open-circuited, shorted or damaged shift control solenoid valve B	Lock in third gear	When code No. 43 or 44 is generated fourth time
85		Open-circuited, shorted or damage pressure control solenoid valve	Locked in third (D) or second (2, L)	When code No. 45 or 46 is generated fourth time.
86		Gear shifting does not match the engine speed	Locked in third (D) or second (2, L)	When either code No. 51, 52, 53 or 54 is generated fourth time.
-	Constant output (or 0V)	Defective transaxle control module (TCM)	Fixed for third speed	-

89446G26

Infiniti

ENGINE CODES

General Information

The Infiniti Electronic Concentrated Control System (ECCS) is an air flow controlled, sequential port fuel injection and engine control system. It is used on all models equipped with 2.0L, 3.0L and 4.5L engines. The ECCS electronic control unit consists of a microcomputer, an inspection lamp, a diagnostic mode selector and connectors for signal input and output, powers and grounds.

The safety relay prevents electrical damage to the electronic control unit, or ECU, and the injectors in case the battery terminals are accidentally connected in reverse. The safety relay is built into the fuel pump control circuit.

Ignition timing is controlled in response to engine operating conditions. The optimum ignition timing in each driving condition is pre-programmed in the computer. The signal from the control unit is transmitted to the power transistor and this signal controls when the transistor turns the ignition coil primary circuit on and off (hence, the ignition timing). The idle speed is also controlled according to engine operating conditions, temperature and gear position. On manual transmission models, if battery voltage is less than 12 volts for a few seconds, a higher idle speed will be maintained by the control unit to improve charging function.

There is a fail-safe system built into the ECCS control unit. This system makes engine starting possible if a portion of the ECU's central processing unit circuit fails. Also, if a major component such as the crank angle sensor or the air flow meter were to malfunction, the ECU substitutes or borrows data to compensate for the fault. For example, if the output voltage of the air flow meter is extremely low, the ECU will substitute a pre-programmed value for the air flow meter signal and allows the vehicle to be driven as long as the engine speed is kept below 2000 rpm. Or, if the cylinder head temperature sensor circuit is open, the control unit clamps the warm-up enrichment at a certain amount. This amount is almost the same as that when the cylinder head temperature is between 68–176°F (20–80°C).

If the fuel pump circuit malfunctions, the fuel pump relay comes on until the engine stops. This allows the fuel pump to receive power from the relay.

The electronic control unit controls the following functions:
- Injector pulse width
- Ignition timing
- Intake valve timing control (045)
- Air regulator control (G20)
- Exhaust gas recirculation (EGR) solenoid valve operation
- Exhaust gas sensor heater operation
- Idle speed
- FICD solenoid valve operation (G20 and M30)
- Fuel pump relay operation
- Fuel pump voltage (M30 and 045)
- Fuel pressure regulator control (M30)
- AIV control (G20)
- Carbon canister control solenoid valve operation
- Air conditioner relay operation (During early wide-open throttle)
- Radiator fan operation (G20)
- Traction control system (TCS) operation (045, if equipped)
- Self-diagnosis
- Fail-safe mode operation

Service Precautions

- Do not disconnect the injector harness connectors with the engine running.
- Do not apply battery power directly to the injectors.
- Do not disconnect the ECU harness connectors before the battery ground cable has been disconnected.
- Make sure all ECU connectors are fastened securely. A poor connection can cause an extremely high surge voltage in the coil and condenser and result in damage to integrated circuits.
- When testing the ECU with a DVOM make sure that the probes of the tester never touch each other as this will result in damage to a transistor in the ECU.
- Keep the ECCS harness at least 4 in. away from adjacent harnesses to prevent an ECCS system malfunction due to external electronic noise.
- Keep all parts and harnesses dry during service.
- Before attempting to remove any parts, turn OFF the ignition switch and disconnect the battery ground cable.
- Always use a 12 volt battery as a power source.
- Do not attempt to disconnect the battery cables with the engine running or the ignition key ON.
- Do not clean the air flow meter with any type of detergent.
- Do not attempt to disassemble the ECCS control unit under any circumstances.
- Avoid static electricity build-up by properly grounding yourself prior to handling any ECU or related parts.

Reading Codes

➡**Diagnostic codes may be retrieved by observing code flashes through the LED lights located on the Electronic Control Module (ECM). A special Nissan Consult monitor tool can be used, but is not required- When using special diagnostic equipment, always observe the tool manufacturer's instructions.**

1990–95 VEHICLES—2-MODE DIAGNOSTIC SYSTEM

Infiniti vehicles use a 2-mode diagnostic system incorporated in the ECU which uses inputs from various sensors to determine the correct air/fuel ratio. If any of the sensors malfunction the ECU will store the code in memory.

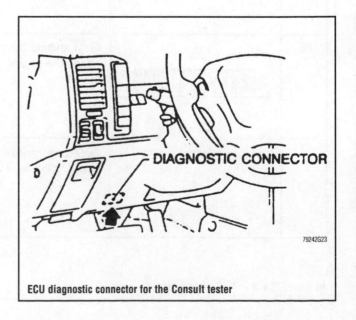

79242G23

ECU diagnostic connector for the Consult tester

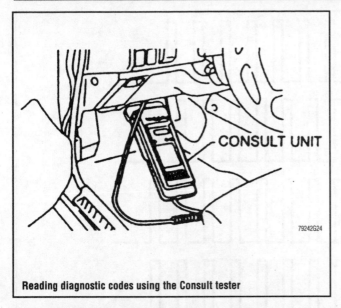

Reading diagnostic codes using the Consult tester

An Infiniti/Nissan Consult monitor may be used to retrieve these codes by simply connecting the monitor to the diagnosis connector located on the driver's side near the hood release.

Turn the ignition switch **ON** and press START, ENGINE and then SELF-DIAG RESULTS, the results will then be output to the monitor.

The conventional CHECK ENGINE or red LED ECU light may be used for self-diagnostics. The conventional 2-mode diagnostic system is broken into 2 separate modes each capable of 2 tests, an ignition switch **ON** or engine running test as outlined below:

1990–95 VEHICLES—MODE I—BULB CHECK

In this mode the RED indicator light on the ECU and the CHECK ENGINE light should be ON. To enter this mode simply turn the ignition switch **ON** and observe the light.

1990–95 VEHICLES—MODE 1—MALFUNCTION WARNING

In this mode the ECU is acknowledging if there is a malfunction by illuminating the RED indicator light on the ECU and the CHECK ENGINE light. If the light turns OFF, the system is normal. To enter this mode, simply start the engine and observe the light.

1990–95 VEHICLES—MODE 2- SELF-DIAGNOSTIC CODES

In this mode the ECU will output all malfunctions via the CHECK ENGINE light or the red LED on the ECU. The code may be retrieved by counting the number of flashes. The longer flashes indicate the first digit and the shorter flashes indicate the second digit. To enter this mode proceed as follows:

1. Turn the ignition switch **ON**, but do not start the vehicle.
2. Turn the ECU diagnostic mode selector fully clockwise for 2 seconds, then turn it back fully counterclockwise.
3. Observe the red LED on the ECU or CHECK ENGINE light for stored codes.

1990–95 VEHICLES—MODE 2- EXHAUST GAS SENSOR MONITOR

In this mode the red LED on the ECU or CHECK ENGINE light will display the condition of the fuel mixture and whether the system is in closed loop or open loop. When the light flashes ON, the exhaust gas sensor is indicating a lean mixture. When the light stays OFF, the sensor is indicating a rich mixture. If the light remains ON or OFF, it is indicating an open loop system. If the sys-

tem is equipped with 2 exhaust gas sensors, the left side will operate first. If already in Mode 2, proceed to Step 3 for exhaust gas sensor monitor.

1. Turn the ignition switch **ON**.
2. Turn the diagnostic switch ON, by turning the switch fully clockwise for 2 seconds and then fully counterclockwise.
3. Start the engine and run until thoroughly warm. Raise the idle to 2,000 rpm and hold for approximately 2 minutes. Ensure the red LED or CHECK ENGINE light flash ON and OFF more than 5 times every 10 seconds with the engine speed at 2,000 rpm.

➡**If equipped with 2 exhaust gas sensors, switch to the right sensor by turning the ECU mode selector fully clockwise for 2 seconds and then fully counterclockwise with the engine running.**

Clearing Codes

1990–95 VEHICLES

All control unit diagnostic codes may be cleared by disconnecting the negative battery for a period of 15 seconds. The codes will be cleared when mode 1 is re-entered from mode 2. The Nissan Consult Monitor or equivalent can also be used to clear codes.

Diagnostic Trouble Codes

1990–95 VEHICLES

Code 16 TCS Signal
Code 21 Ignition signal missing in primary coil
Code 31 ECM (engine ECCS control unit)
Code 32 EGR circuit
Code 33 Heated oxygen sensor circuit
Code 34 Knock Sensor (KS) circuit
Code 35 EGR temperature sensor circuit
Code 42 Fuel temperature sensor circuit
Code 43 Throttle sensor circuit
Code 45 Injector leak
Code 46 Secondary throttle sensor circuit
Code 51 Injector circuit
Code 53 Heated oxygen sensor circuit (right bank)
Code 54 NT controller circuit
Code 55 No malfunctioning in the above circuit
Code 11 Crankshaft position sensor
Code 12 Mass Air flow sensor
Code 13 Engine coolant temperature sensor circuit
Code 14 Vehicle speed sensor

TRANSMISSION CODES

Reading and Retrieving Codes

Infiniti vehicles provide diagnostic trouble codes through one of four different lights, depending on the specific model at hand: Power Shift lamp, O/D Off lamp, A/T Check lamp (J30 models), digital readout in the diagnostic information display (Q45 models). The vehicle will flash the applicable lamp 16 times whenever the ignition key is turned **ON** when the system has stored DTC's.

M30 Infiniti models display DTC's using an 11-flash sequence. The light flashes 11 times in a row. The sequence always starts with a long start flash (approximately two seconds in duration), and is followed by 10 shorter flashes (approximately ¼ second in duration). If there are no malfunctions detected by the system, all ten flashes will be short, but if the computer recognizes a fault in the

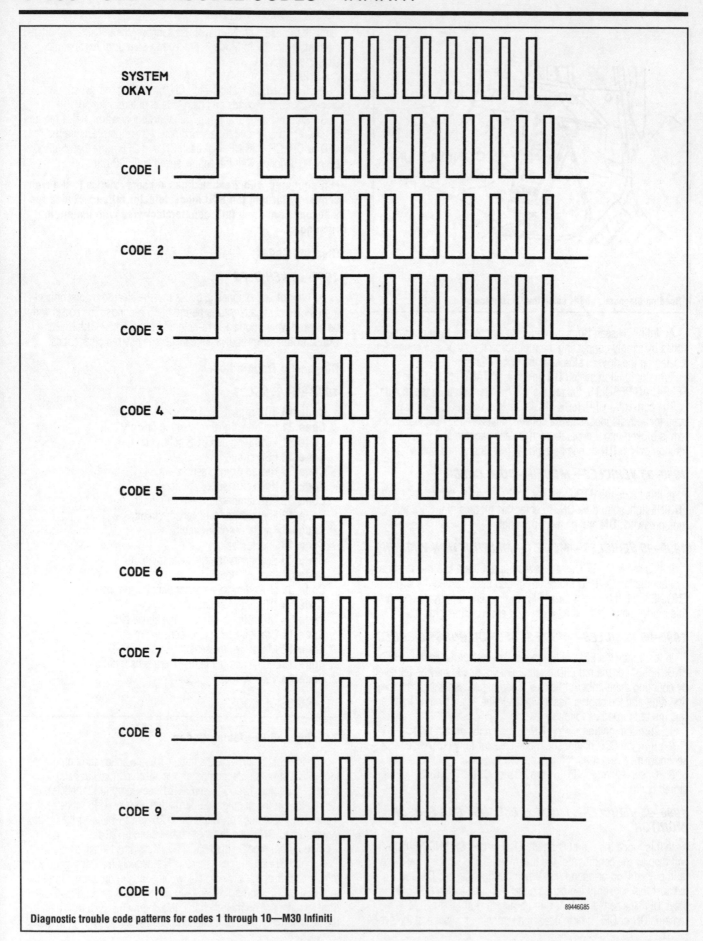

Diagnostic trouble code patterns for codes 1 through 10—M30 Infiniti

89446G85

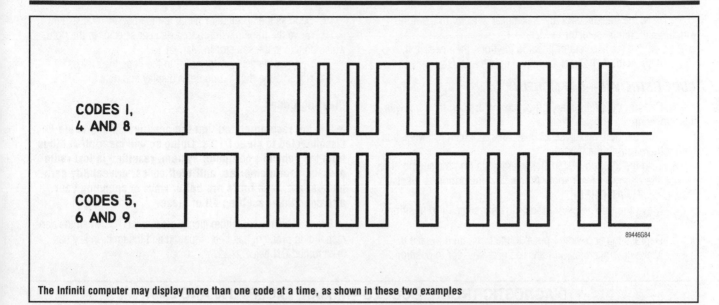

CODES 1, 4 AND 8

CODES 5, 6 AND 9

89446G84

The Infiniti computer may display more than one code at a time, as shown in these two examples

system, one of the 10 flashes will be longer (approximately one second in duration)—the long flash identifies the code stored in memory. For example, if the third flash after the two second start flash is the long one, you're looking at a Code 3.

If the computer has detected more than one code, it displays all of the codes in the same pass. For an example, please refer to the accompanying illustration. After the computer displays the code(s), the light will remain off for approximately 2½ seconds, then repeats the code(s). If the light flashes on and off, in regular, one-second intervals, it indicates the battery is low or was disconnected long enough to interfere with computer memory.

J30 models display diagnostic trouble codes in the same manner as with the other models, the only difference is that they use a 13-flash sequence.

➡**If the light remains on or off, try performing the sequence again. You may have missed one of the steps in the preliminary or retrieval procedure.**

Q45 models display DTC's in a hexadecimal format through a digital display, which also serves as the odometer display. Therefore, it is possible to read codes 1 through 9 and A through D.

➡**For Q45 models, if there are no codes in memory, the odometer will display OK.**

PRELIMINARY PROCEDURE

To prepare the vehicle for code retrieval, perform the following:

➡**It is important to perform all of the following steps; some of the steps are a preliminary inspection, which prepares the system for fault code retrieval.**

1. Start the engine and allow it to reach normal operating temperature.
2. Turn the ignition key **OFF** and apply the parking brake.
3. Without starting the engine, turn the ignition key to the **ON** position.

➡**The next 7 steps do not apply for Q45 models.**

4. For vehicles equipped with an O/D Off button, activate the switch so that the O/D Off light illuminates. Then, deactivate the switch so that the light turns off.

5. For vehicles equipped with a Mode switch, activate the switch so that the Power Shift lamp illuminates. Then, deactivate the switch so that the light turns off.
6. Turn the ignition key **OFF**, and wait a few seconds.
7. Once again turn the ignition switch **ON**, without starting the engine.

The indicator light on the instrument panel should illuminate for a few seconds (lamp bulb check), then extinguish. This is to check the light circuit, to make sure it's working satisfactorily for code flashing. If the lamp does not illuminate, inspect the bulb and circuit for a malfunction.

8. Turn the ignition switch **OFF**.
9. Move the transmission/transaxle gear selector to the D position.
10. If equipped, turn the O/D Off switch off.

CODE RETRIEVAL—EXCEPT J30 AND Q45 MODELS

1. Perform the preliminary procedure before commencing with this procedure.
2. Turn the ignition key **ON**, without starting the engine, and wait for a few seconds.
3. Move the transmission/transaxle gear selector to position 2.
4. Turn the O/D switch on (lamp off).
5. Move the transmission/transaxle gear selector to position 1.
6. Turn the O/D switch off (lamp on).
7. Depress the accelerator pedal to the floor and release it.
8. On models which use the O/D Off light for code display, turn the O/D switch on.
9. Observe the appropriate light in the instrument cluster and note the DTC's.

CODE RETREIVAL—J30 MODELS

1. Perform the preliminary procedure before commencing with this procedure.
2. Turn the ignition key **ON**, without starting the engine.
3. Wait a few seconds, then move the transmission/transaxle gear selector to position 3.
4. Depress the accelerator pedal to the floor, then release it.
5. Move the transmission/transaxle gear selector to position 2.
6. Depress the accelerator pedal to the floor, then release it.

7. Move the transmission/transaxle gear selector to the right, which positions the selector in Manual 1.

8. Depress the accelerator pedal to the floor, then release it.

9. Observe the A/T Check lamp and note the DTC's.

CODE RETREIVAL—Q45 MODELS

1. Perform the preliminary procedure before commencing with this procedure.

2. Turn the odometer reset counter knob counterclockwise, and hold it there during the next step.

3. Turn the ignition switch **ON**, without starting the engine, then release the odometer reset knob. At this time the odometer display should display **AT CHECK**.

4. Move the transmission/transaxle gear selector to position 3.

5. Depress the accelerator pedal to the floor, then release it.

6. Move the transmission/transaxle gear selector to position 2.

7. Depress the accelerator pedal to the floor, then release it.

8. Move the transmission/transaxle gear selector to the right, which positions the selector in Manual 1.

9. Depress the accelerator pedal to the floor, then release it.

10. Observe the digital odometer display and note the DTC's.

Clearing Codes

➡**It is not recommended that the negative battery cable be disconnected to clear DTC's. Doing so will clear all settings from the vehicle's computer system, resulting in lost radio presets, seat memories, anti-theft codes, driveability parameters, etc., and there are better ways of spending your afternoon than resetting all of these.**

Infiniti vehicles will automatically clear any DTC's when the corresponding problem has been repaired and the ignition key has been turned **ON** twice.

DIAGNOSTIC TROUBLE CODES—NISSAN, EXCEPT J30 AND Q45

Code	Component/Circuit Fault
1	Vehicle Speed Sensor (VSS) or circuit
2	Speedometer VSS or circuit
3	Throttle Position (TP) sensor or circuit
4	Shift solenoid A or circuit
5	Shift solenoid B or circuit
6	Timing solenoid or circuit, or over-run clutch solenoid or circuit
7	Lock-up solenoid or circuit
8	Transaxle fluid temperature sensor or circuit, or computer power insufficient
9	Engine speed signal circuit
10	Line pressure solenoid or circuit
Regular Flashing	Battery voltage low, or power was disconnected from the computer

89446G12

Infiniti Transmission related Diagnostic Trouble Codes—except J30 and Q45 models

DIAGNOSTIC TROUBLE CODES—NISSAN J30 AND Q45

Codes J30	Q45	Component/Circuit Fault
1	1	Vehicle Speed Sensor (VSS) or circuit
2	2	Speedometer VSS or circuit
3	3	Throttle Position (TP) sensor or circuit
4	4	Shift solenoid A or circuit
5	5	Shift solenoid B or circuit
6	6	Timing solenoid or circuit, or over-run clutch solenoid or circuit
7	7	Lock-up solenoid or circuit
8	8	Transaxle fluid temperature sensor or circuit, or computer power insufficient
9	9	Engine speed signal circuit
10	A	Turbine shaft speed sensor or circuit
11	B	Line pressure solenoid or circuit
12	C	Engine/Transmission communication circuit
Regular Flashing	D	Battery voltage low, or power was disconnected from the computer

89446G13

Infiniti Transmission related Diagnostic Trouble Codes—J30 and Q45 models

Isuzu

ENGINE CODES

General Information

Isuzu vehicles may be fitted with either a Feedback Carburetor (FBC), a Throttle Body Fuel Injection (TBI) system or a Multi-port Fuel Injection System (MFI).

The Feedback Carburetor System (FBC) is primarily used on 1985 and earlier normally aspirated engines, although some engines may use this system up to 1993. The system Electronic Control Module (ECM) constantly monitors and controls engine operation by reading data from various sensors and outputting signals to the carburetor. This helps lower emissions while maintaining the fuel economy, driveability and performance of the vehicle.

The Throttle Body (TBI) fuel injection system was put into production in 1989. It is used on 2.8L and 3.1L engines. The system functions much the same as a multi-port fuel injection system but with one exception—fuel is injected into the intake manifold rather than into each individual cylinder. This system may also control the ignition system.

The 1-TEC Multi-port Fuel Injection (MFI) system was first used in 1985 and continues to be used today. The system constantly monitors and controls engine operation through the use of data sensors, an Electronic Control Module (ECM) and other components. Individual fuel injectors are mounted at each cylinder and provide a metered amount of fuel as required by current operating conditions. This system may also control the ignition system and, as equipped, the turbocharger system.

All vehicles covered in this section have self-diagnostic capabilities. The ECM diagnostics are in the form of trouble codes stored in the system's memory. When a trouble code is detected by the control module, it will turn the malfunction indicator lamp ON until the code is cleared. An intermittent problem will set a code. The lamp will turn OFF if the problem goes away, but the trouble code will stay in memory until ECM power is interrupted.

Service Precautions

• Keep all ECM parts and harnesses dry during service. Protect the ECM and all solid-state components from rough handling or extremes of temperature.

• Use extreme care when working around the ECM or other solid-state components. Do not allow any open circuit to short or ground in the ECM circuit. Voltage spikes may cause damage to solid-state components.

• Before attempting to remove any parts, turn the ignition switch OFF or disconnect the negative battery cable.

• Remove the ECM before any arc welding is performed to the vehicle.

• Electronic components are very susceptible to damage caused by electrostatic discharge (static electricity). To prevent electronic component damage, do not touch the control module connector pins or soldered components on the control module circuit board.

Reading Codes

➡Diagnostic codes may be retrieve through the use of the **CHECK ENGINE light or Malfunction Indicator Lamp (MIL). A** special Scan tool can be used, but is not required. When using special diagnostic equipment, always observe the tool manufacturer's instructions.

1982–86 VEHICLES

The trouble code system is actuated by connecting a diagnostic lead to ground. The location of the diagnostic lead differs from model-to-model and, in some cases from year-to-year within the same model.

Trooper II for 1985; connect the diagnostic lead terminals together (1 male and 1 female). The terminals are located under dash, on the passenger's side, behind the radio. The terminal leads for 1986 models are located near the ALDL connector, under dash, on the driver's side behind the cigarette lighter.

Pick-up truck for 1982; connect the diagnostic lead terminals together (1 male and 1 female). The terminals are branched from the harness near the ECM, under the dash on the driver's side behind the hood release.

The trouble code is determined by counting the flashes of the 'Check Engine lamp. Trouble Code 12 will flash first, indicating that the self-diagnostic system is working. Code 12 consists of 1 flash, short pause, then 2 flashes. There will be a longer pause and Code 12 will repeat 2 more times. Each code flashes 3 times. The cycle will then repeat itself until the engine is started or the ignition switch is turned OFF. In most cases, the codes will be checked with the engine running since no codes other than 12 or 51 will be present on initial key ON.

1987–94 VEHICLES—USING A SCAN TOOL

1. Turn the ignition key to the **OFF** position.
2. Connect the scan tool to the Assembly Line Diagnostic Link (ALDL).
3. Turn **ON** the ignition for scan tool to access engine computer.

1987–89 VEHICLES—USING THE CHECK ENGINE LAMP

1. With the ignition turned **ON**, and the engine stopped, the CHECK ENGINE lamp should be ON. This is a bulb check to indicate the light is working properly.
2. For the Trooper and Pickup; connect the trouble code TEST lead (white cable) to the ground lead (black cable). It is located 8 in. from the ECM connector next to the clutch pedal or center console.

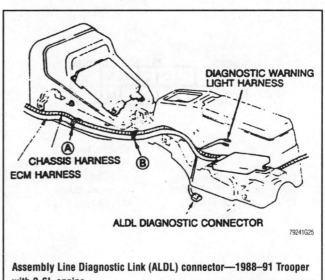

Assembly Line Diagnostic Link (ALDL) connector—1988–91 Trooper with 2.6L engine

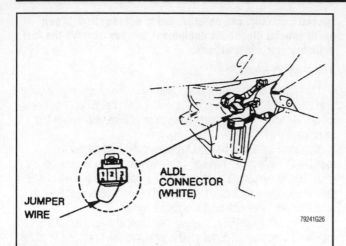

JUMPER WIRE

ALDL CONNECTOR (WHITE)

79241G26

Assembly Line Diagnostic Link (ALDL) connector—1992–94 Trooper with 2.6L engine

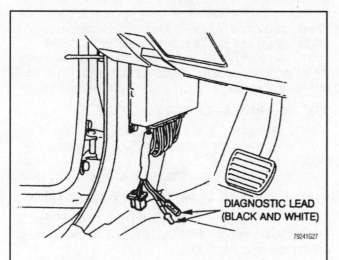

DIAGNOSTIC LEAD (BLACK AND WHITE)

79241G27

Assembly Line Diagnostic link (ALDL) connector—1990–94 Amigo, Pick-up and Rodeo (4 cylinder engines)

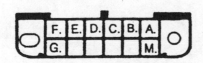

A. Ground
B. Diagnostic terminal
C. A.I.R system (if used)
D. Check engine light
E. Serial data
F. Torque Converter Clutch (TCC)
G. Fuel pump
M. Serial data

79241G28

Assembly Line Diagnostic link (ALDL) connector—1990–94 Amigo, Pick-up and Rodeo (6 cylinder engines)

3. The CHECK ENGINE light will begin to flash a trouble Code 12. Code 12 consists of 1 flash, short pause and then 2 more flashes. There will be a longer pause and then a Code 12 will repeat 2 more times. The check indicates that the self-diagnostic system is working. This cycle will repeat itself until the engine is started or the ignition switch is turned **OFF**. If more than a single fault code is stored in the memory, the lowest number code will flash 3 times followed by the next highest code number until all the codes have been flashed. The faults will then repeat in the same order. In most cases, codes will be checked with the engine running since no codes other than Codes 12 and 51 will be present on the initial key ON. Remove the jumper wire from the test terminal before starting the engine.

➡️**The fault indicated by trouble Code 15 takes 5 minutes of engine operation before it will display.**

1990–94 VEHICLES

1. With the ignition turned **ON** and the engine stopped, the CHECK ENGINE lamp should be ON. This is a bulb check to indicate the light is working properly.

2. Enter the diagnostic modes as follows:

   a. For the Trooper; jumper the 1 and 3 terminals (outer terminals) of the white Assembly Line Diagnostic Link (ALDL). The connector for Impulse and Stylus is located behind the kick panel on the passenger side of the vehicle. On Trooper, the ALDL connector is located behind the left side of the center console.

   b. For the Amigo, Pickup, and Rodeo with 4 cylinder engine; connect the trouble code TEST lead (white cable) and a ground lead (black cable) together. It is located 8 in. from the ECM connector (next to the clutch pedal or brake pedal).

   c. For the Amigo, Pickup, and Rodeo with 6 cylinder engine; jumper wire the A and B terminals together of the Assembly Line Diagnostic Link (ALDL). The ALDL is located in the center console and is sometimes covered by a plastic cover labeled DIAGNOSTIC CONNECTOR. Read the trouble codes with the ignition switch **ON** and the engine OFF.

3. The CHECK ENGINE light will begin to flash a trouble Code 12. Code 12 consists of 1 flash, a short pause and then 2 more flashes. There will be a longer pause and a Code 12 will repeat 2 more times. Code 12 indicates that the self-diagnostic system is working. If any other faults are present, the faults will be displayed 3 times each in the same fashion. Fault codes are flashed from lowest to highest after the Code 12. Remember to remove the jumper wire from the ALDL connector before starting the engine. After all codes have been displayed, the cycle will repeat itself until the engine is started or the ignition switch is turned OFF.

➡️**The fault indicated by trouble Code 15 takes 5 minutes of engine operation before it will display (4 cylinder engine only).**

Clearing Codes

1982–86 VEHICLES

The trouble code memory is fed a continuous 12 volts even with the ignition switch in the **OFF** position. After a fault has been corrected, it will be necessary to remove the voltage for 10 seconds to clear any stored codes. Voltage can be removed by disconnecting the 14 pin ECM connector or by removing the fuse marked 'ECM' or fuse No. 4 on some models. Since all memory will be lost when removing the fuse, it will be necessary to reset the clock and other electrical equipment.

1987–94 VEHICLES

The trouble code memory is fed a continuous 12 volts even with the ignition switch in the **OFF** position. After a fault has been corrected, it will be necessary to remove the voltage for 30 seconds to clear any stored codes. The quickest way to remove the voltage is to remove the ECM fuse from the fuse block or the MAIN 60A fuse for 10 seconds. The voltage can also be removed by disconnecting the negative battery cable. This will mean electronic instrumentation, such as a clock and radio, would have to be reset.

1987–89 Pickup and Trooper; turn the ignition switch OFF and disconnect the ECM 13-pin connector or remove the No. 4 fuse from the fuse block for 30 seconds.

The 60 amp slow blow fuse may be removed from the fuse block in the engine compartment. However, the electronic functions with memory have to be reset after removing the No. 4 fuse for 30 seconds.

1992–94 Amigo, Pickup, Rodeo and Trooper; To clear the trouble codes; turn the ignition switch OFF and remove the ECM fuse from the under-dash fuse block for 30 seconds. Removing the number 3 fuse from the under dash fuse panel will result in having to reset all the electronic functions with memory in the vehicle. This applies to trucks with 4-cylinder engines.

Removing the 60 amp slow blow fuse from the fuse block in the engine compartment will also erase codes.

Diagnostic Trouble Codes

1982 1.8L FBC TRUCK AND 1982–84 1.8L FBC CAR ENGINES

Code 12 Idle switch is not turned ON
Code 13 Idle switch is not turned OFF
Code 14 Wide Open Throttle (WOT) switch is not turned ON
Code 15 Wide Open Throttle (WOT) switch is not turned ON
Code 21 Output transistor is not turned ON
Code 22 Output transistor is not turned OFF
Code 23 Abnormal oxygen sensor
Code 24 Abnormal Water Temperature Sensor (WTS) switch
Code 25 Abnormal Random Access Memory (RAM)
Code 12, 13, 14 and 15 Check Engine lamp not ON
Code 21, 22, 23, 24 and 25 Check Engine lamp ON

1985–89 1.5L FBC (VIN 7), 1983 1.8L FBC TRUCK, 1983–86 2.0L FBC (VIN A), AND 1986–94 2.3L FBC (VIN L) ENGINES

Code 12 Normal
Code 13 Oxygen sensor circuit
Code 14 Coolant Temperature Sensor (CTS)—circuit shorted
Code 15 Coolant Temperature Sensor (CTS)—circuit open
Code 21 Idle switch—circuit open or Wide Open Throttle (WOT) switch—circuit shorted
Code 22 Fuel Cut Solenoid (FCS)—circuit open or grounded
Code 23 Mixture Control (M/C) solenoid—circuit open or grounded, or Vacuum Control Solenoid (VCS)—circuit open or grounded (1983 1.8L Truck, 1983–86 2.0L Truck, 1986–88 2.3L Truck)
Code 24 Vehicle Speed Sensor (VSS) circuit
Code 25 Air Switching Solenoid (ASS)—circuit open or grounded
Code 26 Vacuum Switching Valve (VSV) system for canister purge—circuit open or grounded
Code 27 Vacuum Switching Valve (VSV)-constant high voltage to ECM

Code 31 No ignition reference pulses to ECM
Code 32 EGR temperature sensor—system malfunction
Code 34 EGR temperature sensor—circuit failure electronic idle control
Code 42 Fuel Cut Relay and/or circuit shorted
Code 44 Oxygen Sensor circuit—lean indication
Code 45 Oxygen Sensor circuit—rich indication
Code 52 Faulty Electronic Control Module (ECM)—Random Access Memory (RAM) problem in ECM
Code 53 Shorted Air Switching Solenoid (ASS) or Air Injection System and/or faulty Electronic Control Module (ECM)
Code 54 Shorted Vacuum Control Solenoid (VCS) and/or faulty Electronic Control Module (ECM)
Code 55 Faulty Electronic Control Module (ECM)

1985–87 2.0L TURBO EFI (VIN F), 1983–89 2.0L EFI (VIN A), 1988–89 2.3L EFI (VIN L), AND 1988–94 2.6L EFI (VIN E) ENGINES

Code 12 Normal
Code 13 Oxygen sensor circuit
Code 14 Engine Coolant Temperature (ECT) sensor -grounded
Code 15 Engine Coolant Temperature (ECT) sensor—incorrect signal (open circuit on 1988-94 2.6L)
Code 16 Engine Coolant Temperature (ECT) sensor -open circuit
Code 21 Throttle Valve Switch (TVS) system—idle contact and full contact made simultaneously
Code 22 Starter—no signal input
Code 23 Ignition power transistor—output terminal grounded
Code 25 Vacuum Switching Valve (VSV)—output terminal grounded or open
Code 26 Canister purge Vacuum Switching Valve (VSV)—open or grounded
Code 27 Canister purge Vacuum Switching Valve (VSV)—faulty transistor or bad ground circuit
Code 32 EGR temperature sensor—faulty sensor or harness
Code 33 Fuel injector system—output terminal grounded or open
Code 34 EGR Vacuum switching valve—output terminal grounded or open
Code 35 Ignition power transistor—open circuit
Code 41 Crank Angle sensor (CAS)—no signal or faulty signal
Code 43 Throttle Valve Switch—idle contact closed continuously
Code 44 Fuel metering system—lean signal (Oxygen sensor-low voltage)
Code 45 Fuel metering system—rich signal (Oxygen sensor-high voltage)
Code 51 Faulty ECM
Code 52 Faulty ECM
Code 53 Vacuum Switching Valve (VSV)—grounded or faulty power transistor
Code 54 Ignition power transistor—grounded or faulty power transistor
Code 55 Faulty ECM
Code 61 Air Flow Sensor (AFS)—grounded, shorted, open or broken HOT wire
Code 62 Air Flow Sensor (AFS)—broken COLD wire
Code 63 Vehicle Speed Sensor (VSS)—no signal input
Code 64 Fuel injector system—grounded or faulty transistor
Code 65 Throttle Valve Switch (TVS)—full contact closed continuously

Code 66 Knock sensor—grounded or open circuit
Code 71 Throttle Position Sensor (TPS)—turbo control system—abnormal signal
Code 72 EGR Vacuum switching valve—output terminal grounded or open
Code 73 EGR Vacuum switching valve—faulty transistor or grounded system

1987–89 1.5L TURBO EFI (VIN 9), 1989 1.6L EFI (VIN 5), 1991–92 1.6L TURBO EFI (VIN 4), 1989–91 2.8L TBI (VIN R), AND 1991–94 3.1L TBI (VIN Z) ENGINES

Code 12 Normal
Code 13 Oxygen sensor circuit
Code 14 Engine Coolant Temperature (ECT) sensor—high temperature indicated
Code 15 Engine Coolant Temperature (ECT) sensor—low temperature indicated
Code 21 Throttle Position Sensor (TPS)—voltage high
Code 22 Throttle Position Sensor (TPS)—voltage low
Code 23 Intake Air Temperature (IAT)—low temperature indicated
Code 24 Vehicle Speed Sensor (VSS)—no input signal Code 25—Intake Air Temperature (IAT)—high temperature indicated
Code 31 Turbocharger wastegate control
Code 32 EGR system fault
Code 33 Manifold Absolute Pressure (MAP) sensor—voltage high
Code 34 Manifold Absolute Pressure (MAP) sensor—voltage low
Code 42 Electronic Spark Timing (EST) circuit fault
Code 43 Electronic Spark Control (ESC)—knock failure circuit
Code 44 Oxygen sensor circuit—lean exhaust
Code 45 Oxygen sensor circuit—rich exhaust
Code 51 PROM error—faulty or incorrect PROM
Code 52 CALPAK error—faulty or incorrect CALPAK
Code 54 Fuel Pump Circuit—low voltage
Code 55 ECM error

1990–91 1.6L EFI (VIN 5), 1992–94 1.8L EFI (VIN 8), 1991–94 2.3L EFI (VIN 5/6), AND 1992–94 3.2L EFI (VIN V/W) ENGINES

Code 13 Oxygen sensor circuit
Code 14 Engine Coolant Temperature (ECT) sensor -out of range
Code 21 Throttle Position Sensor (TPS)—out of range
Code 23 Intake Air Temperature (IAT)—out of range
Code 24 Vehicle Speed Sensor (VSS)—no input signal
Code 32 EGR system fault
Code 33 Manifold Absolute Pressure (MAP) sensor—out of range
Code 44 Oxygen sensor circuit—lean exhaust
Code 45 Oxygen sensor circuit—rich exhaust
Code 51 ECM failure

TRANSMISSION CODES

Reading and Retrieving Codes

IMPULSE MODELS

The Isuzu Impulse utilizes two-digit DTC's to inform the technician in which circuit and/or component a fault was detected. The two-digit codes are displayed by the ECONOMY light, located in the instrument cluster. The light flashes the digits out by first flashing the tens digit, then after a slight pause, the ones digit. Therefore, if the ECONOMY light flashes twice, pauses slightly, then flashes once, a Code 21 was displayed. If there is more than one DTC stored in the computer's memory, the light will display the lowest number first and move upward through all of the stored codes in numerical order. After the progression reaches the highest code number, the cycle will repeat itself until the ignition switch is turned **OFF**.

To determine if the vehicle's computer has stored any DTC's, switch the transmission/transaxle mode switch to the NORMAL position and turn the ignition switch **ON**. The ECONOMY light should illuminate for three seconds, then extinguish. If the ECONOMY light stays ON in the NORMAL mode, the computer has stored DTC's.

To retrieve the stored DTC's, perform the following:
1. Perform the preliminary inspection, located earlier in this section. This is very important, since a loose or disconnected wire, or corroded connector terminals can cause a whole slew of unrelated

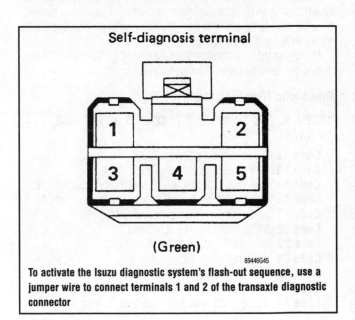

Self-diagnosis terminal

(Green)

89446G45

To activate the Isuzu diagnostic system's flash-out sequence, use a jumper wire to connect terminals 1 and 2 of the transaxle diagnostic connector

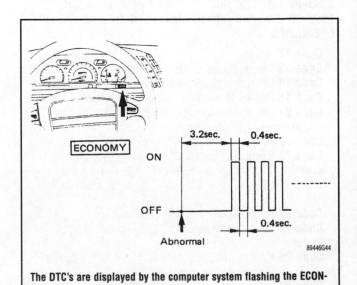

89446G44

The DTC's are displayed by the computer system flashing the ECONOMY light—Isuzu Impulse models

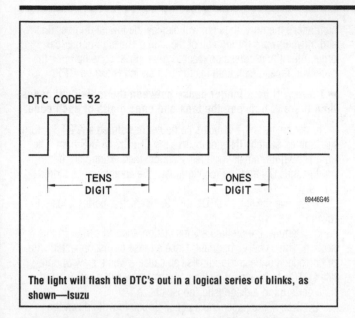

The light will flash the DTC's out in a logical series of blinks, as shown—Isuzu

DTC's to be stored by the computer; you will waste a lot of time performing a diagnostic "goose chase."

2. Grab some paper and a pencil or pen to write down the DTC's when they are flashed out.

3. Locate the five terminal, green transaxle diagnostic connector, which is usually located behind the passenger's kickpanel.

4. Turn the ignition switch **OFF**.

5. Using a jumper wire or a bent paper clip, connect terminals 1 and 2 of the transaxle diagnostic connector. A black wire with a white tracer stripe is usually attached to terminal 1, and a black wire with a blue tracer stripe is usually attached to terminal 2 of the transaxle diagnostic connector.

6. Set the transaxle mode selector in the NORMAL position, then turn the ignition switch to the **ON** position **without starting the engine**.

7. Observe the ECONOMY light and note any stored DTC's flashed out by it.

8. After all of the codes have been noted, turn the ignition switch **OFF**, and remove the jumper wire/paper clip from the transaxle diagnostic connector.

1988–90 TROOPER MODELS

The 1988–90 Isuzu Trooper models utilize two-digit DTC's to inform the technician in which circuit and/or component a fault was detected. The two-digit codes are displayed by the O/D OFF light, located in the center console shifter display. The light flashes the digits out by first flashing the tens digit, then after a slight pause, the ones digit. Therefore, if the O/D OFF light flashes twice, pauses slightly, then flashes once, a Code 21 was displayed. If there is more than one DTC stored in the computer's memory, the light will display the lowest number first and move upward through all of the stored codes in numerical order. After the progression reaches the highest code number, the cycle will repeat itself until the ignition switch is turned **OFF**.

➡**There will be a longer pause between the individual codes than there is between the tens and ones digits of each code.**

To determine if the vehicle's computer has stored any DTC's, switch the transmission/transaxle mode switch so that transmission/transaxle overdrive is enabled (the O/D OFF switch in the out position) and turn the ignition switch **ON**. The O/D OFF light should

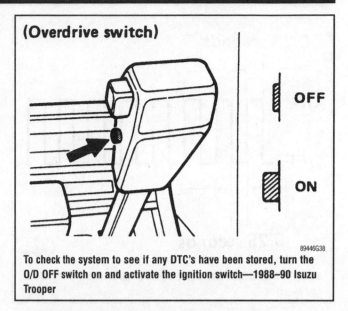

(Overdrive switch)

To check the system to see if any DTC's have been stored, turn the O/D OFF switch on and activate the ignition switch—1988–90 Isuzu Trooper

illuminate for three seconds, then extinguish. If the light flashes ON and OFF, the computer has stored DTC's.

➡**The O/D OFF light should stay illuminated when the overdrive is turned off.**

To retrieve the stored DTC's, perform the following:

1. Perform the preliminary inspection, located earlier in this section. This is very important, since a loose or disconnected wire, or corroded connector terminals can cause a whole slew of unrelated DTC's to be stored by the computer; you will waste a lot of time performing a diagnostic "goose chase."

2. Grab some paper and a pencil or pen to write down the DTC's when they are flashed out.

3. Locate the two terminal diagnostic request connector, which is usually located up and behind the left-hand side of the instrument panel (near the hood latch release handle).

4. Turn the ignition switch **OFF**.

5. Using a jumper wire or a bent paper clip, connect the two terminals of the diagnostic request connector.

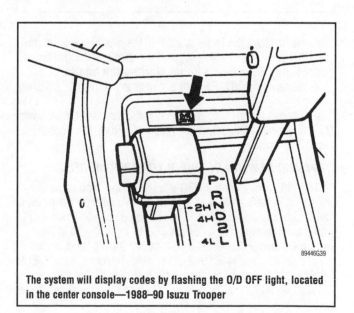

The system will display codes by flashing the O/D OFF light, located in the center console—1988–90 Isuzu Trooper

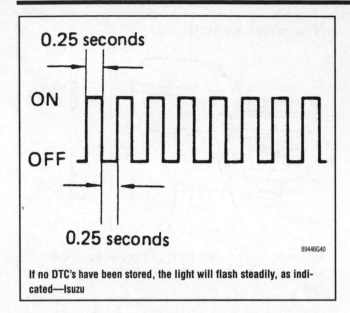

If no DTC's have been stored, the light will flash steadily, as indicated—Isuzu

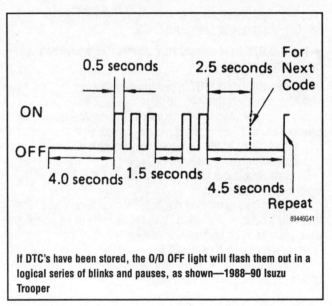

If DTC's have been stored, the O/D OFF light will flash them out in a logical series of blinks and pauses, as shown—1988–90 Isuzu Trooper

6. Set the transaxle mode selector to the overdrive enabled position (O/D OFF button in the out position), then turn the ignition switch to the **ON** position **without starting the engine**.

7. Observe the O/D OFF light and note any stored DTC's flashed out by it.

8. After all of the codes have been noted, turn the ignition switch **OFF**, and remove the jumper wire/paper clip from the diagnostic request connector.

1991 AND NEWER PRE-OBD II TROOPER MODELS

The 1991 and newer pre-OBD II Isuzu Trooper models, as with the earlier models, utilize two-digit DTC's to inform the technician in which circuit and/or component a fault was detected. The two-digit codes are displayed by the CHECK TRANS light, located in the instrument cluster display. The light flashes the digits out in the same manner as with the 1988–90 models. Therefore, if the CHECK TRANS light flashes twice, pauses slightly, then flashes once, a Code 21 was displayed. If there is more than one DTC stored in the

computer's memory, the light will display the lowest number first and move upward through all of the stored codes in numerical order. After the progression reaches the highest code number, the cycle will repeat itself until the ignition switch is turned **OFF**.

➡**There will be a longer pause between the individual codes than there is between the tens and ones digits of each code.**

To determine if the vehicle's computer has stored any DTC's, turn the ignition switch **ON** and observe the CHECK TRANS light. The light should illuminate for three seconds, then extinguish. If the light flashes ON and OFF continuously, the computer has stored DTC's.

To retrieve the stored DTC's for 1991 vehicles, perform the following:

1. Perform the preliminary inspection, located earlier in this section. This is very important, since a loose or disconnected wire, or corroded connector terminals can cause a whole slew of unrelated DTC's to be stored by the computer; you will waste a lot of time performing a diagnostic "goose chase."

2. Grab some paper and a pencil or pen to write down the DTC's when they are flashed out.

3. Locate the **two terminal** diagnostic request connector (not the three terminal connector) attached to the transmission/transaxle computer connector harness, which is usually located up and behind the left-hand side of the instrument panel (near the hood latch release handle). If the diagnostic request connector cannot be found, either call your local Isuzu dealer or refer to the applicable Chilton's Total Car Care manual for your vehicle.

4. Turn the ignition switch **OFF**.

5. Using a jumper wire or a bent paper clip, connect the two terminals of the diagnostic request connector.

6. Turn the ignition switch to the **ON** position **without starting the engine**.

7. Observe the CHECK TRANS light and note any stored DTC's flashed out by it.

8. After all of the codes have been noted, turn the ignition switch **OFF**, and remove the jumper wire/paper clip from the diagnostic request connector.

To retrieve the stored DTC's for 1992 and newer models, perform the following:

9. Perform the preliminary inspection, located earlier in this section. This is very important, since a loose or disconnected wire, or corroded connector terminals can cause a whole slew of unrelated DTC's to be stored by the computer; you will waste a lot of time performing a diagnostic "goose chase."

10. Grab some paper and a pencil or pen to write down the DTC's when they are flashed out.

11. Locate the **three terminal** diagnostic request connector, which may be located either in the same spot as the 1991 models (up and behind the left-hand side of the instrument panel, near the hood latch release handle), or behind the left-hand side of the center console. If the diagnostic request connector cannot be found, either call your local Isuzu dealer or refer to the applicable Chilton's Total Car Care manual for your vehicle.

12. Turn the ignition switch **OFF**.

13. Using a jumper wire or a bent paper clip, connect the two OUTER terminals of the diagnostic request connector.

14. Turn the ignition switch to the **ON** position **without starting the engine**.

15. Observe the CHECK TRANS light and note any stored DTC's flashed out by it.

16. After all of the codes have been noted, turn the ignition switch **OFF**, and remove the jumper wire/paper clip from the diagnostic request connector.

Clearing Codes

IMPULSE MODELS

➡It is not recommended that the negative battery cable be disconnected to clear DTC's. Doing so will clear all settings from the vehicle's computer system, resulting in lost radio presets, seat memories, anti-theft codes, driveability parameters, etc., and there are better ways of spending your afternoon than resetting all of these.

To clear the stored DTC's from the vehicle's computer memory, remove the 4AT fuse from the fuse block for a minimum of one minute. Reinstall the 4AT fuse and all of the stored DTC's should be deleted. If there are still codes present, either the codes were not properly cleared (Are the codes identical to those flashed out previously?) or the underlying problem is still there (Are only some of the codes the same as previously?)

TROOPER MODELS

➡It is not recommended that the negative battery cable be disconnected to clear DTC's. Doing so will clear all settings from the vehicle's computer system, resulting in lost radio presets, seat memories, anti-theft codes, driveability parameters, etc., and there are better ways of spending your afternoon than resetting all of these.

To clear the stored DTC's from the vehicle's computer memory, remove the CLOCK fuse (1988–90 Trooper II models), the STOP/AT CONT fuse (1988–90 Trooper models) or the ROOM LAMP fuse (1991 and newer pre-OBD II Trooper models) from the fuse block for a minimum of one minute. Reinstall the applicable fuse and all of the stored DTC's should be deleted. If there are still codes present, either the codes were not properly cleared (Are the codes identical to those flashed out previously?) or the underlying problem is still there (Are only some of the codes the same as previously?)

Code	Faulty parts	Pattern display
11	Vehicle speed sensor 1 (transmission mounting)	ON OFF
13	Engine speed sensor	
15	ATF temperature sensor ECU power voltage	
21	Engine throttle duty signal	
24	Vehicle speed sensor 2 (speedometer built-in)	
31	Shift solenoid A	
32	Shift solenoid B	
33	Overrun clutch solenoid	
34	Lock-up duty solenoid	
35	Line pressure duty solenoid	

89446G43

Isuzu transmission related Diagnostic Trouble Codes—Impulse

Code No.	Light Pattern	Diagnosis System
21	⎍⎍⎍	Defective No. 1 speed sensor (in combination meter) — severed wire harness or short circuit
22	⎍⎍⎍⎍	Defective No. 2 speed sensor (in Automatic transmission) — severed wire harness or short circuit
23	⎍⎍⎍⎍	Severed throttle sensor or short circuit — Severed wire harness or short circuit
31	⎍⎍⎍⎍	Severed No. 1 solenoid or short circuit — severed wire harness or short circuit
32	⎍⎍⎍⎍	Severed No. 2 solenoid or short circuit — severed wire harness or short circuit
33	⎍⎍⎍⎍⎍	Severed No. 3 solenoid or short circuit — severed wire harness or short circuit
34	⎍⎍⎍⎍⎍⎍	Severed No. 4 solenoid or short circuit — severed wire harness or short circuit

89446G42

Isuzu transmission related Diagnostic Trouble Codes—1988–90 Trooper

DIAGNOSTIC TROUBLE CODES—ISUZU 1991 AND NEWER TROOPER

Code	Component/Circuit Fault
17	1-2/3-4 shift solenoid or circuit
21	Throttle Position (TP) sensor or circuit
22	Throttle Position (TP) sensor or circuit
23	Engine Coolant Temperature (ECT) sensor or circuit
25	1-2/3-4 shift solenoid or circuit
26	2-3 shift solenoid or circuit
28	2-3 shift solenoid or circuit
29	TCC solenoid or circuit
31	No engine speed signal from engine speed sensor
32	Line pressure solenoid or circuit
33	Line pressure solenoid or circuit
34	Band apply solenoid or circuit
35	Band apply solenoid or circuit
36	TCC solenoid or circuit
39	Vehicle Speed Sensor (VSS) or circuit
41	Gear error
43	Computer ground control circuit turned off, because of fault in a solenoid circuit
46	Downshift error
48	Low voltage or bad computer ground
55	Computer failure
56	Shift lever position switch or TP sensor circuit
65	ATF temperature sensor or circuit
66	ATF temperature sensor or circuit
77	Kickdown switch or circuit, or TP sensor or circuit
82	Shift lever position switch

89446G47

Isuzu transmission related Diagnostic Trouble Codes—1991 and newer pre-OBD II Trooper models

ENGINE CODES

General Information

Fuel metering is controlled by regulating the time that the injectors are open during the engine operating cycle. Constant fuel pressure is maintained within the fuel rail; injector duration or operating time controls the volume of fuel admitted to the cylinders.

The injection system is managed by a digital Engine Control Unit (ECU). This micro-processor based unit controls the electrical signals to the injectors, triggering them for the correct time period. The ECU relies primarily on the manifold pressure and rpm signals from the engine. Once these signals, indicating engine speed and load, are received, the controller uses them to choose proper injector operating periods. This basic pulse length will be slightly modified by the signals from other engine sensors. These secondary control factors include engine coolant temperature, inlet air temperature, throttle position and battery condition.

The injectors are triggered 6 times per engine cycle; on V-12 engines, the injectors of both banks are operated 6 times per cycle.

Service Precautions

• Keep all PCME parts and harnesses dry during service. Protect the PCME and all solid-state components from rough handling or extremes of temperature.

• Before attempting to remove any parts, turn the ignition switch **OFF** and disconnect the battery ground cable.

• Make sure all harness connectors are fastened securely. A poor connection can cause an extremely high voltage surge, resulting in damage to integrated circuits.

• Always use a 12 volt battery as a power source.

• Do not disconnect the battery cables with the engine running; never run the engine with battery cables loose or disconnected.

• Do not attempt to disassemble the PCME unit under any circumstances.

• When performing PCME input/output signal diagnosis, remove the water-proofing rubber plug, if equipped, from the connectors to make it easier to insert tester probes into the connector. Always reinstall it after testing.

• When connecting or disconnecting pin connectors from the PCME, take care not to bend or break any pin terminals. Check that there are no bends or breaks on PCME pin terminals before attempting any connections.

• When measuring supply voltage of PCME-controlled components, keep the tester probes separated from each other and from accidental grounding. If the tester probes accidentally make contact with each other during measurement, a short circuit will damage the PCME.

• Use great care when working on or around air bag systems. Due to back-up circuitry, the system may stay armed for a period of time without battery power.

• When working on or around air bag wiring or components, always disarm the system using the correct procedures. Once disarmed, attach a flag or note to the steering wheel. Re-arm the system when repairs are completed and remove the wheel marker.

• Never attempt to measure the resistance of the air bag squib; detonation may occur.

Reading Codes

1987–94 VEHICLES

For 1987–91 vehicles, only the 6 cylinder engines provide fault codes. For 1992–94 vehicles, all engines will provide fault codes. The codes may be read on the trip computer display on the dashboard. Generally, codes will be held in memory and the dash warning lamp will be lit when a fault is sensed by the PCME.

➡**Not every stored code will cause the warning lamp to light after the first occurrence. Always check for codes during diagnosis, including cases where the dash warning lamp was not reported lit.**

To read the stored codes on the XJ6 models, bring the vehicle to a complete stop. Switch the ignition **OFF** and wait at least 10 seconds. Turn the ignition to the II or **ON** position but do not start or crank the engine. Press the trip computer button (VCM). After a short period of time, the stored codes will be displayed graphically on the trip computer panel. The display may include the designation FF, an abbreviation for FUEL system FAILURE. The codes on XJS models will be displayed 5 seconds after the ignition key is turned to the II or **ON** position but do not start or crank the engine.

On both XJ6 and XJS models, the fault codes are displayed in order of priority, one at a time. The next is only displayed when the preceding one is cleared. To retrieve next code, turn the ignition key to the **OFF** position; now repeat the read and clear procedure again for as many times as necessary.

Clearing Codes

1987–91 VEHICLES

Codes will not be cleared from memory until all stored codes have been displayed. Interrupting the display cycle will allow some codes to be retained and the memory will not clear.

To clear stored codes, drive the vehicle in excess of 19 mph (20 km/h); the memories will be cleared electronically. Each system will perform its self-check function and any new codes will be set. If no faults are present in a particular system, the failure memory will remain cleared.

1992–94 VEHICLES

To clear engine codes on XJ6 & XJS models find the Diagnostic Trouble Code (DTC) reset connector. On XJ6 models the DTC connector is a red round econo seal connector with a pink/red wire. The connector is located behind the passenger side under dash panel next to the PCME. On XJS models the DTC connector is a purple male PM5 connector with a yellow/green wire. The connector is located behind the passenger side center console footwell panel. To clear the codes use a jumper wire to connect the DTC connector to ground for 3 seconds. This will clear one code at a time so you can proceed to the next trouble code. Repeat this procedure until no more trouble codes are present.

Diagnostic Trouble Codes

1987–89 3.6L ENGINE

Code 1 Cranking signal failure—crankshaft signal missing after cranking for 6 seconds or cranking signal line from Li 2-8 is active above 2000 rpm

Code 2 Airflow meter circuit—open or shorted to ground

Code 3 Coolant temperature sensor failure

Code 4 Feedback failure (where applicable)

Code 5 Airflow meter/throttle potentiometer failure—low throttle potentiometer voltage with high airflow meter voltage

Code 6 Airflow meter/throttle potentiometer failure—high throttle potentiometer voltage with low airflow meter voltage

Code 7 Idle fuel adjustment potentiometer failure

Code 8 Not allocated—If this code appears, a 6.8 kilo-ohm resistor installed in place of a hot start sensor is faulty

1989–94 4.0L, 1992–94 5.3L ENGINE

Code 11 Idle potentiometer TPS out of range

Code 12 Airflow meter circuit signal out range

Code 13 PCME pressure sensor loss of vacuum signal, incorrect fuel pressure or faulty PCME

Code 14 Engine Coolant Temperature (ECT)—sensor signal out of range or static during engine warm-up

Code 16 Intake Air Temperature (IAT)—sensor resistance out of range or faulty thermistor

Code 17 Throttle Position Sensor (TPS), open or short

Code 18 Throttle Position Sensor (TPS)/Mass Air Flow Meter (MAFS) calibration TPS voltage signal is low with high air flow

Code 19 Throttle Position Sensor (TPS)/Mass Air Flow Meter (MAFS) calibration TPS voltage signal is high with low air flow

Code 22 PCME output to fuel pump relay

Code 23 Poor feedback control in rich direction

Code 23 Fuel supply (rich or lean)—open or short in fuel supply circuit, faulty or restricted fuel line or injectors (5.3L)

Code 24 Ignition drive—PCME output to ignition amplifier module

Code 26 Air leak—poor feedback control in lean direction

Code 29 PCME self check problem

Code 33 Injector drive fault—PCME output to injectors

Code 34 Injector drive circuit—injector leakage

Code 34 Bank A (Right) injectors—open or short circuit / faulty or restricted fuel injectors (5.3L)

Code 36 Bank B (Left) injectors—open or short circuit / faulty or restricted fuel injectors (5.3L)

Code 37 EGR drive—PCME output to EGR switch valve

Code 39 EGR check sensor—checks EGR operation

Code 44 Oxygen (Lambda) sensor—feedback out of control, fuel mixture rich or lean

Code 44 Right Oxygen (Lambda) sensor—resistance out of range (5.3L)

Code 45 Left Oxygen (Lambda) sensor—resistance out of range (5.3L)

Code 46 Idle speed control coil 1 drive—PCME output to idle speed control stepper motor

Code 47 Idle speed control coil 2 drive—PCME output to idle speed control stepper motor

Code 48 Idle speed control motor/valve—stepper motor out of position, temperature less than 860F (300C)

Code 49 Fuel injection ballast resistor—open circuit or faulty resistor (5.3L)

Code 66 Secondary Air Injection Relay—voltage out of operating range

Code 68 Road speed sensor—PCME senses vehicle travel at greater than 5 km/h with high engine air flow

Code 69 Park/Neutral/Park Switch

Code 89 Purge valve drive

ENGINE CODES

General Information

This system is broken down into 3 major systems: the Fuel System, Air Induction System and the Electronic Control System.

The air induction system provides sufficient clean air for the engine operation. This system includes the throttle body, air intake ducting and cleaner and idle control components. Lexus equips the E5300, 5C300 and 5C400 with a system of induction tuning that changes the induction path length. The intake Air Control Valve (IACV) changes the length of the induction path to broaden the power curve by matching the resonance characteristics of the intake charge with the engine speed. When the IACV opens, the effective length of the intake tract is shortened, boosting top end power. With the IACV closed, the intake tract is long, boosting the low end torque.

Lexus engines are equipped with a computer which centrally controls the electronic fuel injection, electronic spark advance and the exhaust gas recirculation valve. The systems can be diagnosed by means of an Electronic Control Unit (ECU).

The ECU receives signals from the various sensors indicating changing engine operations conditions such as:

- Intake air flow
- Intake air temperature
- Coolant temperature sensor
- Engine rpm
- Acceleration/deceleration
- Exhaust oxygen content

These signals are utilized by the ECU to determine the injection duration necessary for an optimum air/fuel ratio.

Service Precautions

• Keep all ECU parts and harnesses dry during service. Protect the ECU and all solid-state components from rough handling or extremes of temperature.

• Before attempting to remove any parts, turn the ignition switch **OFF** and disconnect the battery ground cable.

• Make sure all harness connectors are fastened securely. A poor connection can cause an extremely high voltage surge, resulting in damage to integrated circuits.

• Always use a 12 volt battery as a power source.

• Do not attempt to disconnect the battery cables with the engine running.

• Do not attempt to disassemble the ECU unit under any circumstances.

• If installing a 2-way or CB radio, mobile phone or other radio equipment, keep the antenna as far as possible away from the electronic control unit. Keep the antenna feeder line at least 8 in. away from the EEI harness and do not run the lines parallel for a long distance. Be sure to ground the radio to the vehicle body.

• When performing ECU input/output signal diagnosis, remove the water-proofing rubber plug, if equipped, from the connectors to make it easier to insert tester probes into the connector. Always reinstall it after testing.

• Always insert test probes into a connector from the wiring side when checking continuity, amperage or voltage.

• When connecting or disconnecting pin connectors from the ECU, take care not to bend or break any pin terminals. Check that there are no bends or breaks on ECU pin terminals before attempting any connections.

• When measuring supply voltage of ECU-controlled components, keep the tester probes separated from each other and from accidental grounding. If the tester probes accidentally make contact with each other during measurement, a short circuit will damage the ECU.

• Use great care when working on or around air bag systems. Wait at least 20 seconds after turning the ignition switch to LOCK and disconnecting the negative battery cable before performing any other work. The air bag system is equipped with a back-up power system which will keep the system functional for 20 seconds without battery power.

• All air bag connectors are a standard yellow color. The related wiring is encased in standard yellow sheathing. Testing and diagnostic procedures must be followed exactly when performing diagnosis on this system. Improper procedures may cause accidental deployment or disable the system when needed.

• Never attempt to measure the resistance of the air bag squib. Detonation may occur.

Reading Codes

1990–95 MODELS

All models contain a self-diagnostic system. Stored fault codes are transmitted through the blinking of the CHECK ENGINE warning lamp. This occurs when the system is placed in normal diagnostic mode or in test mode. Normal diagnostic mode is used to read stored codes while the vehicle is stopped. The test mode is used after the vehicle is driven under certain conditions. In test mode, while the ECU monitors, the technician will simulate conditions of the suspected fault in an attempt to cause the malfunction. When a malfunction is found, the CHECK ENGINE lamp will illuminate to alert the technician that the fault is presently occurring.

When troubleshooting 1995 E5300 and L5400 models, an OBD II scan tool or LEXUS hand held tester is necessary to access codes and read data output from the ECM. This is special tool, some aftermarket tools may be available from your auto parts store for purchase or rent.

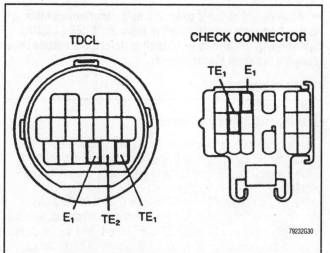

TDCL   **CHECK CONNECTOR**

TE₁ E₁

E₁ TE₂ TE₁

79232G30

Jump terminals TE₁ and E₁ to enter normal diagnostic mode, jump terminals TE₂ and E₁ to enter test mode

Without the OBD II System:

To read the fault codes, the following initial conditions must be met:

1. Battery voltage at or above 11 volts.
2. Throttle fully closed.
3. Transmission in N.
4. All electrical systems and accessories OFF.

Normal Diagnostic Mode:

1. Turn the ignition **ON** but do not start the engine.
2. Use a jumper wire to connect terminals TE1 and E1 of the check connector in the engine compartment or of the TDCL connector below the left side of the dash, if so equipped.
3. Fault codes will be transmitted through the controlled flashing of the CHECK ENGINE warning lamp.
4. If no malfunction was found or no code was stored, the lamp will flash 2 times per second with no other pauses or patterns. This confirms that the diagnostic system is working but has nothing to report. This light pattern may be referred to as the system normal signal. It should be present when no other codes are stored.
5. If faults are present, the CHECK lamp will blink the number of the code(s). All codes are 2 digits; the pulsing of the light represents the digits, not the count. For example, Code 25 is displayed as 2 flashes, a pause and 5 flashes.
6. If more than 1 code is stored, the next will be transmitted after a 2½ second pause.

➡ **If multiple codes are stored, they will be transmitted in numerical order from lowest to highest. This does not indicate the order of fault occurrence.**

7. When all codes have been transmitted, the entire pattern will repeat after a 4½ second pause. The cycle will continue as long as the diagnostic terminals are connected.
8. After recording the codes, disconnect the jumper at the diagnostic connector and turn the ignition **OFF**.

Test Mode:

1. Turn the ignition switch **OFF**.
2. Use a jumper wire to connect the TE2 and E1 terminals of the check connector or TDCL. The test mode cannot be initiated if the connection between TE2 and E1 is made with the key in the **ON** position.
3. Turn the ignition switch **ON**, but do not start the engine. The CHECK ENGINE light should flash. If the light does not flash check the TE1 terminal circuit.
4. Start the engine and simulate the conditions of the problem or malfunction.
5. When the road test is complete, connect the TE1 and E1 terminals of the TDCL or check connector and read trouble codes.
6. After the codes are read and noted, disconnect the jumpers from the connector. When the engine is not cranked, a code for starter signal and cam position sensors will be set, but this is not abnormal. If any of the sensed switches are used, the transmission shift lever, throttle or air conditioner, the switch condition code will be set, but this is not abnormal either.

Clearing Codes

1990–95 VEHICLES

Although the CHECK ENGINE lamp will reset itself after a repair is made, the original fault code will still be stored in memory. It is therefore necessary to clear the code after repairs are completed.

1. Turn the ignition **OFF**.
2. Remove the 20 amp EFI fuse from junction fuse box No. 2

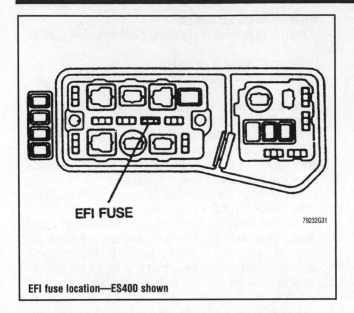

EFI FUSE

79232G31

EFI fuse location—ES400 shown

3. Wait at least 10 seconds before reinstalling the EFI fuse.

4. Road test vehicle and check to see that no fault codes are present.

Diagnostic Trouble Codes

1990–95 VEHICLES WITHOUT THE OBD II SYSTEM

Constant blinking of CHECK ENGINE light: Normal system operation

Code 12 Rpm NE, G1 or G2 signal to ECU—missing for 2 seconds or more after STA turns ON

Code 13 Rpm NE signal to ECU—missing for 50 milliseconds or more with engine speed above 1000 rpm, between 2 pulses of the G signal, NE signal of other than 12 pulses to ECU, or deviance of G1, G2 and NE signal continues for 1 second with engine warm and idling.

Code 14 Igniter IGF1 signal to ECU—missing for 8 successive ignitions

Code 15 Igniter IGF2 signal to ECU—missing for 8 successive ignitions

Code 16 ECT control signal—normal signal missing from ECT CPU (1990–94)

Code 16 NT control system—normal signal missing from between the engine CPU and NT CPU in the ECM (1995)

Code 17 No. 1 cam position sensor—G1 signal to ECU missing

Code 18 No. 2 cam position sensor—G2 signal to ECU—missing

Code 21* Left bank main oxygen sensor signal—signal voltage is remains between 0.35–0.70 V for 60 seconds or more at driving speed between 40–50 mph, NC ON and ECT in 4th gear or open/short sensor heater circuit

Code 22 Engine Coolant Temperature sensor circuit -open/short for 0.5 seconds or more

Code 24 Intake air temperature sensor circuit -open/short for 0.5 seconds or more

Code 25*—Air/fuel ratio LEAN malfunction—voltage output from oxygen sensor is less than 0.45 V for 90 seconds with engine racing at 2000 rpm, feedback frequency 5 Hz or more with main oxygen sensor signal centered at 0.45 V and idle switch ON, or feedback value of right and left banks differs by more than a certain percentage.

Code 26*—Air/fuel ratio RICH malfunction—feedback frequency 5 Hz or more with main oxygen sensor signal centered at 0.45 V and idle switch ON, or feedback value of right and left banks differs by more than a certain percentage.

Code 27*—Left bank sub-oxygen sensor signal—output of main oxygen sensor is 0.45 V or more and output of sub-oxygen sensor is 0.45 V or less with engine at wide open throttle for 4 seconds or more and sensors warmed.

Code 28*—Right bank main oxygen sensor signal—signal voltage is remains between 0.35-0.70 V for 60 seconds or more at driving speed between 40–50 mph, NC ON and ECT in 4th gear or open/short sensor heater circuit

Code 29* Right bank sub-oxygen sensor signal—output of main oxygen sensor is 0.45 V or more and output of sub-oxygen sensor is 0.45 V or less with engine at wide open throttle for 4 seconds or more and sensors warmed.

Code 31 Air flow meter circuit signal to ECU—missing for 2 seconds when engine speed is above 300 rpm

Code 32 Air flow meter circuit—E2 circuit open or VC and VS shorted

Code 35 HAC sensor circuit—open/short for 0.5 seconds or more/Baro sensor

Code 41 Throttle Position Sensor signal—open or short in the throttle position sensor circuit.

Code 42 Vehicle speed sensor circuit—engine RPM over 2350, and VSS shows zero miles per hour.

Code 43 Starter signal to ECU—missing

Code 47 Sub-throttle position sensor signal (VTA2) -open/short for at least 0.5 seconds or signal outputs exceed 1.45 V with idle contacts ON.

Code 51 NC signal ON, IDL contacts off, or shift in R, D, 2 or 1 range—during check mode

Code 52 No. 1 knock sensor signal—missing from ECU for 3 revolutions when engine speed is between 1600-5200 rpm

Code 53 Knock control signal—ECU knock control malfunction detected with engine speed between 650–5200 rpm

Code 55 No. 2 knock sensor signal—missing from ECU for 3 revolutions when engine speed is between 1600-5200 rpm

Code 71* EGR gas temperature below 149°F (65°C) for 90 seconds or more during EGR control

Code 78* Fuel pump control signal—open or short in the fuel pump control circuit

➡***2 trip detection logic code: A single occurrence of this fault will be temporarily stored in memory. CHECK ENGINE light will NOT illuminate until fault is detected a second time (during a separate ignition cycle).**

TRANSMISSION CODES

Reading and Retrieving Codes

Like most other manufacturers, Toyota vehicles built since 1985 are capable of displaying DTC's. The Lexus systems display the DTC's via the O/D OFF lamp (as with many other import manufacturers).

The DTC's displayed by Toyota models consist of two digits. They are flashed out so that first the tens digit is displayed, then the ones digit. And, since all of the O/D OFF lamp blinks are the same length of duration, Code 38 would be displayed as three flashes, then a short pause followed by eight more flashes. After which, the O/D OFF light would remain off for a few seconds, then the code

Model	Year	Design	Location
Camry	1985-94	2	Left Front Shock Tower
	1992-94	3	Under the Left Side of the Instrument Panel
Celica	1986-94	2	Left Front Shock Tower
Corolla	1987	1	Right Front Shock Tower
	1988-94	2	Left Front Shock Tower
Cressida	1985-91	2	Left Front Shock Tower
	1989-91	3	Under Dash; Left of Steering
MR2	1985-89	2	Left Front of Engine Compartment
	1991	2	Left Rear of the Engine Compartment
	1992-94	2	Right Rear Shock Tower
Paseo	1992-94	2	Left Side of the Engine Compartment, Near the No. 2 Fuse Box
Pick-up and 4Runner	1985-89	1	Left Front Shock Tower
	1990-94	2	Left Side of the Engine Compartment, Near the No. 2 Fuse Box
Previa Van	1991-94	2	Under Front of the Driver's Side Seat
Supra	1985-86	1	Left Side of the Engine Compartment
	1986-92	2	Left Front Shock Tower
	1993-94	2	Right Side of the Firewall

89446G59

Before starting the Lexus diagnostic retrieval procedure, locate the connector

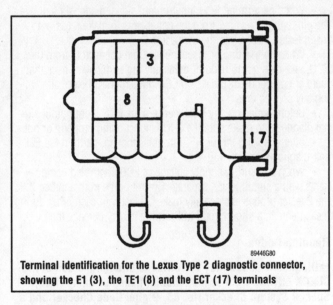

89446G80

Terminal identification for the Lexus Type 2 diagnostic connector, showing the E1 (3), the TE1 (8) and the ECT (17) terminals

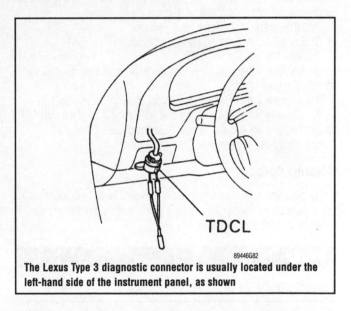

89446G82

The Lexus Type 3 diagnostic connector is usually located under the left-hand side of the instrument panel, as shown

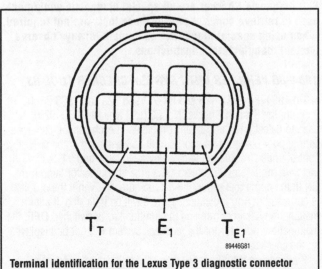

89446G81

Terminal identification for the Lexus Type 3 diagnostic connector

would be repeated. The system repeats each DTC three times, then goes on to the next code stored in memory. After the system finishes displaying all of the codes stored in memory, it returns to the first DTC and starts all over. The system will continue to display the DTC's until the ignition key is turned **OFF**.

Before the codes can be retrieved, the diagnostic connector link must be located. Over the years, however, there have been three different configurations of diagnostic connector links. The earliest connector appears in 1985–88 vehicles. It is a single terminal (ECT terminal) connector. The next design was introduced on some models in 1985. It is an irregularly-shaped, multi-terminal connector and usually always appears in the engine compartment. On vehicles with a single, integrated (engine & transmission) Powertrain Control Module (PCM) the ECT terminal is changed to the T_T terminal. The third configuration is referred to as the Toyota Diagnostic Communication Link (TDCL). This connector is used on 1992 and newer pre-OBD II models that use a single, integrated (engine & transmission) Powertrain Control Module (PCM). Models equipped with this

DIAGNOSTIC TROUBLE CODES—TOYOTA

Code	Component/Circuit Fault
38	ATF temperature sensor or circuit
42	Speedometer Vehicle Speed Sensor (VSS) or circuit
44	Rear transfer case speed sensor or circuit
61	Transmission/transaxle VSS or circuit
62	No. 1 shift solenoid or circuit
63	No. 2 shift solenoid or circuit
64	Lock-up solenoid or circuit
65	Transfer case solenoid or circuit
73	No. 1 center differential control solenoid or circuit
74	No. 2 center differential control solenoid or circuit

89446G58

Lexus transmission related Diagnostic Trouble Codes

third type of connector also have the second design connector in the engine compartment. The third type connector is usually located under the driver's side of the instrument panel.

To retrieve DTC's, perform the following:

1. Locate the diagnostic connector location.
2. Turn the ignition key **OFF**.
3. Using a jumper wire, ground the ECT or TE1 terminal in the diagnostic connector.
4. Turn the ignition key **ON**, without starting the engine.
5. Observe the O/D OFF light and note all DTC's. If the O/D OFF keeps flashing off and on, in even, regular cycles, there are no codes stored in memory.

Clearing Codes

To clear the codes on these vehicles, you'll have to disconnect the computer connector and leave it disconnected for one minute. Then reconnect the computer.

Mazda

ENGINE CODES

General Information

Mazda utilizes 2 types of fuel systems between the years 1984–94. The Feedback Carburetor (FBC) system and Electronic Gas Injection (EGI) system, or fuel injection system. The feedback carburetor system was used in the GLC, 323, 626 and RX-7 between years 1984–87.

It was also used in the B2000, B2200 and B2600 pickup trucks between years 1984–92.

Electronic Gas Injection (EGI) was first available in the 1984 RX-7. 626 picked it up in 1986, 323 in 1987. In 1988 all models except B2200 and B2600 pickup trucks came equipped with fuel injection. Mazda uses various variations of EGI. Navajo uses the Ford EEC-IV system. However, the EEC-IV system will not be covered in this section.

Service Precautions

• Before connecting or disconnecting the ECU harness connectors, make sure the ignition switch is **OFF** and the negative battery

cable is disconnected to avoid the possibility of damage to the control unit.

• When performing ECU input/output signal diagnosis, remove the pin terminal retainer from the connectors to make it easier to insert tester probes into the connector.

• When connecting or disconnecting pin connectors from the ECU, take care not to bend or break any pin terminals. Check that there are no bends or breaks on ECU pin terminals before attempting any connections.

• Before replacing any ECU, perform the ECU input/output signal diagnosis to make sure the ECU is functioning properly or not.

• After checking through EGI troubleshooting, perform the EFI self-diagnosis and driving test.

• When measuring supply voltage of ECU controlled components with a circuit tester, separate 1 tester probe from another. If the 2 tester probes accidentally make contact with each other during measurement, a short circuit will result and may damage the ECU.

Reading Codes

➡**Diagnostic codes may be retrieved through the use of the CHECK ENGINE light or Malfunction Indicator Lamp (MIL). Special System Checker No. 83, Digital Code Checker and a Self-diagnosis Checker are all special diagnostic equipment used to retrieve codes, however these tools are not required. When using special diagnostic equipment, always observe the tool manufacturer's instructions.**

1984–86 VEHICLES WITH SYSTEM CHECKER TOOL 83

On 1984–85 GLC, 626 and RX-7, 1986 323 and 1986 B2000 Pick-up, the System Checker No. 83 (tool No. 49-G030–920), is used to detect and indicate any problems of each sensor, damaged wiring, poor contact or a short circuit between each of the sensor control units. Trouble is indicated by a red lamp and a buzzer. If there are more than 2 problems at a time, the indicator lamp turns ON in the numerical order of the code number. Even if the problem is corrected during indication, 1 cycle will be indicated, If after a malfunction has occurred and the ignition key is switched **OFF**, the malfunction indicator for the feedback system will not be displayed on the checker.

Read engine trouble codes using the following procedures:

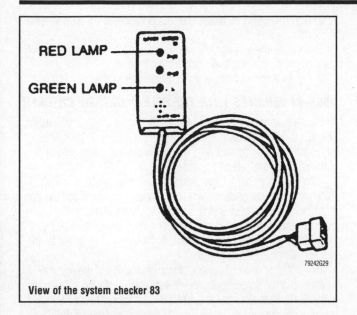

View of the system checker 83

1984–85 626 and B2200:

1. Operate the engine until normal operating temperatures are reached. Allow the engine to run at idle.

2. Connect System Checker tool No. 83 (49-G030-920) to the check connector, located near the ECU.

3. Check whether the trouble indication light turns ON.

If there is more than 2 problems at the same time, the indicator lamp lights on in the numerical order of the code number. Even if the problem is corrected during indication, 1 cycle will be Indicated. If after a malfunction has occurred the Ignition key is switched off, the malfunction indicator for the feedback system will not be displayed on the checker. The control unit has a built In fall-safe mechanism. If a malfunction occurs during driving, the control unit will on its own initiative, send out a command and driving performance will be affected. The commands are as follows:

a. Water Thermo-Sensor—the control unit outputs a constant 176°F (80°C) command.

b. Feed-Back Sensor—the control unit holds air/fuel solenoid to dwell meter reading 27° (duty 30%) for B2200.

c. Vacuum Sensor—the control unit prevents operation of the EGR valve, and holds the air/fuel solenoid to a duty of 0%.

d. EGR Position Sensor—the control unit prevents operation of the EGR valve.

➡️ If the trouble code is code number 3 (feedback system), proceed as follows:

4. Start the engine, letting it run until it reaches normal operating temperature. Connect a tachometer to the engine.

5. Connect a dwell meter (90 degrees, 4 cylinder) to the yellow wire in the service (check) connector of the air/fuel solenoid valve.

6. Run the engine at idle and note the reading on the dwell meter.

7. If the dwell meter reading is 0° degrees, the probable causes are as follows:

a. The wiring harness from the IG to the check connector BrY terminal is open.

b. The wiring harness from the check connector Y terminal to the control unit (F) terminal is grounded.

c. The transistor in the control unit for the air/fuel solenoid is open.

8. If the dwell meter reading is 90°, the probable causes are as follows:

a. The wiring harness from the IG to the check connector BrY terminal is open.

b. The wiring harness from the check connector BrY terminal to the control unit (F) terminal is grounded.

c. The transistor in the control unit for the air/fuel solenoid is short circuited.

9. If the dwell meter reading is 18°, check whether the green lamp (feedback signal) illuminates or does not illuminate.

10. If the oxygen sensor signal lamp does not illuminate, proceed as follows:

a. If the green lamp does not illuminate, the air is sucked from the intake system or the air is sucked from the exhaust manifold.

b. Carburetor jets are clogged.

c. The valve of the air/fuel solenoid is stuck to the lower position, giving a lean air/fuel mixture condition.

11. If the oxygen sensor signal lamp illuminates, proceed as follows:

a. If green lamp turns ON, the mixture is richer than stoichiometric air/fuel ratio.

b. If the green lamp turns ON and OFF, the O_2 sensor signal is fed to the control unit.

c. If the green lamp turns OFF, the mixture is leaner than stoichiometric.

1984–85 RX-7 and 1986 323:

1. Operate the engine until normal temperatures are reached.
2. Allow the engine to run at idle.
3. Check whether the trouble indication light turns ON.

➡️ Trouble is indicated by a red light and a buzzer.

4. If the light turns ON, check the ECM code problems indicated.

WITH DIGITAL CODE CHECKER AND SELF-DIAGNOSIS CHECKER

The Digital Code Checker tool No. 49-G01829A0 for 1986 or Self-Diagnosis Checker tool No. 49-H018-9A1 are used to retrieve code numbers of malfunctions which have happened and were memorized or are continuing. The malfunction is indicated by the code number and buzzer.

If there is more 1 malfunction, the code numbers will display on the self diagnosis checker 1 by 1 in numerical order. In the case of

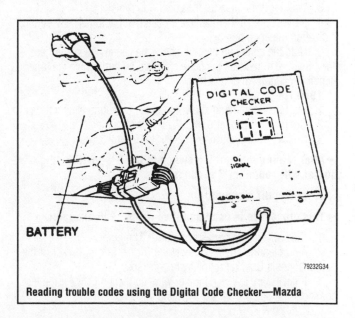

Reading trouble codes using the Digital Code Checker—Mazda

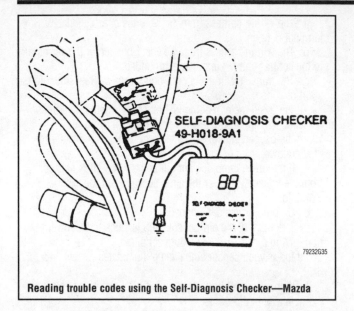

SELF-DIAGNOSIS CHECKER
49-H018-9A1

88

79232G35

Reading trouble codes using the Self-Diagnosis Checker—Mazda

malfunctions, 09, 13 and 01, the code numbers are displayed in order of 01, 09 and then 13.

The ECU has a built in fail-safe mechanism for the main input sensors. If a malfunction occurs, the emission control unit will substitute values; this will slightly effect the driving performance but the vehicle may still be driven.

The ECU continuously checks for malfunctions of the input devices within 2 seconds after turning the ignition switch to the **ON** position and the test connector is grounded.

The malfunction indicator light indicates a pattern the same as the buzzer of the self-diagnosis checker when the self-diagnosis check connector is grounded. When the self-diagnosis check connector is not grounded, the lamp illuminates steady while the malfunction recovers. However, the malfunction code is memorized in the emission control unit.

Read engine trouble codes using the following procedures:
1986 Vehicles Except RX-7:
1. Warm the engine to normal operating temperatures, by keeping the engine speed below 400 rpm.
2. Connect the Digital Code Checker No.
3. Wait for 3 minutes for the code(s) to register.
4. If the code number flashes, a buzzer will automatically sound, indicating the code number.
5. Note the code number and check the causes, repair as necessary. Be sure to recheck the code numbers by performing the 'After Repair Procedure' after repairing.
1986 RX-7:
1. Start the engine and allow it to reach operating temperature.
2. Connect the Digital Code Checker for trouble codes.
3. Check the Digital Code Checker for trouble codes.

➡**After turning the ignition switch to the ON position, the buzzer will sound for 3 seconds.**

After Repair Procedure:

➡**This procedure is used on all vehicles 1986 and later.**

1. Clear all trouble codes from the ECU memory.
2. After clearing codes, connect the Digital Code Checker or the Self-Diagnosis Checker to the test connector.
3. If necessary to use a jumper wire, connect it between the test connector (green: pin 1) and a ground.

4. Turn the ignition switch **ON**, but do not start the engine for 6 seconds.
5. Operate the engine until normal operating temperatures are reached, then, run it at 2000 rpm for 2 minutes.
6. Verify that no code numbers are displayed.

1987–94 VEHICLES WITH THE SELF-DIAGNOSIS CHECKER

The self-diagnosis checker (49-HOI 8-9A1) and System Selector (49-BOI 9-9A0), are used to retrieve code numbers of malfunctions which have happened and were memorized or are continuing. The malfunction is indicated by a code number.

If there is more than 1 malfunction, the code numbers will display on the self-diagnosis checker in numerical order. The ECU has a built in fail-safe mechanism for the main input sensors. If a malfunction occurs, the emission control unit will substitute values. This will affect driving performance, but the vehicle may still be driven.

The ECU continuously checks for malfunctions of the input devices. But the ECU checks for malfunctions of the output devices within 3 seconds after the green (1 pin) test connector or TEN terminal of the diagnosis connector is grounded and the ignition switch is turned to the **ON** position. Read engine trouble codes using the following procedures:
1987–91 323, Miata and Protegé:
1. Connect the tester to the check connector at the rear of the left side wheel housing and to the negative battery cable.
2. Set the tester select switch to the A setting.
3. With a jumper wire, ground the 1-pin test connector.
4. Turn the ignition switch **ON**.
5. Make sure that 88 flashes on the monitor and that the audible buzzer sounds for 3 seconds after turning the ignition switch **ON**.
6. If 88 does not flash, check the main relay, power supply circuit and the check connector wiring.
7. If 88 flashes and the buzzer sounds for more than 20 seconds, replace the engine control unit and repeat Steps 3 and 4.
8. Note any other code numbers that are present and refer to the code chart. Repair if necessary.
1992 626, MX-6, MPV, B2200 and B2600i:
The check connector is located at the rear of the left side wheel house on 626/MX-6, front of the left side wheel house on MPV, above the right side wheel house on B2200 and near the fuel filter on B2600i.
1. Connect the tester to the check connector and to ground.
2. Set the tester select switch to the A setting.
3. With a jumper wire, ground the 1 pin test connector.
4. Turn the ignition switch ON.
5. Make sure that 88 flashes on the monitor and that the audible buzzer sounds for 3 seconds after turning the ignition switch ON.
6. If 88 does not flash, check the main relay, power supply circuit and the check connector wiring.
7. If 88 flashes and the buzzer sounds for more than 20 seconds, replace the engine control unit and perform steps number 1 through 6 again.

➡**Before replacing the ECU on the MPV or B2600i, check for a short circuit between ECU terminal IB for JE engine and 1F for G6 engine and the 6 pin check connector.**

8. Note and record any other code numbers that are present.
1992–94 323, Protegé, 929, Miata (MX-5), MX-3 and 1993–94 RX-7:
1. Connect the system selector to the diagnosis connector at the rear of the left side wheel housing.

2. Set the SYSTEM SELECT switch to the 1 setting.

3. Set the TEST switch to the SELF—TEST position.

4. Connect the self-diagnosis checker, the system selector and ground.

5. Set the self-diagnosis checker SELECT switch to the A position.

6. Turn the ignition switch **ON**.

7. Make sure that 88 flashes on the monitor and that the audible buzzer sounds for 3 seconds after turning the ignition switch **ON**.

8. If 88 does not flash, check the main relay, power supply circuit and the diagnosis connector wiring.

9. If 88 flashes and the buzzer sounds for more than 20 seconds, check for a short between ECU terminal 1F and the FEN terminal of the diagnosis connector. Replace the engine control unit if necessary and perform Steps 1 through 7 again.

10. Note and record any other code numbers that are present.

1993–94 626 and MX-6:

1. Connect the system selector to the diagnosis connector at the rear of the left side wheel housing.

2. Set the SYSTEM SELECT switch to the 1 setting.

3. Set the TEST switch to the SELF—TEST position.

4. Connect the self-diagnosis checker the system selector and ground.

5. Set the self-diagnosis checker SELECT switch to the A position.

6. Turn the ignition switch **ON**.

7. Make sure that 88 flashes on the monitor and that the audible buzzer sounds for 3 seconds after turning the ignition switch **ON**.

8. If 88 does not flash, check the main relay, power supply circuit and the diagnosis connector wiring.

9. If 88 flashes and the buzzer sounds for more than 20 seconds, check for a short between PCM terminal 1F (manual trans.), (1G auto trans.) and the FEN terminal of the diagnosis connector. Replace the engine control unit if necessary and perform Steps 1 through 7 again.

10. Note and record any other code numbers that are present.

1984–94 VEHICLES WITHOUT THE SELF-DIAGNOSIS CHECKER

The malfunction indicator light indicates a pattern the same as the buzzer of the self-diagnosis checker when the green (1 pin) test connector or FEN terminal of the diagnosis connector is grounded.

Clearing Codes

1984–86 VEHICLES

1. Turn the ignition switch **OFF**.
2. Disconnect the negative battery cable.
3. Depress the brake pedal for at least 5 seconds.
4. Reconnect the negative battery cable.

1987–91 VEHICLES

1. Cancel the memory of the malfunction by disconnecting the negative battery cable and depressing the brake pedal for at least 20 seconds, then reconnect the negative battery cable.

2. Except Miata, MX-3, 323 and Protegé, connect the Self-Diagnosis Checker 49-H018–9A1 to the check connector. Ground the test connector (green: 1 pin) using a jumper wire.

3. On Miata, MX-3, 323 and Protegé, connect Self-Diagnosis Checker (49-BOI 9–9A0) to the diagnosis connector.

4. Turn the ignition switch **ON**, but do not start the engine for approximately 6 seconds.

5. Start the engine and allow it to reach normal operating temperature. Then run the engine at 2000 rpm for 2 minutes. Check that no code numbers are displayed.

1992–94 VEHICLES

323, MX-3 (B6 engine), and MX-5/Miata:

1. Disconnect the negative battery cable.
2. Press the brake pedal for at least 20 seconds.
3. Connect the negative battery cable.
4. Connect the self-diagnosis tester to the diagnosis connector.
5. Turn the ignition switch **ON**.
6. Start and warm-up the engine.
7. Run engine at 2,000 rpm for 3 minutes.
8. Verify that no more codes are stored.

1992 626, MX-6, B2200 (EGI) and B2600i:

1. Disconnect the negative battery cable.
2. Press the brake pedal for at least 5 seconds.
3. Connect the negative battery cable.
4. Connect the self-diagnosis tester and ground the test connector.
5. Turn the ignition switch to the **ON** position for 6 seconds.
6. Start and warm-up the engine.
7. Run engine at 2,000 rpm for 2 minutes (3 minutes on truck).
8. Verify that no more codes are stored.

1993–94 626 and MX-6 (FS Engine):

1. Disconnect the negative battery cable.
2. Press the brake pedal for at least 20 seconds.
3. Connect the negative battery cable.
4. Connect the self-diagnosis tester to the diagnosis connector.
5. Turn the ignition switch **ON**.
6. Start and warm-up the engine.
7. Run engine at 2,000 rpm for 2 minutes.
8. Verify that no more codes are stored.

1993–94 626 and MX-6 (KL Engine), MX-3 (K8 engine) and 1993–94 RX-7:

1. Disconnect the negative battery cable.
2. Press the brake pedal for at least 20 seconds.
3. Connect the negative battery cable.
4. Connect the self-diagnosis tester to the diagnosis connector.
5. Turn the ignition switch **ON**.
6. Verify that no more codes are stored.

929:

1. Turn the ignition switch **OFF**.
2. Disconnect the negative battery cable for 20 seconds.

MPV (G6 engine):

1. Disconnect the negative battery cable for at least 20 seconds.
2. Connect the negative battery cable.
3. Connect the self-diagnosis tester and ground the test connector.
4. Turn the ignition switch to the **ON** position for 6 seconds.
5. Start and warm-up the engine.
6. Run engine at 2,000 rpm for 3 minutes.
7. Verify that no more codes are stored.

B2200 (FOC engine):

For the pickup with the feedback carburetor fuel system disconnect the negative battery cable for at least 5 seconds.

Diagnostic Trouble Codes

1984–94 EXCEPT RX-7

1984–85 Vehicles With The 2.0L (FE) Engine:
Code 01 Engine speed
Code 02 Water thermosensor

Code 03 Oxygen sensor
Code 04 Vacuum sensor
Code 05 EGR position sensor
1986–87 Vehicles With the 1.6L, 2.0L and 2.2L Engines:
Code 01 Ignition pulse
Code 02 Air flow meter
Code 03 Water thermosensor
Code 04 Intake air thermo or Temperature sensor
Code 05 Feedback system
Code 06 Atmospheric pressure sensor (1986 1 .6L)
Code 08 EGR position sensor
Code 09 Atmospheric pressure sensor
Code 22 No. 1 Cylinder sensor (2.2L turbocharged)
1988–94 Vehicles With The 1.6L, 1.8L, 2.0L, 2.2L, 2.5L, 2.6L and 3.0L Engines:
Code 01 Ignition pulse
Code 02 Ne signal—distributor
Code 02 NE 2 signal—crankshaft (1992–93 1.8L V6, 1994 2.0L, 1992–94 3.0L)
Code 03 Gi signal—distributor (2.2L turbo, 1988–91 3.0L)
Code 03 G signal—distributor
Code 04 G2 signal—distributor (2.2L turbo, 1988–91 3.0L); NE 1 signal—distributor (1992–94 1.8L V6, 1994 2.0L, 1992–93 3.0L)
Code 05 Knock sensor and control unit (Left side on 1992–943.0L)
Code 06 Speed signal
Code 07 Knock sensor; right side (1992–94 3.0L)
Code 08 Air flow meter
Code 09 Engine coolant temperature sensor (C IS)
Code 10 Intake air temperature sensor
Code 11 Intake air thermosensor—dynamic chamber (3.0L, 2.6L)
Code 13 Intake manifold pressure sensor (1 .3L)
Code 14 Atmospheric pressure sensor (in ECU on 2.6L and 1994 2.5L)
Code 15 Oxygen sensor
Code 15 Oxygen sensor; left side on 1992–94 1.8L V6, 1994 2.5L, 1990–94 3.0L
Code 16 EGR position sensor
Code 17 Closed loop system
Code 17 Closed loop system; left side on 1992–94 1.8L V6, 1993–94 2.5L 1990–94 3.0L
Code 23 Heated oxygen sensor; right side on 1992–94 1.8L V6, 1994 2.5L 1990–91 3.0L
Code 24 Closed loop system; right side on 1992–94 1.8L V6, 1993 2.5L 1990–91 3.0L
Code 25 Solenoid valve—pressure regulator
Code 26 Solenoid valve—purge control
Code 26 Solenoid valve—purge control No. 2 (1988–89 3.0L)
Code 27 Solenoid valve—purge control No. 1 (1988–89 3.0L)
Code 27 Solenoid valve—No. 2 purge control (1989 1.6L)
Code 28 Solenoid valve—EGR vacuum
Code 29 Solenoid valve—EGR vent
Code 30 Relay (cold start injector 3.0L)
Code 34 ISC valve
Code 34 Idle air control valve (1993–94 2.0L and 2.5L, 1.6L, 1.8L, 2.6L, 3.1L)
Code 36 Oxygen sensor heater relay (1990 3.0L)
Code 36 Right side oxygen sensor heater (1 992–94 3.0L)
Code 37 Left side oxygen sensor heater (1 992–94 3.0L)
Code 37 Coolant fan relay
Code 40 Oxygen sensor heater relay (1991 3.0L)

Code 40 Solenoid (triple induction system) and oxygen sensor relay (1 988–89 3.0L)
Code 41 Solenoid valve—VRIS (1989–94 MPV 3.0L)
Code 41 Solenoid valve—VRIS 1 (1992–94 1.8L V6, 1993 2.5L)
Code 41 Solenoid valve—VICS (3.0L)
Code 42 Solenoid valve—Waste gate (turbocharged)
Code 46 Solenoid valve—VRIS 2 (1992–94 1.8L V6, 1993 2.5L)
Code 65 NC signal—PCMT (1 992–94 3.0L)
Code 67 Coolant fan relay No. 1 (1993 2.5L)
Code 67 Coolant fan relay No. 2 (1992–94 1.8L V6)
Code 68 Coolant fan relay No. 2, No.3 with ATX (1993 2.5L)
Code 69 Engine coolant temperature sensor—fan (1992–94 1.8L V6, 1993 2.0L and 2.5L)

1984–94 RX-7

1.3L Rotary Engine:
Code 01 Crank angle sensor (1984–87)
Code 01 Ignition coil—trailing (1988–91)
Code 02 Air flow meter (1984–87)
Code 02 Ne signal—crank angle sensor (1988–91)
Code 03 Water thermosensor (1984–87)
Code 03 G signal—crank angle sensor (1988–91)
Code 04 Intake air temperature sensor—in the air flow meter (1984–87)
Code 05 Oxygen sensor (1984–87)
Code 05 Knock sensor (1993)
Code 06 Throttle sensor (1984–87)
Code 06 Speedometer sensor (1993)
Code 07 Boost sensor / Pressure sensor (1984–87 turbo)
Code 08 Air flow meter (1988–91)
Code 09 Atmospheric pressure sensor (1984–87)
Code 09 Water thermosensor (1988–94)
Code 10 Intake air thermosensor—in air flow meter (1988–91)
Code 11 Intake air thermosensor (1988–93)
Code 12 Coil with igniter—trailing (1984–87)
Code 12 Throttle sensor—wide open throttle (1988–94)
Code 13 Intake manifold pressure sensor (1988–94)
Code 14 Atmospheric pressure sensor (1988–94, in ECU on 1993)
Code 15 Intake air temperature sensor—in dynamic chamber (1984–87)
Code 15 Oxygen sensor (1988–94)
Code 16 EGR switch (1993 California)
Code 17 Closed loop system (1988–93)
Code 18 Throttle sensor—closed or narrow throttle (1988–94)
Code 20 Metering oil pump position sensor (1988–94)
Code 23 Fuel thermosensor (1993)
Code 25 Solenoid valve—pressure regulator control (1993)
Code 26 Metering oil pump stepping motor (1993)
Code 27 Step motor—metering oil pump (1988–91)
Code 27 Metering oil pump (1993)
Code 28 Solenoid valve—EGR (1993)
Code 30 Solenoid valve—split air bypass (1988–94)
Code 31 Solenoid valve—relief No. 1 (1988–94)
Code 32 Solenoid valve—switching (1988–94)
Code 33 Solenoid valve—port air bypass (1988–94)
Code 34 Solenoid valve—bypass air control (1988–91)
Code 34 Solenoid valve—idle speed control (1993)
Code 37 Metering oil pump (1988–93)
Code 38 Solenoid valve—accelerated warm-up system (1988–94)

Code 39 Solenoid valve—relief No. 2 (1993)
Code 40 Auxiliary port valve (1988–91)
Code 40 Solenoid valve—purge control (1993)
Code 41 Solenoid valve—variable dynamic effect Intake control (1988–91)
Code 42 Solenoid valve—turbo boost pressure regulator (1988–91)
Code 42 Solenoid valve—turbo pre-control (1993)
Code 43 Solenoid valve—wastegate control (1993)
Code 44 Solenoid valve—turbo control (1993)
Code 45 Solenoid valve—charge control (1993)
Code 46 Solenoid valve—charge relief (1993)
Code 50 Solenoid valve—double throttle control (1993)
Code 51 Fuel pump relay (1988–94)
Code 54 Air pump relay (1993)
Code 71 Injector—front secondary (1988–94)
Code 73 Injector—rear secondary (1988–94)
Code 76 Slip lockup off signal—EC-AT CU (1993)
Code 77 Torque reduced—EC-AT CU (1993)

TRANSMISSION CODES

Reading and Retrieving Codes

Generally, the transaxle related DTC's are easy to retrieve and read on all pre-OBD II Mazda vehicles.

➡️**The 1994 and newer pre-OBD II 626 and MX6 models with the 2.0L engine use a Ford transaxle. Please refer to the Ford procedures in this section for these models.**

All Mazda models utilize two-digit DTC's to inform the technician in which circuit and/or component a fault was detected. The two-digit codes are displayed by either the HOLD light or the CHECK ENGINE light, both of which are located in the instrument cluster display. The light flashes the digits out by first flashing the tens digits as long flashes, then after a slight pause, the ones digits as shorter flashes. Therefore, if the HOLD or CHECK ENGINE light flashes once (long flash), pauses slightly, then flashes twice (short flashes), a Code 12 was displayed. If there is more than one DTC stored in the computer's memory, the light will display the lowest number first and move upward through all of the stored codes in numerical order. After the progression reaches the highest code number, the cycle will repeat itself until the ignition switch is turned **OFF**.

➡️**There will be a longer pause between the individual codes than there is between the tens and ones digits of each code.**

To determine if the vehicle's computer has stored any DTC's, turn the ignition switch **ON** and observe the HOLD or CHECK ENGINE light. The light should illuminate for three seconds, then extinguish. If the light flashes ON and OFF continuously, the computer has stored DTC's.

To retrieve the stored DTC's, perform the following:

1. Perform the preliminary inspection, located earlier in this section. This is very important, since a loose or disconnected wire, or corroded connector terminals can cause a whole slew of unrelated DTC's to be stored by the computer; you will waste a lot of time performing a diagnostic "goose chase."

2. Grab some paper and a pencil or pen to write down the DTC's when they are flashed out.

3. Locate the Diagnostic Request Connector (DRC). The location and configuration of the DRC depends largely on the model of the

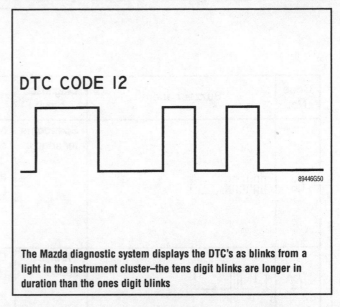

DTC CODE 12

89446G50

The Mazda diagnostic system displays the DTC's as blinks from a light in the instrument cluster–the tens digit blinks are longer in duration than the ones digit blinks

particular vehicle at hand. The DRC will either be a green or blue two-terminal connector, or will be one terminal as a part of the large diagnostic connector in the engine compartment.

➡️**If the DRC terminal for the vehicle at hand is designated as the TAT terminal of the large diagnostic connector in the engine compartment, but the connector does not utilize the TAT terminal, ground the TEN terminal instead. If the TEN terminal is to be used, observe the CHECK ENGINE light, not the HOLD light.**

To locate the DRC in the specific vehicle at hand, refer to the following general list:

- 1990–91 323 and MX3—TAT terminal, located in the diagnostic connector near the battery, actuates the HOLD light
- 1992–95 323 and MX3—TAT terminal, located in the diagnostic connector near the battery, actuates the CHECK ENGINE light
- 1988–89 626 and MX6—blue, single connector, located under the far left-hand side of the instrument panel, near the transaxle computer, actuates the HOLD light
- 1990–92 323 and MX3 non-Turbo—green, single connector, located near the left, front shock tower, actuates the CHECK ENGINE light
- 1990–92 626 and MX6 Turbo—blue, single connector, located under the far left-hand side of the instrument panel, near the transaxle computer, actuates the HOLD light
- 1993 626 and MX6—TAT terminal, located in the diagnostic connector near the left, front shock tower, actuates the HOLD light
- 1994–95 626 and MX6 models with 2.5L engine—TAT terminal, located in the diagnostic connector near the left, front shock tower, actuates the HOLD light
- 1988–91 929—blue, single connector, located near the right, front shock tower, actuates the HOLD light
- 1992–94 929—TAT terminal, located in the diagnostic connector near the left, front shock tower, actuates the HOLD light
- 1989–91 RX7—blue, single connector, located near the left, front shock tower, actuates the HOLD light
- 1992–93 RX7—TAT terminal, located in the diagnostic connector near the left, front shock tower, actuates the HOLD light
- All MPV—blue, single connector, located under the far left-hand side of the instrument panel, near the transaxle computer, actuates the HOLD light

Code No.	Buzzer pattern	Diagnosed circuit	Condition	Point
06		Speedometer sensor	No input signal from speedometer sensor while driving at drum speed above 600 rpm in D, S, or L ranges	• Speedometer sensor connector • Wiring from speedometer sensor to instrument cluster • Wiring from instrument cluster to EC-AT CU • Speedometer sensor resistance
12		Throttle sensor	Open or short circuit	• Throttle sensor connector • Wiring from throttle sensor to EC-AT CU • Throttle sensor resistance.
55		Pulse generator	No input signal from pulse generator while driving at 40 km/h (25 mph) or higher in D, S, and L ranges	• Pulse generator connector • Wiring from pulse generator to EC-AT CU • Pulse generator connector
57		Reduce torque signal 1	Open or short circiut of reduce torque signal 1 wire harness	• Wiring from ECU to EC-AT CU
58		Reduce torque signal 2	Open or short circuit of reduce torque signal 2 and/or torque reduced signal/water thermo signal wire harness	• Wiring from ECU to EC-AT CU

89446G48

A few examples of some of the possible DTC's displayed by Mazda vehicles

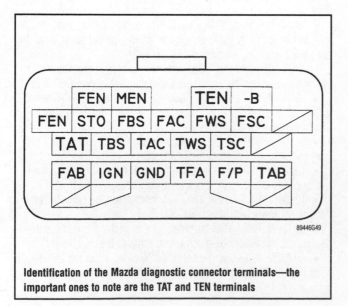

89446G49

Identification of the Mazda diagnostic connector terminals—the important ones to note are the TAT and TEN terminals

89446P06

The DRC for Mazda MPV models is a single wire connector, located up, under the left side of the instrument panel

4. Ensure the ignition switch is in the **OFF** position.

5. Using a jumper wire, attach the applicable DRC terminal to a good engine or chassis ground. If the DRC is located in the diagnostic connector, you can even attach it to the GND terminal of the connector.

6. Without starting the engine, turn the ignition to the **ON** position.

7. Observe the applicable light and note the DTC's.

8. After all of the codes have been noted, turn the ignition switch **OFF** and remove the jumper wire.

Clearing Codes

To clear the DTC's from the vehicle's computer memory, disconnect the negative battery cable for at least one minute, then reattach it. Reattach the negative battery cable and all of the stored DTC's should be deleted. If there are still codes present, either the codes were not properly cleared (Are the codes identical to those flashed out previously?) or the underlying problem is still there (Are only some of the codes the same as previously?)

Mitsubishi

ENGINE CODES

General Information

Mitsubishi uses 2 types of fuel systems. Feedback carburetor system and fuel injection. The type of fuel injection system is known as Electronic Controlled Injection (ECI).

Mitsubishi uses a conventional downdraft two-barrel compound type carburetor which incorporates an automatic choke, accelerator pump, and enrichment system. In addition, a deceleration device is provided.

The Electronic Fuel Injection (EFI) system, used on Mitsubishi vehicles, is classified as a Multi-Point Injection (MPI) system. The MPI system controls the fuel flow, idle speed, and ignition timing. The basic function of the MPI system is to control the air/fuel ratio in accordance with all engine operating conditions. An Electronic Control Unit (ECU) is the heart of the MPI system. Based on data from various sensors, the ECU computes the desired air/fuel ratio.

DIAGNOSTIC TROUBLE CODES—MAZDA

Code		Component/Circuit Fault
1		No engine speed signal from ECM
6		Vehicle Speed Sensor (VSS) or circuit
7		Output speed sensor or circuit
12		Throttle Position (TP) sensor or circuit
55		Input speed sensor or circuit
56		ATF temperature sensor or circuit
57	①	Lost communication between engine computer terminal 1S and transmission and computer terminal 1J
58	①	Lost communication between engine computer terminal 1B and transmission and computer terminal 1L
59	①	Lost communication between engine and transmission computer terminals 1K
60		Shift solenoid A (1-2) or circuit
61		Shift solenoid B (2-3) or circuit
62		Shift solenoid C (3-4) or circuit
63		Lockup solenoid or circuit
64	②	3-2 control solenoid or circuit
64	③	Line pressure control solenoid or circuit
65		Lock-up engagement control solenoid or circuit
66	②	Line pressure control solenoid or circuit
46		Downshift error
48		Low voltage or bad computer ground
55		Computer failure
56		Shift lever position switch or TP sensor circuit
65		ATF temperature sensor or circuit
66		ATF temperature sensor or circuit
77		Kickdown switch or circuit, or TP sensor or circuit
82		Shift lever position switch

① GF4AEL only
② Except R4AEL
③ R4AEL

89446G51

Mazda transmission related Diagnostic Trouble Codes

Service Precautions

• Before connecting or disconnecting the ECU harness connectors, make sure the ignition switch is **OFF** and the negative battery cable is disconnected to avoid the possibility of damage to the control unit.

• When performing ECU input/output signal diagnosis, remove the pin terminal retainer from the connectors to make it easier to insert tester probes into the connector.

• When connecting or disconnecting pin connectors from the ECU, take care not to bend or break any pin terminals. Check that there are no bends or breaks on ECU pin terminals before attempting any connections.

• Before replacing any ECU, perform the ECU input/output signal diagnosis to make sure the ECU is functioning properly.

• When measuring supply voltage of ECU-controlled components with a circuit tester, separate 1 tester probe from another. If the 2 tester probes accidentally make contact with each other during measurement, a short circuit will result and damage the ECU.

Reading Codes

➡**All though the CHECK ENGINE light or Malfunction Indicator Lamp (MIL) will illuminate when there is trouble detected, diagnostic codes can only be retrieved with the use of either a analog voltmeter or a Multi-use Tester. When using diagnostic equipment, always observe the tool manufacturer's instructions.**

1984–86 VEHICLES—USING THE ECI/MPI TESTER

Refer to manufacturer's tester manual regarding diagnosis with this tester.

1985–94 VEHICLES—UISNG AN ANALOG VOLTMETER

The voltmeter can be used to retrieve code numbers of malfunctions which have happened and were memorized or are continuing to happen. On the voltmeter, the malfunction is indicated by a sweep of the needle. The voltmeter should be connected to the data link connector located under the driver side dashboard. Connect the voltmeter between the Multi-Point Injection (MPI) terminal and the ground terminal. Turn the ignition switch **ON** if the normal condi-

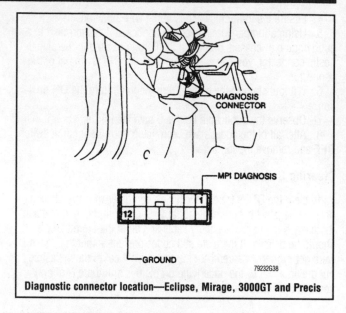

Diagnostic connector location—Eclipse, Mirage, 3000GT and Precis

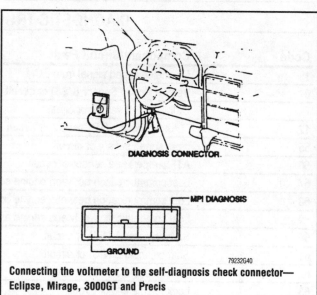

Connecting the voltmeter to the self-diagnosis check connector— Eclipse, Mirage, 3000GT and Precis

tion exists, the voltmeter pointer will indicate a normal pattern. A normal pattern is indicated by constant needle sweeps. If a problem exists in the system the voltmeter pointer will indicate it in a series of pointer sweeps. For example, a Code 3 would be 3 consecutive short sweeps of the voltmeter needle.

If there is more than 1 malfunction, the low code numbers will first be indicated and after a 2 second pause (no code indication) the higher code will be indicated.

1985–94 VEHICLES—USING THE MULTI-USE TESTER

To read the trouble codes using the Multi-Use Tester (MB991341 or equivalent) follow the steps below:

1. Turn the ignition switch **OFF**.
2. Insert the power supply terminal to the cigarette lighter socket.
3. Connect the tester connector to the diagnosis connector in the glove compartment, under the 'hood or under the driver side dashboard.
4. Turn the ignition switch **ON** and push the DIAG key.
5. Observe the trouble code and make the necessary repairs.

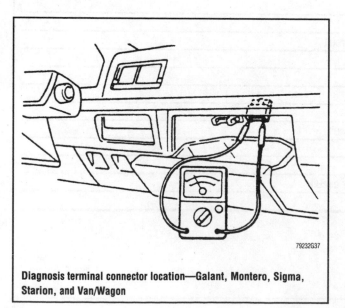

Diagnosis terminal connector location—Galant, Montero, Sigma, Starion, and Van/Wagon

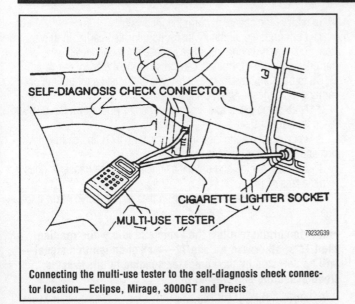

Connecting the multi-use tester to the self-diagnosis check connector location—Eclipse, Mirage, 3000GT and Precis

On most models the CHECK ENGINE malfunction indicator light will light up and remain illuminated to indicate that there is a problem in the system. After this light has been reported to be ON, the system should be checked for malfunction codes.

Clearing Codes

WITHOUT MULTI-USE TESTER

1984–86 vehicles—engine codes can be cleared by disconnecting the negative battery terminal or by disconnecting ECU connector for 15 seconds or longer.

1987–94 vehicles—engine codes can be cleared by disconnecting the negative battery terminal for 10 seconds or longer.

WITH MULTI-USE TESTER

Engine codes may also be cleared by setting the ignition switch to the **ON** position and using the malfunction code ERASE signal.

Diagnostic Trouble Codes

1989–94 VEHICLES

Diamante, Eclipse, 3000gt, Galant, Mirage, Montero, Precis, Sigma, Starion and Expo:
Code 11 Oxygen sensor
Code 12 Air flow sensor
Code 13 Intake Air Temperature Sensor
Code 14 Throttle Position Sensor (TPS)
Code 15 SC Motor Position Sensor (MPS)
Code 21 Engine Coolant Temperature Sensor
Code 22 Crank angle sensor
Code 23 No. 1 cylinder TDC (Camshaft position) Sensor
Code 24 Vehicle speed sensor
Code 25 Barometric pressure sensor
Code 31 Knock (KS) sensor
Code 32 Manifold pressure sensor
Code 36 Ignition timing adjustment signal
Code 39 Oxygen sensor (rear—turbocharged)
Code 41 Injector
Code 42 Fuel pump
Code 43 EGR-California
Code 44 Ignition Coil; power transistor unit (No. 1 and No. 4 cylinders) on 3.0L

Code 52 Ignition Coil; power transistor unit (No. 2 and No. 5 cylinders) on 3.0L
Code 53 Ignition Coil; power transistor unit (No. 3 and No. 6 cylinders)
Code 55 AC valve position sensor
Code 59 Heated oxygen sensor
Code 61 Transaxle control unit cable (automatic transmission)
Code 62 Warm-up control valve position sensor (non-turbo)

TRANSMISSION CODES

Reading and Retrieving Codes

GENERAL INFORMATION

Most Mitsubishi computerized transaxle control systems have no way of indicating a problem in the system. If the computer detects a problem, it begins pulsing a code at the diagnostic link connector. The only way you'll know there's a code is to look for it; nothing lights up to tell you there's a problem.

Mitsubishi vehicles use Diagnostic Link Connectors (DLC) in a possible three different configurations, four different sets of diagnostic codes (types 1 through 4), and three methods for displaying DTC's. Mitsubishi transaxles appear in Mitsubishi (of course), Hyundai and some Chrysler vehicles.

The Diagnostic Link Connector (DLC) must be located to retrieve any DTC's. Luckily, unlike some of the other import models covered in this manual, there are only three various locations for the DLC. The three locations are as follows: 1985–88 Galant, inside the glove box above the compartment itself; 1989 and newer pre-OBD II Mitsubishi and Chrysler vehicles, either in the fuse box, or under the far left-hand side of the instrument panel; all OBD II vehicles, under the left-hand side of the instrument panel.

➡**Remember that only OBD II models can let the driver know that codes have been stored; all other models do not alert the driver that any faults have been detected. Therefore, the only way to ascertain whether codes have been stored is to actually perform the retrieval procedure.**

Over the years, the transaxle computer control system used with Mitsubishi vehicles has been designed to display the transaxle DTC's in three different formats. All of these formats consist of changes to the on- and off-time of the signal. Also, three different sets of DTC's have been utilized to inform the person servicing the transaxle of any detected problems. The Mitsubishi models and their corresponding code types are presented in the following list:

➡**The precise date when models switched from one code type to another wasn't always consistent, and using this list as a Bible for the breaking points between code types could have you chasing down the wrong component for the wrong code.**

Type 1 codes:
• 1985–88 Galant models
Type 2 codes:
• 1989–90½ Galant models
• 1989–93 Precis models
• 1989–90 except Galant and Precis models
• 1989–90 Chrysler models equipped with Mitsubishi transaxles
Type 3 codes:
• 1990½ and newer Galant models
• 1994 and newer Precis models

- 1991 and newer Galant and Precis models
- 1991 and newer Chrysler models equipped with Mitsubishi transaxles

Since the precise date when models switched from one code type to another listed in the preceding list is not set in stone, record any codes in memory, clear the codes, then create a fault in the transaxle computer system. For example, if the shift solenoid code was not stored by the transaxle computer, disconnect the shift solenoid transmission/transaxle connector and restart the engine, after clearing the previous codes. The diagnostic system should store the appropriate code for a shift solenoid/circuit fault, which will allow you to ascertain precisely what set of codes the system being worked on uses.

TYPE 1 CODES

The first format only appeared in 1985–88 models. The Type 1 system was designed with what is known as the "Type 1" computer system. The Type 1 DTC's are made up of a series of signal pulses that varied in on-time and off-time. There is no regular pattern to these signals, except for a 4 second start and end signal, to indicate when each DTC starts and finishes. To identify the code, you must compare the signal to the accompanying code chart.

The Type 1 Mitsubishi transaxle diagnostic system functions so that once the system detects a malfunction, it pulses the DTC continuously through the DLC. There's no wire to ground, no special procedure to perform; just connect a DVOM to the respective DLC terminals and read the DTC's.

Because of the lack of a regular pattern, Type 1 codes can be difficult to read and interpret. One way to make these early codes easier to read is to use an oscilloscope or Digital Multi-Meter (DMM) with a graphic display instead of a voltmeter. By capturing the signal as a single waveform, you can compare the waveform to the accompanying code chart.

The Type 1 code system utilized volatile code memory, which means that whenever the ignition key is turned OFF, the codes are cleared. Also, because of the simplistic design of the system, any code will force the vehicle to run in a failsafe mode.

➡The Type 1 code system can only display one DTC at a time. Therefore, you must repair the problem for the first DTC before the second DTC can be read. So do not be surprised if you must fix several problems before there are no DTC's left.

To retrieve the codes, perform the following:
1. Perform the preliminary inspection, located earlier in this section. This is very important, since a loose or disconnected wire, or corroded connector terminals can cause a whole slew of unrelated DTC's to be stored by the computer; you will waste a lot of time performing a diagnostic "goose chase."
2. Grab some paper and a pencil or pen to write down the DTC's when they are flashed out.
3. Locate the Diagnostic Link Connector (DLC), inside the glove box above the compartment.
4. Start the engine and drive the vehicle until the transaxle goes into the failsafe mode.
5. Park the vehicle, but do not turn the ignition OFF. Allow it to idle.

➡It is important to allow the vehicle to idle while reading the DTC's, otherwise a Code 12—lack of an ignition signal—will be detected by the computer, leading to misdirected troubleshooting and wasted time.

6. Attach a Digital Volt-Ohmmeter (DVOM), a Digital Multi-Meter (DMM)—expensive versions are often equipped with a scope dis-

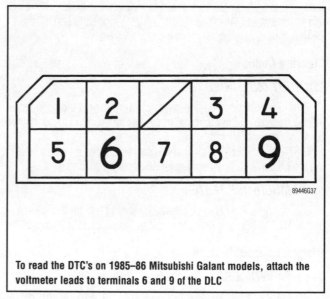

To read the DTC's on 1985–86 Mitsubishi Galant models, attach the voltmeter leads to terminals 6 and 9 of the DLC

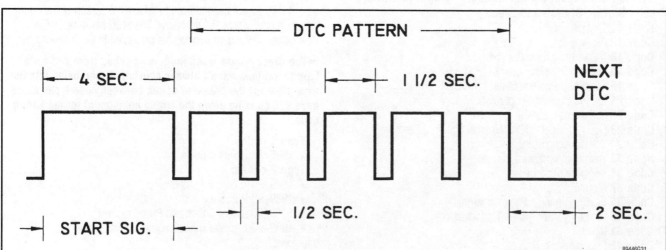

The Type 1 Mitsubishi codes are comprised of one 4 second start pulse followed by five individual pulses, lasting either 1½ seconds (ON) or ½ second (OFF) in duration

play function capable of showing waveforms—or an oscilloscope to the test terminals on the Diagnostic Link Connector (DLC). For 1985–86 Galant models, attach the instrument's (DVOM/DMM/oscilloscope) negative lead to terminal 9 of the DLC and the positive lead to terminal 6. For 1987–88 Galant models, attach the negative lead to terminal 6 of the DLC and the positive lead to terminal 12.

7. Observe the DVOM/DMM/oscilloscope and note the wave pattern. Compare the pattern to the accompanying chart to determine which code it corresponds to.

➡**Be sure of which system the vehicle at hand uses, since Type 1 Code 12 looks the same as Types 2 and 3 Code 15.**

8. After all of the DTC(s) have been retrieved, fix the applicable problems, clear the codes, drive the vehicle, and perform the retrieval procedure again to ensure that all of the codes are gone.

TYPE 2 AND 3 CODES

➡**Reading and retrieving Type 2 and Type 3 codes is identical; the only difference between Type 2 and 3 codes is what each of the codes stands for.**

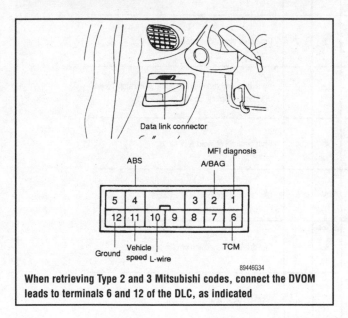

When retrieving Type 2 and 3 Mitsubishi codes, connect the DVOM leads to terminals 6 and 12 of the DLC, as indicated

Unlike the Type 1 diagnostic system, the system used for Type 2 and 3 codes was designed with the capability of storing up to 10 DTC's. Therefore, you will not have to fix each problem one at a time. However, unlike other import manufacturers, the codes are not displayed from lowest number to highest number. Rather, they are flashed out in the order in which they detected. This means that if the same fault was detected numerous times, the same code will be flashed out the same number of times as detected. Also, since only 10 codes can be stored at one time, if the computer detects an eleventh code, the first code stored will be deleted to make room for the latest code. This makes it important to check for codes whenever you suspect a problem with the transaxle.

With the introduction of the Type 2 code system, Mitsubishi also developed Failsafe Codes. Failsafe codes are DTC's which, if detected, will switch the vehicle's management system into a failsafe mode (thus the name). The failsafe mode sets the control computer to run (albeit not as efficiently) on predetermined parameters, so that erroneous information from the vehicle's sensors will be ignored until the problem at hand can be repaired. Other manufacturers call the failsafe mode the Limp In mode.

➡**Remember that since many codes are redundant, the system may store a non-failsafe code and a failsafe code for the same malfunction. In which case, the system will go into failsafe mode.**

Mitsubishi also implemented an easier way for reading the DTC's in Type 2 and 3 codes. The DTC's are 2-digit codes represented by a logical series of flashes (a DVOM, DMM or voltmeter can be used). The system displays the codes on the voltmeter by first flashing out the tens digit, which is a series of long flashes, followed by a short pause, then flashing out the ones digits, which is a series of short flashes. Therefore, two long flashes followed by three short flashes would indicate a Code 23. Between each complete DTC is a longer separator code.

To retrieve the codes, perform the following:

1. Perform the preliminary inspection, located earlier in this section. This is very important, since a loose or disconnected wire, or corroded connector terminals can cause a whole slew of unrelated DTC's to be stored by the computer; you will waste a lot of time performing a diagnostic "goose chase."

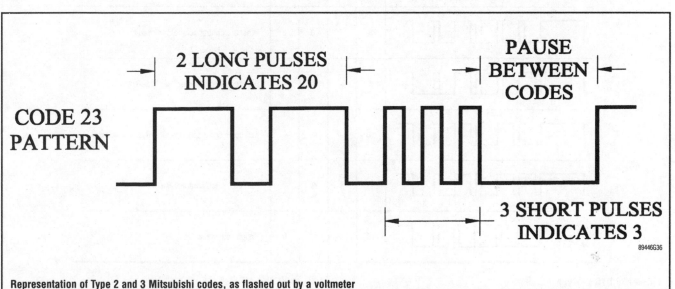

Representation of Type 2 and 3 Mitsubishi codes, as flashed out by a voltmeter

2. Grab some paper and a pencil or pen to write down the DTC's when they are flashed out.

3. Locate the Diagnostic Link Connector (DLC), either in the fuse box or under the far left-hand side of the instrument panel for 1989 and newer pre-OBD II Mitsubishi and Chrysler vehicles.

4. Start the engine and drive the vehicle until the transaxle goes into the failsafe mode.

5. Park the vehicle, but do not turn the ignition **OFF**. Allow it to idle.

6. Attach a voltmeter (analog or digital) to the test terminals on the Diagnostic Link Connector (DLC). The negative lead should be attached to terminal 6 and the positive lead to terminal 12.

7. Observe the voltmeter and count the flashes (or arm sweeps if using an analog voltmeter); note the applicable codes.

8. After all of the DTC(s) have been retrieved, fix the applicable problems, clear the codes, drive the vehicle, and perform the retrieval procedure again to ensure that all of the codes are gone.

Clearing Codes

TYPE 1 CODES

It is not necessary to clear Type 1 codes, because they are stored in volatile memory and automatically erase whenever the ignition key is turned **OFF**.

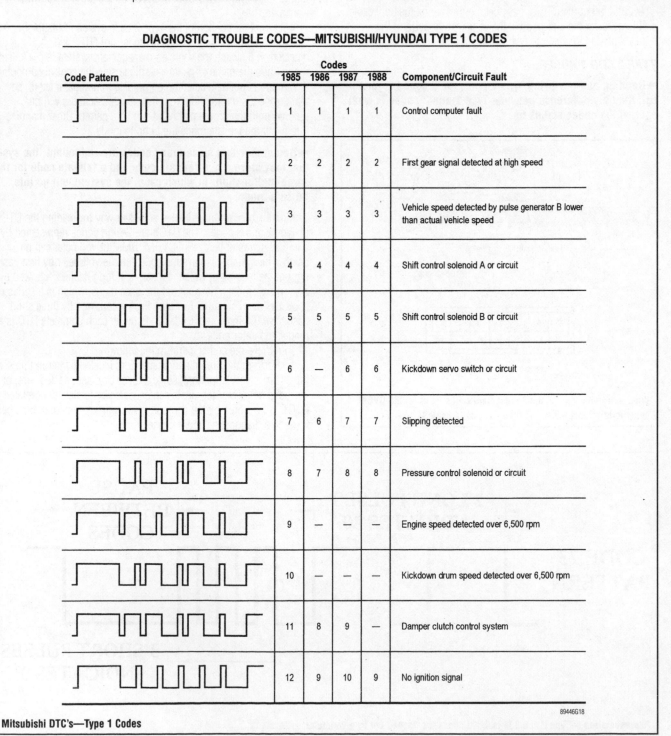

DIAGNOSTIC TROUBLE CODES—MITSUBISHI/HYUNDAI TYPE 1 CODES

Code Pattern	1985	1986	1987	1988	Component/Circuit Fault
	1	1	1	1	Control computer fault
	2	2	2	2	First gear signal detected at high speed
	3	3	3	3	Vehicle speed detected by pulse generator B lower than actual vehicle speed
	4	4	4	4	Shift control solenoid A or circuit
	5	5	5	5	Shift control solenoid B or circuit
	6	—	6	6	Kickdown servo switch or circuit
	7	6	7	7	Slipping detected
	8	7	8	8	Pressure control solenoid or circuit
	9	—	—	—	Engine speed detected over 6,500 rpm
	10	—	—	—	Kickdown drum speed detected over 6,500 rpm
	11	8	9	—	Damper clutch control system
	12	9	10	9	No ignition signal

Mitsubishi DTC's—Type 1 Codes

89446G18

Fault code	Fault code (for voltmeter)	Cause	Remedy
21	5V - - - - - - - - ⎍⎍⎍ 0V - - - - - ⎍⎍⎍	Abnormal increase of TPS output	o Check the throttle position sensor connector. o Check the throttle position sensor itself.
22	⎍⎍⎍⎍	Abnormal decrease of TPS output	o Adjust the throttle position sensor. o Check the accelerator switch (No.28: output or not).
23	⎍⎍⎍⎍	Incorrect adjustment of the throttle-position sensor system	o Check the throttle position sensor output circuit harness.
24	⎍⎍⎍⎍⎍	Damaged or disconnected wiring of the oil temperature sensor system	o Check the oil temperature sensor circuit harness. o Check the oil temperature sensor connector. o Check the oil temperature sensor itself.
25	⎍⎍⎍⎍⎍⎍	Damaged or disconnected wiring of the kickdown servo switch system, or improper contact	o Check the kickdown servo switch output circuit harness. o Check the kickdown servo switch connector.
26	⎍⎍⎍⎍⎍⎍⎍	Short circuit of the kickdown servo switch system	o Check the kickdown servo switch itself.
27	⎍⎍⎍⎍⎍⎍⎍⎍	Damaged or disconnected wiring of the ignition pulse pick-up cable system	o Check the ignition pulse signal line.
28	⎍⎍⎍⎍⎍⎍⎍⎍⎍	Short circuit of the accelerator switch system or improper adjustment	o Check the accelerator switch output circuit harness. o Check the accelerator switch connector. o Check the accelerator switch itself. o Adjust the accelerator switch.
31	⎍⎍⎍⎍	Malfunction of the microprocessor	o Replace the control unit.

89446G19

Mitsubishi DTC's, 1 of 4—Type 2 Codes

Fault code	Fault code (for voltmeter)	Cause	Remedy
32		First gear command during high-speed driving	o Replace the control unit
33		Damaged or disconnected wiring of the pulse generator B system	o Check the pulse generator B output circuit harness. o Check pulse generator B itself. o Check the vehicle speed reed switch (for chattering).
41		Damaged or disconnected wiring of the shift control solenoid valve A system	o Check the solenoid valve connector. o Check shift control solenoid valve A itself.
42		Short circuit of the shift-control solenoid valve A system	o Check the shift control solenoid valve A drive circuit harness
43		Damaged or disconnected wiring of the shift control solenoid valve B system	o Check the solenoid valve connector. o Check shift control solenoid valve B itself.
44		Short circuit of the shift control solenoid valve B system	o Check the shift control solenoid valve B drive circuit harness.
45		Damaged or disconnected wiring of the pressure control solenoid valve system	o Check the solenoid valve connector. o Check the pressure control solenoid valve itself.
46		Short circuit of the pressure control solenoid valve system	o Check the pressure control solenoid valve drive circuit harness.
47		Damaged or disconnected wiring of the damper clutch control solenoid valve system	o Check the solenoid valve connector. o Check the damper clutch control solenoid valve itself.
48		Short circuit of the damper clutch control solenoid valve system	o Check the damper clutch control solenoid valve drive circuit harness.

Mitsubishi DTC's, 2 of 4—Type 2 Codes

89446G20

FAIL-SAFE ITEM

Code No.	Output code — Output pattern (for voltmeter)	Description	Fail-safe	Note (relation to fault code)
11	5V / 0V pattern	Malfunction of the microprocessor	3rd gear hold	When code No. 31 is generated a 4th time.
12	pattern	First gear command during high speed driving	3rd gear (D) or 2nd gear (2, L) hold	When code No. 32 is generated a 4th time.
13	pattern	Damaged or disconnected wiring of the pulse generator B system	3rd gear (D) or 2nd gear (2, L) hold	When code No. 33 is generated a 4th time.
14	pattern	Damaged or disconnected wiring, or short circuit, of shift control solenoid valve A	3rd gear hold	When code No. 41 or 42 is generated a 4th time.
15	pattern	Damaged or disconnected wiring, or short circuit, of shift control solenoid valve B	3rd gear hold	When code No. 43 or 44 is generated a 4th time.
16	pattern	Damaged or disconnected wiring, or short circuit, of the pressure control solenoid valve	3rd gear (D) or 2nd gear (2, L) hold	When code No. 45 or 46 is generated a 4th time.
17	pattern	Shift steps non-synchronous	3rd gear (D) or 2nd gear (2, L) hold	When code No. 51, 52 53 or 54 is generated a 4th time.

89446G21

Fault code	Fault code (for voltmeter)	Cause	Remedy
49		Malfunction of the damper clutch system	o Check the damper clutch control solenoid valve drive circuit harness. o Check the damper clutch hydraulic pressure system. o Check the damper clutch control solenoid valve itself. o Replace the control unit.
51		First gear non-synchronous	o Check the pulse generator output circuit harness. o Check the pulse generator connector. o Check pulse generator A and pulse generator B themselves. o Kickdown brake slippage.
52		Second gear non-synchronous	o Check the pulse generator A output circuit harness. o Check the pulse generator A connector. o Check pulse generator A itself. o Kickdown brake slippage.
53		Third gear non-synchronous	o Check the pulse generator A output circuit harness. o Check the pulse generator connector. o Check pulse generator A and pulse generator B themselves. o Front clutch slippage. o Rear clutch slippage.
54		Fourth gear non-synchronous	o Check the pulse generator A output circuit harness. o Check the pulse generator A connector. o Check pulse generator A itself. o Kickdown brake slippage.

89446G22

code	Diagnostic trouble code (for voltmeter)	Cause	Remedy
11		o Throttle position sensor malfunction o Open throttle position sensor circuit o Shorted throttle position sensor circuit	o Check the throttle position sensor connector. o Check the throttle position sensor itself. o Check the idle switch o Check the throttle position sensor wiring harness o Check the wiring between ECM and throttle position sensor
12			
13			
14			
15		Oil temperature sensor damaged or disconnected wiring	o Oil temperature sensor connector inspection o Oil temperature sensor inspection o Oil temperature sensor wiring harness inspection
21		Open kickdown servo switch circuit.	o Check the kickdown servo switch connector. o Check the kickdown servo switch. o Check the kickdown serro switch wiring harness
22		Shorted kickdown servo switch circuit.	
23		Open ignition pulse pickup cable circuit	o Check the ignition pulse signal line. o Check the wiring between ECM and ignition system
24		Open-circuited or improperly adjusted idle switch	o Check the idle switch connector o Check the idle switch itself o Adjust the idle switch. o Check the idle switch wiring harness

89446G23

code	Diagnostic trouble code (for voltmeter)	Cause	Remedy
31		Pulse generator A damaged or disconnectorted wiring	o Check the pulse generator A and pulse generator B.
32		Pluse generator B damaged or disconnected wiring	o Check the vehicle speed reed switch (for chattering)
			o Check the pulse generator A and B wiring harness
41		Open shift control solenoid valve A circuit	o Check the solenoid valve connector
42		Shorted shift control solenoid valve A circuit	o Check the shift control solenoid valve A.
			o Check the shift control solenoid valve A wiring harness
43		Open shift control solenoid valve B circuit	o Check the solenoid valve connector.
			o Check the shift control solenoid valve B wiring harness
44		Shorted shift control solenoid valve B circuit	o Check the solenoid valve connector.
45		Open pressure control solenoid valve circuit	o Check the pressure control solenoid valve.
46		Shorted pressure control solenoid valve circuit	o Check the pressure control solenoid valve wiring harness
47		Open circuit in damper clutch control control solenoid valve	o Inspection of solenoid valve connector.
		Short circuit in damper clutch control solenoid valve	o Individual inspection of damper clutch control solenoid valve
48		Defect in the damper clutch system	o Check the damper clutch control solenoid valve wiring harness
			o Check the TCM
49			o Inspection of damper clutch hydraulic system
51		Shifting to first gear does not match the engine speed.	o Check the pulse generator A and pulse generator B connector
			o Check the pulse generator A and pulse generator B
			o Check the one way clutch or rear clutch
			o Check the pulse generator wiring harness

89446G24

code	Diagnostic trouble code (for voltmeter)	Cause	Remedy
52		Shifting to second gear does not match the engine speed.	o Check the pulse generator A and pulse generator B connector o Check the pulse generator A and pulse generator B. o Check the one way clutch or rear clutch o Check the pulse generator wiring harness
53		Shifting to third gear does not match the engine speed	o Check the rear clutch or control system o Check the pulse generator A and pulse generator B connector o Check the pulse generator A and pulse generator B. o Check the front clutch slippage or control system o Check the rear clutch slippage or control system o Check the pulse generator wiring harness
54		Shifting to fourth gear does not match the engine speed.	o Check the pulse generator A and B connector o Check the pulse generator A and B o Kickdown brake slippage. o Check the end clutch or control system o Check the pulse generator wiring harness
-		Normal	-
		Defective transaxle control module (TCM)	o TCM power supply inspection o TCM earth inspection o TCM replacement

89446G25

FAIL-SAFE ITEM

Output code		Description	Fail-safe	Note (relation to diagnostic trouble code)
Code No.	Output pattern (for voltmeter)			
81		Open-circuited or damaged pulse generator A	Locked in third (D) or second (2,L)	When code No. 31 is generated fourth time
82		Open-circuited or damaged pulse generator B	Locked in third (D) or second (2, L)	When code No. 32 is generated fourth time
83		Open-circuited, shorted or damaged shift control solenoid valve A	Lock in third	When code No. 41 or 42 is generated fourth time
84		Open-circuited, shorted or damaged shift control solenoid valve B	Lock in third gear	When code No. 43 or 44 is generated fourth time
85		Open-circuited, shorted or damage pressure control solenoid valve	Locked in third (D) or second (2, L)	When code No. 45 or 46 is generated fourth time.
86		Gear shifting does not match the engine speed	Locked in third (D) or second (2, L)	When either code No. 51, 52, 53 or 54 is generated fourth time.
-	Constant output (or 0V)	Defective transaxle control module (TCM)	Fixed for third speed	-

89446G26

TYPE 2 AND 3 CODES

→It is not recommended that the negative battery cable be disconnected to clear DTC's. Doing so will clear all settings from the vehicle's computer system, resulting in lost radio presets, seat memories, anti-theft codes, driveability parameters, etc., and there are better ways of spending your afternoon than resetting all of these.

1. Turn the ignition switch **OFF**.
2. Disconnect the wiring harness connector from the control system computer.
3. Wait at least one minute, then reconnect the computer. The codes should now be cleared.

If there are still codes present, either the codes were not properly cleared (Are the codes identical to those flashed out previously?) or the underlying problem is still there (Are only some of the codes the same as previously?)

Nissan

ENGINE CODES

General Information

Nissan uses 2 types of fuel systems. Electronic Control Carburetor (ECC) system and Electronic Concentrated Control System (ECCS). The ECC system is a Feedback carburetor system. The ECCS is a fuel injected system which may be either throttle body injection or Multi-port injection. Both ECC and ECCS systems were available as of 1984.

Service Precautions

- Do not disconnect the injector harness connectors with the engine running.
- Do not apply battery power directly to the injectors.
- Do not disconnect the ECU harness connectors before the battery ground cable has been disconnected.
- Make sure all ECU connectors are fastened securely. A poor connection can cause an extremely high surge voltage in the coil and condenser and result in damage to integrated circuits.
- When testing the ECU with a DVOM make sure that the probes of the tester never touch each other as this will result in damage to a transistor in the ECU.
- Keep the ECCS harness at least 4 in. away from adjacent harnesses to prevent an ECCS system malfunction due to external electronic noise.
- Keep all parts and harnesses dry during service.
- Before attempting to remove any parts, turn OFF the ignition switch and disconnect the battery ground cable.
- Always use a 12 volt battery as a power source.
- Do not attempt to disconnect the battery cables with the engine running or the ignition key **ON**.
- Do not clean the air flow meter with any type of detergent.
- Do not attempt to disassemble the ECCS control unit under any circumstances.
- Avoid static electricity build-up by properly grounding yourself prior to handling any ECU or related parts.

Reading Codes

→Diagnostic codes may be retrieved by observing the code flashes through the LED lights located on the Electronic Con-

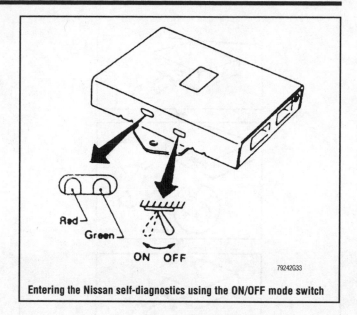

Entering the Nissan self-diagnostics using the ON/OFF mode switch

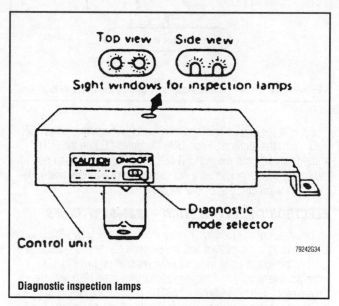

Diagnostic inspection lamps

trol Module (ECM). A special Nissan Consult monitor tool can be used, but is not required. When using special diagnostic equipment, always observe the tool manufacturer's instructions.

ELECTRONIC CONTROLLED CARBURETOR— 1987–88 VEHICLES

The 1 .6L (EI6S) carbureted engine utilizes a duty-controlled solenoid valve for fuel enrichment and an Idle Speed Control (ISC) actuator for basic controls instead of the conventional choke valve plate and fast idle cam. There are several other inputs which further affect the air/fuel ratio. The system is controlled in 2 ways: open or closed loop. To inspect the system for malfunctions, proceed as follows:

1. Position the ECU so the red and green LED's are visible.
2. Run the engine until it is at normal operating temperature.
3. Verify the diagnosis switch on the ECU is OFF.
4. Run the engine 2000 rpm for 5 minutes. After 5 minutes, observe the green LED light while maintaining 2000 rpm. The light should be blinking ON and OFF at least 5 times in 10 seconds. If not as specified, inspect the exhaust gas sensor.

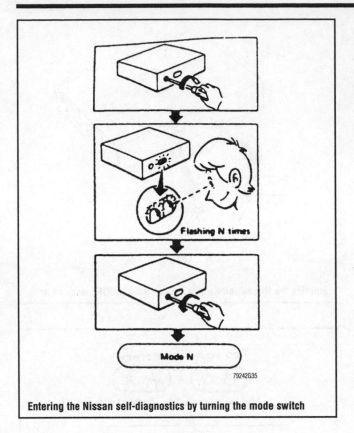

Entering the Nissan self-diagnostics by turning the mode switch

5. Turn the engine OFF and turn the ECU diagnosis switch ON.

6. Turn the ignition switch **ON**. The green LED on the ECU should stay ON and the red LED will either flash for a short period indicating a malfunctioning input sensor or for a longer time indicating a malfunctioning output sensor.

ELECTRONIC FUEL INJECTION—1984–94 VEHICLES

Two types of diagnostic systems are used in Nissan vehicles: the 2-mode diagnostic system and the 5-mode diagnostic system. The 2 mode system is used in some vehicles starting in 1990, ultimately, all vehicles used the 2-mode system after 1991 with the exception of 1991-94 Maxima (VG3OE engine), Pathfinder and Truck. These vehicles continued to use the 5-mode system. The 5-mode system began in 1984.

The 5-mode diagnostic system is incorporated in the ECU which uses inputs from various sensors to determine the correct air/fuel ratio. If any of the sensors malfunction the ECU will store the code in memory. The 5-mode diagnostic system is capable of various tests as outlined below. When using these modes, the ECM may have to be removed from its mounting bracket to better access the mode selector switch.

➡**Vehicles are equipped with a CHECK ENGINE light on the instrument panel. If any systems are malfunctioning, the light will illuminate the same time as the red lamp while the engine is running and the system is in Mode 1.**

Mode 1—Heated Oxygen Sensor:

During closed loop operation the green lamp turns ON when a lean condition is detected and turns OFF under a rich condition. During open loop the green lamp remains ON or OFF. This mode is used to check Heated Oxygen sensor functions for correct operation. To enter Mode 1, proceed as follows:

1. Turn the ignition switch ON.

2. Turn the diagnostic switch located on the side of the ECU ON by either flipping the switch to the ON position or turning the screw switch fully clockwise.

3. Turn the diagnostic switch OFF or fully counterclockwise as soon as the inspection lamps flash 1 time.

4. The self-diagnostic system is now in Mode 1.

Mode 2—Mixture Ratio Feedback Control Monitor:

The green inspection lamp is operating in the same manner as in Mode 1. During closed loop operation the red inspection lamp turns ON and OFF simultaneously with the green lamp when the mixture ratio is controlled within the specified value. During open loop the red lamp remains ON or OFF. Mode 2 is used for checking that optimum control of the fuel mixture is obtained. To enter Mode 2, proceed as follows:

1. Turn the ignition switch ON.

2. Turn the diagnostic switch ON, by either flipping the switch to the ON position or use a screwdriver and turn the switch fully clockwise.

3. Turn the diagnostic switch OFF or fully counterclockwise as soon as the inspection lamps flash 2 times.

4. The self-diagnostic system is now in Mode 2.

Mode 3—Self-Diagnosis System:

This mode of the self-diagnostics is for stored code retrieval. To enter Mode 3, proceed as follows:

1. Thoroughly warm the engine before proceeding. With the engine OFF, turn the ignition switch ON.

2. Turn the diagnostic switch located on the side of the ECU ON by either flipping the switch to the ON position or using a screwdriver, turn the switch fully clockwise.

2. Turn the diagnostic switch OFF or fully counterclockwise as soon as the inspection lamps flash 3 times.

3. The self-diagnostic system is now in Mode 3.

➡**When the battery is disconnected or self-diagnostic Mode 4 is selected after using Mode 3, all stored codes will be cleared. However if the ignition key is turned OFF and then the procedure is followed to enter Mode 4 directly, the stored codes will not be cleared.**

4. The codes will now be displayed by the red and green inspection lamps flashing. The red lamp will flash first and the green lamp will follow. The red lamp is the tens and the green lamp is the units, that is, the red lamp flashes 1 time and the green lamp flashes 2 times, this would indicate a Code 12.

Mode 4—On/Off Switches:

This mode checks the operation of the Vehicle Speed Sensor (VSS), Closed Throttle Position (CTP) and starter switches. Entering this mode will also clear all stored codes in the ECU. To enter Mode 4, proceed as follows:

1. Turn the ignition switch **ON**.

2. Turn the diagnostic switch located on the side of the ECU ON by either flipping the switch to the ON position or turning the mode switch fully clockwise.

3. Turn the diagnostic switch OFF or fully counterclockwise as soon as the inspection lamps flash 4 times.

4. The self-diagnostic system is now in Mode 4.

5. Turn the ignition switch to the START position and verify the red inspection lamp illuminates. This verifies that the starter switch is working.

6. Depress the accelerator and verify the red inspection lamp goes OFF. This verifies that the CTP switch is working.

7. Raise and properly support the vehicle and verify the lamp goes ON when the vehicle speed is above 12 mph (20 km/h). This verifies that the VSS is working

8. Turn the ignition switch OFF.

Mode 5—Real Time Diagnostics:

In this mode the ECU is capable of detecting and alerting the technician the instant a malfunction in the crank angle sensor, air flow meter, ignition signal or the fuel pump occurs while operating/driving the vehicle. Items which are noted to be malfunctioning are not stored in the ECU's memory. To enter Mode 5, proceed as follows:

1. Turn the ignition switch **ON**.
2. Turn the diagnostic switch located on the side of the ECU ON by either flipping the switch to the ON position or by turning the switch fully clockwise.
3. Turn the diagnostic switch OFF or fully counterclockwise as soon as the inspection lamps flash 5 times.
4. The self-diagnostic system is now in Mode 5.
5. Ensure the inspection lamps are not flashing. If they are, count the number of flashes within a 3.2 second period:

- 1 Flash = Crank angle sensor
- 2 Flashes = Air flow meter
- 3 Flashes = Fuel pump
- 4 Flashes = Ignition signal

2-Mode Diagnostic System:

The 1992–94 300ZX, Stanza, 240ZX, Sentra/NX Coupe, Maxima (VE30DE engine), and the 1993–94 Altima and Quest uses a 2-mode diagnostic system incorporated in the ECU which uses inputs from various sensors to determine the correct air/fuel ratio. If any of the sensors malfunction the ECU will store the code in memory.

A Nissan Consult monitor, or equivalent may be used to retrieve these codes by simply connecting the monitor to the diagnostic connector located on the driver's side near the hood release. Turn the ignition switch to **ON** and press START, ENGINE and then SELF-DIAG RESULTS, the results will then be output to the monitor.

The conventional CHECK ENGINE or red LED ECU light may also be used for self-diagnostics. The conventional 2-Mode diagnostic system is broken into 2 separate modes each capable of 2 tests, an ignition switch **ON** or engine running test as outlined below:

Mode 1—Bulb Check:

In this mode the RED indicator light on the ECU and the CHECK ENGINE light should be ON. To enter this mode simply turn the ignition switch **ON** and observe the light.

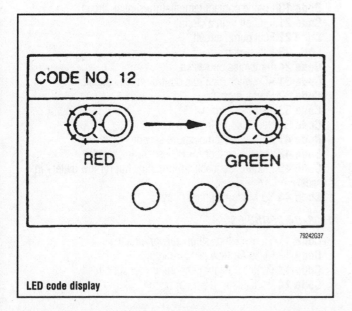

Nissan Rear time code 1—crank angle sensor

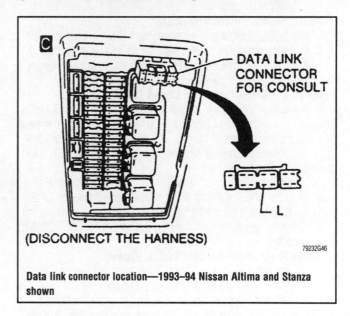

Data link connector location—1993–94 Nissan Altima and Stanza shown

LED code display

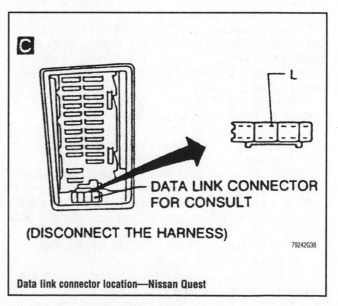

Data link connector location—Nissan Quest

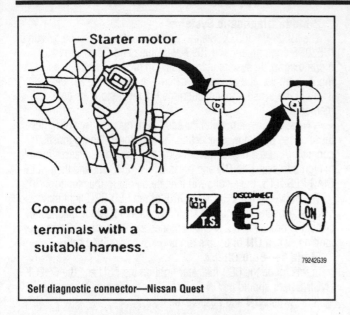

Connect ⓐ and ⓑ terminals with a suitable harness.

79242G39

Self diagnostic connector—Nissan Quest

Mode 1—Malfunction Warning:

In this mode the ECU is acknowledging if there is a malfunction by illuminating the RED indicator light on the ECU and the CHECK ENGINE light. If the light turns OFF, the system is normal. To enter this mode, simply start the engine and observe the light.

Mode 2—Self-Diagnostic Codes (Except Quest):

In this mode the ECU will output all malfunctions via the CHECK ENGINE light or the red LED on the ECU. The code may be retrieved by counting the number of flashes. The longer flashes indicate the first digit and the shorter flashes indicate the second digit. To enter this mode proceed as follows:

1. Turn the ignition switch **ON**, but do not start the vehicle.
2. Turn the ECU diagnosis mode selector fully clockwise for 2 seconds, then turn it back fully counterclockwise.
3. Observe the red LED on the ECU or CHECK ENGINE light for stored codes.

Mode 2—Self-Diagnostic Codes (Quest):

In this mode the ECU will output all malfunctions via the CHECK ENGINE light or the red LED on the ECU. The code may be retrieved by counting the number of flashes. The longer flashes indicate the first digit and the shorter flashes indicate the second digit. To enter this mode proceed as follows:

1. Turn the ignition switch **ON**, but do not start the vehicle.
2. Disconnect harness connectors and connect terminals A and B with a jumper wire.
3. Wait 2 seconds, remove the jumper wire and reconnect the harness connector.
4. Observe the CHECK ENGINE light for stored codes.

Mode 2—Exhaust Gas Sensor Monitor:

In this mode the red LED on the ECU or CHECK ENGINE light will display the condition of the fuel mixture and whether the system is in closed loop or open loop. When the light flashes ON, the exhaust gas sensor is indicating a lean mixture. When the light stays OFF, the sensor is indicating a rich mixture. If the light remains ON or OFF, it is indicating an open loop system. If the system is equipped with 2 exhaust gas sensors, the left side will operate first. If already in Mode 2, proceed to Step C to enter the exhaust gas sensor monitor.

1. On all models except Quest, perform the following steps:
   a. Turn the ignition switch **ON**.
   b. Turn the diagnostic switch ON, by turning the switch fully clockwise for 2 seconds and then fully counterclockwise.

c. Start the engine and run until thoroughly warm. Raise the idle to 2,000 rpm and hold for approximately 2 minutes. Ensure the red LED or CHECK ENGINE light flashes ON and OFF more than 5 times every 10 seconds with the engine speed at 2,000 rpm.

➡ **If equipped with 2 exhaust gas sensors, switch to the right sensor by turning the ECU mode selector fully clockwise for 2 seconds and then fully counterclockwise with the engine running.**

2. On Quest models, perform the following steps:
   a. Turn the ignition switch **ON**.
   b. Disconnect harness connectors and connect terminals A and B with a jumper wire.
   c. Wait 2 seconds, remove the jumper wire and reconnect the harness connectors.
   d. Start the engine and run until thoroughly warm. Raise the idle to 2,000 rpm and hold for approximately 2 minutes. Ensure the red LED or CHECK ENGINE light flashes ON and OFF more than 5 times every 10 seconds with the engine speed at 2,000 rpm.

Clearing Codes

ENGINE CODES—EXCEPT MODE 2, 3 AND 5 SYSTEMS

All control unit diagnostic codes may be cleared by disconnecting the negative battery cable for a period of 15 seconds. Entering Mode 4 of the Electronic Fuel Injection system diagnostics will also clear stored ECM engine codes.

ENGINE CODES—MODE 2, 3 AND 5 SYSTEMS

On 5-mode systems, enter mode 4 immediately after using mode 3 and the codes will be cleared. On 2-mode systems, the codes will be cleared when mode 1 is re-entered from mode 2. The Nissan Consult Monitor or equivalent can also be used to clear codes on 2-mode systems.

Diagnostic Trouble Codes

1984–87 VEHICLES

Code 11 Crankshaft position sensor circuit
Code 12 Mass Air flow sensor circuit
Code 13 Engine coolant temperature sensor circuit
Code 21 Ignition signal circuit
Code 22 Fuel pump circuit
Code 23 Idle switch circuit
Code 24 Transmission switch
Code 31 AC switch, fast idle control of load signal
Code 32 Starter signal
Code 33 EGR gas sensor
Code 34 Detonation (Knock) sensor
Code 41 Air or Fuel temperature sensor
Code 42 Throttle sensor (or BP sensor in Canada)
Code 43 Mixture feedback control slips out (or low battery in Canada)
Code 44 No Malfunctioning circuits

1988–94 VEHICLES

Code 11 Crankshaft position sensor circuit
Code 12 Mass Air flow sensor circuit
Code 13 Engine coolant temperature sensor circuit
Code 14 Vehicle speed sensor circuit

Code 15 Mixture ratio feedback control slips out (1988)
Code 21 Ignition signal circuit
Code 22 Fuel pump circuit (to 1991)
Code 23 Idle switch circuit (to 1991)
Code 24 Fuel Switch circuit or OD. switch circuit (to 1990)
Code 25 AAC valve circuit (to 1991)
Code 31 Electronic Control Module (ECM) or A/C circuit
Code 32 Exhaust Gas Recirculation (EGR) function
Code 33 Oxygen sensor circuit (left side, if two)
Code 34 Knock sensor circuit
Code 35 Exhaust gas temperature sensor circuit
Code 41 Air temperature sensor circuit
Code 42 Fuel temperature sensor circuit
Code 43 Throttle position sensor circuit

Code 44 No malfunctioning circuits
Code 45 Injector leak
Code 51 Injector circuit
Code 53 Heated oxygen sensor circuit (right side)
Code 54 Signal circuit from NT control unit to ECM
Code 55 No malfunctioning in the above circuits

TRANSMISSION CODES

Reading and Retrieving Codes

Nissan vehicles provide diagnostic trouble codes through one of four different lights, depending on the specific model at hand: Power Shift lamp, O/D Off lamp, A/T Check lamp (300ZX and J30

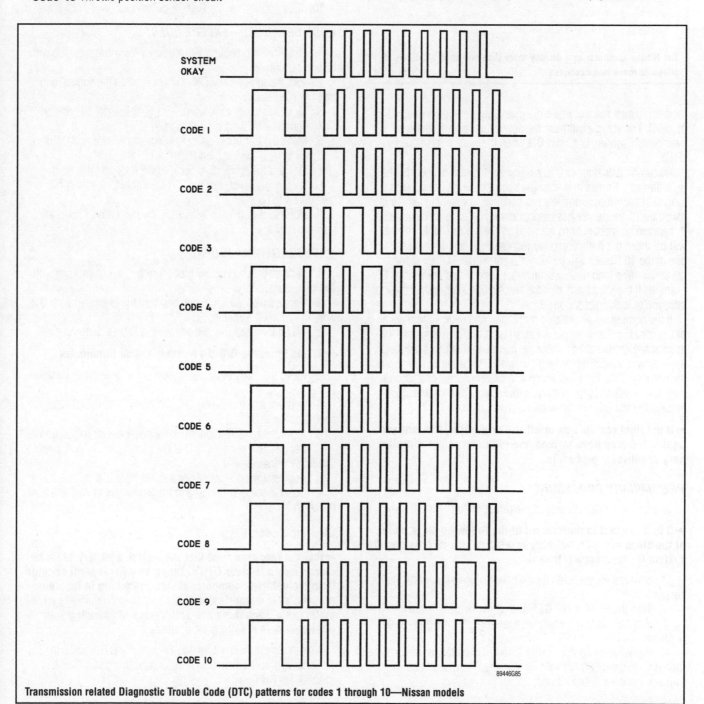

89446G85

Transmission related Diagnostic Trouble Code (DTC) patterns for codes 1 through 10—Nissan models

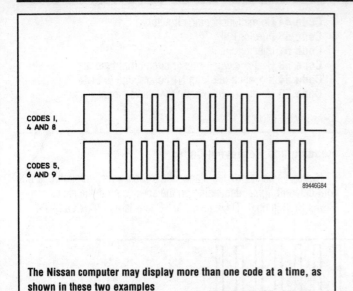

The Nissan computer may display more than one code at a time, as shown in these two examples

models), digital readout in the diagnostic information display (Q45 models). The vehicle will flash the applicable lamp 16 times whenever the ignition key is turned **ON** when the system has stored DTC's.

Nissan models display DTC's using an 11-flash sequence. The light flashes 11 times in a row. The sequence always starts with a long start flash (approximately two seconds in duration), and is followed by 10 shorter flashes (approximately ¼ second in duration). If there are no malfunctions detected by the system, all ten flashes will be short, but if the computer recognizes a fault in the system, one of the 10 flashes will be longer (approximately one second in duration)—the long flash identifies the code stored in memory. For example, if the third flash after the two second start flash is the long one, you're looking at a Code 3.

If the computer has detected more than one code, it displays all of the codes in the same pass. For an example, please refer to the accompanying illustration. After the computer displays the code(s), the light will remain off for approximately 2½ seconds, then repeats the code(s). If the light flashes on and off, in regular, one-second intervals, it indicates the battery is low or was disconnected long enough to interfere with computer memory.

➡**If the light remains on or off, try performing the sequence again. You may have missed one of the steps in the preliminary or retrieval procedure.**

PRELIMINARY PROCEDURE

To prepare the vehicle for code retrieval, perform the following:

➡**It is important to perform all of the following steps; some of the steps are a preliminary inspection, which prepares the system for fault code retrieval.**

1. Start the engine and allow it to reach normal operating temperature.
2. Turn the ignition key **OFF** and apply the parking brake.
3. Without starting the engine, turn the ignition key to the **ON** position.
4. For vehicles equipped with an O/D Off button, activate the switch so that the O/D Off light illuminates. Then, deactivate the switch so that the light turns off.

5. For vehicles equipped with a Mode switch, activate the switch so that the Power Shift lamp illuminates. Then, deactivate the switch so that the light turns off.
6. Turn the ignition key **OFF**, and wait a few seconds.
7. Once again turn the ignition switch **ON**, without starting the engine.

The indicator light on the instrument panel should illuminate for a few seconds (lamp bulb check), then extinguish. This is to check the light circuit, to make sure it's working satisfactorily for code flashing. If the lamp does not illuminate, inspect the bulb and circuit for a malfunction.

8. Turn the ignition switch **OFF**.
9. Move the transmission/transaxle gear selector to the D position.
10. If equipped, turn the O/D Off switch off.

CODE RETREIVAL—EXCEPT QUEST

1. Perform the preliminary procedure before commencing with this procedure.
2. Turn the ignition key **ON**, without starting the engine, and wait for a few seconds.
3. Move the transmission/transaxle gear selector to position 2.
4. Turn the O/D switch on (lamp off).
5. Move the transmission/transaxle gear selector to position 1.
6. Turn the O/D switch off (lamp on).
7. Depress the accelerator pedal to the floor and release it.
8. On models which use the O/D Off light for code display, turn the O/D switch on.
9. Observe the appropriate light in the instrument cluster and note the DTC's.

CODE RETREIVAL—QUEST

1. Perform the preliminary procedure before commencing with this procedure.
2. Depress the O/D Off button and turn the ignition switch **ON**, without starting the engine.
3. Wait a few seconds, and release the O/D Off button.

➡**At this time, the O/D Off light should be illuminated.**

4. Move the transmission/transaxle gear selector to position 2.
5. Depress and release the O/D Off switch (the O/D Off lamp should extinguish).
6. Move the transmission/transaxle gear selector to position 1.
7. Depress and release the O/D Off switch (the O/D Off lamp should illuminate again).
8. Depress the accelerator pedal to the floor and release it.
9. Observe the O/D Off lamp in the instrument cluster and note the DTC's.

Clearing Codes

➡**It is not recommended that the negative battery cable be disconnected to clear DTC's. Doing so will clear all settings from the vehicle's computer system, resulting in lost radio presets, seat memories, anti-theft codes, driveability parameters, etc., and there are better ways of spending your afternoon than resetting all of these.**

Nissan vehicles will automatically clear any DTC's when the corresponding problem has been repaired and the ignition key has been turned **ON** twice.

DIAGNOSTIC TROUBLE CODES—NISSAN, EXCEPT J30 AND Q45

Code	Component/Circuit Fault
1	Vehicle Speed Sensor (VSS) or circuit
2	Speedometer VSS or circuit
3	Throttle Position (TP) sensor or circuit
4	Shift solenoid A or circuit
5	Shift solenoid B or circuit
6	Timing solenoid or circuit, or over-run clutch solenoid or circuit
7	Lock-up solenoid or circuit
8	Transaxle fluid temperature sensor or circuit, or computer power insufficient
9	Engine speed signal circuit
10	Line pressure solenoid or circuit
Regular Flashing	Battery voltage low, or power was disconnected from the computer

89446G12

Nissan transmission related Diagnostic Trouble Codes—except J30 and Q45 models

DIAGNOSTIC TROUBLE CODES—NISSAN J30 AND Q45

Codes		
J30	**Q45**	**Component/Circuit Fault**
1	1	Vehicle Speed Sensor (VSS) or circuit
2	2	Speedometer VSS or circuit
3	3	Throttle Position (TP) sensor or circuit
4	4	Shift solenoid A or circuit
5	5	Shift solenoid B or circuit
6	6	Timing solenoid or circuit, or over-run clutch solenoid or circuit
7	7	Lock-up solenoid or circuit
8	8	Transaxle fluid temperature sensor or circuit, or computer power insufficient
9	9	Engine speed signal circuit
10	A	Turbine shaft speed sensor or circuit
11	B	Line pressure solenoid or circuit
12	C	Engine/Transmission communication circuit
Regular Flashing	D	Battery voltage low, or power was disconnected from the computer

89446G13

Nissan transmission related Diagnostic Trouble Codes—J30 and Q45 models

Porsche

ENGINE CODES

General Information

EXCEPT 911 TURBO

Porsche vehicle use 2 forms of electronic fuel injection. The 928 uses LH-Jetronic fuel injection. The 911, 944 and the 968 use Digital Motor Electronics (DME) fuel injection. Both systems are advanced versions of their fuel injection systems and provide excellent control of emissions, fuel economy and performance.

911 TURBO

The Porsche 911 Turbo has always been a specialized vehicle. The latest version is a hybrid of the 911 Carrera 2 body and the proven 3.3L turbocharged engine. While the 3.6L engine of the Carrera 2 uses Digital Motor Electronics as the fuel injection system, the 911 Turbo has used since its introduction the traditional Continuous Injection System (CIS).

Service Precautions

• Do not disconnect the battery or power to the control unit before reading the fault codes. Fault code memory is erased when power is interrupted.

• Make sure the ignition switch is **OFF** before disconnecting any wiring.

• Before removing or installing a control unit, disconnect the negative battery cable. The unit receives power through the main connector at all times and will be permanently damaged if improperly powered up or down.

• Keep all parts and harnesses dry during service. Protect the control unit and all solid-state components from rough handling or extremes of temperature.

• All air bag system wiring is in a yellow harness. Use extreme care when working around this wiring. Do not test these circuits without first disconnecting the air bag units.

Reading Codes

Vehicles prior to 1991 with self diagnostic capability require the use of special tool 9288 Tester or 9268 flash tester to retrieve diagnostics codes. On 1991 and later models, the flash codes can be read on the CHECK ENGINE light on the instrument panel. When using special diagnostic equipment, always observe the tool manufacturer's instructions.

➡ **1987 vehicles with LH and EZK systems did not have OBD capability.**

1987–93 VEHICLES—USING A 9288 DIAGNOSTIC TESTER

When the tester is attached to the diagnostic connector, the control units will present country and application codes to the tester for display on the screen. The tester will then provide a menu with instructions for retrieving fault codes from the control unit memory. If the tester display shows fault not present, this indicates the fault is intermittent or the conditions under which the fault occurs do not exist at this time. The necessary conditions will be displayed on the screen. If the display shows signal not plausible, the input or output signal does exist but is out of the correct operating range.

1987–93 VEHICLES—NOT USING A 9288 DIAGNOSTIC TESTER

The control unit is equipped with a self diagnostic program that will detect emissions related malfunctions and turn the CHECK ENGINE light ON while the engine is running. Emissions related fault codes stored in the control unit can be read with the 9268 flash tester. On 1991 and later models, the flash codes can be read on the CHECK ENGINE light on the instrument panel. Only faults that may effect exhaust emissions are reported as flash codes. All other codes are only accessible with the 9288 Diagnostic Tester.

1. If required, connect the flash tester to the diagnostic connector using the adapter connector.
2. Turn the ignition switch **ON** without starting the engine.
3. Fully press the accelerator pedal to close the full load switch. After about 3 seconds the CHECK ENGINE light or tester will flash.
4. When the pedal is released, flash codes will be reported. All codes are 4 digits. Each digit will be flashed with about 2.5 seconds between digits. When the whole code has been displayed, the light will stay ON or OFF. Count the flashes and write the numbers down. If Code 1500 or 2500 is displayed, no codes are stored in memory.
5. Repeat Steps 3 and 4 until Code 1000 appears, indicating all codes have been reported. On 928 models, the EZK unit will display Code 2000 when all codes have been reported.

➡ **On all 928 models, if the first digit is 2, the fault is in the EZK ignition system control unit.**

If the second digit is 1, the detected fault is current. If the second digit is 2, the detected fault has not occurred during the last running of the vehicle but did occur within the last 50 engine starts. The remaining 2 digits indicate which component or circuit is at fault.

Clearing Codes

1987–93 VEHICLES

The fault code memory should be cleared before returning the vehicle to service. All Codes in memory and the idle control adaptation are lost when the control unit or the battery is disconnected. To avoid the loss of the learned idle program, use the instructions on the 9288 tester to clear the memory. If this tester is not available, disconnect the battery or control unit to clear the memory. It will be necessary to drive the vehicle for at least 6 minutes and run the engine at idle for about 10 minutes so the control unit can learn idle speed, timing and mixture parameters. Make sure the engine is at operating temperature and that all accessories are OFF. The throttle-at-idle position switch must be closed and functioning or system adaptation will not take place.

Diagnostic Trouble Codes

1223 Coolant temperature sensor
1224 Air temperature sensor
1231 Battery voltage
1232 Throttle idle switch
1233 Throttle full load switch
1251 Fuel injector group 1, even numbers
1252 Fuel injector group 2, odd numbers
1261 Fuel pump relay
1262 Idle speed control actuator
1263 Carbon canister purge valve
1264 Oxygen sensor heater relay
1221 Control unit self test
1215 Airflow sensor
1221 Oxygen sensor
1222 Oxygen regulation

Saab

ENGINE CODES

General Information

The LH-Jetronic fuel injection system was introduced in the Saab 900 in 1985. The 9000 picked it up in 1986. The 1990 models and in some markets 1991–94 models are equipped with an LH 2.4 fuel system Electronic Control Unit (ECU). Most 1991–94 models, are equipped with an LH 2.4.2 fuel system ECU, except the 1991 9000 with the B234 engine, which has an LH 2.4.1 fuel system ECU.

Turbocharged 900 models are also equipped with an Automatic Performance Control (APC) ECU and 9000 models are equipped with an integrated Direct Ignition-Automatic Performance Control (DI/APC) control unit, which control ignition functions and turbocharger-related functions.

The LH-system ECU has the particular fuel system identification marked on it for identification. Visually, the LH 2.4 fuel system can be differentiated from the LH 2.4.2 fuel system by the pins on the Automatic Idle Control (AIC) valve. The LH 2.4 fuel system AIC valve has 2 pins and the LH 2.4.2 fuel system AIC valve has 3 pins.

The LH 2.4.1 fuel system differs from the LH 2.4 fuel system in that the cold start injector has been discontinued. The direct ignition system has taken over this function. Also, the vehicle speed sensor is used in the control of the AIC function to tell the fuel system ECU whether the car is moving or at a standstill.

The LH 2.4 (some markets) and LH 2.4.2 fuel systems with the Electronic Throttle System (ETS), used on the 1992–94 9000 with traction control, differ from the systems without ETS in the following ways: The electronically controlled throttle eliminates the need for automatic idling and load control throughout the load range, the throttle angle transmitter is located in the actuator motor, which is integrated with the throttle housing and the electronically controlled

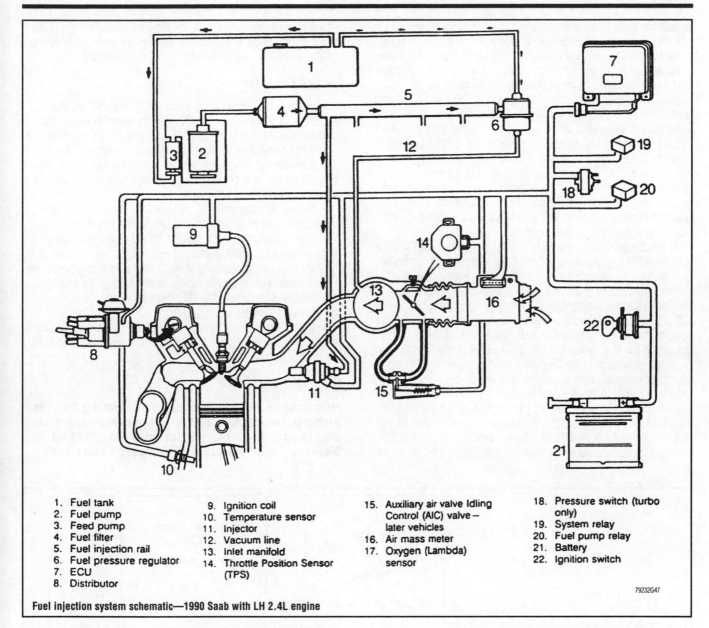

1. Fuel tank
2. Fuel pump
3. Feed pump
4. Fuel filter
5. Fuel injection rail
6. Fuel pressure regulator
7. ECU
8. Distributor

9. Ignition coil
10. Temperature sensor
11. Injector
12. Vacuum line
13. Inlet manifold
14. Throttle Position Sensor (TPS)

15. Auxiliary air valve Idling Control (AIC) valve — later vehicles
16. Air mass meter
17. Oxygen (Lambda) sensor

18. Pressure switch (turbo only)
19. System relay
20. Fuel pump relay
21. Battery
22. Ignition switch

79232G47

Fuel injection system schematic—1990 Saab with LH 2.4L engine

throttle system carries out compensation for the air conditioning. Vehicles with ETS also have an Automatic Slip Reduction (ASR) control unit that carries out other traction control system functions.

Visually, the throttle housing on engines with ETS is larger, is vacuum operated and has an emergency cable, but in other respects this system operates in the same way as the LH 2.4 and 2.4.2 systems without ETS.

The central component of the LH-Jetronic fuel injection system is the air mass meter that measures the mass of air flow instead of the volume. The microprocessor measures how much electrical energy is used when air flow passes an electrically heated platinum wire in the air mass meter. The higher the rate of air flow, the higher the energy necessary to keep the temperature of the wire constant. At the same time, the microprocessor monitors the engine speed and temperature, calculating the exact amount of fuel needed for optimum performance. The microprocessor also incorporates an rpm limiter that ensures that no opening signals will be transmitted to the injectors at engine speeds above 6000 rpm.

The LH-Jetronic fuel injection system provides the air mass meter with a self-cleaning function. During burn-off the platinum

wire in the air mass meter is quickly heated to about 1800°F (1000°C) for a 1 second duration, 4 seconds after the ignition is switched **OFF**. This burns away any deposits on the wire that would be detrimental to efficient operation.

The APC system on turbocharged vehicles enables the engine to achieve optimum performance and good fuel economy, regardless of the grade of fuel being used. A knock sensor, in conjunction with the pressure transducer and ignition system information, detects knocking in the engine and sends an electrical signal to the microprocessor inside the ECU. The ECU processes these signals and sends electrical pulses to a solenoid valve that controls the charging pressure in the intake manifold. The turbocharger is designed to come into operation at fairly low engine speeds, thereby providing a high torque within the speed range of normal driving. It is water cooled and the coolant for the bearing housing is supplied by a pipe connected to the cooling system.

Charging pressure is regulated by a pressure regulator valve (known as a wastegate). The charging pressure regulator is fitted to the exhaust side of the engine and regulates the flow of exhaust gas to the compressor. The valve remains closed when the engine load is low. As the demand on the engine is increased, the wastegate opens.

The DI/APC system was updated in 1991 to include an air temperature sensor, located upstream of the throttle housing. The boost pressure is governed by the position of the throttle valve, but it is subject to temperature compensation based on information supplied by the new air temperature sensor. The separate pressure-switch function is discontinued, with the pressure-sensing function now being regulated by the DI/APC system ECU. The load signal provided by the air mass meter is sent to the LH-system ECU, which wil! keep the DI/APC system ECU informed of boost status.

Starting in 1991 there is also a spark plug burn-off function that occurs when the engine is stopped. The burn-off function, which operates in all cylinders simultaneously, lasts for 5 seconds at a frequency corresponding to 6000 rpm.

Turbocharged California vehicles, except the 1991–94 9000 Turbo with the B234 engine, are equipped with an electronically controlled EGR system. A modulating valve functions as a 3-way valve as it controls the vacuum to the EGR valve. A vacuum regulator is incorporated in the modulating valve to maintain a constant vacuum to the EGR valve. A vacuum storage tank is connected via a vacuum check valve to the intake manifold. The check valve prevents the loss of vacuum in the tank during acceleration. An EGR temperature sensor provides information to the LH-system ECU. If the temperature deviates from a normal range, a problem is indicated due to improper exhaust gas flow in the EGR pipe and a fault code will be set.

The LH-Jetronic system also incorporates an emergency system known as a limp home function. If a malfunction is detected, the limp home feature of the ECU is actuated, enabling the vehicle to continue its journey, but with somewhat diminished performance. If the vehicle is operated in this mode, the Check Engine Light (CEL) on the display panel will be illuminated. An integrated fault-storing capability enables diagnosis to be carried out efficiently.

Service Precautions

- Before connecting or disconnecting ECU harness connectors, make sure the ignition switch is **OFF** and the negative battery cable is disconnected to avoid the possibility of damage to the control unit.
- When connecting or disconnecting pin connectors from the ECU, take care not to bend or break any pin terminals. Check that there are no bends or breaks on ECU pin terminals before attempting any connections.
- Before replacing any ECU, perform ECU input/output signal diagnosis to determine if the ECU is functioning properly.
- When measuring supply voltage of ECU controlled components with a circuit tester, separate 1 tester probe from another. If the 2 tester probes accidentally make contact with each other during measurement, a short circuit may result and damage the ECU.
- Always disarm the airbag (SRS) system when working on the airbag or ABS system.
- Always verify the ignition is switched **OFF** before connecting or disconnecting any electrical connections, especially connections to a control unit.

Reading Codes

➡**Diagnostic codes may be retrieved by observing the code flashes through the CHECK ENGINE light or Malfunction Indicator Lamp (MIL) only on vehicles listed under 'Without Diagnostic Tester' procedure. With this procedure a basic**

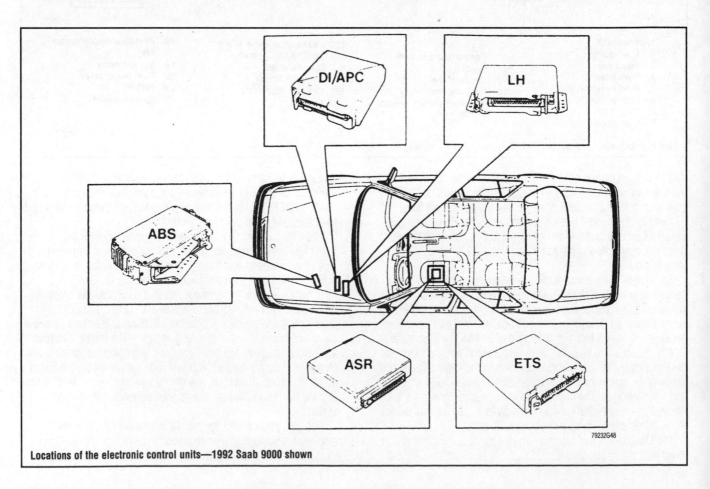

Locations of the electronic control units—1992 Saab 9000 shown

79232G48

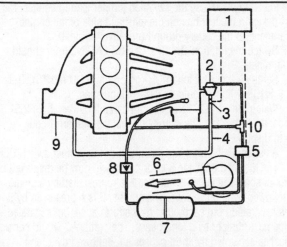

1. LH-system ECU
2. EGR valve
3. Thermostatic switch
4. EGR pipe
5. Modulating valve with vacuum regulator
6. Turbocharger delivery pipe
7. Vacuum tank
8. Check valve
9. Exhaust manifold
10. Overflow valve

79232G49

Electronically controlled EGR system—Saab California vehicles

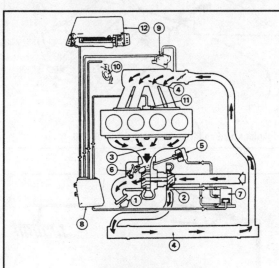

1. Turbine, pressure side
2. Turbine, delivery side
3. Exhaust manifold
4. Intake manifold
5. Wastegate valve diaphragm unit
6. Wastegate valve
7. Solenoid valve
8. DI/APC electronic control unit
9. Pressure sensor
10. Crankshaft sensor
11. Knock sensor
12. LH-system electronic control unit

79232G50

Automatic performance control system schematic—Saab Turbo shown

jumper switch (momentary type) is required to activate the computer. Other vehicles would require the use of special diagnostic tools: LH System tester or a ISAT Tester to retrieve codes. Read all procedures before attempting to perform checks. When using special diagnostic tools always observe the tool manufacturer's instructions.

1985–94 VEHICLES—WITH THE LH SYSTEM TESTER

The Saab LH system tester 8394223 has been developed to simplify service and fault diagnosis work on the LH fuel injection system. The tester consists of a test unit, power supply lead, test lead incorporating a 2-way 35-pin connector and a pressure sensor with magnetic base.

The tester is equipped with an automatic program for diagnosing faults, both permanent and intermittent, while the vehicle is operating. Faults detected are then stored in memo for recall after the vehicle is road-tested.

Connect the diagnostic tester, or equivalent, as follows:

1. Insert the power supply lead between the door and body where there is a break in the seal, then run it under the back of the hood on the left hand side.

2. Clean the battery terminals to ensure proper contact with lead clips.

3. Connect the power supply lead to the tester first, then connect the lead clips to the battery.

4. Remove the cover on the left side over the space behind the false bulkhead panel.

5. Remove the ABS system ECU and bracket, if equipped.

6. Remove the LH system ECU connector. Connect the test lead between the LH-system ECU and the vehicle's wiring loom. Fit a couple of ties around the connector and ECU to hold them tightly together.

The tester, designed to perform 3 basic functions, monitor mode, test mode and fuel mode, is now in start mode. If at any time the operation in progress must be interrupted, simultaneously press all 3 control buttons and the tester will revert to the starting point.

Monitor Mode:

The monitor mode can be selected either by switching the ignition **ON**, or if the ignition is **OFF**, by pressing the START TEST button when MON appears on the tester display.

When in monitor mode, the tester is used to manually control parameter and functional checks.

Test Mode:

Test mode can only be selected from the monitor mode. Once the LH system version has been selected, the test mode can be activated. Press the START TEST button to select the test mode. TEST will now appear on the display.

In the test mode, the program instructs by way of prompts in the upper part of the display.

Fuel Mode:

To select the fuel mode, the ignition must be **OFF**. Initially, MON will be displayed on the tester for approximately 5 seconds. During this time, if required, the monitor mode can be selected. If none of the tester buttons are activated, FUEL will appear on the display for approximately 2 seconds. To activate the fuel mode, press the START button when FUEL is displayed.

When the tester is in fuel mode the following checks can be performed:

- Fuel pump delivery flow
- Fuel pressure and fuel-pressure regulator
- Residual pressure
- Fuel pump delivery pressure Delivery flow from injectors

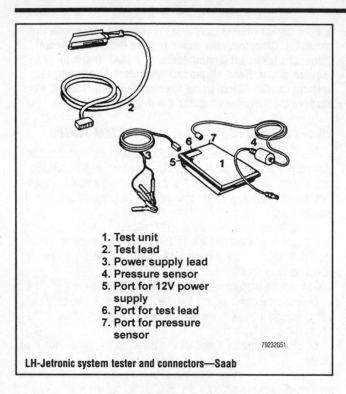

1. Test unit
2. Test lead
3. Power supply lead
4. Pressure sensor
5. Port for 12V power supply
6. Port for test lead
7. Port for pressure sensor

79232G51

LH-Jetronic system tester and connectors—Saab

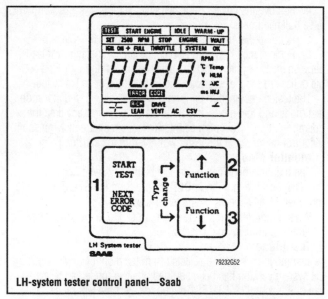

79232G52

LH-system tester control panel—Saab

1985–94 VEHICLES—WITHOUT DIAGNOSTIC TESTER

This method can be used to retrieve fault codes from 1988–94 Saab models equipped with LH 2.4, LH 2.4.1 and LH 2.4.2 fuel injection systems. These systems are capable of an internal self-diagnostic checks and have the ability to store up to 3 intermittent faults at a time. Serious malfunctions are always given priority and must be rectified before the memory can store information on minor faults. The built in diagnosis function also has the capability to manually test the components and signals of the LH system.

The Saab switched jumper lead 8393886, or equivalent, is necessary to conduct these tests.

Connect the switched jumper lead for as follows:

For the Saab 900, Use the switched jumper lead to connect the No. 3 pin in the 3-pin test socket, on the right-hand side in the engine compartment, to the battery ground (negative terminal).

For the Saab 9000, use the switched jumper lead to connect the 3-pin socket, in the test box on the left-hand side of the engine compartment, to the battery ground (negative terminal).

1. Switch the ignition **ON**. The CHECK ENGINE light should now illuminate.

2. Set the jumper switch to ON (grounding ECU pin 16). The CHECK ENGINE light should now be extinguished.

3. Watch the CHECK ENGINE light carefully. After about 2.5 seconds, it will flash briefly, signifying that the first error code is about to display.

4. As soon as the light has flashed, turn the jumper switch OFF.

5. The first of a possible 3 error codes will now be displayed by a series of short flashes. The number 1 is represented by a single flash followed by a long pause. The number 2 is represented by 2 flashes separated from each other by a short pause, but separated from the next number by a long pause. The number 3 would consist of 3 flashes separated by short pauses and followed by a long pause, and so on. For example Code 12112 would consist of: flash-long pause, flash-short pause, flash-long pause, flash-long pause,

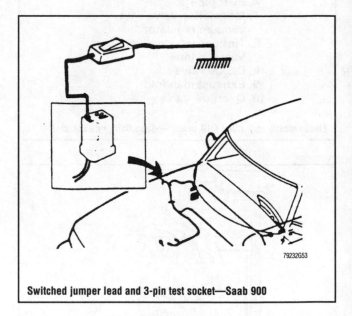

79232G53

Switched jumper lead and 3-pin test socket—Saab 900

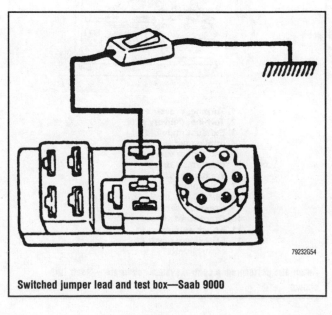

79232G54

Switched jumper lead and test box—Saab 9000

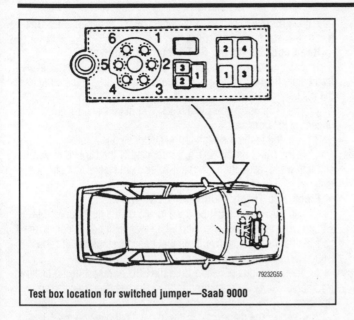

Test box location for switched jumper—Saab 9000

flash-long pause, flash-short pause, flash-long pause. The code will be displayed repeatedly until the next test step is taken.

6. To check for any additional error codes, turn the jumper switch ON.

7. Watch the CHECK ENGINE light carefully. After a short flash, turn the jumper switch OFF.

8. If present, the next error code will now display in the same fashion as the first.

9. If there are no more faults stored or all faults have been remedied, an uninterrupted series of flashes will be displayed.

10. Follow the same procedure until all faults have been identified and corrected.

11. To restart the test procedure (return to the first fault), set the jumper switch to ON.

12. After 2 short flashes, turn the jumper switch OFF. The fault code for the first fault should now be displayed.

13. Proceed with the test from Step 5.

TESTING COMPONENTS AND SIGNALS

1. Connect the jumper lead in the same manner as for reading fault codes.

2. Set the jumper switch to ON.

3. Turn the ignition switch **ON** and wait for a short flash of the CHECK ENGINE light.

4. Immediately following the flash, turn the jumper switch OFF.

5. The moment the CHECK ENGINE light begins flashing, the fuel pump should begin running for about 1 second (if it is not faulty). There will be no identification codes sent during this test.

6. To move on to the next test, set the jumper switch to ON.

7. After a short flash, set the jumper switch to OFF. A test code (NOT FAULT CODE) will be displayed and the corresponding component will activate.

8. Continue through the remaining items in the test sequence in the same method—set switch to ON, wait for a short flash, set the switch to OFF. Components and signals are checked in the following order:

Fuel pump (no code displayed).
Injection valves (1.5 ms-10 Hz).
AIC valve (switches between open and closed positions).
EVAP Canister Purge valve (switches between closed and open)—CHECK ENGINE flashing stops.

EGR valve operates—CHECK ENGINE flashing stops. Drive signal (changes when shifting from D to N)—CHECK ENGINE flashing stops. Air conditioning operates—CHECK ENGINE flashing stops.

Throttle position switch position (changes as accelerator is depressed)—CHECK ENGINE flashing stops. Throttle position switch WOT. position (changes as accelerator is pressed down to the floor)—CHECK ENGINE flashing stops.

Fuel pump operates—CHECK ENGINE flashing stops.

1990–94 VEHICLES—WITH SAAB ISAT TEST

The Saab SAT tester is also available to extract fault codes, both constant and intermittent, or to issue command codes.

Read codes with the ISAT tester, or equivalent, as follows:

1. Never unplug connector from the ECU or disconnect a battery lead before the fault data stored in the ECU has been transferred to the tester.

2. Connect the diagnostic tester to the diagnostic socket. The diagnosis socket is a black 10-pin connector located under the RH front seat.

3. Turn the ignition to the **ON** position.

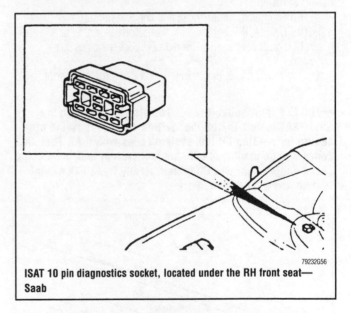

ISAT 10 pin diagnostics socket, located under the RH front seat— Saab

ISAT tester—Saab

4. Press the No. 1 on the SAT to identify that you are checking the LH system.

The Trionic system engine fault codes can be read in the same manner, but the Saab adapter # 8611188 must be used with SAT and current EPROM update.

EZK Ignition Codes:

The EZK ignition system is capable of self diagnostics only with the aid of the system tester 8394058, or equivalent. The test should be performed only in the event that a malfunction is suspected or when adjusting ignition timing.

Read fault codes as follows:

1. With the ignition switch **OFF**, connect the tester to the test box on the left-hand side of the engine compartment on the Saab 9000 or to the test socket located forward of the electrical distribution box on the Saab 900.

2. Turn the ignition switch **ON** and start engine. The fault indication LED (green) on the tester should illuminate for about 2 seconds while the starter motor is running.

3. Warm the engine to normal operating temperature, making sure that the engine is briefly run above 2300 rpm at some point during warm-up.

4. Run engine at idling speed and check tester LEDs for flashing. (CHECK ENGINE light will flash at a corresponding rate to any green LED fault indication.) The red LED light indicates spark knocking.

5. The fault code is determined by counting the number of LED flashes.

➡The EZK ignition fault codes can also be read with the Saab ISAT tester. To pull the ignition codes, follow the procedure for reading LH fuel system codes with ISAT. The EZK codes will be displayed with the fuel system fault codes. The check engine light will illuminate steady if the EZK or fuel system has a fault in memory.

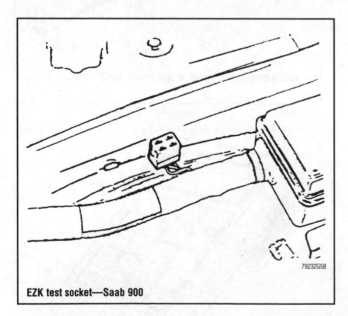

EZK test socket—Saab 900

DL/APC Ignition Codes:

The Saab combined Direct Ignition and Automatic Performance Control (DI/APC) system is controlled by the LH fuel injection ECU. The DI/APC system is an adaptive system which compensates for engine wear and other conditions which would adversely affect engine performance. With the aid of the Saab SAT tester, or equiva-

lent, it is possible to extract fault codes, both constant and intermittent or to issue command codes.

Read codes with the SAT tester as follows:

1. Never unplug connector from the ECU or disconnect a battery lead before the fault data stored in the ECU has been transferred to the ISAT.

2. The diagnosis socket is a black 10-pin connector located under the RH front seat.

3. Turn the ignition switch to the **ON** position.

4. Press No. 2 on the SAT tester to identify the DI/APC system and follow the instructions in the ISAT manual that accompany the tester to read fault codes or issue command codes.

Electronic Throttle Control (ETS) Codes:

A Saab ISAT tester must be used to access the fault codes. Make sure to use the correct code chart when diagnosing fault codes, the codes differ for automatic and manual transmissions. Read codes with SAT tester as follows:

1. Never unplug connector from the ECU or disconnect a battery lead before the fault data stored in the ECU has been transferred to the tester.

2. Connect the diagnostic tester to the diagnostic socket. The diagnosis socket is a black 10-pin diagnostic connector located under the RH front seat.

3. Turn the ignition to the **ON** position.

4. Press the No. 3 on the ISAT to identify that you are checking the ETS system.

Clearing Codes

1985–94 VEHICLES—WITH LH SYSTEM TESTER

After completing repairs, reset the tester to the start mode by pressing all 3 control buttons at the same time. If fault codes are still found, disconnect the battery for at least 10 minutes or road test vehicle for 10 minutes or until CHECK ENGINE light extinguishes. The fault code should erase itself after extended operation with the repaired component.

1985–94 VEHICLES—WITH SAAB ISAT TESTER

After repairs are completed, the SAT command code 900 is used to clear diagnostic memory.

1985–94 VEHICLES—WITHOUT SYSTEM TESTER

1. Set the jumper switch to ON.

2. After 3 short flashes, turn the jumper switch OFF.

3. The CHECK ENGINE light will now either flash in a continuous series of long flashes (this represents Code 00000) or display the Code 12444, indicating that the contents of the memory have been erased.

EZK IGNITION CODES

The EZK ignition does not store intermittent fault codes. All codes should clear when repair procedures are carried out.

DI/APC IGNITION CODES

After repairs are completed, the ISAT command Code 900 is used to clear diagnostic memory.

ELECTRONIC THROTTLE CONTROL (ETS) SYSTEM CODES

After repairs are completed, the SAT command Code 900 is used to clear diagnostic memory. A confirmation code '11111' will be displayed after the codes have been successfully erased. If this display is not received, repeat the code erasure procedure.

Diagnostic Trouble Codes

1988–94 FLASH CODES

Fault codes on 1988–94 vehicles using LH 2.4, 2.4.1 and LH 2.4.2 fuel injected systems may be read without the use of a diagnostic tester.

Code 00000 No more faults or faults not detected

Code 12111 Oxygen sensor adaptation fault; air/fuel mixture during idling

Code 12112 Oxygen sensor adaptation fault; air/fuel mixture with engine running

Code 12113 Idling control (IAC) adaptation fault; pulse ratio too low

Code 12114 Idling control (IAC) adaptation fault; pulse ratio too high

Code 12211 Incorrect battery voltage with engine running (below 10V or over 16V)

Code 12212 Throttle Position Sensor; faulty idling contacts (grounding when throttle open)

Code 12213 Throttle Position Sensor; faulty full-throttle contacts (grounding when engine idling)

Code 12214 Engine Coolant Temperature sensor signal; faulty (signal below -90 degrees or above 160 degrees Centigrade)

Code 12221 Mass Air Flow (MAF) sensor signal; missing (engine in limp-home mode)

Code 12222 Idling adjustment (IAC); faulty

Code 12223 Air/Fuel mixture; lean

Code 12224 Air/Fuel mixture; rich

Code 12225 Heated Oxygen sensor; faulty or preheating defective (engine temperature must be 80 degrees Centigrade

Code 12231 No ignition signal; (always occurs with the engine switched off)

Code 12232 Memory voltage greater than 1 V

Code 12233 Change made in EPROM (ROM fault 1992 and newer)

Code 12241 Fuel injector malfunction (1992 and newer)

Code 12242—Mass Air Flow (MAF) sensor; No filament burn-off (1992 and newer)

Code 12243 Vehicle Speed Sensor (VSS) signal; missing

Code 12244—No drive signal to pin 30 in ECM (automatic transmission, 1992 and newer)

Code 12245 EGR function faulty

Code 12251 Throttle Position (TP) sensor is faulty (1992 and newer)

Code 12252 EVAP canister purge valve not working (1992 and newer)

Code 12253 PRE-Ignition signal lasts more than 20 seconds (1992 and newer)

Code 12254 Engine RPM signal is missing (1992 and newer)

1985–94 LH-TESTER

Fault codes on 1985–94 vehicles using LH 2.2 and LH 2.4 fuel injection systems may be read by using an LH system tester.

Code E001 No ignition pulse

Code E002 No signal from Coolant Temperature Sensor (CTS) (LH 2.2) or Throttle Position Sensor (TPS); idling contacts not closing on idling (LH 2.4)

Code E003 Throttle Position Sensor (TPS); idling contacts not closing on idling (LH 2.2) or Throttle Position Sensor (TPS); full load contacts constantly open (LH 2.4)

Code E004 Battery voltage to Electronic Control Unit (ECU) memory; missing

Code E005 Electronic Control Unit (ECU) pin 5 not grounding

Code E006 Air Mass Meter (AMM) not grounding

Code E007 No signal from Air Mass Meter (AMM)

Code E008 Air Mass Meter (AMM); no filament burn-off function

Code E009 No power to system relay

Code E010 No signal from Electronic Control Unit (ECU) pin 10 to Automatic Idle Control (AIC) valve

Code E011 Electronic Control Unit (ECU) pin 11 not grounding

Code E012 Throttle Position sensor (TPS)—full throttle contacts constantly open

Code E013 No injection pulse (LH 2.2) or No signal from temperature sensor (LH 2.4)

Code E014 Air Mass Meter (AMM)—break in CO—adjusting circuit

Code E017 Fuel pump relay—control circuit faulty (LH 2.2) or break in ground circuit continuity (LH 2.4)

Code E018 No power at + 15 supply terminal

Code E020 Faulty signal from Oxygen sensor (LH 2.2) or Fuel pump relay; faulty control circuit (LH 2.4)

Code E021 System relay; faulty control circuit

Code E023 No signal from Automatic Idle Control (AIC) valve

Code E024 No load signal (LH 2.2) or Lambda sensor; faulty signal (LH 2.4)

Code E025 Electronic Control Unit (ECU) pin 25 not grounding

Code E033 Signal to Automatic Idle Control (AIC) valve from Electronic Control Unit (ECU); missing

Code E035 No power at +15 supply terminal

Code EI0I Starter motor revolutions too low

Code E102 Short in Coolant Temperature Sensor (CTS) circuit (LH 2.2) or Throttle Position Sensor (TPS); idling contacts not opening on increase from idling to 2500 rpm (LH 2.4)

Code E103 Throttle Position Sensor (TPS); idling contacts not opening on increase from idling to 2500 rpm (LH 2.2) or Throttle Position Sensor (TPS); full load contacts constantly closed (LH 2.4)

Code E107 Low signal from Air Mass Meter (AMM)

Code E108—Air Mass Meter (AMM); filament burn-off function constantly actuated

Code E109 Low voltage from system relay

Code E112 Throttle Position Sensor (TPS)—full load contacts constantly closed

Code E113 Erratic or No Injection Pulse

Code E120 Lambda sensor—signal too low

Code E207 High signal from Air Mass Meter (AMM)

Code E213 Continuous pulses to injectors

Code E218 Continuous pulses from injectors

Code E220 Lambda sensor—signal too high

Code E320 DI/APC system Electronic Control Unit (ECU)—pre-ignition signal constantly actuated

Code E328 Pre-ignition signal constantly grounded Code AICO—Automatic Idle Control (AIC) valve pulse ratio—faulty

Code GLOU (Glow) Air Mass Meter (AMM) filament burn-off function operating

Code C1 Turbo

Code C2 Turbo with AIC

Code C3 Turbo with AIC and catalytic converter

Code C4 Turbo with AIC and Saab DI

Code CS Non-Turbo with AIC

Code FPU Fuel pump relay and system relay operating

Code FUEL Fuel mode

Code OFF Starting point for injection valve test

Code FIn Injection valve open

Code MON Monitor mode

Subaru

ENGINE CODES

General Information

FEEDBACK CARBURETOR

The DFC328 feedback carburetor is a 2 barrel, downdraft type which consists of the following systems:

Float system—Provided with the fuel return system. Primary side—Which consists of a slow system, main system, accelerating pump system and a choke system.

Secondary side—Which consists of step system and main system.

The primary and secondary side use the same float system. Fuel in the fuel tank is routed through the fuel pump and the needle valve, into the float chamber. Fuel level in the float chamber is maintained constant by the function of the needle and float. Fuel level height is adjusted by adjusting the float seat.

The float system consists of a Float Chamber Ventilation (FCV) system. When the engine is started and the coolant temperature is above 680F (200C), the FCV solenoid valve turns on. This allows air from the air filter to flow through the float chamber to the air vent. This ventilates the float chamber.

The choke system consists of a choke valve linked to a bimetal through a choke lever, so that the choke valve is kept opened at a suitable angle relative to ambient temperature by means of the bimetal force. When the engine is started, the main vacuum diaphragm is operated by the vacuum sensed in the downstream portion of the secondary throttle valve, so that the choke valve is opened through a vacuum piston and a connecting rod. This allows an appropriate amount of air to be inducted, and the over-choke is prevented.

The auxiliary vacuum diaphragm is also operated by the vacuum, and allows the setting angle of the bi-metal force through a setting piston and a connecting rod. This operation allows the choke valve to be kept open at a moderate position to prevent an over-rich mixture. After the engine is started, the heater warms the bi-metal which adjusts the opening of the choke valve automatically.

A Coasting Fuel Cut (CFC) system is used to activate the anti-dieseling switch on the carburetor during deceleration. This closes the slow system passage for improved fuel economy. The control unit detects deceleration when the following conditions are met:

1. When intake manifold pressure is below -21.65 in. Hg, and the NC system is OFF; or -17.72 in. Hg and the air conditioning system is ON.
2. When the vehicle is operated at 25 mph.
3. Clutch pedal is released.
4. When engine reaches 2500 rpm.

When the control unit determines that the vehicle is decelerating, current to the anti-dieseling switch is interrupted. This closes the slow system passage so that the fuel flow is shut off. However, when coolant temperature is below 176°F (80°C), the fuel flow will not be shut off.

SINGLE POINT FUEL INJECTION

The SPFI is used on the Loyale 1.8L engine only. The system electronically controls the amount of injection from the fuel injector, and supplies the optimum air/fuel mixture under all operating conditions of the engine. Features of the SPFI system are as follows:

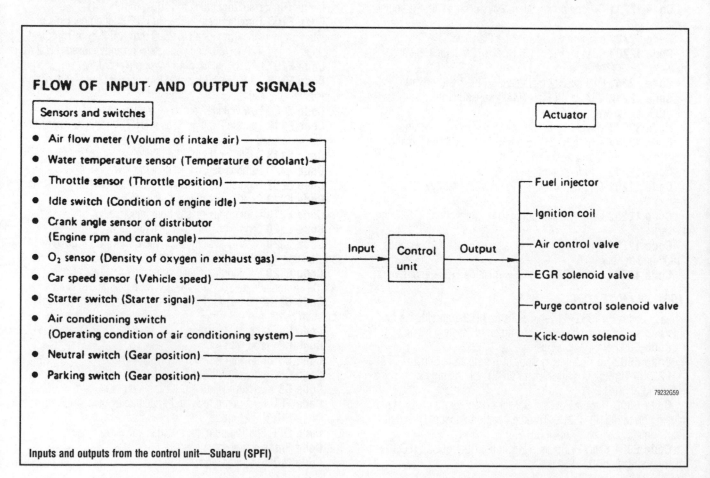

FLOW OF INPUT AND OUTPUT SIGNALS

Sensors and switches

- Air flow meter (Volume of intake air)
- Water temperature sensor (Temperature of coolant)
- Throttle sensor (Throttle position)
- Idle switch (Condition of engine idle)
- Crank angle sensor of distributor (Engine rpm and crank angle)
- O₂ sensor (Density of oxygen in exhaust gas)
- Car speed sensor (Vehicle speed)
- Starter switch (Starter signal)
- Air conditioning switch (Operating condition of air conditioning system)
- Neutral switch (Gear position)
- Parking switch (Gear position)

Input → Control unit → Output

Actuator

- Fuel injector
- Ignition coil
- Air control valve
- EGR solenoid valve
- Purge control solenoid valve
- Kick-down solenoid

79232G59

Inputs and outputs from the control unit—Subaru (SPFI)

1. Precise control of the air/fuel mixture is accomplished by an increased number of input signals transmitting engine operating conditions to the control unit.

2. The use of hot wire type air flow meter not only eliminates the need for high altitude compensation, but improves driving performance at high altitudes.

3. The air control valve automatically regulates the idle speed to the set value under all engine operating conditions.

4. Ignition timing is electrically controlled, thereby allowing the use of complicated spark advances characteristics.

5. Wear of the air flow meter and fuel injector is automatically corrected so that they maintain their original performance.

6. Troubleshooting can easily be accomplished by the built-in self-diagnosis function.

MULTI-POINT FUEL INJECTION

The MPFI system supplies the optimum air/fuel mixture to the engine under all various operating conditions.

System fuel, which is pressurized at a constant pressure, is injected into the intake air passage of the cylinder head. The amount of fuel injected is controlled by the intermittent injection system where the electro-magnetic injection valve (fuel injector) opens only for a short period of time, depending on the amount of fuel required for 1 cycle of operation. During system operation, the amount of injection is determined by the duration of an electric pulse sent to the fuel injector, which permits precise metering of the fuel.

Each of the operating conditions of the engine are converted into electric signals, resulting in additional features of the system, such as improved adaptability and easier addition of compensating element. The MPFI system also incorporates the following features:

- Reduced emission of exhaust gases
- Reduction in fuel consumption
- Increased engine output
- Superior acceleration and deceleration
- Superior starting and warm-up performance in cold weather since compensation is made for coolant and intake air temperature
- Good performance with turbocharger, if equipped.

Service Precautions

- Before connecting or disconnecting ECU harness connectors, make sure the ignition switch is **OFF** and the negative battery cable is disconnected to avoid the possibility of damage to the control unit.

- When connecting or disconnecting pin connectors from the ECU, take care not to bend or break any pin terminals. Check that there are no bends or breaks on ECU pin terminals before attempting any connections.

- Before replacing any ECU, perform ECU input/output signal diagnosis to determine if the ECU is functioning properly.

- When measuring supply voltage of ECU-controlled components with a circuit tester, separate 1 tester probe from another. If the 2 tester probes accidentally make contact with each other during measurement, a short circuit may result and damage the EC U.

Reading Codes

1987–92 FEEDBACK CARBURETOR VEHICLES

The self-diagnosis system has 4 modes: U-check mode, Read memory mode, D-check mode and Clear memory mode. Two connectors, Read memory and Test mode, are used. Also, the CHECK ENGINE light is utilized. Connectors are used in various combina-

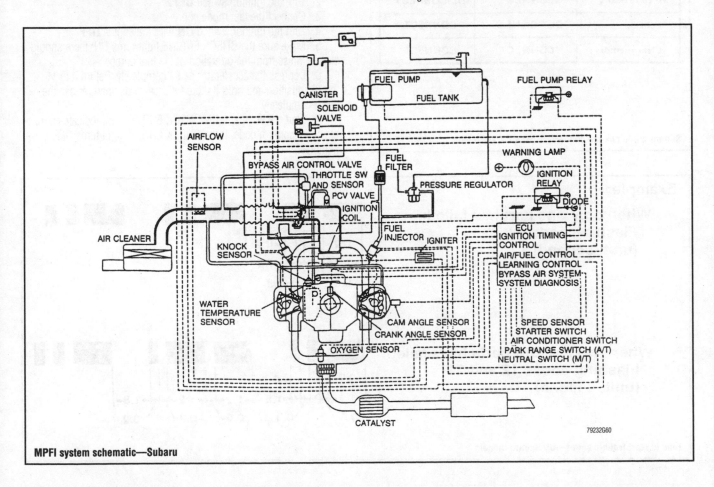

MPFI system schematic—Subaru

79232G60

tions to select the proper test mode and the lamps are used to read codes. No scan tool is necessary to extract codes.

➡**The engine should be running when in the D-check or clear memory modes.**

U-Check Mode:

The U-check is a user-oriented mode in which only the components necessary for start-up and drive are diagnosed. On occurrence of a fault the CHECK ENGINE light is turned ON to indicate that system inspection is necessary. The diagnosis of less significant components which do not adversely effect start-up and driving are excluded from this mode.

Read Memory Mode:

The Read memory mode is used to detect faults which recently occurred but are not currently present.

1. Turn the ignition switch **OFF**.
2. Connect the Read memory connector.
3. Turn the ignition switch **ON** with the engine **OFF**.
4. If the CHECK ENGINE light turns ON, trouble code(s) are present.

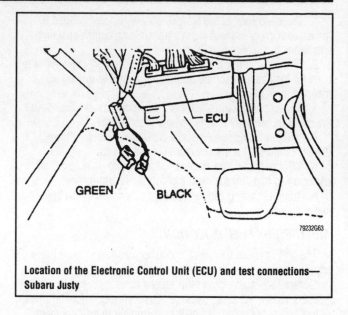

Location of the Electronic Control Unit (ECU) and test connections—Subaru Justy

5. If the oxygen monitor lamp turns ON, trouble code(s) are being produced; confirm the trouble code(s).
6. Disconnect the read memory connector.
7. Perform the D-check mode.

D-Check Mode:

The D-check mode is used to check the current status of the entire system.

1. Start the engine and warm it to normal operating temperatures.
2. Turn the ignition switch **OFF**.
3. Connect the test mode connector.
4. Turn the ignition switch **ON** with the engine **OFF**.
5. Make sure the CHECK ENGINE light turns ON; there should also be noise from the operation of the fuel pump.
6. Depress the accelerator pedal completely. Return it to ½ throttle position and hold it there for 2 seconds, then release the pedal completely.
7. Start the engine: If the CHECK ENGINE light indicates a trouble code, confirm code. If the CHECK ENGINE light turns OFF, continue test.

Mode	Read memory connector	Test mode connector
U-check	DISCONNECT	DISCONNECT
Read memory	CONNECT	DISCONNECT
D-check	DISCONNECT	CONNECT
(Clear memory)	CONNECT	CONNECT

Subaru mode change connectors for different modes

Example:

**When only one part has failed:
Flashing code 12
(unit: second)**

0.2
1.2
0.3
0.2
1.8
0.3

**When two or more parts have failed:
Flashing codes 12 and 21
(unit: second)**

0.2
1.2
0.3
0.2
1.8
0.3
1.2
0.3
1.2
0.3
1.8
0.2

How to read trouble codes—all Subaru models

8. Race the engine briefly with the throttle fully opened.

9. Drive the vehicle above 5 mph, at engine speeds above 1500 rpm, for at least 1 minute.

10. If the CHECK ENGINE light blinks, there are no trouble codes. If the CHECK ENGINE light stays ON, trouble codes are present and must be read.

1987–92 FUEL INJECTED VEHICLES

The self-diagnosis system has 4 modes: U-check mode, read memory mode, D-check mode and clear memory mode. Two connectors, Read memory and Test mode, are used. Also, the CHECK ENGINE light is utilized. Connectors are used in various combinations to select the proper test mode and the lamps are used to read codes. No scan tool is necessary to extract codes.

➡ **The engine should be running when in the D-check or clear memory modes.**

U-Check Mode:

The U-check is a user-oriented mode in which only the components necessary for start-up and drive are diagnosed. On occur-

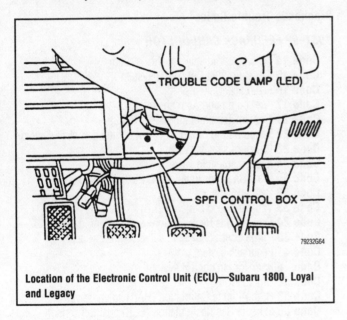

Location of the Electronic Control Unit (ECU)—Subaru 1800, Loyal and Legacy

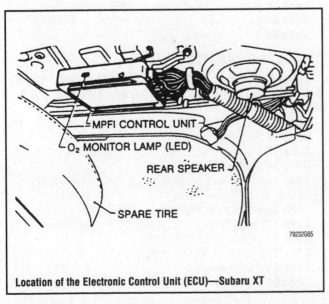

Location of the Electronic Control Unit (ECU)—Subaru XT

rence of a fault, the CHECK ENGINE light is turned ON to indicate that system inspection is necessary. The diagnosis of less significant components which do not adversely effect start-up and driving are excluded from this mode.

Read Memory Mode:

The Read memory mode is used to detect faults which recently occurred but are not currently present.

1. Turn the ignition switch **OFF**.

2. Connect the Read memory connector.

3. Turn the ignition switch **ON** with the engine **OFF**.

4. If the CHECK ENGINE light turns ON, trouble code(s) are present.

5. If the oxygen monitor lamp turns ON, trouble code(s) are being produced; confirm the trouble code(s).

6. Disconnect the read memory connector.

7. Perform the D-check mode.

D-Check Mode:

The D-check mode is used to check the current status of the entire system.

1. Start the engine and warm it to normal operating temperatures.

2. Turn the ignition switch **OFF**.

3. Connect the test mode connector.

4. Turn the ignition switch **ON** with the engine OFF.

5. Make sure the CHECK ENGINE light turns ON; there should also be noise from the operation of the fuel pump.

6. Depress the accelerator pedal completely. Return it to ½ throttle position and hold it there for 2 seconds, then release the pedal completely.

7. Start the engine: If the CHECK ENGINE light indicates a trouble code, confirm code. If the CHECK ENGINE light turns OFF, continue test.

8. Race the engine briefly with the throttle fully opened.

9. Drive the vehicle above 5 mph, at engine speeds above 1500 rpm, for at least 1 minute.

10. If the CHECK ENGINE light blinks, there are no trouble codes. If the CHECK ENGINE light stays ON, trouble codes are present and must be read.

Clearing Codes

1987–92 FEEDBACK CARBURETOR VEHICLES

1. Start the engine and warm it to normal operating temperatures.

2. Turn the ignition switch **OFF**.

3. Connect the test mode connector and the read memory connector.

4. Turn the ignition switch **ON** with the engine **OFF**.

5. Make sure the CHECK ENGINE light turns ON.

6. Depress the accelerator pedal completely. Return it to ½ throttle position and hold it there for 2 seconds, then release the pedal completely.

7. Start the engine; the CHECK ENGINE light should turn OFF.

8. Race the engine with the throttle fully opened for a second or two.

9. Drive the vehicle above 5 mph, at engine speeds above 1500 rpm, for at least 1 minute.

10. The CHECK ENGINE light should blink showing that there are no trouble codes. If the CHECK ENGINE light stays ON, read the trouble codes and re-perform the D-check mode.

1984–94 FUEL INJECTED VEHICLES

1. Start the engine and warm it to normal operating temperatures.

2. Turn the ignition switch **OFF**.

3. Connect the test mode connector and the read memory connector.

4. Turn the ignition switch **ON** with the engine **OFF**.

5. Make sure the CHECK ENGINE light turns ON.

6. Depress the accelerator pedal completely. Return it to ½ throttle position and hold it there for 2 seconds then release the pedal completely.

7. Start the engine; the CHECK ENGINE light should turn OFF.

8. Race the engine with the throttle fully opened for a second or two.

9. Drive the vehicle above 5 mph, at engine speeds above 1500 rpm, for at least 1 minute.

10. The CHECK ENGINE light should blink showing that there are no trouble codes. If the CHECK ENGINE light stays ON, read the trouble codes and re-perform the D-check mode.

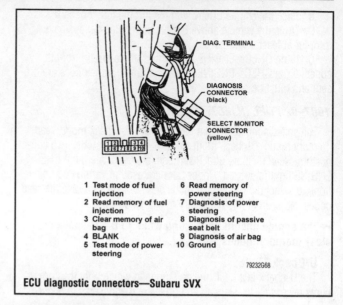

1 Test mode of fuel injection
2 Read memory of fuel injection
3 Clear memory of air bag
4 BLANK
5 Test mode of power steering
6 Read memory of power steering
7 Diagnosis of power steering
8 Diagnosis of passive seat belt
9 Diagnosis of air bag
10 Ground

ECU diagnostic connectors—Subaru SVX

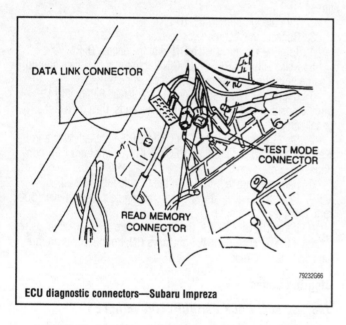

ECU diagnostic connectors—Subaru Impreza

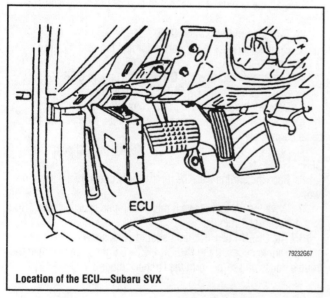

Location of the ECU—Subaru SVX

Diagnostic Trouble Codes

1987–92 FEEDBACK CARBURETOR

Code 14 Duty solenoid valve control system
Code 15 Coasting Fuel Cut (CFC) system
Code 16 Feedback system
Code 17 Fuel pump and Auto choke
Code 21 Coolant Temperature Sensor (CTS)
Code 22 Vacuum Line Charging (VLC) solenoid control system
Code 23 Pressure sensor system
Code 24 Idle-up solenoid valve
Code 25 Float Chamber Ventilation (FCV) solenoid valve
Code 32 Oxygen Sensor
Code 33 Vehicle Speed Sensor (VSS)
Code 34 Exhaust Gas Recirculation (EGR) solenoid valve
Code 35 Purge control solenoid valve control system
Code 41 Feedback system
Code 46 Radiator fan control system
Code 52 Clutch switch
Code 53 High Altitude Calibration (HAC) solenoid valve
Code 55 Exhaust Gas Recirculation (EGR) position sensor
Code 56 Exhaust Gas Recirculation (EGR) system
Code 62 Idle-up system 1
Code 63 Idle-up system 2

1984–94 FUEL INJECTED

1984–86 Vehicles With 1.6L (VIN 2 SPFI), 1.8L (VIN 4 & 5 SPFI), 18L (VIN 4, 5 & 7 MPFI) and 2.7L (VIN 8 & 9 MPFI) Engines

Code 11 No ignition pulse
Code 12 Starter switch; continuously in OFF position
Code 13 Starter switch; continuously in ON position
Code 14 Air flow meter
Code 15 Atmospheric pressure switch (1.8L VIN 4 & 5 SPFI 1986 only)
Code 16 Crank angle sensor (1.8L VIN 4 & 5 SPFI 1986 only)
Code 17 Starter switch (1.8L VIN 4 & 5 SPFI 1986 only)
Code 21 Air flow meter flap seized
Code 22 Pressure (Vacuum) switch
Code 23 Throttle sensor
Code 24 Wide Open Throttle (WOT) sensor

Code 25 Throttle sensor (1.8L VIN 4 & 5 SPFI 1986 only)
Code 31 Vehicle Speed Sensor (VSS)
Code 32 Oxygen sensor
Code 33 Coolant Temperature Sensor (CTS)
Code 34 Intake Air Thermosensor (IAT)
Code 35 EGR solenoid or -Air flow meter (1.8L VIN 4 & 5 SPFI 1986 only)
Code 41 Open or ground in sensor
Code 42 Fuel injector
Code 43 Kickdown Low Hold (KDLH) relay
Code 46 Neutral safety switch (1.8L VIN 4 & 5 SPFI 1986 only)
Code 53 Fuel pump (1.8L VIN 4 & 5 SPFI 1986 only)
Code 55 Kickdown Low Hold (KDLH) relay (1.8L VIN 4 & 5 SPFI 1986 only)
Code 57 Canister purge control (1.8L VIN 4 & 5 SPFI 1986 only)
Code 58 Air control valve (1.8L VIN 4 & 5 SPFI 1986 only)
Code 62 EGR control system (1.8L VIN 4 & 5 SPFI 1986 only)
Code 88 Faulty ECU (1.8L VIN 4 & 5 SPFI 1986 only)
1987–94 Vehicles With 1.2L (VIN 7 & 8 MPFI), 1.8L (VIN 4 & 5 SPFI), 1.8L (VIN 2, 4, 5 & 7 MPFI), 2.2L (VIN 6 MPFI), 2.7L (VIN 8 & 9 MPFI) and 3.3L (VIN 3) Engines:
Code 11 Crank angle sensor
Code 12 Starter switch
Code 13 Crank angle (Cam or Cylinder distinction) sensor
Code 14 Fuel injector (1.8L VIN 4 & 5 SPFI) or—Fuel injector No. 1 (1.2L VIN 7 & 8 MPFI), (1.8L VIN 2 MPFI), (2.2L VIN 6 MPFI) and (3.3L VIN 3 MPFI) or—Fuel injectors No. 1 and No. 2 (1.8L VIN 4 & 5 MPFI) or—Fuel injectors No. 5 and No. 6 (2.7L VIN 8 & 9 MPFI)
Code 15 Fuel injector No. 2 (1.2L VIN 7 & 8 MPFI), (1.8L VIN 2 MPFI), (2.2L VIN 6 MPFI) and (3.3L VIN 3 MPFI) or—Fuel injectors No. 3 and No. 4 (1.8L VIN 4 & 5 MPFI) or—Fuel injectors No. 1 and No. 2 (2.7L VIN 8 & 9 MPFI)
Code 16 Fuel injector No. 3
Code 17 Fuel injector No. 4
Code 18 Fuel injector No. 5
Code 19 Fuel injector No. 6
Code 21 Coolant Temperature Sensor (CTS)
Code 22 Knock sensor (1.8L VIN 2 MPFI), (2.2L VIN 6 MPFI), (2.7L VIN 8 & 9 MPFI)
Code 22 Knock sensor 1; right (3.3L VIN 3 MPFI)
Code 23 Air flow meter
Code 24 Air control valve
Code 25 Fuel injectors No. 3 and No. 4; abnormal injector output (2.7L VIN 8 & 9 MPFI)
Code 26 Air temperature sensor; abnormal signal (1.2L VIN 7 & 8 MPFI)
Code 28 Knock sensor 2; left (3.3L VIN 3 MPFI)
Code 29 Crank angle sensor 2 (3.3L VIN 3 MPFI)
Code 31 Throttle sensor
Code 32 Oxygen sensor 1; right (3.3L VIN 3 MPFI)
Code 33 Vehicle speed sensor 2 (1.8L VIN 2 MPFI), (3.3L VIN 3 MPFI)
Code 34 EGR solenoid valve or (California) clogged EGR line
Code 35 Purge control solenoid valve
Code 36 Air suction valve; faulty valve function or—Igniter; abnormal signal
Code 37 Oxygen sensor 2; left (3.3L VIN 3 MPFI)
Code 38 Engine torque control (3.3L VIN 3 MPFI)

Code 41 AF (Air/Fuel) learning control—or—System too lean (1.8 VIN 4 & 5 MPFI), (2.7L VIN 8 & 9 MPFI)
Code 42 Idle switch
Code 43 Power switch
Code 44 Duty solenoid valve (wastegate control); valve inoperative
Code 45 Kickdown control relay (1.8L VIN 4 & 5 SPFI)
Code 45 A: Atmospheric pressure sensor; faulty sensor (2.2L VIN 6 MPFI 1991 only)
Code 45 B: Pressure exchange solenoid valve; valve inoperative (2.2L VIN 6 MPFI 1991 only)
Code 45 Atmospheric pressure sensor (1.2L VIN 7 & 8 MPFI), (2.2L VIN 6 MPFI 1992 1993), (3.3L VIN 3 MPFI)
Code 49 Air flow sensor
Code 51 Neutral switch
Code 52 Clutch switch; signal remains ON or OFF (Front Wheel Drive/Manual Transaxle only) (1.2L VIN 7 & 8 MPFI)
Code 52 Parking switch (2.2L VIN 6 MPFI), (3.3L VIN 3 MPFI)
Code 55 EGR gas temperature sensor
Code 56 EGR System; faulty EGR function (1.8L VIN 2 MPFI)
Code 56 EGR system (California) (3.3L VIN 3 MPFI)
Code 61 Parking switch; continuously in ON position
Code 62 Electric load signal; headlight HI/LO signal or rear defogger signal remains ON or OFF
Code 63 Blower fan switch; signal remains ON or OFF
Code 65 Vacuum pressure sensor; abnormal signal

➡**If more than one definition is listed for a code or the code is not listed here, consult your 'Chilton Total Car Care' manual to obtain the specific meaning for your vehicle. This list is for reference and does not mean that a component is defective. The code identifies the circuit and component that require further testing.**

TRANSMISSION CODES

Reading and Retrieving Codes

EXCEPT JUSTY

Subaru indicates malfunctions in the transmission/transaxle system and displays DTC's via the POWER lamp in the instrument cluster display. This lamp should illuminate when the ignition key is turned **ON**, then extinguish when the system is functioning normally. However, when the system detects a malfunction, the POWER light will blink on and off for a couple of seconds during start up. This isn't a trouble code; it's the computer's way of letting you know that there's a problem in the system. To retrieve the DTC's themselves, you must perform the retrieval procedure.

Subaru models, except Justy, Legacy and SVX, display DTC's using an 12-flash sequence. The light flashes 12 times in a row. The sequence always starts with a long start flash (approximately two seconds in duration), and is followed by 11 shorter flashes (approximately ¼ second in duration). If there are no malfunctions detected by the system, all 11 flashes will be short, but if the computer recognizes a fault in the system, one of the 11 flashes will be longer (approximately one second in duration)—the long flash identifies the code stored in memory. For example, if the third flash after the two second start flash is the long one, you're looking at a Code 3.

If the computer has detected more than one code, it displays all of the codes in the same pass. For an example, please refer to the accompanying illustration. After the computer displays the code(s), the light will remain off for approximately 2½ seconds, then repeats the code(s). If the light flashes on and off, in regular, one-second intervals, it indicates the battery is low or was disconnected long enough to interfere with computer memory.

Legacy and SVX models display diagnostic trouble codes as two digit codes, also via the POWER lamp. They display the tens digit as long flashes, then the ones digit as short flashes after a brief pause. Therefore, a Code 13 would consist of one long flash, a short pause, then three short flashes. There is a longer pause between each individual DTC. The system repeats each code three times, starting with the lowest number code in memory. It then moves on to the next code stored in memory. After the system finishes displaying all of the DTC's, the process repeats itself. The system continues displaying DTC's until the ignition key is turned **OFF**.

If there are no DTC's stored in memory, the POWER light will flash on and off, in short flashes. The POWER light flashing very rapidly indicates low battery voltage.

If the lamp worked prior to the diagnostic procedure, but will not display any DTC's, there may be a problem in one or more of the following:

- Throttle position idle switch
- Shift lever position switch
- Hold switch

To retrieve codes on Subaru models except Justy, Legacy or SVX, perform the following:

1. If equipped with 4WD, lock the four-wheel drive system into two wheel drive, as follows:

   a. Locate the special plug, marked FWD. It is usually located around the left front shock tower.

➡**This is the 4WD lockout connector.**

   b. Open the connector cover.

   c. Insert a 15A spade-type fuse into the connector, thereby locking the system out of 4WD.

2. Start the engine and allow it reach normal operating temperature.
3. Turn the ignition switch **OFF**.
4. Without starting the engine, turn the ignition key to the **ON** position. At this time the POWER lamp should come on.
5. Turn the ignition key **OFF** and apply the parking brake.

➡**From this point on, do not disturb or touch either the accelerator or brake pedal unless specifically instructed to.**

6. Move the transmission/transaxle gear selector to D.
7. Turn the 1-Hold switch on (down position).
8. Turn the ignition key **ON**, without starting the engine.
9. Move the transmission/transaxle gear selector to position 3.
10. Turn the 1-Hold switch off (up position).
11. Move the transmission/transaxle gear selector to position 2.
12. Turn the 1-Hold switch on (down position).
13. Depress the accelerator pedal halfway to the floor.
14. Observe the POWER lamp and note the DTC's.

Legacy and SVX models are designed with two types of DTC's: existing codes and history codes. Existing codes indicate problems that are currently detected by the diagnostic system, also known as hard codes. History codes indicate problems that were there in the past; you won't find the problem there now. Intermittent problems will often set history codes. If a problem sets a history code, but not

an existing code, look for conditions like loose connections or bad grounds; anything that can cause a failure that comes and goes. The procedure for setting the system to display existing codes is different from the procedure for displaying history codes.

To retrieve existing codes on Subaru Legacy and SVX models, perform the following:

1. Start the engine and allow it reach normal operating temperature.
2. If the transmission/transaxle is functioning well enough to road test the car, drive it at least 12 mph (20 km/h).
3. Apply the parking brake, then turn the ignition switch **OFF**.
4. Move the transmission/transaxle gear selector to D.
5. Turn the Hold or Manual switch on (down or in position).
6. Turn the ignition key **ON**, without starting the engine.
7. Move the transmission/transaxle gear selector to position 3.
8. Turn the Hold or Manual switch off (up or out position).
9. Move the transmission/transaxle gear selector to position 2.
10. Turn the Hold or Manual switch on (down or in position).
11. Move the transmission/transaxle gear selector to position 1.
12. Turn the Hold or Manual switch off (up or out position).
13. Depress the accelerator pedal halfway to the floor.
14. Observe the POWER lamp and note the DTC's.

To retrieve history codes on Subaru Legacy and SVX models, perform the following:

1. Apply the parking brake, then turn the ignition switch **OFF**.
2. Move the transmission/transaxle gear selector to 1.
3. Turn the Hold or Manual switch on (down or in position).
4. Turn the ignition key **ON**, without starting the engine.
5. Move the transmission/transaxle gear selector to position 2.
6. Turn the Hold or Manual switch off (up or out position).
7. Move the transmission/transaxle gear selector to position 3.
8. Turn the Hold or Manual switch on (down or in position).
9. Move the transmission/transaxle gear selector to position D.
10. Turn the Hold or Manual switch off (up or out position).
11. Depress the accelerator pedal halfway to the floor.
12. Observe the POWER lamp and note the DTC's.

JUSTY

The Subaru Justy diagnostic system indicates detected faults in the transmission/transaxle system and displays DTC's CHECK ECVT lamp, located in the instrument cluster display. This lamp should illuminate when the ignition key is turned **ON**, then extinguish when the system is functioning normally. However, when the system detects a malfunction, the CHECK ECVT light will illuminate during vehicle operation. This isn't a trouble code; it's the computer's way of letting you know that there's a problem in the system. To retrieve the DTC's themselves, you must perform the retrieval procedure.

Justy models display diagnostic trouble codes as two digit codes, also via the POWER lamp. They display the tens digit as long flashes, then the ones digit as short flashes after a brief pause. Therefore, a Code 13 would consist of one long flash, a short pause, then three short flashes. There is a longer pause between each individual DTC. The system repeats each code three times, starting with the lowest number code in memory. It then moves on to the next code stored in memory. After the system finishes displaying all of the DTC's, the process repeats itself. The system continues displaying DTC's until the ignition key is turned **OFF**.

If there are no DTC's stored in memory, the POWER light will flash on and off, in short flashes. The POWER light flashing very rapidly indicates low battery voltage.

To retrieve the codes, perform the following:

1. Perform the preliminary inspection, located earlier in this section. This is very important, since a loose or disconnected wire, or corroded connector terminals can cause a whole slew of unrelated DTC's to be stored by the computer; you will waste a lot of time performing a diagnostic "goose chase."

2. Grab some paper and a pencil or pen to write down the DTC's when they are flashed out.

3. Locate the Transaxle Check Mode Connectors and the Engine Check Connectors. The Transaxle Check Mode Connectors are single terminal, white, disconnected connector halves hanging from the transaxle computer harness, just before the computer connector. Don't confuse this with the two-terminal memory connector that's already connected. The Engine Check Connectors are usually located near the Transaxle Check Mode Connectors, and are disconnected, green single-terminal connector halves. If neither of these can be found, either call your local Subaru dealer or refer to the applicable Chilton's Total Car Care manual for your vehicle.

➡**The code retrieval procedure must be performed on the road. Take the vehicle to an open stretch of roadway, and pull off to the side to begin. You will need to accelerate the vehicle for 1/8 to 1/4 mile (0.2–0.4 km) during the procedure. Be sure there is enough clear road to complete the diagnostic procedure.**

4. Start the engine and allow it reach normal operating temperature with the A/C off.

5. Turn the ignition key **OFF**.

6. Connect the two Transaxle Check Mode connector halves together.

7. Start the engine, with the transaxle in P.

➡**From this point on, do not disturb or touch either the accelerator or brake pedal unless specifically instructed to. If you must use one of these pedals to avoid an accident, after the problem is dealt with shut the ignition key OFF and start over, beginning with Step 4.**

8. Engage the two green Engine Check Connector halves for 10 seconds, then separate them.

➡**If the CHECK ECVT light starts to blink or display codes, ignore it for now.**

9. Apply the parking brake.
10. Move the transmission/transaxle gear selector to R.
11. Move the transmission/transaxle gear selector to N.
12. Move the transmission/transaxle gear selector to D.
13. Move the transmission/transaxle gear selector to Ds.
14. Move the transmission/transaxle gear selector to D.
15. When the road is clear, disengage the parking brake.
16. Floor the throttle, and accelerate to approximately 25–30 mph (40–48 km/h), and hold the vehicle at this speed constantly for at least 5 seconds.
17. Release the throttle, and COAST to a stop. Pull off the road where it's safe to sit for a few minutes.

➡**Don't touch the brake pedal to stop; If necessary, use the parking brake to help stop the vehicle.**

18. Apply the parking brake.
19. Depress and release the brake pedal three times.

➡**The CHECK ECVT light will blink three sets of short flashes. These are normal system identification flashes, not DTC's.**

20. Observe the CHECK ECVT lamp and note the DTC's.

Clearing Codes

EXCEPT JUSTY

To clear the codes from memory, disconnect the negative battery cable for at least one minute. Then, reconnect the cable. The codes should now be cleared. If there are still codes present, either the codes were not properly cleared (Are the codes identical to those flashed out previously?) or the underlying problem is still there (Are only some of the codes the same as previously?)

JUSTY

To clear the DTC's from the system's memory, detach the transmission/transaxle Check Mode connector.

DIAGNOSTIC TROUBLE CODES—SUBARU EXCEPT JUSTY, LEGACY & SVX	
Code	**Component/Circuit Fault**
1	No. 1 speed sensor
2	No. 2 speed sensor
3	Throttle Position (TP) sensor
4	No. 1 shift solenoid
5	No. 2 shift solenoid
6	No. 3 shift solenoid
7	Duty cycle solenoid B
8	Duty cycle solenoid C
9	ATF temperature sensor
10	No ignition signal
11	Duty cycle solenoid A

89446GAA

Subaru transmission related Diagnostic Trouble Codes—except Justy, Legacy and SVX

DIAGNOSTIC TROUBLE CODES—SUBARU LEGACY & SVX

Code	Component/Circuit Fault
11	Duty solenoid A or circuit
12	Duty solenoid B or circuit
13	Shift solenoid No. 3 or circuit
14	Shift solenoid No. 2 or circuit
15	Shift solenoid No. 1 or circuit
21	ATF temperature sensor or circuit
22	Atmospheric pressure (Baro) sensor
23	Engine speed signal circuit
24	Duty solenoid C or circuit
31	Throttle Position (TP) sensor or circuit
32	No. 1 Vehicle Speed Sensor (VSS) or circuit
33	No. 2 Vehicle Speed Sensor (VSS) or circuit

89446GAB

Subaru transmission related Diagnostic Trouble Codes—Legacy and SVX

DIAGNOSTIC TROUBLE CODES—SUBARU JUSTY

Code	Component/Circuit Fault
13	D-range switch or circuit
14	Ds-range switch or circuit
15	R-range switch or circuit
21	ECM communication circuit
22	Coolant temperature signal from ECM
25 ①	Slow Cut solenoid, CFC solenoid or circuit
31	Throttle pedal switch or circuit
32	Throttle position sensor or circuit
33	No vehicle speed sensor signal
34	Clutch coil system current out of specifications
35	Line pressure solenoid or circuit
41, 42	Improper high altitude signal from ECM
45	Brake switch or circuit

① Coolant temperature sensor problems may also set a Code 25.

89446GAC

Subaru transmission related Diagnostic Trouble Codes—Justy

Suzuki

ENGINE CODES

General Information

Suzuki used both feedback carburetor and fuel injection systems.

Suzuki uses the Hitachi 2-barrel, downdraft type carburetor, which has both a primary and secondary system. A feedback system is provided to maintain the air/fuel ratio, to reduce emission levels and to improve fuel economy simultaneously.

The primary system operates under normal driving conditions and the secondary system operates under high speed-high load driving conditions. A choke valve is provided in the primary system.

The primary system is equipped with a choke system. The choke system is a fully automatic type using a thermo-wax. A mixture con-trol solenoid valve is also incorporated which is operated by an electrical signal from the Electronic Control Module (ECM). The acceleration pump system and a fuel cutoff solenoid are also part of the primary system.

The secondary system is equipped with a secondary diaphragm through which vacuum is supplied from the primary side, via a Vacuum Switching Valve (VSV) and a Vacuum Transmitting Valve (VTV), to operate the secondary throttle valve. The VSV and VTV are used only on the Suzuki Samurai and have been eliminated on the Suzuki Sidekick 1300cc.

A Switch vent solenoid valve is provided on the top of the float chamber. Its purpose is to reduce the evaporative emissions. The 2-barrel, downdraft type carburetor is also equipped with a idle-up system. This system operates at idle and compensates the idle speed when any of the following conditions exist:

• When any electrical load (lights, rear defogger, heater fan, etc.) is operating.

• When the vehicle is at a high altitude.

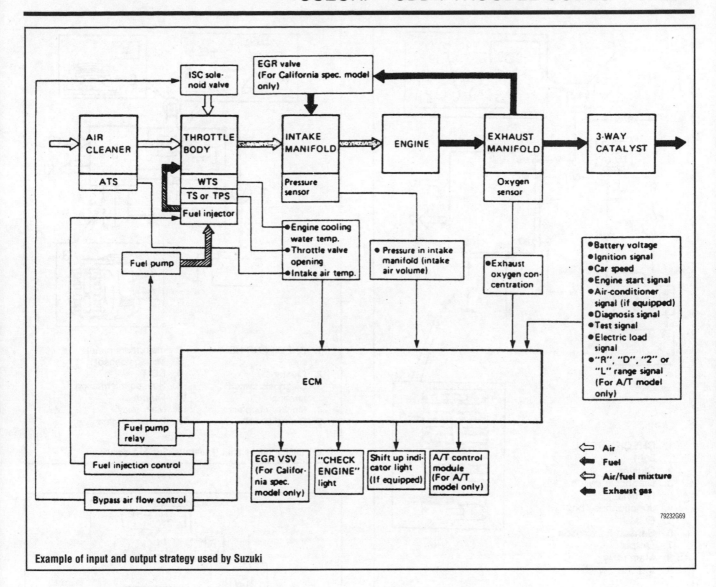

Example of input and output strategy used by Suzuki

- When the engine temperature is below 44°F (7°C).
- When the engine speed is lower than 1500 rpm after the engine is started.

The Electronic Fuel Injection (EFI) system supplies the vehicle's combustion chambers with air/fuel mixture of optimized ratio under varying driving conditions. Fuel delivery through the injector is controlled electrically by the Electronic Control Module (ECM).

Service Precautions

- Keep the ECM parts and harnesses dry during service. Protect the ECM and all solid-state components from rough handling or temperature extremes.
- Use extreme care when working around the ECM or other components.
- Disconnect the negative battery cable before attempting to disconnect or remove any parts.

a Disconnect the negative battery cable and ECM connector before performing arc welding on the vehicle.

- Disconnect and remove the ECM from the vehicle before subjecting the vehicle to the temperatures experienced in a heated paint booth.

Reading Codes

1989–95 VEHICLES

On Swift and Samurai, the ECM memory is activated by connecting the spare fuse to the diagnosis switch terminal and turning the ignition switch ON. The fuse panel is located under the instrument panel, near the driver's side kick panel. On Sidekick models the diagnostic terminals B and C must be grounded. The diagnostic terminal is located under the hood, on the right rear side.

The diagnostic codes are flashed by the CHECK ENGINE light or Malfunction Indicator Lamp (MIL) on the dash. The memory displays the codes in numerical order from lowest to highest. The order in which the codes are displayed does not necessarily indicate the order in which the malfunction occurred. The ECM displays each code 3 times, then moves on the next code in numerical

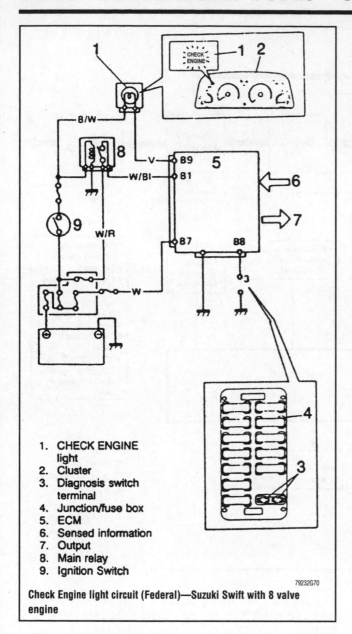

1. CHECK ENGINE
   light
2. Cluster
3. Diagnosis switch
   terminal
4. Junction/fuse box
5. ECM
6. Sensed information
7. Output
8. Main relay
9. Ignition Switch

79232G70

**Check Engine light circuit (Federal)—Suzuki Swift with 8 valve
engine**

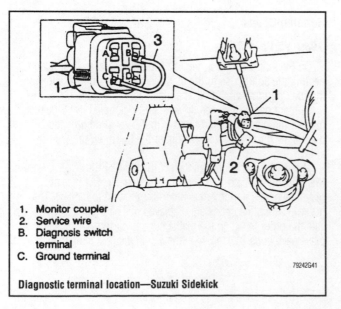

1. Monitor coupler
2. Service wire
B. Diagnosis switch
   terminal
C. Ground terminal

79242G41

Diagnostic terminal location—Suzuki Sidekick

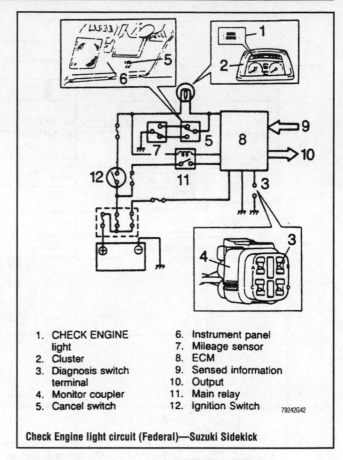

1. CHECK ENGINE
   light
2. Cluster
3. Diagnosis switch
   terminal
4. Monitor coupler
5. Cancel switch
6. Instrument panel
7. Mileage sensor
8. ECM
9. Sensed information
10. Output
11. Main relay
12. Ignition Switch

79242G42

Check Engine light circuit (Federal)—Suzuki Sidekick

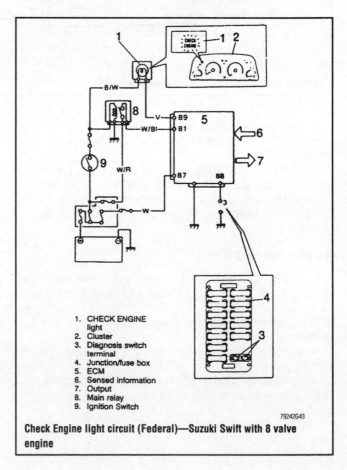

1. CHECK ENGINE
   light
2. Cluster
3. Diagnosis switch
   terminal
4. Junction/fuse box
5. ECM
6. Sensed information
7. Output
8. Main relay
9. Ignition Switch

79242G43

**Check Engine light circuit (Federal)—Suzuki Swift with 8 valve
engine**

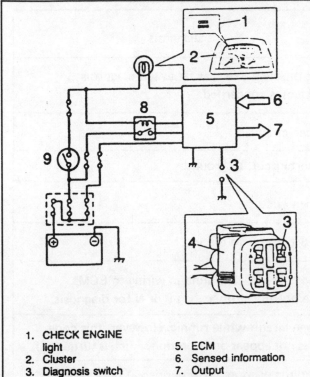

1. CHECK ENGINE light
2. Cluster
3. Diagnosis switch terminal
4. Fuse box
5. ECM
6. Sensed information
7. Output
8. Main relay
9. Ignition switch

79242G44

Check Engine light circuit (California and Canada)—Suzuki Sidekick

order. The entire sequence is repeated as long as the diagnosis switch terminal is grounded and the ignition switch is in the **ON** position.

Clearing Codes

1989–95 VEHICLES

When repairs have been completed, erase the ECM back-up memory by disconnecting the negative battery cable or the ECM harness connector for 30 seconds or longer.

Diagnostic Trouble Codes

1986–95 VEHICLES

Code 12 Normal
Code 13 Oxygen sensor circuit
Code 14 Engine Coolant Temperature (ECT) Sensor circuit—low temperature indicated, signal voltage high
Code 15 Engine Coolant Temperature (ECT) Sensor circuit—high temperature indicated, signal voltage low
Code 21 Throttle Position Sensor (TPS) circuit—signal voltage high
Code 22 Throttle Position Sensor (TPS) circuit—signal voltage low
Code 23 Air Temperature Sensor (ATS) circuit—low temperature indicated, signal voltage high
Code 24 Vehicle Speed Sensor (VSS) circuit
Code 25 Air Temperature Sensor (ATS) circuit—high temperature indicated, signal voltage low
Code 31 Pressure Sensor (PS) circuit—high pressure indicated, signal voltage high

Code 32 Pressure Sensor (PS) circuit—low pressure indicated, signal voltage low
Code 33 Mass Air Flow Sensor (MAS) circuit—signal voltage high
Code 34 Mass Air Flow Sensor (MAS) circuit—signal voltage low
Code 41 Ignition signal
Code 42 Crank Angle Sensor (CAS) circuit (except 1989–90 Sidekick) or Fifth switch circuit, Lock-up signal circuit (1989–90 Sidekick)
Code 44 Idle switch of Throttle Position Sensor (TPS) -open circuit
Code 45 Idle switch of Throttle Position Sensor (TPS) -shorted circuit
Code 51 Exhaust Gas Recirculation (EGR) system and/or Recirculated Exhaust Gas Temperature Sensor (REGTS) system—California vehicle
Code 52 Fuel Injector—California vehicle
Code 53 Ground circuit—California vehicle
Code 54 Fifth gear switch circuit
Code 71 Test switch circuit

TRANSMISSION CODES

Reading Codes

On Suzuki models equipped with the Electric Shift Control System (ESCS), retrieve the transmission codes as follows:

1. Run the engine until it reaches normal operating temperature.

➡**Leave the engine running during this procedure.**

2. Position the transmission shifter in the **P** (park) position and apply the parking brake.

3. Attach the leads of an analog voltmeter to the terminals of the green monitor connector, which should be located under the left-hand side of the instrument panel near the steering column.

4. Using a jumper wire or Suzuki's DIAG switch coupler lead, connect the two terminals of the black DIAG SW connector. This connector is located near the monitor connector under the instrument panel.

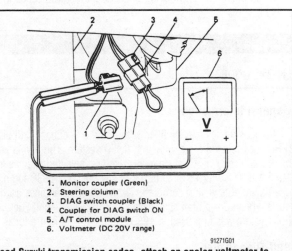

1. Monitor coupler (Green)
2. Steering column
3. DIAG switch coupler (Black)
4. Coupler for DIAG switch ON
5. A/T control module
6. Voltmeter (DC 20V range)

91271G01

To read Suzuki transmission codes, attach an analog voltmeter to the monitor connector and jumper the DIAG SW terminals while the engine is running

Diagnostic Code		Diagnostic Area	Diagnosis
No.	Mode		
12	⎍⎍	Normal	No problem exists as far as self-diagnosis system is concerned.
21	⎍⎍⎍	Direct clutch solenoid	Open circuit.
22	⎍⎍⎍		Short circuit to ground.
23	⎍⎍⎍	2nd brake solenoid	Open circuit.
24	⎍⎍⎍		Short circuit to ground.
25	⎍⎍⎍	R, D, 2 and L range signal	Short circuit to ground in wiring to ECM. Use positions other than P or N for diagnosis.
31	⎍⎍	Speed sensor	Open circuit while runnig. However, this code does not appear once ignition switch is turned off.
32	⎍⎍⎍	Shift lever switch	2 points or more are grounded at once or all points are open.
33	⎍⎍⎍	Ignition signal	No ignition signal for more than 9 seconds while running at 30 km/h (19 mile/h) or more with throttle position sensor opened more than 28%.

91271G02

Suzuki transmission related Diagnostic Trouble Codes

5. Read the voltmeter needle pulses and compare the code(s) with the accompanying code chart.

Clearing Codes

When the ignition switch is turned **OFF** all stored codes will be erased.

Toyota

ENGINE CODES

General Information

Toyota vehicles may be fitted with either a Feedback Carburetor (FBC) or Multi-port Fuel Injection (MEI) system. The Toyota Feedback Carburetor (FBC) system was used on selected engines from 1983 1990. Two types of carburetors were used. The 3E engine used a variable venturi carburetor, while all other engines used a more typical down draft style carburetor. The Multi-port Fuel Injection system was first used in 1980 and continues in use today.

Self-Diagnostics

As the engine control computers became capable of more functions, self-diagnostic and memory circuits were added.

These systems allow the ECU to note a fault, assign an identity code and store the code in memory for later retrieval.

All fuel injected control engine units possess the ability to provide fault codes during diagnosis. The number, type and meaning of engine codes vary by year and model.

While most fault codes are held in an electronic memory and are retained even after the ignition is switched **OFF**, certain codes are only held or displayed as long as the ignition is **ON**. If the fault is present at the next restart, the code will reset.

When a controller or ECU notes a fault, the dash warning lamp for the appropriate system will be lit to advise the operator. If the dash lamp is normally lit during system operation, as in the case of cruise control, the lamp will flash when a fault is found. The illumination or flashing of the dash lamp indicates that the controller has detected a fault and placed itself into the back-up or default mode.

Beginning in 1995 some models were equipped with an on board diagnostic system known as OBD II. To diagnose this system an OBD II scan tool, complying with SAE J1978 or TOYOTA hand held tester is necessary to access codes and read data output from the ECM. This is a rather expensive tool and not cost effective for the general public. The following model and engine applications are equipped with the OBD II system:

- 1995 Tercel
- 1995 Camry with a 1MZ-FE engine
- 1995 Avalon

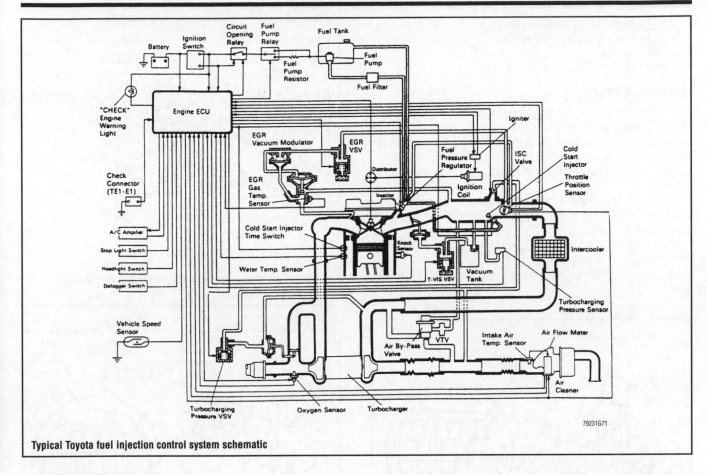

Typical Toyota fuel injection control system schematic

79231G71

- 1995 Previa with a 2TZ-FZE engine
- 1995 Tacoma
- 1995 T100

Service Precautions

- Keep all ECU parts and harnesses dry during service. Protect the ECU and all solid-state components from rough handling or extremes of temperature.
- Before attempting to remove any parts, turn the ignition switch **OFF** and disconnect the battery ground cable.
- Make sure all harness connectors are fastened securely. A poor connection can cause an extremely high voltage surge, resulting in damage to integrated circuits.
- Always use a 12 volt battery as a power source.
- Do not attempt to disconnect the battery cables with the engine running.
- Do not attempt to disassemble the ECU unit under any circumstances.
- If installing a 2-way or CB radio, mobile phone or other radio equipment, keep the antenna as far as possible away from the electronic control unit. Keep the antenna feeder line at least 8 in. away from the EEI harness and do not run the lines parallel for a long distance. Be sure to ground the radio to the vehicle body.
- When performing ECU input/output signal diagnosis, remove the water-proofing rubber plug, if equipped, from the connectors to make it easier to insert tester probes into the connector. Always reinstall it after testing.
- When connecting or disconnecting pin connectors from the ECU, take care not to bend or break any pin terminals. Check that there are no bends or breaks on ECU pin terminals before attempting any connections.
- When measuring supply voltage of ECU-controlled components, keep the tester probes separated from each other and from accidental grounding. If the tester probes accidentally make contact with each other during measurement, a short circuit will damage the ECU.
- Use great care when working on or around air bag systems. Wait at least 20 seconds after turning the ignition switch to LOCK and disconnecting the negative battery cable before performing any other work. The air bag system is equipped with a back-up power system which will keep the system functional for 20 seconds without battery power.
- All air bag connectors are a standard yellow color; the related wiring is encased in standard yellow sheathing. Testing and diagnostic procedures must be followed exactly when performing diagnosis on this system. Improper procedures may cause accidental deployment or disable the system when needed.
- Never attempt to measure the resistance of the air bag squib; detonation may occur.

Reading Codes

The following procedures are for all vehicles except those equipped with the OBD II system. Accessing OBD II system codes can only be accomplished with the use of a OBD II scan tool, complying with SAE J1978 or TOYOTA hand held tester. This is a rather expensive tool and not cost effective for the general public. The following models are equipped with the OBD II system:

- 1995 Tercel
- 1995 Camry with a 1 MZ-FE engine

- 1995 Avalon
- 1995 Previa with a 2TZ-FZE engine
- 1995 Tacoma
- 1995 T100

1983–86 VEHICLES

The diagnostic codes can be read by the number of blinks of the 'Check Engine' warning light when the proper terminals of the check connector are short-circuited. If the vehicle is equipped with a super monitor display, the diagnostic code is indicated on the display screen. The initial conditions for entering the self-diagnostics are as follows:

1. The battery voltage of the vehicle should be above 11 volts. The throttle valve must be in a fully closed position (throttle position sensor IDL points closed).

2. If equipped with an automatic transmission, place it in P or N.

3. Turn the air conditioning switch OFF.

4. Start the engine and allow it reach normal operating temperature.

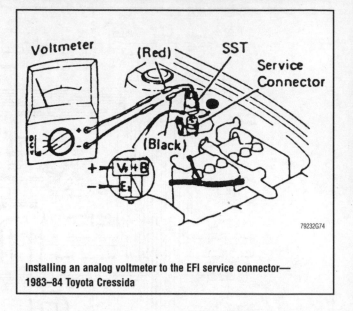

Installing an analog voltmeter to the EFI service connector—1983–84 Toyota Cressida

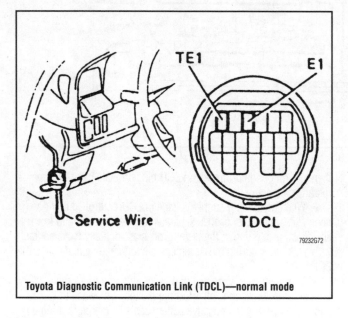

Toyota Diagnostic Communication Link (TDCL)—normal mode

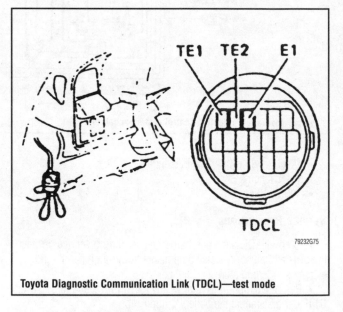

Toyota Diagnostic Communication Link (TDCL)—test mode

Except Super Monitor Display—Normal Mode:

1. Turn the ignition switch to the **ON** position. Do not start the engine. Remove the protective rubber cap and, with a jumper wire connect the terminals of the check connector.

2. Read the diagnostic code as indicated by the number of flashes of the 'Check Engine' warning light.

➡**On some early models, install an analog voltmeter to the EFI service connector. Read diagnostic codes by voltmeter needle deflection between 0V–2.5V–5V. The voltmeter needle will fluctuate between 5V and 2.5V every 0.6 seconds.**

3. If the system is operating normally (no malfunction), the light will blink once every ¼ second. On single digit code number systems, the light will blink once every 3 or 4.5 seconds.

4. In the event of a malfunction, the light will blink once every ½ second (on some models it may be 1, 2 or 3 seconds). The 1st number of blinks will equal the 1st digit of a 2-digit diagnostic code. After a 1.5 second pause, the 2nd number of blinks will equal the 2nd number of a 2-digit diagnostic code. If there are 2 or more

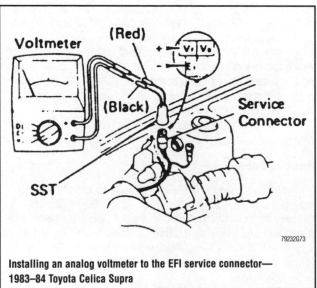

Installing an analog voltmeter to the EFI service connector—1983–84 Toyota Celica Supra

codes, there will be a 2.5 second pause between each. On single digit code number systems the light will blink a number of times equal to the malfunction code indication every 2 or 4.5 seconds.

5. After all the codes have been output, there will be a 4.5 second pause and they will be repeated as long as the terminals of the check connector are shorted.

➡**In event of multiple trouble codes, indication will begin from the smaller value and continue to the larger in order.**

6. After the diagnosis check, remove the jumper wire from the check connector and install the protective rubber cap.

Test Mode:

1. Using a jumper wire, connect the TE2 and E1 terminals of the Toyota Diagnostic Communication Link (TDCL), then turn the ignition switch **ON** to begin the diagnostic test mode.

2. Start the engine and drive the vehicle at a speed of 10 mph or more. Simulate the conditions where the malfunction has been reported to happen.

3. Using a jumper wire, connect the TE2 and E1 terminals of the TDCL connector.

4. Read the diagnosis code as indicated by the number of 'Check Engine' light flashes.

5. After diagnosis check remove the jumper wires.

Super Monitor Display:

The super monitor display system was offered as an option on some late model Toyota vehicles.

1. Turn the ignition switch **ON** but do not start the engine.

2. Simultaneously push and hold in the SELECT and INPUT M keys for at least 3 seconds. The letters DIAG will appear on the screen.

3. After a short pause, hold the SET key in for at least 3 seconds. If the system is normal (no malfunctions), ENG-OK will appear on the screen.

4. If there is a malfunction, the code number for it will appear on the screen. In event of 2 or more numbers, there will be a 3 second pause between each (example: EN-42).

1987–89 VEHICLES

Stored fault codes are transmitted through the blinking of the CHECK engine warning lamp. This occurs only when the system is placed into the diagnostic mode; it does not occur while the vehicle is being driven.

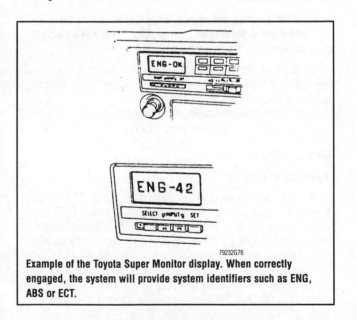

Example of the Toyota Super Monitor display. When correctly engaged, the system will provide system identifiers such as ENG, ABS or ECT.

To read the engine codes, the computer must be put into the diagnostic mode by connecting the proper terminals

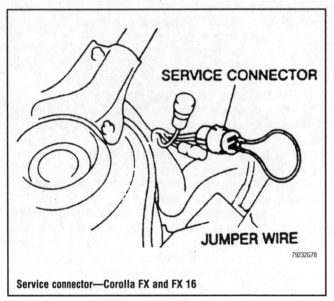

Service connector—Corolla FX and FX 16

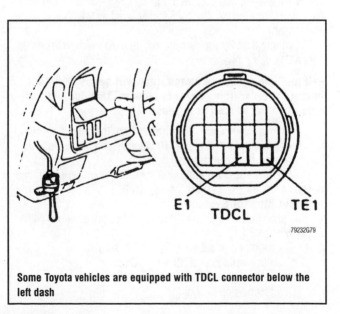

Some Toyota vehicles are equipped with TDCL connector below the left dash

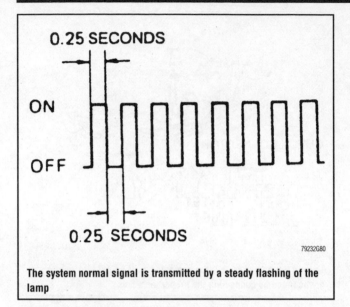

The system normal signal is transmitted by a steady flashing of the lamp

To read the fault codes:
1. The following initial conditions must be met:
   a. Battery voltage at or above 11 volts.
   b. Throttle fully closed.
   c. Transmission in N or P.
   d. All electrical systems and accessories OFF.
2. Turn the ignition **ON** but do not start the engine.
3. Use a jumper wire to connect terminals T and E1 at the diagnostic connector. On all 1989 vehicles except Corolla, MR2 and Tercel, connect terminals TE1 and Et. For 1988–89 Vans, jumper the 2 pins of the service connector. On 1989 Corolla, MR2 and Tercel, connect terminals T and E1.
4. The fault codes will be transmitted through the controlled flashing of the CHECK engine warning lamp.
5. If no malfunction was found or no code was stored, the lamp will flash 2 times per second with no other pauses or patterns. This confirms that the diagnostic system is working but has nothing to report. This light pattern may be referred to as the System Normal signal; it should be present when no other codes are stored.
6. The CHECK lamp will blink the number of the code(s). All codes are 2 digits; the pulsing of the light represents the digits, not the count. For example, Code 25 is displayed as 2 flashes a pause and 5 flashes.
7. If more than 2 codes are stored, the next will be transmitted after a 2½ second pause.

➡️**If multiple codes are stored, they will be transmitted in numerical order from lowest to highest. This does not indicate the order of fault occurrence.**

8. When all codes have been transmitted, the entire pattern will repeat after a 4½ second pause. The repeats continue as long as the diagnostic terminals are connected.
9. After recording the codes, disconnect the jumper at the diagnostic connector and turn the ignition **OFF**.

Super Monitor System:
This procedure is used on Cressida and Supra equipped with Super Monitor.
1. The following initial conditions must be met:
   a. Battery voltage at or above 11 volts.
   b. Throttle fully closed.
   c. Transmission in N or P.

d. All electrical systems and accessories OFF.
2. Turn the ignition **ON** but do not start the engine.
3. Simultaneously press and hold the SELECT and INPUT M keys for at least 2 seconds. The letters DIAG will appear on the screen, showing that the system is in the diagnostic mode.
4. After a short pause, hold in the SET key for at least 2 seconds.
5. If the system is normal, with no faults stored, the message ENG OK will appear on the screen. If faults are stored, the code number will appear on the screen with a system designator; for example, ENG-42. If 2 or more codes are stored, each will appear after a 3 second pause.

1990–95 VEHICLES

Stored fault codes are transmitted through the blinking of the CHECK engine warning lamp. This occurs only when the system is placed into the diagnostic mode; it does not occur while the vehicle is being driven.

To read the fault codes:
1. The following initial conditions must be met:
   a. Battery voltage at or above 11 volts.
   b. Throttle fully closed.
   c. Transmission in N or P.
   d. All electrical systems and accessories OFF.
2. Turn the ignition **ON** but do not start the engine.
3. Use a jumper wire to connect terminals TE1 and E1 at the diagnostic connector in the engine compartment or at the TDCL connector below the left dashboard if so equipped.
4. The fault codes will be transmitted through the controlled flashing of the CHECK ENGINE warning lamp.
5. If no malfunction was found or no code was stored, the lamp will flash 2 times per second with no other pauses or patterns. This confirms that the diagnostic system is working but has nothing to report. This light pattern may be referred to as the System Normal signal; it should be present when no other codes are stored.
6. The CHECK lamp will blink the number of the code(s). All codes are 2-digit; the pulsing of the light represents the digits, not the count. For example, Code 25 is displayed as 2 flashes, a pause and 5 flashes.
7. If more than 1 code is stored, the next will be transmitted after a 2½ second pause.

➡️**If multiple codes are stored, they will be transmitted in numerical order from lowest to highest. This does not indicate the order of fault occurrence.**

8. When all codes have been transmitted, the entire pattern will repeat after a 4½ second pause. The repeats continue as long as the diagnostic terminals are connected.
9. After recording the codes, disconnect the jumper at the diagnostic connector and turn the ignition **OFF**.

Clearing Codes

1986–95 VEHICLES

Stored codes will remain in memory until cleared. The correct method of clearing codes is to turn the ignition switch **OFF**, then remove the proper fuse. On all vehicles except as noted below, remove the EEI fuse. Each fuse must be removed for at least 10 seconds. The time required may be longer in cold weather.

Disconnecting the negative battery cable will also clear the memory but is not recommended due to other on-board memories being cleared as well. Once the system power is restored, re-check for stored codes. Only the System Normal indication should be present. If any other code is stored, the clearing procedure must be repeated or additional repairs performed; the old code will remain stored along with any new ones.

After repairs, it is recommended to clear the memory before test driving the vehicle. Upon returning from the drive, interrogate the memory; if the original code is again present, the repair was unsuccessful.

Except EFI Fuse:
- Corolla—15A Stop fuse
- 1990–94 MR2—7.5 AMP fuse
- Starlet—yellow fusible link
- 1990–94 Tercel—Stop fuse
- 1990–94 Corolla—15A Stop fuse

Diagnostic Trouble Codes

1983–1984 ENGINES

Code 1 Normal operation

Code 2 Open or shorted air flow meter circuit—defective air flow meter or Electronic Control Unit (ECU)

Code 3 Open or shorted air flow meter circuit—defective air flow meter or Electronic Control Unit (ECU)

Code 4 Open Water Thermosensor (THW) circuit—defective Water Thermosensor (THW) or Electronic Control Unit (ECU)

Code 5 Open or shorted oxygen sensor circuit—lean or rich indication—defective oxygen sensor or Electronic Control Unit (ECU)

Code 6 No ignition signal—defective ignition system circuit, Integrated Ignition Assembly (IIA) or Electronic Control Unit (ECU)

Code 7 Defective Throttle Position Sensor (TPS) circuit, Throttle Position Sensor (TPS) or Electronic Control Unit (ECU)

1985–1987 ENGINES

➡ **The 1985 2.0L (25-E and 3Y-EC) engines use 1984 Codes**

Code 1 Normal operation

Code 2 Open or shorted air flow meter circuit—defective air flow meter or Electronic Control Unit (ECU)

Code 3 No signal from igniter 4 times in succession -defective igniter or main relay circuit, igniter or Electronic Control Unit (ECU)

Code 4 Open Water Thermosensor (THW) circuit—defective Water Thermosensor (THW) or Electronic Control Unit (ECU)

Code 5 Open or shorted oxygen sensor circuit—lean or rich indication—defective oxygen sensor or Electronic Control Unit (ECU)

Code 6 No engine revolution sensor (Ne) signal to Electronic Control Unit (ECU) or Ne value being over 1000 rpm in spite of no Ne signal to ECU—defective igniter circuit, igniter, distributor or Electronic Control Unit (ECU)

Code 7 Open or shorted Throttle Position Sensor (TPS) circuit, Throttle Position Sensor (TPS) or Electronic Control Unit (ECU)

Code 8 Open or shorted intake air thermosensor circuit -defective intake air thermosensor circuit or Electronic Control Unit (ECU)

Code 10 No starter switch signal to Electronic Control Unit (ECU) with vehicle speed at 0 and engine speed over 800 rpm—defective speed sensor circuit, main relay circuit, igniter switch to starter circuit, igniter switch or Electronic Control Unit (ECU)

Code 11 Short circuit in check connector terminal T with the air conditioning switch ON or throttle switch (IDL) contact point OFF—defective air conditioner switch, Throttle Position Sensor (TPS) circuit, Throttle Position Sensor (TPS) or Electronic Control Unit (ECU)

Code 12 Knock control sensor signal has not reached judgment level in succession—defective knock control sensor circuit, knock control sensor or Electronic Control Unit (ECU)

Code 13 Knock CPU faulty

1988–95 ENGINES

Constant blinking of indicator light: No faults detected

Code 11 Momentary interruption in power supply to ECU; up to 1991

Code 12 Engine revolution (NE or G) signal to ECU; missing within several seconds after engine is cranked

Code 13 Rpm NE signal to ECU; missing when engine speed is above 1000 rpm

Code 14 Igniter (IGE) signal to ECU; missing 4–11 times in succession

Code 16 ECT control signal—normal signal missing from ECT CPU (1990–94)

Code 16 A/T control system—normal signal missing from between the engine CPU and A/T CPU in the ECM (1995)

Code 21 Main oxygen sensor signal; voltage output does not exceed a set value on the lean and rich sides continuously for a certain period of time or open/short sensor heater circuit

Code 22 Water temperature sensor circuit (THW); open/short for 500 milliseconds. or more

Code 23 Intake air temperature signal (THA)

Code 24 Intake air temperature sensor circuit (THA); open/short for 500 milliseconds. or more

Code 25 Air/fuel ratio LEAN malfunction; Oxygen sensor output is less than 0.45 V for at least 90 seconds when oxygen sensor is warmed up (engine racing at 2000 rpm). California only: air/fuel ratio feedback compensation/adaptive control: feedback value continues at upper (LEAN) limit, or is not renewed, for a certain period of time.

Code 26 Air/fuel ratio RICH malfunction; California only: Air/fuel ratio feedback compensation/adaptive control: feedback value continues at lower (RICH) limit, or is not renewed, for a certain period of time.

Code 27 Sub-oxygen sensor signal; detection of sensor/signal deterioration or open/short sensor heater circuit (California only)

Code 28 No. 2 oxygen sensor signal/heater signal

Code 31 Air flow meter circuit; open or shorted when idle contacts are closed

Code 31 Vacuum (Manifold absolute pressure) sensor signal; open/short circuit

Code 32 Air flow meter circuit; circuit open or shorted when idling

Code 34 Turbocharging pressure signal; excessive pressure

Code 35 Altitude compensation (HAC) sensor signal; open/short

Code 35 Turbocharging pressure sensor signal; open/short

Code 36 Turbocharging pressure sensor signal; open or short detected for 0.5 sec or more in the turbocharging pressure sensor signal circuit; 1992–94

Code 41 Throttle position sensor circuit (VTA); open/short

Code 42 Vehicle speed sensor circuit

Code 43 No starter switch (STA) signal to ECU until engine speed reaches 800 rpm when cranking

Code 51 NC signal ON, DL contact OFF, or shift position in R, D, 2 or 1 range; with check terminals T and El connected

Code 52 Knock sensor signal (KNK); open/short

Code 53 Knock control signal in ECU; ECU knock control faulty

Code 55 Knock sensor (rear side) signal in ECU; ECU knock control faulty

Code 71 EGR system malfunction; EGR gas temperature signal (THG) is below water temperature sensor signal or below intake air temperature sensor signal plus 86°F (30°C), after driving for 240 seconds in EGR operation range (California only)

Code 72 Fuel cut solenoid signal circuit (FCS) open; up to 1991

Code 78 Fuel pump control signal input circuit to pump (FPC) open

Code 81 TCM communication; open detected in ECT1 circuit for 2 or more seconds

Code 83 TCM communication; open detected in ESA1 circuit 0.5 sec after idle

Code 84 TCM communication; open in ESA2 circuit for 0.5 seconds after idle

Code 85 TCM communication; open in ESA3 circuit for more than 0.5 seconds after idle

TRANSMISSION CODES

Reading and Retrieving Codes

Like most other manufacturers, Toyota vehicles built since 1985 are capable of displaying DTC's. The Toyota systems display the DTC's via the O/D OFF lamp (as with many other import manufacturers).

The DTC's displayed by Toyota models consist of two digits. They are flashed out so that first the tens digit is displayed, then the ones digit. And, since all of the O/D OFF lamp blinks are the same length of duration, Code 38 would be displayed as three flashes, then a short pause followed by eight more flashes. After which, the O/D OFF light would remain off for a few seconds, then the code would be repeated. The system repeats each DTC three times, then goes on to the next code stored in memory. After the system finishes displaying all of the codes stored in memory, it returns to the first DTC and starts all over. The system will continue to display the DTC's until the ignition key is turned **OFF**.

Before the codes can be retrieved, the diagnostic connector link must be located. Over the years, however, there have been three dif-

ferent configurations of diagnostic connector links. The earliest connector appears in 1985–88 vehicles. It is a single terminal (ECT terminal) connector. The next design was introduced on some models in 1985. It is an irregularly-shaped, multi-terminal connector and usually always appears in the engine compartment. On vehicles with a single, integrated (engine & transmission) Powertrain Control Module (PCM) the ECT terminal is changed to the T$_T$ terminal. The third configuration is referred to as the Toyota Diagnostic Communication Link (TDCL). This connector is used on 1989–91 Cressida and 1992 and newer pre-OBD II Camry that use a single, integrated (engine & transmission) Powertrain Control Module (PCM). Model equipped with this third type of connector also have the second design connector in the engine compartment. The third type connector is usually located under the driver's side of the instrument panel.

To retrieve DTC's, perform the following:
1. Locate the diagnostic connector location.
2. Turn the ignition key **OFF**.

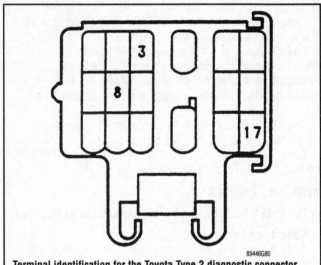

89446G80

Terminal identification for the Toyota Type 2 diagnostic connector, showing the E1 (3), the TE1 (8) and the ECT (17) terminals

Model	Year	Design	Location
Camry	1985-94	2	Left Front Shock Tower
	1992-94	3	Under the Left Side of the Instrument Panel
Celica	1986-94	2	Left Front Shock Tower
Corolla	1987	1	Right Front Shock Tower
	1988-94	2	Left Front Shock Tower
Cressida	1985-91	2	Left Front Shock Tower
	1989-91	3	Under Dash; Left of Steering
MR2	1985-89	2	Left Front of Engine Compartment
	1991	2	Left Rear of the Engine Compartment
	1992-94	2	Right Rear Shock Tower
Paseo	1992-94	2	Left Side of the Engine Compartment, Near the No. 2 Fuse Box
Pick-up and 4Runner	1985-89	1	Left Front Shock Tower
	1990-94	2	Left Side of the Engine Compartment, Near the No. 2 Fuse Box
Previa Van	1991-94	2	Under Front of the Driver's Side Seat
Supra	1985-86	1	Left Side of the Engine Compartment
	1986-92	2	Left Front Shock Tower
	1993-94	2	Right Side of the Firewall

89446G59

Before starting the Toyota diagnostic retrieval procedure, locate the connector

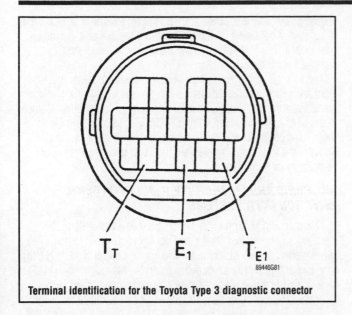

Terminal identification for the Toyota Type 3 diagnostic connector

T_T E_1 T_{E1}

89446G81

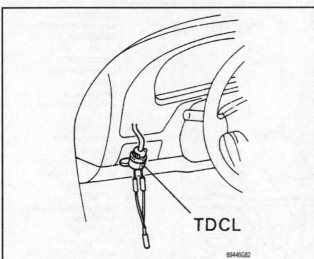

TDCL

89446G82

The Toyota Type 3 diagnostic connector is usually located under the left-hand side of the instrument panel, as shown

3. Using a jumper wire, ground the ECT or TE1 terminal in the diagnostic connector.

4. Turn the ignition key **ON**, without starting the engine.

5. Observe the O/D OFF light and note all DTC's. If the O/D OFF keeps flashing off and on, in even, regular cycles, there are no codes stored in memory.

Clearing Codes

➡**It is not recommended that the negative battery cable be disconnected to clear DTC's. Doing so will clear all settings from the vehicle's computer system, resulting in lost radio presets, seat memories, anti-theft codes, driveability parameters, etc., and there are better ways of spending your afternoon than resetting all of these.**

To clear the codes from memory, remove the appropriate fuse from the fuse block for at least one minute, as follows:

1985–86 Camry—Turn the ignition switch **OFF**
1986 Camry—Fuse: ECU B 15A
1987–88 Camry with A140E—Fuse: radio No. 1 15A
1989 Camry except A540E—Fuse: dome 20A
1988–89 Camry with A540E—Fuse: EFI 15A
1986–87 Celica—Fuse: No. 1 radio 15A
1988–89 Celica—Fuse: dome 20A
1985–86 Cressida—Fuse: No. 1 radio 15A
1987–88 Cressida—Fuse: EFI 15A
1989–92 Cressida—Fuse: dome 7.5A
1987–88 FX—Fuse: stop lamp 15A
1986–89 MR2—Fuse: AM 7.5A
1985–86 Supra with A43DE—Fuse: ECU B 15A
1986–89 Supra with A340E—Fuse: No. 1 radio 15A
1985–88 Truck—Fuse: stop lamp 15A
1989–92 Truck—Fuse: EFI 15A

Then reinstall the fuse. The codes should now be cleared. If there are still codes present, either the codes were not properly cleared (Are the codes identical to those flashed out previously?) or the underlying problem is still there (Are only some of the codes the same as previously?).

To clear codes on vehicles that don't appear in the accompanying chart, you'll have to disconnect the computer connector, and leave it disconnected for one minute. Then reconnect the computer.

DIAGNOSTIC TROUBLE CODES—TOYOTA

Code	Component/Circuit Fault
38	ATF temperature sensor or circuit
42	Speedometer Vehicle Speed Sensor (VSS) or circuit
44	Rear transfer case speed sensor or circuit
61	Transmission/transaxle VSS or circuit
62	No. 1 shift solenoid or circuit
63	No. 2 shift solenoid or circuit
64	Lock-up solenoid or circuit
65	Transfer case solenoid or circuit
73	No. 1 center differential control solenoid or circuit
74	No. 2 center differential control solenoid or circuit

89446G58

Toyota transmission related Diagnostic Trouble Codes

Volkswagen

ENGINE CODES

General Information

CIS-E FUEL INJECTION

The CIS-E Motronic system is the latest development of the electronically controlled mechanical Continuous Injection System (CIS). This system uses injectors, fuel pump and air flow sensor that are similar to those on earlier systems. The fuel distributor is equipped with an electronically controlled differential pressure regulator. This is operated by the ECU to control the fuel pressure in the lower chamber of the fuel distributor, which controls air/fuel mixture.

The ECU is now equipped with an adaptive learning program which allows it to learn and remember the normal operating range of the mixture control output signal. This gives the system the capability to compensate for changes in altitude, slight vacuum leaks or other changes due to things such as engine wear. Cold

engine driveability and emissions are improved. The new ECU also is capable of cold start enrichment without the use of a thermo-time switch. The Fox still uses the thermo-time switch on CIS equipped vehicles.

The fuel injector pressure has been increased for better fuel atomization and residual pressure. The threads on the new injectors are different so they cannot be interchanged with older units. Some of the other components used on the CIS-E system are similar to those used on the fully electronic engine management systems. Some of the testing procedures are the same but the parts are not necessarily interchangeable.

MOTRONIC AND MOTRONIC 2.9 MULTIPORT FUEL INJECTION (MFI) SYSTEMS

Motronic and Motronic 2.9 systems are developments of the electronically controlled Multiport Fuel Injection (MFI) system. The two systems are almost identical. The engine control module (ECM) monitors engine intake air quantity using the Mass Air Flow (MAF) sensor. This is a true mass air measurement system. Using input from the MAE and other sensors, the ECM can calculate the length

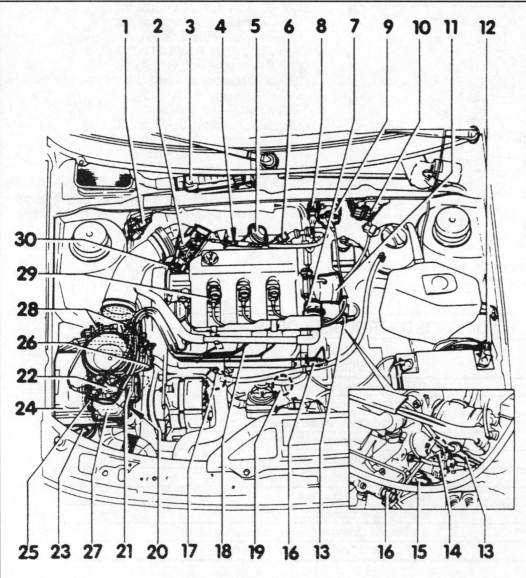

1. Oxygen sensor harness connector on right engine mount
2. Throttle body
3. Control module
4. Intake air temperature sensor—California only
5. EGR valve—California only
6. Exhaust tap
7. Idle stabiiizer valve
8. Ignition coil power output stage
9. Ignition coil
10. 6-pin wiring harness connector
11. Distributor
12. EGR vacuum amplifier—California only
13. Ignition timing sensor or plug wire 4
14. EGR vacuum valve—California only
15. Coolant temperature sensor
16. Cold start valve
17. Knock sensor I
18. Fuel injector
19. Knock sensor II
20. Heated air intake control door
21. Differential pressure regulator
22. Fuel distributor
23. Charcoal canister below air cleaner
24. Air filter
25. Potentiometer
26. Fuel pressure regulator
27. Air flow sensor
28. Charcoal canister solenoid valves
29. Spark plug
30. Throttle switch harness connectors

Engine compartment layout—Volkswagen GTI 16V and Passat shown

79232G81

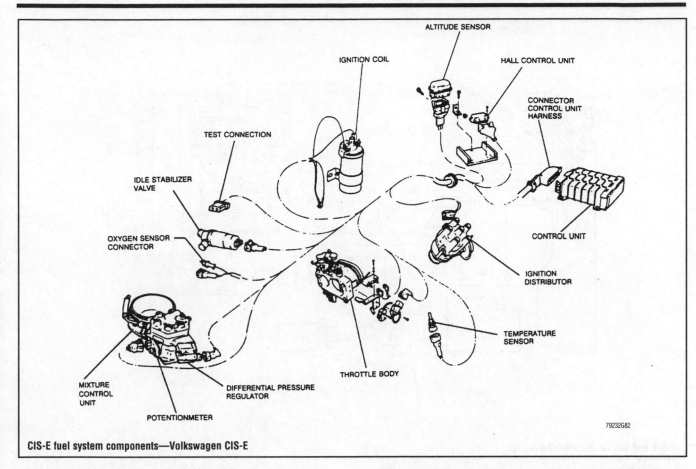

CIS-E fuel system components—Volkswagen CIS-E

of time the injectors should be opened, and also controls the ignition system timing.

The ECM is equipped with an adaptive learning program which allows it to learn and remember the normal operating range of the mixture control output signal. This allows the system to compensate for changes in altitude, slight vacuum leaks or other changes due to other things such as engine wear. Cold engine driveability and emissions are improved.

The ECM is also equipped with a fault memory. If the sensor signal or output solenoid feedback signal is outside preprogrammed parameters, the ECM will store a fault code representing the fault and sensor involved. The ECM will also illuminate the Malfunction Indicator Lamp (MIL) to inform the vehicle operator that the vehicle requires service.

MONO-MOTRONIC THROTTLE BODY FUEL INJECTION (TBI) SYSTEM

The Mono-Motronic system is development of the electronically controlled Throttle Body Fuel Injection (TBI) system. Based on outputs from the Throttle Position (TP), Engine Coolant Temperature (ECT) and Intake Air Temperature/Fuel Injector Temperature (IAT/FIT) sensors, the Engine Control Module (ECM) can infer intake air flow by air temperature and throttle position. This is a speed-density type control system. The ECM calculates the length of time the injector(s) should be opened, and with input from other sensors, also controls ignition timing.

The ECM is equipped with an adaptive learning program which allows it to learn and remember the normal operating range of the mixture control output signal. This allows the system to compensate for changes in altitude, slight vacuum leaks or other changes due to other things such as engine wear. Cold engine driveability and

emissions are improved. The ECM is also equipped with a fault memory. If the sensor signal or output solenoid feedback signal is outside preprogrammed parameters, the ECM will store a fault code representing the fault and sensor involved. The ECM will also illuminate the Malfunction Indicator Lamp (MIL) to inform the vehicle operator that the vehicle requires service.

DIGIFANT MULTIPORT FUEL INJECTION (MFI) SYSTEM

The Digifant Motronic system is development of the electronically controlled Multiport Fuel Injection (MFI) system. This system is quite similar on all models, but there is one significant difference among the three engines, intake air flow is measured using one of two systems. The 1.8L NA engine is equipped with a Vane Air Flow (VAF) sensor. This is a true mass air measurement system. The 1.8L SC and 2.5L NA engines use a Manifold Absolute Pressure (MAP) sensor.

Along with output from the Intake Air Temperature (IAT) sensor, the ECM can infer intake air flow by air temperature and pressure. This is a speed-density type control system. Using either measurement system, the ECM calculates the length of time the injectors should be opened, and with input from other sensors, also controls ignition timing.

The ECM is equipped with an adaptive learning program which allows it to learn and remember the normal operating range of the mixture control output signal. This allows the system to compensate for changes in altitude, slight vacuum leaks or other changes due to other things such as engine wear. Cold engine driveability and emissions are improved.

The ECM is also equipped with a fault memory. If the sensor signal or output solenoid feedback signal is outside preprogrammed parameters, the ECM will store a fault code representing the fault

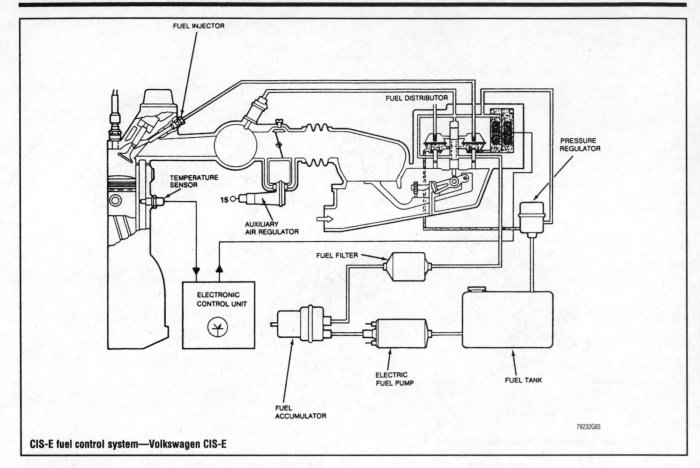

FUEL INJECTOR

FUEL DISTRIBUTOR

PRESSURE REGULATOR

TEMPERATURE SENSOR

15

AUXILIARY AIR REGULATOR

ELECTRONIC CONTROL UNIT

FUEL FILTER

ELECTRIC FUEL PUMP

FUEL TANK

FUEL ACCUMULATOR

79232G83

CIS-E fuel control system—Volkswagen CIS-E

and sensor involved. The ECM will also illuminate the Malfunction Indicator Lamp (MIL) to inform the vehicle operator that the vehicle requires service. Some vehicles equipped with both the Digifant Motronic system and California specification emissions equipment also have the capability of flashing diagnostic codes.

Service Precautions

- Do not disconnect the battery or the control unit before reading the fault codes. On the Motronic system, fault code memory is erased when power is interrupted.
- Make sure the ignition switch is **OFF** before disconnecting any wiring.
- Before removing or installing a control unit, disconnect the negative battery cable. The unit receives power through the main connector at all times and will be permanently damaged if improperly powered up or down.
- Keep all parts and harnesses dry during service. Protect the control unit and all solid-state components from rough handling or extremes of temperature.

Reading Codes

Only California vehicles with the Digifant II and Digifant I systems were equipped with the capability of flashing diagnostic codes.

On the Digifant II system codes were viewed through a combination rocker switch/indicator light. The following California vehicles were equipped with the Digifant II system:

1988–90 Golf, Jetta and GTI with 2.0L 16 valve engine
1990 Cabriolet with engine code 2H

On the Digifant I system a jumper cable would have to be connected and then the codes would flash from the CHECK engine light on the dash. The following California vehicles were equipped with the Digifant I system:

1990–93 Fox—Digifant I
1991–93 Cabriolet—Digifant I
1991–92 Corrado—Digifant I

On all other systems, codes can only be retrieved with the use of a special diagnostic tester, the VAG 1551. This tester is available at car dealerships. The VAG 1551 tester can be used on all vehicles that have code capability.

➡**Some diagnostic codes may be retrieved by connecting special jumper cable 357 971 514E or an equivalent to the check connectors. Others for the most part are going to require the use of a special VAG 1551 tester and adapter to retrieve any remaining codes.**

DIGIFANT II SYSTEM—ROCKER SWITCH METHOD

California models are the only vehicles equipped with the On-Board Diagnostic (OBD) lamp. On these models codes may be accessed by 2 methods. The first is the use of a combination rocker switch/lamp located on the instrument panel. The second is used by the dealers, they use a special tool called the VAG 1551.

An indicator light labeled CHECK is located in a rocker switch on the instrument panel. Each time the engine is started, the indicator light will flash once to inform the operator the bulb is working.

The light will come on and stay on if a fault develops in the engine management system. It will also display diagnostic codes to assist in trouble diagnosis.

A diagnostic code consists of 4 groups of flashes. There is a 2.5 second pause (light OFF), between each group of flashes.

The indicator light will come on for two and half seconds prior to displaying a fault code when the diagnostic procedure has been activated. The fault code will continue repeating while the ignition is **ON**.

If the fault is not repaired the indicator light will come on and stay on when the ignition is turned **ON** to signify the fault still exists.

The following California vehicles were equipped with the Digifant II system:

1988–90 Golf, Jetta and GTI with 2.0L 16 valve engine

1990 Cabriolet with engine code 2H

Prior to checking for codes, drive the vehicle for 10 minutes or more.

1. Turn the ignition to the **ON** position, do not start the engine.

2. Press and hold down the rocker switch for 4–6 seconds then release the switch. The CHECK indicator lamp will begin flashing a diagnostic code.

3. Press and hold down the rocker switch again for 4–6 seconds then release it. The indicator lamp will flash the next diagnostic code and will continue until all codes have been displayed.

When all diagnostic codes have been displayed, the indicator lamp will flash a series of 2.5 second flashes ON and 2.5 seconds OFF. This is an 'End Of Fault Sequence' code. If there are no faults stored in the control unit memory, the indicator lamp will flash Code 4444.

➡**Occasionally the control unit will sense various deviations or changes in the air/fuel mixture. Because of the sensitivity of this system a fault code may set without any apparent problem showing up. This is a normal function with systems of this type.**

DIGIFANT I SYSTEM—JUMPER CABLE METHOD

California models are the only vehicles equipped with the On-Board Diagnostic (OBD) lamp. On these models codes may be access by 2 methods. The first is the use of a special jumper cable connected to the diagnostic connector. The second is used by the dealers, they use a special tool called the VAG 1551.

The following California vehicles were equipped with the Digifant I system:

1990–93 Fox—Digifant I

1991–93 Cabriolet—Digifant I

1991–92 Corrado—Digifant I

1. Verify that all the fuses and grounds in the engine compartment are good.

2. Turn the ignition key **ON**.

3. Connect jumper cable 357 971 514E or equivalent to the connectors located in the center console. The black end of jumper wire connects to the black diagnostic connector in the console. The white end of the jumper wire connects to white diagnostic connector in the console.

4. Connect the jumper wire for about S seconds. When the OBD light begins flashing, remove the jumper wire.

5. Count the flashes of the light to get codes, each separate flash, in one code, will be have a short interval in between. Each interval between codes will be about 2.5 seconds. Count flashes until either Code 4444 or 0000 appears. To end this procedure turn the ignition switch **OFF**.

6. Output checks can not be performed without a diagnostic tester.

DIGIFANT I SYSTEM—WITH DIAGNOSTIC TESTER

This method is used by the Dealers and requires the use of the special tester VAG 1551. The following vehicles can only be accessed using this tester:

1991–92 Golf, Jetta—Digifant I system

1993–94 Corrado—with VR6 Engine—Motronic system

All CIS-E Motronic and Motronic systems

1. Make sure ignition switch is **OFF** and that all fuses are good. Make sure all grounds in engine compartment are good, especially those for the battery and control module.

2. Make sure the air conditioning system is OFF.

3. Connect the VAG 1551 diagnostic tester using VAG 1551/1 Adapter cable or equivalent. The diagnostic connectors are located in the center console, under the shifter. The shifter knob and console cover must be removed to access diagnostic connectors. Connect diagnostic connector 1 (black) to the black connector on the scan tool. Connect diagnostic connector 2 (white) to the white connector on the scan tool. The blue connector is not required.

4. Turn the ignition switch **ON**; now start the vehicle and let it idle. If vehicle will not start, crank engine for 6 seconds and leave ignition **ON**.

5. Turn the tester ON and make sure it is receiving power. The screen will display 2 menu options; Rapid Data Transfer and Blink Code Output:

6. If you choose to use mode 02, Blink Code Output, skip to Step 10.

7. Select mode 01, Rapid Data Transfer, and address word 01. Now press the 0 button to enter your selection. The tester will display a control unit part number, the system it controls and an application (country) code.

a. If the information is displayed and is correct, press the (run) key to continue. The display 'Select function XX' will appear.

b. If 'Control unit does not answer' is displayed, use the Help key to display a list of possible causes. When the problem is repaired, return to step 1 and start over again.

8. When function 02 is selected, the control module will report fault codes to the diagnostic tester.

9. When all codes have been reported, proceed to the Output Check diagnosis or select function 06 to exit the fault code memory without erasing the codes. Repair and erase the faults, then check and see if all faults have been corrected.

10. The following steps will retrieve engine codes by using Blink Code Output.

11. To operate the VAG 1551 tester in Blink Code Output, select menu option # 2. An asterisk will appear and flash the codes, which the tester will count and report on the screen as numbers. If Code 4444 or 0000 is displayed, no faults are found in memory.

➡**On vehicles that use 4444 for no codes present, the 0000 will stand for output ended.**

12. If the engine is not running, some codes may be displayed. These can be ignored if the engine was intentionally stalled, but should be investigated if the engine will not start.

13. Press the (run) key to advance to the next code. Read through entire code list before starting repairs.

14. When the 0000 (output ended) code is displayed, pressing the (run) key again will proceed to another control module. If no other control modules are to be tested, the following display will appear: Blink Code Output is ended. To stop the program without

erasing the codes, turn the ignition key **OFF** and press the clear C button once.

15. Repair and erase the faults, then check to see if all faults have been corrected.

Output Check Diagnosis

Only the Motronic system is equipped with this program. It allows testing most of the engine output devices without running the engine. The program cannot be run without the VAG 1551 Diagnostic Tester or equivalent. During the test, four output devices are activated in the following order:

- Differential pressure regulator
- Carbon canister frequency valve
- Idle stabilizer valve
- Cold start valve

Testing the differential pressure regulator requires a multi-meter that will read milliamps. The other items can be checked with a voltmeter, test light or by listening and feeling for valve activation. The cold start valve is activated for a limited time to avoid flooding the engine.

1. Connect the diagnostic tester, turn the ignition switch **ON** and confirm that the tester will communicate with the control unit. See the procedure for retrieving fault codes.

2. Select Rapid Data Transfer and Function 03. When the test is started by pressing the Q button (enter), the first output signal is generated.

3. Each time the Run button is pressed, the tester will send an output signal to the next device on the list.

4. When the last item has been tested, select Function 06 to exit the program. To repeat the test, turn the ignition switch **OFF** and **ON** again.

➡**Leave the ignition OFF for approximately 20 seconds, before selecting Output Check diagnosis again.**

Clearing Codes

1988–95 VEHICLES—WITHOUT DIAGNOSTIC TESTER

1. To erase codes, wait until Code 4444 or 0000 is displayed.
2. Turn ignition switch **OFF** and connect jumper wire to diagnostic connectors again.
3. Turn the ignition switch **ON** and leave connectors jumpered for about 5 seconds, When Code 4444 or 0000 appears the codes will be erased.
4. Turn the ignition switch **OFF** and remove jumper wire.

1988–95 VEHICLES—WITH DIAGNOSTIC TESTER

For both engine and automatic transaxle, after all fault codes have been retrieved, select Function 05 and press the Q button to enter the selection. The memory will be erased only if all fault codes have been retrieved. Test drive the vehicle for at least 10 minutes, including at least 1 full throttle application above 3000 rpm. Check the fault code memory again to make sure all faults have been repaired.

Diagnostic Trouble Codes

1988–95 VEHICLES

➡**The 5 digit code groups are used with a diagnostic tester. The 4 digit code groups are the flashing codes.**

00000 or 4444 No faults in memory
00281 or 1231 Vehicle Speed Sensor (VSS) signal is missing
00282 or 1232 Throttle actuator solenoid or wiring harness
00513 or 2111 Engine RPM sensor signal is missing
00514 or 2112 Ignition reference sensor signal is missing
00515 or 2113 Hall sender signal is missing
00516 or 2121 Idle switch has open short in circuit
00517 or 2123 Full throttle switch
00518 or 2212 Throttle position sensor
00519 or 2222 Manifold absolute pressure (MAP) sensor
00520 or 2232 Air flow sensor signal is missing
00521 or 2242 CO potentiometer
00522 or 2312 Engine coolant temperature (ECT) sensor
00523 or 2322 Intake air temperature (IAT) sensor
00524 or 2142 Knock sensor 1 signal is missing
00525 or 2342 Oxygen sensor signal missing
00527 or 2412 Intake air temperature (IAT) sensor has open/short in circuit
00532 or 2234 Supply voltage is too high
00533 or 2231 Idle speed regulation out of limit
00535 or Both 2141/2142 Knock sensor or control program
00537 or 2341 Oxygen sensor signal out of limit
00540 or 2144 Knock sensor 2 signal is missing
00543 or 2214 RPM exceeds maximum limit
00545 or 2314 Engine/Transmission electrical connection
00549 or 2314 Fuel consumption signal
00552 or 2323 Air flow sensor signal missing
00553 or 2324 Mass Air Flow (MAE) sensor signal is out of range
00558 or NA Adaptive mixture control lean (Fuel injector leak, EVAP purge system)
00559 or NA Adaptive mixture control rich (vacuum leak)
00560 or 2411 EGR temperature sensor circuit
00561 or 2413 Mixture adaptation limits are out of range
00585 or 2411 EGR temperature sensor circuit (2.8L AAA engine only)
00586 EGR controlling system, EGR valve is sticking or false signals
00587 Adjustment limit mixture regulator is lean
00609 Ignition output 1 circuit
00624 A/C compressor engagement circuit has mechanical or electrical malfunction
00640 or 3434 Heated Oxygen sensor relay has open/short circuit
01025—Malfunction indicator lamp (MIL) circuit
01242 or 4332 Output stages in engine control module (ECM)
01247 or 4343 EVAP frequency valve 1 has open/short in circuit
01249 or 4411 Fuel injector #1 circuit has open/short
01250 or 4412 Fuel injector #2 circuit has open/short
01251 or 4413 Fuel injector #3 circuit has open/short
01252 or 4414 Fuel injector #4 circuit has open/short
01253 or 4421 Fuel injector #5 circuit has open/short
01254 or 4422 Fuel injector #6 circuit has open/short
01257 or 4431 Idle Air Control (IAC) valve has open/short in circuit or a mechanical malfunction
01259 or 4433 Fuel pump relay is faulty or short circuited

01265 or 4312 EGR frequency valve has open/short in circuit
65535 or 1111 Engine Control Module (ECM) is defective
0000 End of output
NA Not Available

TRANSMISSION CODES

Reading Codes

It is necessary to use Volkswagen's VAG 1551 scan tool/tester to retrieve/read transmission related Diagnostic Trouble Codes (DTC's). Follow Audi's instructions to properly connect the tester/scan tool, then to retrieve the transmission codes.

Clearing Codes

It is necessary to use Volkswagen's VAG 1551 scan tool/tester to clear transmission related Diagnostic Trouble Codes (DTC's) from the Transmission Control Module's (TCM's) memory. Follow Audi's instructions to properly connect the tester/scan tool, then to retrieve the transmission codes.

Diagnostic Trouble Codes

Code 0000 or 4444 No fault recognized
Code 00258 or 1113 Solenoid valve 1, or open circuit or short to ground
Code 00260 or 1121 Solenoid valve 2, or open circuit or short to ground
Code 00262 or 1123 Solenoid valve 3, or open circuit or short to ground
Code 00263 or 1124 Mechanical or hydraulic transmission fault
Code 00264 or 1131 Solenoid valve 4, or open circuit or short to ground
Code 00266 or 1133 Solenoid valve 5, or open circuit or short to ground
Code 00268 or 1141 Solenoid valve 6, or open circuit or short to ground
Code 00270 or 1143 Solenoid valve 7, or open circuit or short to ground
Code 00281 or 1231 Vehicle Speed Sensor (VSS) and/or circuit
Code 00293 or 1314 Multi-Function switch and/or circuit
Code 00296 or 1323 Kickdown switch and/or circuit
Code 00299 or 1332 Trans-Program switch and/or circuit
Code 00300 or 1333 Trans oil temperature sender and/or circuit
Code 00518 or 2212 Throttle valve potentiometer and/or circuit
Code 00526 or 2131 Brake light switch and/or circuit
Code 00529 or 2122 Engine speed signal missing
Code 00532 or 2234 Supply voltage, voltage for all values too low
Code 00545 or 2314 Engine/transmission electrical connection
Code 00596 Short between solenoid wires
Code 00638 Engine/transmission electrical connection
Code 00641 Trans oil temperature and/or circuit
Code 00652 Gear monitoring implausible signal—electric/hydraulic fault
Code 00660 Kickdown switch/Throttle valve potentiometer and/or circuits

Code 01236 or 4314 Selector lever lock solenoid and/or circuit
Code 65535 or 1111 Control unit—electrical influences from outside sources, bad ground connection and/or defective Transmission Control Module (TCM)

Volvo

ENGINE CODES

General Information

The LH-Jetronic 2.2 was used on models through 1990. The LH-Jetronic 2.4 and 3.1 fuel injection systems are used on 1990 and newer 240, 700 and 900 series vehicles. The LHJetronic 3.2 fuel injection system is used on the 1993 and later 850. The Motronic 1.8 was used on the 1992 and later 960. The Motronic 4.3 is used on 1994 and later 850 Turbo. On all fuel systems except the LH 2.2 system are monitored by a self-diagnostic system that lights up a warning lamp on the instrument panel. The LH-Jetronic 2.2 does not have self-diagnostic ability. Many different fault codes can be set, however only three can be stored at any one time. Fault tracing can be carried out by utilizing the diagnostic unit.

Service Precautions

• Do not operate the fuel pump when the fuel lines are empty.
• Do not operate the fuel pump when removed from the fuel tank.
• Do not reuse fuel hose clamps.
• The washer(s) below any fuel system bolt (banjo fittings, service bolt, fuel filter, etc.) must be replaced whenever the bolt is loosened. Do not reuse the washers; a high-pressure fuel leak may result.
• Make sure all ECU harness connectors are fastened securely. A poor connection can cause an extremely high voltage surge and result in damage to integrated circuits.
• Keep all ECU parts and harnesses dry during service. Protect the ECU and all solid-state components from rough handling or extremes of temperature.
• Use extreme care when working around the ECU or other components; the airbag or SRS wiring may be in the vicinity. On these vehicles, the SRS wiring and connectors are yellow; do not cut or test these circuits.
• Before attempting to remove any parts, turn the ignition switch **OFF** and disconnect the battery ground cable.
• Always use a 12 volt battery as a power source for the engine, never a booster or high-voltage charging unit.
• Do not disconnect the battery cables with the engine running.
• Do not disconnect any wiring connector with the engine running or the ignition **ON** unless specifically instructed to do so.
• Do not apply battery power directly to injectors.
• Whenever possible, use a flashlight instead of a drop light.
• Keep all open flame and smoking material out of the area. a Use a shop cloth or similar to catch fuel when opening a fuel system. Consider the fuel-soaked rag to be a flammable solid and dispose of it in the proper manner.
• Relieve fuel system pressure before servicing any fuel system component.
• Always use eye or full-face protection when working around fuel lines, fittings or components.
• Always keep a dry chemical (class B-C) fire extinguisher near the area.

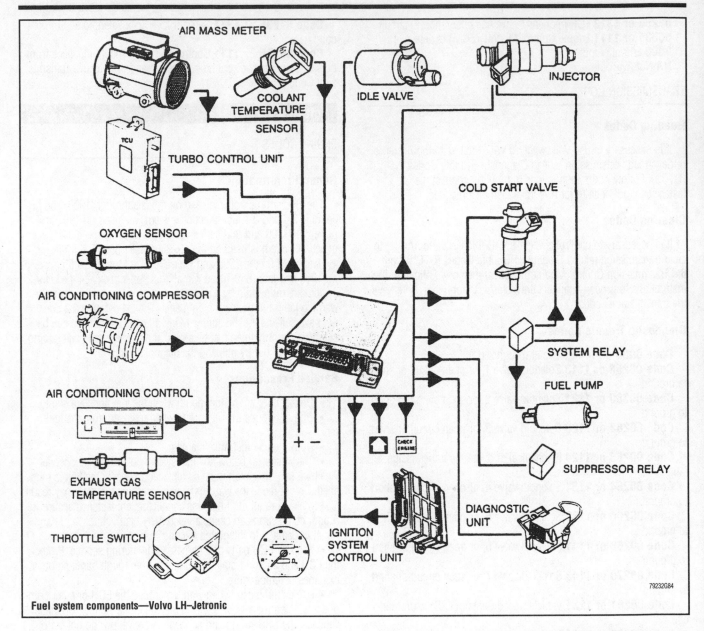

Fuel system components—Volvo LH–Jetronic

Reading Codes

➡**On-board engine diagnostics were not available on vehicles prior to 1988.**

1. Open diagnostic socket cover and install selector cable into socket No. 2 for fuel injection codes or socket No. 6 (except Motronic systems) for ignition codes.

2. Turn ignition to the **ON** position.

3. Enter control system 1 by pressing the button once. Hold the button for at least 1 second, but not more than 3.

4. Watch the diode light and count the number of flashes in the 3 flash series indicating a fault code. The flash series are separated by 3 second intervals. Note fault codes.

➡**If there are no fault codes in the diagnostic unit, the diode will flash 1-1-1 and the fuel system is operating correctly.**

5. If diode light does not flash when button is pressed, or no code is flashed there is a problem with the soft-diagnostic system, proceed as follows:

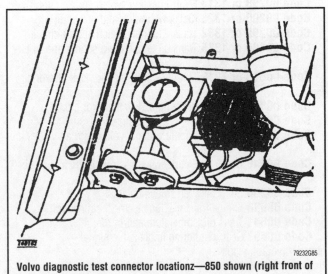

Volvo diagnostic test connector locationz—850 shown (right front of engine compartment)

a. Check ground connections on the intake manifold, and the ground connection for the Lambda-sond at the right front mud-guard.

b. Check the fuses for the pump relay and the primary pump. On 240 models, fuses are located inside the engine compartment on the left side wheel well housing. On 760/780 models, fuses are located in the center console, just below the radio. On 740/940 models, fuses are located behind the ashtray. Access can be gained by removing the ashtray, and pressing upward on the tab marked 'electrical fuses press'. On 850 models, the fuses are located on the left side of the engine compartment behind the strut mount plate. The fuses on 960 models are located on the far left side of the dashboard. The driver's door must be open to gain access to the fuses.

c. Remove glove compartment, and check control unit ground connections.

d. Turn ignition switch to the **OFF** position. Remove control unit connector and connector protective sleeve.

e. Check diagnostic socket, (Steps 5e–5j), by connecting a voltmeter between ground and No. 4 connection on the control unit connector. Reading should be 12 volts. If no voltage is present, check lead between control unit connector and fuse No. 1 in the fuse/relay box.

f. Turn the ignition to the **ON** position, and install selector cable into the No. 2 socket on the diagnostic socket. Connect a voltmeter between ground and No. 12 connection on the control unit connector. Reading should be 12 volts. Press the button on diagnostic socket and note reading. Reading on the voltmeter should be 0 volts. If no voltage at the control unit is present, take reading at the diagnostic socket connector. If reading remains at 12 volts when button is pressed, check diagnostic socket.

g. Connect a voltmeter between ground and the red/black lead on the diagnostic socket connector. Reading should be 12 volts.

h. Connect a suitable ohmmeter between ground and the brown/black lead in the diagnostic socket connector. Reading should be 0 ohms.

i. Turn ignition to the **OFF** position. Connect ohmmeter between diagnostic socket selector cable and the pin under selector button. Ohmmeter should read infinity. Press button, and note reading. Reading should be 0 ohms.

j. Connect a suitable diode/multimeter tester, or equivalent, between the diagnostic socket diode light and the selector cable. Connect red test pin from the tester to pin under diode light and black test pin from tester to selector cable. A reading on the tester indicates correct diode light function. With no reading on tester, replace diagnostic socket.

k. Check the system relay/primary relay by connecting a volt-meter between ground and the No. 9 connection on the control unit connector, then connect a jumper wire between ground and No. 21 connection on the control unit connector. The relay should activate and the reading should be 12 volts.

6. Press the diagnostic socket button. Note any additional fault codes.

➡ **The diagnostic system memory is full when it contains 3 fault codes. Until those codes are corrected and the memory erased, the system cannot give information on any other problems.**

7. Press the diagnostic socket button for the third time to see if a third fault code is stored in the memory. If the diode light flashes the same Code 1-1-1, there are no other codes in the memory.

Clearing Codes

1989–94 VEHICLES

1. Turn the ignition switch to the **ON** position.
2. Read fault codes.
3. Press diagnostic socket button 1 time and hold for approximately 5 seconds. Release button. After 3 seconds the diode light should light up. While the light is still lit, press the button again and hold for approximately 5 seconds. After releasing the button, the diode light should go off.
4. To ensure that the memory is erased, press the button 1 time, for 1 second but not more than 3 seconds. The diode light should flash Code 1-1-1.
5. Start and run engine. If engine will not start, correct the problem before proceeding and start over with step 1.

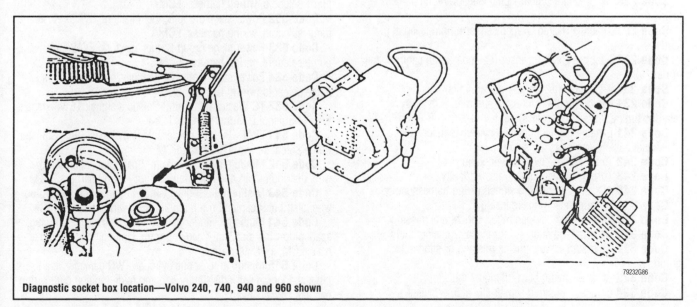

Diagnostic socket box location—Volvo 240, 740, 940 and 960 shown

79232G86

6. Check to see if new fault codes have been stored in the memory by pressing the diagnostic socket button 1 time, for 1 second but not more than 3 seconds.

7. If fault Code 1-1-1 flashes, it indicates that there are no additional fault codes stored in its memory.

Diagnostic Trouble Codes

1989–94 VEHICLES

Code 111 No fault

Code 112 Fault in control module

Code 113 Heated oxygen sensor at maximum enrichment limit; injector clogged, break in lead, etc.

Code 115 Fuel injector 1; wiring harness, ECM

Code 121 Air Mass Meter (MAF) or air pressure sensor signal; missing/faulty

Code 122 Air temperature sensor signal; missing/faulty

Code 123 Engine temperature sensor signal faulty or missing

Code 125 Fuel injector 2; wiring harness, ECM

Code 131 Engine Speed Signal

Code 132 Battery voltage; too low or too high

Code 133 Throttle (Shutter) switch; idle setting faulty

Code 135 Fuel injector 3; wiring harness, ECM

Code 142 Fault in control module

Code 143 Front knock sensor signal missing/faulty

Code 144 Load signal missing from fuel system control module

Code 145 Fuel injector 4; wiring harness, ECM

Code 153 Rear HO2S signal; wiring harness, ECM

Code 154 EGR system leakage; EGR valve, EGR transfer pipes

Code 155 Fuel injector 5; wiring harness, ECM

Code 212 Oxygen sensor (Lambda-sond) signal; missing/faulty

Code 213 Throttle switch; full load setting faulty

Code 214 Timing pick-up signal; missing intermittently

Code 214 Vehicle speed sensor signal intermittent

Code 221 Adaptive Heated Oxygen sensor running rich at part load

Code 222 System relay; signal is missing/faulty

Code 223 Idling valve signal; missing/faulty

Code 224 Missing/faulty temperature sensor circuit; signal missing/faulty

Code 225 NC pressure sensor signal; sensor/wiring harness faulty

Code 231 Adaptive Heated Oxygen sensor running lean at part load

Code 232 Adaptive Oxygen sensor (Lambda-sond) control; lean or rich, idle

Code 233 Adaptive idling control out of limits

Code 234 Faulty throttle control; engine runs with safety retarded timing (about 10 degrees); up to 1991

Code 241 Exhaust gas recirculation system; sensor senses flow of exhaust back to engine is too small

Code 242 Turbo control valve; not operating

Code 243 Throttle switch signal; missing/faulty

Code 245 AC solenoid closing signal; wiring harness/faulty

Code 311 Speedometer signal missing

Code 312 Signal for knock-controlled enrichment missing

Code 314 Camshaft position sensor signal; missing/faulty

Code 321 Cold start valve signal is missing or shorted to ground

Code 322 Air mass meter burn-off signal missing

Code 324 Camshaft position sensor signal intermittent

Code 325 Memory failure; ECM wiring harness

Code 335 Request to illuminate MIL from TCM; TCM, wiring harness

Code 342 NC blocking relay; current too high

Code 411 Throttle switch signal; missing/faulty

Code 416 Boost pressure reduction request from TCM; TCM wiring harness/faulty

Code 413 EGR temperature sensor signal; missing or incorrect

Code 421 Boost pressure sensor in control module

Code 423 Throttle position sensor signal; missing/faulty

Code 424 Load signal from fuel system; RPM too low for boost

Code 431 Coolant temperature sensor signal; missing

Code 432 High temperature warning in control box (temp. above 185°F (85°C))

Code 433 Rear knock sensor signal; missing/faulty

Code 435 Front HO2S slow response; front HO2S

Code 436 Rear HO2S compensation; rear HO2S

Code 443 TWC efficiency; TWC converter

Code 444 Acceleration sensor signal; acceleration sensor wiring harness

Code 451 Misfire, cylinder 1; Spark plug, spark plug wire, distributor, ignition coil, wiring harness

Code 452 Misfire, cylinder 2; Spark plug, spark plug wire, distributor, ignition coil, wiring harness

Code 453 Misfire, cylinder 3; Spark plug, spark plug wire, distributor, ignition coil, wiring harness

Code 454 Misfire, cylinder 4; Spark plug, spark plug wire, distributor, ignition coil, wiring harness

Code 455 Misfire, cylinder 5; Spark plug, spark plug wire, distributor, ignition coil, wiring harness

Code 512 Heated oxygen sensor at maximum lean running limit

Code 513 High temperature warning in control box (temp. above 203°F (95°C))

Code 514 Engine cooling fan, Low speed signal; engine cooling fan relay wiring harness

Code 521 Front HO2S preheating; Front HO2S, wiring harness

Code 522 Rear HO2S preheating; Rear HO2S, wiring harness

Code 531 Power stage group A; fuel injectors, EVAP canister purge solenoid, wiring harness, ECM

Code 532 Power stage group B; fuel injectors, EVAP canister purge solenoid, wiring harness, ECM

Code 533 Power stage group C; fuel injectors, EVAP canister purge solenoid, wiring harness, ECM

Code 534 Power stage group D; fuel injectors, EVAP canister purge solenoid, wiring harness, ECM

Code 535 TC Wastegate Control Solenoid signal; TC Wastegate Control Solenoid, wiring harness, ECM

Code 541 EVAP Canister purge Solenoid signal; EVAP Canister Purge Solenoid, wiring harness, ECM

Code 542 Multiple cylinder misfire; Spark plug, spark plug wire, distributor cap and rotor, ignition coil, wiring harness, ECM

Code 543 Misfire at least one cylinder; Spark plug, spark plug wire, distributor cap and rotor, ignition coil, wiring harness, ECM

Code 544 Multiple cylinder misfire TWC damage; Spark plug, spark plug wire, distributor cap and rotor, ignition coil, wiring harness, ECM, TWC converter

Code 545 Misfire at least one cylinder, TWC damage; Spark plug, spark plug wire, distributor cap and rotor, ignition coil, wiring harness, TWC converter

Code 551 Misfire in cylinder 1, TWC damage; Cylinder 1 spark plug, spark plug wire, distributor cap, ignition coil, wiring harness, TWC converter

Code 552 Misfire in cylinder 2, TWC damage; Cylinder 1 spark plug, spark plug wire, distributor cap, ignition coil, wiring harness, TWC converter

Code 553 Misfire in cylinder 3, TWC damage; Cylinder 1 spark plug, spark plug wire, distributor cap, ignition coil, wiring harness, TWC converter

➡**Code combination explanations are as follows:**

Code 113, 221 & 232 In the part load range; Air/Fuel mixture is lean and on idle

Code 113 & 221 Air/Fuel mixture is lean in the part load range

Code 113 & 231 Air/Fuel mixture is probably rich in the part load range

TRANSMISSION CODES

Reading Codes

On Volvo models equipped with transmission related Diagnostic Trouble Codes (DTC's), it is necessary to use the Volvo Scan Tool (VST), connected to the Diagnostic Link Connector (DLC), to retrieve/read the transmission DTC's. Follow Volvo's instructions to properly connect the tester/scan tool, then to retrieve the transmission codes.

Clearing Codes

It is necessary to use the VST to clear transmission related Diagnostic Trouble Codes (DTC's) from the Transmission Control Module's (TCM's) memory. Follow Audi's instructions to properly connect the tester/scan tool, then to retrieve the transmission codes.

Diagnostic Trouble Codes

MODE 1 CODES

Code 111 No faults found
Code 112 Solenoid S1 and/or circuit
Code 113 Transmission Control Module (TCM) fault
Code 114 Program selector open circuit or short circuit
Code 121 Solenoid S1 and/or circuit; control module fault
Code 122 Solenoid S1 and/or circuit
Code 123 Solenoid STH and/or circuit
Code 124 Program selector faulty and/or circuit
Code 131 Solenoid STH and/or circuit; control module fault
Code 132 TCM fault
Code 134 Faulty load signal from fuel sys.
Code 141 Oil temperature sensor and/or circuit
Code 142 Oil temperature sensor and/or circuit
Code 143 Kickdown switch and/or circuit
Code 211 TCM fault
Code 212 Solenoid S2 and/or circuit
Code 213 Throttle Position (TP) signal and/or circuit
Code 221 Solenoid S2 and/or circuit
Code 222 Solenoid S2 and/or circuit
Code 223 Throttle Position (TP) signal and/or circuit
Code 231 Throttle Position (TP) signal and/or circuit
Code 232 Speedometer signal missing
Code 233 Speedometer signal faulty
Code 235 Oil temperature signal too high

Code 245 Torque limiting signal and/or circuit
Code 311 Engine speed signal from gearbox missing
Code 312 Engine speed signal from gearbox faulty
Code 313 Gear position sensor and/or circuit
Code 321 Gear changing too slow
Code 322 Incorrect gear change ratio
Code 323 Lock-up slipping or not engaged
Code 331 Solenoid SL and/or circuit
Code 332 Solenoid SL and/or circuit; control module fault
Code 333 Solenoid SL and/or circuit; control module fault

MODE 2 CODES

Code 144 Gear selector in position R OK
Code 214 Gear selector in position D OK
Code 224 Gear selector in position 3 OK
Code 234 Gear selector in position L OK
Code 241 Gear selector in position N OK
Code 242 Gear selector in position P OK
Code 243 Undefined signal from the program selector
Code 244 Economy position OK
Code 314 Sport position OK
Code 324 Winter position OK
Code 334 Brake light switch OK
Code 341 Kickdown switch OK

MODE 3 CODES

Solenoid S1 operating.
Solenoid S2 operating.
Solenoid SL operating.
Solenoid STH operating.
Indicating LED in combination instr. flashes.
Malfunction Indicator Lamp (MIL) flashes.
Drive compensation and torque control signals are tested with engine idling.
Engine speed changes.

MODE 4 CODES

Code 311 Normal speed
Code 312 x2 speed
Code 313 x10 speed
Code 342 Solenoid S1
Code 343 Solenoid S2
Code 344 Solenoid SL
Code 411 Solenoid STH
Code 412 Indicating lamp
Code 414 Drive compensation (P&N)
Code 422 Torque control (TC2)
Code 423 Torque control (TC1)
Code 431 Malfunction Indicator Lamp (MIL)
Code 432 Torque control (TCT)

MODE 5 CODES

Code 115 Oil temperature
Code 125 Throttle signal status
Code 135 Engagement time status
Code 424 Speed
Code 434 Gearbox rpm
Code 444 Throttle opening

MODE 6 CODES

Code 125 Reset throttle signal
Code 135 Reset engagement time

OBD II POWERTRAIN DIAGNOSTIC TROUBLE CODES

Acura

READING CODES

Reading the control module memory is on of the first steps in OBD II system diagnostics. This step should be initially performed to determine the general nature of the fault. Subsequent readings will determine if the fault has been cleared.

Reading codes can be performed by any of the methods below:

- Read the control module memory with the Generic Scan Tool (GST)
- Read the control module memory with the vehicle manufacturer's specific tester

To read the fault codes, connect the scan tool or tester according to the manufacturer's instructions. Follow the manufacturer's specified procedure for reading the codes.

CLEARING CODES

Control module reset procedures are a very important part of OBD II system diagnostics. This step should be done at the end of any fault code repair and at the end of any driveability repair.

Clearing codes can be performed by any of the methods below:

- Clear the control module memory with the Generic Scan Tool (GST)
- Clear the control module memory with the vehicle manufacturer's specific tester
- Turn the ignition **OFF** and remove the negative battery cable for at least 1 minute.

Removing the negative battery cable may cause other systems in the vehicle to lose their memory. Prior to removing the cable, ensure you have the proper reset codes for radios and alarms.

➡**The MIL may also be de-activated for some codes if the vehicle completes three consecutive trips without a fault detected with vehicle conditions similar to those present during the fault.**

DIAGNOSTIC TROUBLE CODES

P0000 No Failures
P0100 Mass or Volume Air Flow Circuit Malfunction
P0101 Mass or Volume Air Flow Circuit Range/Performance Problem
P0102 Mass or Volume Air Flow Circuit Low Input
P0103 Mass or Volume Air Flow Circuit High Input
P0104 Mass or Volume Air Flow Circuit Intermittent
P0105 Manifold Absolute Pressure/Barometric Pressure Circuit Malfunction
P0106 Manifold Absolute Pressure/Barometric Pressure Circuit Range/Performance Problem
P0107 Manifold Absolute Pressure/Barometric Pressure Circuit Low Input
P0108 Manifold Absolute Pressure/Barometric Pressure Circuit High Input

P0109 Manifold Absolute Pressure/Barometric Pressure Circuit Intermittent
P0110 Intake Air Temperature Circuit Malfunction
P0111 Intake Air Temperature Circuit Range/Performance Problem
P0112 Intake Air Temperature Circuit Low Input
P0113 Intake Air Temperature Circuit High Input
P0114 Intake Air Temperature Circuit Intermittent
P0115 Engine Coolant Temperature Circuit Malfunction
P0116 Engine Coolant Temperature Circuit Range/Performance Problem
P0117 Engine Coolant Temperature Circuit Low Input
P0118 Engine Coolant Temperature Circuit High Input
P0119 Engine Coolant Temperature Circuit Intermittent
P0120 Throttle/Pedal Position Sensor/Switch "A" Circuit Malfunction
P0121 Throttle/Pedal Position Sensor/Switch "A" Circuit Range/Performance Problem
P0122 Throttle/Pedal Position Sensor/Switch "A" Circuit Low Input
P0123 Throttle/Pedal Position Sensor/Switch "A" Circuit High Input
P0124 Throttle/Pedal Position Sensor/Switch "A" Circuit Intermittent
P0125 Insufficient Coolant Temperature For Closed Loop Fuel Control
P0126 Insufficient Coolant Temperature For Stable Operation
P0130 O_2 Circuit Malfunction (Bank no. 1 Sensor no. 1)
P0131 O_2 Sensor Circuit Low Voltage (Bank no. 1 Sensor no. 1)
P0132 O_2 Sensor Circuit High Voltage (Bank no. 1 Sensor no. 1)
P0133 O_2 Sensor Circuit Slow Response (Bank no. 1 Sensor no. 1)
P0134 O_2 Sensor Circuit No Activity Detected (Bank no. 1 Sensor no. 1)
P0135 O_2 Sensor Heater Circuit Malfunction (Bank no. 1 Sensor no. 1)
P0136 O_2 Sensor Circuit Malfunction (Bank no. 1 Sensor no. 2)
P0137 O_2 Sensor Circuit Low Voltage (Bank no. 1 Sensor no. 2)
P0138 O_2 Sensor Circuit High Voltage (Bank no. 1 Sensor no. 2)
P0139 O_2 Sensor Circuit Slow Response (Bank no. 1 Sensor no. 2)
P0140 O_2 Sensor Circuit No Activity Detected (Bank no. 1 Sensor no. 2)
P0141 O_2 Sensor Heater Circuit Malfunction (Bank no. 1 Sensor no. 2)
P0142 O_2 Sensor Circuit Malfunction (Bank no. 1 Sensor no. 3)
P0143 O_2 Sensor Circuit Low Voltage (Bank no. 1 Sensor no. 3)
P0144 O_2 Sensor Circuit High Voltage (Bank no. 1 Sensor no. 3)
P0145 O_2 Sensor Circuit Slow Response (Bank no. 1 Sensor no. 3)
P0146 O_2 Sensor Circuit No Activity Detected (Bank no. 1 Sensor no. 3)
P0147 O_2 Sensor Heater Circuit Malfunction (Bank no. 1 Sensor no. 3)

P0150 O_2 Sensor Circuit Malfunction (Bank no. 2 Sensor no. 1)
P0151 O_2 Sensor Circuit Low Voltage (Bank no. 2 Sensor no. 1)
P0152 O_2 Sensor Circuit High Voltage (Bank no. 2 Sensor no. 1)
P0153 O_2 Sensor Circuit Slow Response (Bank no. 2 Sensor no. 1)
P0154 O_2 Sensor Circuit No Activity Detected (Bank no. 2 Sensor no. 1)
P0155 O_2 Sensor Heater Circuit Malfunction (Bank no. 2 Sensor no. 1)
P0156 O_2 Sensor Circuit Malfunction (Bank no. 2 Sensor no. 2)
P0157 O_2 Sensor Circuit Low Voltage (Bank no. 2 Sensor no. 2)
P0158 O_2 Sensor Circuit High Voltage (Bank no. 2 Sensor no. 2)
P0159 O_2 Sensor Circuit Slow Response (Bank no. 2 Sensor no. 2)
P0160 O_2 Sensor Circuit No Activity Detected (Bank no. 2 Sensor no. 2)
P0161 O_2 Sensor Heater Circuit Malfunction (Bank no. 2 Sensor no. 2)
P0162 O_2 Sensor Circuit Malfunction (Bank no. 2 Sensor no. 3)
P0163 O_2 Sensor Circuit Low Voltage (Bank no. 2 Sensor no. 3)
P0164 O_2 Sensor Circuit High Voltage (Bank no. 2 Sensor no. 3)
P0165 O_2 Sensor Circuit Slow Response (Bank no. 2 Sensor no. 3)
P0166 O_2 Sensor Circuit No Activity Detected (Bank no. 2 Sensor no. 3)
P0167 O_2 Sensor Heater Circuit Malfunction (Bank no. 2 Sensor no. 3)
P0170 Fuel Trim Malfunction (Bank no. 1)
P0171 System Too Lean (Bank no. 1)
P0172 System Too Rich (Bank no. 1)
P0173 Fuel Trim Malfunction (Bank no. 2)
P0174 System Too Lean (Bank no. 2)
P0175 System Too Rich (Bank no. 2)
P0176 Fuel Composition Sensor Circuit Malfunction
P0177 Fuel Composition Sensor Circuit Range/Performance
P0178 Fuel Composition Sensor Circuit Low Input
P0179 Fuel Composition Sensor Circuit High Input
P0180 Fuel Temperature Sensor "A" Circuit Malfunction
P0181 Fuel Temperature Sensor "A" Circuit Range/Performance
P0182 Fuel Temperature Sensor "A" Circuit Low Input
P0183 Fuel Temperature Sensor "A" Circuit High Input
P0184 Fuel Temperature Sensor "A" Circuit Intermittent
P0185 Fuel Temperature Sensor "B" Circuit Malfunction
P0186 Fuel Temperature Sensor "B" Circuit Range/Performance
P0187 Fuel Temperature Sensor "B" Circuit Low Input
P0188 Fuel Temperature Sensor "B" Circuit High Input
P0189 Fuel Temperature Sensor "B" Circuit Intermittent
P0190 Fuel Rail Pressure Sensor Circuit Malfunction
P0191 Fuel Rail Pressure Sensor Circuit Range/Performance
P0192 Fuel Rail Pressure Sensor Circuit Low Input
P0193 Fuel Rail Pressure Sensor Circuit High Input
P0194 Fuel Rail Pressure Sensor Circuit Intermittent
P0195 Engine Oil Temperature Sensor Malfunction
P0196 Engine Oil Temperature Sensor Range/Performance
P0197 Engine Oil Temperature Sensor Low

P0198 Engine Oil Temperature Sensor High
P0199 Engine Oil Temperature Sensor Intermittent
P0200 Injector Circuit Malfunction
P0201 Injector Circuit Malfunction—Cylinder no. 1
P0202 Injector Circuit Malfunction—Cylinder no. 2
P0203 Injector Circuit Malfunction—Cylinder no. 3
P0204 Injector Circuit Malfunction—Cylinder no. 4
P0205 Injector Circuit Malfunction—Cylinder no. 5
P0206 Injector Circuit Malfunction—Cylinder no. 6
P0207 Injector Circuit Malfunction—Cylinder no. 7
P0208 Injector Circuit Malfunction—Cylinder no. 8
P0209 Injector Circuit Malfunction—Cylinder no. 9
P0210 Injector Circuit Malfunction—Cylinder no. 10
P0211 Injector Circuit Malfunction—Cylinder no. 11
P0212 Injector Circuit Malfunction—Cylinder no. 12
P0213 Cold Start Injector no. 1 Malfunction
P0214 Cold Start Injector no. 2 Malfunction
P0215 Engine Shutoff Solenoid Malfunction
P0216 Injection Timing Control Circuit Malfunction
P0217 Engine Over Temperature Condition
P0218 Transmission Over Temperature Condition
P0219 Engine Over Speed Condition
P0220 Throttle/Pedal Position Sensor/Switch "B" Circuit Malfunction
P0221 Throttle/Pedal Position Sensor/Switch "B" Circuit Range/Performance Problem
P0222 Throttle/Pedal Position Sensor/Switch "B" Circuit Low Input
P0223 Throttle/Pedal Position Sensor/Switch "B" Circuit High Input
P0224 Throttle/Pedal Position Sensor/Switch "B" Circuit Intermittent
P0225 Throttle/Pedal Position Sensor/Switch "C" Circuit Malfunction
P0226 Throttle/Pedal Position Sensor/Switch "C" Circuit Range/Performance Problem
P0227 Throttle/Pedal Position Sensor/Switch "C" Circuit Low Input
P0228 Throttle/Pedal Position Sensor/Switch "C" Circuit High Input
P0229 Throttle/Pedal Position Sensor/Switch "C" Circuit Intermittent
P0230 Fuel Pump Primary Circuit Malfunction
P0231 Fuel Pump Secondary Circuit Low
P0232 Fuel Pump Secondary Circuit High
P0233 Fuel Pump Secondary Circuit Intermittent
P0234 Engine Over Boost Condition
P0261 Cylinder no. 1 Injector Circuit Low
P0262 Cylinder no. 1 Injector Circuit High
P0263 Cylinder no. 1 Contribution/Balance Fault
P0264 Cylinder no. 2 Injector Circuit Low
P0265 Cylinder no. 2 Injector Circuit High
P0266 Cylinder no. 2 Contribution/Balance Fault
P0267 Cylinder no. 3 Injector Circuit Low
P0268 Cylinder no. 3 Injector Circuit High
P0269 Cylinder no. 3 Contribution/Balance Fault
P0270 Cylinder no. 4 Injector Circuit Low
P0271 Cylinder no. 4 Injector Circuit High
P0272 Cylinder no. 4 Contribution/Balance Fault
P0273 Cylinder no. 5 Injector Circuit Low
P0274 Cylinder no. 5 Injector Circuit High

P0275 Cylinder no. 5 Contribution/Balance Fault
P0276 Cylinder no. 6 Injector Circuit Low
P0277 Cylinder no. 6 Injector Circuit High
P0278 Cylinder no. 6 Contribution/Balance Fault
P0279 Cylinder no. 7 Injector Circuit Low
P0280 Cylinder no. 7 Injector Circuit High
P0281 Cylinder no. 7 Contribution/Balance Fault
P0282 Cylinder no. 8 Injector Circuit Low
P0283 Cylinder no. 8 Injector Circuit High
P0284 Cylinder no. 8 Contribution/Balance Fault
P0285 Cylinder no. 9 Injector Circuit Low
P0286 Cylinder no. 9 Injector Circuit High
P0287 Cylinder no. 9 Contribution/Balance Fault
P0288 Cylinder no. 10 Injector Circuit Low
P0289 Cylinder no. 10 Injector Circuit High
P0290 Cylinder no. 10 Contribution/Balance Fault
P0291 Cylinder no. 11 Injector Circuit Low
P0292 Cylinder no. 11 Injector Circuit High
P0293 Cylinder no. 11 Contribution/Balance Fault
P0294 Cylinder no. 12 Injector Circuit Low
P0295 Cylinder no. 12 Injector Circuit High
P0296 Cylinder no. 12 Contribution/Balance Fault
P0300 Random/Multiple Cylinder Misfire Detected
P0301 Cylinder no. 1—Misfire Detected
P0302 Cylinder no. 2—Misfire Detected
P0303 Cylinder no. 3—Misfire Detected
P0304 Cylinder no. 4—Misfire Detected
P0305 Cylinder no. 5—Misfire Detected
P0306 Cylinder no. 6—Misfire Detected
P0307 Cylinder no. 7—Misfire Detected
P0308 Cylinder no. 8—Misfire Detected
P0309 Cylinder no. 9—Misfire Detected
P0310 Cylinder no. 10—Misfire Detected
P0311 Cylinder no. 11—Misfire Detected
P0312 Cylinder no. 12—Misfire Detected
P0320 Ignition/Distributor Engine Speed Input Circuit Malfunction
P0321 Ignition/Distributor Engine Speed Input Circuit Range/Performance
P0322 Ignition/Distributor Engine Speed Input Circuit No Signal
P0323 Ignition/Distributor Engine Speed Input Circuit Intermittent
P0325 Knock Sensor no. 1—Circuit Malfunction (Bank no. 1 or Single Sensor)
P0326 Knock Sensor no. 1—Circuit Range/Performance (Bank no. 1 or Single Sensor)
P0327 Knock Sensor no. 1—Circuit Low Input (Bank no. 1 or Single Sensor)
P0328 Knock Sensor no. 1—Circuit High Input (Bank no. 1 or Single Sensor)
P0329 Knock Sensor no. 1—Circuit Input Intermittent (Bank no. 1 or Single Sensor)
P0330 Knock Sensor no. 2—Circuit Malfunction (Bank no. 2)
P0331 Knock Sensor no. 2—Circuit Range/Performance (Bank no. 2)
P0332 Knock Sensor no. 2—Circuit Low Input (Bank no. 2)
P0333 Knock Sensor no. 2—Circuit High Input (Bank no. 2)
P0334 Knock Sensor no. 2—Circuit Input Intermittent (Bank no. 2)
P0335 Crankshaft Position Sensor "A" Circuit Malfunction

P0336 Crankshaft Position Sensor "A" Circuit Range/Performance
P0337 Crankshaft Position Sensor "A" Circuit Low Input
P0338 Crankshaft Position Sensor "A" Circuit High Input
P0339 Crankshaft Position Sensor "A" Circuit Intermittent
P0340 Camshaft Position Sensor Circuit Malfunction
P0341 Camshaft Position Sensor Circuit Range/Performance
P0342 Camshaft Position Sensor Circuit Low Input
P0343 Camshaft Position Sensor Circuit High Input
P0344 Camshaft Position Sensor Circuit Intermittent
P0350 Ignition Coil Primary/Secondary Circuit Malfunction
P0351 Ignition Coil "A" Primary/Secondary Circuit Malfunction
P0352 Ignition Coil "B" Primary/Secondary Circuit Malfunction
P0353 Ignition Coil "C" Primary/Secondary Circuit Malfunction
P0354 Ignition Coil "D" Primary/Secondary Circuit Malfunction
P0355 Ignition Coil "E" Primary/Secondary Circuit Malfunction
P0356 Ignition Coil "F" Primary/Secondary Circuit Malfunction
P0357 Ignition Coil "G" Primary/Secondary Circuit Malfunction
P0358 Ignition Coil "H" Primary/Secondary Circuit Malfunction
P0359 Ignition Coil "I" Primary/Secondary Circuit Malfunction
P0360 Ignition Coil "J" Primary/Secondary Circuit Malfunction
P0361 Ignition Coil "K" Primary/Secondary Circuit Malfunction
P0362 Ignition Coil "L" Primary/Secondary Circuit Malfunction
P0370 Timing Reference High Resolution Signal "A" Malfunction
P0371 Timing Reference High Resolution Signal "A" Too Many Pulses
P0372 Timing Reference High Resolution Signal "A" Too Few Pulses
P0373 Timing Reference High Resolution Signal "A" Intermittent/Erratic Pulses
P0374 Timing Reference High Resolution Signal "A" No Pulses
P0375 Timing Reference High Resolution Signal "B" Malfunction
P0376 Timing Reference High Resolution Signal "B" Too Many Pulses
P0377 Timing Reference High Resolution Signal "B" Too Few Pulses
P0378 Timing Reference High Resolution Signal "B" Intermittent/Erratic Pulses
P0379 Timing Reference High Resolution Signal "B" No Pulses
P0380 Glow Plug/Heater Circuit "A" Malfunction
P0381 Glow Plug/Heater Indicator Circuit Malfunction
P0382 Glow Plug/Heater Circuit "B" Malfunction
P0385 Crankshaft Position Sensor "B" Circuit Malfunction
P0386 Crankshaft Position Sensor "B" Circuit Range/Performance
P0387 Crankshaft Position Sensor "B" Circuit Low Input
P0388 Crankshaft Position Sensor "B" Circuit High Input

P0389 Crankshaft Position Sensor "B" Circuit Intermittent
P0400 Exhaust Gas Recirculation Flow Malfunction
P0401 Exhaust Gas Recirculation Flow Insufficient Detected
P0402 Exhaust Gas Recirculation Flow Excessive Detected
P0403 Exhaust Gas Recirculation Circuit Malfunction
P0404 Exhaust Gas Recirculation Circuit Range/Performance
P0405 Exhaust Gas Recirculation Sensor "A" Circuit Low
P0406 Exhaust Gas Recirculation Sensor "A" Circuit High
P0407 Exhaust Gas Recirculation Sensor "B" Circuit Low
P0408 Exhaust Gas Recirculation Sensor "B" Circuit High
P0410 Secondary Air Injection System Malfunction
P0411 Secondary Air Injection System Incorrect Flow Detected
P0412 Secondary Air Injection System Switching Valve "A" Circuit Malfunction
P0413 Secondary Air Injection System Switching Valve "A" Circuit Open
P0414 Secondary Air Injection System Switching Valve "A" Circuit Shorted
P0415 Secondary Air Injection System Switching Valve "B" Circuit Malfunction
P0416 Secondary Air Injection System Switching Valve "B" Circuit Open
P0417 Secondary Air Injection System Switching Valve "B" Circuit Shorted
P0418 Secondary Air Injection System Relay "A" Circuit Malfunction
P0419 Secondary Air Injection System Relay "B" Circuit Malfunction
P0420 Catalyst System Efficiency Below Threshold (Bank no. 1)
P0421 Warm Up Catalyst Efficiency Below Threshold (Bank no. 1)
P0422 Main Catalyst Efficiency Below Threshold (Bank no. 1)
P0423 Heated Catalyst Efficiency Below Threshold (Bank no. 1)
P0424 Heated Catalyst Temperature Below Threshold (Bank no. 1)
P0430 Catalyst System Efficiency Below Threshold (Bank no. 2)
P0431 Warm Up Catalyst Efficiency Below Threshold (Bank no. 2)
P0432 Main Catalyst Efficiency Below Threshold (Bank no. 2)
P0433 Heated Catalyst Efficiency Below Threshold (Bank no. 2)
P0434 Heated Catalyst Temperature Below Threshold (Bank no. 2)
P0440 Evaporative Emission Control System Malfunction
P0441 Evaporative Emission Control System Incorrect Purge Flow
P0442 Evaporative Emission Control System Leak Detected (Small Leak)
P0443 Evaporative Emission Control System Purge Control Valve Circuit Malfunction
P0444 Evaporative Emission Control System Purge Control Valve Circuit Open
P0445 Evaporative Emission Control System Purge Control Valve Circuit Shorted
P0446 Evaporative Emission Control System Vent Control Circuit Malfunction
P0447 Evaporative Emission Control System Vent Control Circuit Open
P0448 Evaporative Emission Control System Vent Control Circuit Shorted
P0449 Evaporative Emission Control System Vent Valve/Solenoid Circuit Malfunction

P0450 Evaporative Emission Control System Pressure Sensor Malfunction
P0451 Evaporative Emission Control System Pressure Sensor Range/Performance
P0452 Evaporative Emission Control System Pressure Sensor Low Input
P0453 Evaporative Emission Control System Pressure Sensor High Input
P0454 Evaporative Emission Control System Pressure Sensor Intermittent
P0455 Evaporative Emission Control System Leak Detected (Gross Leak)
P0460 Fuel Level Sensor Circuit Malfunction
P0461 Fuel Level Sensor Circuit Range/Performance
P0462 Fuel Level Sensor Circuit Low Input
P0463 Fuel Level Sensor Circuit High Input
P0464 Fuel Level Sensor Circuit Intermittent
P0465 Purge Flow Sensor Circuit Malfunction
P0466 Purge Flow Sensor Circuit Range/Performance
P0467 Purge Flow Sensor Circuit Low Input
P0468 Purge Flow Sensor Circuit High Input
P0469 Purge Flow Sensor Circuit Intermittent
P0470 Exhaust Pressure Sensor Malfunction
P0471 Exhaust Pressure Sensor Range/Performance
P0472 Exhaust Pressure Sensor Low
P0473 Exhaust Pressure Sensor High
P0474 Exhaust Pressure Sensor Intermittent
P0475 Exhaust Pressure Control Valve Malfunction
P0476 Exhaust Pressure Control Valve Range/Performance
P0477 Exhaust Pressure Control Valve Low
P0478 Exhaust Pressure Control Valve High
P0479 Exhaust Pressure Control Valve Intermittent
P0480 Cooling Fan no. 1 Control Circuit Malfunction
P0481 Cooling Fan no. 2 Control Circuit Malfunction
P0482 Cooling Fan no. 3 Control Circuit Malfunction
P0483 Cooling Fan Rationality Check Malfunction
P0484 Cooling Fan Circuit Over Current
P0485 Cooling Fan Power/Ground Circuit Malfunction
P0500 Vehicle Speed Sensor Malfunction
P0501 Vehicle Speed Sensor Range/Performance
P0502 Vehicle Speed Sensor Circuit Low Input
P0503 Vehicle Speed Sensor Intermittent/Erratic/High
P0505 Idle Control System Malfunction
P0506 Idle Control System RPM Lower Than Expected
P0507 Idle Control System RPM Higher Than Expected
P0510 Closed Throttle Position Switch Malfunction
P0520 Engine Oil Pressure Sensor/Switch Circuit Malfunction
P0521 Engine Oil Pressure Sensor/Switch Range/Performance
P0522 Engine Oil Pressure Sensor/Switch Low Voltage
P0523 Engine Oil Pressure Sensor/Switch High Voltage
P0530 A/C Refrigerant Pressure Sensor Circuit Malfunction
P0531 A/C Refrigerant Pressure Sensor Circuit Range/Performance
P0532 A/C Refrigerant Pressure Sensor Circuit Low Input
P0533 A/C Refrigerant Pressure Sensor Circuit High Input
P0534 A/C Refrigerant Charge Loss
P0550 Power Steering Pressure Sensor Circuit Malfunction
P0551 Power Steering Pressure Sensor Circuit Range/Performance
P0552 Power Steering Pressure Sensor Circuit Low Input
P0553 Power Steering Pressure Sensor Circuit High Input

P0554 Power Steering Pressure Sensor Circuit Intermittent
P0560 System Voltage Malfunction
P0561 System Voltage Unstable
P0562 System Voltage Low
P0563 System Voltage High
P0565 Cruise Control On Signal Malfunction
P0566 Cruise Control Off Signal Malfunction
P0567 Cruise Control Resume Signal Malfunction
P0568 Cruise Control Set Signal Malfunction
P0569 Cruise Control Coast Signal Malfunction
P0570 Cruise Control Accel Signal Malfunction
P0571 Cruise Control/Brake Switch "A" Circuit Malfunction
P0572 Cruise Control/Brake Switch "A" Circuit Low
P0573 Cruise Control/Brake Switch "A" Circuit High
P0574 Through P0580 Reserved for Cruise Codes
P0600 Serial Communication Link Malfunction
P0601 Internal Control Module Memory Check Sum Error
P0602 Control Module Programming Error
P0603 Internal Control Module Keep Alive Memory (KAM) Error
P0604 Internal Control Module Random Access Memory (RAM) Error
P0605 Internal Control Module Read Only Memory (ROM) Error
P0606 PCM Processor Fault
P0608 Control Module VSS Output "A" Malfunction
P0609 Control Module VSS Output "B" Malfunction
P0620 Generator Control Circuit Malfunction
P0621 Generator Lamp "L" Control Circuit Malfunction
P0622 Generator Field "F" Control Circuit Malfunction
P0650 Malfunction Indicator Lamp (MIL) Control Circuit Malfunction
P0654 Engine RPM Output Circuit Malfunction
P0655 Engine Hot Lamp Output Control Circuit Malfunction
P0656 Fuel Level Output Circuit Malfunction
P0700 Transmission Control System Malfunction
P0701 Transmission Control System Range/Performance
P0702 Transmission Control System Electrical
P0703 Torque Converter/Brake Switch "B" Circuit Malfunction
P0704 Clutch Switch Input Circuit Malfunction
P0705 Transmission Range Sensor Circuit Malfunction (PRNDL Input)
P0706 Transmission Range Sensor Circuit Range/Performance
P0707 Transmission Range Sensor Circuit Low Input
P0708 Transmission Range Sensor Circuit High Input
P0709 Transmission Range Sensor Circuit Intermittent
P0710 Transmission Fluid Temperature Sensor Circuit Malfunction
P0711 Transmission Fluid Temperature Sensor Circuit Range/Performance
P0712 Transmission Fluid Temperature Sensor Circuit Low Input
P0713 Transmission Fluid Temperature Sensor Circuit High Input
P0714 Transmission Fluid Temperature Sensor Circuit Intermittent
P0715 Input/Turbine Speed Sensor Circuit Malfunction
P0716 Input/Turbine Speed Sensor Circuit Range/Performance
P0717 Input/Turbine Speed Sensor Circuit No Signal
P0718 Input/Turbine Speed Sensor Circuit Intermittent
P0719 Torque Converter/Brake Switch "B" Circuit Low
P0720 Output Speed Sensor Circuit Malfunction
P0721 Output Speed Sensor Circuit Range/Performance

P0722 Output Speed Sensor Circuit No Signal
P0723 Output Speed Sensor Circuit Intermittent
P0724 Torque Converter/Brake Switch "B" Circuit High
P0725 Engine Speed Input Circuit Malfunction
P0726 Engine Speed Input Circuit Range/Performance
P0727 Engine Speed Input Circuit No Signal
P0728 Engine Speed Input Circuit Intermittent
P0730 Incorrect Gear Ratio
P0731 Gear no. 1 Incorrect Ratio
P0732 Gear no. 2 Incorrect Ratio
P0733 Gear no. 3 Incorrect Ratio
P0734 Gear no. 4 Incorrect Ratio
P0735 Gear no. 5 Incorrect Ratio
P0736 Reverse Incorrect Ratio
P0740 Torque Converter Clutch Circuit Malfunction
P0741 Torque Converter Clutch Circuit Performance or Stuck Off
P0742 Torque Converter Clutch Circuit Stuck On
P0743 Torque Converter Clutch Circuit Electrical
P0744 Torque Converter Clutch Circuit Intermittent
P0745 Pressure Control Solenoid Malfunction
P0746 Pressure Control Solenoid Performance or Stuck Off
P0747 Pressure Control Solenoid Stuck On
P0748 Pressure Control Solenoid Electrical
P0749 Pressure Control Solenoid Intermittent
P0750 Shift Solenoid "A" Malfunction
P0751 Shift Solenoid "A" Performance or Stuck Off
P0752 Shift Solenoid "A" Stuck On
P0753 Shift Solenoid "A" Electrical
P0754 Shift Solenoid "A" Intermittent
P0755 Shift Solenoid "B" Malfunction
P0756 Shift Solenoid "B" Performance or Stuck Off
P0757 Shift Solenoid "B" Stuck On
P0758 Shift Solenoid "B" Electrical
P0759 Shift Solenoid "B" Intermittent
P0760 Shift Solenoid "C" Malfunction
P0761 Shift Solenoid "C" Performance Or Stuck Off
P0762 Shift Solenoid "C" Stuck On
P0763 Shift Solenoid "C" Electrical
P0764 Shift Solenoid "C" Intermittent
P0765 Shift Solenoid "D" Malfunction
P0766 Shift Solenoid "D" Performance Or Stuck Off
P0767 Shift Solenoid "D" Stuck On
P0768 Shift Solenoid "D" Electrical
P0769 Shift Solenoid "D" Intermittent
P0770 Shift Solenoid "E" Malfunction
P0771 Shift Solenoid "E" Performance Or Stuck Off
P0772 Shift Solenoid "E" Stuck On
P0773 Shift Solenoid "E" Electrical
P0774 Shift Solenoid "E" Intermittent
P0780 Shift Malfunction
P0781 1–2 Shift Malfunction
P0782 2–3 Shift Malfunction
P0783 3–4 Shift Malfunction
P0784 4–5 Shift Malfunction
P0785 Shift/Timing Solenoid Malfunction
P0786 Shift/Timing Solenoid Range/Performance
P0787 Shift/Timing Solenoid Low
P0788 Shift/Timing Solenoid High
P0789 Shift/Timing Solenoid Intermittent
P0790 Normal/Performance Switch Circuit Malfunction

P0801 Reverse Inhibit Control Circuit Malfunction
P0803 1–4 Upshift (Skip Shift) Solenoid Control Circuit Malfunction
P0804 1–4 Upshift (Skip Shift) Lamp Control Circuit Malfunction
P1106 Map Sensor Circuit Intermittent High Voltage
P1107 MAP Sensor Circuit Intermittent Low Voltage
P1111 IAT Sensor Circuit Intermittent High Voltage
P1112 IAT Sensor Circuit Intermittent Low Voltage
P1114 ECT Sensor Circuit Intermittent Low Voltage
P1115 ECT Sensor Circuit Intermittent High Voltage
P1121 TP Sensor Circuit Intermittent High Voltage
P1122 TP Sensor Circuit Intermittent Low Voltage
P1133 HO_2 S-11 Insufficient Switching (Bank 1 Sensor 1)
P1134 HO_2 S-11 Transition Time Ratio (Bank 1 Sensor 1)
P1153 HO_2 S-21 Insufficient Switching (Bank 2 Sensor I)
P1154 HO_2 S-21 Transition Time Ratio (Bank 2 Sensor 1)
P1171 Fuel System Lean During Acceleration
P1391 G-Acceleration Sensor Intermittent Low Voltage
P1390 G-Acceleration (Low G) Sensor Performance
P1392 Rough Road G-Sensor Circuit Low Voltage
P1393 Rough Road G-Sensor Circuit High Voltage
P1394 G-Acceleration Sensor Intermittent High Voltage
P1406 EGR Valve Pintle Position Sensor Circuit Fault
P1441 EVAP System Flow During Non-Purge
P1442 EVAP System Flow During Non-Purge
P1508 Idle Speed Control System-Low
P1509 Idle Speed Control System-High
P1618 Serial Peripheral Interface Communication Error
P1640 Output Driver Module 'A' Fault
P1790 PCM ROM (Transmission Side) Check Sum Error
P1792 PCM EEPROM (Transmission Side) Check Sum Error
P1835 Kick Down Switch Always On
P1850 Brake Band Apply Solenoid Electrical Fault
P1860 TCC PWM Solenoid Electrical Fault
P1870 Transmission Component Slipping

Audi

READING CODES

Reading the control module memory is one of the first steps in OBD II system diagnostics. This step should be initially performed to determine the general nature of the fault. Subsequent readings will determine if the fault has been cleared.

Reading codes can be performed by any of the methods below:

• Read the control module memory with the Generic Scan Tool (GST)

• Read the control module memory with Audi's VAG 1551 scan tool/ tester—if the VAG 1551 is used, refer to the trouble code equivalent list for the proper code

To read the fault codes, connect the scan tool or tester according to the manufacturer's instructions. Follow the manufacturer's specified procedure for reading the codes.

CLEARING CODES

Control module reset procedures are a very important part of OBD II System diagnostics. This step should be done at the end of any fault code repair and at the end of any driveability repair.

Clearing codes can be performed by any of the methods below:

• Clear the control module memory with the Generic Scan Tool (GST)

• Clear the control module memory with the vehicle manufacturer's specific tester

• Turn the ignition **OFF** and remove the negative battery cable for at least 1 minute.

Removing the negative battery cable may cause other systems in the vehicle to lose their memory. Prior to removing the cable, ensure you have the proper reset codes for radios and alarms.

➡**The MIL will may also be de-activated for some codes if the vehicle completes three consecutive trips without a fault detected with vehicle conditions similar to those present during the fault.**

DIAGNOSTIC TROUBLE CODES

P0000 No Failures
P0100 Mass or Volume Air Flow Circuit Malfunction
P0101 Mass or Volume Air Flow Circuit Range/Performance Problem
P0102 Mass or Volume Air Flow Circuit Low Input
P0103 Mass or Volume Air Flow Circuit High Input
P0104 Mass or Volume Air Flow Circuit Intermittent
P0105 Manifold Absolute Pressure/Barometric Pressure Circuit Malfunction
P0106 Manifold Absolute Pressure/Barometric Pressure Circuit Range/Performance Problem
P0107 Manifold Absolute Pressure/Barometric Pressure Circuit Low Input
P0108 Manifold Absolute Pressure/Barometric Pressure Circuit High Input
P0109 Manifold Absolute Pressure/Barometric Pressure Circuit Intermittent
P0110 Intake Air Temperature Circuit Malfunction
P0111 Intake Air Temperature Circuit Range/Performance Problem
P0112 Intake Air Temperature Circuit Low Input
P0113 Intake Air Temperature Circuit High Input
P0114 Intake Air Temperature Circuit Intermittent
P0115 Engine Coolant Temperature Circuit Malfunction
P0116 Engine Coolant Temperature Circuit Range/Performance Problem
P0117 Engine Coolant Temperature Circuit Low Input
P0118 Engine Coolant Temperature Circuit High Input
P0119 Engine Coolant Temperature Circuit Intermittent
P0120 Throttle/Pedal Position Sensor/Switch "A" Circuit Malfunction
P0121 Throttle/Pedal Position Sensor/Switch "A" Circuit Range/Performance Problem
P0122 Throttle/Pedal Position Sensor/Switch "A" Circuit Low Input
P0123 Throttle/Pedal Position Sensor/Switch "A" Circuit High Input
P0124 Throttle/Pedal Position Sensor/Switch "A" Circuit Intermittent
P0125 Insufficient Coolant Temperature For Closed Loop Fuel Control
P0126 Insufficient Coolant Temperature For Stable Operation
P0130 O_2 Circuit Malfunction (Bank no. 1 Sensor no. 1)

P0131 O_2 Sensor Circuit Low Voltage (Bank no. 1 Sensor no. 1)

P0132 O_2 Sensor Circuit High Voltage (Bank no. 1 Sensor no. 1)

P0133 O_2 Sensor Circuit Slow Response (Bank no. 1 Sensor no. 1)

P0134 O_2 Sensor Circuit No Activity Detected (Bank no. 1 Sensor no. 1)

P0135 O_2 Sensor Heater Circuit Malfunction (Bank no. 1 Sensor no. 1)

P0136 O_2 Sensor Circuit Malfunction (Bank no. 1 Sensor no. 2)

P0137 O_2 Sensor Circuit Low Voltage (Bank no. 1 Sensor no. 2)

P0138 O_2 Sensor Circuit High Voltage (Bank no. 1 Sensor no. 2)

P0139 O_2 Sensor Circuit Slow Response (Bank no. 1 Sensor no. 2)

P0140 O_2 Sensor Circuit No Activity Detected (Bank no. 1 Sensor no. 2)

P0141 O_2 Sensor Heater Circuit Malfunction (Bank no. 1 Sensor no. 2)

P0142 O_2 Sensor Circuit Malfunction (Bank no. 1 Sensor no. 3)

P0143 O_2 Sensor Circuit Low Voltage (Bank no. 1 Sensor no. 3)

P0144 O_2 Sensor Circuit High Voltage (Bank no. 1 Sensor no. 3)

P0145 O_2 Sensor Circuit Slow Response (Bank no. 1 Sensor no. 3)

P0146 O_2 Sensor Circuit No Activity Detected (Bank no. 1 Sensor no. 3)

P0147 O_2 Sensor Heater Circuit Malfunction (Bank no. 1 Sensor no. 3)

P0150 O_2 Sensor Circuit Malfunction (Bank no. 2 Sensor no. 1)

P0151 O_2 Sensor Circuit Low Voltage (Bank no. 2 Sensor no. 1)

P0152 O_2 Sensor Circuit High Voltage (Bank no. 2 Sensor no. 1)

P0153 O_2 Sensor Circuit Slow Response (Bank no. 2 Sensor no. 1)

P0154 O_2 Sensor Circuit No Activity Detected (Bank no. 2 Sensor no. 1)

P0155 O_2 Sensor Heater Circuit Malfunction (Bank no. 2 Sensor no. 1)

P0156 O_2 Sensor Circuit Malfunction (Bank no. 2 Sensor no. 2)

P0157 O_2 Sensor Circuit Low Voltage (Bank no. 2 Sensor no. 2)

P0158 O_2 Sensor Circuit High Voltage (Bank no. 2 Sensor no. 2)

P0159 O_2 Sensor Circuit Slow Response (Bank no. 2 Sensor no. 2)

P0160 O_2 Sensor Circuit No Activity Detected (Bank no. 2 Sensor no. 2)

P0161 O_2 Sensor Heater Circuit Malfunction (Bank no. 2 Sensor no. 2)

P0162 O_2 Sensor Circuit Malfunction (Bank no. 2 Sensor no. 3)

P0163 O_2 Sensor Circuit Low Voltage (Bank no. 2 Sensor no. 3)

P0164 O_2 Sensor Circuit High Voltage (Bank no. 2 Sensor no. 3)

P0165 O_2 Sensor Circuit Slow Response (Bank no. 2 Sensor no. 3)

P0166 O_2 Sensor Circuit No Activity Detected (Bank no. 2 Sensor no. 3)

P0167 O_2 Sensor Heater Circuit Malfunction (Bank no. 2 Sensor no. 3)

P0170 Fuel Trim Malfunction (Bank no. 1)

P0171 System Too Lean (Bank no. 1)

P0172 System Too Rich (Bank no. 1)

P0173 Fuel Trim Malfunction (Bank no. 2)

P0174 System Too Lean (Bank no. 2)

P0175 System Too Rich (Bank no. 2)

P0176 Fuel Composition Sensor Circuit Malfunction

P0177 Fuel Composition Sensor Circuit Range/Performance

P0178 Fuel Composition Sensor Circuit Low Input

P0179 Fuel Composition Sensor Circuit High Input

P0180 Fuel Temperature Sensor "A" Circuit Malfunction

P0181 Fuel Temperature Sensor "A" Circuit Range/Performance

P0182 Fuel Temperature Sensor "A" Circuit Low Input

P0183 Fuel Temperature Sensor "A" Circuit High Input

P0184 Fuel Temperature Sensor "A" Circuit Intermittent

P0185 Fuel Temperature Sensor "B" Circuit Malfunction

P0186 Fuel Temperature Sensor "B" Circuit Range/Performance

P0187 Fuel Temperature Sensor "B" Circuit Low Input

P0188 Fuel Temperature Sensor "B" Circuit High Input

P0189 Fuel Temperature Sensor "B" Circuit Intermittent

P0190 Fuel Rail Pressure Sensor Circuit Malfunction

P0191 Fuel Rail Pressure Sensor Circuit Range/Performance

P0192 Fuel Rail Pressure Sensor Circuit Low Input

P0193 Fuel Rail Pressure Sensor Circuit High Input

P0194 Fuel Rail Pressure Sensor Circuit Intermittent

P0195 Engine Oil Temperature Sensor Malfunction

P0196 Engine Oil Temperature Sensor Range/Performance

P0197 Engine Oil Temperature Sensor Low

P0198 Engine Oil Temperature Sensor High

P0199 Engine Oil Temperature Sensor Intermittent

P0200 Injector Circuit Malfunction

P0201 Injector Circuit Malfunction—Cylinder no. 1

P0202 Injector Circuit Malfunction—Cylinder no. 2

P0203 Injector Circuit Malfunction—Cylinder no. 3

P0204 Injector Circuit Malfunction—Cylinder no. 4

P0205 Injector Circuit Malfunction—Cylinder no. 5

P0206 Injector Circuit Malfunction—Cylinder no. 6

P0207 Injector Circuit Malfunction—Cylinder no. 7

P0208 Injector Circuit Malfunction—Cylinder no. 8

P0209 Injector Circuit Malfunction—Cylinder no. 9

P0210 Injector Circuit Malfunction—Cylinder no. 10

P0211 Injector Circuit Malfunction—Cylinder no. 11

P0212 Injector Circuit Malfunction—Cylinder no. 12

P0213 Cold Start Injector no. 1 Malfunction

P0214 Cold Start Injector no. 2 Malfunction

P0215 Engine Shutoff Solenoid Malfunction

P0216 Injection Timing Control Circuit Malfunction

P0217 Engine Over Temperature Condition

P0218 Transmission Over Temperature Condition

P0219 Engine Over Speed Condition

P0220 Throttle/Pedal Position Sensor/Switch "B" Circuit Malfunction

P0221 Throttle/Pedal Position Sensor/Switch "B" Circuit Range/Performance Problem

P0222 Throttle/Pedal Position Sensor/Switch "B" Circuit Low Input

P0223 Throttle/Pedal Position Sensor/Switch "B" Circuit High Input

P0224 Throttle/Pedal Position Sensor/Switch "B" Circuit Intermittent

P0225 Throttle/Pedal Position Sensor/Switch "C" Circuit Malfunction

P0226 Throttle/Pedal Position Sensor/Switch "C" Circuit Range/Performance Problem

P0227 Throttle/Pedal Position Sensor/Switch "C" Circuit Low Input

P0228 Throttle/Pedal Position Sensor/Switch "C" Circuit High Input

P0229 Throttle/Pedal Position Sensor/Switch "C" Circuit Intermittent

P0230 Fuel Pump Primary Circuit Malfunction

P0231 Fuel Pump Secondary Circuit Low

P0232 Fuel Pump Secondary Circuit High

P0233 Fuel Pump Secondary Circuit Intermittent

P0234 Engine Over Boost Condition

P0261 Cylinder no. 1 Injector Circuit Low

P0262 Cylinder no. 1 Injector Circuit High

P0263 Cylinder no. 1 Contribution/Balance Fault

P0264 Cylinder no. 2 Injector Circuit Low

P0265 Cylinder no. 2 Injector Circuit High

P0266 Cylinder no. 2 Contribution/Balance Fault

P0267 Cylinder no. 3 Injector Circuit Low

P0268 Cylinder no. 3 Injector Circuit High

P0269 Cylinder no. 3 Contribution/Balance Fault

P0270 Cylinder no. 4 Injector Circuit Low

P0271 Cylinder no. 4 Injector Circuit High

P0272 Cylinder no. 4 Contribution/Balance Fault

P0273 Cylinder no. 5 Injector Circuit Low

P0274 Cylinder no. 5 Injector Circuit High

P0275 Cylinder no. 5 Contribution/Balance Fault

P0276 Cylinder no. 6 Injector Circuit Low

P0277 Cylinder no. 6 Injector Circuit High

P0278 Cylinder no. 6 Contribution/Balance Fault

P0279 Cylinder no. 7 Injector Circuit Low

P0280 Cylinder no. 7 Injector Circuit High

P0281 Cylinder no. 7 Contribution/Balance Fault

P0282 Cylinder no. 8 Injector Circuit Low

P0283 Cylinder no. 8 Injector Circuit High

P0284 Cylinder no. 8 Contribution/Balance Fault

P0285 Cylinder no. 9 Injector Circuit Low

P0286 Cylinder no. 9 Injector Circuit High

P0287 Cylinder no. 9 Contribution/Balance Fault

P0288 Cylinder no. 10 Injector Circuit Low

P0289 Cylinder no. 10 Injector Circuit High

P0290 Cylinder no. 10 Contribution/Balance Fault

P0291 Cylinder no. 11 Injector Circuit Low

P0292 Cylinder no. 11 Injector Circuit High

P0293 Cylinder no. 11 Contribution/Balance Fault

P0294 Cylinder no. 12 Injector Circuit Low

P0295 Cylinder no. 12 Injector Circuit High

P0296 Cylinder no. 12 Contribution/Balance Fault

P0300 Random/Multiple Cylinder Misfire Detected

P0301 Cylinder no. 1—Misfire Detected

P0302 Cylinder no. 2—Misfire Detected

P0303 Cylinder no. 3—Misfire Detected

P0304 Cylinder no. 4—Misfire Detected

P0305 Cylinder no. 5—Misfire Detected

P0306 Cylinder no. 6—Misfire Detected

P0307 Cylinder no. 7—Misfire Detected

P0308 Cylinder no. 8—Misfire Detected

P0309 Cylinder no. 9—Misfire Detected

P0310 Cylinder no. 10—Misfire Detected

P0311 Cylinder no. 11—Misfire Detected

P0312 Cylinder no. 12—Misfire Detected

P0320 Ignition/Distributor Engine Speed Input Circuit Malfunction

P0321 Ignition/Distributor Engine Speed Input Circuit Range/Performance

P0322 Ignition/Distributor Engine Speed Input Circuit No Signal

P0323 Ignition/Distributor Engine Speed Input Circuit Intermittent

P0325 Knock Sensor no. 1—Circuit Malfunction (Bank no. 1 or Single Sensor)

P0326 Knock Sensor no. 1—Circuit Range/Performance (Bank no. 1 or Single Sensor)

P0327 Knock Sensor no. 1—Circuit Low Input (Bank no. 1 or Single Sensor)

P0328 Knock Sensor no. 1—Circuit High Input (Bank no. 1 or Single Sensor)

P0329 Knock Sensor no. 1—Circuit Input Intermittent (Bank no. 1 or Single Sensor)

P0330 Knock Sensor no. 2—Circuit Malfunction (Bank no. 2)

P0331 Knock Sensor no. 2—Circuit Range/Performance (Bank no. 2)

P0332 Knock Sensor no. 2—Circuit Low Input (Bank no. 2)

P0333 Knock Sensor no. 2—Circuit High Input (Bank no. 2)

P0334 Knock Sensor no. 2—Circuit Input Intermittent (Bank no. 2)

P0335 Crankshaft Position Sensor "A" Circuit Malfunction

P0336 Crankshaft Position Sensor "A" Circuit Range/Performance

P0337 Crankshaft Position Sensor "A" Circuit Low Input

P0338 Crankshaft Position Sensor "A" Circuit High Input

P0339 Crankshaft Position Sensor "A" Circuit Intermittent

P0340 Camshaft Position Sensor Circuit Malfunction

P0341 Camshaft Position Sensor Circuit Range/Performance

P0342 Camshaft Position Sensor Circuit Low Input

P0343 Camshaft Position Sensor Circuit High Input

P0344 Camshaft Position Sensor Circuit Intermittent

P0350 Ignition Coil Primary/Secondary Circuit Malfunction

P0351 Ignition Coil "A" Primary/Secondary Circuit Malfunction

P0352 Ignition Coil "B" Primary/Secondary Circuit Malfunction

P0353 Ignition Coil "C" Primary/Secondary Circuit Malfunction

P0354 Ignition Coil "D" Primary/Secondary Circuit Malfunction

P0355 Ignition Coil "E" Primary/Secondary Circuit Malfunction

P0356 Ignition Coil "F" Primary/Secondary Circuit Malfunction

P0357 Ignition Coil "G" Primary/Secondary Circuit Malfunction

P0358 Ignition Coil "H" Primary/Secondary Circuit Malfunction

P0359 Ignition Coil "I" Primary/Secondary Circuit Malfunction

P0360 Ignition Coil "J" Primary/Secondary Circuit Malfunction

P0361 Ignition Coil "K" Primary/Secondary Circuit Malfunction

P0362 Ignition Coil "L" Primary/Secondary Circuit Malfunction

P0370 Timing Reference High Resolution Signal "A" Malfunction

P0371 Timing Reference High Resolution Signal "A" Too Many Pulses

P0372 Timing Reference High Resolution Signal "A" Too Few Pulses

P0373 Timing Reference High Resolution Signal "A" Intermittent/Erratic Pulses

P0374 Timing Reference High Resolution Signal "A" No Pulses

P0375 Timing Reference High Resolution Signal "B" Malfunction

P0376 Timing Reference High Resolution Signal "B" Too Many Pulses

P0377 Timing Reference High Resolution Signal "B" Too Few Pulses

P0378 Timing Reference High Resolution Signal "B" Intermittent/Erratic Pulses

P0379 Timing Reference High Resolution Signal "B" No Pulses

P0380 Glow Plug/Heater Circuit "A" Malfunction

P0381 Glow Plug/Heater Indicator Circuit Malfunction

P0382 Glow Plug/Heater Circuit "B" Malfunction

P0385 Crankshaft Position Sensor "B" Circuit Malfunction

P0386 Crankshaft Position Sensor "B" Circuit Range/Performance

P0387 Crankshaft Position Sensor "B" Circuit Low Input

P0388 Crankshaft Position Sensor "B" Circuit High Input

P0389 Crankshaft Position Sensor "B" Circuit Intermittent

P0400 Exhaust Gas Recirculation Flow Malfunction

P0401 Exhaust Gas Recirculation Flow Insufficient Detected

P0402 Exhaust Gas Recirculation Flow Excessive Detected

P0403 Exhaust Gas Recirculation Circuit Malfunction

P0404 Exhaust Gas Recirculation Circuit Range/Performance

P0405 Exhaust Gas Recirculation Sensor "A" Circuit Low

P0406 Exhaust Gas Recirculation Sensor "A" Circuit High

P0407 Exhaust Gas Recirculation Sensor "B" Circuit Low

P0408 Exhaust Gas Recirculation Sensor "B" Circuit High

P0410 Secondary Air Injection System Malfunction

P0411 Secondary Air Injection System Incorrect Flow Detected

P0412 Secondary Air Injection System Switching Valve "A" Circuit Malfunction

P0413 Secondary Air Injection System Switching Valve "A" Circuit Open

P0414 Secondary Air Injection System Switching Valve "A" Circuit Shorted

P0415 Secondary Air Injection System Switching Valve "B" Circuit Malfunction

P0416 Secondary Air Injection System Switching Valve "B" Circuit Open

P0417 Secondary Air Injection System Switching Valve "B" Circuit Shorted

P0418 Secondary Air Injection System Relay "A" Circuit Malfunction

P0419 Secondary Air Injection System Relay "B" Circuit Malfunction

P0420 Catalyst System Efficiency Below Threshold (Bank no. 1)

P0421 Warm Up Catalyst Efficiency Below Threshold (Bank no. 1)

P0422 Main Catalyst Efficiency Below Threshold (Bank no. 1)

P0423 Heated Catalyst Efficiency Below Threshold (Bank no. 1)

P0424 Heated Catalyst Temperature Below Threshold (Bank no. 1)

P0430 Catalyst System Efficiency Below Threshold (Bank no. 2)

P0431 Warm Up Catalyst Efficiency Below Threshold (Bank no. 2)

P0432 Main Catalyst Efficiency Below Threshold (Bank no. 2)

P0433 Heated Catalyst Efficiency Below Threshold (Bank no. 2)

P0434 Heated Catalyst Temperature Below Threshold (Bank no. 2)

P0440 Evaporative Emission Control System Malfunction

P0441 Evaporative Emission Control System Incorrect Purge Flow

P0442 Evaporative Emission Control System Leak Detected (Small Leak)

P0443 Evaporative Emission Control System Purge Control Valve Circuit Malfunction

P0444 Evaporative Emission Control System Purge Control Valve Circuit Open

P0445 Evaporative Emission Control System Purge Control Valve Circuit Shorted

P0446 Evaporative Emission Control System Vent Control Circuit Malfunction

P0447 Evaporative Emission Control System Vent Control Circuit Open

P0448 Evaporative Emission Control System Vent Control Circuit Shorted

P0449 Evaporative Emission Control System Vent Valve/Solenoid Circuit Malfunction

P0450 Evaporative Emission Control System Pressure Sensor Malfunction

P0451 Evaporative Emission Control System Pressure Sensor Range/Performance

P0452 Evaporative Emission Control System Pressure Sensor Low Input

P0453 Evaporative Emission Control System Pressure Sensor High Input

P0454 Evaporative Emission Control System Pressure Sensor Intermittent

P0455 Evaporative Emission Control System Leak Detected (Gross Leak)

P0460 Fuel Level Sensor Circuit Malfunction

P0461 Fuel Level Sensor Circuit Range/Performance

P0462 Fuel Level Sensor Circuit Low Input

P0463 Fuel Level Sensor Circuit High Input

P0464 Fuel Level Sensor Circuit Intermittent

P0465 Purge Flow Sensor Circuit Malfunction

P0466 Purge Flow Sensor Circuit Range/Performance

P0467 Purge Flow Sensor Circuit Low Input

P0468 Purge Flow Sensor Circuit High Input

P0469 Purge Flow Sensor Circuit Intermittent
P0470 Exhaust Pressure Sensor Malfunction
P0471 Exhaust Pressure Sensor Range/Performance
P0472 Exhaust Pressure Sensor Low
P0473 Exhaust Pressure Sensor High
P0474 Exhaust Pressure Sensor Intermittent
P0475 Exhaust Pressure Control Valve Malfunction
P0476 Exhaust Pressure Control Valve Range/Performance
P0477 Exhaust Pressure Control Valve Low
P0478 Exhaust Pressure Control Valve High
P0479 Exhaust Pressure Control Valve Intermittent
P0480 Cooling Fan no. 1 Control Circuit Malfunction
P0481 Cooling Fan no. 2 Control Circuit Malfunction
P0482 Cooling Fan no. 3 Control Circuit Malfunction
P0483 Cooling Fan Rationality Check Malfunction
P0484 Cooling Fan Circuit Over Current
P0485 Cooling Fan Power/Ground Circuit Malfunction
P0500 Vehicle Speed Sensor Malfunction
P0501 Vehicle Speed Sensor Range/Performance
P0502 Vehicle Speed Sensor Circuit Low Input
P0503 Vehicle Speed Sensor Intermittent/Erratic/High
P0505 Idle Control System Malfunction
P0506 Idle Control System RPM Lower Than Expected
P0507 Idle Control System RPM Higher Than Expected
P0510 Closed Throttle Position Switch Malfunction
P0520 Engine Oil Pressure Sensor/Switch Circuit Malfunction
P0521 Engine Oil Pressure Sensor/Switch Range/Performance
P0522 Engine Oil Pressure Sensor/Switch Low Voltage
P0523 Engine Oil Pressure Sensor/Switch High Voltage
P0530 A/C Refrigerant Pressure Sensor Circuit Malfunction
P0531 A/C Refrigerant Pressure Sensor Circuit Range/Performance
P0532 A/C Refrigerant Pressure Sensor Circuit Low Input
P0533 A/C Refrigerant Pressure Sensor Circuit High Input
P0534 A/C Refrigerant Charge Loss
P0550 Power Steering Pressure Sensor Circuit Malfunction
P0551 Power Steering Pressure Sensor Circuit Range/Performance
P0552 Power Steering Pressure Sensor Circuit Low Input
P0553 Power Steering Pressure Sensor Circuit High Input
P0554 Power Steering Pressure Sensor Circuit Intermittent
P0560 System Voltage Malfunction
P0561 System Voltage Unstable
P0562 System Voltage Low
P0563 System Voltage High
P0565 Cruise Control On Signal Malfunction
P0566 Cruise Control Off Signal Malfunction
P0567 Cruise Control Resume Signal Malfunction
P0568 Cruise Control Set Signal Malfunction
P0569 Cruise Control Coast Signal Malfunction
P0570 Cruise Control Accel Signal Malfunction
P0571 Cruise Control/Brake Switch "A" Circuit Malfunction
P0572 Cruise Control/Brake Switch "A" Circuit Low
P0573 Cruise Control/Brake Switch "A" Circuit High
P0574 **Through P0580** Reserved for Cruise Codes
P0600 Serial Communication Link Malfunction
P0601 Internal Control Module Memory Check Sum Error
P0602 Control Module Programming Error
P0603 Internal Control Module Keep Alive Memory (KAM) Error

P0604 Internal Control Module Random Access Memory (RAM) Error
P0605 Internal Control Module Read Only Memory (ROM) Error
P0606 PCM Processor Fault
P0608 Control Module VSS Output "A" Malfunction
P0609 Control Module VSS Output "B" Malfunction
P0620 Generator Control Circuit Malfunction
P0621 Generator Lamp "L" Control Circuit Malfunction
P0622 Generator Field "F" Control Circuit Malfunction
P0650 Malfunction Indicator Lamp (MIL) Control Circuit Malfunction
P0654 Engine RPM Output Circuit Malfunction
P0655 Engine Hot Lamp Output Control Circuit Malfunction
P0656 Fuel Level Output Circuit Malfunction
P0700 Transmission Control System Malfunction
P0701 Transmission Control System Range/Performance
P0702 Transmission Control System Electrical
P0703 Torque Converter/Brake Switch "B" Circuit Malfunction
P0704 Clutch Switch Input Circuit Malfunction
P0705 Transmission Range Sensor Circuit Malfunction (PRNDL Input)
P0706 Transmission Range Sensor Circuit Range/Performance
P0707 Transmission Range Sensor Circuit Low Input
P0708 Transmission Range Sensor Circuit High Input
P0709 Transmission Range Sensor Circuit Intermittent
P0710 Transmission Fluid Temperature Sensor Circuit Malfunction
P0711 Transmission Fluid Temperature Sensor Circuit Range/Performance
P0712 Transmission Fluid Temperature Sensor Circuit Low Input
P0713 Transmission Fluid Temperature Sensor Circuit High Input
P0714 Transmission Fluid Temperature Sensor Circuit Intermittent
P0715 Input/Turbine Speed Sensor Circuit Malfunction
P0716 Input/Turbine Speed Sensor Circuit Range/Performance
P0717 Input/Turbine Speed Sensor Circuit No Signal
P0718 Input/Turbine Speed Sensor Circuit Intermittent
P0719 Torque Converter/Brake Switch "B" Circuit Low
P0720 Output Speed Sensor Circuit Malfunction
P0721 Output Speed Sensor Circuit Range/Performance
P0722 Output Speed Sensor Circuit No Signal
P0723 Output Speed Sensor Circuit Intermittent
P0724 Torque Converter/Brake Switch "B" Circuit High
P0725 Engine Speed Input Circuit Malfunction
P0726 Engine Speed Input Circuit Range/Performance
P0727 Engine Speed Input Circuit No Signal
P0728 Engine Speed Input Circuit Intermittent
P0730 Incorrect Gear Ratio
P0731 Gear no. 1 Incorrect Ratio
P0732 Gear no. 2 Incorrect Ratio
P0733 Gear no. 3 Incorrect Ratio
P0734 Gear no. 4 Incorrect Ratio
P0735 Gear no. 5 Incorrect Ratio
P0736 Reverse Incorrect Ratio
P0740 Torque Converter Clutch Circuit Malfunction
P0741 Torque Converter Clutch Circuit Performance or Stuck Off

P0742 Torque Converter Clutch Circuit Stuck On
P0743 Torque Converter Clutch Circuit Electrical
P0744 Torque Converter Clutch Circuit Intermittent
P0745 Pressure Control Solenoid Malfunction
P0746 Pressure Control Solenoid Performance or Stuck Off
P0747 Pressure Control Solenoid Stuck On
P0748 Pressure Control Solenoid Electrical
P0749 Pressure Control Solenoid Intermittent
P0750 Shift Solenoid "A" Malfunction
P0751 Shift Solenoid "A" Performance or Stuck Off
P0752 Shift Solenoid "A" Stuck On
P0753 Shift Solenoid "A" Electrical
P0754 Shift Solenoid "A" Intermittent
P0755 Shift Solenoid "B" Malfunction
P0756 Shift Solenoid "B" Performance or Stuck Off
P0757 Shift Solenoid "B" Stuck On
P0758 Shift Solenoid "B" Electrical
P0759 Shift Solenoid "B" Intermittent
P0760 Shift Solenoid "C" Malfunction
P0761 Shift Solenoid "C" Performance Or Stuck Off
P0762 Shift Solenoid "C" Stuck On
P0763 Shift Solenoid "C" Electrical
P0764 Shift Solenoid "C" Intermittent
P0765 Shift Solenoid "D" Malfunction
P0766 Shift Solenoid "D" Performance Or Stuck Off
P0767 Shift Solenoid "D" Stuck On
P0768 Shift Solenoid "D" Electrical
P0769 Shift Solenoid "D" Intermittent
P0770 Shift Solenoid "E" Malfunction
P0771 Shift Solenoid "E" Performance Or Stuck Off
P0772 Shift Solenoid "E" Stuck On
P0773 Shift Solenoid "E" Electrical
P0774 Shift Solenoid "E" Intermittent
P0780 Shift Malfunction
P0781 1–2 Shift Malfunction
P0782 2–3 Shift Malfunction
P0783 3–4 Shift Malfunction
P0784 4–5 Shift Malfunction
P0785 Shift/Timing Solenoid Malfunction
P0786 Shift/Timing Solenoid Range/Performance
P0787 Shift/Timing Solenoid Low
P0788 Shift/Timing Solenoid High
P0789 Shift/Timing Solenoid Intermittent
P0790 Normal/Performance Switch Circuit Malfunction
P0801 Reverse Inhibit Control Circuit Malfunction
P0803 1–4 Upshift (Skip Shift) Solenoid Control Circuit Malfunction
P0804 1–4 Upshift (Skip Shift) Lamp Control Circuit Malfunction
P1102 Oxygen Sensor Heating Circuit, Bank 1-Sensor 1 Short to B+
P1105 Oxygen Sensor Heating Circuit, Bank 1-Sensor 2 Short to B+
P1107 Oxygen Sensor Heating Circuit, Bank 2-Sensor I Short to B+
P1110 Oxygen Sensor Heating Circuit, Bank 2-Sensor 2 Short to B+
P1127 Long Term Fuel Trim Multiplicative, Bank 1 System Too Rich
P1128 Long Term Fuel Trim Multiplicative, Bank 1 System Too Lean

P1129 Long Term Fuel Trim Multiplicative, Bank2 System too Rich
P1130 Long Term Fuel Trim Multiplicative, Bank2 System too Lean
P1136 Long Term Fuel Trim Additive, Bank 1 System Too Lean
P1137 Long Term Fuel Trim Additive, Bank 1 System Too Rich
P1138 Long Term Fuel Trim Additive Fuel, Bank I System too Lean
P1139 Long Term Fuel Trim Additive Fuel, Bank 1 System too Rich
P1141 Load Calculation Cross Check Range/Performance
P1176 Oxygen Correction Behind Catalyst, B1 Limit Attained
P1177 Oxygen Correction Behind Catalyst. 82 Limit Attained
P1196 Oxygen Sensor Heater Circuit, Bank 1-Sensor 1 Electrical Malfunction
P1197 Oxygen Sensor Heater Circuit, Bank 2-Sensor I Electrical Malfunction
P1198 Oxygen Sensor Heater Circuit, Bankl-Sensor2 Electrical Malfunction
P1198 Oxygen Sensor Heater Circuit, Bank 1-Sensor 2 Electrical Malfunction
P1199 Oxygen Sensor Heater Circuit, Bank 2-Sensor 2 Electrical Malfunction
P1201 Cylinder 1, Fuel Injection Circuit Electrical Malfunction
P1202 Cylinder 2, Fuel Injection Circuit Electrical Malfunction
P1203 Cylinder 3, Fuel Injection Circuit Electrical Malfunction
P1204 Cylinder 4, Fuel Injection Circuit Electrical Malfunction
P1205 Cylinder 5, Fuel Injection Circuit Electrical Malfunction
P1206 Cylinder 6, Fuel Injection Circuit Electrical Malfunction
P1207 Cylinder 7, Fuel Injection Circuit Electrical Malfunction
P1208 Cylinder 8, Fuel Injection Circuit Electrical Malfunction
P1213 Cylinder I-Fuel Injection Circuit Short to B+
P1214 Cylinder 2-Fuel Injection Circuit Short to B+
P1215 Cylinder 3 Fuel Injection Circuit Short to B+
P1216 Cylinder 4 Fuel Injection Circuit Short to B+
P1217 Cylinder 5 Fuel Injection Circuit Short to B+
P1218 Cylinder 6 Fuel Injection Circuit Short to B+
P1219 Cylinder 7, Fuel Injection Circuit Short to B+
P1219 Cylinder 8, Fuel Injection Circuit Short to B+
P1225 Cylinder I Fuel Injection Circuit Short to Ground
P1226 Cylinder 2 Fuel Injection Circuit Short to Ground
P1227 Cylinder 3 Fuel Injection Circuit Short to Ground
P1228 Cylinder 4 Fuel Injection Circuit Short to Ground
P1229 Cylinder 5 Fuel Injection Circuit Short to Ground
P1230 Cylinder 6 Fuel Injection Circuit Short to Ground
P1237 Cylinder I Fuel Injection Circuit Open Circuit
P1238 Cylinder 2 Fuel Injection Circuit Open Circuit
P1239 Cylinder 3 Fuel Injection Circuit Open Circuit
P1240 Cylinder 4 Fuel Injection Circuit Open Circuit
P1241 Cylinder 5 Fuel Injection Circuit Open Circuit
P1242 Cylinder 6 Fuel Injection Circuit Open Circuit

P1250 Fuel Level Too Low
P1280 Fuel Injection Air Control Valve Circuit Flow too Low
P1283 Fuel Injection Air Control Valve Circuit Electrical Malfunction
P1325 Cylinder 1 Knock Control Limit Attained
P1326 Cylinder 2 Knock Control Limit Attained
P1327 Cylinder 3 Knock Control Limit Attained
P1328 Cylinder 4 Knock Control Limit Attained
P1329 Cylinder 5 Knock Control Limit Attained
P1330 Cylinder 6 Knock Control Limit Attained
P1331 Cylinder 7, Knock Control Limit Attained
P1332 Cylinder 8, Knock Control Limit Attained
P1337 Camshaft Position Sensor, Bank 1 Short to Ground
P1338 Camshaft Position Sensor, Bank 1 Open Circuit/Short to B+
P1340 Boost Pressure Control Valve Short to B+
P1386 Internal Control Module Knock Control Circuit Error
P1391 Camshaft Position Sensor, Bank 2 Short to Ground
P1392 Camshaft Position Sensor, Bank 2 Open Circuit/Short to B+
P1410 Tank Ventilation Valve Circuit Short to B+
P1420 Secondary Air Injection Module Short To B+
P1421 Secondary Air Injection Module Short To Ground
P1421 Secondary Air Injection Valve Circuit Short to Ground
P1422 Secondary Air Injection System Control Valve Circuit Short To B+
P1422 Secondary Air Injection Valve Circuit Short to B+
P1425 Tank Vent Valve Short To Ground
P1425 Tank Vent Valve Short to Ground
P1426 Tank Vent Valve Open
P1426 Tank Vent Valve Open
P1432 Secondary Air Injection Valve Open
P1433 Secondary Air Injection System Pump Relay Circuit Open
P1434 Secondary Air Injection System Pump Relay Circuit Short to B+
P1435 Secondary Air Injection System Pump Relay Circuit Short to Ground
P1436 Secondary Air Injection System Pump Relay Circuit Electrical Malfunction
P1450 Secondary Air Injection System Circuit Short To B+
P1451 Secondary Air Injection System Circuit Short To Ground
P1452 Secondary Air Injection System Open Circuit
P1471 EVAP Emission Control LDP Circuit Short to B+
P1472 EVAP Emission Control LDP Circuit Short to Ground
P1473 EVAP Emission Control LDP Circuit Open Circuit
P1475 EVAP Emission Control LDP Circuit Malfunction/Signal Circuit Open
P1476 EVAP Emission Control LDP Circuit Malfunction/Insufficient Vacuum
P1477 EVAP Emission Control LDP Circuit Malfunction
P1500 Fuel Pump Relay Circuit Electrical Malfunction
P1501 Fuel Pump Relay Circuit Short to Ground
P1502 Fuel Pump Relay Circuit Short to B+
P1505 Closed Throttle Position Switch Does Not Close/Open Circuit
P1506 Closed Throttle Position Switch Does Not Open/Short to Ground
P1507 Idle System Learned Value Lower Limit Attained
P1508 Idle System Learned Value Upper Limit Attained
P1512 Intake Manifold Changeover Valve Circuit, Short to B+

P1515 Intake Manifold Changeover Valve Circuit, Short to Ground
P1516 Intake Manifold Changeover Valve Circuit, Open
P1519 Intake Camshaft Control, Bank 1 Malfunction
P1522 Intake Camshaft Control, Bank 2 Malfunction
P1543 Throttle Actuation Potentiometer Signal Too Low
P1544 Throttle Actuation Potentiometer Signal Too High
P1545 Throttle Position Control Malfunction
P1547 Boost Pressure Control Valve Short to Ground
P1548 Boost Pressure Control Valve Open
P1555 Charge Pressure Upper Limit Exceeded
P1556 Charge Pressure Negative Deviation
P1557 Charge Pressure Positive Deviation
P1558 Throttle Actuator Electrical Malfunction
P1559 Idle Speed Control Throttle Position Adaptation Malfunction
P1560 Maximum Engine Speed Exceeded
P1564 Idle Speed Control, Throttle Position Low Voltage During Adaptation
P1580 Throttle Actuator (B1) Malfunction
P1582 Idle Adaptation At Limit
P1602 Power Supply (B+) Terminal 30 Low Voltage
P1606 Rough Road Spec Engine Torque ABS-ECU Electrical Malfunction
P1611 MIL Call-up Circuit/Transmission Control Module Short to Ground
P1612 Electronic Control Module Incorrect Coding
P1613 MIL Call-up Circuit Open/Short to B+
P1624 MIL Request Signal Active
P1625 CAN-Bus Implausible Message from Transmission Control
P1626 CAN-Bus Missing Message from Transmission Control
P1640 Internal Control Module (EEPROM) Error
P1640 Internal Control Module (EEPROM) Error
P1681 Control Unit Programming not Finished
P1690 Malfunction Indicator Light Malfunction
P1693 Malfunction Indicator Light Short to B+
P1778 Solenoid EV7 Electrical Malfunction
P1780 Engine Intervention Readable

TROUBLE CODE EQUIVALENT LIST

The following codes represent those displayed by Audi's VAG 1551 tester/scan tool.

16486 Mass or Volume Air Flow Circuit Low Input
16487 Mass or Volume Air Flow Circuit High Input
16491 Manifold Absolute Pressure or Barometric Pressure Low Input
16492 Manifold Absolute Pressure or Barometric Pressure High Input
16496 Intake Air Temperature Circuit Low Input
16497 Intake Air Temperature Circuit High Input
16500 Engine Coolant Temperature Circuit Range/Performance
16501 Engine Coolant Temperature Circuit Low Input
16502 Oxygen Engine Coolant Temperature Circuit High Input
16504 Throttle Position Sensor A Circuit Malfunction
16505 Throttle/Pedal Position Sensor A Circuit Range/Performance
16506 Throttle/Pedal Position Sensor A Circuit Low Input
16507 Throttle/Pedal Position Sensor A Circuit High Input

16509 Insufficient Coolant Temperature For Closed Loop Fuel Control

16514 Oxygen Sensor Circuit, Bank 1-Sensor 1 Malfunction

16515 Oxygen Sensor Circuit, Bank 1-Sensor 1 Low Voltage

16516 Oxygen Sensor Circuit, Bank 1-Sensor 1 High Voltage

16517 Oxygen Sensor Circuit, Bank 1-Sensor 1 Slow Response

16518 Oxygen Sensor Circuit, Bank 1-Sensor 1 No Activity Detected

16519 Oxygen Sensor Heater Circuit, Bank 1-Sensor 1 Malfunction

16520 Oxygen Sensor Circuit, Bank 1-Sensor 2 Malfunction

16521 Oxygen Sensor Circuit, Bank 1-Sensor 2 Low Voltage

16522 Oxygen Sensor Circuit, Bank 1-Sensor 2 High Voltage

16524 Oxygen Sensor Circuit, Bank 1-Sensor 2 No Activity Detected

16525 Oxygen Sensor Heater Circuit, Bank 1-Sensor 2 Malfunction

16534 Oxygen Sensor Circuit, Bank 2-Sensor 1 Malfunction

16535 Oxygen Sensor Circuit, Bank 2-Sensor 1 Low Voltage

16536 Oxygen Sensor Circuit, Bank 2-Sensor 1 High Voltage

16537 Oxygen Sensor Circuit, Bank 2-Sensor 1 Slow Response

16538 Oxygen Sensor Circuit, Bank 2-Sensor 1 No Activity Detected

16540 Oxygen Sensor Circuit, Bank 2-Sensor 2 Malfunction

16541 Oxygen Sensor Circuit, Bank 2-Sensor 2 Low Voltage

16542 Oxygen Sensor Circuit, Bank 2-Sensor 2 High Voltage

16544 Oxygen Sensor Circuit, Bank 2-Sensor 2 No Activity Detected

16555 Oxygen System Too Lean, Bank 1

16556 Oxygen System Too Rich, Bank 1

16684 Random Multiple Misfire Detected

16685 Cylinder 1 Misfire Detected

16686 Cylinder 2 Misfire Detected

16687 Cylinder 3 Misfire Detected

16688 Cylinder 4 Misfire Detected

16689 Cylinder 5 Misfire Detected

16690 Cylinder 6 Misfire Detected

16705 Ignition Distributor Engine Speed Input Circuit Range/Performance

16706 Ignition /Distributor Engine Speed Input Circuit No Signal

16711 Knock sensor 1 Circuit Low Input

16712 Knock Sensor Circuit, High Input

16716 Knock sensor 2 Circuit Low Input

16716 Knock Sensor Circuit, Low Input

16717 Knock Sensor 2 Circuit, High Input

16725 Camshaft Position Sensor Circuit Range/Performance

16795 Secondary Air Injection System Incorrect Flow Detected

16806 Main Catalyst Efficiency Below Threshold (Bank 1)

16806 Main Catalyst, Bank I Efficiency Below Threshold

16824 Evaporative Emission Control System Malfunction

16825 EVAP Emission Contr. Sys. Incorrect Purge Flow

16826 EVAP Emission Contr. Sys. (Small Leak) Leak Detected

16839 EVAP Emission Contr. Sys. (Gross Leak) Leak Detected

16885 Vehicle Speed Sensor Range/Performance

16890 Idle Control System RPM Lower Than Expected

16891 Idle Control System RPM Higher Than Expected

16894 Closed Throttle Position Switch Malfunction

16944 System Voltage Malfunction

16946 System Voltage Low Voltage

16947 System Voltage High Voltage

16985 Internal Contr. Module Memory Check Sum Error

16988 Internal Contr. Module Random Access Memory (RAM) Error

16989 Internal Control Module Read Only Memory (ROM) Error

17091 Transmission Range Sensor Circuit Low Input

17092 Transmission Range Sensor Circuit High Input

17099 Input Turbine Speed Sensor Circuit Malfunction

17106 Output Speed Sensor Circuit No Signal

17109 Engine Speed Input Circuit Malfunction

17132 Pressure Control Solenoid Electrical

17137 Shift Solenoid A Electrical

17142 Shift Solenoid B Electrical

17147 Shift Solenoid C Electrical

17152 Shift Solenoid D Electrical

17157 Shift Solenoid E Electrical

16684 Random/Multiple Cylinder Misfire Detected

16685 Cylinder I Misfire Detected

16686 Cylinder 2 Misfire Detected

17510 Oxygen Sensor Heating Circuit, Bank 1-Sensor 1 Short to B+

17513 Oxygen Sensor Heating Circuit, Bank 1-Sensor 2 Short to B+

17515 Oxygen Sensor Heating Circuit, Bank 2-Sensor I Short to B+

17518 Oxygen Sensor Heating Circuit, Bank 2-Sensor 2 Short to B+

17535 Long Term Fuel Trim Multiplicative, Bank 1 System Too Rich

17536 Long Term Fuel Trim Multiplicative, Bank 1 System Too Lean

17537 Long Term Fuel Trim Multiplicative, Bank2 System too Rich

17538 Long Term Fuel Trim Multiplicative, Bank2 System too Lean

17544 Long Term Fuel Trim Additive, Bank 1 System Too Lean

17545 Long Term Fuel Trim Additive, Bank 1 System Too Rich

17546 Long Term Fuel Trim Additive Fuel, Bank I System too Lean

17547 Long Term Fuel Trim Additive Fuel, Bank 1 System too Rich

17549 Load Calculation Cross Check Range/Performance

17584 Oxygen Correction Behind Catalyst, B1 Limit Attained

17585 Oxygen Correction Behind Catalyst. 82 Limit Attained

17604 Oxygen Sensor Heater Circuit, Bank 1-Sensor 1 Electrical Malfunction

17605 Oxygen Sensor Heater Circuit, Bank 2-Sensor I Electrical Malfunction

17606 Oxygen Sensor Heater Circuit, BankI-Sensor2 Electrical Malfunction

17606 Oxygen Sensor Heater Circuit, Bank 1-Sensor 2 Electrical Malfunction

17607 Oxygen Sensor Heater Circuit, Bank 2-Sensor 2 Electrical Malfunction

17609 Cylinder 1, Fuel Injection Circuit Electrical Malfunction

17610 Cylinder 2, Fuel Injection Circuit Electrical Malfunction

17611 Cylinder 3, Fuel Injection Circuit Electrical Malfunction

17612 Cylinder 4, Fuel Injection Circuit Electrical Malfunction

17613 Cylinder 5, Fuel Injection Circuit Electrical Malfunction

17614 Cylinder 6, Fuel Injection Circuit Electrical Malfunction

17615 Cylinder 7, Fuel Injection Circuit Electrical Malfunction

17616 Cylinder 8, Fuel Injection Circuit Electrical Malfunction

17621 Cylinder I-Fuel Injection Circuit Short to B+

17622 Cylinder 2-Fuel Injection Circuit Short to B+

17623 Cylinder 3 Fuel Injection Circuit Short to B+

17624 Cylinder 4 Fuel Injection Circuit Short to B+

17625 Cylinder 5 Fuel Injection Circuit Short to B+

17626 Cylinder 6 Fuel Injection Circuit Short to B+

17627 Cylinder 7, Fuel Injection Circuit Short to B+

17628 Cylinder 8, Fuel Injection Circuit Short to B+

17633 Cylinder I Fuel Injection Circuit Short to Ground

17634 Cylinder 2 Fuel Injection Circuit Short to Ground

17635 Cylinder 3 Fuel Injection Circuit Short to Ground

17636 Cylinder 4 Fuel Injection Circuit Short to Ground

17637 Cylinder 5 Fuel Injection Circuit Short to Ground

17638 Cylinder 6 Fuel Injection Circuit Short to Ground

17645 Cylinder I Fuel Injection Circuit Open Circuit

17646 Cylinder 2 Fuel Injection Circuit Open Circuit

17647 Cylinder 3 Fuel Injection Circuit Open Circuit

17648 Cylinder 4 Fuel Injection Circuit Open Circuit

17649 Cylinder 5 Fuel Injection Circuit Open Circuit

17650 Cylinder 6 Fuel Injection Circuit Open Circuit

17658 Fuel Level Too Low

17688 Fuel Injection Air Control Valve Circuit Flow too Low

17691 Fuel Injection Air Control Valve Circuit Electrical Malfunction

17733 Cylinder I Knock Control Limit Attained

17734 Cylinder 2 Knock Control Limit Attained

17735 Cylinder 3 Knock Control Limit Attained

17736 Cylinder 4 Knock Control Limit Attained

17737 Cylinder 5 Knock Control Limit Attained

17738 Cylinder 6 Knock Control Limit Attained

17739 Cylinder 7, Knock Control Limit Attained

17740 Cylinder 8, Knock Control Limit Attained

17745 Camshaft Position Sensor, Bank 1 Short to Ground

17746 Camshaft Position Sensor, Bank 1 Open Circuit/Short to B+

17954 Boost Pressure Control Valve Short to B+

17794 Internal Control Module Knock Control Circuit Error

17799 Camshaft Position Sensor, Bank 2 Short to Ground

17800 Camshaft Position Sensor, Bank 2 Open Circuit/Short to B+

17818 Tank Ventilation Valve Circuit Short to B+

17818 Tank Ventilation Valve Circuit Short to B+

17828 Secondary Air Injection Module Short To B+

17829 Secondary Air Injection Module Short To Ground

17829 Secondary Air Injection Valve Circuit Short to Ground

17830 Secondary Air Injection System Control Valve Circuit Short To B+

17830 Secondary Air Injection Valve Circuit Short to B+

17833 Tank Vent Valve Short To Ground

17833 Tank Vent Valve Short to Ground

17834 Tank Vent Valve Open

17834 Tank Vent Valve Open

17840 Secondary Air Injection Valve Open

17841 Secondary Air Injection System Pump Relay Circuit Open

17842 Secondary Air Injection System Pump Relay Circuit Short to B+

17843 Secondary Air Injection System Pump Relay Circuit Short to Ground

17844 Secondary Air Injection System Pump Relay Circuit Electrical Malfunction

17858 Secondary Air Injection System Circuit Short To B+

17859 Secondary Air Injection System Circuit Short To Ground

17860 Secondary Air Injection System Open Circuit

17879 EVAP Emission Control LDP Circuit Short to B+

17880 EVAP Emission Control LDP Circuit Short to Ground

17881 EVAP Emission Control LDP Circuit Open Circuit

17883 EVAP Emission Control LDP Circuit Malfunction/Signal Circuit Open

17884 EVAP Emission Control LDP Circuit Malfunction/Insufficient Vacuum

17885 EVAP Emission Control LDP Circuit Malfunction

17908 Fuel Pump Relay Circuit Electrical Malfunction

17909 Fuel Pump Relay Circuit Short to Ground

17910 Fuel Pump Relay Circuit Short to B+

17913 Closed Throttle Position Switch Does Not Close/Open Circuit

17914 Closed Throttle Position Switch Does Not Open/Short to Ground

17915 Idle System Learned Value Lower Limit Attained

17916 Idle System Learned Value Upper Limit Attained

17920 Intake Manifold Changeover Valve Circuit, Short to B+

17923 Intake Manifold Changeover Valve Circuit, Short to Ground

17924 Intake Manifold Changeover Valve Circuit, Open

17927 Intake Camshaft Control, Bank 2 Malfunction

17951 Throttle Actuation Potentiometer Signal Too Low

17952 Throttle Actuation Potentiometer Signal Too High

17953 Throttle Position Control Malfunction

17955 Boost Pressure Control Valve Short to Ground

17956 Boost Pressure Control Valve Open

17963 Charge Pressure Upper Limit Exceeded

17964 Charge Pressure Negative Deviation

17965 Charge Pressure Positive Deviation

17966 Throttle Actuator Electrical Malfunction

17967 Idle Speed Control Throttle Position Adaptation Malfunction

17968 Maximum Engine Speed Exceeded

17972 Idle Speed Control, Throttle Position Low Voltage During Adaptation

17988 Throttle Actuator (B1) Malfunction

17990 Idle Adaptation At Limit

18010 Power Supply (B+) Terminal 30 Low Voltage

18014 Rough Road Spec Engine Torque ABS-ECU Electrical Malfunction

18019 MIL Call-up Circuit/Transmission Control Module Short to Ground

18020 Electronic Control Module Incorrect Coding

18021 MIL Call-up Circuit Open/Short to B+

18032 MIL Request Signal Active

18033 CAN-Bus Implausible Message from Transmission Control

18034 CAN-Bus Missing Message from Transmission Control

18048 Internal Control Module (EEPROM) Error

18048 Internal Control Module (EEPROM) Error

18089 Control Unit Programming not Finished
18098 Malfunction Indicator Light Malfunction
18101 Malfunction Indicator Light Short to B+
18186 Solenoid EV7 Electrical Malfunction
18188 Engine Intervention Readable

Chrysler Corporation

READING CODES

With Scan Tool

Reading the control module memory is on of the first steps in OBD II system diagnostics. This step should be initially performed to determine the general nature of the fault. Subsequent readings will determine if the fault has been cleared.

Reading codes can be performed by any of the methods below:
- Read the control module memory with the Generic Scan Tool (GST)
- Read the control module memory with the vehicle manufacturer's specific tester

To read the fault codes, connect the scan tool or tester according to the manufacturer's instructions. Follow the manufacturer's specified procedure for reading the codes.

Without Scan Tool

On all 1995–97 vehicles, as well as the 1998 Sebring coupe and Avenger models with a 2.0L engine, DTC's can be accessed by observing the 2-digit number (referred to as an OBD II Equivalent code) displayed by the Malfunction Indicator Lamp (MIL). The MIL is shown on the instrument panel as the Check Engine lamp. This method should be used as a "quick test" only. You should always use a scan tool to get the most detailed information.

➡**Be advised that the MIL can only perform a limited number of functions, and it is a good idea to have the system checked with a scan tool to double check the circuit function.**

Within a period of 5 seconds, cycle the ignition key **ON—OFF—ON—OFF—ON**.

1. Count the number of times the MIL (check engine lamp) on the instrument panel flashes on and off.

The number of flashes represents the trouble code. There is a short pause between the flashes representing the 1st and 2nd digits of the code. Longer pauses are used to separate individual 2-digit trouble codes.

An example of a flashed DTC is as follows:
- Lamp flashes 4 times, pauses, then flashes 6 more times. This denotes a DTC number 46.
- Lamp flashes 5 times, pauses, then flashes 5 more times. This indicates a DTC number 55.

DTC 55 will always be the last code to be displayed.

CLEARING CODES

With Scan Tool

Control module reset procedures are a very important part of OBD II System diagnostics. This step should be done at the end of any fault code repair and at the end of any driveability repair.

Clearing codes can be performed by any of the methods below:

- Clear the control module memory with the Generic Scan Tool (GST)
- Clear the control module memory with the vehicle manufacturer's specific tester
- Turn the ignition off and remove the negative battery cable for at least 1 minute.

Removing the negative battery cable may cause other systems in the vehicle to lose their memory. Prior to removing the cable, ensure you have the proper reset codes for radios and alarms.

➡**The MIL will may also be de-activated for some codes if the vehicle completes three consecutive trips without a fault detected with vehicle conditions similar to those present during the fault.**

Without Scan Tool

Control module reset procedures are a very important part of OBD II System diagnostics. This step should be done at the end of any fault code repair and at the end of any driveability repair.

Clearing codes can be performed by turning the ignition off and removing the negative battery cable for at least 1 minute. Removing the negative battery cable may cause other systems in the vehicle to loose their memory. Prior to removing the cable, ensure you have the proper reset codes for radios and alarms.

➡**The MIL will may also be de-activated for some codes if the vehicle completes three consecutive trips without a fault detected with vehicle conditions similar to those present during the fault.**

DIAGNOSTIC TROUBLE CODES

P0000 No Failures
P0100 Mass or Volume Air Flow Circuit Malfunction
P0101 Mass or Volume Air Flow Circuit Range/Performance Problem
P0102 Mass or Volume Air Flow Circuit Low Input
P0103 Mass or Volume Air Flow Circuit High Input
P0104 Mass or Volume Air Flow Circuit Intermittent
P0105 Manifold Absolute Pressure/Barometric Pressure Circuit Malfunction
P0106 Manifold Absolute Pressure/Barometric Pressure Circuit Range/Performance Problem
P0107 Manifold Absolute Pressure/Barometric Pressure Circuit Low Input
P0108 Manifold Absolute Pressure/Barometric Pressure Circuit High Input
P0109 Manifold Absolute Pressure/Barometric Pressure Circuit Intermittent
P0110 Intake Air Temperature Circuit Malfunction
P0111 Intake Air Temperature Circuit Range/Performance Problem
P0112 Intake Air Temperature Circuit Low Input
P0113 Intake Air Temperature Circuit High Input
P0114 Intake Air Temperature Circuit Intermittent
P0115 Engine Coolant Temperature Circuit Malfunction
P0116 Engine Coolant Temperature Circuit Range/Performance Problem
P0117 Engine Coolant Temperature Circuit Low Input
P0118 Engine Coolant Temperature Circuit High Input
P0119 Engine Coolant Temperature Circuit Intermittent

P0120 Throttle/Pedal Position Sensor/Switch "A" Circuit Malfunction

P0121 Throttle/Pedal Position Sensor/Switch "A" Circuit Range/Performance Problem

P0122 Throttle/Pedal Position Sensor/Switch "A" Circuit Low Input

P0123 Throttle/Pedal Position Sensor/Switch "A" Circuit High Input

P0124 Throttle/Pedal Position Sensor/Switch "A" Circuit Intermittent

P0125 Insufficient Coolant Temperature For Closed Loop Fuel Control

P0126 Insufficient Coolant Temperature For Stable Operation

P0130 O_2 Circuit Malfunction (Bank no. 1 Sensor no. 1)

P0131 O_2 Sensor Circuit Low Voltage (Bank no. 1 Sensor no. 1)

P0132 O_2 Sensor Circuit High Voltage (Bank no. 1 Sensor no. 1)

P0133 O_2 Sensor Circuit Slow Response (Bank no. 1 Sensor no. 1)

P0134 O_2 Sensor Circuit No Activity Detected (Bank no. 1 Sensor no. 1)

P0135 O_2 Sensor Heater Circuit Malfunction (Bank no. 1 Sensor no. 1)

P0136 O_2 Sensor Circuit Malfunction (Bank no. 1 Sensor no. 2)

P0137 O_2 Sensor Circuit Low Voltage (Bank no. 1 Sensor no. 2)

P0138 O_2 Sensor Circuit High Voltage (Bank no. 1 Sensor no. 2)

P0139 O_2 Sensor Circuit Slow Response (Bank no. 1 Sensor no. 2)

P0140 O_2 Sensor Circuit No Activity Detected (Bank no. 1 Sensor no. 2)

P0141 O_2 Sensor Heater Circuit Malfunction (Bank no. 1 Sensor no. 2)

P0142 O_2 Sensor Circuit Malfunction (Bank no. 1 Sensor no. 3)

P0143 O_2 Sensor Circuit Low Voltage (Bank no. 1 Sensor no. 3)

P0144 O_2 Sensor Circuit High Voltage (Bank no. 1 Sensor no. 3)

P0145 O_2 Sensor Circuit Slow Response (Bank no. 1 Sensor no. 3)

P0146 O_2 Sensor Circuit No Activity Detected (Bank no. 1 Sensor no. 3)

P0147 O_2 Sensor Heater Circuit Malfunction (Bank no. 1 Sensor no. 3)

P0150 O_2 Sensor Circuit Malfunction (Bank no. 2 Sensor no. 1)

P0151 O_2 Sensor Circuit Low Voltage (Bank no. 2 Sensor no. 1)

P0152 O_2 Sensor Circuit High Voltage (Bank no. 2 Sensor no. 1)

P0153 O_2 Sensor Circuit Slow Response (Bank no. 2 Sensor no. 1)

P0154 O_2 Sensor Circuit No Activity Detected (Bank no. 2 Sensor no. 1)

P0155 O_2 Sensor Heater Circuit Malfunction (Bank no. 2 Sensor no. 1)

P0156 O_2 Sensor Circuit Malfunction (Bank no. 2 Sensor no. 2)

P0157 O_2 Sensor Circuit Low Voltage (Bank no. 2 Sensor no. 2)

P0158 O_2 Sensor Circuit High Voltage (Bank no. 2 Sensor no. 2)

P0159 O_2 Sensor Circuit Slow Response (Bank no. 2 Sensor no. 2)

P0160 O_2 Sensor Circuit No Activity Detected (Bank no. 2 Sensor no. 2)

P0161 O_2 Sensor Heater Circuit Malfunction (Bank no. 2 Sensor no. 2)

P0162 O_2 Sensor Circuit Malfunction (Bank no. 2 Sensor no. 3)

P0163 O_2 Sensor Circuit Low Voltage (Bank no. 2 Sensor no. 3)

P0164 O_2 Sensor Circuit High Voltage (Bank no. 2 Sensor no. 3)

P0165 O_2 Sensor Circuit Slow Response (Bank no. 2 Sensor no. 3)

P0166 O_2 Sensor Circuit No Activity Detected (Bank no. 2 Sensor no. 3)

P0167 O_2 Sensor Heater Circuit Malfunction (Bank no. 2 Sensor no. 3)

P0170 Fuel Trim Malfunction (Bank no. 1)

P0171 System Too Lean (Bank no. 1)

P0172 System Too Rich (Bank no. 1)

P0173 Fuel Trim Malfunction (Bank no. 2)

P0174 System Too Lean (Bank no. 2)

P0175 System Too Rich (Bank no. 2)

P0176 Fuel Composition Sensor Circuit Malfunction

P0177 Fuel Composition Sensor Circuit Range/Performance

P0178 Fuel Composition Sensor Circuit Low Input

P0179 Fuel Composition Sensor Circuit High Input

P0180 Fuel Temperature Sensor "A" Circuit Malfunction

P0181 Fuel Temperature Sensor "A" Circuit Range/Performance

P0182 Fuel Temperature Sensor "A" Circuit Low Input

P0183 Fuel Temperature Sensor "A" Circuit High Input

P0184 Fuel Temperature Sensor "A" Circuit Intermittent

P0185 Fuel Temperature Sensor "B" Circuit Malfunction

P0186 Fuel Temperature Sensor "B" Circuit Range/Performance

P0187 Fuel Temperature Sensor "B" Circuit Low Input

P0188 Fuel Temperature Sensor "B" Circuit High Input

P0189 Fuel Temperature Sensor "B" Circuit Intermittent

P0190 Fuel Rail Pressure Sensor Circuit Malfunction

P0191 Fuel Rail Pressure Sensor Circuit Range/Performance

P0192 Fuel Rail Pressure Sensor Circuit Low Input

P0193 Fuel Rail Pressure Sensor Circuit High Input

P0194 Fuel Rail Pressure Sensor Circuit Intermittent

P0195 Engine Oil Temperature Sensor Malfunction

P0196 Engine Oil Temperature Sensor Range/Performance

P0197 Engine Oil Temperature Sensor Low

P0198 Engine Oil Temperature Sensor High

P0199 Engine Oil Temperature Sensor Intermittent

P0200 Injector Circuit Malfunction

P0201 Injector Circuit Malfunction—Cylinder no. 1

P0202 Injector Circuit Malfunction—Cylinder no. 2

P0203 Injector Circuit Malfunction—Cylinder no. 3

P0204 Injector Circuit Malfunction—Cylinder no. 4

P0205 Injector Circuit Malfunction—Cylinder no. 5

P0206 Injector Circuit Malfunction—Cylinder no. 6

P0207 Injector Circuit Malfunction—Cylinder no. 7

P0208 Injector Circuit Malfunction—Cylinder no. 8

P0209 Injector Circuit Malfunction—Cylinder no. 9

P0210 Injector Circuit Malfunction—Cylinder no. 10

P0211 Injector Circuit Malfunction—Cylinder no. 11

P0212 Injector Circuit Malfunction—Cylinder no. 12

P0213 Cold Start Injector no. 1 Malfunction

P0214 Cold Start Injector no. 2 Malfunction

P0215 Engine Shutoff Solenoid Malfunction
P0216 Injection Timing Control Circuit Malfunction
P0217 Engine Over Temperature Condition
P0218 Transmission Over Temperature Condition
P0219 Engine Over Speed Condition
P0220 Throttle/Pedal Position Sensor/Switch "B" Circuit Malfunction
P0221 Throttle/Pedal Position Sensor/Switch "B" Circuit Range/Performance Problem
P0222 Throttle/Pedal Position Sensor/Switch "B" Circuit Low Input
P0223 Throttle/Pedal Position Sensor/Switch "B" Circuit High Input
P0224 Throttle/Pedal Position Sensor/Switch "B" Circuit Intermittent
P0225 Throttle/Pedal Position Sensor/Switch "C" Circuit Malfunction
P0226 Throttle/Pedal Position Sensor/Switch "C" Circuit Range/Performance Problem
P0227 Throttle/Pedal Position Sensor/Switch "C" Circuit Low Input
P0228 Throttle/Pedal Position Sensor/Switch "C" Circuit High Input
P0229 Throttle/Pedal Position Sensor/Switch "C" Circuit Intermittent
P0230 Fuel Pump Primary Circuit Malfunction
P0231 Fuel Pump Secondary Circuit Low
P0232 Fuel Pump Secondary Circuit High
P0233 Fuel Pump Secondary Circuit Intermittent
P0234 Engine Over Boost Condition
P0261 Cylinder no. 1 Injector Circuit Low
P0262 Cylinder no. 1 Injector Circuit High
P0263 Cylinder no. 1 Contribution/Balance Fault
P0264 Cylinder no. 2 Injector Circuit Low
P0265 Cylinder no. 2 Injector Circuit High
P0266 Cylinder no. 2 Contribution/Balance Fault
P0267 Cylinder no. 3 Injector Circuit Low
P0268 Cylinder no. 3 Injector Circuit High
P0269 Cylinder no. 3 Contribution/Balance Fault
P0270 Cylinder no. 4 Injector Circuit Low
P0271 Cylinder no. 4 Injector Circuit High
P0272 Cylinder no. 4 Contribution/Balance Fault
P0273 Cylinder no. 5 Injector Circuit Low
P0274 Cylinder no. 5 Injector Circuit High
P0275 Cylinder no. 5 Contribution/Balance Fault
P0276 Cylinder no. 6 Injector Circuit Low
P0277 Cylinder no. 6 Injector Circuit High
P0278 Cylinder no. 6 Contribution/Balance Fault
P0279 Cylinder no. 7 Injector Circuit Low
P0280 Cylinder no. 7 Injector Circuit High
P0281 Cylinder no. 7 Contribution/Balance Fault
P0282 Cylinder no. 8 Injector Circuit Low
P0283 Cylinder no. 8 Injector Circuit High
P0284 Cylinder no. 8 Contribution/Balance Fault
P0285 Cylinder no. 9 Injector Circuit Low
P0286 Cylinder no. 9 Injector Circuit High
P0287 Cylinder no. 9 Contribution/Balance Fault
P0288 Cylinder no. 10 Injector Circuit Low
P0289 Cylinder no. 10 Injector Circuit High
P0290 Cylinder no. 10 Contribution/Balance Fault
P0291 Cylinder no. 11 Injector Circuit Low
P0292 Cylinder no. 11 Injector Circuit High

P0293 Cylinder no. 11 Contribution/Balance Fault
P0294 Cylinder no. 12 Injector Circuit Low
P0295 Cylinder no. 12 Injector Circuit High
P0296 Cylinder no. 12 Contribution/Balance Fault
P0300 Random/Multiple Cylinder Misfire Detected
P0301 Cylinder no. 1—Misfire Detected
P0302 Cylinder no. 2—Misfire Detected
P0303 Cylinder no. 3—Misfire Detected
P0304 Cylinder no. 4—Misfire Detected
P0305 Cylinder no. 5—Misfire Detected
P0306 Cylinder no. 6—Misfire Detected
P0307 Cylinder no. 7—Misfire Detected
P0308 Cylinder no. 8—Misfire Detected
P0309 Cylinder no. 9—Misfire Detected
P0310 Cylinder no. 10—Misfire Detected
P0311 Cylinder no. 11—Misfire Detected
P0312 Cylinder no. 12—Misfire Detected
P0320 Ignition/Distributor Engine Speed Input Circuit Malfunction
P0321 Ignition/Distributor Engine Speed Input Circuit Range/Performance
P0322 Ignition/Distributor Engine Speed Input Circuit No Signal
P0323 Ignition/Distributor Engine Speed Input Circuit Intermittent
P0325 Knock Sensor no. 1—Circuit Malfunction (Bank no. 1 or Single Sensor)
P0326 Knock Sensor no. 1—Circuit Range/Performance (Bank no. 1 or Single Sensor)
P0327 Knock Sensor no. 1—Circuit Low Input (Bank no. 1 or Single Sensor)
P0328 Knock Sensor no. 1—Circuit High Input (Bank no. 1 or Single Sensor)
P0329 Knock Sensor no. 1—Circuit Input Intermittent (Bank no. 1 or Single Sensor)
P0330 Knock Sensor no. 2—Circuit Malfunction (Bank no. 2)
P0331 Knock Sensor no. 2—Circuit Range/Performance (Bank no. 2)
P0332 Knock Sensor no. 2—Circuit Low Input (Bank no. 2)
P0333 Knock Sensor no. 2—Circuit High Input (Bank no. 2)
P0334 Knock Sensor no. 2—Circuit Input Intermittent (Bank no. 2)
P0335 Crankshaft Position Sensor "A" Circuit Malfunction
P0336 Crankshaft Position Sensor "A" Circuit Range/Performance
P0337 Crankshaft Position Sensor "A" Circuit Low Input
P0338 Crankshaft Position Sensor "A" Circuit High Input
P0339 Crankshaft Position Sensor "A" Circuit Intermittent
P0340 Camshaft Position Sensor Circuit Malfunction
P0341 Camshaft Position Sensor Circuit Range/Performance
P0342 Camshaft Position Sensor Circuit Low Input
P0343 Camshaft Position Sensor Circuit High Input
P0344 Camshaft Position Sensor Circuit Intermittent
P0350 Ignition Coil Primary/Secondary Circuit Malfunction
P0351 Ignition Coil "A" Primary/Secondary Circuit Malfunction
P0352 Ignition Coil "B" Primary/Secondary Circuit Malfunction
P0353 Ignition Coil "C" Primary/Secondary Circuit Malfunction
P0354 Ignition Coil "D" Primary/Secondary Circuit Malfunction

P0355 Ignition Coil "E" Primary/Secondary Circuit Malfunction

P0356 Ignition Coil "F" Primary/Secondary Circuit Malfunction

P0357 Ignition Coil "G" Primary/Secondary Circuit Malfunction

P0358 Ignition Coil "H" Primary/Secondary Circuit Malfunction

P0359 Ignition Coil "I" Primary/Secondary Circuit Malfunction

P0360 Ignition Coil "J" Primary/Secondary Circuit Malfunction

P0361 Ignition Coil "K" Primary/Secondary Circuit Malfunction

P0362 Ignition Coil "L" Primary/Secondary Circuit Malfunction

P0370 Timing Reference High Resolution Signal "A" Malfunction

P0371 Timing Reference High Resolution Signal "A" Too Many Pulses

P0372 Timing Reference High Resolution Signal "A" Too Few Pulses

P0373 Timing Reference High Resolution Signal "A" Intermittent/Erratic Pulses

P0374 Timing Reference High Resolution Signal "A" No Pulses

P0375 Timing Reference High Resolution Signal "B" Malfunction

P0376 Timing Reference High Resolution Signal "B" Too Many Pulses

P0377 Timing Reference High Resolution Signal "B" Too Few Pulses

P0378 Timing Reference High Resolution Signal "B" Intermittent/Erratic Pulses

P0379 Timing Reference High Resolution Signal "B" No Pulses

P0380 Glow Plug/Heater Circuit "A" Malfunction

P0381 Glow Plug/Heater Indicator Circuit Malfunction

P0382 Glow Plug/Heater Circuit "B" Malfunction

P0385 Crankshaft Position Sensor "B" Circuit Malfunction

P0386 Crankshaft Position Sensor "B" Circuit Range/Performance

P0387 Crankshaft Position Sensor "B" Circuit Low Input

P0388 Crankshaft Position Sensor "B" Circuit High Input

P0389 Crankshaft Position Sensor "B" Circuit Intermittent

P0400 Exhaust Gas Recirculation Flow Malfunction

P0401 Exhaust Gas Recirculation Flow Insufficient Detected

P0402 Exhaust Gas Recirculation Flow Excessive Detected

P0403 Exhaust Gas Recirculation Circuit Malfunction

P0404 Exhaust Gas Recirculation Circuit Range/Performance

P0405 Exhaust Gas Recirculation Sensor "A" Circuit Low

P0406 Exhaust Gas Recirculation Sensor "A" Circuit High

P0407 Exhaust Gas Recirculation Sensor "B" Circuit Low

P0408 Exhaust Gas Recirculation Sensor "B" Circuit High

P0410 Secondary Air Injection System Malfunction

P0411 Secondary Air Injection System Incorrect Flow Detected

P0412 Secondary Air Injection System Switching Valve "A" Circuit Malfunction

P0413 Secondary Air Injection System Switching Valve "A" Circuit Open

P0414 Secondary Air Injection System Switching Valve "A" Circuit Shorted

P0415 Secondary Air Injection System Switching Valve "B" Circuit Malfunction

P0416 Secondary Air Injection System Switching Valve "B" Circuit Open

P0417 Secondary Air Injection System Switching Valve "B" Circuit Shorted

P0418 Secondary Air Injection System Relay "A" Circuit Malfunction

P0419 Secondary Air Injection System Relay "B" Circuit Malfunction

P0420 Catalyst System Efficiency Below Threshold (Bank no. 1)

P0421 Warm Up Catalyst Efficiency Below Threshold (Bank no. 1)

P0422 Main Catalyst Efficiency Below Threshold (Bank no. 1)

P0423 Heated Catalyst Efficiency Below Threshold (Bank no. 1)

P0424 Heated Catalyst Temperature Below Threshold (Bank no. 1)

P0430 Catalyst System Efficiency Below Threshold (Bank no. 2)

P0431 Warm Up Catalyst Efficiency Below Threshold (Bank no. 2)

P0432 Main Catalyst Efficiency Below Threshold (Bank no. 2)

P0433 Heated Catalyst Efficiency Below Threshold (Bank no. 2)

P0434 Heated Catalyst Temperature Below Threshold (Bank no. 2)

P0440 Evaporative Emission Control System Malfunction

P0441 Evaporative Emission Control System Incorrect Purge Flow

P0442 Evaporative Emission Control System Leak Detected (Small Leak)

P0443 Evaporative Emission Control System Purge Control Valve Circuit Malfunction

P0444 Evaporative Emission Control System Purge Control Valve Circuit Open

P0445 Evaporative Emission Control System Purge Control Valve Circuit Shorted

P0446 Evaporative Emission Control System Vent Control Circuit Malfunction

P0447 Evaporative Emission Control System Vent Control Circuit Open

P0448 Evaporative Emission Control System Vent Control Circuit Shorted

P0449 Evaporative Emission Control System Vent Valve/Solenoid Circuit Malfunction

P0450 Evaporative Emission Control System Pressure Sensor Malfunction

P0451 Evaporative Emission Control System Pressure Sensor Range/Performance

P0452 Evaporative Emission Control System Pressure Sensor Low Input

P0453 Evaporative Emission Control System Pressure Sensor High Input

P0454 Evaporative Emission Control System Pressure Sensor Intermittent

P0455 Evaporative Emission Control System Leak Detected (Gross Leak)

P0460 Fuel Level Sensor Circuit Malfunction

P0461 Fuel Level Sensor Circuit Range/Performance

P0462 Fuel Level Sensor Circuit Low Input

P0463 Fuel Level Sensor Circuit High Input

P0464 Fuel Level Sensor Circuit Intermittent

P0465 Purge Flow Sensor Circuit Malfunction

P0466 Purge Flow Sensor Circuit Range/Performance

P0467 Purge Flow Sensor Circuit Low Input

P0468 Purge Flow Sensor Circuit High Input

P0469 Purge Flow Sensor Circuit Intermittent

P0470 Exhaust Pressure Sensor Malfunction

P0471 Exhaust Pressure Sensor Range/Performance
P0472 Exhaust Pressure Sensor Low
P0473 Exhaust Pressure Sensor High
P0474 Exhaust Pressure Sensor Intermittent
P0475 Exhaust Pressure Control Valve Malfunction
P0476 Exhaust Pressure Control Valve Range/Performance
P0477 Exhaust Pressure Control Valve Low
P0478 Exhaust Pressure Control Valve High
P0479 Exhaust Pressure Control Valve Intermittent
P0480 Cooling Fan no. 1 Control Circuit Malfunction
P0481 Cooling Fan no. 2 Control Circuit Malfunction
P0482 Cooling Fan no. 3 Control Circuit Malfunction
P0483 Cooling Fan Rationality Check Malfunction
P0484 Cooling Fan Circuit Over Current
P0485 Cooling Fan Power/Ground Circuit Malfunction
P0500 Vehicle Speed Sensor Malfunction
P0501 Vehicle Speed Sensor Range/Performance
P0502 Vehicle Speed Sensor Circuit Low Input
P0503 Vehicle Speed Sensor Intermittent/Erratic/High
P0505 Idle Control System Malfunction
P0506 Idle Control System RPM Lower Than Expected
P0507 Idle Control System RPM Higher Than Expected
P0510 Closed Throttle Position Switch Malfunction
P0520 Engine Oil Pressure Sensor/Switch Circuit Malfunction
P0521 Engine Oil Pressure Sensor/Switch Range/Performance
P0522 Engine Oil Pressure Sensor/Switch Low Voltage
P0523 Engine Oil Pressure Sensor/Switch High Voltage
P0530 A/C Refrigerant Pressure Sensor Circuit Malfunction
P0531 A/C Refrigerant Pressure Sensor Circuit Range/Performance
P0532 A/C Refrigerant Pressure Sensor Circuit Low Input
P0533 A/C Refrigerant Pressure Sensor Circuit High Input
P0534 A/C Refrigerant Charge Loss
P0550 Power Steering Pressure Sensor Circuit Malfunction
P0551 Power Steering Pressure Sensor Circuit Range/Performance
P0552 Power Steering Pressure Sensor Circuit Low Input
P0553 Power Steering Pressure Sensor Circuit High Input
P0554 Power Steering Pressure Sensor Circuit Intermittent
P0560 System Voltage Malfunction
P0561 System Voltage Unstable
P0562 System Voltage Low
P0563 System Voltage High
P0565 Cruise Control On Signal Malfunction
P0566 Cruise Control Off Signal Malfunction
P0567 Cruise Control Resume Signal Malfunction
P0568 Cruise Control Set Signal Malfunction
P0569 Cruise Control Coast Signal Malfunction
P0570 Cruise Control Accel Signal Malfunction
P0571 Cruise Control/Brake Switch "A" Circuit Malfunction
P0572 Cruise Control/Brake Switch "A" Circuit Low
P0573 Cruise Control/Brake Switch "A" Circuit High
P0574 **Through P0580** Reserved for Cruise Codes
P0600 Serial Communication Link Malfunction
P0601 Internal Control Module Memory Check Sum Error
P0602 Control Module Programming Error
P0603 Internal Control Module Keep Alive Memory (KAM) Error
P0604 Internal Control Module Random Access Memory (RAM) Error
P0605 Internal Control Module Read Only Memory (ROM) Error
P0606 PCM Processor Fault
P0608 Control Module VSS Output "A" Malfunction

P0609 Control Module VSS Output "B" Malfunction
P0620 Generator Control Circuit Malfunction
P0621 Generator Lamp "L" Control Circuit Malfunction
P0622 Generator Field "F" Control Circuit Malfunction
P0650 Malfunction Indicator Lamp (MIL) Control Circuit Malfunction
P0654 Engine RPM Output Circuit Malfunction
P0655 Engine Hot Lamp Output Control Circuit Malfunction
P0656 Fuel Level Output Circuit Malfunction
P0700 Transmission Control System Malfunction
P0701 Transmission Control System Range/Performance
P0702 Transmission Control System Electrical
P0703 Torque Converter/Brake Switch "B" Circuit Malfunction
P0704 Clutch Switch Input Circuit Malfunction
P0705 Transmission Range Sensor Circuit Malfunction (PRNDL Input)
P0706 Transmission Range Sensor Circuit Range/Performance
P0707 Transmission Range Sensor Circuit Low Input
P0708 Transmission Range Sensor Circuit High Input
P0709 Transmission Range Sensor Circuit Intermittent
P0710 Transmission Fluid Temperature Sensor Circuit Malfunction
P0711 Transmission Fluid Temperature Sensor Circuit Range/Performance
P0712 Transmission Fluid Temperature Sensor Circuit Low Input
P0713 Transmission Fluid Temperature Sensor Circuit High Input
P0714 Transmission Fluid Temperature Sensor Circuit Intermittent
P0715 Input/Turbine Speed Sensor Circuit Malfunction
P0716 Input/Turbine Speed Sensor Circuit Range/Performance
P0717 Input/Turbine Speed Sensor Circuit No Signal
P0718 Input/Turbine Speed Sensor Circuit Intermittent
P0719 Torque Converter/Brake Switch "B" Circuit Low
P0720 Output Speed Sensor Circuit Malfunction
P0721 Output Speed Sensor Circuit Range/Performance
P0722 Output Speed Sensor Circuit No Signal
P0723 Output Speed Sensor Circuit Intermittent
P0724 Torque Converter/Brake Switch "B" Circuit High
P0725 Engine Speed Input Circuit Malfunction
P0726 Engine Speed Input Circuit Range/Performance
P0727 Engine Speed Input Circuit No Signal
P0728 Engine Speed Input Circuit Intermittent
P0730 Incorrect Gear Ratio
P0731 Gear no. 1 Incorrect Ratio
P0732 Gear no. 2 Incorrect Ratio
P0733 Gear no. 3 Incorrect Ratio
P0734 Gear no. 4 Incorrect Ratio
P0735 Gear no. 5 Incorrect Ratio
P0736 Reverse Incorrect Ratio
P0740 Torque Converter Clutch Circuit Malfunction
P0741 Torque Converter Clutch Circuit Performance or Stuck Off
P0742 Torque Converter Clutch Circuit Stuck On
P0743 Torque Converter Clutch Circuit Electrical
P0744 Torque Converter Clutch Circuit Intermittent
P0745 Pressure Control Solenoid Malfunction
P0746 Pressure Control Solenoid Performance or Stuck Off
P0747 Pressure Control Solenoid Stuck On
P0748 Pressure Control Solenoid Electrical
P0749 Pressure Control Solenoid Intermittent

P0750 Shift Solenoid "A" Malfunction
P0751 Shift Solenoid "A" Performance or Stuck Off
P0752 Shift Solenoid "A" Stuck On
P0753 Shift Solenoid "A" Electrical
P0754 Shift Solenoid "A" Intermittent
P0755 Shift Solenoid "B" Malfunction
P0756 Shift Solenoid "B" Performance or Stuck Off
P0757 Shift Solenoid "B" Stuck On
P0758 Shift Solenoid "B" Electrical
P0759 Shift Solenoid "B" Intermittent
P0760 Shift Solenoid "C" Malfunction
P0761 Shift Solenoid "C" Performance Or Stuck Off
P0762 Shift Solenoid "C" Stuck On
P0763 Shift Solenoid "C" Electrical
P0764 Shift Solenoid "C" Intermittent
P0765 Shift Solenoid "D" Malfunction
P0766 Shift Solenoid "D" Performance Or Stuck Off
P0767 Shift Solenoid "D" Stuck On
P0768 Shift Solenoid "D" Electrical
P0769 Shift Solenoid "D" Intermittent
P0770 Shift Solenoid "E" Malfunction
P0771 Shift Solenoid "E" Performance Or Stuck Off
P0772 Shift Solenoid "E" Stuck On
P0773 Shift Solenoid "E" Electrical
P0774 Shift Solenoid "E" Intermittent
P0780 Shift Malfunction
P0781 1–2 Shift Malfunction
P0782 2–3 Shift Malfunction
P0783 3–4 Shift Malfunction
P0784 4–5 Shift Malfunction
P0785 Shift/Timing Solenoid Malfunction
P0786 Shift/Timing Solenoid Range/Performance
P0787 Shift/Timing Solenoid Low
P0788 Shift/Timing Solenoid High
P0789 Shift/Timing Solenoid Intermittent
P0790 Normal/Performance Switch Circuit Malfunction
P0801 Reverse Inhibit Control Circuit Malfunction
P0803 1–4 Upshift (Skip Shift) Solenoid Control Circuit Malfunction
P0804 1–4 Upshift (Skip Shift) Lamp Control Circuit Malfunction
P1290 CNG Fuel System Pressure Too High—3.3L CNG vehicles only
P1291 No Temp Rise Seen From Intake Air Heaters
P1292 CNG Pressure Sensor Voltage Too High—3.3L CNG vehicles only
P1293 CNG Pressure Sensor Voltage Too Low—3.3L CNG vehicles only
P1294 Target Idle Not Reached
P1296 No 5-Volts To MAP Sensor
P1297 No Change In MAP From Start To Run
P1391 Intermittent Loss Of CMP Or CKP
P1398 Misfire Adaptive Numerator At Limit
P1486 EVAP Leak Monitor Pinched Hose Or Obstruction Found
P1491 Radiator Fan Control Relay Circuit
P1492 Battery Temp Sensor Voltage Too High
P1493 Battery Temp Sensor Voltage Too Low
P1494 Leak Detection Pump Pressure Switch Or Mechanical Fault
P1495 Leak Detection Pump Solenoid Circuit
P1498 Auxiliary 5-Volt Supply Output Too Low
P1697 PCM Failure SRI Mile Not Stored

P1698 PCM Failure EEPROM Write Denied
P1756 Governor Pressure Not Equal To Target @ 15–20 PSI
P1757 Governor Pressure Above 3 PSI In Gear With 0 MPH
P1762 Governor Pressure Sensor Offset Volts Too Low Or High
P1763 Governor Pressure Sensor Volts Too High
P1764 Governor Pressure Sensor Volts Too Low
P1765 Trans. 12-Volt Supply Relay Control Circuit
P1899 P/N Switch Stuck In Park Or In Gear

TROUBLE CODE EQUIVALENT LIST

11 No crank reference signal at PCM
11 Timing belt skipped t tooth or more
11 Intermittent loss of CMP or CKP
11 Misfire adaptive numerator at limit
12 Battery disconnect
13 Slow change in idle MAP sensor signal (VIN N engine)
13 No change in MAP from start to run
14 MAP sensor voltage too low
14 MAP sensor voltage too high
14 No 5 volts to MAP sensor
15 5 volt supply output too low
16 No vehicle speed sensor signal
16 Knock sensor signal
17 Engine cold too long
17 Closed loop temperature not reached
21 Front 02S shorted to voltage
21 Front 02S stays at center
21 Rear 02S shorted to voltage
21 Rear 02S stays at center
21 Upstream 02S shorted to ground
21 Upstream 02S shorted to voltage
21 Upstream 02S response
21 Upstream 02S stays at center
21 Upstream 02S heater failure
21 Downstream 02S shorted to ground
21 Downstream 02S shorted to voltage
21 Downstream 025 response
21 Downstream 02S signal inactive
21 Downstream 02S heater failure
21 Front bank upstream 02S shorted to ground (6 cylinder)
21 Front bank upstream 02S shorted to voltage (6 cylinder)
21 Front bank upstream 02S slow response 6 cylinder)
21 Front bank upstream 02S stays at center (6 cylinder)
21 Front bank upstream 025 heater failure (6 cylinder)
21 Front bank downstream 025 shorted to ground (6 cylinder)
21 Front bank downstream 02S shorted to voltage (6 cylinder)
21 Front bank downstream 02S stays at center (6 cylinder)
21 Front bank downstream 02S heater failure (6 cylinder)
22 ECT sensor voltage too low
22 ECT sensor voltage too high
23 Intake air temperature voltage low
23 Intake air temperature voltage high
24 TPS voltage does not agree with MAP
24 Throttle position sensor voltage low
24 Throttle position sensor voltage high
24 No 5 volts to TPS
25 Idle air control motor circuits
25 Target idle not reached
25 Vacuum leak found (IAC fully seated)
27 Injector #I control circuit

27 Injector #2 control circuit
27 Injector #3 control circuit
27 Injector #4 control circuit
27 Injector #5 control circuit (6 cylinder)
27 Injector #6 control circuit (6 cylinder)
31 EVAP purge flow monitor failure
31 EVAP system small leak
31 EVAP solenoid circuit
31 EVAP system large leak
31 EVAP leak monitor pinched hose
31 Leak detection pump pressure switch
31 EVAP emission vent solenoid switch or mechanical failure
31 Leak detection pump solenoid circuit
31 EVAP emission vent solenoid circuit
31 High speed radiator fan ground control relay circuit
32 EGR system failure
32 EGR solenoid circuit
33 A/C pressure sensor volts too high
33 A/C pressure sensor volts too low
33 A/C clutch relay circuit
34 Speed control switch always low
34 Speed control switch always high
31 Speed control solenoid circuit
35 High speed condenser fan control relay circuit
35 High fan and high fan ground control relay circuit
35 High speed radiator fan control relay circuit
35 High speed fan control relay circuit
35 Low speed fan control relay circuit
37 Park/Neutral switch failure
41 Alternator field not switching properly
42 Auto shutdown relay circuit
42 No ASD relay output voltage at PCM
42 Fuel level sending unit volts too low
42 Fuel level sending unit volts too high
42 Fuel level unit no change over miles
42 Fuel pump relay control circuit
43 Multiple cylinder misfire
43 Cylinder #1 misfire
43 Cylinder #2 misfire
43 Cylinder #3 mislire
43 Cylinder #4 misfire
43 Cylinder #5 misfire
43 Cylinder #6 misfire
43 Ignition coil #1 primary circuit
43 Ignition coil #2 primary circuit
44 Ambient temperature sensor
44 Battery temperature sensor volts out of limit
44 Battery temperature sensor voltage too high
44 Battery temperature sensor voltage too low
45 Transaxle fault present
46 Charging system voltage too high
47 Charging system voltage too low
51 Fuel system lean (4 cylinder)
51 Rear bank fuel system lean (6 cylinder)
51 Front bank fuel system lean (6 cylinder)
52 Fuel system rich (4 cylinder)
52 Rear bank fuel system rich (6 cylinder)
52 Front bank fuel system rich (6 cylinder)
53 Internal controller failure
53 PCM failure SPI communications
53 Internal controller failure
53 PCM failure SPI communications

54 No cam signal at PCM
55 Completion or fault code display on Check Engine Lamp
62 PCM failure SRI mile not stared
63 PCM failure EEPROM write denied
64 Catalytic converter efficiency failure
64 Rear bank catalytic converter efficiency failure
65 Power steering switch failure
65 Brake switch performance circuit
66 No CCD message from body controller
66 No CCD message from TCM
71 5 volt output low speed control power circuit
72 Catalytic Converter efficiency failure
72 Front bank catalytic converter efficiency failure
77 Malfunction detected with power feed to speed control servo

Ford Motor Company

➡ **The Mercury Villager is covered under the Nissan section since it shares a platform with the Nissan Quest.**

READING CODES

Reading the control module memory is on of the first steps in OBD II system diagnostics. This step should be initially performed to determine the general nature of the fault. Subsequent readings will determine if the fault has been cleared.

Reading codes can be performed by any of the methods below:

• Read the control module memory with the Generic Scan Tool (GST)

• Read the control module memory with the vehicle manufacturer's specific tester

To read the fault codes, connect the scan tool or tester according to the manufacturer's instructions. Follow the manufacturer's specified procedure for reading the codes.

CLEARING CODES

Control module reset procedures are a very important part of OBD II System diagnostics. This step should be done at the end of any fault code repair and at the end of any driveability repair.

Clearing codes can be performed by any of the methods below:

• Clear the control module memory with the Generic Scan Tool (GST)

• Clear the control module memory with the vehicle manufacturer's specific tester

• Turn the ignition off and remove the negative battery cable for at least 1 minute.

Removing the negative battery cable may cause other systems in the vehicle to lose their memory. Prior to removing the cable, ensure you have the proper reset codes for radios and alarms.

➡ **The MIL will may also be de-activated for some codes if the vehicle completes three consecutive trips without a fault detected with vehicle conditions similar to those present during the fault.**

DIAGNOSTIC TROUBLE CODES

1995 Models

P0000 No Failures
P0100 Mass or Volume Air Flow Circuit Malfunction

P0101 Mass or Volume Air Flow Circuit Range/Performance Problem

P0102 Mass or Volume Air Flow Circuit Low Input

P0103 Mass or Volume Air Flow Circuit High Input

P0104 Mass or Volume Air Flow Circuit Intermittent

P0105 Manifold Absolute Pressure/Barometric Pressure Circuit Malfunction

P0106 Manifold Absolute Pressure/Barometric Pressure Circuit Range/Performance Problem

P0107 Manifold Absolute Pressure/Barometric Pressure Circuit Low Input

P0108 Manifold Absolute Pressure/Barometric Pressure Circuit High Input

P0109 Manifold Absolute Pressure/Barometric Pressure Circuit Intermittent

P0110 Intake Air Temperature Circuit Malfunction

P0111 Intake Air Temperature Circuit Range/Performance Problem

P0112 Intake Air Temperature Circuit Low Input

P0113 Intake Air Temperature Circuit High Input

P0114 Intake Air Temperature Circuit Intermittent

P0115 Engine Coolant Temperature Circuit Malfunction

P0116 Engine Coolant Temperature Circuit Range/Performance Problem

P0117 Engine Coolant Temperature Circuit Low Input

P0118 Engine Coolant Temperature Circuit High Input

P0119 Engine Coolant Temperature Circuit Intermittent

P0120 Throttle/Pedal Position Sensor/Switch "A" Circuit Malfunction

P0121 Throttle/Pedal Position Sensor/Switch "A" Circuit Range/Performance Problem

P0122 Throttle/Pedal Position Sensor/Switch "A" Circuit Low Input

P0123 Throttle/Pedal Position Sensor/Switch "A" Circuit High Input

P0124 Throttle/Pedal Position Sensor/Switch "A" Circuit Intermittent

P0125 Insufficient Coolant Temperature For Closed Loop Fuel Control

P0126 Insufficient Coolant Temperature For Stable Operation

P0130 O_2 Circuit Malfunction (Bank no. 1 Sensor no. 1)

P0131 O_2 Sensor Circuit Low Voltage (Bank no. 1 Sensor no. 1)

P0132 O_2 Sensor Circuit High Voltage (Bank no. 1 Sensor no. 1)

P0133 O_2 Sensor Circuit Slow Response (Bank no. 1 Sensor no. 1)

P0134 O_2 Sensor Circuit No Activity Detected (Bank no. 1 Sensor no. 1)

P0135 O_2 Sensor Heater Circuit Malfunction (Bank no. 1 Sensor no. 1)

P0136 O_2 Sensor Circuit Malfunction (Bank no. 1 Sensor no. 2)

P0137 O_2 Sensor Circuit Low Voltage (Bank no. 1 Sensor no. 2)

P0138 O_2 Sensor Circuit High Voltage (Bank no. 1 Sensor no. 2)

P0139 O_2 Sensor Circuit Slow Response (Bank no. 1 Sensor no. 2)

P0140 O_2 Sensor Circuit No Activity Detected (Bank no. 1 Sensor no. 2)

P0141 O_2 Sensor Heater Circuit Malfunction (Bank no. 1 Sensor no. 2)

P0142 O_2 Sensor Circuit Malfunction (Bank no. 1 Sensor no. 3)

P0143 O_2 Sensor Circuit Low Voltage (Bank no. 1 Sensor no. 3)

P0144 O_2 Sensor Circuit High Voltage (Bank no. 1 Sensor no. 3)

P0145 O_2 Sensor Circuit Slow Response (Bank no. 1 Sensor no. 3)

P0146 O_2 Sensor Circuit No Activity Detected (Bank no. 1 Sensor no. 3)

P0147 O_2 Sensor Heater Circuit Malfunction (Bank no. 1 Sensor no. 3)

P0150 O_2 Sensor Circuit Malfunction (Bank no. 2 Sensor no. 1)

P0151 O_2 Sensor Circuit Low Voltage (Bank no. 2 Sensor no. 1)

P0152 O_2 Sensor Circuit High Voltage (Bank no. 2 Sensor no. 1)

P0153 O_2 Sensor Circuit Slow Response (Bank no. 2 Sensor no. 1)

P0154 O_2 Sensor Circuit No Activity Detected (Bank no. 2 Sensor no. 1)

P0155 O_2 Sensor Heater Circuit Malfunction (Bank no. 2 Sensor no. 1)

P0156 O_2 Sensor Circuit Malfunction (Bank no. 2 Sensor no. 2)

P0157 O_2 Sensor Circuit Low Voltage (Bank no. 2 Sensor no. 2)

P0158 O_2 Sensor Circuit High Voltage (Bank no. 2 Sensor no. 2)

P0159 O_2 Sensor Circuit Slow Response (Bank no. 2 Sensor no. 2)

P0160 O_2 Sensor Circuit No Activity Detected (Bank no. 2 Sensor no. 2)

P0161 O_2 Sensor Heater Circuit Malfunction (Bank no. 2 Sensor no. 2)

P0162 O_2 Sensor Circuit Malfunction (Bank no. 2 Sensor no. 3)

P0163 O_2 Sensor Circuit Low Voltage (Bank no. 2 Sensor no. 3)

P0164 O_2 Sensor Circuit High Voltage (Bank no. 2 Sensor no. 3)

P0165 O_2 Sensor Circuit Slow Response (Bank no. 2 Sensor no. 3)

P0166 O_2 Sensor Circuit No Activity Detected (Bank no. 2 Sensor no. 3)

P0167 O_2 Sensor Heater Circuit Malfunction (Bank no. 2 Sensor no. 3)

P0170 Fuel Trim Malfunction (Bank no. 1)

P0171 System Too Lean (Bank no. 1)

P0172 System Too Rich (Bank no. 1)

P0173 Fuel Trim Malfunction (Bank no. 2)

P0174 System Too Lean (Bank no. 2)

P0175 System Too Rich (Bank no. 2)

P0176 Fuel Composition Sensor Circuit Malfunction

P0177 Fuel Composition Sensor Circuit Range/Performance

P0178 Fuel Composition Sensor Circuit Low Input

P0179 Fuel Composition Sensor Circuit High Input

P0180 Fuel Temperature Sensor "A" Circuit Malfunction

P0181 Fuel Temperature Sensor "A" Circuit Range/Performance

P0182 Fuel Temperature Sensor "A" Circuit Low Input

P0183 Fuel Temperature Sensor "A" Circuit High Input

P0184 Fuel Temperature Sensor "A" Circuit Intermittent

P0185 Fuel Temperature Sensor "B" Circuit Malfunction

P0186 Fuel Temprature Sensor "B" Circuit Range/Performance

P0187 Fuel Temperature Sensor "B" Circuit Low Input
P0188 Fuel Temperature Sensor "B" Circuit High Input
P0189 Fuel Temperature Sensor "B" Circuit Intermittent
P0190 Fuel Rail Pressure Sensor Circuit Malfunction
P0191 Fuel Rail Pressure Sensor Circuit Range/Performance
P0192 Fuel Rail Pressure Sensor Circuit Low Input
P0193 Fuel Rail Pressure Sensor Circuit High Input
P0194 Fuel Rail Pressure Sensor Circuit Intermittent
P0195 Engine Oil Temperature Sensor Malfunction
P0196 Engine Oil Temperature Sensor Range/Performance
P0197 Engine Oil Temperature Sensor Low
P0198 Engine Oil Temperature Sensor High
P0199 Engine Oil Temperature Sensor Intermittent
P0200 Injector Circuit Malfunction
P0201 Injector Circuit Malfunction—Cylinder no. 1
P0202 Injector Circuit Malfunction—Cylinder no. 2
P0203 Injector Circuit Malfunction—Cylinder no. 3
P0204 Injector Circuit Malfunction—Cylinder no. 4
P0205 Injector Circuit Malfunction—Cylinder no. 5
P0206 Injector Circuit Malfunction—Cylinder no. 6
P0207 Injector Circuit Malfunction—Cylinder no. 7
P0208 Injector Circuit Malfunction—Cylinder no. 8
P0209 Injector Circuit Malfunction—Cylinder no. 9
P0210 Injector Circuit Malfunction—Cylinder no. 10
P0211 Injector Circuit Malfunction—Cylinder no. 11
P0212 Injector Circuit Malfunction—Cylinder no. 12
P0213 Cold Start Injector no. 1 Malfunction
P0214 Cold Start Injector no. 2 Malfunction
P0215 Engine Shutoff Solenoid Malfunction
P0216 Injection Timing Control Circuit Malfunction
P0217 Engine Over Temperature Condition
P0218 Transmission Over Temperature Condition
P0219 Engine Over Speed Condition
P0220 Throttle/Pedal Position Sensor/Switch "B" Circuit Malfunction
P0221 Throttle/Pedal Position Sensor/Switch "B" Circuit Range/Performance Problem
P0222 Throttle/Pedal Position Sensor/Switch "B" Circuit Low Input
P0223 Throttle/Pedal Position Sensor/Switch "B" Circuit High Input
P0224 Throttle/Pedal Position Sensor/Switch "B" Circuit Intermittent
P0225 Throttle/Pedal Position Sensor/Switch "C" Circuit Malfunction
P0226 Throttle/Pedal Position Sensor/Switch "C" Circuit Range/Performance Problem
P0227 Throttle/Pedal Position Sensor/Switch "C" Circuit Low Input
P0228 Throttle/Pedal Position Sensor/Switch "C" Circuit High Input
P0229 Throttle/Pedal Position Sensor/Switch "C" Circuit Intermittent
P0230 Fuel Pump Primary Circuit Malfunction
P0231 Fuel Pump Secondary Circuit Low
P0232 Fuel Pump Secondary Circuit High
P0233 Fuel Pump Secondary Circuit Intermittent
P0234 Engine Over Boost Condition
P0261 Cylinder no. 1 Injector Circuit Low
P0262 Cylinder no. 1 Injector Circuit High
P0263 Cylinder no. 1 Contribution/Balance Fault
P0264 Cylinder no. 2 Injector Circuit Low

P0265 Cylinder no. 2 Injector Circuit High
P0266 Cylinder no. 2 Contribution/Balance Fault
P0267 Cylinder no. 3 Injector Circuit Low
P0268 Cylinder no. 3 Injector Circuit High
P0269 Cylinder no. 3 Contribution/Balance Fault
P0270 Cylinder no. 4 Injector Circuit Low
P0271 Cylinder no. 4 Injector Circuit High
P0272 Cylinder no. 4 Contribution/Balance Fault
P0273 Cylinder no. 5 Injector Circuit Low
P0274 Cylinder no. 5 Injector Circuit High
P0275 Cylinder no. 5 Contribution/Balance Fault
P0276 Cylinder no. 6 Injector Circuit Low
P0277 Cylinder no. 6 Injector Circuit High
P0278 Cylinder no. 6 Contribution/Balance Fault
P0279 Cylinder no. 7 Injector Circuit Low
P0280 Cylinder no. 7 Injector Circuit High
P0281 Cylinder no. 7 Contribution/Balance Fault
P0282 Cylinder no. 8 Injector Circuit Low
P0283 Cylinder no. 8 Injector Circuit High
P0284 Cylinder no. 8 Contribution/Balance Fault
P0285 Cylinder no. 9 Injector Circuit Low
P0286 Cylinder no. 9 Injector Circuit High
P0287 Cylinder no. 9 Contribution/Balance Fault
P0288 Cylinder no. 10 Injector Circuit Low
P0289 Cylinder no. 10 Injector Circuit High
P0290 Cylinder no. 10 Contribution/Balance Fault
P0291 Cylinder no. 11 Injector Circuit Low
P0292 Cylinder no. 11 Injector Circuit High
P0293 Cylinder no. 11 Contribution/Balance Fault
P0294 Cylinder no. 12 Injector Circuit Low
P0295 Cylinder no. 12 Injector Circuit High
P0296 Cylinder no. 12 Contribution/Balance Fault
P0300 Random/Multiple Cylinder Misfire Detected
P0301 Cylinder no. 1—Misfire Detected
P0302 Cylinder no. 2—Misfire Detected
P0303 Cylinder no. 3—Misfire Detected
P0304 Cylinder no. 4—Misfire Detected
P0305 Cylinder no. 5—Misfire Detected
P0306 Cylinder no. 6—Misfire Detected
P0307 Cylinder no. 7—Misfire Detected
P0308 Cylinder no. 8—Misfire Detected
P0309 Cylinder no. 9—Misfire Detected
P0310 Cylinder no. 10—Misfire Detected
P0311 Cylinder no. 11—Misfire Detected
P0312 Cylinder no. 12—Misfire Detected
P0320 Ignition/Distributor Engine Speed Input Circuit Malfunction
P0321 Ignition/Distributor Engine Speed Input Circuit Range/Performance
P0322 Ignition/Distributor Engine Speed Input Circuit No Signal
P0323 Ignition/Distributor Engine Speed Input Circuit Intermittent
P0325 Knock Sensor no. 1—Circuit Malfunction (Bank no. 1 or Single Sensor)
P0326 Knock Sensor no. 1—Circuit Range/Performance (Bank no. 1 or Single Sensor)
P0327 Knock Sensor no. 1—Circuit Low Input (Bank no. 1 or Single Sensor)
P0328 Knock Sensor no. 1—Circuit High Input (Bank no. 1 or Single Sensor)
P0329 Knock Sensor no. 1—Circuit Input Intermittent (Bank no. 1 or Single Sensor)

P0330 Knock Sensor no. 2—Circuit Malfunction (Bank no. 2)
P0331 Knock Sensor no. 2—Circuit Range/Performance (Bank no. 2)
P0332 Knock Sensor no. 2—Circuit Low Input (Bank no. 2)
P0333 Knock Sensor no. 2—Circuit High Input (Bank no. 2)
P0334 Knock Sensor no. 2—Circuit Input Intermittent (Bank no. 2)
P0335 Crankshaft Position Sensor "A" Circuit Malfunction
P0336 Crankshaft Position Sensor "A" Circuit Range/Performance
P0337 Crankshaft Position Sensor "A" Circuit Low Input
P0338 Crankshaft Position Sensor "A" Circuit High Input
P0339 Crankshaft Position Sensor "A" Circuit Intermittent
P0340 Camshaft Position Sensor Circuit Malfunction
P0341 Camshaft Position Sensor Circuit Range/Performance
P0342 Camshaft Position Sensor Circuit Low Input
P0343 Camshaft Position Sensor Circuit High Input
P0344 Camshaft Position Sensor Circuit Intermittent
P0350 Ignition Coil Primary/Secondary Circuit Malfunction
P0351 Ignition Coil "A" Primary/Secondary Circuit Malfunction
P0352 Ignition Coil "B" Primary/Secondary Circuit Malfunction
P0353 Ignition Coil "C" Primary/Secondary Circuit Malfunction
P0354 Ignition Coil "D" Primary/Secondary Circuit Malfunction
P0355 Ignition Coil "E" Primary/Secondary Circuit Malfunction
P0356 Ignition Coil "F" Primary/Secondary Circuit Malfunction
P0357 Ignition Coil "G" Primary/Secondary Circuit Malfunction
P0358 Ignition Coil "H" Primary/Secondary Circuit Malfunction
P0359 Ignition Coil "I" Primary/Secondary Circuit Malfunction
P0360 Ignition Coil "J" Primary/Secondary Circuit Malfunction
P0361 Ignition Coil "K" Primary/Secondary Circuit Malfunction
P0362 Ignition Coil "L" Primary/Secondary Circuit Malfunction
P0370 Timing Reference High Resolution Signal "A" Malfunction
P0371 Timing Reference High Resolution Signal "A" Too Many Pulses
P0372 Timing Reference High Resolution Signal "A" Too Few Pulses
P0373 Timing Reference High Resolution Signal "A" Intermittent/Erratic Pulses
P0374 Timing Reference High Resolution Signal "A" No Pulses
P0375 Timing Reference High Resolution Signal "B" Malfunction
P0376 Timing Reference High Resolution Signal "B" Too Many Pulses
P0377 Timing Reference High Resolution Signal "B" Too Few Pulses
P0378 Timing Reference High Resolution Signal "B" Intermittent/Erratic Pulses
P0379 Timing Reference High Resolution Signal "B" No Pulses
P0380 Glow Plug/Heater Circuit "A" Malfunction
P0381 Glow Plug/Heater Indicator Circuit Malfunction
P0382 Glow Plug/Heater Circuit "B" Malfunction
P0385 Crankshaft Position Sensor "B" Circuit Malfunction
P0386 Crankshaft Position Sensor "B" Circuit Range/Performance
P0387 Crankshaft Position Sensor "B" Circuit Low Input
P0388 Crankshaft Position Sensor "B" Circuit High Input
P0389 Crankshaft Position Sensor "B" Circuit Intermittent
P0400 Exhaust Gas Recirculation Flow Malfunction
P0401 Exhaust Gas Recirculation Flow Insufficient Detected
P0402 Exhaust Gas Recirculation Flow Excessive Detected
P0403 Exhaust Gas Recirculation Circuit Malfunction

P0404 Exhaust Gas Recirculation Circuit Range/Performance
P0405 Exhaust Gas Recirculation Sensor "A" Circuit Low
P0406 Exhaust Gas Recirculation Sensor "A" Circuit High
P0407 Exhaust Gas Recirculation Sensor "B" Circuit Low
P0408 Exhaust Gas Recirculation Sensor "B" Circuit High
P0410 Secondary Air Injection System Malfunction
P0411 Secondary Air Injection System Incorrect Flow Detected
P0412 Secondary Air Injection System Switching Valve "A" Circuit Malfunction
P0413 Secondary Air Injection System Switching Valve "A" Circuit Open
P0414 Secondary Air Injection System Switching Valve "A" Circuit Shorted
P0415 Secondary Air Injection System Switching Valve "B" Circuit Malfunction
P0416 Secondary Air Injection System Switching Valve "B" Circuit Open
P0417 Secondary Air Injection System Switching Valve "B" Circuit Shorted
P0418 Secondary Air Injection System Relay "A" Circuit Malfunction
P0419 Secondary Air Injection System Relay "B" Circuit Malfunction
P0420 Catalyst System Efficiency Below Threshold (Bank no. 1)
P0421 Warm Up Catalyst Efficiency Below Threshold (Bank no. 1)
P0422 Main Catalyst Efficiency Below Threshold (Bank no. 1)
P0423 Heated Catalyst Efficiency Below Threshold (Bank no. 1)
P0424 Heated Catalyst Temperature Below Threshold (Bank no. 1)
P0430 Catalyst System Efficiency Below Threshold (Bank no. 2)
P0431 Warm Up Catalyst Efficiency Below Threshold (Bank no. 2)
P0432 Main Catalyst Efficiency Below Threshold (Bank no. 2)
P0433 Heated Catalyst Efficiency Below Threshold (Bank no. 2)
P0434 Heated Catalyst Temperature Below Threshold (Bank no. 2)
P0440 Evaporative Emission Control System Malfunction
P0441 Evaporative Emission Control System Incorrect Purge Flow
P0442 Evaporative Emission Control System Leak Detected (Small Leak)
P0443 Evaporative Emission Control System Purge Control Valve Circuit Malfunction
P0444 Evaporative Emission Control System Purge Control Valve Circuit Open
P0445 Evaporative Emission Control System Purge Control Valve Circuit Shorted
P0446 Evaporative Emission Control System Vent Control Circuit Malfunction
P0447 Evaporative Emission Control System Vent Control Circuit Open
P0448 Evaporative Emission Control System Vent Control Circuit Shorted
P0449 Evaporative Emission Control System Vent Valve/Solenoid Circuit Malfunction
P0450 Evaporative Emission Control System Pressure Sensor Malfunction
P0451 Evaporative Emission Control System Pressure Sensor Range/Performance
P0452 Evaporative Emission Control System Pressure Sensor Low Input

P0453 Evaporative Emission Control System Pressure Sensor High Input

P0454 Evaporative Emission Control System Pressure Sensor Intermittent

P0455 Evaporative Emission Control System Leak Detected (Gross Leak)

P0460 Fuel Level Sensor Circuit Malfunction

P0461 Fuel Level Sensor Circuit Range/Performance

P0462 Fuel Level Sensor Circuit Low Input

P0463 Fuel Level Sensor Circuit High Input

P0464 Fuel Level Sensor Circuit Intermittent

P0465 Purge Flow Sensor Circuit Malfunction

P0466 Purge Flow Sensor Circuit Range/Performance

P0467 Purge Flow Sensor Circuit Low Input

P0468 Purge Flow Sensor Circuit High Input

P0469 Purge Flow Sensor Circuit Intermittent

P0470 Exhaust Pressure Sensor Malfunction

P0471 Exhaust Pressure Sensor Range/Performance

P0472 Exhaust Pressure Sensor Low

P0473 Exhaust Pressure Sensor High

P0474 Exhaust Pressure Sensor Intermittent

P0475 Exhaust Pressure Control Valve Malfunction

P0476 Exhaust Pressure Control Valve Range/Performance

P0477 Exhaust Pressure Control Valve Low

P0478 Exhaust Pressure Control Valve High

P0479 Exhaust Pressure Control Valve Intermittent

P0480 Cooling Fan no. 1 Control Circuit Malfunction

P0481 Cooling Fan no. 2 Control Circuit Malfunction

P0482 Cooling Fan no. 3 Control Circuit Malfunction

P0483 Cooling Fan Rationality Check Malfunction

P0484 Cooling Fan Circuit Over Current

P0485 Cooling Fan Power/Ground Circuit Malfunction

P0500 Vehicle Speed Sensor Malfunction

P0501 Vehicle Speed Sensor Range/Performance

P0502 Vehicle Speed Sensor Circuit Low Input

P0503 Vehicle Speed Sensor Intermittent/Erratic/High

P0505 Idle Control System Malfunction

P0506 Idle Control System RPM Lower Than Expected

P0507 Idle Control System RPM Higher Than Expected

P0510 Closed Throttle Position Switch Malfunction

P0520 Engine Oil Pressure Sensor/Switch Circuit Malfunction

P0521 Engine Oil Pressure Sensor/Switch Range/Performance

P0522 Engine Oil Pressure Sensor/Switch Low Voltage

P0523 Engine Oil Pressure Sensor/Switch High Voltage

P0530 A/C Refrigerant Pressure Sensor Circuit Malfunction

P0531 A/C Refrigerant Pressure Sensor Circuit Range/Performance

P0532 A/C Refrigerant Pressure Sensor Circuit Low Input

P0533 A/C Refrigerant Pressure Sensor Circuit High Input

P0534 A/C Refrigerant Charge Loss

P0550 Power Steering Pressure Sensor Circuit Malfunction

P0551 Power Steering Pressure Sensor Circuit Range/Performance

P0552 Power Steering Pressure Sensor Circuit Low Input

P0553 Power Steering Pressure Sensor Circuit High Input

P0554 Power Steering Pressure Sensor Circuit Intermittent

P0560 System Voltage Malfunction

P0561 System Voltage Unstable

P0562 System Voltage Low

P0563 System Voltage High

P0565 Cruise Control On Signal Malfunction

P0566 Cruise Control Off Signal Malfunction

P0567 Cruise Control Resume Signal Malfunction

P0568 Cruise Control Set Signal Malfunction

P0569 Cruise Control Coast Signal Malfunction

P0570 Cruise Control Accel Signal Malfunction

P0571 Cruise Control/Brake Switch "A" Circuit Malfunction

P0572 Cruise Control/Brake Switch "A" Circuit Low

P0573 Cruise Control/Brake Switch "A" Circuit High

P0574 Through P0580 Reserved for Cruise Codes

P0600 Serial Communication Link Malfunction

P0601 Internal Control Module Memory Check Sum Error

P0602 Control Module Programming Error

P0603 Internal Control Module Keep Alive Memory (KAM) Error

P0604 Internal Control Module Random Access Memory (RAM) Error

P0605 Internal Control Module Read Only Memory (ROM) Error

P0606 PCM Processor Fault

P0608 Control Module VSS Output "A" Malfunction

P0609 Control Module VSS Output "B" Malfunction

P0620 Generator Control Circuit Malfunction

P0621 Generator Lamp "L" Control Circuit Malfunction

P0622 Generator Field "F" Control Circuit Malfunction

P0650 Malfunction Indicator Lamp (MIL) Control Circuit Malfunction

P0654 Engine RPM Output Circuit Malfunction

P0655 Engine Hot Lamp Output Control Circuit Malfunction

P0656 Fuel Level Output Circuit Malfunction

P0700 Transmission Control System Malfunction

P0701 Transmission Control System Range/Performance

P0702 Transmission Control System Electrical

P0703 Torque Converter/Brake Switch "B" Circuit Malfunction

P0704 Clutch Switch Input Circuit Malfunction

P0705 Transmission Range Sensor Circuit Malfunction (PRNDL Input)

P0706 Transmission Range Sensor Circuit Range/Performance

P0707 Transmission Range Sensor Circuit Low Input

P0708 Transmission Range Sensor Circuit High Input

P0709 Transmission Range Sensor Circuit Intermittent

P0710 Transmission Fluid Temperature Sensor Circuit Malfunction

P0711 Transmission Fluid Temperature Sensor Circuit Range/Performance

P0712 Transmission Fluid Temperature Sensor Circuit Low Input

P0713 Transmission Fluid Temperature Sensor Circuit High Input

P0714 Transmission Fluid Temperature Sensor Circuit Intermittent

P0715 Input/Turbine Speed Sensor Circuit Malfunction

P0716 Input/Turbine Speed Sensor Circuit Range/Performance

P0717 Input/Turbine Speed Sensor Circuit No Signal

P0718 Input/Turbine Speed Sensor Circuit Intermittent

P0719 Torque Converter/Brake Switch "B" Circuit Low

P0720 Output Speed Sensor Circuit Malfunction

P0721 Output Speed Sensor Circuit Range/Performance

P0722 Output Speed Sensor Circuit No Signal

P0723 Output Speed Sensor Circuit Intermittent

P0724 Torque Converter/Brake Switch "B" Circuit High

P0725 Engine Speed Input Circuit Malfunction

P0726 Engine Speed Input Circuit Range/Performance

P0727 Engine Speed Input Circuit No Signal

P0728 Engine Speed Input Circuit Intermittent

P0730 Incorrect Gear Ratio

P0731 Gear no. 1 Incorrect Ratio
P0732 Gear no. 2 Incorrect Ratio
P0733 Gear no. 3 Incorrect Ratio
P0734 Gear no. 4 Incorrect Ratio
P0735 Gear no. 5 Incorrect Ratio
P0736 Reverse Incorrect Ratio
P0740 Torque Converter Clutch Circuit Malfunction
P0741 Torque Converter Clutch Circuit Performance or Stuck Off
P0742 Torque Converter Clutch Circuit Stuck On
P0743 Torque Converter Clutch Circuit Electrical
P0744 Torque Converter Clutch Circuit Intermittent
P0745 Pressure Control Solenoid Malfunction
P0746 Pressure Control Solenoid Performance or Stuck Off
P0747 Pressure Control Solenoid Stuck On
P0748 Pressure Control Solenoid Electrical
P0749 Pressure Control Solenoid Intermittent
P0750 Shift Solenoid "A" Malfunction
P0751 Shift Solenoid "A" Performance or Stuck Off
P0752 Shift Solenoid "A" Stuck On
P0753 Shift Solenoid "A" Electrical
P0754 Shift Solenoid "A" Intermittent
P0755 Shift Solenoid "B" Malfunction
P0756 Shift Solenoid "B" Performance or Stuck Off
P0757 Shift Solenoid "B" Stuck On
P0758 Shift Solenoid "B" Electrical
P0759 Shift Solenoid "B" Intermittent
P0760 Shift Solenoid "C" Malfunction
P0761 Shift Solenoid "C" Performance Or Stuck Off
P0762 Shift Solenoid "C" Stuck On
P0763 Shift Solenoid "C" Electrical
P0764 Shift Solenoid "C" Intermittent
P0765 Shift Solenoid "D" Malfunction
P0766 Shift Solenoid "D" Performance Or Stuck Off
P0767 Shift Solenoid "D" Stuck On
P0768 Shift Solenoid "D" Electrical
P0769 Shift Solenoid "D" Intermittent
P0770 Shift Solenoid "E" Malfunction
P0771 Shift Solenoid "E" Performance Or Stuck Off
P0772 Shift Solenoid "E" Stuck On
P0773 Shift Solenoid "E" Electrical
P0774 Shift Solenoid "E" Intermittent
P0780 Shift Malfunction
P0781 1–2 Shift Malfunction
P0782 2–3 Shift Malfunction
P0783 3–4 Shift Malfunction
P0784 4–5 Shift Malfunction
P0785 Shift/Timing Solenoid Malfunction
P0786 Shift/Timing Solenoid Range/Performance
P0787 Shift/Timing Solenoid Low
P0788 Shift/Timing Solenoid High
P0789 Shift/Timing Solenoid Intermittent
P0790 Normal/Performance Switch Circuit Malfunction
P0801 Reverse Inhibit Control Circuit Malfunction
P0803 1–4 Upshift (Skip Shift) Solenoid Control Circuit Malfunction
P0804 1–4 Upshift (Skip Shift) Lamp Control Circuit Malfunction
P1000 OBD II Monitor Testing not complete
P1039 OBD II Monitor not complete
P1051 Brake switch signal missing or incorrect
P1100 Mass Air Flow (MAF) sensor intermittent

P1101 Mass Air Flow (MAF) sensor out of Self-Test range
P1112 Intake Air Temperature (IAT) sensor intermittent
P1116 Engine Coolant Temperature (ECT) sensor out of Self-Test range
P1117 Engine Coolant Temperature (ECT) sensor intermittent
P1120 Throttle Position (TP) sensor out of range low
P1121 Throttle Position (TP) sensor inconsistent with MAF sensor
P1124 Throttle Position (TP) sensor out of Self-Test range
P1125 Throttle Position (TP) sensor circuit intermittent
P1130 Lack of HO_2S 11 switch, adaptive fuel at limit
P1131 Lack of HO_2S 11 switch, sensor indicates lean (Bank #1)
P1132 Lack of HO_2S 11 switch, sensor indicates rich (Bank #1)
U1135 Ignition switch signal missing or incorrect
P1137 Lack of HO_2S 12 switch, sensor indicates lean (Bank #1)
P1138 Lack of HO_2S 12 switch, sensor indicates rich (Bank #1)
P1150 Lack of HO_2S 21 switch, adaptive fuel at limit
P1151 Lack of HO_2S 21 switch, sensor indicates lean (Bank #2)
P1152 Lack of HO_2S 21 switch, sensor indicates rich (Bank #2)
P1157 Lack of HO_2S 22 switch, sensor indicates lean (Bank #2)
P1158 Lack of HO_2S 22 switch, sensor indicates rich (Bank #2)
P1220 Series Throttle Control malfunction
P1224 Throttle Position Sensor (TP-B) out of Self-test range
P1233 Fuel Pump driver Module off-line
P1234 Fuel Pump driver Module off-line
P1235 Fuel Pump control out of range
P1236 Fuel Pump control out of range
P1237 Fuel Pump secondary circuit malfunction
P1238 Fuel Pump secondary circuit malfunction
P1260 THEFT detected—engine disabled
P1270 Engine RPM or vehicle speed limiter reached
P1351 Ignition Diagnostic Monitor (IDM) circuit input malfunction
P1352 Ignition coil A primary circuit malfunction
P1353 Ignition coil B primary circuit malfunction
P1354 Ignition coil C primary circuit malfunction
P1355 Ignition coil D primary circuit malfunction
P1358 Ignition Diagnostic Monitor (IDM) signal out of Self-Test range
P1359 Spark output circuit malfunction
P1364 Ignition coil primary circuit malfunction
P1390 Octane Adjust (OCT ADJ) out of Self-Test range
P1400 Differential Pressure Feedback Electronic (DPFE) sensor circuit low voltage detected
P1401 Differential Pressure Feedback Electronic (DPFE) sensor circuit high voltage detected
P1403 Differential Pressure Feedback Electronic (DPFE) sensor hoses reversed
P1405 Differential Pressure Feedback Electronic (DPFE) sensor upstream hose off or plugged
P1406 Differential Pressure Feedback Electronic (DPFE) sensor downstream hose off or plugged
P1407 Exhaust Gas Recirculation (EGR) no flow detected (valve stuck closed or inoperative)
P1408 Exhaust Gas Recirculation (EGR) flow out of Self-Test range
P1409 Electronic Vacuum Regulator (EVR) control circuit malfunction
P1414 Secondary Air Injection system monitor circuit high voltage
P1443 Evaporative emission control system—vacuum system purge control solenoid or purge control valve malfunction

P1444 Purge Flow Sensor (PFS) circuit low input

P1445 Purge Flow Sensor (PFS) circuit high input

U1451 Lack of response from Passive Anti-Theft system (PATS) module—engine disabled

P1460 Wide Open Throttle Air Conditioning Cut-off (WAC) circuit malfunction

P1461 Air Conditioning Pressure (ACP) sensor circuit low input

P1462 Air Conditioning Pressure (ACP) sensor circuit high input

P1463 Air Conditioning Pressure (ACP) sensor insufficient pressure change

P1469 Low air conditioning cycling period

P1473 Fan Secondary High with fan(s) off

P1474 Low Fan Control primary circuit malfunction

P1479 High Fan Control primary circuit malfunction

P1480 Fan Secondary low with low fan on

P1481 Fan Secondary low with high fan on

P1500 Vehicle Speed Sensor (VSS) circuit intermittent

P1505 Idle Air Control (IAC) system at adaptive clip

P1506 Idle Air control (IAC) overspeed error

P1518 Intake Manifold Runner Control (IMRC) malfunction (stuck open)

P1519 Intake Manifold Runner Control (IMRC) malfunction (stuck closed)

P1520 Intake Manifold Runner Control (IMRC) circuit malfunction

P1507 Idle Air control (IAC) under speed error

P1605 Powertrain Control Module (POM)—Keep Alive Memory (KAM) test error

P1650 Power steering Pressure (PSP) switch out of Self-Test range

P1651 Power steering Pressure (PSP) switch input malfunction

P1701 Reverse engagement error

P1703 Brake On/Off (BOO) switch out of Self-Test range

P1705 Manual Lever Position (MLP) sensor out of Self-Test range

P1709 Park or Neutral Position (PNP) switch out of Self-test range

P1729 4X4 Low switch error

P1711 Transmission Fluid Temperature (TFT) sensor out of Self-Test range

P1741 Torque Converter Clutch (TCC) control error

P1742 Torque Converter Clutch (TCC) solenoid mechanically failed (turns MIL on)

P1743 Torque Converter Clutch (TCC) solenoid mechanically failed (turns TOIL on)

P1744 Torque Converter Clutch (TCC) system mechanically stuck in off position

P1748 Electronic Pressure Control (EPC) solenoid circuit low input (open circuit)

P1747 Electronic Pressure Control (EPC) solenoid circuit high input (short circuit)

P1749 Electric Pressure Control (EPC) solenoid failed low

P1751 Shift Solenoid #1(SS1) performance

P1756 Shift Solenoid #2 (SS2) performance

P1780 Transmission Control Switch (TCS) circuit out of Self-Test range

1996–99 Models

P0000 No Failures

P0100 Mass or Volume Air Flow Circuit Malfunction

P0101 Mass or Volume Air Flow Circuit Range/Performance Problem

P0102 Mass or Volume Air Flow Circuit Low Input

P0103 Mass or Volume Air Flow Circuit High Input

P0104 Mass or Volume Air Flow Circuit Intermittent

P0105 Manifold Absolute Pressure/Barometric Pressure Circuit Malfunction

P0106 Manifold Absolute Pressure/Barometric Pressure Circuit Range/Performance Problem

P0107 Manifold Absolute Pressure/Barometric Pressure Circuit Low Input

P0108 Manifold Absolute Pressure/Barometric Pressure Circuit High Input

P0109 Manifold Absolute Pressure/Barometric Pressure Circuit Intermittent

P0110 Intake Air Temperature Circuit Malfunction

P0111 Intake Air Temperature Circuit Range/Performance Problem

P0112 Intake Air Temperature Circuit Low Input

P0113 Intake Air Temperature Circuit High Input

P0114 Intake Air Temperature Circuit Intermittent

P0115 Engine Coolant Temperature Circuit Malfunction

P0116 Engine Coolant Temperature Circuit Range/Performance Problem

P0117 Engine Coolant Temperature Circuit Low Input

P0118 Engine Coolant Temperature Circuit High Input

P0119 Engine Coolant Temperature Circuit Intermittent

P0120 Throttle/Pedal Position Sensor/Switch "A" Circuit Malfunction

P0121 Throttle/Pedal Position Sensor/Switch "A" Circuit Range/Performance Problem

P0122 Throttle/Pedal Position Sensor/Switch "A" Circuit Low Input

P0123 Throttle/Pedal Position Sensor/Switch "A" Circuit High Input

P0124 Throttle/Pedal Position Sensor/Switch "A" Circuit Intermittent

P0125 Insufficient Coolant Temperature For Closed Loop Fuel Control

P0126 Insufficient Coolant Temperature For Stable Operation

P0130 O_2 Circuit Malfunction (Bank no. 1 Sensor no. 1)

P0131 O_2 Sensor Circuit Low Voltage (Bank no. 1 Sensor no. 1)

P0132 O_2 Sensor Circuit High Voltage (Bank no. 1 Sensor no. 1)

P0133 O_2 Sensor Circuit Slow Response (Bank no. 1 Sensor no. 1)

P0134 O_2 Sensor Circuit No Activity Detected (Bank no. 1 Sensor no. 1)

P0135 O_2 Sensor Heater Circuit Malfunction (Bank no. 1 Sensor no. 1)

P0136 O_2 Sensor Circuit Malfunction (Bank no. 1 Sensor no. 2)

P0137 O_2 Sensor Circuit Low Voltage (Bank no. 1 Sensor no. 2)

P0138 O_2 Sensor Circuit High Voltage (Bank no. 1 Sensor no. 2)

P0139 O_2 Sensor Circuit Slow Response (Bank no. 1 Sensor no. 2)

P0140 O_2 Sensor Circuit No Activity Detected (Bank no. 1 Sensor no. 2)

P0141 O_2 Sensor Heater Circuit Malfunction (Bank no. 1 Sensor no. 2)

P0142 O_2 Sensor Circuit Malfunction (Bank no. 1 Sensor no. 3)
P0143 O_2 Sensor Circuit Low Voltage (Bank no. 1 Sensor no. 3)
P0144 O_2 Sensor Circuit High Voltage (Bank no. 1 Sensor no. 3)
P0145 O_2 Sensor Circuit Slow Response (Bank no. 1 Sensor no. 3)
P0146 O_2 Sensor Circuit No Activity Detected (Bank no. 1 Sensor no. 3)
P0147 O_2 Sensor Heater Circuit Malfunction (Bank no. 1 Sensor no. 3)
P0150 O_2 Sensor Circuit Malfunction (Bank no. 2 Sensor no. 1)
P0151 O_2 Sensor Circuit Low Voltage (Bank no. 2 Sensor no. 1)
P0152 O_2 Sensor Circuit High Voltage (Bank no. 2 Sensor no. 1)
P0153 O_2 Sensor Circuit Slow Response (Bank no. 2 Sensor no. 1)
P0154 O_2 Sensor Circuit No Activity Detected (Bank no. 2 Sensor no. 1)
P0155 O_2 Sensor Heater Circuit Malfunction (Bank no. 2 Sensor no. 1)
P0156 O_2 Sensor Circuit Malfunction (Bank no. 2 Sensor no. 2)
P0157 O_2 Sensor Circuit Low Voltage (Bank no. 2 Sensor no. 2)
P0158 O_2 Sensor Circuit High Voltage (Bank no. 2 Sensor no. 2)
P0159 O_2 Sensor Circuit Slow Response (Bank no. 2 Sensor no. 2)
P0160 O_2 Sensor Circuit No Activity Detected (Bank no. 2 Sensor no. 2)
P0161 O_2 Sensor Heater Circuit Malfunction (Bank no. 2 Sensor no. 2)
P0162 O_2 Sensor Circuit Malfunction (Bank no. 2 Sensor no. 3)
P0163 O_2 Sensor Circuit Low Voltage (Bank no. 2 Sensor no. 3)
P0164 O_2 Sensor Circuit High Voltage (Bank no. 2 Sensor no. 3)
P0165 O_2 Sensor Circuit Slow Response (Bank no. 2 Sensor no. 3)
P0166 O_2 Sensor Circuit No Activity Detected (Bank no. 2 Sensor no. 3)
P0167 O_2 Sensor Heater Circuit Malfunction (Bank no. 2 Sensor no. 3)
P0170 Fuel Trim Malfunction (Bank no. 1)
P0171 System Too Lean (Bank no. 1)
P0172 System Too Rich (Bank no. 1)
P0173 Fuel Trim Malfunction (Bank no. 2)
P0174 System Too Lean (Bank no. 2)
P0175 System Too Rich (Bank no. 2)
P0176 Fuel Composition Sensor Circuit Malfunction
P0177 Fuel Composition Sensor Circuit Range/Performance
P0178 Fuel Composition Sensor Circuit Low Input
P0179 Fuel Composition Sensor Circuit High Input
P0180 Fuel Temperature Sensor "A" Circuit Malfunction
P0181 Fuel Temperature Sensor "A" Circuit Range/Performance
P0182 Fuel Temperature Sensor "A" Circuit Low Input
P0183 Fuel Temperature Sensor "A" Circuit High Input
P0184 Fuel Temperature Sensor "A" Circuit Intermittent
P0185 Fuel Temperature Sensor "B" Circuit Malfunction
P0186 Fuel Temperature Sensor "B" Circuit Range/Performance
P0187 Fuel Temperature Sensor "B" Circuit Low Input

P0188 Fuel Temperature Sensor "B" Circuit High Input
P0189 Fuel Temperature Sensor "B" Circuit Intermittent
P0190 Fuel Rail Pressure Sensor Circuit Malfunction
P0191 Fuel Rail Pressure Sensor Circuit Range/Performance
P0192 Fuel Rail Pressure Sensor Circuit Low Input
P0193 Fuel Rail Pressure Sensor Circuit High Input
P0194 Fuel Rail Pressure Sensor Circuit Intermittent
P0195 Engine Oil Temperature Sensor Malfunction
P0196 Engine Oil Temperature Sensor Range/Performance
P0197 Engine Oil Temperature Sensor Low
P0198 Engine Oil Temperature Sensor High
P0199 Engine Oil Temperature Sensor Intermittent
P0200 Injector Circuit Malfunction
P0201 Injector Circuit Malfunction—Cylinder no. 1
P0202 Injector Circuit Malfunction—Cylinder no. 2
P0203 Injector Circuit Malfunction—Cylinder no. 3
P0204 Injector Circuit Malfunction—Cylinder no. 4
P0205 Injector Circuit Malfunction—Cylinder no. 5
P0206 Injector Circuit Malfunction—Cylinder no. 6
P0207 Injector Circuit Malfunction—Cylinder no. 7
P0208 Injector Circuit Malfunction—Cylinder no. 8
P0209 Injector Circuit Malfunction—Cylinder no. 9
P0210 Injector Circuit Malfunction—Cylinder no. 10
P0211 Injector Circuit Malfunction—Cylinder no. 11
P0212 Injector Circuit Malfunction—Cylinder no. 12
P0213 Cold Start Injector no. 1 Malfunction
P0214 Cold Start Injector no. 2 Malfunction
P0215 Engine Shutoff Solenoid Malfunction
P0216 Injection Timing Control Circuit Malfunction
P0217 Engine Over Temperature Condition
P0218 Transmission Over Temperature Condition
P0219 Engine Over Speed Condition
P0220 Throttle/Pedal Position Sensor/Switch "B" Circuit Malfunction
P0221 Throttle/Pedal Position Sensor/Switch "B" Circuit Range/Performance Problem
P0222 Throttle/Pedal Position Sensor/Switch "B" Circuit Low Input
P0223 Throttle/Pedal Position Sensor/Switch "B" Circuit High Input
P0224 Throttle/Pedal Position Sensor/Switch "B" Circuit Intermittent
P0225 Throttle/Pedal Position Sensor/Switch "C" Circuit Malfunction
P0226 Throttle/Pedal Position Sensor/Switch "C" Circuit Range/Performance Problem
P0227 Throttle/Pedal Position Sensor/Switch "C" Circuit Low Input
P0228 Throttle/Pedal Position Sensor/Switch "C" Circuit High Input
P0229 Throttle/Pedal Position Sensor/Switch "C" Circuit Intermittent
P0230 Fuel Pump Primary Circuit Malfunction
P0231 Fuel Pump Secondary Circuit Low
P0232 Fuel Pump Secondary Circuit High
P0233 Fuel Pump Secondary Circuit Intermittent
P0234 Engine Over Boost Condition
P0261 Cylinder no. 1 Injector Circuit Low
P0262 Cylinder no. 1 Injector Circuit High
P0263 Cylinder no. 1 Contribution/Balance Fault
P0264 Cylinder no. 2 Injector Circuit Low
P0265 Cylinder no. 2 Injector Circuit High

P0266 Cylinder no. 2 Contribution/Balance Fault
P0267 Cylinder no. 3 Injector Circuit Low
P0268 Cylinder no. 3 Injector Circuit High
P0269 Cylinder no. 3 Contribution/Balance Fault
P0270 Cylinder no. 4 Injector Circuit Low
P0271 Cylinder no. 4 Injector Circuit High
P0272 Cylinder no. 4 Contribution/Balance Fault
P0273 Cylinder no. 5 Injector Circuit Low
P0274 Cylinder no. 5 Injector Circuit High
P0275 Cylinder no. 5 Contribution/Balance Fault
P0276 Cylinder no. 6 Injector Circuit Low
P0277 Cylinder no. 6 Injector Circuit High
P0278 Cylinder no. 6 Contribution/Balance Fault
P0279 Cylinder no. 7 Injector Circuit Low
P0280 Cylinder no. 7 Injector Circuit High
P0281 Cylinder no. 7 Contribution/Balance Fault
P0282 Cylinder no. 8 Injector Circuit Low
P0283 Cylinder no. 8 Injector Circuit High
P0284 Cylinder no. 8 Contribution/Balance Fault
P0285 Cylinder no. 9 Injector Circuit Low
P0286 Cylinder no. 9 Injector Circuit High
P0287 Cylinder no. 9 Contribution/Balance Fault
P0288 Cylinder no. 10 Injector Circuit Low
P0289 Cylinder no. 10 Injector Circuit High
P0290 Cylinder no. 10 Contribution/Balance Fault
P0291 Cylinder no. 11 Injector Circuit Low
P0292 Cylinder no. 11 Injector Circuit High
P0293 Cylinder no. 11 Contribution/Balance Fault
P0294 Cylinder no. 12 Injector Circuit Low
P0295 Cylinder no. 12 Injector Circuit High
P0296 Cylinder no. 12 Contribution/Balance Fault
P0300 Random/Multiple Cylinder Misfire Detected
P0301 Cylinder no. 1—Misfire Detected
P0302 Cylinder no. 2—Misfire Detected
P0303 Cylinder no. 3—Misfire Detected
P0304 Cylinder no. 4—Misfire Detected
P0305 Cylinder no. 5—Misfire Detected
P0306 Cylinder no. 6—Misfire Detected
P0307 Cylinder no. 7—Misfire Detected
P0308 Cylinder no. 8—Misfire Detected
P0309 Cylinder no. 9—Misfire Detected
P0310 Cylinder no. 10—Misfire Detected
P0311 Cylinder no. 11—Misfire Detected
P0312 Cylinder no. 12—Misfire Detected
P0320 Ignition/Distributor Engine Speed Input Circuit Malfunction
P0321 Ignition/Distributor Engine Speed Input Circuit Range/Performance
P0322 Ignition/Distributor Engine Speed Input Circuit No Signal
P0323 Ignition/Distributor Engine Speed Input Circuit Intermittent
P0325 Knock Sensor no. 1—Circuit Malfunction (Bank no. 1 or Single Sensor)
P0326 Knock Sensor no. 1—Circuit Range/Performance (Bank no. 1 or Single Sensor)
P0327 Knock Sensor no. 1—Circuit Low Input (Bank no. 1 or Single Sensor)
P0328 Knock Sensor no. 1—Circuit High Input (Bank no. 1 or Single Sensor)
P0329 Knock Sensor no. 1—Circuit Input Intermittent (Bank no. 1 or Single Sensor)

P0330 Knock Sensor no. 2—Circuit Malfunction (Bank no. 2)
P0331 Knock Sensor no. 2—Circuit Range/Performance (Bank no. 2)
P0332 Knock Sensor no. 2—Circuit Low Input (Bank no. 2)
P0333 Knock Sensor no. 2—Circuit High Input (Bank no. 2)
P0334 Knock Sensor no. 2—Circuit Input Intermittent (Bank no. 2)
P0335 Crankshaft Position Sensor "A" Circuit Malfunction
P0336 Crankshaft Position Sensor "A" Circuit Range/Performance
P0337 Crankshaft Position Sensor "A" Circuit Low Input
P0338 Crankshaft Position Sensor "A" Circuit High Input
P0339 Crankshaft Position Sensor "A" Circuit Intermittent
P0340 Camshaft Position Sensor Circuit Malfunction
P0341 Camshaft Position Sensor Circuit Range/Performance
P0342 Camshaft Position Sensor Circuit Low Input
P0343 Camshaft Position Sensor Circuit High Input
P0344 Camshaft Position Sensor Circuit Intermittent
P0350 Ignition Coil Primary/Secondary Circuit Malfunction
P0351 Ignition Coil "A" Primary/Secondary Circuit Malfunction
P0352 Ignition Coil "B" Primary/Secondary Circuit Malfunction
P0353 Ignition Coil "C" Primary/Secondary Circuit Malfunction
P0354 Ignition Coil "D" Primary/Secondary Circuit Malfunction
P0355 Ignition Coil "E" Primary/Secondary Circuit Malfunction
P0356 Ignition Coil "F" Primary/Secondary Circuit Malfunction
P0357 Ignition Coil "G" Primary/Secondary Circuit Malfunction
P0358 Ignition Coil "H" Primary/Secondary Circuit Malfunction
P0359 Ignition Coil "I" Primary/Secondary Circuit Malfunction
P0360 Ignition Coil "J" Primary/Secondary Circuit Malfunction
P0361 Ignition Coil "K" Primary/Secondary Circuit Malfunction
P0362 Ignition Coil "L" Primary/Secondary Circuit Malfunction
P0370 Timing Reference High Resolution Signal "A" Malfunction
P0371 Timing Reference High Resolution Signal "A" Too Many Pulses
P0372 Timing Reference High Resolution Signal "A" Too Few Pulses
P0373 Timing Reference High Resolution Signal "A" Intermittent/Erratic Pulses
P0374 Timing Reference High Resolution Signal "A" No Pulses
P0375 Timing Reference High Resolution Signal "B" Malfunction
P0376 Timing Reference High Resolution Signal "B" Too Many Pulses
P0377 Timing Reference High Resolution Signal "B" Too Few Pulses
P0378 Timing Reference High Resolution Signal "B" Intermittent/Erratic Pulses
P0379 Timing Reference High Resolution Signal "B" No Pulses
P0380 Glow Plug/Heater Circuit "A" Malfunction
P0381 Glow Plug/Heater Indicator Circuit Malfunction
P0382 Glow Plug/Heater Circuit "B" Malfunction
P0385 Crankshaft Position Sensor "B" Circuit Malfunction
P0386 Crankshaft Position Sensor "B" Circuit Range/Performance
P0387 Crankshaft Position Sensor "B" Circuit Low Input
P0388 Crankshaft Position Sensor "B" Circuit High Input
P0389 Crankshaft Position Sensor "B" Circuit Intermittent
P0400 Exhaust Gas Recirculation Flow Malfunction
P0401 Exhaust Gas Recirculation Flow Insufficient Detected
P0402 Exhaust Gas Recirculation Flow Excessive Detected
P0403 Exhaust Gas Recirculation Circuit Malfunction

P0404 Exhaust Gas Recirculation Circuit Range/Performance
P0405 Exhaust Gas Recirculation Sensor "A" Circuit Low
P0406 Exhaust Gas Recirculation Sensor "A" Circuit High
P0407 Exhaust Gas Recirculation Sensor "B" Circuit Low
P0408 Exhaust Gas Recirculation Sensor "B" Circuit High
P0410 Secondary Air Injection System Malfunction
P0411 Secondary Air Injection System Incorrect Flow Detected
P0412 Secondary Air Injection System Switching Valve "A" Circuit Malfunction
P0413 Secondary Air Injection System Switching Valve "A" Circuit Open
P0414 Secondary Air Injection System Switching Valve "A" Circuit Shorted
P0415 Secondary Air Injection System Switching Valve "B" Circuit Malfunction
P0416 Secondary Air Injection System Switching Valve "B" Circuit Open
P0417 Secondary Air Injection System Switching Valve "B" Circuit Shorted
P0418 Secondary Air Injection System Relay "A" Circuit Malfunction
P0419 Secondary Air Injection System Relay "B" Circuit Malfunction
P0420 Catalyst System Efficiency Below Threshold (Bank no. 1)
P0421 Warm Up Catalyst Efficiency Below Threshold (Bank no. 1)
P0422 Main Catalyst Efficiency Below Threshold (Bank no. 1)
P0423 Heated Catalyst Efficiency Below Threshold (Bank no. 1)
P0424 Heated Catalyst Temperature Below Threshold (Bank no. 1)
P0430 Catalyst System Efficiency Below Threshold (Bank no. 2)
P0431 Warm Up Catalyst Efficiency Below Threshold (Bank no. 2)
P0432 Main Catalyst Efficiency Below Threshold (Bank no. 2)
P0433 Heated Catalyst Efficiency Below Threshold (Bank no. 2)
P0434 Heated Catalyst Temperature Below Threshold (Bank no. 2)
P0440 Evaporative Emission Control System Malfunction
P0441 Evaporative Emission Control System Incorrect Purge Flow
P0442 Evaporative Emission Control System Leak Detected (Small Leak)
P0443 Evaporative Emission Control System Purge Control Valve Circuit Malfunction
P0444 Evaporative Emission Control System Purge Control Valve Circuit Open
P0445 Evaporative Emission Control System Purge Control Valve Circuit Shorted
P0446 Evaporative Emission Control System Vent Control Circuit Malfunction
P0447 Evaporative Emission Control System Vent Control Circuit Open
P0448 Evaporative Emission Control System Vent Control Circuit Shorted
P0449 Evaporative Emission Control System Vent Valve/Solenoid Circuit Malfunction
P0450 Evaporative Emission Control System Pressure Sensor Malfunction
P0451 Evaporative Emission Control System Pressure Sensor Range/Performance
P0452 Evaporative Emission Control System Pressure Sensor Low Input

P0453 Evaporative Emission Control System Pressure Sensor High Input
P0454 Evaporative Emission Control System Pressure Sensor Intermittent
P0455 Evaporative Emission Control System Leak Detected (Gross Leak)
P0460 Fuel Level Sensor Circuit Malfunction
P0461 Fuel Level Sensor Circuit Range/Performance
P0462 Fuel Level Sensor Circuit Low Input
P0463 Fuel Level Sensor Circuit High Input
P0464 Fuel Level Sensor Circuit Intermittent
P0465 Purge Flow Sensor Circuit Malfunction
P0466 Purge Flow Sensor Circuit Range/Performance
P0467 Purge Flow Sensor Circuit Low Input
P0468 Purge Flow Sensor Circuit High Input
P0469 Purge Flow Sensor Circuit Intermittent
P0470 Exhaust Pressure Sensor Malfunction
P0471 Exhaust Pressure Sensor Range/Performance
P0472 Exhaust Pressure Sensor Low
P0473 Exhaust Pressure Sensor High
P0474 Exhaust Pressure Sensor Intermittent
P0475 Exhaust Pressure Control Valve Malfunction
P0476 Exhaust Pressure Control Valve Range/Performance
P0477 Exhaust Pressure Control Valve Low
P0478 Exhaust Pressure Control Valve High
P0479 Exhaust Pressure Control Valve Intermittent
P0480 Cooling Fan no. 1 Control Circuit Malfunction
P0481 Cooling Fan no. 2 Control Circuit Malfunction
P0482 Cooling Fan no. 3 Control Circuit Malfunction
P0483 Cooling Fan Rationality Check Malfunction
P0484 Cooling Fan Circuit Over Current
P0485 Cooling Fan Power/Ground Circuit Malfunction
P0500 Vehicle Speed Sensor Malfunction
P0501 Vehicle Speed Sensor Range/Performance
P0502 Vehicle Speed Sensor Circuit Low Input
P0503 Vehicle Speed Sensor Intermittent/Erratic/High
P0505 Idle Control System Malfunction
P0506 Idle Control System RPM Lower Than Expected
P0507 Idle Control System RPM Higher Than Expected
P0510 Closed Throttle Position Switch Malfunction
P0520 Engine Oil Pressure Sensor/Switch Circuit Malfunction
P0521 Engine Oil Pressure Sensor/Switch Range/Performance
P0522 Engine Oil Pressure Sensor/Switch Low Voltage
P0523 Engine Oil Pressure Sensor/Switch High Voltage
P0530 A/C Refrigerant Pressure Sensor Circuit Malfunction
P0531 A/C Refrigerant Pressure Sensor Circuit Range/Performance
P0532 A/C Refrigerant Pressure Sensor Circuit Low Input
P0533 A/C Refrigerant Pressure Sensor Circuit High Input
P0534 A/C Refrigerant Charge Loss
P0550 Power Steering Pressure Sensor Circuit Malfunction
P0551 Power Steering Pressure Sensor Circuit Range/Performance
P0552 Power Steering Pressure Sensor Circuit Low Input
P0553 Power Steering Pressure Sensor Circuit High Input
P0554 Power Steering Pressure Sensor Circuit Intermittent
P0560 System Voltage Malfunction
P0561 System Voltage Unstable
P0562 System Voltage Low
P0563 System Voltage High
P0565 Cruise Control On Signal Malfunction
P0566 Cruise Control Off Signal Malfunction

P0567 Cruise Control Resume Signal Malfunction
P0568 Cruise Control Set Signal Malfunction
P0569 Cruise Control Coast Signal Malfunction
P0570 Cruise Control Accel Signal Malfunction
P0571 Cruise Control/Brake Switch "A" Circuit Malfunction
P0572 Cruise Control/Brake Switch "A" Circuit Low
P0573 Cruise Control/Brake Switch "A" Circuit High
P0574 Through P0580 Reserved for Cruise Codes
P0600 Serial Communication Link Malfunction
P0601 Internal Control Module Memory Check Sum Error
P0602 Control Module Programming Error
P0603 Internal Control Module Keep Alive Memory (KAM) Error
P0604 Internal Control Module Random Access Memory (RAM) Error
P0605 Internal Control Module Read Only Memory (ROM) Error
P0606 PCM Processor Fault
P0608 Control Module VSS Output "A" Malfunction
P0609 Control Module VSS Output "B" Malfunction
P0620 Generator Control Circuit Malfunction
P0621 Generator Lamp "L" Control Circuit Malfunction
P0622 Generator Field "F" Control Circuit Malfunction
P0650 Malfunction Indicator Lamp (MIL) Control Circuit Malfunction
P0654 Engine RPM Output Circuit Malfunction
P0655 Engine Hot Lamp Output Control Circuit Malfunction
P0656 Fuel Level Output Circuit Malfunction
P0700 Transmission Control System Malfunction
P0701 Transmission Control System Range/Performance
P0702 Transmission Control System Electrical
P0703 Torque Converter/Brake Switch "B" Circuit Malfunction
P0704 Clutch Switch Input Circuit Malfunction
P0705 Transmission Range Sensor Circuit Malfunction (PRNDL Input)
P0706 Transmission Range Sensor Circuit Range/Performance
P0707 Transmission Range Sensor Circuit Low Input
P0708 Transmission Range Sensor Circuit High Input
P0709 Transmission Range Sensor Circuit Intermittent
P0710 Transmission Fluid Temperature Sensor Circuit Malfunction
P0711 Transmission Fluid Temperature Sensor Circuit Range/Performance
P0712 Transmission Fluid Temperature Sensor Circuit Low Input
P0713 Transmission Fluid Temperature Sensor Circuit High Input
P0714 Transmission Fluid Temperature Sensor Circuit Intermittent
P0715 Input/Turbine Speed Sensor Circuit Malfunction
P0716 Input/Turbine Speed Sensor Circuit Range/Performance
P0717 Input/Turbine Speed Sensor Circuit No Signal
P0718 Input/Turbine Speed Sensor Circuit Intermittent
P0719 Torque Converter/Brake Switch "B" Circuit Low
P0720 Output Speed Sensor Circuit Malfunction
P0721 Output Speed Sensor Circuit Range/Performance
P0722 Output Speed Sensor Circuit No Signal
P0723 Output Speed Sensor Circuit Intermittent
P0724 Torque Converter/Brake Switch "B" Circuit High
P0725 Engine Speed Input Circuit Malfunction
P0726 Engine Speed Input Circuit Range/Performance
P0727 Engine Speed Input Circuit No Signal
P0728 Engine Speed Input Circuit Intermittent
P0730 Incorrect Gear Ratio

P0731 Gear no. 1 Incorrect Ratio
P0732 Gear no. 2 Incorrect Ratio
P0733 Gear no. 3 Incorrect Ratio
P0734 Gear no. 4 Incorrect Ratio
P0735 Gear no. 5 Incorrect Ratio
P0736 Reverse Incorrect Ratio
P0740 Torque Converter Clutch Circuit Malfunction
P0741 Torque Converter Clutch Circuit Performance or Stuck Off
P0742 Torque Converter Clutch Circuit Stuck On
P0743 Torque Converter Clutch Circuit Electrical
P0744 Torque Converter Clutch Circuit Intermittent
P0745 Pressure Control Solenoid Malfunction
P0746 Pressure Control Solenoid Performance or Stuck Off
P0747 Pressure Control Solenoid Stuck On
P0748 Pressure Control Solenoid Electrical
P0749 Pressure Control Solenoid Intermittent
P0750 Shift Solenoid "A" Malfunction
P0751 Shift Solenoid "A" Performance or Stuck Off
P0752 Shift Solenoid "A" Stuck On
P0753 Shift Solenoid "A" Electrical
P0754 Shift Solenoid "A" Intermittent
P0755 Shift Solenoid "B" Malfunction
P0756 Shift Solenoid "B" Performance or Stuck Off
P0757 Shift Solenoid "B" Stuck On
P0758 Shift Solenoid "B" Electrical
P0759 Shift Solenoid "B" Intermittent
P0760 Shift Solenoid "C" Malfunction
P0761 Shift Solenoid "C" Performance Or Stuck Off
P0762 Shift Solenoid "C" Stuck On
P0763 Shift Solenoid "C" Electrical
P0764 Shift Solenoid "C" Intermittent
P0765 Shift Solenoid "D" Malfunction
P0766 Shift Solenoid "D" Performance Or Stuck Off
P0767 Shift Solenoid "D" Stuck On
P0768 Shift Solenoid "D" Electrical
P0769 Shift Solenoid "D" Intermittent
P0770 Shift Solenoid "E" Malfunction
P0771 Shift Solenoid "E" Performance Or Stuck Off
P0772 Shift Solenoid "E" Stuck On
P0773 Shift Solenoid "E" Electrical
P0774 Shift Solenoid "E" Intermittent
P0780 Shift Malfunction
P0781 1–2 Shift Malfunction
P0782 2–3 Shift Malfunction
P0783 3–4 Shift Malfunction
P0784 4–5 Shift Malfunction
P0785 Shift/Timing Solenoid Malfunction
P0786 Shift/Timing Solenoid Range/Performance
P0787 Shift/Timing Solenoid Low
P0788 Shift/Timing Solenoid High
P0789 Shift/Timing Solenoid Intermittent
P0790 Normal/Performance Switch Circuit Malfunction
P0801 Reverse Inhibit Control Circuit Malfunction
P0803 1–4 Upshift (Skip Shift) Solenoid Control Circuit Malfunction
P0804 1–4 Upshift (Skip Shift) Lamp Control Circuit Malfunction
P1000 OBD II Monitor Testing Not Complete More Driving Required
P1001 Key On Engine Running (KOER) Self-Test Not Able To Complete, KOER Aborted

P1100 Mass Air Flow (MAF) Sensor Intermittent
P1101 Mass Air Flow (MAF) Sensor Out Of Self-Test Range
P1111 System Pass 49 State Except Econoline
P1112 Intake Air Temperature (IAT) Sensor Intermittent
P1116 Engine Coolant Temperature (ECT) Sensor Out Of Self-Test Range
P1117 Engine Coolant Temperature (ECT) Sensor Intermittent
P1120 Throttle Position (TP) Sensor Out Of Range (Low)
P1121 Throttle Position (TP) Sensor Inconsistent With MAF Sensor
P1124 Throttle Position (TP) Sensor Out Of Self-Test Range
P1125 Throttle Position (TP) Sensor Circuit Intermittent
P1127 Exhaust Not Warm Enough, Downstream Heated Oxygen Sensors (HO2S) Not Tested
P1128 Upstream Heated Oxygen Sensors (HO2S) Swapped From Bank To Bank
P1129 Downstream Heated Oxygen Sensors (HO2S) Swapped From Bank To Bank
P1130 Lack Of Upstream Heated Oxygen Sensor (HO2S 11) Switch, Adaptive Fuel At Limit (Bank #1)
P1131 Lack Of Upstream Heated Oxygen Sensor (HO2S 11) Switch, Sensor Indicates Lean (Bank #1)
P1132 Lack Of Upstream Heated Oxygen Sensor (HO2S 11) Switch, Sensor Indicates Rich (Bank#1)
P1137 Lack Of Downstream Heated Oxygen Sensor (HO2S 12) Switch, Sensor Indicates Lean (Bank#1)
P1138 Lack Of Downstream Heated Oxygen Sensor (HO2S 12) Switch, Sensor Indicates Rich (Bank#1)
P1150 Lack Of Upstream Heated Oxygen Sensor (HO2S 21) Switch, Adaptive Fuel At Limit (Bank #2)
P1151 Lack Of Upstream Heated Oxygen Sensor (HO2S 21) Switch, Sensor Indicates Lean (Bank#2)
P1152 Lack Of Upstream Heated Oxygen Sensor (HO2S 21) Switch, Sensor Indicates Rich (Bank #2)
P1157 Lack Of Downstream Heated Oxygen Sensor (HO2S 22) Switch, Sensor Indicates Lean (Bank #2)
P1158 Lack Of Downstream Heated Oxygen Sensor (HO2S 22) Switch, Sensor Indicates Rich (Bank#2)
P1169 (HO2S 12) Signal Remained Unchanged For More Than 20 Seconds After Closed Loop
P1170 (HO2S 11) Signal Remained Unchanged For More Than 20 Seconds After Closed Loop
P1173 Feedback A/F Mixture Control (HO2S 21) Signal Remained Unchanged For More Than 20 Seconds After Closed Loop
P1184 Engine Oil Temp Sensor Circuit Performance
P1195 Barometric (BARO) Pressure Sensor Circuit Malfunction (Signal Is From EGR Boost Sensor)
P1196 Starter Switch Circuit Malfunction
P1209 Injection Control Pressure (ICP) Peak Fault
P1210 Injection Control Pressure (ICP) Above Expected Level
P1211 Injection Control Pressure (ICP) Not Controllable—Pressure Above/Below Desired
P1212 Injection Control Pressure (ICP) Voltage Not At Expected Level
P1218 Cylinder Identification (CID) Stuck High
P1219 Cylinder Identification (CID) Stuck Low
P1220 Series Throttle Control Malfunction (Traction Control System)
P1224 Throttle Position Sensor "B" (TP-B) Out Of Self-Test Range (Traction Control System)
P1230 Fuel Pump Low Speed Malfunction

P1231 Fuel Pump Secondary Circuit Low With High Speed Pump On
P1232 Low Speed Fuel Pump Primary Circuit Malfunction
P1233 Fuel Pump Driver Module Off-line (MIL DTC)
P1234 Fuel Pump Driver Module Disabled Or Off-line (No MIL)
P1235 Fuel Pump Control Out Of Range (MIL DTC)
P1236 Fuel Pump Control Out Of Range (No MIL)
P1237 Fuel Pump Secondary Circuit Malfunction (MIL DTC)
P1238 Fuel Pump Secondary Circuit Malfunction (No DMIL)
P1250 Fuel Pressure Regulator Control (FPRC) Solenoid Malfunction
P1260 THEFT Detected—Engine Disabled
P1261 High To Low Side Short—Cylinder #1 (Indicates Low side Circuit Is Shorted To B+ Or To The High Side Between The IDM And The Injector)
P1262 High To Low Side Short—Cylinder #2 (Indicates Low side Circuit Is Shorted To B+ Or To The High Side Between The IDM And The Injector)
P1263 High To Low Side Short—Cylinder #3 (Indicates Low side Circuit Is Shorted To B+ Or To The High Side Between The IDM And The Injector)
P1264 High To Low Side Short—Cylinder #4 (Indicates Low side Circuit Is Shorted To B+ Or To The High Side Between The IDM And The Injector)
P1265 High To Low Side Short—Cylinder #5 (Indicates Low side Circuit Is Shorted To B+ Or To The High Side Between The IDM And The Injector)
P1266 High To Low Side Short—Cylinder #6 (Indicates Low side Circuit Is Shorted To B+ Or To The High Side Between The IDM And The Injector)
P1267 High To Low Side Short—Cylinder #7 (Indicates Low side Circuit Is Shorted To B+ Or To The High Side Between The IDM And The Injector)
P1268 High To Low Side Short—Cylinder #8 (Indicates Low side Circuit Is Shorted To B+ Or To The High Side Between The IDM And The Injector)
P1270 Engine RPM Or Vehicle Speed Limiter Reached
P1271 High To Low Side Open—Cylinder #1 (Indicates A High To Low Side Open Between The Injector And The IDM)
P1272 High To Low Side Open—Cylinder #2 (Indicates A High To Low Side Open Between The Injector And The IDM)
P1273 High To Low Side Open—Cylinder #3 (Indicates A High To Low Side Open Between The Injector And The IDM)
P1274 High To Low Side Open—Cylinder #4 (Indicates A High To Low Side Open Between The Injector And The IDM)
P1275 High To Low Side Open—Cylinder #5 (Indicates A High To Low Side Open Between The Injector And The IDM)
P1276 High To Low Side Open—Cylinder #6 (Indicates A High To Low Side Open Between The Injector And The IDM)
P1277 High To Low Side Open—Cylinder #7 (Indicates A High To Low Side Open Between The Injector And The IDM)
P1278 High To Low Side Open—Cylinder #8 (Indicates A High To Low Side Open Between The Injector And The IDM)
P1280 Injection Control Pressure (ICP) Circuit Out Of Range Low
P1281 Injection Control Pressure (ICP) Circuit Out Of Range High
P1282 Injection Control Pressure (ICP) Excessive
P1283 Injection Pressure Regulator (IPR) Circuit Failure
P1284 Injection Control Pressure (ICP) Failure—Aborts KOER Or CCT Test

P1285 Cylinder Head Temperature (CHT) Over Temperature Sensed

P1288 Cylinder Head Temperature (CHT) Sensor Out Of Self-Test Range

P1289 Cylinder Head Temperature (CHT) Sensor Circuit Low Input

P1290 Cylinder Head Temperature (CHT) Sensor Circuit High Input

P1291 IDM To Injector High Side Circuit #1 (Right Bank) Short To GND Or B+

P1292 IDM To Injector High Side Circuit #2 (Right Bank) Short To GND Or B+

P1293 IDM To Injector High Side Circuit Open Bank #1 (Right Bank)

P1294 IDM To Injector High Side Circuit Open Bank #2 (Left Bank)

P1295 Multiple IDM/Injector Circuit Faults On Bank #1 (Right)

P1296 Multiple IDM/Injector Circuit Faults On Bank#2 (Left)

P1297 High Sides Shorted Together

P1298 IDM Failure

P1299 Engine Over Temperature Condition

P1309 Misfire Detection Monitor Is Not Enabled

P1316 Injector Circuit/IDM Codes Detected

P1320 Distributor Signal Interrupt

P1336 Crankshaft Position Sensor (Gear)

P1345 No Camshaft Position Sensor Signal

P1351 Ignition Diagnostic Monitor (IDM) Circuit Input Malfunction

P1351 Indicates Ignition System Malfunction

P1352 Indicates Ignition System Malfunction

P1353 Indicates Ignition System Malfunction

P1354 Indicates Ignition System Malfunction

P1355 Indicates Ignition System Malfunction

P1356 PIPs Occurred While IDM Pulse width Indicates Engine Not Turning

P1357 Ignition Diagnostic Monitor (IDM) Pulse width Not Defined

P1358 Ignition Diagnostic Monitor (IDM) Signal Out Of Self-Test Range

P1359 Spark Output Circuit Malfunction

P1364 Spark Output Circuit Malfunction

P1390 Octane Adjust (OCT ADJ) Out Of Self-Test Range

P1391 Glow Plug Circuit Low Input Bank #1 (Right)

P1392 Glow Plug Circuit High Input Bank #1 (Right)

P1393 Glow Plug Circuit Low Input Bank #2 (Left)

P1394 Glow Plug Circuit High Input Bank #2 (Left)

P1395 Glow Plug Monitor Fault Bank #1

P1396 Glow Plug Monitor Fault Bank #2

P1397 System Voltage Out Of Self Test Range

P1400 Differential Pressure Feedback EGR (DPFE) Sensor Circuit Low Voltage Detected

P1401 Differential Pressure Feedback EGR (DPFE) Sensor Circuit High Voltage Detected/EGR Temperature Sensor

P1402 EGR Valve Position Sensor Open Or Short

P1403 Differential Pressure Feedback EGR (DPFE) Sensor Hoses Reversed

P1405 Differential Pressure Feedback EGR (DPFE) Sensor Upstream Hose Off Or Plugged

P1406 Differential Pressure Feedback EGR (DPFE) Sensor Downstream Hose Off Or Plugged

P1407 Exhaust Gas Recirculation (EGR) No Flow Detected (Valve Stuck Closed Or Inoperative)

P1408 Exhaust Gas Recirculation (EGR) Flow Out Of Self-Test Range

P1409 Electronic Vacuum Regulator (EVR) Control Circuit Malfunction

P1410 Check That Fuel Pressure Regulator Control Solenoid And The EGR Check Solenoid Connectors Are Not Swapped

P1411 Secondary Air Injection System Incorrect Downstream Flow Detected

P1413 Secondary Air Injection System Monitor Circuit Low Voltage

P1414 Secondary Air Injection System Monitor Circuit High Voltage

P1442 Evaporative Emission Control System Small Leak Detected

P1443 Evaporative Emission Control System—Vacuum System, Purge Control Solenoid Or Purge Control Valve Malfunction

P1444 Purge Flow Sensor (PFS) Circuit Low Input

P1445 Purge Flow Sensor (PFS) Circuit High Input

P1449 Evaporative Emission Control System Unable To Hold Vacuum

P1450 Unable To Bleed Up Fuel Tank Vacuum

P1455 Evaporative Emission Control System Control Leak Detected (Gross Leak)

P1460 Wide Open Throttle Air Conditioning Cut-Off Circuit Malfunction

P1461 Air Conditioning Pressure (ACP) Sensor Circuit Low Input

P1462 Air Conditioning Pressure (ACP) Sensor Circuit High Input

P1463 Air Conditioning Pressure (ACP) Sensor Insufficient Pressure Change

P1464 Air Conditioning (A/C) Demand Out Of Self-Test Range/A/C On During KOER Or CCT Test

P1469 Low Air Conditioning Cycling Period

P1473 Fan Secondary High, With Fan(s) Off

P1474 Low Fan Control Primary Circuit Malfunction

P1479 High Fan Control Primary Circuit Malfunction

P1480 Fan Secondary Low, With Low Fan On

P1481 Fan Secondary Low, With High Fan On

P1483 Power To Fan Circuit Over current

P1484 Open Power/Ground To Variable Load Control Module (VLCM)

P1485 EGR Control Solenoid Open Or Short

P1486 EGR Vent Solenoid Open Or Short

P1487 EGR Boost Check Solenoid Open Or Short

P1500 Vehicle Speed Sensor (VSS) Circuit Intermittent

P1501 Vehicle Speed Sensor (VSS) Out Of Self-Test Range/Vehicle Moved During Test

P1502 Invalid Self Test—Auxiliary Powertrain Control Module (APCM) Functioning

P1504 Idle Air Control (IAC) Circuit Malfunction

P1505 Idle Air Control (IAC) System At Adaptive Clip

P1506 Idle Air Control (IAC) Overspeed Error

P1507 Idle Air Control (IAC) Underspeed Error

P1512 Intake Manifold Runner Control (IMRC) Malfunction (Bank#1 Stuck Closed)

P1513 Intake Manifold Runner Control (IMRC) Malfunction (Bank#2 Stuck Closed)

P1516 Intake Manifold Runner Control (IMRC) Input Error (Bank #1)

P1517 Intake Manifold Runner Control (IMRC) Input Error (Bank #2)

P1518 Intake Manifold Runner Control (IMRC) Malfunction (Stuck Open)

P1519 Intake Manifold Runner Control (IMRC) Malfunction (Stuck Closed)

P1520 Intake Manifold Runner Control (IMRC) Circuit Malfunction

P1521 Variable Resonance Induction System (VRIS) Solenoid #1 Open Or Short

P1522 Variable Resonance Induction System (VRIS) Solenoid#2 Open Or Short

P1523 High Speed Inlet Air (HSIA) Solenoid Open Or Short

P1530 Air Condition (A/C) Clutch Circuit Malfunction

P1531 Invalid Test—Accelerator Pedal Movement

P1536 Parking Brake Applied Failure

P1537 Intake Manifold Runner Control (IMRC) Malfunction (Bank#1 Stuck Open)

P1538 Intake Manifold Runner Control (IMRC) Malfunction (Bank#2 Stuck Open)

P1539 Power To Air Condition (A/C) Clutch Circuit Overcurrent

P1549 Problem In Intake Manifold Tuning (IMT) Valve System

P1550 Power Steering Pressure (PSP) Sensor Out Of Self-Test Range

P1601 Serial Communication Error

P1605 Powertrain Control Module (PCM)—Keep Alive Memory (KAM) Test Error

P1608 PCM Internal Circuit Malfunction

P1609 PCM Internal Circuit Malfunction (2.5L Only)

P1625 B+ Supply To Variable Load Control Module (VLCM) Fan Circuit Malfunction

P1626 B+ Supply To Variable Load Control Module (VLCM) Air Conditioning (A/C) Circuit

P1650 Power Steering Pressure (PSP) Switch Out Of Self-Test Range

P1651 Power Steering Pressure (PSP) Switch Input Malfunction

P1660 Output Circuit Check Signal High

P1661 Output Circuit Check Signal Low

P1662 Injection Driver Module Enable (IDM EN) Circuit Failure

P1663 Fuel Delivery Command Signal (FDCS) Circuit Failure

P1667 Cylinder Identification (CID) Circuit Failure

P1668 PCM—IDM Diagnostic Communication Error

P1670 EF Feedback Signal Not Detected

P1701 Reverse Engagement Error

P1701 Fuel Trim Malfunction (Villager)

P1703 Brake On/Off (BOO) Switch Out Of Self-Test Range

P1704 Digital Transmission Range (TR) Sensor Failed To Transition State

P1705 Transmission Range (TR) Sensor Out Of Self-Test Range

P1705 TP Sensor (AT) Villager

P1705 Clutch Pedal Position (CPP) Or Park Neutral Position (PNP) Problem

P1706 High Vehicle Speed In Park

P1709 Park Or Neutral Position (PNP) Or Clutch Pedal Position (CPP) Switch Out Of Self-Test Range

P1709 Throttle Position (TP) Sensor Malfunction (Aspire 1.3L, Escort/ Tracer 1.8L, Probe 2.5L)

P1711 Transmission Fluid Temperature (TFT) Sensor Out Of Self-Test Range

P1714 Shift Solenoid "A" Inductive Signature Malfunction

P1715 Shift Solenoid "B" Inductive Signature Malfunction

P1716 Transmission Malfunction

P1717 Transmission Malfunction

P1719 Transmission Malfunction

P1720 Vehicle Speed Sensor (VSS) Circuit Malfunction

P1727 Coast Clutch Solenoid Inductive Signature Malfunction

P1728 Transmission Slip Error—Converter Clutch Failed

P1729 4x4 Low Switch Error

P1731 Improper 1–2 Shift

P1732 Improper 2–3 Shift

P1733 Improper 3–4 Shift

P1734 Improper 4–5 Shift

P1740 Torque Converter Clutch (TCC) Inductive Signature Malfunction

P1741 Torque Converter Clutch (TCC) Control Error

P1742 Torque Converter Clutch (TCC) Solenoid Failed On (Turns On MIL)

P1743 Torque Converter Clutch (TCC) Solenoid Failed On (Turns On TCIL)

P1744 Torque Converter Clutch (TCC) System Mechanically Stuck In Off Position

P1744 Torque Converter Clutch (TCC) Solenoid Malfunction (2.5L Only)

P1746 Electronic Pressure Control (EPC) Solenoid Open Circuit (Low Input)

P1747 Electronic Pressure Control (EPC) Solenoid Short Circuit (High Input)

P1748 Electronic Pressure Control (EPC) Malfunction

P1749 Electronic Pressure Control (EPC) Solenoid Failed Low

P1751 Shift Solenoid#1 (SS1) Performance

P1754 Coast Clutch Solenoid (CCS) Circuit Malfunction

P1756 Shift Solenoid#2 (SS2) Performance

P1760 Overrun Clutch SN

P1761 Shift Solenoid #(SS2) Performance

P1762 Transmission Malfunction

P1765 3–2 Timing Solenoid Malfunction (2.5L Only)

P1779 TCIL Circuit Malfunction

P1780 Transmission Control Switch (TCS) Circuit Out Of Self-Test Range

P1781 4x4 Low Switch, Out Of Self-Test Range

P1783 Transmission Over Temperature Condition

P1784 Transmission Malfunction

P1785 Transmission Malfunction

P1786 Transmission Malfunction

P1787 Transmission Malfunction

P1788 3–2 Timing/Coast Clutch Solenoid (3–2/CCS) Circuit Open

P1789 3–2 Timing/Coast Clutch Solenoid (3–2/CCS) Circuit Shorted

P1792 Idle (IDL) Switch (Closed Throttle Position Switch) Malfunction

P1794 Loss Of Battery Voltage Input

P1795 EGR Boost Sensor Malfunction

P1797 Clutch Pedal Position (CPP) Switch Or Neutral Switch Circuit Malfunction

P1900 Cooling Fan

U1021 SCP Indicating The Lack Of Air Conditioning (A/C) Clutch Status Response

U1039 Vehicle Speed Signal (VSS) Missing Or Incorrect

U1051 Brake Switch Signal Missing Or Incorrect

U1073 SCP Indicating The Lack Of Engine Coolant Fan Status Response

U1131 SCP Indicating The Lack Of Fuel Pump Status Response

U1135 SCP Indicating The Ignition Switch Signal Missing Or Incorrect

U1256 SCP Indicating A Communications Error

U1451 Lack Of Response From Passive Anti-Theft System (PATS) Module—Engine Disabled

General Motors Corporation

READING CODES

Reading the control module memory is one of the first steps in OBD II system diagnostics.

This step should be initially performed to determine the general nature of the fault. Subsequent readings will determine if the fault has been cleared.

Reading codes can be performed by any of the methods below:
- Read the control module memory with the Generic Scan Tool (GST)
- Read the control module memory with the vehicle manufacturer's specific tester

To read the fault codes, connect the scan tool or tester according to the manufacturer's instructions. Follow the manufacturer's specified procedure for reading the codes.

CLEARING CODES

Control module reset procedures are a very important part of OBD II System diagnostics. This step should be done at the end of any fault code repair and at the end of any driveability repair.

Clearing codes can be performed by any of the methods below:
- Clear the control module memory with the Generic Scan Tool (GST)
- Clear the control module memory with the vehicle manufacturer's specific tester
- Turn the ignition off and remove the negative battery cable for at least 1 minute.

Removing the negative battery cable may cause other systems in the vehicle to lose their memory. Prior to removing the cable, ensure you have the proper reset codes for radios and alarms.

➡**The MIL will may also be de-activated for some codes if the vehicle completes three consecutive trips without a fault detected with vehicle conditions similar to those present during the fault.**

DIAGNOSTIC TROUBLE CODES

Except Geo Models

P0000 No Failures

P0100 Mass or Volume Air Flow Circuit Malfunction

P0101 Mass or Volume Air Flow Circuit Range/Performance Problem

P0102 Mass or Volume Air Flow Circuit Low Input

P0103 Mass or Volume Air Flow Circuit High Input

P0104 Mass or Volume Air Flow Circuit Intermittent

P0105 Manifold Absolute Pressure/Barometric Pressure Circuit Malfunction

P0106 Manifold Absolute Pressure/Barometric Pressure Circuit Range/Performance Problem

P0107 Manifold Absolute Pressure/Barometric Pressure Circuit Low Input

P0108 Manifold Absolute Pressure/Barometric Pressure Circuit High Input

P0109 Manifold Absolute Pressure/Barometric Pressure Circuit Intermittent

P0110 Intake Air Temperature Circuit Malfunction

P0111 Intake Air Temperature Circuit Range/Performance Problem

P0112 Intake Air Temperature Circuit Low Input

P0113 Intake Air Temperature Circuit High Input

P0114 Intake Air Temperature Circuit Intermittent

P0115 Engine Coolant Temperature Circuit Malfunction

P0116 Engine Coolant Temperature Circuit Range/Performance Problem

P0117 Engine Coolant Temperature Circuit Low Input

P0118 Engine Coolant Temperature Circuit High Input

P0119 Engine Coolant Temperature Circuit Intermittent

P0120 Throttle/Pedal Position Sensor/Switch "A" Circuit Malfunction

P0121 Throttle/Pedal Position Sensor/Switch "A" Circuit Range/Performance Problem

P0122 Throttle/Pedal Position Sensor/Switch "A" Circuit Low Input

P0123 Throttle/Pedal Position Sensor/Switch "A" Circuit High Input

P0124 Throttle/Pedal Position Sensor/Switch "A" Circuit Intermittent

P0125 Insufficient Coolant Temperature For Closed Loop Fuel Control

P0126 Insufficient Coolant Temperature For Stable Operation

P0130 O_2 Circuit Malfunction (Bank no. 1 Sensor no. 1)

P0131 O_2 Sensor Circuit Low Voltage (Bank no. 1 Sensor no. 1)

P0132 O_2 Sensor Circuit High Voltage (Bank no. 1 Sensor no. 1)

P0133 O_2 Sensor Circuit Slow Response (Bank no. 1 Sensor no. 1)

P0134 O_2 Sensor Circuit No Activity Detected (Bank no. 1 Sensor no. 1)

P0135 O_2 Sensor Heater Circuit Malfunction (Bank no. 1 Sensor no. 1)

P0136 O_2 Sensor Circuit Malfunction (Bank no. 1 Sensor no. 2)

P0137 O_2 Sensor Circuit Low Voltage (Bank no. 1 Sensor no. 2)

P0138 O_2 Sensor Circuit High Voltage (Bank no. 1 Sensor no. 2)

P0139 O_2 Sensor Circuit Slow Response (Bank no. 1 Sensor no. 2)

P0140 O_2 Sensor Circuit No Activity Detected (Bank no. 1 Sensor no. 2)

P0141 O_2 Sensor Heater Circuit Malfunction (Bank no. 1 Sensor no. 2)

P0142 O_2 Sensor Circuit Malfunction (Bank no. 1 Sensor no. 3)

P0143 O_2 Sensor Circuit Low Voltage (Bank no. 1 Sensor no. 3)

P0144 O_2 Sensor Circuit High Voltage (Bank no. 1 Sensor no. 3)

P0145 O_2 Sensor Circuit Slow Response (Bank no. 1 Sensor no. 3)

P0146 O_2 Sensor Circuit No Activity Detected (Bank no. 1 Sensor no. 3)

P0147 O_2 Sensor Heater Circuit Malfunction (Bank no. 1 Sensor no. 3)

P0150 O_2 Sensor Circuit Malfunction (Bank no. 2 Sensor no. 1)

P0151 O_2 Sensor Circuit Low Voltage (Bank no. 2 Sensor no. 1)

P0152 O_2 Sensor Circuit High Voltage (Bank no. 2 Sensor no. 1)

P0153 O_2 Sensor Circuit Slow Response (Bank no. 2 Sensor no. 1)

P0154 O_2 Sensor Circuit No Activity Detected (Bank no. 2 Sensor no. 1)

P0155 O_2 Sensor Heater Circuit Malfunction (Bank no. 2 Sensor no. 1)

P0156 O_2 Sensor Circuit Malfunction (Bank no. 2 Sensor no. 2)

P0157 O_2 Sensor Circuit Low Voltage (Bank no. 2 Sensor no. 2)

P0158 O_2 Sensor Circuit High Voltage (Bank no. 2 Sensor no. 2)

P0159 O_2 Sensor Circuit Slow Response (Bank no. 2 Sensor no. 2)

P0160 O_2 Sensor Circuit No Activity Detected (Bank no. 2 Sensor no. 2)

P0161 O_2 Sensor Heater Circuit Malfunction (Bank no. 2 Sensor no. 2)

P0162 O_2 Sensor Circuit Malfunction (Bank no. 2 Sensor no. 3)

P0163 O_2 Sensor Circuit Low Voltage (Bank no. 2 Sensor no. 3)

P0164 O_2 Sensor Circuit High Voltage (Bank no. 2 Sensor no. 3)

P0165 O_2 Sensor Circuit Slow Response (Bank no. 2 Sensor no. 3)

P0166 O_2 Sensor Circuit No Activity Detected (Bank no. 2 Sensor no. 3)

P0167 O_2 Sensor Heater Circuit Malfunction (Bank no. 2 Sensor no. 3)

P0170 Fuel Trim Malfunction (Bank no. 1)

P0171 System Too Lean (Bank no. 1)

P0172 System Too Rich (Bank no. 1)

P0173 Fuel Trim Malfunction (Bank no. 2)

P0174 System Too Lean (Bank no. 2)

P0175 System Too Rich (Bank no. 2)

P0176 Fuel Composition Sensor Circuit Malfunction

P0177 Fuel Composition Sensor Circuit Range/Performance

P0178 Fuel Composition Sensor Circuit Low Input

P0179 Fuel Composition Sensor Circuit High Input

P0180 Fuel Temperature Sensor "A" Circuit Malfunction

P0181 Fuel Temperature Sensor "A" Circuit Range/Performance

P0182 Fuel Temperature Sensor "A" Circuit Low Input

P0183 Fuel Temperature Sensor "A" Circuit High Input

P0184 Fuel Temperature Sensor "A" Circuit Intermittent

P0185 Fuel Temperature Sensor "B" Circuit Malfunction

P0186 Fuel Temperature Sensor "B" Circuit Range/Performance

P0187 Fuel Temperature Sensor "B" Circuit Low Input

P0188 Fuel Temperature Sensor "B" Circuit High Input

P0189 Fuel Temperature Sensor "B" Circuit Intermittent

P0190 Fuel Rail Pressure Sensor Circuit Malfunction

P0191 Fuel Rail Pressure Sensor Circuit Range/Performance

P0192 Fuel Rail Pressure Sensor Circuit Low Input

P0193 Fuel Rail Pressure Sensor Circuit High Input

P0194 Fuel Rail Pressure Sensor Circuit Intermittent

P0195 Engine Oil Temperature Sensor Malfunction

P0196 Engine Oil Temperature Sensor Range/Performance

P0197 Engine Oil Temperature Sensor Low

P0198 Engine Oil Temperature Sensor High

P0199 Engine Oil Temperature Sensor Intermittent

P0200 Injector Circuit Malfunction

P0201 Injector Circuit Malfunction—Cylinder no. 1

P0202 Injector Circuit Malfunction—Cylinder no. 2

P0203 Injector Circuit Malfunction—Cylinder no. 3

P0204 Injector Circuit Malfunction—Cylinder no. 4

P0205 Injector Circuit Malfunction—Cylinder no. 5

P0206 Injector Circuit Malfunction—Cylinder no. 6

P0207 Injector Circuit Malfunction—Cylinder no. 7

P0208 Injector Circuit Malfunction—Cylinder no. 8

P0209 Injector Circuit Malfunction—Cylinder no. 9

P0210 Injector Circuit Malfunction—Cylinder no. 10

P0211 Injector Circuit Malfunction—Cylinder no. 11

P0212 Injector Circuit Malfunction—Cylinder no. 12

P0213 Cold Start Injector no. 1 Malfunction

P0214 Cold Start Injector no. 2 Malfunction

P0215 Engine Shutoff Solenoid Malfunction

P0216 Injection Timing Control Circuit Malfunction

P0217 Engine Over Temperature Condition

P0218 Transmission Over Temperature Condition

P0219 Engine Over Speed Condition

P0220 Throttle/Pedal Position Sensor/Switch "B" Circuit Malfunction

P0221 Throttle/Pedal Position Sensor/Switch "B" Circuit Range/Performance Problem

P0222 Throttle/Pedal Position Sensor/Switch "B" Circuit Low Input

P0223 Throttle/Pedal Position Sensor/Switch "B" Circuit High Input

P0224 Throttle/Pedal Position Sensor/Switch "B" Circuit Intermittent

P0225 Throttle/Pedal Position Sensor/Switch "C" Circuit Malfunction

P0226 Throttle/Pedal Position Sensor/Switch "C" Circuit Range/Performance Problem

P0227 Throttle/Pedal Position Sensor/Switch "C" Circuit Low Input

P0228 Throttle/Pedal Position Sensor/Switch "C" Circuit High Input

P0229 Throttle/Pedal Position Sensor/Switch "C" Circuit Intermittent

P0230 Fuel Pump Primary Circuit Malfunction

P0231 Fuel Pump Secondary Circuit Low

P0232 Fuel Pump Secondary Circuit High

P0233 Fuel Pump Secondary Circuit Intermittent

P0234 Engine Over Boost Condition

P0261 Cylinder no. 1 Injector Circuit Low

P0262 Cylinder no. 1 Injector Circuit High

P0263 Cylinder no. 1 Contribution/Balance Fault

P0264 Cylinder no. 2 Injector Circuit Low

P0265 Cylinder no. 2 Injector Circuit High

P0266 Cylinder no. 2 Contribution/Balance Fault

P0267 Cylinder no. 3 Injector Circuit Low

P0268 Cylinder no. 3 Injector Circuit High

P0269 Cylinder no. 3 Contribution/Balance Fault

P0270 Cylinder no. 4 Injector Circuit Low

P0271 Cylinder no. 4 Injector Circuit High

P0272 Cylinder no. 4 Contribution/Balance Fault

P0273 Cylinder no. 5 Injector Circuit Low

P0274 Cylinder no. 5 Injector Circuit High

P0275 Cylinder no. 5 Contribution/Balance Fault

P0276 Cylinder no. 6 Injector Circuit Low

P0277 Cylinder no. 6 Injector Circuit High
P0278 Cylinder no. 6 Contribution/Balance Fault
P0279 Cylinder no. 7 Injector Circuit Low
P0280 Cylinder no. 7 Injector Circuit High
P0281 Cylinder no. 7 Contribution/Balance Fault
P0282 Cylinder no. 8 Injector Circuit Low
P0283 Cylinder no. 8 Injector Circuit High
P0284 Cylinder no. 8 Contribution/Balance Fault
P0285 Cylinder no. 9 Injector Circuit Low
P0286 Cylinder no. 9 Injector Circuit High
P0287 Cylinder no. 9 Contribution/Balance Fault
P0288 Cylinder no. 10 Injector Circuit Low
P0289 Cylinder no. 10 Injector Circuit High
P0290 Cylinder no. 10 Contribution/Balance Fault
P0291 Cylinder no. 11 Injector Circuit Low
P0292 Cylinder no. 11 Injector Circuit High
P0293 Cylinder no. 11 Contribution/Balance Fault
P0294 Cylinder no. 12 Injector Circuit Low
P0295 Cylinder no. 12 Injector Circuit High
P0296 Cylinder no. 12 Contribution/Balance Fault
P0300 Random/Multiple Cylinder Misfire Detected
P0301 Cylinder no. 1—Misfire Detected
P0302 Cylinder no. 2—Misfire Detected
P0303 Cylinder no. 3—Misfire Detected
P0304 Cylinder no. 4—Misfire Detected
P0305 Cylinder no. 5—Misfire Detected
P0306 Cylinder no. 6—Misfire Detected
P0307 Cylinder no. 7—Misfire Detected
P0308 Cylinder no. 8—Misfire Detected
P0309 Cylinder no. 9—Misfire Detected
P0310 Cylinder no. 10—Misfire Detected
P0311 Cylinder no. 11—Misfire Detected
P0312 Cylinder no. 12—Misfire Detected
P0320 Ignition/Distributor Engine Speed Input Circuit Malfunction
P0321 Ignition/Distributor Engine Speed Input Circuit Range/Performance
P0322 Ignition/Distributor Engine Speed Input Circuit No Signal
P0323 Ignition/Distributor Engine Speed Input Circuit Intermittent
P0325 Knock Sensor no. 1—Circuit Malfunction (Bank no. 1 or Single Sensor)
P0326 Knock Sensor no. 1—Circuit Range/Performance (Bank no. 1 or Single Sensor)
P0327 Knock Sensor no. 1—Circuit Low Input (Bank no. 1 or Single Sensor)
P0328 Knock Sensor no. 1—Circuit High Input (Bank no. 1 or Single Sensor)
P0329 Knock Sensor no. 1—Circuit Input Intermittent (Bank no. 1 or Single Sensor)
P0330 Knock Sensor no. 2—Circuit Malfunction (Bank no. 2)
P0331 Knock Sensor no. 2—Circuit Range/Performance (Bank no. 2)
P0332 Knock Sensor no. 2—Circuit Low Input (Bank no. 2)
P0333 Knock Sensor no. 2—Circuit High Input (Bank no. 2)
P0334 Knock Sensor no. 2—Circuit Input Intermittent (Bank no. 2)
P0335 Crankshaft Position Sensor "A" Circuit Malfunction
P0336 Crankshaft Position Sensor "A" Circuit Range/Performance
P0337 Crankshaft Position Sensor "A" Circuit Low Input

P0338 Crankshaft Position Sensor "A" Circuit High Input
P0339 Crankshaft Position Sensor "A" Circuit Intermittent
P0340 Camshaft Position Sensor Circuit Malfunction
P0341 Camshaft Position Sensor Circuit Range/Performance
P0342 Camshaft Position Sensor Circuit Low Input
P0343 Camshaft Position Sensor Circuit High Input
P0344 Camshaft Position Sensor Circuit Intermittent
P0350 Ignition Coil Primary/Secondary Circuit Malfunction
P0351 Ignition Coil "A" Primary/Secondary Circuit Malfunction
P0352 Ignition Coil "B" Primary/Secondary Circuit Malfunction
P0353 Ignition Coil "C" Primary/Secondary Circuit Malfunction
P0354 Ignition Coil "D" Primary/Secondary Circuit Malfunction
P0355 Ignition Coil "E" Primary/Secondary Circuit Malfunction
P0356 Ignition Coil "F" Primary/Secondary Circuit Malfunction
P0357 Ignition Coil "G" Primary/Secondary Circuit Malfunction
P0358 Ignition Coil "H" Primary/Secondary Circuit Malfunction
P0359 Ignition Coil "I" Primary/Secondary Circuit Malfunction
P0360 Ignition Coil "J" Primary/Secondary Circuit Malfunction
P0361 Ignition Coil "K" Primary/Secondary Circuit Malfunction
P0362 Ignition Coil "L" Primary/Secondary Circuit Malfunction
P0370 Timing Reference High Resolution Signal "A" Malfunction
P0371 Timing Reference High Resolution Signal "A" Too Many Pulses
P0372 Timing Reference High Resolution Signal "A" Too Few Pulses
P0373 Timing Reference High Resolution Signal "A" Intermittent/Erratic Pulses
P0374 Timing Reference High Resolution Signal "A" No Pulses
P0375 Timing Reference High Resolution Signal "B" Malfunction
P0376 Timing Reference High Resolution Signal "B" Too Many Pulses
P0377 Timing Reference High Resolution Signal "B" Too Few Pulses
P0378 Timing Reference High Resolution Signal "B" Intermittent/Erratic Pulses
P0379 Timing Reference High Resolution Signal "B" No Pulses
P0380 Glow Plug/Heater Circuit "A" Malfunction
P0381 Glow Plug/Heater Indicator Circuit Malfunction
P0382 Glow Plug/Heater Circuit "B" Malfunction
P0385 Crankshaft Position Sensor "B" Circuit Malfunction
P0386 Crankshaft Position Sensor "B" Circuit Range/Performance
P0387 Crankshaft Position Sensor "B" Circuit Low Input
P0388 Crankshaft Position Sensor "B" Circuit High Input
P0389 Crankshaft Position Sensor "B" Circuit Intermittent
P0400 Exhaust Gas Recirculation Flow Malfunction
P0401 Exhaust Gas Recirculation Flow Insufficient Detected
P0402 Exhaust Gas Recirculation Flow Excessive Detected
P0403 Exhaust Gas Recirculation Circuit Malfunction
P0404 Exhaust Gas Recirculation Circuit Range/Performance
P0405 Exhaust Gas Recirculation Sensor "A" Circuit Low
P0406 Exhaust Gas Recirculation Sensor "A" Circuit High
P0407 Exhaust Gas Recirculation Sensor "B" Circuit Low
P0408 Exhaust Gas Recirculation Sensor "B" Circuit High
P0410 Secondary Air Injection System Malfunction
P0411 Secondary Air Injection System Incorrect Flow Detected
P0412 Secondary Air Injection System Switching Valve "A" Circuit Malfunction
P0413 Secondary Air Injection System Switching Valve "A" Circuit Open

P0414 Secondary Air Injection System Switching Valve "A" Circuit Shorted

P0415 Secondary Air Injection System Switching Valve "B" Circuit Malfunction

P0416 Secondary Air Injection System Switching Valve "B" Circuit Open

P0417 Secondary Air Injection System Switching Valve "B" Circuit Shorted

P0418 Secondary Air Injection System Relay "A" Circuit Malfunction

P0419 Secondary Air Injection System Relay "B" Circuit Malfunction

P0420 Catalyst System Efficiency Below Threshold (Bank no. 1)

P0421 Warm Up Catalyst Efficiency Below Threshold (Bank no. 1)

P0422 Main Catalyst Efficiency Below Threshold (Bank no. 1)

P0423 Heated Catalyst Efficiency Below Threshold (Bank no. 1)

P0424 Heated Catalyst Temperature Below Threshold (Bank no. 1)

P0430 Catalyst System Efficiency Below Threshold (Bank no. 2)

P0431 Warm Up Catalyst Efficiency Below Threshold (Bank no. 2)

P0432 Main Catalyst Efficiency Below Threshold (Bank no. 2)

P0433 Heated Catalyst Efficiency Below Threshold (Bank no. 2)

P0434 Heated Catalyst Temperature Below Threshold (Bank no. 2)

P0440 Evaporative Emission Control System Malfunction

P0441 Evaporative Emission Control System Incorrect Purge Flow

P0442 Evaporative Emission Control System Leak Detected (Small Leak)

P0443 Evaporative Emission Control System Purge Control Valve Circuit Malfunction

P0444 Evaporative Emission Control System Purge Control Valve Circuit Open

P0445 Evaporative Emission Control System Purge Control Valve Circuit Shorted

P0446 Evaporative Emission Control System Vent Control Circuit Malfunction

P0447 Evaporative Emission Control System Vent Control Circuit Open

P0448 Evaporative Emission Control System Vent Control Circuit Shorted

P0449 Evaporative Emission Control System Vent Valve/Solenoid Circuit Malfunction

P0450 Evaporative Emission Control System Pressure Sensor Malfunction

P0451 Evaporative Emission Control System Pressure Sensor Range/Performance

P0452 Evaporative Emission Control System Pressure Sensor Low Input

P0453 Evaporative Emission Control System Pressure Sensor High Input

P0454 Evaporative Emission Control System Pressure Sensor Intermittent

P0455 Evaporative Emission Control System Leak Detected (Gross Leak)

P0460 Fuel Level Sensor Circuit Malfunction

P0461 Fuel Level Sensor Circuit Range/Performance

P0462 Fuel Level Sensor Circuit Low Input

P0463 Fuel Level Sensor Circuit High Input

P0464 Fuel Level Sensor Circuit Intermittent

P0465 Purge Flow Sensor Circuit Malfunction

P0466 Purge Flow Sensor Circuit Range/Performance

P0467 Purge Flow Sensor Circuit Low Input

P0468 Purge Flow Sensor Circuit High Input

P0469 Purge Flow Sensor Circuit Intermittent

P0470 Exhaust Pressure Sensor Malfunction

P0471 Exhaust Pressure Sensor Range/Performance

P0472 Exhaust Pressure Sensor Low

P0473 Exhaust Pressure Sensor High

P0474 Exhaust Pressure Sensor Intermittent

P0475 Exhaust Pressure Control Valve Malfunction

P0476 Exhaust Pressure Control Valve Range/Performance

P0477 Exhaust Pressure Control Valve Low

P0478 Exhaust Pressure Control Valve High

P0479 Exhaust Pressure Control Valve Intermittent

P0480 Cooling Fan no. 1 Control Circuit Malfunction

P0481 Cooling Fan no. 2 Control Circuit Malfunction

P0482 Cooling Fan no. 3 Control Circuit Malfunction

P0483 Cooling Fan Rationality Check Malfunction

P0484 Cooling Fan Circuit Over Current

P0485 Cooling Fan Power/Ground Circuit Malfunction

P0500 Vehicle Speed Sensor Malfunction

P0501 Vehicle Speed Sensor Range/Performance

P0502 Vehicle Speed Sensor Circuit Low Input

P0503 Vehicle Speed Sensor Intermittent/Erratic/High

P0505 Idle Control System Malfunction

P0506 Idle Control System RPM Lower Than Expected

P0507 Idle Control System RPM Higher Than Expected

P0510 Closed Throttle Position Switch Malfunction

P0520 Engine Oil Pressure Sensor/Switch Circuit Malfunction

P0521 Engine Oil Pressure Sensor/Switch Range/Performance

P0522 Engine Oil Pressure Sensor/Switch Low Voltage

P0523 Engine Oil Pressure Sensor/Switch High Voltage

P0530 A/C Refrigerant Pressure Sensor Circuit Malfunction

P0531 A/C Refrigerant Pressure Sensor Circuit Range/Performance

P0532 A/C Refrigerant Pressure Sensor Circuit Low Input

P0533 A/C Refrigerant Pressure Sensor Circuit High Input

P0534 A/C Refrigerant Charge Loss

P0550 Power Steering Pressure Sensor Circuit Malfunction

P0551 Power Steering Pressure Sensor Circuit Range/Performance

P0552 Power Steering Pressure Sensor Circuit Low Input

P0553 Power Steering Pressure Sensor Circuit High Input

P0554 Power Steering Pressure Sensor Circuit Intermittent

P0560 System Voltage Malfunction

P0561 System Voltage Unstable

P0562 System Voltage Low

P0563 System Voltage High

P0565 Cruise Control On Signal Malfunction

P0566 Cruise Control Off Signal Malfunction

P0567 Cruise Control Resume Signal Malfunction

P0568 Cruise Control Set Signal Malfunction

P0569 Cruise Control Coast Signal Malfunction

P0570 Cruise Control Accel Signal Malfunction

P0571 Cruise Control/Brake Switch "A" Circuit Malfunction

P0572 Cruise Control/Brake Switch "A" Circuit Low

P0573 Cruise Control/Brake Switch "A" Circuit High

P0574 Through P0580 Reserved for Cruise Codes

P0600 Serial Communication Link Malfunction

P0601 Internal Control Module Memory Check Sum Error

P0602 Control Module Programming Error

P0603 Internal Control Module Keep Alive Memory (KAM) Error

P0604 Internal Control Module Random Access Memory (RAM) Error

P0605 Internal Control Module Read Only Memory (ROM) Error

P0606 PCM Processor Fault

P0608 Control Module VSS Output "A" Malfunction

P0609 Control Module VSS Output "B" Malfunction

P0620 Generator Control Circuit Malfunction

P0621 Generator Lamp "L" Control Circuit Malfunction

P0622 Generator Field "F" Control Circuit Malfunction

P0650 Malfunction Indicator Lamp (MIL) Control Circuit Malfunction

P0654 Engine RPM Output Circuit Malfunction

P0655 Engine Hot Lamp Output Control Circuit Malfunction

P0656 Fuel Level Output Circuit Malfunction

P0700 Transmission Control System Malfunction

P0701 Transmission Control System Range/Performance

P0702 Transmission Control System Electrical

P0703 Torque Converter/Brake Switch "B" Circuit Malfunction

P0704 Clutch Switch Input Circuit Malfunction

P0705 Transmission Range Sensor Circuit Malfunction (PRNDL Input)

P0706 Transmission Range Sensor Circuit Range/Performance

P0707 Transmission Range Sensor Circuit Low Input

P0708 Transmission Range Sensor Circuit High Input

P0709 Transmission Range Sensor Circuit Intermittent

P0710 Transmission Fluid Temperature Sensor Circuit Malfunction

P0711 Transmission Fluid Temperature Sensor Circuit Range/Performance

P0712 Transmission Fluid Temperature Sensor Circuit Low Input

P0713 Transmission Fluid Temperature Sensor Circuit High Input

P0714 Transmission Fluid Temperature Sensor Circuit Intermittent

P0715 Input/Turbine Speed Sensor Circuit Malfunction

P0716 Input/Turbine Speed Sensor Circuit Range/Performance

P0717 Input/Turbine Speed Sensor Circuit No Signal

P0718 Input/Turbine Speed Sensor Circuit Intermittent

P0719 Torque Converter/Brake Switch "B" Circuit Low

P0720 Output Speed Sensor Circuit Malfunction

P0721 Output Speed Sensor Circuit Range/Performance

P0722 Output Speed Sensor Circuit No Signal

P0723 Output Speed Sensor Circuit Intermittent

P0724 Torque Converter/Brake Switch "B" Circuit High

P0725 Engine Speed Input Circuit Malfunction

P0726 Engine Speed Input Circuit Range/Performance

P0727 Engine Speed Input Circuit No Signal

P0728 Engine Speed Input Circuit Intermittent

P0730 Incorrect Gear Ratio

P0731 Gear no. 1 Incorrect Ratio

P0732 Gear no. 2 Incorrect Ratio

P0733 Gear no. 3 Incorrect Ratio

P0734 Gear no. 4 Incorrect Ratio

P0735 Gear no. 5 Incorrect Ratio

P0736 Reverse Incorrect Ratio

P0740 Torque Converter Clutch Circuit Malfunction

P0741 Torque Converter Clutch Circuit Performance or Stuck Off

P0742 Torque Converter Clutch Circuit Stuck On

P0743 Torque Converter Clutch Circuit Electrical

P0744 Torque Converter Clutch Circuit Intermittent

P0745 Pressure Control Solenoid Malfunction

P0746 Pressure Control Solenoid Performance or Stuck Off

P0747 Pressure Control Solenoid Stuck On

P0748 Pressure Control Solenoid Electrical

P0749 Pressure Control Solenoid Intermittent

P0750 Shift Solenoid "A" Malfunction

P0751 Shift Solenoid "A" Performance or Stuck Off

P0752 Shift Solenoid "A" Stuck On

P0753 Shift Solenoid "A" Electrical

P0754 Shift Solenoid "A" Intermittent

P0755 Shift Solenoid "B" Malfunction

P0756 Shift Solenoid "B" Performance or Stuck Off

P0757 Shift Solenoid "B" Stuck On

P0758 Shift Solenoid "B" Electrical

P0759 Shift Solenoid "B" Intermittent

P0760 Shift Solenoid "C" Malfunction

P0761 Shift Solenoid "C" Performance Or Stuck Off

P0762 Shift Solenoid "C" Stuck On

P0763 Shift Solenoid "C" Electrical

P0764 Shift Solenoid "C" Intermittent

P0765 Shift Solenoid "D" Malfunction

P0766 Shift Solenoid "D" Performance Or Stuck Off

P0767 Shift Solenoid "D" Stuck On

P0768 Shift Solenoid "D" Electrical

P0769 Shift Solenoid "D" Intermittent

P0770 Shift Solenoid "E" Malfunction

P0771 Shift Solenoid "E" Performance Or Stuck Off

P0772 Shift Solenoid "E" Stuck On

P0773 Shift Solenoid "E" Electrical

P0774 Shift Solenoid "E" Intermittent

P0780 Shift Malfunction

P0781 1–2 Shift Malfunction

P0782 2–3 Shift Malfunction

P0783 3–4 Shift Malfunction

P0784 4–5 Shift Malfunction

P0785 Shift/Timing Solenoid Malfunction

P0786 Shift/Timing Solenoid Range/Performance

P0787 Shift/Timing Solenoid Low

P0788 Shift/Timing Solenoid High

P0789 Shift/Timing Solenoid Intermittent

P0790 Normal/Performance Switch Circuit Malfunction

P0801 Reverse Inhibit Control Circuit Malfunction

P0803 1–4 Upshift (Skip Shift) Solenoid Control Circuit Malfunction

P0804 1–4 Upshift (Skip Shift) Lamp Control Circuit Malfunction

P1106 MAP Sensor Voltage Intermittently High (Except 2.2L)

P1107 MAP Sensor Voltage Intermittently Low (Except 2.2L)

P1111 IAT Sensor Circuit Intermittent High Voltage (Except 2.2L)

P1112 IAT Sensor Circuit Intermittent Low Voltage (Except 2.2L)

P1114 ECT Sensor Circuit Intermittent Low Voltage (Except 2.2L)

P1115 ECT Sensor Circuit Intermittent High Voltage (Except 2.2L)

P1121 TP Sensor Voltage Intermittently High (Except 2.2L)

P1122 TP Sensor Voltage Intermittently Low (Except 2.2L)

P1133 HO2S Insufficient Switching Sensor (3.4L)

P1133 HO2S Insufficient Switching Bank #1, Sensor #1 (Except 3.4L & 4.3L)

P1134 HO_2 #1 Transition Time Ratio (3.4L)
P1134 HO_2 Transition Time Ratio Bank #1, Sensor #1 (4.3, 5.0L, 5.7L & 7.4L)
P1153 HO_2 Insufficient Switching Sensor Bank #2, Sensor #1 (4.3L, 5.0L, 5.7L & 7.4L)
P1154 HO_2 Transition Time Ratio Bank #2, Sensor #1 (4.3L, 5.0L, 5.7L & 7.4L)
P1345 Crankshaft/Camshaft (CKP/CMP) Correlation (4.3L, 5.0L, 5.7L & 7.4L)
P1350 Ignition Control (IC) Circuit Malfunction (3.4L)
P1351 Ignition Control (IC) Circuit High Voltage (4.3L, 5.0L, 5.7L & 7.4L)
P1361 Ignition Control (IC) Circuit Not Toggling (3.4L)
P1361 Ignition Control (IC) Circuit Low Voltage (4.3L, 5.0L, 5.7L & 7.4L)
P1380 Electronic Brake Control Module (EBCM) DTC Detected Rough Road Data Unusable
P1381 Misfire Detected, No EBCM/PCM/VCM Serial Data (Except "P" Series)
P1406 EGR Pintle Position Circuit Fault (Except "P" Series)
P1415 AIR System Bank #1 (Except "P" Series)
P1416 AIR System Bank #2 (Except "P" Series)
P1441 EVAP Control System Flow During Non-Purge
P1442 EVAP Vacuum Switch Circuit (3.4L)
P1508 IAC System Low RPM (4.3L, 5.0L, 5.7L & 7.4L)
P1509 IAC System High RPM (4.3L, 5.7L & 7.4L)
P1520 PNP Circuit (2.2L)
P1635 5-Volt Reference "A" Circuit (3.4L)
P1639 5-Volt Reference "B" Circuit (3.4L)
P1641 MIL Control Circuit (3.4L)
P1651 Fan #1 Relay Control Circuit (3.4L)
P1652 Fan #2 Relay Control Circuit (3.4L)
P1654 A/C Relay Control (3.4L)
P1655 EVAP Purge Solenoid Control Circuit (3.4L)
P1672 Low Engine Oil Level Light Control Circuit (3.4L)

Geo Models

P0000 No Failures
P0100 Mass or Volume Air Flow Circuit Malfunction
P0101 Mass or Volume Air Flow Circuit Range/Performance Problem
P0102 Mass or Volume Air Flow Circuit Low Input
P0103 Mass or Volume Air Flow Circuit High Input
P0104 Mass or Volume Air Flow Circuit Intermittent
P0105 Manifold Absolute Pressure/Barometric Pressure Circuit Malfunction
P0106 Manifold Absolute Pressure/Barometric Pressure Circuit Range/Performance Problem
P0107 Manifold Absolute Pressure/Barometric Pressure Circuit Low Input
P0108 Manifold Absolute Pressure/Barometric Pressure Circuit High Input
P0109 Manifold Absolute Pressure/Barometric Pressure Circuit Intermittent
P0110 Intake Air Temperature Circuit Malfunction
P0111 Intake Air Temperature Circuit Range/Performance Problem
P0112 Intake Air Temperature Circuit Low Input
P0113 Intake Air Temperature Circuit High Input
P0114 Intake Air Temperature Circuit Intermittent
P0115 Engine Coolant Temperature Circuit Malfunction

P0116 Engine Coolant Temperature Circuit Range/Performance Problem
P0117 Engine Coolant Temperature Circuit Low Input
P0118 Engine Coolant Temperature Circuit High Input
P0119 Engine Coolant Temperature Circuit Intermittent
P0120 Throttle/Pedal Position Sensor/Switch "A" Circuit Malfunction
P0121 Throttle/Pedal Position Sensor/Switch "A" Circuit Range/Performance Problem
P0122 Throttle/Pedal Position Sensor/Switch "A" Circuit Low Input
P0123 Throttle/Pedal Position Sensor/Switch "A" Circuit High Input
P0124 Throttle/Pedal Position Sensor/Switch "A" Circuit Intermittent
P0125 Insufficient Coolant Temperature For Closed Loop Fuel Control
P0126 Insufficient Coolant Temperature For Stable Operation
P0130 O_2 Circuit Malfunction (Bank no. 1 Sensor no. 1)
P0131 O_2 Sensor Circuit Low Voltage (Bank no. 1 Sensor no. 1)
P0132 O_2 Sensor Circuit High Voltage (Bank no. 1 Sensor no. 1)
P0133 O_2 Sensor Circuit Slow Response (Bank no. 1 Sensor no. 1)
P0134 O_2 Sensor Circuit No Activity Detected (Bank no. 1 Sensor no. 1)
P0135 O_2 Sensor Heater Circuit Malfunction (Bank no. 1 Sensor no. 1)
P0136 O_2 Sensor Circuit Malfunction (Bank no. 1 Sensor no. 2)
P0137 O_2 Sensor Circuit Low Voltage (Bank no. 1 Sensor no. 2)
P0138 O_2 Sensor Circuit High Voltage (Bank no. 1 Sensor no. 2)
P0139 O_2 Sensor Circuit Slow Response (Bank no. 1 Sensor no. 2)
P0140 O_2 Sensor Circuit No Activity Detected (Bank no. 1 Sensor no. 2)
P0141 O_2 Sensor Heater Circuit Malfunction (Bank no. 1 Sensor no. 2)
P0142 O_2 Sensor Circuit Malfunction (Bank no. 1 Sensor no. 3)
P0143 O_2 Sensor Circuit Low Voltage (Bank no. 1 Sensor no. 3)
P0144 O_2 Sensor Circuit High Voltage (Bank no. 1 Sensor no. 3)
P0145 O_2 Sensor Circuit Slow Response (Bank no. 1 Sensor no. 3)
P0146 O_2 Sensor Circuit No Activity Detected (Bank no. 1 Sensor no. 3)
P0147 O_2 Sensor Heater Circuit Malfunction (Bank no. 1 Sensor no. 3)
P0150 O_2 Sensor Circuit Malfunction (Bank no. 2 Sensor no. 1)
P0151 O_2 Sensor Circuit Low Voltage (Bank no. 2 Sensor no. 1)
P0152 O_2 Sensor Circuit High Voltage (Bank no. 2 Sensor no. 1)
P0153 O_2 Sensor Circuit Slow Response (Bank no. 2 Sensor no. 1)
P0154 O_2 Sensor Circuit No Activity Detected (Bank no. 2 Sensor no. 1)
P0155 O_2 Sensor Heater Circuit Malfunction (Bank no. 2 Sensor no. 1)

P0156 O_2 Sensor Circuit Malfunction (Bank no. 2 Sensor no. 2)
P0157 O_2 Sensor Circuit Low Voltage (Bank no. 2 Sensor no. 2)
P0158 O_2 Sensor Circuit High Voltage (Bank no. 2 Sensor no. 2)
P0159 O_2 Sensor Circuit Slow Response (Bank no. 2 Sensor no. 2)
P0160 O_2 Sensor Circuit No Activity Detected (Bank no. 2 Sensor no. 2)
P0161 O_2 Sensor Heater Circuit Malfunction (Bank no. 2 Sensor no. 2)
P0162 O_2 Sensor Circuit Malfunction (Bank no. 2 Sensor no. 3)
P0163 O_2 Sensor Circuit Low Voltage (Bank no. 2 Sensor no. 3)
P0164 O_2 Sensor Circuit High Voltage (Bank no. 2 Sensor no. 3)
P0165 O_2 Sensor Circuit Slow Response (Bank no. 2 Sensor no. 3)
P0166 O_2 Sensor Circuit No Activity Detected (Bank no. 2 Sensor no. 3)
P0167 O_2 Sensor Heater Circuit Malfunction (Bank no. 2 Sensor no. 3)
P0170 Fuel Trim Malfunction (Bank no. 1)
P0171 System Too Lean (Bank no. 1)
P0172 System Too Rich (Bank no. 1)
P0173 Fuel Trim Malfunction (Bank no. 2)
P0174 System Too Lean (Bank no. 2)
P0175 System Too Rich (Bank no. 2)
P0176 Fuel Composition Sensor Circuit Malfunction
P0177 Fuel Composition Sensor Circuit Range/Performance
P0178 Fuel Composition Sensor Circuit Low Input
P0179 Fuel Composition Sensor Circuit High Input
P0180 Fuel Temperature Sensor "A" Circuit Malfunction
P0181 Fuel Temperature Sensor "A" Circuit Range/Performance
P0182 Fuel Temperature Sensor "A" Circuit Low Input
P0183 Fuel Temperature Sensor "A" Circuit High Input
P0184 Fuel Temperature Sensor "A" Circuit Intermittent
P0185 Fuel Temperature Sensor "B" Circuit Malfunction
P0186 Fuel Temperature Sensor "B" Circuit Range/Performance
P0187 Fuel Temperature Sensor "B" Circuit Low Input
P0188 Fuel Temperature Sensor "B" Circuit High Input
P0189 Fuel Temperature Sensor "B" Circuit Intermittent
P0190 Fuel Rail Pressure Sensor Circuit Malfunction
P0191 Fuel Rail Pressure Sensor Circuit Range/Performance
P0192 Fuel Rail Pressure Sensor Circuit Low Input
P0193 Fuel Rail Pressure Sensor Circuit High Input
P0194 Fuel Rail Pressure Sensor Circuit Intermittent
P0195 Engine Oil Temperature Sensor Malfunction
P0196 Engine Oil Temperature Sensor Range/Performance
P0197 Engine Oil Temperature Sensor Low
P0198 Engine Oil Temperature Sensor High
P0199 Engine Oil Temperature Sensor Intermittent
P0200 Injector Circuit Malfunction
P0201 Injector Circuit Malfunction—Cylinder no. 1
P0202 Injector Circuit Malfunction—Cylinder no. 2
P0203 Injector Circuit Malfunction—Cylinder no. 3
P0204 Injector Circuit Malfunction—Cylinder no. 4
P0205 Injector Circuit Malfunction—Cylinder no. 5
P0206 Injector Circuit Malfunction—Cylinder no. 6
P0207 Injector Circuit Malfunction—Cylinder no. 7
P0208 Injector Obcd Malfunction—Cylinder no. 8
P0209 Injector Circuit Malfunction—Cylinder no. 9

P0210 Injector Circuit Malfunction—Cylinder no. 10
P0211 Injector Circuit Malfunction—Cylinder no. 11
P0212 Injector Circuit Malfunction—Cylinder no. 12
P0213 Cold Start Injector no. 1 Malfunction
P0214 Cold Start Injector no. 2 Malfunction
P0215 Engine Shutoff Solenoid Malfunction
P0216 Injection Timing Control Circuit Malfunction
P0217 Engine Over Temperature Condition
P0218 Transmission Over Temperature Condition
P0219 Engine Over Speed Condition
P0220 Throttle/Pedal Position Sensor/Switch "B" Circuit Malfunction
P0221 Throttle/Pedal Position Sensor/Switch "B" Circuit Range/Performance Problem
P0222 Throttle/Pedal Position Sensor/Switch "B" Circuit Low Input
P0223 Throttle/Pedal Position Sensor/Switch "B" Circuit High Input
P0224 Throttle/Pedal Position Sensor/Switch "B" Circuit Intermittent
P0225 Throttle/Pedal Position Sensor/Switch "C" Circuit Malfunction
P0226 Throttle/Pedal Position Sensor/Switch "C" Circuit Range/Performance Problem
P0227 Throttle/Pedal Position Sensor/Switch "C" Circuit Low Input
P0228 Throttle/Pedal Position Sensor/Switch "C" Circuit High Input
P0229 Throttle/Pedal Position Sensor/Switch "C" Circuit Intermittent
P0230 Fuel Pump Primary Circuit Malfunction
P0231 Fuel Pump Secondary Circuit Low
P0232 Fuel Pump Secondary Circuit High
P0233 Fuel Pump Secondary Circuit Intermittent
P0234 Engine Over Boost Condition
P0261 Cylinder no. 1 Injector Circuit Low
P0262 Cylinder no. 1 Injector Circuit High
P0263 Cylinder no. 1 Contribution/Balance Fault
P0264 Cylinder no. 2 Injector Circuit Low
P0265 Cylinder no. 2 Injector Circuit High
P0266 Cylinder no. 2 Contribution/Balance Fault
P0267 Cylinder no. 3 Injector Circuit Low
P0268 Cylinder no. 3 Injector Circuit High
P0269 Cylinder no. 3 Contribution/Balance Fault
P0270 Cylinder no. 4 Injector Circuit Low
P0271 Cylinder no. 4 Injector Circuit High
P0272 Cylinder no. 4 Contribution/Balance Fault
P0273 Cylinder no. 5 Injector Circuit Low
P0274 Cylinder no. 5 Injector Circuit High
P0275 Cylinder no. 5 Contribution/Balance Fault
P0276 Cylinder no. 6 Injector Circuit Low
P0277 Cylinder no. 6 Injector Circuit High
P0278 Cylinder no. 6 Contribution/Balance Fault
P0279 Cylinder no. 7 Injector Circuit Low
P0280 Cylinder no. 7 Injector Circuit High
P0281 Cylinder no. 7 Contribution/Balance Fault
P0282 Cylinder no. 8 Injector Circuit Low
P0283 Cylinder no. 8 Injector Circuit High
P0284 Cylinder no. 8 Contribution/Balance Fault
P0285 Cylinder no. 9 Injector Circuit Low
P0286 Cylinder no. 9 Injector Circuit High
P0287 Cylinder no. 9 Contribution/Balance Fault

P0288 Cylinder no. 10 Injector Circuit Low
P0289 Cylinder no. 10 Injector Circuit High
P0290 Cylinder no. 10 Contribution/Balance Fault
P0291 Cylinder no. 11 Injector Circuit Low
P0292 Cylinder no. 11 Injector Circuit High
P0293 Cylinder no. 11 Contribution/Balance Fault
P0294 Cylinder no. 12 Injector Circuit Low
P0295 Cylinder no. 12 Injector Circuit High
P0296 Cylinder no. 12 Contribution/Balance Fault
P0300 Random/Multiple Cylinder Misfire Detected
P0301 Cylinder no. 1—Misfire Detected
P0302 Cylinder no. 2—Misfire Detected
P0303 Cylinder no. 3—Misfire Detected
P0304 Cylinder no. 4—Misfire Detected
P0305 Cylinder no. 5—Misfire Detected
P0306 Cylinder no. 6—Misfire Detected
P0307 Cylinder no. 7—Misfire Detected
P0308 Cylinder no. 8—Misfire Detected
P0309 Cylinder no. 9—Misfire Detected
P0310 Cylinder no. 10—Misfire Detected
P0311 Cylinder no. 11—Misfire Detected
P0312 Cylinder no. 12—Misfire Detected
P0320 Ignition/Distributor Engine Speed Input Circuit Malfunction
P0321 Ignition/Distributor Engine Speed Input Circuit Range/Performance
P0322 Ignition/Distributor Engine Speed Input Circuit No Signal
P0323 Ignition/Distributor Engine Speed Input Circuit Intermittent
P0325 Knock Sensor no. 1—Circuit Malfunction (Bank no. 1 or Single Sensor)
P0326 Knock Sensor no. 1—Circuit Range/Performance (Bank no. 1 or Single Sensor)
P0327 Knock Sensor no. 1—Circuit Low Input (Bank no. 1 or Single Sensor)
P0328 Knock Sensor no. 1—Circuit High Input (Bank no. 1 or Single Sensor)
P0329 Knock Sensor no. 1—Circuit Input Intermittent (Bank no. 1 or Single Sensor)
P0330 Knock Sensor no. 2—Circuit Malfunction (Bank no. 2)
P0331 Knock Sensor no. 2—Circuit Range/Performance (Bank no. 2)
P0332 Knock Sensor no. 2—Circuit Low Input (Bank no. 2)
P0333 Knock Sensor no. 2—Circuit High Input (Bank no. 2)
P0334 Knock Sensor no. 2—Circuit Input Intermittent (Bank no. 2)
P0335 Crankshaft Position Sensor "A" Circuit Malfunction
P0336 Crankshaft Position Sensor "A" Circuit Range/Performance
P0337 Crankshaft Position Sensor "A" Circuit Low Input
P0338 Crankshaft Position Sensor "A" Circuit High Input
P0339 Crankshaft Position Sensor "A" Circuit Intermittent
P0340 Camshaft Position Sensor Circuit Malfunction
P0341 Camshaft Position Sensor Circuit Range/Performance
P0342 Camshaft Position Sensor Circuit Low Input
P0343 Camshaft Position Sensor Circuit High Input
P0344 Camshaft Position Sensor Circuit Intermittent
P0350 Ignition Coil Primary/Secondary Circuit Malfunction
P0351 Ignition Coil "A" Primary/Secondary Circuit Malfunction
P0352 Ignition Coil "B" Primary/Secondary Circuit Malfunction
P0353 Ignition Coil "C" Primary/Secondary Circuit Malfunction
P0354 Ignition Coil "D" Primary/Secondary Circuit Malfunction

P0355 Ignition Coil "E" Primary/Secondary Circuit Malfunction
P0356 Ignition Coil "F" Primary/Secondary Circuit Malfunction
P0357 Ignition Coil "G" Primary/Secondary Circuit Malfunction
P0358 Ignition Coil "H" Primary/Secondary Circuit Malfunction
P0359 Ignition Coil "I" Primary/Secondary Circuit Malfunction
P0360 Ignition Coil "J" Primary/Secondary Circuit Malfunction
P0361 Ignition Coil "K" Primary/Secondary Circuit Malfunction
P0362 Ignition Coil "L" Primary/Secondary Circuit Malfunction
P0370 Timing Reference High Resolution Signal "A" Malfunction
P0371 Timing Reference High Resolution Signal "A" Too Many Pulses
P0372 Timing Reference High Resolution Signal "A" Too Few Pulses
P0373 Timing Reference High Resolution Signal "A" Intermittent/Erratic Pulses
P0374 Timing Reference High Resolution Signal "A" No Pulses
P0375 Timing Reference High Resolution Signal "B" Malfunction
P0376 Timing Reference High Resolution Signal "B" Too Many Pulses
P0377 Timing Reference High Resolution Signal "B" Too Few Pulses
P0378 Timing Reference High Resolution Signal "B" Intermittent/Erratic Pulses
P0379 Timing Reference High Resolution Signal "B" No Pulses
P0380 Glow Plug/Heater Circuit "A" Malfunction
P0381 Glow Plug/Heater Indicator Circuit Malfunction
P0382 Glow Plug/Heater Circuit "B" Malfunction
P0385 Crankshaft Position Sensor "B" Circuit Malfunction
P0386 Crankshaft Position Sensor "B" Circuit Range/Performance
P0387 Crankshaft Position Sensor "B" Circuit Low Input
P0388 Crankshaft Position Sensor "B" Circuit High Input
P0389 Crankshaft Position Sensor "B" Circuit Intermittent
P0400 Exhaust Gas Recirculation Flow Malfunction
P0401 Exhaust Gas Recirculation Flow Insufficient Detected
P0402 Exhaust Gas Recirculation Flow Excessive Detected
P0403 Exhaust Gas Recirculation Circuit Malfunction
P0404 Exhaust Gas Recirculation Circuit Range/Performance
P0405 Exhaust Gas Recirculation Sensor "A" Circuit Low
P0406 Exhaust Gas Recirculation Sensor "A" Circuit High
P0407 Exhaust Gas Recirculation Sensor "B" Circuit Low
P0408 Exhaust Gas Recirculation Sensor "B" Circuit High
P0410 Secondary Air Injection System Malfunction
P0411 Secondary Air Injection System Incorrect Flow Detected
P0412 Secondary Air Injection System Switching Valve "A" Circuit Malfunction
P0413 Secondary Air Injection System Switching Valve "A" Circuit Open
P0414 Secondary Air Injection System Switching Valve "A" Circuit Shorted
P0415 Secondary Air Injection System Switching Valve "B" Circuit Malfunction
P0416 Secondary Air Injection System Switching Valve "B" Circuit Open
P0417 Secondary Air Injection System Switching Valve "B" Circuit Shorted
P0418 Secondary Air Injection System Relay "A" Circuit Malfunction
P0419 Secondary Air Injection System Relay "B" Circuit Malfunction

P0420 Catalyst System Efficiency Below Threshold (Bank no. 1)

P0421 Warm Up Catalyst Efficiency Below Threshold (Bank no. 1)

P0422 Main Catalyst Efficiency Below Threshold (Bank no. 1)

P0423 Heated Catalyst Efficiency Below Threshold (Bank no. 1)

P0424 Heated Catalyst Temperature Below Threshold (Bank no. 1)

P0430 Catalyst System Efficiency Below Threshold (Bank no. 2)

P0431 Warm Up Catalyst Efficiency Below Threshold (Bank no. 2)

P0432 Main Catalyst Efficiency Below Threshold (Bank no. 2)

P0433 Heated Catalyst Efficiency Below Threshold (Bank no. 2)

P0434 Heated Catalyst Temperature Below Threshold (Bank no. 2)

P0440 Evaporative Emission Control System Malfunction

P0441 Evaporative Emission Control System Incorrect Purge Flow

P0442 Evaporative Emission Control System Leak Detected (Small Leak)

P0443 Evaporative Emission Control System Purge Control Valve Circuit Malfunction

P0444 Evaporative Emission Control System Purge Control Valve Circuit Open

P0445 Evaporative Emission Control System Purge Control Valve Circuit Shorted

P0446 Evaporative Emission Control System Vent Control Circuit Malfunction

P0447 Evaporative Emission Control System Vent Control Circuit Open

P0448 Evaporative Emission Control System Vent Control Circuit Shorted

P0449 Evaporative Emission Control System Vent Valve/Solenoid Circuit Malfunction

P0450 Evaporative Emission Control System Pressure Sensor Malfunction

P0451 Evaporative Emission Control System Pressure Sensor Range/Performance

P0452 Evaporative Emission Control System Pressure Sensor Low Input

P0453 Evaporative Emission Control System Pressure Sensor High Input

P0454 Evaporative Emission Control System Pressure Sensor Intermittent

P0455 Evaporative Emission Control System Leak Detected (Gross Leak)

P0460 Fuel Level Sensor Circuit Malfunction

P0461 Fuel Level Sensor Circuit Range/Performance

P0462 Fuel Level Sensor Circuit Low Input

P0463 Fuel Level Sensor Circuit High Input

P0464 Fuel Level Sensor Circuit Intermittent

P0465 Purge Flow Sensor Circuit Malfunction

P0466 Purge Flow Sensor Circuit Range/Performance

P0467 Purge Flow Sensor Circuit Low Input

P0468 Purge Flow Sensor Circuit High Input

P0469 Purge Flow Sensor Circuit Intermittent

P0470 Exhaust Pressure Sensor Malfunction

P0471 Exhaust Pressure Sensor Range/Performance

P0472 Exhaust Pressure Sensor Low

P0473 Exhaust Pressure Sensor High

P0474 Exhaust Pressure Sensor Intermittent

P0475 Exhaust Pressure Control Valve Malfunction

P0476 Exhaust Pressure Control Valve Range/Performance

P0477 Exhaust Pressure Control Valve Low

P0478 Exhaust Pressure Control Valve High

P0479 Exhaust Pressure Control Valve Intermittent

P0480 Cooling Fan no. 1 Control Circuit Malfunction

P0481 Cooling Fan no. 2 Control Circuit Malfunction

P0482 Cooling Fan no. 3 Control Circuit Malfunction

P0483 Cooling Fan Rationality Check Malfunction

P0484 Cooling Fan Circuit Over Current

P0485 Cooling Fan Power/Ground Circuit Malfunction

P0500 Vehicle Speed Sensor Malfunction

P0501 Vehicle Speed Sensor Range/Performance

P0502 Vehicle Speed Sensor Circuit Low Input

P0503 Vehicle Speed Sensor Intermittent/Erratic/High

P0505 Idle Control System Malfunction

P0506 Idle Control System RPM Lower Than Expected

P0507 Idle Control System RPM Higher Than Expected

P0510 Closed Throttle Position Switch Malfunction

P0520 Engine Oil Pressure Sensor/Switch Circuit Malfunction

P0521 Engine Oil Pressure Sensor/Switch Range/Performance

P0522 Engine Oil Pressure Sensor/Switch Low Voltage

P0523 Engine Oil Pressure Sensor/Switch High Voltage

P0530 A/C Refrigerant Pressure Sensor Circuit Malfunction

P0531 A/C Refrigerant Pressure Sensor Circuit Range/Performance

P0532 A/C Refrigerant Pressure Sensor Circuit Low Input

P0533 A/C Refrigerant Pressure Sensor Circuit High Input

P0534 A/C Refrigerant Charge Loss

P0550 Power Steering Pressure Sensor Circuit Malfunction

P0551 Power Steering Pressure Sensor Circuit Range/Performance

P0552 Power Steering Pressure Sensor Circuit Low Input

P0553 Power Steering Pressure Sensor Circuit High Input

P0554 Power Steering Pressure Sensor Circuit Intermittent

P0560 System Voltage Malfunction

P0561 System Voltage Unstable

P0562 System Voltage Low

P0563 System Voltage High

P0565 Cruise Control On Signal Malfunction

P0566 Cruise Control Off Signal Malfunction

P0567 Cruise Control Resume Signal Malfunction

P0568 Cruise Control Set Signal Malfunction

P0569 Cruise Control Coast Signal Malfunction

P0570 Cruise Control Accel Signal Malfunction

P0571 Cruise Control/Brake Switch "A" Circuit Malfunction

P0572 Cruise Control/Brake Switch "A" Circuit Low

P0573 Cruise Control/Brake Switch "A" Circuit High

P0574 Through P0580 Reserved for Cruise Codes

P0600 Serial Communication Link Malfunction

P0601 Internal Control Module Memory Check Sum Error

P0602 Control Module Programming Error

P0603 Internal Control Module Keep Alive Memory (KAM) Error

P0604 Internal Control Module Random Access Memory (RAM) Error

P0605 Internal Control Module Read Only Memory (ROM) Error

P0606 PCM Processor Fault

P0608 Control Module VSS Output "A" Malfunction

P0609 Control Module VSS Output "B" Malfunction

P0620 Generator Control Circuit Malfunction

P0621 Generator Lamp "L" Control Circuit Malfunction

P0622 Generator Field "F" Control Circuit Malfunction

P0650 Malfunction Indicator Lamp (MIL) Control Circuit Malfunction

P0654 Engine RPM Output Circuit Malfunction
P0655 Engine Hot Lamp Output Control Circuit Malfunction
P0656 Fuel Level Output Circuit Malfunction
P0700 Transmission Control System Malfunction
P0701 Transmission Control System Range/Performance
P0702 Transmission Control System Electrical
P0703 Torque Converter/Brake Switch "B" Circuit Malfunction
P0704 Clutch Switch Input Circuit Malfunction
P0705 Transmission Range Sensor Circuit Malfunction (PRNDL Input)
P0706 Transmission Range Sensor Circuit Range/Performance
P0707 Transmission Range Sensor Circuit Low Input
P0708 Transmission Range Sensor Circuit High Input
P0709 Transmission Range Sensor Circuit Intermittent
P0710 Transmission Fluid Temperature Sensor Circuit Malfunction
P0711 Transmission Fluid Temperature Sensor Circuit Range/Performance
P0712 Transmission Fluid Temperature Sensor Circuit Low Input
P0713 Transmission Fluid Temperature Sensor Circuit High Input
P0714 Transmission Fluid Temperature Sensor Circuit Intermittent
P0715 Input/Turbine Speed Sensor Circuit Malfunction
P0716 Input/Turbine Speed Sensor Circuit Range/Performance
P0717 Input/Turbine Speed Sensor Circuit No Signal
P0718 Input/Turbine Speed Sensor Circuit Intermittent
P0719 Torque Converter/Brake Switch "B" Circuit Low
P0720 Output Speed Sensor Circuit Malfunction
P0721 Output Speed Sensor Circuit Range/Performance
P0722 Output Speed Sensor Circuit No Signal
P0723 Output Speed Sensor Circuit Intermittent
P0724 Torque Converter/Brake Switch "B" Circuit High
P0725 Engine Speed Input Circuit Malfunction
P0726 Engine Speed Input Circuit Range/Performance
P0727 Engine Speed Input Circuit No Signal
P0728 Engine Speed Input Circuit Intermittent
P0730 Incorrect Gear Ratio
P0731 Gear no. 1 Incorrect Ratio
P0732 Gear no. 2 Incorrect Ratio
P0733 Gear no. 3 Incorrect Ratio
P0734 Gear no. 4 Incorrect Ratio
P0735 Gear no. 5 Incorrect Ratio
P0736 Reverse Incorrect Ratio
P0740 Torque Converter Clutch Circuit Malfunction
P0741 Torque Converter Clutch Circuit Performance or Stuck Off
P0742 Torque Converter Clutch Circuit Stuck On
P0743 Torque Converter Clutch Circuit Electrical
P0744 Torque Converter Clutch Circuit Intermittent
P0745 Pressure Control Solenoid Malfunction
P0746 Pressure Control Solenoid Performance or Stuck Off
P0747 Pressure Control Solenoid Stuck On
P0748 Pressure Control Solenoid Electrical
P0749 Pressure Control Solenoid Intermittent
P0750 Shift Solenoid "A" Malfunction
P0751 Shift Solenoid "A" Performance or Stuck Off
P0752 Shift Solenoid "A" Stuck On
P0753 Shift Solenoid "A" Electrical
P0754 Shift Solenoid "A" Intermittent
P0755 Shift Solenoid "B" Malfunction

P0756 Shift Solenoid "B" Performance or Stuck Off
P0757 Shift Solenoid "B" Stuck On
P0758 Shift Solenoid "B" Electrical
P0759 Shift Solenoid "B" Intermittent
P0760 Shift Solenoid "C" Malfunction
P0761 Shift Solenoid "C" Performance Or Stuck Off
P0762 Shift Solenoid "C" Stuck On
P0763 Shift Solenoid "C" Electrical
P0764 Shift Solenoid "C" Intermittent
P0765 Shift Solenoid "D" Malfunction
P0766 Shift Solenoid "D" Performance Or Stuck Off
P0767 Shift Solenoid "D" Stuck On
P0768 Shift Solenoid "D" Electrical
P0769 Shift Solenoid "D" Intermittent
P0770 Shift Solenoid "E" Malfunction
P0771 Shift Solenoid "E" Performance Or Stuck Off
P0772 Shift Solenoid "E" Stuck On
P0773 Shift Solenoid "E" Electrical
P0774 Shift Solenoid "E" Intermittent
P0780 Shift Malfunction
P0781 1–2 Shift Malfunction
P0782 2–3 Shift Malfunction
P0783 3–4 Shift Malfunction
P0784 4–5 Shift Malfunction
P0785 Shift/Timing Solenoid Malfunction
P0786 Shift/Timing Solenoid Range/Performance
P0787 Shift/Timing Solenoid Low
P0788 Shift/Timing Solenoid High
P0789 Shift/Timing Solenoid Intermittent
P0790 Normal/Performance Switch Circuit Malfunction
P0801 Reverse Inhibit Control Circuit Malfunction
P0803 1–4 Upshift (Skip Shift) Solenoid Control Circuit Malfunction
P0804 1–4 Upshift (Skip Shift) Lamp Control Circuit Malfunction
P1450 Barometric Pressure Sensor Circuit Fault
P1451 Barometric Pressure Sensor Performance
P1460 Cooling Fan Control System Fault
P1500 Starter Signal Circuit Fault
P1510 Back-up Power Supply Fault
P1530 Ignition Timing Adjustment Switch Circuit
P1600 PCM Battery Circuit Fault

Honda

➡**The Honda Passport is covered in the Isuzu section since it shares a platform with the Isuzu Rodeo.**

READING CODES

With Scan Tool

Reading the control module memory is one of the first steps in OBD II system diagnostics. This step should be initially performed to determine the general nature of the fault. Subsequent readings will determine if the fault has been cleared.

Reading codes can be performed by any of the methods below:

• Read the control module memory with the Generic Scan Tool (GST)

• Read the control module memory with the vehicle manufacturer's specific tester

To read the fault codes, connect the scan tool or tester according to the manufacturer's instructions. Follow the manufacturer's specified procedure for reading the codes.

Without Scan Tool

Honda also provides a way of reading OBD II trouble code equivalents using a service connector and viewing the MIL. This method is similar to the flash codes from non-OBD II vehicles.

To read codes, plug the service connector into the service check connector and turn the ignition on. The MIL will flash any stored trouble codes.

CLEARING CODES

Control module reset procedures are a very important part of OBD II System diagnostics.

This step should be done at the end of any fault code repair and at the end of any driveability repair.

Clearing codes can be performed by any of the methods below:
• Clear the control module memory with the Generic Scan Tool (GST)
• Clear the control module memory with the vehicle manufacturer's specific tester
• Turn the ignition off and remove the negative battery cable for at least 1 minute.

Removing the negative battery cable may cause other systems in the vehicle to lose their memory. Prior to removing the cable, ensure you have the proper reset codes for radios and alarms.

➡The MIL will may also be de-activated for some codes if the vehicle completes three consecutive trips without a fault detected with vehicle conditions similar to those present during the fault.

DIAGNOSTIC TROUBLE CODES

P0000 No Failures
P0100 Mass or Volume Air Flow Circuit Malfunction
P0101 Mass or Volume Air Flow Circuit Range/Performance Problem
P0102 Mass or Volume Air Flow Circuit Low Input
P0103 Mass or Volume Air Flow Circuit High Input
P0104 Mass or Volume Air Flow Circuit Intermittent
P0105 Manifold Absolute Pressure/Barometric Pressure Circuit Malfunction
P0106 Manifold Absolute Pressure/Barometric Pressure Circuit Range/Performance Problem
P0107 Manifold Absolute Pressure/Barometric Pressure Circuit Low Input
P0108 Manifold Absolute Pressure/Barometric Pressure Circuit High Input
P0109 Manifold Absolute Pressure/Barometric Pressure Circuit Intermittent
P0110 Intake Air Temperature Circuit Malfunction
P0111 Intake Air Temperature Circuit Range/Performance Problem
P0112 Intake Air Temperature Circuit Low Input
P0113 Intake Air Temperature Circuit High Input
P0114 Intake Air Temperature Circuit Intermittent
P0115 Engine Coolant Temperature Circuit Malfunction
P0116 Engine Coolant Temprature Circuit Range/Performance Problem

P0117 Engine Coolant Temperature Circuit Low Input
P0118 Engine Coolant Temperature Circuit High Input
P0119 Engine Coolant Temperature Circuit Intermittent
P0120 Throttle/Pedal Position Sensor/Switch "A" Circuit Malfunction
P0121 Throttle/Pedal Position Sensor/Switch "A" Circuit Range/Performance Problem
P0122 Throttle/Pedal Position Sensor/Switch "A" Circuit Low Input
P0123 Throttle/Pedal Position Sensor/Switch "A" Circuit High Input
P0124 Throttle/Pedal Position Sensor/Switch "A" Circuit Intermittent
P0125 Insufficient Coolant Temperature For Closed Loop Fuel Control
P0126 Insufficient Coolant Temperature For Stable Operation
P0130 O_2 Circuit Malfunction (Bank no. 1 Sensor no. 1)
P0131 O_2 Sensor Circuit Low Voltage (Bank no. 1 Sensor no. 1)
P0132 O_2 Sensor Circuit High Voltage (Bank no. 1 Sensor no. 1)
P0133 O_2 Sensor Circuit Slow Response (Bank no. 1 Sensor no. 1)
P0134 O_2 Sensor Circuit No Activity Detected (Bank no. 1 Sensor no. 1)
P0135 O_2 Sensor Heater Circuit Malfunction (Bank no. 1 Sensor no. 1)
P0136 O_2 Sensor Circuit Malfunction (Bank no. 1 Sensor no. 2)
P0137 O_2 Sensor Circuit Low Voltage (Bank no. 1 Sensor no. 2)
P0138 O_2 Sensor Circuit High Voltage (Bank no. 1 Sensor no. 2)
P0139 O_2 Sensor Circuit Slow Response (Bank no. 1 Sensor no. 2)
P0140 O_2 Sensor Circuit No Activity Detected (Bank no. 1 Sensor no. 2)
P0141 O_2 Sensor Heater Circuit Malfunction (Bank no. 1 Sensor no. 2)
P0142 O_2 Sensor Circuit Malfunction (Bank no. 1 Sensor no. 3)
P0143 O_2 Sensor Circuit Low Voltage (Bank no. 1 Sensor no. 3)
P0144 O_2 Sensor Circuit High Voltage (Bank no. 1 Sensor no. 3)
P0145 O_2 Sensor Circuit Slow Response (Bank no. 1 Sensor no. 3)
P0146 O_2 Sensor Circuit No Activity Detected (Bank no. 1 Sensor no. 3)
P0147 O_2 Sensor Heater Circuit Malfunction (Bank no. 1 Sensor no. 3)
P0150 O_2 Sensor Circuit Malfunction (Bank no. 2 Sensor no. 1)
P0151 O_2 Sensor Circuit Low Voltage (Bank no. 2 Sensor no. 1)
P0152 O_2 Sensor Circuit High Voltage (Bank no. 2 Sensor no. 1)
P0153 O_2 Sensor Circuit Slow Response (Bank no. 2 Sensor no. 1)
P0154 O_2 Sensor Circuit No Activity Detected (Bank no. 2 Sensor no. 1)
P0155 O_2 Sensor Heater Circuit Malfunction (Bank no. 2 Sensor no. 1)
P0156 O_2 Sensor Circuit Malfunction (Bank no. 2 Sensor no. 2)

P0157 O₂ Sensor Circuit Low Voltage (Bank no. 2 Sensor no. 2)

P0158 O₂ Sensor Circuit High Voltage (Bank no. 2 Sensor no. 2)

P0159 O₂ Sensor Circuit Slow Response (Bank no. 2 Sensor no. 2)

P0160 O₂ Sensor Circuit No Activity Detected (Bank no. 2 Sensor no. 2)

P0161 O₂ Sensor Heater Circuit Malfunction (Bank no. 2 Sensor no. 2)

P0162 O₂ Sensor Circuit Malfunction (Bank no. 2 Sensor no. 3)

P0163 O₂ Sensor Circuit Low Voltage (Bank no. 2 Sensor no. 3)

P0164 O₂ Sensor Circuit High Voltage (Bank no. 2 Sensor no. 3)

P0165 O₂ Sensor Circuit Slow Response (Bank no. 2 Sensor no. 3)

P0166 O₂ Sensor Circuit No Activity Detected (Bank no. 2 Sensor no. 3)

P0167 O₂ Sensor Heater Circuit Malfunction (Bank no. 2 Sensor no. 3)

P0170 Fuel Trim Malfunction (Bank no. 1)

P0171 System Too Lean (Bank no. 1)

P0172 System Too Rich (Bank no. 1)

P0173 Fuel Trim Malfunction (Bank no. 2)

P0174 System Too Lean (Bank no. 2)

P0175 System Too Rich (Bank no. 2)

P0176 Fuel Composition Sensor Circuit Malfunction

P0177 Fuel Composition Sensor Circuit Range/Performance

P0178 Fuel Composition Sensor Circuit Low Input

P0179 Fuel Composition Sensor Circuit High Input

P0180 Fuel Temperature Sensor "A" Circuit Malfunction

P0181 Fuel Temperature Sensor "A" Circuit Range/Performance

P0182 Fuel Temperature Sensor "A" Circuit Low Input

P0183 Fuel Temperature Sensor "A" Circuit High Input

P0184 Fuel Temperature Sensor "A" Circuit Intermittent

P0185 Fuel Temperature Sensor "B" Circuit Malfunction

P0186 Fuel Temperature Sensor "B" Circuit Range/Performance

P0187 Fuel Temperature Sensor "B" Circuit Low Input

P0188 Fuel Temperature Sensor "B" Circuit High Input

P0189 Fuel Temperature Sensor "B" Circuit Intermittent

P0190 Fuel Rail Pressure Sensor Circuit Malfunction

P0191 Fuel Rail Pressure Sensor Circuit Range/Performance

P0192 Fuel Rail Pressure Sensor Circuit Low Input

P0193 Fuel Rail Pressure Sensor Circuit High Input

P0194 Fuel Rail Pressure Sensor Circuit Intermittent

P0195 Engine Oil Temperature Sensor Malfunction

P0196 Engine Oil Temperature Sensor Range/Performance

P0197 Engine Oil Temperature Sensor Low

P0198 Engine Oil Temperature Sensor High

P0199 Engine Oil Temperature Sensor Intermittent

P0200 Injector Circuit Malfunction

P0201 Injector Circuit Malfunction—Cylinder no. 1

P0202 Injector Circuit Malfunction—Cylinder no. 2

P0203 Injector Circuit Malfunction—Cylinder no. 3

P0204 Injector Circuit Malfunction—Cylinder no. 4

P0205 Injector Circuit Malfunction—Cylinder no. 5

P0206 Injector Circuit Malfunction—Cylinder no. 6

P0207 Injector Circuit Malfunction—Cylinder no. 7

P0208 Injector Circuit Malfunction—Cylinder no. 8

P0209 Injector Circuit Malfunction—Cylinder no. 9

P0210 Injector Circuit Malfunction—Cylinder no. 10

P0211 Injector Circuit Malfunction—Cylinder no. 11

P0212 Injector Circuit Malfunction—Cylinder no. 12

P0213 Cold Start Injector no. 1 Malfunction

P0214 Cold Start Injector no. 2 Malfunction

P0215 Engine Shutoff Solenoid Malfunction

P0216 Injection Timing Control Circuit Malfunction

P0217 Engine Over Temperature Condition

P0218 Transmission Over Temperature Condition

P0219 Engine Over Speed Condition

P0220 Throttle/Pedal Position Sensor/Switch "B" Circuit Malfunction

P0221 Throttle/Pedal Position Sensor/Switch "B" Circuit Range/Performance Problem

P0222 Throttle/Pedal Position Sensor/Switch "B" Circuit Low Input

P0223 Throttle/Pedal Position Sensor/Switch "B" Circuit High Input

P0224 Throttle/Pedal Position Sensor/Switch "B" Circuit Intermittent

P0225 Throttle/Pedal Position Sensor/Switch "C" Circuit Malfunction

P0226 Throttle/Pedal Position Sensor/Switch "C" Circuit Range/Performance Problem

P0227 Throttle/Pedal Position Sensor/Switch "C" Circuit Low Input

P0228 Throttle/Pedal Position Sensor/Switch "C" Circuit High Input

P0229 Throttle/Pedal Position Sensor/Switch "C" Circuit Intermittent

P0230 Fuel Pump Primary Circuit Malfunction

P0231 Fuel Pump Secondary Circuit Low

P0232 Fuel Pump Secondary Circuit High

P0233 Fuel Pump Secondary Circuit Intermittent

P0234 Engine Over Boost Condition

P0261 Cylinder no. 1 Injector Circuit Low

P0262 Cylinder no. 1 Injector Circuit High

P0263 Cylinder no. 1 Contribution/Balance Fault

P0264 Cylinder no. 2 Injector Circuit Low

P0265 Cylinder no. 2 Injector Circuit High

P0266 Cylinder no. 2 Contribution/Balance Fault

P0267 Cylinder no. 3 Injector Circuit Low

P0268 Cylinder no. 3 Injector Circuit High

P0269 Cylinder no. 3 Contribution/Balance Fault

P0270 Cylinder no. 4 Injector Circuit Low

P0271 Cylinder no. 4 Injector Circuit High

P0272 Cylinder no. 4 Contribution/Balance Fault

P0273 Cylinder no. 5 Injector Circuit Low

P0274 Cylinder no. 5 Injector Circuit High

P0275 Cylinder no. 5 Contribution/Balance Fault

P0276 Cylinder no. 6 Injector Circuit Low

P0277 Cylinder no. 6 Injector Circuit High

P0278 Cylinder no. 6 Contribution/Balance Fault

P0279 Cylinder no. 7 Injector Circuit Low

P0280 Cylinder no. 7 Injector Circuit High

P0281 Cylinder no. 7 Contribution/Balance Fault

P0282 Cylinder no. 8 Injector Circuit Low

P0283 Cylinder no. 8 Injector Circuit High

P0284 Cylinder no. 8 Contribution/Balance Fault

P0285 Cylinder no. 9 Injector Circuit Low

P0286 Cylinder no. 9 Injector Circuit High

P0287 Cylinder no. 9 Contribution/Balance Fault

P0288 Cylinder no. 10 Injector Circuit Low

P0289 Cylinder no. 10 Injector Circuit High
P0290 Cylinder no. 10 Contribution/Balance Fault
P0291 Cylinder no. 11 Injector Circuit Low
P0292 Cylinder no. 11 Injector Circuit High
P0293 Cylinder no. 11 Contribution/Balance Fault
P0294 Cylinder no. 12 Injector Circuit Low
P0295 Cylinder no. 12 Injector Circuit High
P0296 Cylinder no. 12 Contribution/Balance Fault
P0300 Random/Multiple Cylinder Misfire Detected
P0301 Cylinder no. 1—Misfire Detected
P0302 Cylinder no. 2—Misfire Detected
P0303 Cylinder no. 3—Misfire Detected
P0304 Cylinder no. 4—Misfire Detected
P0305 Cylinder no. 5—Misfire Detected
P0306 Cylinder no. 6—Misfire Detected
P0307 Cylinder no. 7—Misfire Detected
P0308 Cylinder no. 8—Misfire Detected
P0309 Cylinder no. 9—Misfire Detected
P0310 Cylinder no. 10—Misfire Detected
P0311 Cylinder no. 11—Misfire Detected
P0312 Cylinder no. 12—Misfire Detected
P0320 Ignition/Distributor Engine Speed Input Circuit Malfunction
P0321 Ignition/Distributor Engine Speed Input Circuit Range/Performance
P0322 Ignition/Distributor Engine Speed Input Circuit No Signal
P0323 Ignition/Distributor Engine Speed Input Circuit Intermittent
P0325 Knock Sensor no. 1—Circuit Malfunction (Bank no. 1 or Single Sensor)
P0326 Knock Sensor no. 1—Circuit Range/Performance (Bank no. 1 or Single Sensor)
P0327 Knock Sensor no. 1—Circuit Low Input (Bank no. 1 or Single Sensor)
P0328 Knock Sensor no. 1—Circuit High Input (Bank no. 1 or Single Sensor)
P0329 Knock Sensor no. 1—Circuit Input Intermittent (Bank no. 1 or Single Sensor)
P0330 Knock Sensor no. 2—Circuit Malfunction (Bank no. 2)
P0331 Knock Sensor no. 2—Circuit Range/Performance (Bank no. 2)
P0332 Knock Sensor no. 2—Circuit Low Input (Bank no. 2)
P0333 Knock Sensor no. 2—Circuit High Input (Bank no. 2)
P0334 Knock Sensor no. 2—Circuit Input Intermittent (Bank no. 2)
P0335 Crankshaft Position Sensor "A" Circuit Malfunction
P0336 Crankshaft Position Sensor "A" Circuit Range/Performance
P0337 Crankshaft Position Sensor "A" Circuit Low Input
P0338 Crankshaft Position Sensor "A" Circuit High Input
P0339 Crankshaft Position Sensor "A" Circuit Intermittent
P0340 Camshaft Position Sensor Circuit Malfunction
P0341 Camshaft Position Sensor Circuit Range/Performance
P0342 Camshaft Position Sensor Circuit Low Input
P0343 Camshaft Position Sensor Circuit High Input
P0344 Camshaft Position Sensor Circuit Intermittent
P0350 Ignition Coil Primary/Secondary Circuit Malfunction
P0351 Ignition Coil "A" Primary/Secondary Circuit Malfunction
P0352 Ignition Coil "B" Primary/Secondary Circuit Malfunction
P0353 Ignition Coil "C" Primary/Secondary Circuit Malfunction
P0354 Ignition Coil "D" Primary/Secondary Circuit Malfunction
P0355 Ignition Coil "E" Primary/Secondary Circuit Malfunction

P0356 Ignition Coil "F" Primary/Secondary Circuit Malfunction
P0357 Ignition Coil "G" Primary/Secondary Circuit Malfunction
P0358 Ignition Coil "H" Primary/Secondary Circuit Malfunction
P0359 Ignition Coil "I" Primary/Secondary Circuit Malfunction
P0360 Ignition Coil "J" Primary/Secondary Circuit Malfunction
P0361 Ignition Coil "K" Primary/Secondary Circuit Malfunction
P0362 Ignition Coil "L" Primary/Secondary Circuit Malfunction
P0370 Timing Reference High Resolution Signal "A" Malfunction
P0371 Timing Reference High Resolution Signal "A" Too Many Pulses
P0372 Timing Reference High Resolution Signal "A" Too Few Pulses
P0373 Timing Reference High Resolution Signal "A" Intermittent/Erratic Pulses
P0374 Timing Reference High Resolution Signal "A" No Pulses
P0375 Timing Reference High Resolution Signal "B" Malfunction
P0376 Timing Reference High Resolution Signal "B" Too Many Pulses
P0377 Timing Reference High Resolution Signal "B" Too Few Pulses
P0378 Timing Reference High Resolution Signal "B" Intermittent/Erratic Pulses
P0379 Timing Reference High Resolution Signal "B" No Pulses
P0380 Glow Plug/Heater Circuit "A" Malfunction
P0381 Glow Plug/Heater Indicator Circuit Malfunction
P0382 Glow Plug/Heater Circuit "B" Malfunction
P0385 Crankshaft Position Sensor "B" Circuit Malfunction
P0386 Crankshaft Position Sensor "B" Circuit Range/Performance
P0387 Crankshaft Position Sensor "B" Circuit Low Input
P0388 Crankshaft Position Sensor "B" Circuit High Input
P0389 Crankshaft Position Sensor "B" Circuit Intermittent
P0400 Exhaust Gas Recirculation Flow Malfunction
P0401 Exhaust Gas Recirculation Flow Insufficient Detected
P0402 Exhaust Gas Recirculation Flow Excessive Detected
P0403 Exhaust Gas Recirculation Circuit Malfunction
P0404 Exhaust Gas Recirculation Circuit Range/Performance
P0405 Exhaust Gas Recirculation Sensor "A" Circuit Low
P0406 Exhaust Gas Recirculation Sensor "A" Circuit High
P0407 Exhaust Gas Recirculation Sensor "B" Circuit Low
P0408 Exhaust Gas Recirculation Sensor "B" Circuit High
P0410 Secondary Air Injection System Malfunction
P0411 Secondary Air Injection System Incorrect Flow Detected
P0412 Secondary Air Injection System Switching Valve "A" Circuit Malfunction
P0413 Secondary Air Injection System Switching Valve "A" Circuit Open
P0414 Secondary Air Injection System Switching Valve "A" Circuit Shorted
P0415 Secondary Air Injection System Switching Valve "B" Circuit Malfunction
P0416 Secondary Air Injection System Switching Valve "B" Circuit Open
P0417 Secondary Air Injection System Switching Valve "B" Circuit Shorted
P0418 Secondary Air Injection System Relay "A" Circuit Malfunction
P0419 Secondary Air Injection System Relay "B" Circuit Malfunction
P0420 Catalyst System Efficiency Below Threshold (Bank no. 1)

P0421 Warm Up Catalyst Efficiency Below Threshold (Bank no. 1)

P0422 Main Catalyst Efficiency Below Threshold (Bank no. 1)

P0423 Heated Catalyst Efficiency Below Threshold (Bank no. 1)

P0424 Heated Catalyst Temperature Below Threshold (Bank no. 1)

P0430 Catalyst System Efficiency Below Threshold (Bank no. 2)

P0431 Warm Up Catalyst Efficiency Below Threshold (Bank no. 2)

P0432 Main Catalyst Efficiency Below Threshold (Bank no. 2)

P0433 Heated Catalyst Efficiency Below Threshold (Bank no. 2)

P0434 Heated Catalyst Temperature Below Threshold (Bank no. 2)

P0440 Evaporative Emission Control System Malfunction

P0441 Evaporative Emission Control System Incorrect Purge Flow

P0442 Evaporative Emission Control System Leak Detected (Small Leak)

P0443 Evaporative Emission Control System Purge Control Valve Circuit Malfunction

P0444 Evaporative Emission Control System Purge Control Valve Circuit Open

P0445 Evaporative Emission Control System Purge Control Valve Circuit Shorted

P0446 Evaporative Emission Control System Vent Control Circuit Malfunction

P0447 Evaporative Emission Control System Vent Control Circuit Open

P0448 Evaporative Emission Control System Vent Control Circuit Shorted

P0449 Evaporative Emission Control System Vent Valve/Solenoid Circuit Malfunction

P0450 Evaporative Emission Control System Pressure Sensor Malfunction

P0451 Evaporative Emission Control System Pressure Sensor Range/Performance

P0452 Evaporative Emission Control System Pressure Sensor Low Input

P0453 Evaporative Emission Control System Pressure Sensor High Input

P0454 Evaporative Emission Control System Pressure Sensor Intermittent

P0455 Evaporative Emission Control System Leak Detected (Gross Leak)

P0460 Fuel Level Sensor Circuit Malfunction

P0461 Fuel Level Sensor Circuit Range/Performance

P0462 Fuel Level Sensor Circuit Low Input

P0463 Fuel Level Sensor Circuit High Input

P0464 Fuel Level Sensor Circuit Intermittent

P0465 Purge Flow Sensor Circuit Malfunction

P0466 Purge Flow Sensor Circuit Range/Performance

P0467 Purge Flow Sensor Circuit Low Input

P0468 Purge Flow Sensor Circuit High Input

P0469 Purge Flow Sensor Circuit Intermittent

P0470 Exhaust Pressure Sensor Malfunction

P0471 Exhaust Pressure Sensor Range/Performance

P0472 Exhaust Pressure Sensor Low

P0473 Exhaust Pressure Sensor High

P0474 Exhaust Pressure Sensor Intermittent

P0475 Exhaust Pressure Control Valve Malfunction

P0476 Exhaust Pressure Control Valve Range/Performance

P0477 Exhaust Pressure Control Valve Low

P0478 Exhaust Pressure Control Valve High

P0479 Exhaust Pressure Control Valve Intermittent

P0480 Cooling Fan no. 1 Control Circuit Malfunction

P0481 Cooling Fan no. 2 Control Circuit Malfunction

P0482 Cooling Fan no. 3 Control Circuit Malfunction

P0483 Cooling Fan Rationality Check Malfunction

P0484 Cooling Fan Circuit Over Current

P0485 Cooling Fan Power/Ground Circuit Malfunction

P0500 Vehicle Speed Sensor Malfunction

P0501 Vehicle Speed Sensor Range/Performance

P0502 Vehicle Speed Sensor Circuit Low Input

P0503 Vehicle Speed Sensor Intermittent/Erratic/High

P0505 Idle Control System Malfunction

P0506 Idle Control System RPM Lower Than Expected

P0507 Idle Control System RPM Higher Than Expected

P0510 Closed Throttle Position Switch Malfunction

P0520 Engine Oil Pressure Sensor/Switch Circuit Malfunction

P0521 Engine Oil Pressure Sensor/Switch Range/Performance

P0522 Engine Oil Pressure Sensor/Switch Low Voltage

P0523 Engine Oil Pressure Sensor/Switch High Voltage

P0530 A/C Refrigerant Pressure Sensor Circuit Malfunction

P0531 A/C Refrigerant Pressure Sensor Circuit Range/Performance

P0532 A/C Refrigerant Pressure Sensor Circuit Low Input

P0533 A/C Refrigerant Pressure Sensor Circuit High Input

P0534 A/C Refrigerant Charge Loss

P0550 Power Steering Pressure Sensor Circuit Malfunction

P0551 Power Steering Pressure Sensor Circuit Range/Performance

P0552 Power Steering Pressure Sensor Circuit Low Input

P0553 Power Steering Pressure Sensor Circuit High Input

P0554 Power Steering Pressure Sensor Circuit Intermittent

P0560 System Voltage Malfunction

P0561 System Voltage Unstable

P0562 System Voltage Low

P0563 System Voltage High

P0565 Cruise Control On Signal Malfunction

P0566 Cruise Control Off Signal Malfunction

P0567 Cruise Control Resume Signal Malfunction

P0568 Cruise Control Set Signal Malfunction

P0569 Cruise Control Coast Signal Malfunction

P0570 Cruise Control Accel Signal Malfunction

P0571 Cruise Control/Brake Switch "A" Circuit Malfunction

P0572 Cruise Control/Brake Switch "A" Circuit Low

P0573 Cruise Control/Brake Switch "A" Circuit High

P0574 **Through P0580** Reserved for Cruise Codes

P0600 Serial Communication Link Malfunction

P0601 Internal Control Module Memory Check Sum Error

P0602 Control Module Programming Error

P0603 Internal Control Module Keep Alive Memory (KAM) Error

P0604 Internal Control Module Random Access Memory (RAM) Error

P0605 Internal Control Module Read Only Memory (ROM) Error

P0606 PCM Processor Fault

P0608 Control Module VSS Output "A" Malfunction

P0609 Control Module VSS Output "B" Malfunction

P0620 Generator Control Circuit Malfunction

P0621 Generator Lamp "L" Control Circuit Malfunction

P0622 Generator Field "F" Control Circuit Malfunction

P0650 Malfunction Indicator Lamp (MIL) Control Circuit Malfunction

P0654 Engine RPM Output Circuit Malfunction

P0655 Engine Hot Lamp Output Control Circuit Malfunction
P0656 Fuel Level Output Circuit Malfunction
P0700 Transmission Control System Malfunction
P0701 Transmission Control System Range/Performance
P0702 Transmission Control System Electrical
P0703 Torque Converter/Brake Switch "B" Circuit Malfunction
P0704 Clutch Switch Input Circuit Malfunction
P0705 Transmission Range Sensor Circuit Malfunction (PRNDL Input)
P0706 Transmission Range Sensor Circuit Range/Performance
P0707 Transmission Range Sensor Circuit Low Input
P0708 Transmission Range Sensor Circuit High Input
P0709 Transmission Range Sensor Circuit Intermittent
P0710 Transmission Fluid Temperature Sensor Circuit Malfunction
P0711 Transmission Fluid Temperature Sensor Circuit Range/Performance
P0712 Transmission Fluid Temperature Sensor Circuit Low Input
P0713 Transmission Fluid Temperature Sensor Circuit High Input
P0714 Transmission Fluid Temperature Sensor Circuit Intermittent
P0715 Input/Turbine Speed Sensor Circuit Malfunction
P0716 Input/Turbine Speed Sensor Circuit Range/Performance
P0717 Input/Turbine Speed Sensor Circuit No Signal
P0718 Input/Turbine Speed Sensor Circuit Intermittent
P0719 Torque Converter/Brake Switch "B" Circuit Low
P0720 Output Speed Sensor Circuit Malfunction
P0721 Output Speed Sensor Circuit Range/Performance
P0722 Output Speed Sensor Circuit No Signal
P0723 Output Speed Sensor Circuit Intermittent
P0724 Torque Converter/Brake Switch "B" Circuit High
P0725 Engine Speed Input Circuit Malfunction
P0726 Engine Speed Input Circuit Range/Performance
P0727 Engine Speed Input Circuit No Signal
P0728 Engine Speed Input Circuit Intermittent
P0730 Incorrect Gear Ratio
P0731 Gear no. 1 Incorrect Ratio
P0732 Gear no. 2 Incorrect Ratio
P0733 Gear no. 3 Incorrect Ratio
P0734 Gear no. 4 Incorrect Ratio
P0735 Gear no. 5 Incorrect Ratio
P0736 Reverse Incorrect Ratio
P0740 Torque Converter Clutch Circuit Malfunction
P0741 Torque Converter Clutch Circuit Performance or Stuck Off
P0742 Torque Converter Clutch Circuit Stuck On
P0743 Torque Converter Clutch Circuit Electrical
P0744 Torque Converter Clutch Circuit Intermittent
P0745 Pressure Control Solenoid Malfunction
P0746 Pressure Control Solenoid Performance or Stuck Off
P0747 Pressure Control Solenoid Stuck On
P0748 Pressure Control Solenoid Electrical
P0749 Pressure Control Solenoid Intermittent
P0750 Shift Solenoid "A" Malfunction
P0751 Shift Solenoid "A" Performance or Stuck Off
P0752 Shift Solenoid "A" Stuck On
P0753 Shift Solenoid "A" Electrical
P0754 Shift Solenoid "A" Intermittent
P0755 Shift Solenoid "B" Malfunction
P0756 Shift Solenoid "B" Performance or Stuck Off

P0757 Shift Solenoid "B" Stuck On
P0758 Shift Solenoid "B" Electrical
P0759 Shift Solenoid "B" Intermittent
P0760 Shift Solenoid "C" Malfunction
P0761 Shift Solenoid "C" Performance Or Stuck Off
P0762 Shift Solenoid "C" Stuck On
P0763 Shift Solenoid "C" Electrical
P0764 Shift Solenoid "C" Intermittent
P0765 Shift Solenoid "D" Malfunction
P0766 Shift Solenoid "D" Performance Or Stuck Off
P0767 Shift Solenoid "D" Stuck On
P0768 Shift Solenoid "D" Electrical
P0769 Shift Solenoid "D" Intermittent
P0770 Shift Solenoid "E" Malfunction
P0771 Shift Solenoid "E" Performance Or Stuck Off
P0772 Shift Solenoid "E" Stuck On
P0773 Shift Solenoid "E" Electrical
P0774 Shift Solenoid "E" Intermittent
P0780 Shift Malfunction
P0781 1–2 Shift Malfunction
P0782 2–3 Shift Malfunction
P0783 3–4 Shift Malfunction
P0784 4–5 Shift Malfunction
P0785 Shift/Timing Solenoid Malfunction
P0786 Shift/Timing Solenoid Range/Performance
P0787 Shift/Timing Solenoid Low
P0788 Shift/Timing Solenoid High
P0789 Shift/Timing Solenoid Intermittent
P0790 Normal/Performance Switch Circuit Malfunction
P0801 Reverse Inhibit Control Circuit Malfunction
P0803 1–4 Upshift (Skip Shift) Solenoid Control Circuit Malfunction
P0804 1–4 Upshift (Skip Shift) Lamp Control Circuit Malfunction
P1106 Barometric Pressure Circuit Range/Performance Problem
P1107 Barometric Pressure Circuit Low Input
P1108 Barometric Pressure Circuit High Input
P1121 Throttle Position Lower Than Expected
P1122 Throttle Position Higher Than Expected
P1128 Manifold Absolute Pressure Lower Than Expected
P1129 Manifold Absolute Pressure Higher Than Expected
P1259 VTEC System Malfunction
P1297 Electrical Load Detector Circuit Low Input
P1298 Electrical Load Detector Circuit High Input
P1297 Electrical Load Detector Circuit Low Input
P1298 Electrical Load Detector Circuit High Input
P1336 Crankshaft Speed Fluctuation Sensor Intermittent Interruption
P1337 Crankshaft Speed Fluctuation Sensor No Signal
P1359 Crankshaft Position Top Dead Center Sensor/Cylinder Position Connector Disconnection
P1361 Top Dead Center Sensor Intermittent Interruption
P1362 Top Dead Center Sensor No Signal
P1381 Cylinder Position Sensor Intermittent Interruption
P1382 Cylinder Position Sensor No Signal
P1456 Evaporative Emission Control System Leak Detected (Fuel Tank System)
P1457 Evaporative Emission Control System Leak Detected (EVAP Control Canister Leak)
P1491 EGR Valve Lift Insufficient Detected
P1498 EGR Valve Lift Sensor High Voltage
P1519 Idle Air Control Valve Circuit Failure

P1508 Idle Air Control Valve Circuit Failure
P1607 Powertrain Control Module Internal Circuit Failure A
P1705 Automatic Transaxle
P1706 Automatic Transaxle
P1753 Automatic Transaxle
P1768 Automatic Transaxle
P1790 Automatic Transaxle
P1791 Automatic Transaxle

TROUBLE CODE EQUIVALENT LIST

If a scan tool is not available for code retrieval, the following codes may be retrieved without one.

1 O_2 Sensor Circuit High Voltage (Bank no. 1 Sensor no. 1)
1 O_2 Sensor Circuit Low Voltage (Bank no. 1 Sensor no. 1)
3 Manifold Absolute Pressure/Barometric Pressure Circuit Low Input
3 Manifold Absolute Pressure/Barometric Pressure Circuit High Input
4 Crankshaft Position Sensor "A" Circuit Malfunction
4 Crankshaft Position Sensor "A" Circuit Range/Performance
5 Manifold Absolute Pressure Higher Than Expected
5 Manifold Absolute Pressure Lower Than Expected
6 Engine Coolant Temperature Circuit High Input
6 Engine Coolant Temperature Circuit Low Input
7 Throttle Position Higher Than Expected
7 Throttle Position Lower Than Expected
7 Throttle/Pedal Position Sensor/Switch "A" Circuit High Input
7 Throttle/Pedal Position Sensor/Switch "A" Circuit Low Input
8 Crankshaft Position Top Dead Center Sensor/Cylinder Position Connector Disconnection
8 Top Dead Center Sensor Intermittent Interruption
8 Top Dead Center Sensor No Signal
9 Cylinder Position Sensor Intermittent Interruption
9 Cylinder Position Sensor No Signal
10 Intake Air Temperature Circuit High Input
10 Intake Air Temperature Circuit Low Input
12 EGR Valve Lift Insufficient Detected
12 EGR Valve Lift Sensor High Voltage
13 Barometric Pressure Circuit High Input
13 Barometric Pressure Circuit Low Input
13 Barometric Pressure Circuit Range/Performance Problem
14 Idle Air Control Valve Circuit Failure
14 Idle Air Control Valve Circuit Failure
14 Idle Control System Malfunction
20 Electrical Load Detector Circuit High Input
20 Electrical Load Detector Circuit High Input
20 Electrical Load Detector Circuit Low Input
20 Electrical Load Detector Circuit Low Input
22 VTEC System Malfunction
23 Knock Sensor no. 1—Circuit Malfunction (Bank no. 1 or Single Sensor)
80 Exhaust Gas Recirculation Flow Insufficient Detected
41 O_2 Sensor Heater Circuit Malfunction (Bank no. 1 Sensor no. 1)
45 System Too Lean (Bank no. 1)
45 System Too Rich (Bank no. 1)
54 Crankshaft Speed Fluctuation Sensor Intermittent Interruption
54 Crankshaft Speed Fluctuation Sensor No Signal
61 O_2 Sensor Circuit Slow Response (Bank no. 1 Sensor no. 1)

63 O_2 Sensor Circuit High Voltage (Bank no. 1 Sensor no. 2)
63 O_2 Sensor Circuit Low Voltage (Bank no. 1 Sensor no. 2)
63 O_2 Sensor Circuit Slow Response (Bank no. 1 Sensor no. 2)
65 O_2 Sensor Heater Circuit Malfunction (Bank no. 1 Sensor no. 2)
67 Catalyst System Efficiency Below Threshold (Bank no. 1)
70 Automatic Transaxle
70 Transmission Control System Malfunction
70 Input/Turbine Speed Sensor Circuit Malfunction
70 Output Speed Sensor Circuit Malfunction
70 Incorrect Gear Ratio
70 Torque Converter Clutch Circuit Malfunction
70 Shift Solenoid "A" Electrical
70 Shift Solenoid "B" Electrical
71 Cylinder no. 1—Misfire Detected
72 Cylinder no. 2—Misfire Detected
73 Cylinder no. 3—Misfire Detected
74 Cylinder no. 4—Misfire Detected
86 Engine Coolant Temperature Circuit Range/Performance Problem
90 Evaporative Emission Control System Leak Detected (EVAP Control Canister Leak)
90 Evaporative Emission Control System Leak Detected (Fuel Tank System)
91 Evaporative Emission Control System Pressure Sensor Low Input
91 Evaporative Emission Control System Pressure Sensor High Input

Hyundai

READING CODES

With a Scan Tool

Reading the control module memory is on of the first steps in OBD II system diagnostics. This step should be initially performed to determine the general nature of the fault. Subsequent readings will determine if the fault has been cleared.

Reading codes can be performed by any of the methods below:
• Read the control module memory with the Generic Scan Tool (GST)
• Read the control module memory with the vehicle manufacturer's specific tester

To read the fault codes, connect the scan tool or tester according to the manufacturer's instructions. Follow the manufacturer's specified procedure for reading the codes.

Without a Scan Tool

The Diagnostic Link Connector (DLC), under the left-hand side of the instrument panel, must be located to retrieve any DTC's.

In 1996, all Hyundai (except the Sonata) switched from an arbitrary code listing and format, to the federally regulated On Board Diagnostics 2nd Generation (OBD II) code system. Normally, OBD II equipped vehicles do not have the option of allowing the person servicing the vehicle to flash the codes out with a voltmeter; usually a scan tool is necessary to retrieve OBD II codes. Hyundai, however, does provide this option.

The Federal government decided that it was time to create a standard for vehicle diagnostic systems codes for ease of servicing and to insure that certain of the vehicle's systems were being monitored

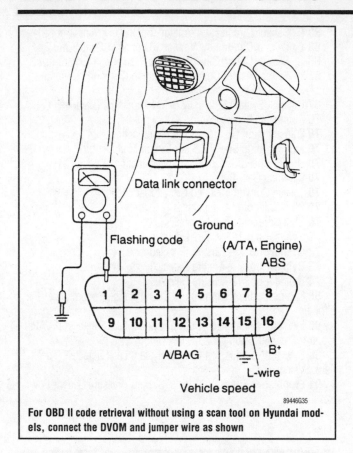

For OBD II code retrieval without using a scan tool on Hyundai models, connect the DVOM and jumper wire as shown

for emissions purposes. Since OBD II codes are standardized (they all contain one letter and four numbers), they are easy to decipher. For a breakdown of a typical transaxle-related OBD II code, refer to the accompanying illustration.

The OBD II system in the Hyundai models is designed so that it will flash the DTC's out on a voltmeter (even though a scan tool is better). However, the first two characters of the code are not used.

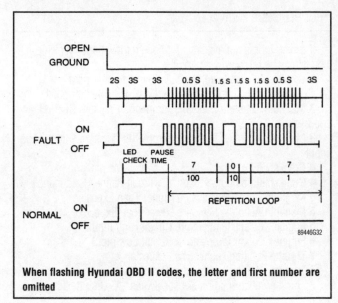

When flashing Hyundai OBD II codes, the letter and first number are omitted

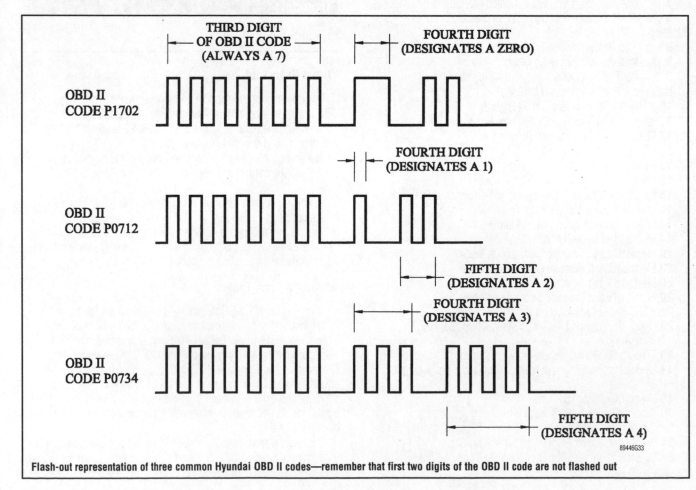

Flash-out representation of three common Hyundai OBD II codes—remember that first two digits of the OBD II code are not flashed out

This is because the transaxle is a part of the powertrain, so all transaxle related codes will begin with a P. Also, since there are no overlapping numbers between SAE and Hyundai codes, the second digit is also not necessary.

The system flashes the codes out in a series of flashes in three groups, each group corresponding to one of the three last digits of the OBD II code. Therefore, Code P0753 would be flashed out in seven flashes, followed five flashes, then by three flashes. Each group of flashes is separated by a brief pause. All of the flashes are of the same duration, with the only exception being zero. Zero is represented by a long flash. Therefore, seven flashes, one long flash, two flashes would indicate a P0702 code (shorted TP sensor circuit).

To retrieve the codes, perform the following:

1. Perform the preliminary inspection, located earlier in this section. This is very important, since a loose or disconnected wire, or corroded connector terminals can cause a whole slew of unrelated DTC's to be stored by the computer; you will waste a lot of time performing a diagnostic "goose chase."

2. Grab some paper and a pencil or pen to write down the DTC's when they are flashed out.

3. Locate the Diagnostic Link Connector (DLC), which is located under the left-hand side of the instrument panel.

4. Start the engine and drive the vehicle until the transaxle goes into the failsafe mode.

5. Park the vehicle, but do not turn the ignition **OFF**. Allow it to idle.

6. Attach a voltmeter (analog or digital) to the test terminals on the Diagnostic Link Connector (DLC). The negative lead should be attached to terminal 4 and the positive lead to terminal 1.

7. Observe the voltmeter and count the flashes (or arm sweeps if using an analog voltmeter); note the applicable codes.

8. After all of the DTC(s) have been retrieved, fix the applicable problems, clear the codes, drive the vehicle, and perform the retrieval procedure again to ensure that all of the codes are gone.

CLEARING CODES

With a Scan Tool

Control module reset procedures are a very important part of OBD II System diagnostics.

This step should be done at the end of any fault code repair and at the end of any driveability repair.

Clearing codes can be performed by any of the methods below:
• Clear the control module memory with the Generic Scan Tool (GST)
• Clear the control module memory with the vehicle manufacturer's specific tester

➡**The MIL will may also be de-activated for some codes if the vehicle completes three consecutive trips without a fault detected with vehicle conditions similar to those present during the fault.**

Without a Scan Tool

➡**It is not recommended that the negative battery cable be disconnected to clear DTC's. Doing so will clear all settings from the vehicle's computer system, resulting in lost radio presets, seat memories, anti-theft codes, driveability parameters, etc., and there are better ways of spending your afternoon than resetting all of these.**

1. Turn the ignition switch **OFF**.
2. Disconnect the wiring harness connector from the control system computer.
3. Wait at least one minute, then reconnect the computer. The codes should now be cleared.

If there are still codes present, either the codes were not properly cleared (Are the codes identical to those flashed out previously?) or the underlying problem is still there (Are only some of the codes the same as previously?).

DIAGNOSTIC TROUBLE CODES

P0000 No Failures
P0100 Mass or Volume Air Flow Circuit Malfunction
P0101 Mass or Volume Air Flow Circuit Range/Performance Problem
P0102 Mass or Volume Air Flow Circuit Low Input
P0103 Mass or Volume Air Flow Circuit High Input
P0104 Mass or Volume Air Flow Circuit Intermittent
P0105 Manifold Absolute Pressure/Barometric Pressure Circuit Malfunction
P0106 Manifold Absolute Pressure/Barometric Pressure Circuit Range/Performance Problem
P0107 Manifold Absolute Pressure/Barometric Pressure Circuit Low Input
P0108 Manifold Absolute Pressure/Barometric Pressure Circuit High Input
P0109 Manifold Absolute Pressure/Barometric Pressure Circuit Intermittent
P0110 Intake Air Temperature Circuit Malfunction
P0111 Intake Air Temperature Circuit Range/Performance Problem
P0112 Intake Air Temperature Circuit Low Input
P0113 Intake Air Temperature Circuit High Input
P0114 Intake Air Temperature Circuit Intermittent
P0115 Engine Coolant Temperature Circuit Malfunction
P0116 Engine Coolant Temperature Circuit Range/Performance Problem
P0117 Engine Coolant Temperature Circuit Low Input
P0118 Engine Coolant Temperature Circuit High Input
P0119 Engine Coolant Temperature Circuit Intermittent
P0120 Throttle/Pedal Position Sensor/Switch "A" Circuit Malfunction
P0121 Throttle/Pedal Position Sensor/Switch "A" Circuit Range/Performance Problem
P0122 Throttle/Pedal Position Sensor/Switch "A" Circuit Low Input
P0123 Throttle/Pedal Position Sensor/Switch "A" Circuit High Input
P0124 Throttle/Pedal Position Sensor/Switch "A" Circuit Intermittent
P0125 Insufficient Coolant Temperature For Closed Loop Fuel Control
P0126 Insufficient Coolant Temperature For Stable Operation
P0130 O_2 Circuit Malfunction (Bank no. 1 Sensor no. 1)
P0131 O_2 Sensor Circuit Low Voltage (Bank no. 1 Sensor no. 1)
P0132 O_2 Sensor Circuit High Voltage (Bank no. 1 Sensor no. 1)
P0133 O_2 Sensor Circuit Slow Response (Bank no. 1 Sensor no. 1)
P0134 O_2 Sensor Circuit No Activity Detected (Bank no. 1 Sensor no. 1)

P0135 O_2 Sensor Heater Circuit Malfunction (Bank no. 1 Sensor no. 1)

P0136 O_2 Sensor Circuit Malfunction (Bank no. 1 Sensor no. 2)

P0137 O_2 Sensor Circuit Low Voltage (Bank no. 1 Sensor no. 2)

P0138 O_2 Sensor Circuit High Voltage (Bank no. 1 Sensor no. 2)

P0139 O_2 Sensor Circuit Slow Response (Bank no. 1 Sensor no. 2)

P0140 O_2 Sensor Circuit No Activity Detected (Bank no. 1 Sensor no. 2)

P0141 O_2 Sensor Heater Circuit Malfunction (Bank no. 1 Sensor no. 2)

P0142 O_2 Sensor Circuit Malfunction (Bank no. 1 Sensor no. 3)

P0143 O_2 Sensor Circuit Low Voltage (Bank no. 1 Sensor no. 3)

P0144 O_2 Sensor Circuit High Voltage (Bank no. 1 Sensor no. 3)

P0145 O_2 Sensor Circuit Slow Response (Bank no. 1 Sensor no. 3)

P0146 O_2 Sensor Circuit No Activity Detected (Bank no. 1 Sensor no. 3)

P0147 O_2 Sensor Heater Circuit Malfunction (Bank no. 1 Sensor no. 3)

P0150 O_2 Sensor Circuit Malfunction (Bank no. 2 Sensor no. 1)

P0151 O_2 Sensor Circuit Low Voltage (Bank no. 2 Sensor no. 1)

P0152 O_2 Sensor Circuit High Voltage (Bank no. 2 Sensor no. 1)

P0153 O_2 Sensor Circuit Slow Response (Bank no. 2 Sensor no. 1)

P0154 O_2 Sensor Circuit No Activity Detected (Bank no. 2 Sensor no. 1)

P0155 O_2 Sensor Heater Circuit Malfunction (Bank no. 2 Sensor no. 1)

P0156 O_2 Sensor Circuit Malfunction (Bank no. 2 Sensor no. 2)

P0157 O_2 Sensor Circuit Low Voltage (Bank no. 2 Sensor no. 2)

P0158 O_2 Sensor Circuit High Voltage (Bank no. 2 Sensor no. 2)

P0159 O_2 Sensor Circuit Slow Response (Bank no. 2 Sensor no. 2)

P0160 O_2 Sensor Circuit No Activity Detected (Bank no. 2 Sensor no. 2)

P0161 O_2 Sensor Heater Circuit Malfunction (Bank no. 2 Sensor no. 2)

P0162 O_2 Sensor Circuit Malfunction (Bank no. 2 Sensor no. 3)

P0163 O_2 Sensor Circuit Low Voltage (Bank no. 2 Sensor no. 3)

P0164 O_2 Sensor Circuit High Voltage (Bank no. 2 Sensor no. 3)

P0165 O_2 Sensor Circuit Slow Response (Bank no. 2 Sensor no. 3)

P0166 O_2 Sensor Circuit No Activity Detected (Bank no. 2 Sensor no. 3)

P0167 O_2 Sensor Heater Circuit Malfunction (Bank no. 2 Sensor no. 3)

P0170 Fuel Trim Malfunction (Bank no. 1)

P0171 System Too Lean (Bank no. 1)

P0172 System Too Rich (Bank no. 1)

P0173 Fuel Trim Malfunction (Bank no. 2)

P0174 System Too Lean (Bank no. 2)

P0175 System Too Rich (Bank no. 2)

P0176 Fuel Composition Sensor Circuit Malfunction

P0177 Fuel Composition Sensor Circuit Range/Performance

P0178 Fuel Composition Sensor Circuit Low Input

P0179 Fuel Composition Sensor Circuit High Input

P0180 Fuel Temperature Sensor "A" Circuit Malfunction

P0181 Fuel Temperature Sensor "A" Circuit Range/Performance

P0182 Fuel Temperature Sensor "A" Circuit Low Input

P0183 Fuel Temperature Sensor "A" Circuit High Input

P0184 Fuel Temperature Sensor "A" Circuit Intermittent

P0185 Fuel Temperature Sensor "B" Circuit Malfunction

P0186 Fuel Temperature Sensor "B" Circuit Range/Performance

P0187 Fuel Temperature Sensor "B" Circuit Low Input

P0188 Fuel Temperature Sensor "B" Circuit High Input

P0189 Fuel Temperature Sensor "B" Circuit Intermittent

P0190 Fuel Rail Pressure Sensor Circuit Malfunction

P0191 Fuel Rail Pressure Sensor Circuit Range/Performance

P0192 Fuel Rail Pressure Sensor Circuit Low Input

P0193 Fuel Rail Pressure Sensor Circuit High Input

P0194 Fuel Rail Pressure Sensor Circuit Intermittent

P0195 Engine Oil Temperature Sensor Malfunction

P0196 Engine Oil Temperature Sensor Range/Performance

P0197 Engine Oil Temperature Sensor Low

P0198 Engine Oil Temperature Sensor High

P0199 Engine Oil Temperature Sensor Intermittent

P0200 Injector Circuit Malfunction

P0201 Injector Circuit Malfunction—Cylinder no. 1

P0202 Injector Circuit Malfunction—Cylinder no. 2

P0203 Injector Circuit Malfunction—Cylinder no. 3

P0204 Injector Circuit Malfunction—Cylinder no. 4

P0205 Injector Circuit Malfunction—Cylinder no. 5

P0206 Injector Circuit Malfunction—Cylinder no. 6

P0207 Injector Circuit Malfunction—Cylinder no. 7

P0208 Injector Circuit Malfunction—Cylinder no. 8

P0209 Injector Circuit Malfunction—Cylinder no. 9

P0210 Injector Circuit Malfunction—Cylinder no. 10

P0211 Injector Circuit Malfunction—Cylinder no. 11

P0212 Injector Circuit Malfunction—Cylinder no. 12

P0213 Cold Start Injector no. 1 Malfunction

P0214 Cold Start Injector no. 2 Malfunction

P0215 Engine Shutoff Solenoid Malfunction

P0216 Injection Timing Control Circuit Malfunction

P0217 Engine Over Temperature Condition

P0218 Transmission Over Temperature Condition

P0219 Engine Over Speed Condition

P0220 Throttle/Pedal Position Sensor/Switch "B" Circuit Malfunction

P0221 Throttle/Pedal Position Sensor/Switch "B" Circuit Range/Performance Problem

P0222 Throttle/Pedal Position Sensor/Switch "B" Circuit Low Input

P0223 Throttle/Pedal Position Sensor/Switch "B" Circuit High Input

P0224 Throttle/Pedal Position Sensor/Switch "B" Circuit Intermittent

P0225 Throttle/Pedal Position Sensor/Switch "C" Circuit Malfunction

P0226 Throttle/Pedal Position Sensor/Switch "C" Circuit Range/Performance Problem

P0227 Throttle/Pedal Position Sensor/Switch "C" Circuit Low Input

P0228 Throttle/Pedal Position Sensor/Switch "C" Circuit High Input

P0229 Throttle/Pedal Position Sensor/Switch "C" Circuit Intermittent

P0230 Fuel Pump Primary Circuit Malfunction

P0231 Fuel Pump Secondary Circuit Low

P0232 Fuel Pump Secondary Circuit High

P0233 Fuel Pump Secondary Circuit Intermittent

P0234 Engine Over Boost Condition

P0261 Cylinder no. 1 Injector Circuit Low

P0262 Cylinder no. 1 Injector Circuit High

P0263 Cylinder no. 1 Contribution/Balance Fault

P0264 Cylinder no. 2 Injector Circuit Low

P0265 Cylinder no. 2 Injector Circuit High

P0266 Cylinder no. 2 Contribution/Balance Fault

P0267 Cylinder no. 3 Injector Circuit Low

P0268 Cylinder no. 3 Injector Circuit High

P0269 Cylinder no. 3 Contribution/Balance Fault

P0270 Cylinder no. 4 Injector Circuit Low

P0271 Cylinder no. 4 Injector Circuit High

P0272 Cylinder no. 4 Contribution/Balance Fault

P0273 Cylinder no. 5 Injector Circuit Low

P0274 Cylinder no. 5 Injector Circuit High

P0275 Cylinder no. 5 Contribution/Balance Fault

P0276 Cylinder no. 6 Injector Circuit Low

P0277 Cylinder no. 6 Injector Circuit High

P0278 Cylinder no. 6 Contribution/Balance Fault

P0279 Cylinder no. 7 Injector Circuit Low

P0280 Cylinder no. 7 Injector Circuit High

P0281 Cylinder no. 7 Contribution/Balance Fault

P0282 Cylinder no. 8 Injector Circuit Low

P0283 Cylinder no. 8 Injector Circuit High

P0284 Cylinder no. 8 Contribution/Balance Fault

P0285 Cylinder no. 9 Injector Circuit Low

P0286 Cylinder no. 9 Injector Circuit High

P0287 Cylinder no. 9 Contribution/Balance Fault

P0288 Cylinder no. 10 Injector Circuit Low

P0289 Cylinder no. 10 Injector Circuit High

P0290 Cylinder no. 10 Contribution/Balance Fault

P0291 Cylinder no. 11 Injector Circuit Low

P0292 Cylinder no. 11 Injector Circuit High

P0293 Cylinder no. 11 Contribution/Balance Fault

P0294 Cylinder no. 12 Injector Circuit Low

P0295 Cylinder no. 12 Injector Circuit High

P0296 Cylinder no. 12 Contribution/Balance Fault

P0300 Random/Multiple Cylinder Misfire Detected

P0301 Cylinder no. 1—Misfire Detected

P0302 Cylinder no. 2—Misfire Detected

P0303 Cylinder no. 3—Misfire Detected

P0304 Cylinder no. 4—Misfire Detected

P0305 Cylinder no. 5—Misfire Detected

P0306 Cylinder no. 6—Misfire Detected

P0307 Cylinder no. 7—Misfire Detected

P0308 Cylinder no. 8—Misfire Detected

P0309 Cylinder no. 9—Misfire Detected

P0310 Cylinder no. 10—Misfire Detected

P0311 Cylinder no. 11—Misfire Detected

P0312 Cylinder no. 12—Misfire Detected

P0320 Ignition/Distributor Engine Speed Input Circuit Malfunction

P0321 Ignition/Distributor Engine Speed Input Circuit Range/Performance

P0322 Ignition/Distributor Engine Speed Input Circuit No Signal

P0323 Ignition/Distributor Engine Speed Input Circuit Intermittent

P0325 Knock Sensor no. 1—Circuit Malfunction (Bank no. 1 or Single Sensor)

P0326 Knock Sensor no. 1—Circuit Range/Performance (Bank no. 1 or Single Sensor)

P0327 Knock Sensor no. 1—Circuit Low Input (Bank no. 1 or Single Sensor)

P0328 Knock Sensor no. 1—Circuit High Input (Bank no. 1 or Single Sensor)

P0329 Knock Sensor no. 1—Circuit Input Intermittent (Bank no. 1 or Single Sensor)

P0330 Knock Sensor no. 2—Circuit Malfunction (Bank no. 2)

P0331 Knock Sensor no. 2—Circuit Range/Performance (Bank no. 2)

P0332 Knock Sensor no. 2—Circuit Low Input (Bank no. 2)

P0333 Knock Sensor no. 2—Circuit High Input (Bank no. 2)

P0334 Knock Sensor no. 2—Circuit Input Intermittent (Bank no. 2)

P0335 Crankshaft Position Sensor "A" Circuit Malfunction

P0336 Crankshaft Position Sensor "A" Circuit Range/Performance

P0337 Crankshaft Position Sensor "A" Circuit Low Input

P0338 Crankshaft Position Sensor "A" Circuit High Input

P0339 Crankshaft Position Sensor "A" Circuit Intermittent

P0340 Camshaft Position Sensor Circuit Malfunction

P0341 Camshaft Position Sensor Circuit Range/Performance

P0342 Camshaft Position Sensor Circuit Low Input

P0343 Camshaft Position Sensor Circuit High Input

P0344 Camshaft Position Sensor Circuit Intermittent

P0350 Ignition Coil Primary/Secondary Circuit Malfunction

P0351 Ignition Coil "A" Primary/Secondary Circuit Malfunction

P0352 Ignition Coil "B" Primary/Secondary Circuit Malfunction

P0353 Ignition Coil "C" Primary/Secondary Circuit Malfunction

P0354 Ignition Coil "D" Primary/Secondary Circuit Malfunction

P0355 Ignition Coil "E" Primary/Secondary Circuit Malfunction

P0356 Ignition Coil "F" Primary/Secondary Circuit Malfunction

P0357 Ignition Coil "G" Primary/Secondary Circuit Malfunction

P0358 Ignition Coil "H" Primary/Secondary Circuit Malfunction

P0359 Ignition Coil "I" Primary/Secondary Circuit Malfunction

P0360 Ignition Coil "J" Primary/Secondary Circuit Malfunction

P0361 Ignition Coil "K" Primary/Secondary Circuit Malfunction

P0362 Ignition Coil "L" Primary/Secondary Circuit Malfunction

P0370 Timing Reference High Resolution Signal "A" Malfunction

P0371 Timing Reference High Resolution Signal "A" Too Many Pulses

P0372 Timing Reference High Resolution Signal "A" Too Few Pulses

P0373 Timing Reference High Resolution Signal "A" Intermittent/Erratic Pulses

P0374 Timing Reference High Resolution Signal "A" No Pulses

P0375 Timing Reference High Resolution Signal "B" Malfunction

P0376 Timing Reference High Resolution Signal "B" Too Many Pulses

P0377 Timing Reference High Resolution Signal "B" Too Few Pulses

P0378 Timing Reference High Resolution Signal "B" Intermittent/Erratic Pulses

P0379 Timing Reference High Resolution Signal "B" No Pulses

P0380 Glow Plug/Heater Circuit "A" Malfunction

P0381 Glow Plug/Heater Indicator Circuit Malfunction

P0382 Glow Plug/Heater Circuit "B" Malfunction

P0385 Crankshaft Position Sensor "B" Circuit Malfunction

P0386 Crankshaft Position Sensor "B" Circuit Range/Performance

P0387 Crankshaft Position Sensor "B" Circuit Low Input

P0388 Crankshaft Position Sensor "B" Circuit High Input

P0389 Crankshaft Position Sensor "B" Circuit Intermittent

P0400 Exhaust Gas Recirculation Flow Malfunction

P0401 Exhaust Gas Recirculation Flow Insufficient Detected

P0402 Exhaust Gas Recirculation Flow Excessive Detected

P0403 Exhaust Gas Recirculation Circuit Malfunction

P0404 Exhaust Gas Recirculation Circuit Range/Performance

P0405 Exhaust Gas Recirculation Sensor "A" Circuit Low

P0406 Exhaust Gas Recirculation Sensor "A" Circuit High

P0407 Exhaust Gas Recirculation Sensor "B" Circuit Low

P0408 Exhaust Gas Recirculation Sensor "B" Circuit High

P0410 Secondary Air Injection System Malfunction

P0411 Secondary Air Injection System Incorrect Flow Detected

P0412 Secondary Air Injection System Switching Valve "A" Circuit Malfunction

P0413 Secondary Air Injection System Switching Valve "A" Circuit Open

P0414 Secondary Air Injection System Switching Valve "A" Circuit Shorted

P0415 Secondary Air Injection System Switching Valve "B" Circuit Malfunction

P0416 Secondary Air Injection System Switching Valve "B" Circuit Open

P0417 Secondary Air Injection System Switching Valve "B" Circuit Shorted

P0418 Secondary Air Injection System Relay "A" Circuit Malfunction

P0419 Secondary Air Injection System Relay "B" Circuit Malfunction

P0420 Catalyst System Efficiency Below Threshold (Bank no. 1)

P0421 Warm Up Catalyst Efficiency Below Threshold (Bank no. 1)

P0422 Main Catalyst Efficiency Below Threshold (Bank no. 1)

P0423 Heated Catalyst Efficrency Below Threshold (Bank no. 1)

P0424 Heated Catalyst Temperature Below Threshold (Bank no. 1)

P0430 Catalyst System Efficiency Below Threshold (Bank no. 2)

P0431 Warm Up Catalyst Efficiency Below Threshold (Bank no. 2)

P0432 Main Catalyst Efficiency Below Threshold (Bank no. 2)

P0433 Heated Catalyst Efficiency Below Threshold (Bank no. 2)

P0434 Heated Catalyst Temperature Below Threshold (Bank no. 2)

P0440 Evaporative Emission Control System Malfunction

P0441 Evaporative Emission Control System Incorrect Purge Flow

P0442 Evaporative Emission Control System Leak Detected (Small Leak)

P0443 Evaporative Emission Control System Purge Control Valve Circuit Malfunction

P0444 Evaporative Emission Control System Purge Control Valve Circuit Open

P0445 Evaporative Emission Control System Purge Control Valve Circuit Shorted

P0446 Evaporative Emission Control System Vent Control Circuit Malfunction

P0447 Evaporative Emission Control System Vent Control Circuit Open

P0448 Evaporative Emission Control System Vent Control Circuit Shorted

P0449 Evaporative Emission Control System Vent Valve/Solenoid Circuit Malfunction

P0450 Evaporative Emission Control System Pressure Sensor Malfunction

P0451 Evaporative Emission Control System Pressure Sensor Range/Performance

P0452 Evaporative Emission Control System Pressure Sensor Low Input

P0453 Evaporative Emission Control System Pressure Sensor High Input

P0454 Evaporative Emission Control System Pressure Sensor Intermittent

P0455 Evaporative Emission Control System Leak Detected (Gross Leak)

P0460 Fuel Level Sensor Circuit Malfunction

P0461 Fuel Level Sensor Circuit Range/Performance

P0462 Fuel Level Sensor Circuit Low Input

P0463 Fuel Level Sensor Circuit High Input

P0464 Fuel Level Sensor Circuit Intermittent

P0465 Purge Flow Sensor Circuit Malfunction

P0466 Purge Flow Sensor Circuit Range/Performance

P0467 Purge Flow Sensor Circuit Low Input

P0468 Purge Flow Sensor Circuit High Input

P0469 Purge Flow Sensor Circuit Intermittent

P0470 Exhaust Pressure Sensor Malfunction

P0471 Exhaust Pressure Sensor Range/Performance

P0472 Exhaust Pressure Sensor Low

P0473 Exhaust Pressure Sensor High

P0474 Exhaust Pressure Sensor Intermittent

P0475 Exhaust Pressure Control Valve Malfunction

P0476 Exhaust Pressure Control Valve Range/Performance

P0477 Exhaust Pressure Control Valve Low

P0478 Exhaust Pressure Control Valve High

P0479 Exhaust Pressure Control Valve Intermittent

P0480 Cooling Fan no. 1 Control Circuit Malfunction

P0481 Cooling Fan no. 2 Control Circuit Malfunction
P0482 Cooling Fan no. 3 Control Circuit Malfunction
P0483 Cooling Fan Rationality Check Malfunction
P0484 Cooling Fan Circuit Over Current
P0485 Cooling Fan Power/Ground Circuit Malfunction
P0500 Vehicle Speed Sensor Malfunction
P0501 Vehicle Speed Sensor Range/Performance
P0502 Vehicle Speed Sensor Circuit Low Input
P0503 Vehicle Speed Sensor Intermittent/Erratic/High
P0505 Idle Control System Malfunction
P0506 Idle Control System RPM Lower Than Expected
P0507 Idle Control System RPM Higher Than Expected
P0510 Closed Throttle Position Switch Malfunction
P0520 Engine Oil Pressure Sensor/Switch Circuit Malfunction
P0521 Engine Oil Pressure Sensor/Switch Range/Performance
P0522 Engine Oil Pressure Sensor/Switch Low Voltage
P0523 Engine Oil Pressure Sensor/Switch High Voltage
P0530 A/C Refrigerant Pressure Sensor Circuit Malfunction
P0531 A/C Refrigerant Pressure Sensor Circuit Range/Performance
P0532 A/C Refrigerant Pressure Sensor Circuit Low Input
P0533 A/C Refrigerant Pressure Sensor Circuit High Input
P0534 A/C Refrigerant Charge Loss
P0550 Power Steering Pressure Sensor Circuit Malfunction
P0551 Power Steering Pressure Sensor Circuit Range/Performance
P0552 Power Steering Pressure Sensor Circuit Low Input
P0553 Power Steering Pressure Sensor Circuit High Input
P0554 Power Steering Pressure Sensor Circuit Intermittent
P0560 System Voltage Malfunction
P0561 System Voltage Unstable
P0562 System Voltage Low
P0563 System Voltage High
P0565 Cruise Control On Signal Malfunction
P0566 Cruise Control Off Signal Malfunction
P0567 Cruise Control Resume Signal Malfunction
P0568 Cruise Control Set Signal Malfunction
P0569 Cruise Control Coast Signal Malfunction
P0570 Cruise Control Accel Signal Malfunction
P0571 Cruise Control/Brake Switch "A" Circuit Malfunction
P0572 Cruise Control/Brake Switch "A" Circuit Low
P0573 Cruise Control/Brake Switch "A" Circuit High
P0574 Through P0580 Reserved for Cruise Codes
P0600 Serial Communication Link Malfunction
P0601 Internal Control Module Memory Check Sum Error
P0602 Control Module Programming Error
P0603 Internal Control Module Keep Alive Memory (KAM) Error
P0604 Internal Control Module Random Access Memory (RAM) Error
P0605 Internal Control Module Read Only Memory (ROM) Error
P0606 PCM Processor Fault
P0608 Control Module VSS Output "A" Malfunction
P0609 Control Module VSS Output "B" Malfunction
P0620 Generator Control Circuit Malfunction
P0621 Generator Lamp "L" Control Circuit Malfunction
P0622 Generator Field "F" Control Circuit Malfunction
P0650 Malfunction Indicator Lamp (MIL) Control Circuit Malfunction

P0654 Engine RPM Output Circuit Malfunction
P0655 Engine Hot Lamp Output Control Circuit Malfunction
P0656 Fuel Level Output Circuit Malfunction
P0700 Transmission Control System Malfunction
P0701 Transmission Control System Range/Performance
P0702 Transmission Control System Electrical
P0703 Torque Converter/Brake Switch "B" Circuit Malfunction
P0704 Clutch Switch Input Circuit Malfunction
P0705 Transmission Range Sensor Circuit Malfunction (PRNDL Input)
P0706 Transmission Range Sensor Circuit Range/Performance
P0707 Transmission Range Sensor Circuit Low Input
P0708 Transmission Range Sensor Circuit High Input
P0709 Transmission Range Sensor Circuit Intermittent
P0710 Transmission Fluid Temperature Sensor Circuit Malfunction
P0711 Transmission Fluid Temperature Sensor Circuit Range/Performance
P0712 Transmission Fluid Temperature Sensor Circuit Low Input
P0713 Transmission Fluid Temperature Sensor Circuit High Input
P0714 Transmission Fluid Temperature Sensor Circuit Intermittent
P0715 Input/Turbine Speed Sensor Circuit Malfunction
P0716 Input/Turbine Speed Sensor Circuit Range/Performance
P0717 Input/Turbine Speed Sensor Circuit No Signal
P0718 Input/Turbine Speed Sensor Circuit Intermittent
P0719 Torque Converter/Brake Switch "B" Circuit Low
P0720 Output Speed Sensor Circuit Malfunction
P0721 Output Speed Sensor Circuit Range/Performance
P0722 Output Speed Sensor Circuit No Signal
P0723 Output Speed Sensor Circuit Intermittent
P0724 Torque Converter/Brake Switch "B" Circuit High
P0725 Engine Speed Input Circuit Malfunction
P0726 Engine Speed Input Circuit Range/Performance
P0727 Engine Speed Input Circuit No Signal
P0728 Engine Speed Input Circuit Intermittent
P0730 Incorrect Gear Ratio
P0731 Gear no. 1 Incorrect Ratio
P0732 Gear no. 2 Incorrect Ratio
P0733 Gear no. 3 Incorrect Ratio
P0734 Gear no. 4 Incorrect Ratio
P0735 Gear no. 5 Incorrect Ratio
P0736 Reverse Incorrect Ratio
P0740 Torque Converter Clutch Circuit Malfunction
P0741 Torque Converter Clutch Circuit Performance or Stuck Off
P0742 Torque Converter Clutch Circuit Stuck On
P0743 Torque Converter Clutch Circuit Electrical
P0744 Torque Converter Clutch Circuit Intermittent
P0745 Pressure Control Solenoid Malfunction
P0746 Pressure Control Solenoid Performance or Stuck Off
P0747 Pressure Control Solenoid Stuck On
P0748 Pressure Control Solenoid Electrical
P0749 Pressure Control Solenoid Intermittent
P0750 Shift Solenoid "A" Malfunction
P0751 Shift Solenoid "A" Performance or Stuck Off
P0752 Shift Solenoid "A" Stuck On
P0753 Shift Solenoid "A" Electrical

P0754 Shift Solenoid "A" Intermittent
P0755 Shift Solenoid "B" Malfunction
P0756 Shift Solenoid "B" Performance or Stuck Off
P0757 Shift Solenoid "B" Stuck On
P0758 Shift Solenoid "B" Electrical
P0759 Shift Solenoid "B" Intermittent
P0760 Shift Solenoid "C" Malfunction
P0761 Shift Solenoid "C" Performance Or Stuck Off
P0762 Shift Solenoid "C" Stuck On
P0763 Shift Solenoid "C" Electrical
P0764 Shift Solenoid "C" Intermittent
P0765 Shift Solenoid "D" Malfunction
P0766 Shift Solenoid "D" Performance Or Stuck Off
P0767 Shift Solenoid "D" Stuck On
P0768 Shift Solenoid "D" Electrical
P0769 Shift Solenoid "D" Intermittent
P0770 Shift Solenoid "E" Malfunction
P0771 Shift Solenoid "E" Performance Or Stuck Off
P0772 Shift Solenoid "E" Stuck On
P0773 Shift Solenoid "E" Electrical

P0774 Shift Solenoid "E" Intermittent
P0780 Shift Malfunction
P0781 1–2 Shift Malfunction
P0782 2–3 Shift Malfunction
P0783 3–4 Shift Malfunction
P0784 4–5 Shift Malfunction
P0785 Shift/Timing Solenoid Malfunction
P0786 Shift/Timing Solenoid Range/Performance
P0787 Shift/Timing Solenoid Low
P0788 Shift/Timing Solenoid High
P0789 Shift/Timing Solenoid Intermittent
P0790 Normal/Performance Switch Circuit Malfunction
P0801 Reverse Inhibit Control Circuit Malfunction
P0803 1–4 Upshift (Skip Shift) Solenoid Control Circuit Malfunction
P0804 1–4 Upshift (Skip Shift) Lamp Control Circuit Malfunction

FLASH OUT CODE LIST

Code	Output pattern (for voltmeter)	Cause	Remedy
P1702	A5AT005E	Shorted throttle position sensor circuit	o Check the throttle position sensor connector o check the throttle position sensor itself o Check the closed throttle position switch o Check the throttle position sensor wiring harness o Check the wiring between ECM and throttle position sensor
P1701	A5AT005F	Open throttle position sensor circuit	
P1704	A5AT005H	Throttle position sensor malfunction Improperly adjusted throttle position sensor	
P0712	A5AT005I	Open fluid temperature sensor circuit	o Fluid temperature sensor connector inspection o Fluid temperature sensor inspection o Fluid temperature sensor wiring harness inspection
P0713	A5AT005J	Shorted fluid temperature sensor circuit	
P1709	A5AT005K	Open kickdown servo switch circuit Shorted kickdown servo switch circuit	o Check the kickdown servo switch connector o Check the kickdown servo switch o Check the kickdown servo switch wiring harness

89446G27

Hyundai flash out DTC's, 1 of 4—Type 4 (OBD II) Codes

Code	Output pattern (for voltmeter)	Cause	Remedy
P0727	A5AT005L	Open ignition pulse pickup cable circuit	o Check the ignition pulse signal line o Check the wiring between ECM and ignisiton system
P1714	A5AT005M	Short-circuited or improperly adjusted closed throttle position switch	o Check the closed throttle position switch connector o Check the closed throttle position switch itself o Adjust the closed throttle position switch o Check the closed throttle position switch wiring harness
P0717	A5AT005N	Open-circuited pulse generator A	o Check the pulse generator A and pulse generator B o Check the vehicle speed reed switch (for chattering) o Check the pulse generator A and B wiring harness
P0722	A5AT005O	Open-circuited pulse generator B	
P0707	A5AT005Z	No input signal	o Check the transaxle range switch o Check the transaxle range wiring harness o Check the manual control cable
P0708	A5ATA05A	More than two input signals	
P0752	A5AT005P	Open shift control solenoid vave A circuit	o Check the solenoid valve connector o Check the shift control solenoid valve A o Check the shift control solenoid valve A wiring harness
P0753	A5AT005Q	Shorted shift control solenoid valve A circuit	
P0757	A5AT005R	Open shift control solenoid valve B circuit	o Check the shift control solenoid valve connector o Check the shift control solenoid valve B wiring harness o Check the shift control solenoid valve B
P0758	A5AT005S	Short shift control solenoid valve B circuit	
P0747	A5AT005T	Open pressure control solenoid valve circuit	o Check the pressure control solenoid valve o Check the pressure control solenoid valve wiring harness
P0748	A5AT005U	Shorted pressure control solenoid valve circuit	

89446G28

Hyundai flash out DTC's, 2 of 4—Type 4 (OBD II) Codes

Code	Output pattern (for voltmeter)	Cause	Remedy
P0743	A5AT005V	Open circuit in damper clutch control solenoid valve	o Inspection of solenoid valve connector
P0742	A5AT005W	Short circuit in damper clutch control solenoid valve	o Individual inspection of damper clutch control solenoid valve
P0740	A5AT005X	Defect in the damper clutch system	o Check the damper clutch control solenoid valve wiring harness
P1744	A5AT005Y		o Chck the TCM o Inspection of damper clutch hydraulic system
P0731	A5ATA05B	Shifting to first gear does not match the engine speed	o Check the pulse generator A and pulse generator B connector o Check the pulse generator A and puls generator B
P0732	A5ATA05C	Shifting to second gear does not match the engine speed	o Check the one way clutch or rear clutch o Check the pulse generator wiring harness o Kickdown brake slippage
P0733	A5ATA05D	Shifting to third gear does not match the engine speed	o Check the rear clutch or control system o Check the pulse generator A and pulse generator B connector o Check the pulse generator A and pulse generator B o Check the pulse generator wiring harness o Check the rear clutch slippage or control system o Check the front clutch slippage or control system
P0734	A5ATA05E	Shifting to fourth gear does not match the engine speed	o Check the pulse generator A and B connector o Check the pulse generator A and B o Kickdown brake slippage o Check the end clutch or control system o Check the pulse generator wiring harness
-		Normal	

Hyundai flash out DTC's, 3 of 4—Type 4 (OBD II) Codes

FAIL-SAFE ITEM

Code No.	Output code — Output pattern (for voltmeter)	Description	Fail-safe	Note (relation to diagnostic trouble code)
P0717	A5AT005N	Open-circuited pulse generator A	Locked in third (D) or second (2,L)	When code No.0717 is generated fourth time
P0722	A5AT005O	Open-circuited pulse generator B	Locked in third (D) or second (2,L)	When code No.0722 is generated fourth time
P0752	A5AT005P	Open-circuited or shorted shift control solenoid valve A	Lock in third	When code No.0752 or 0753 is generated fourth time
P0753	A5AT005Q			
P0757	A5AT005R	Open-circuited or shorted shift control solenoid valve B	Lock in third gear	When code No.0757 or 0758 is generated fourth time
P0758	A5AT005S			
P0747	A5AT005T	Open-circuited or shorted pressure control solenoid valve	Locked in third (D) or second (2,L)	When code No.0747 or 0748 is generated fourth time
P0748	A5AT005U			
P0731	A5ATA05B	Gear shifting does not match the engine speed	Locked in third (D) or second (2,L)	When either code No.0731, 0732, 0733 or 0734 is generated fourth time
P0732	A5ATA05C			
P0733	A5ATA05D			
P0734	A5ATA05E			

89446G30

Hyundai flash out DTC's, 4 of 4—Type 4 (OBD II) Codes

Infiniti

READING CODES

Reading the control module memory is on of the first steps in OBD II system diagnostics. This step should be initially performed to determine the general nature of the fault. Subsequent readings will determine if the fault has been cleared.

Reading codes can be performed by any of the methods below:

• Read the control module memory with the Generic Scan Tool (GST)

• Read the control module memory with the vehicle manufacturer's specific tester

To read the fault codes, connect the scan tool or tester according to the manufacturer's instructions. Follow the manufacturer's specified procedure for reading the codes.

CLEARING CODES

Control module reset procedures are a very important part of OBD II System diagnostics.

This step should be done at the end of any fault code repair and at the end of any driveability repair.

Clearing codes can be performed by any of the methods below:

• Clear the control module memory with the Generic Scan Tool (GST)

• Clear the control module memory with the vehicle manufacturer's specific tester

➡**The MIL will may also be de-activated for some codes if the vehicle completes three consecutive trips without a fault detected with vehicle conditions similar to those present during the fault.**

DIAGNOSTIC TROUBLE CODES

P0000 No Failures
P0100 Mass or Volume Air Flow Circuit Malfunction
P0101 Mass or Volume Air Flow Circuit Range/Performance Problem
P0102 Mass or Volume Air Flow Circuit Low Input
P0103 Mass or Volume Air Flow Circuit High Input
P0104 Mass or Volume Air Flow Circuit Intermittent
P0105 Manifold Absolute Pressure/Barometric Pressure Circuit Malfunction
P0106 Manifold Absolute Pressure/Barometric Pressure Circuit Range/Performance Problem
P0107 Manifold Absolute Pressure/Barometric Pressure Circuit Low Input
P0108 Manifold Absolute Pressure/Barometric Pressure Circuit High Input
P0109 Manifold Absolute Pressure/Barometric Pressure Circuit Intermittent
P0110 Intake Air Temperature Circuit Malfunction
P0111 Intake Air Temperature Circuit Range/Performance Problem
P0112 Intake Air Temperature Circuit Low Input
P0113 Intake Air Temperature Circuit High Input
P0114 Intake Air Temperature Circuit Intermittent
P0115 Engine Coolant Temperature Circuit Malfunction
P0116 Engine Coolant Temperature Circuit Range/Performance Problem

P0117 Engine Coolant Temperature Circuit Low Input
P0118 Engine Coolant Temperature Circuit High Input
P0119 Engine Coolant Temperature Circuit Intermittent
P0120 Throttle/Pedal Position Sensor/Switch "A" Circuit Malfunction
P0121 Throttle/Pedal Position Sensor/Switch "A" Circuit Range/Performance Problem
P0122 Throttle/Pedal Position Sensor/Switch "A" Circuit Low Input
P0123 Throttle/Pedal Position Sensor/Switch "A" Circuit High Input
P0124 Throttle/Pedal Position Sensor/Switch "A" Circuit Intermittent
P0125 Insufficient Coolant Temperature For Closed Loop Fuel Control
P0126 Insufficient Coolant Temperature For Stable Operation
P0130 O_2 Circuit Malfunction (Bank no. 1 Sensor no. 1)
P0131 O_2 Sensor Circuit Low Voltage (Bank no. 1 Sensor no. 1)
P0132 O_2 Sensor Circuit High Voltage (Bank no. 1 Sensor no. 1)
P0133 O_2 Sensor Circuit Slow Response (Bank no. 1 Sensor no. 1)
P0134 O_2 Sensor Circuit No Activity Detected (Bank no. 1 Sensor no. 1)
P0135 O_2 Sensor Heater Circuit Malfunction (Bank no. 1 Sensor no. 1)
P0136 O_2 Sensor Circuit Malfunction (Bank no. 1 Sensor no. 2)
P0137 O_2 Sensor Circuit Low Voltage (Bank no. 1 Sensor no. 2)
P0138 O_2 Sensor Circuit High Voltage (Bank no. 1 Sensor no. 2)
P0139 O_2 Sensor Circuit Slow Response (Bank no. 1 Sensor no. 2)
P0140 O_2 Sensor Circuit No Activity Detected (Bank no. 1 Sensor no. 2)
P0141 O_2 Sensor Heater Circuit Malfunction (Bank no. 1 Sensor no. 2)
P0142 O_2 Sensor Circuit Malfunction (Bank no. 1 Sensor no. 3)
P0143 O_2 Sensor Circuit Low Voltage (Bank no. 1 Sensor no. 3)
P0144 O_2 Sensor Circuit High Voltage (Bank no. 1 Sensor no. 3)
P0145 O_2 Sensor Circuit Slow Response (Bank no. 1 Sensor no. 3)
P0146 O_2 Sensor Circuit No Activity Detected (Bank no. 1 Sensor no. 3)
P0147 O_2 Sensor Heater Circuit Malfunction (Bank no. 1 Sensor no. 3)
P0150 O_2 Sensor Circuit Malfunction (Bank no. 2 Sensor no. 1)
P0151 O_2 Sensor Circuit Low Voltage (Bank no. 2 Sensor no. 1)
P0152 O_2 Sensor Circuit High Voltage (Bank no. 2 Sensor no. 1)
P0153 O_2 Sensor Circuit Slow Response (Bank no. 2 Sensor no. 1)
P0154 O_2 Sensor Circuit No Activity Detected (Bank no. 2 Sensor no. 1)
P0155 O_2 Sensor Heater Circuit Malfunction (Bank no. 2 Sensor no. 1)
P0156 O_2 Sensor Circuit Malfunction (Bank no. 2 Sensor no. 2)

P0157 O$_2$ Sensor Circuit Low Voltage (Bank no. 2 Sensor no. 2)

P0158 O$_2$ Sensor Circuit High Voltage (Bank no. 2 Sensor no. 2)

P0159 O$_2$ Sensor Circuit Slow Response (Bank no. 2 Sensor no. 2)

P0160 O$_2$ Sensor Circuit No Activity Detected (Bank no. 2 Sensor no. 2)

P0161 O$_2$ Sensor Heater Circuit Malfunction (Bank no. 2 Sensor no. 2)

P0162 O$_2$ Sensor Circuit Malfunction (Bank no. 2 Sensor no. 3)

P0163 O$_2$ Sensor Circuit Low Voltage (Bank no. 2 Sensor no. 3)

P0164 O$_2$ Sensor Circuit High Voltage (Bank no. 2 Sensor no. 3)

P0165 O$_2$ Sensor Circuit Slow Response (Bank no. 2 Sensor no. 3)

P0166 O$_2$ Sensor Circuit No Activity Detected (Bank no. 2 Sensor no. 3)

P0167 O$_2$ Sensor Heater Circuit Malfunction (Bank no. 2 Sensor no. 3)

P0170 Fuel Trim Malfunction (Bank no. 1)
P0171 System Too Lean (Bank no. 1)
P0172 System Too Rich (Bank no. 1)
P0173 Fuel Trim Malfunction (Bank no. 2)
P0174 System Too Lean (Bank no. 2)
P0175 System Too Rich (Bank no. 2)
P0176 Fuel Composition Sensor Circuit Malfunction
P0177 Fuel Composition Sensor Circuit Range/Performance
P0178 Fuel Composition Sensor Circuit Low Input
P0179 Fuel Composition Sensor Circuit High Input
P0180 Fuel Temperature Sensor "A" Circuit Malfunction
P0181 Fuel Temperature Sensor "A" Circuit Range/Performance
P0182 Fuel Temperature Sensor "A" Circuit Low Input
P0183 Fuel Temperature Sensor "A" Circuit High Input
P0184 Fuel Temperature Sensor "A" Circuit Intermittent
P0185 Fuel Temperature Sensor "B" Circuit Malfunction
P0186 Fuel Temperature Sensor "B" Circuit Range/Performance
P0187 Fuel Temperature Sensor "B" Circuit Low Input
P0188 Fuel Temperature Sensor "B" Circuit High Input
P0189 Fuel Temperature Sensor "B" Circuit Intermittent
P0190 Fuel Rail Pressure Sensor Circuit Malfunction
P0191 Fuel Rail Pressure Sensor Circuit Range/Performance
P0192 Fuel Rail Pressure Sensor Circuit Low Input
P0193 Fuel Rail Pressure Sensor Circuit High Input
P0194 Fuel Rail Pressure Sensor Circuit Intermittent
P0195 Engine Oil Temperature Sensor Malfunction
P0196 Engine Oil Temperature Sensor Range/Performance
P0197 Engine Oil Temperature Sensor Low
P0198 Engine Oil Temperature Sensor High
P0199 Engine Oil Temperature Sensor Intermittent
P0200 Injector Circuit Malfunction
P0201 Injector Circuit Malfunction—Cylinder no. 1
P0202 Injector Circuit Malfunction—Cylinder no. 2
P0203 Injector Circuit Malfunction—Cylinder no. 3
P0204 Injector Circuit Malfunction—Cylinder no. 4
P0205 Injector Circuit Malfunction—Cylinder no. 5
P0206 Injector Circuit Malfunction—Cylinder no. 6
P0207 Injector Circuit Malfunction—Cylinder no. 7
P0208 Injector Circuit Malfunction—Cylinder no. 8
P0209 Injector Circuit Malfunction—Cylinder no. 9
P0210 Injector Circuit Malfunction—Cylinder no. 10

P0211 Injector Circuit Malfunction—Cylinder no. 11
P0212 Injector Circuit Malfunction—Cylinder no. 12
P0213 Cold Start Injector no. 1 Malfunction
P0214 Cold Start Injector no. 2 Malfunction
P0215 Engine Shutoff Solenoid Malfunction
P0216 Injection Timing Control Circuit Malfunction
P0217 Engine Over Temperature Condition
P0218 Transmission Over Temperature Condition
P0219 Engine Over Speed Condition
P0220 Throttle/Pedal Position Sensor/Switch "B" Circuit Malfunction
P0221 Throttle/Pedal Position Sensor/Switch "B" Circuit Range/Performance Problem
P0222 Throttle/Pedal Position Sensor/Switch "B" Circuit Low Input
P0223 Throttle/Pedal Position Sensor/Switch "B" Circuit High Input
P0224 Throttle/Pedal Position Sensor/Switch "B" Circuit Intermittent
P0225 Throttle/Pedal Position Sensor/Switch "C" Circuit Malfunction
P0226 Throttle/Pedal Position Sensor/Switch "C" Circuit Range/Performance Problem
P0227 Throttle/Pedal Position Sensor/Switch "C" Circuit Low Input
P0228 Throttle/Pedal Position Sensor/Switch "C" Circuit High Input
P0229 Throttle/Pedal Position Sensor/Switch "C" Circuit Intermittent
P0230 Fuel Pump Primary Circuit Malfunction
P0231 Fuel Pump Secondary Circuit Low
P0232 Fuel Pump Secondary Circuit High
P0233 Fuel Pump Secondary Circuit Intermittent
P0234 Engine Over Boost Condition
P0261 Cylinder no. 1 Injector Circuit Low
P0262 Cylinder no. 1 Injector Circuit High
P0263 Cylinder no. 1 Contribution/Balance Fault
P0264 Cylinder no. 2 Injector Circuit Low
P0265 Cylinder no. 2 Injector Circuit High
P0266 Cylinder no. 2 Contribution/Balance Fault
P0267 Cylinder no. 3 Injector Circuit Low
P0268 Cylinder no. 3 Injector Circuit High
P0269 Cylinder no. 3 Contribution/Balance Fault
P0270 Cylinder no. 4 Injector Circuit Low
P0271 Cylinder no. 4 Injector Circuit High
P0272 Cylinder no. 4 Contribution/Balance Fault
P0273 Cylinder no. 5 Injector Circuit Low
P0274 Cylinder no. 5 Injector Circuit High
P0275 Cylinder no. 5 Contribution/Balance Fault
P0276 Cylinder no. 6 Injector Circuit Low
P0277 Cylinder no. 6 Injector Circuit High
P0278 Cylinder no. 6 Contribution/Balance Fault
P0279 Cylinder no. 7 Injector Circuit Low
P0280 Cylinder no. 7 Injector Circuit High
P0281 Cylinder no. 7 Contribution/Balance Fault
P0282 Cylinder no. 8 Injector Circuit Low
P0283 Cylinder no. 8 Injector Circuit High
P0284 Cylinder no. 8 Contribution/Balance Fault
P0285 Cylinder no. 9 Injector Circuit Low
P0286 Cylinder no. 9 Injector Circuit High
P0287 Cylinder no. 9 Contribution/Balance Fault
P0288 Cylinder no. 10 Injector Circuit Low

P0289 Cylinder no. 10 Injector Circuit High
P0290 Cylinder no. 10 Contribution/Balance Fault
P0291 Cylinder no. 11 Injector Circuit Low
P0292 Cylinder no. 11 Injector Circuit High
P0293 Cylinder no. 11 Contribution/Balance Fault
P0294 Cylinder no. 12 Injector Circuit Low
P0295 Cylinder no. 12 Injector Circuit High
P0296 Cylinder no. 12 Contribution/Balance Fault
P0300 Random/Multiple Cylinder Misfire Detected
P0301 Cylinder no. 1—Misfire Detected
P0302 Cylinder no. 2—Misfire Detected
P0303 Cylinder no. 3—Misfire Detected
P0304 Cylinder no. 4—Misfire Detected
P0305 Cylinder no. 5—Misfire Detected
P0306 Cylinder no. 6—Misfire Detected
P0307 Cylinder no. 7—Misfire Detected
P0308 Cylinder no. 8—Misfire Detected
P0309 Cylinder no. 9—Misfire Detected
P0310 Cylinder no. 10—Misfire Detected
P0311 Cylinder no. 11—Misfire Detected
P0312 Cylinder no. 12—Misfire Detected
P0320 Ignition/Distributor Engine Speed Input Circuit Malfunction
P0321 Ignition/Distributor Engine Speed Input Circuit Range/Performance
P0322 Ignition/Distributor Engine Speed Input Circuit No Signal
P0323 Ignition/Distributor Engine Speed Input Circuit Intermittent
P0325 Knock Sensor no. 1—Circuit Malfunction (Bank no. 1 or Single Sensor)
P0326 Knock Sensor no. 1—Circuit Range/Performance (Bank no. 1 or Single Sensor)
P0327 Knock Sensor no. 1—Circuit Low Input (Bank no. 1 or Single Sensor)
P0328 Knock Sensor no. 1—Circuit High Input (Bank no. 1 or Single Sensor)
P0329 Knock Sensor no. 1—Circuit Input Intermittent (Bank no. 1 or Single Sensor)
P0330 Knock Sensor no. 2—Circuit Malfunction (Bank no. 2)
P0331 Knock Sensor no. 2—Circuit Range/Performance (Bank no. 2)
P0332 Knock Sensor no. 2—Circuit Low Input (Bank no. 2)
P0333 Knock Sensor no. 2—Circuit High Input (Bank no. 2)
P0334 Knock Sensor no. 2—Circuit Input Intermittent (Bank no. 2)
P0335 Crankshaft Position Sensor "A" Circuit Malfunction
P0336 Crankshaft Position Sensor "A" Circuit Range/Performance
P0337 Crankshaft Position Sensor "A" Circuit Low Input
P0338 Crankshaft Position Sensor "A" Circuit High Input
P0339 Crankshaft Position Sensor "A" Circuit Intermittent
P0340 Camshaft Position Sensor Circuit Malfunction
P0341 Camshaft Position Sensor Circuit Range/Performance
P0342 Camshaft Position Sensor Circuit Low Input
P0343 Camshaft Position Sensor Circuit High Input
P0344 Camshaft Position Sensor Circuit Intermittent
P0350 Ignition Coil Primary/Secondary Circuit Malfunction
P0351 Ignition Coil "A" Primary/Secondary Circuit Malfunction
P0352 Ignition Coil "B" Primary/Secondary Circuit Malfunction
P0353 Ignition Coil "C" Primary/Secondary Circuit Malfunction
P0354 Ignition Coil "D" Primary/Secondary Circuit Malfunction
P0355 Ignition Coil "E" Primary/Secondary Circuit Malfunction

P0356 Ignition Coil "F" Primary/Secondary Circuit Malfunction
P0357 Ignition Coil "G" Primary/Secondary Circuit Malfunction
P0358 Ignition Coil "H" Primary/Secondary Circuit Malfunction
P0359 Ignition Coil "I" Primary/Secondary Circuit Malfunction
P0360 Ignition Coil "J" Primary/Secondary Circuit Malfunction
P0361 Ignition Coil "K" Primary/Secondary Circuit Malfunction
P0362 Ignition Coil "L" Primary/Secondary Circuit Malfunction
P0370 Timing Reference High Resolution Signal "A" Malfunction
P0371 Timing Reference High Resolution Signal "A" Too Many Pulses
P0372 Timing Reference High Resolution Signal "A" Too Few Pulses
P0373 Timing Reference High Resolution Signal "A" Intermittent/Erratic Pulses
P0374 Timing Reference High Resolution Signal "A" No Pulses
P0375 Timing Reference High Resolution Signal "B" Malfunction
P0376 Timing Reference High Resolution Signal "B" Too Many Pulses
P0377 Timing Reference High Resolution Signal "B" Too Few Pulses
P0378 Timing Reference High Resolution Signal "B" Intermittent/Erratic Pulses
P0379 Timing Reference High Resolution Signal "B" No Pulses
P0380 Glow Plug/Heater Circuit "A" Malfunction
P0381 Glow Plug/Heater Indicator Circuit Malfunction
P0382 Glow Plug/Heater Circuit "B" Malfunction
P0385 Crankshaft Position Sensor "B" Circuit Malfunction
P0386 Crankshaft Position Sensor "B" Circuit Range/Performance
P0387 Crankshaft Position Sensor "B" Circuit Low Input
P0388 Crankshaft Position Sensor "B" Circuit High Input
P0389 Crankshaft Position Sensor "B" Circuit Intermittent
P0400 Exhaust Gas Recirculation Flow Malfunction
P0401 Exhaust Gas Recirculation Flow Insufficient Detected
P0402 Exhaust Gas Recirculation Flow Excessive Detected
P0403 Exhaust Gas Recirculation Circuit Malfunction
P0404 Exhaust Gas Recirculation Circuit Range/Performance
P0405 Exhaust Gas Recirculation Sensor "A" Circuit Low
P0406 Exhaust Gas Recirculation Sensor "A" Circuit High
P0407 Exhaust Gas Recirculation Sensor "B" Circuit Low
P0408 Exhaust Gas Recirculation Sensor "B" Circuit High
P0410 Secondary Air Injection System Malfunction
P0411 Secondary Air Injection System Incorrect Flow Detected
P0412 Secondary Air Injection System Switching Valve "A" Circuit Malfunction
P0413 Secondary Air Injection System Switching Valve "A" Circuit Open
P0414 Secondary Air Injection System Switching Valve "A" Circuit Shorted
P0415 Secondary Air Injection System Switching Valve "B" Circuit Malfunction
P0416 Secondary Air Injection System Switching Valve "B" Circuit Open
P0417 Secondary Air Injection System Switching Valve "B" Circuit Shorted
P0418 Secondary Air Injection System Relay "A" Circuit Malfunction
P0419 Secondary Air Injection System Relay "B" Circuit Malfunction
P0420 Catalyst System Efficiency Below Threshold (Bank no. 1)

P0421 Warm Up Catalyst Efficiency Below Threshold (Bank no. 1)

P0422 Main Catalyst Efficiency Below Threshold (Bank no. 1)

P0423 Heated Catalyst Efficiency Below Threshold (Bank no. 1)

P0424 Heated Catalyst Temperature Below Threshold (Bank no. 1)

P0430 Catalyst System Efficiency Below Threshold (Bank no. 2)

P0431 Warm Up Catalyst Efficiency Below Threshold (Bank no. 2)

P0432 Main Catalyst Efficiency Below Threshold (Bank no. 2)

P0433 Heated Catalyst Efficiency Below Threshold (Bank no. 2)

P0434 Heated Catalyst Temperature Below Threshold (Bank no. 2)

P0440 Evaporative Emission Control System Malfunction

P0441 Evaporative Emission Control System Incorrect Purge Flow

P0442 Evaporative Emission Control System Leak Detected (Small Leak)

P0443 Evaporative Emission Control System Purge Control Valve Circuit Malfunction

P0444 Evaporative Emission Control System Purge Control Valve Circuit Open

P0445 Evaporative Emission Control System Purge Control Valve Circuit Shorted

P0446 Evaporative Emission Control System Vent Control Circuit Malfunction

P0447 Evaporative Emission Control System Vent Control Circuit Open

P0448 Evaporative Emission Control System Vent Control Circuit Shorted

P0449 Evaporative Emission Control System Vent Valve/Solenoid Circuit Malfunction

P0450 Evaporative Emission Control System Pressure Sensor Malfunction

P0451 Evaporative Emission Control System Pressure Sensor Range/Performance

P0452 Evaporative Emission Control System Pressure Sensor Low Input

P0453 Evaporative Emission Control System Pressure Sensor High Input

P0454 Evaporative Emission Control System Pressure Sensor Intermittent

P0455 Evaporative Emission Control System Leak Detected (Gross Leak)

P0460 Fuel Level Sensor Circuit Malfunction

P0461 Fuel Level Sensor Circuit Range/Performance

P0462 Fuel Level Sensor Circuit Low Input

P0463 Fuel Level Sensor Circuit High Input

P0464 Fuel Level Sensor Circuit Intermittent

P0465 Purge Flow Sensor Circuit Malfunction

P0466 Purge Flow Sensor Circuit Range/Performance

P0467 Purge Flow Sensor Circuit Low Input

P0468 Purge Flow Sensor Circuit High Input

P0469 Purge Flow Sensor Circuit Intermittent

P0470 Exhaust Pressure Sensor Malfunction

P0471 Exhaust Pressure Sensor Range/Performance

P0472 Exhaust Pressure Sensor Low

P0473 Exhaust Pressure Sensor High

P0474 Exhaust Pressure Sensor Intermittent

P0475 Exhaust Pressure Control Valve Malfunction

P0476 Exhaust Pressure Control Valve Range/Performance

P0477 Exhaust Pressure Control Valve Low

P0478 Exhaust Pressure Control Valve High

P0479 Exhaust Pressure Control Valve Intermittent

P0480 Cooling Fan no. 1 Control Circuit Malfunction

P0481 Cooling Fan no. 2 Control Circuit Malfunction

P0482 Cooling Fan no. 3 Control Circuit Malfunction

P0483 Cooling Fan Rationality Check Malfunction

P0484 Cooling Fan Circuit Over Current

P0485 Cooling Fan Power/Ground Circuit Malfunction

P0500 Vehicle Speed Sensor Malfunction

P0501 Vehicle Speed Sensor Range/Performance

P0502 Vehicle Speed Sensor Circuit Low Input

P0503 Vehicle Speed Sensor Intermittent/Erratic/High

P0505 Idle Control System Malfunction

P0506 Idle Control System RPM Lower Than Expected

P0507 Idle Control System RPM Higher Than Expected

P0510 Closed Throttle Position Switch Malfunction

P0520 Engine Oil Pressure Sensor/Switch Circuit Malfunction

P0521 Engine Oil Pressure Sensor/Switch Range/Performance

P0522 Engine Oil Pressure Sensor/Switch Low Voltage

P0523 Engine Oil Pressure Sensor/Switch High Voltage

P0530 A/C Refrigerant Pressure Sensor Circuit Malfunction

P0531 A/C Refrigerant Pressure Sensor Circuit Range/Performance

P0532 A/C Refrigerant Pressure Sensor Circuit Low Input

P0533 A/C Refrigerant Pressure Sensor Circuit High Input

P0534 A/C Refrigerant Charge Loss

P0550 Power Steering Pressure Sensor Circuit Malfunction

P0551 Power Steering Pressure Sensor Circuit Range/Performance

P0552 Power Steering Pressure Sensor Circuit Low Input

P0553 Power Steering Pressure Sensor Circuit High Input

P0554 Power Steering Pressure Sensor Circuit Intermittent

P0560 System Voltage Malfunction

P0561 System Voltage Unstable

P0562 System Voltage Low

P0563 System Voltage High

P0565 Cruise Control On Signal Malfunction

P0566 Cruise Control Off Signal Malfunction

P0567 Cruise Control Resume Signal Malfunction

P0568 Cruise Control Set Signal Malfunction

P0569 Cruise Control Coast Signal Malfunction

P0570 Cruise Control Accel Signal Malfunction

P0571 Cruise Control/Brake Switch "A" Circuit Malfunction

P0572 Cruise Control/Brake Switch "A" Circuit Low

P0573 Cruise Control/Brake Switch "A" Circuit High

P0574 **Through P0580** Reserved for Cruise Codes

P0600 Serial Communication Link Malfunction

P0601 Internal Control Module Memory Check Sum Error

P0602 Control Module Programming Error

P0603 Internal Control Module Keep Alive Memory (KAM) Error

P0604 Internal Control Module Random Access Memory (RAM) Error

P0605 Internal Control Module Read Only Memory (ROM) Error

P0606 PCM Processor Fault

P0608 Control Module VSS Output "A" Malfunction

P0609 Control Module VSS Output "B" Malfunction

P0620 Generator Control Circuit Malfunction

P0621 Generator Lamp "L" Control Circuit Malfunction

P0622 Generator Field "F" Control Circuit Malfunction

P0650 Malfunction Indicator Lamp (MIL) Control Circuit Malfunction

P0654 Engine RPM Output Circuit Malfunction

P0655 Engine Hot Lamp Output Control Circuit Malfunction
P0656 Fuel Level Output Circuit Malfunction
P0700 Transmission Control System Malfunction
P0701 Transmission Control System Range/Performance
P0702 Transmission Control System Electrical
P0703 Torque Converter/Brake Switch "B" Circuit Malfunction
P0704 Clutch Switch Input Circuit Malfunction
P0705 Transmission Range Sensor Circuit Malfunction (PRNDL Input)
P0706 Transmission Range Sensor Circuit Range/Performance
P0707 Transmission Range Sensor Circuit Low Input
P0708 Transmission Range Sensor Circuit High Input
P0709 Transmission Range Sensor Circuit Intermittent
P0710 Transmission Fluid Temperature Sensor Circuit Malfunction
P0711 Transmission Fluid Temperature Sensor Circuit Range/Performance
P0712 Transmission Fluid Temperature Sensor Circuit Low Input
P0713 Transmission Fluid Temperature Sensor Circuit High Input
P0714 Transmission Fluid Temperature Sensor Circuit Intermittent
P0715 Input/Turbine Speed Sensor Circuit Malfunction
P0716 Input/Turbine Speed Sensor Circuit Range/Performance
P0717 Input/Turbine Speed Sensor Circuit No Signal
P0718 Input/Turbine Speed Sensor Circuit Intermittent
P0719 Torque Converter/Brake Switch "B" Circuit Low
P0720 Output Speed Sensor Circuit Malfunction
P0721 Output Speed Sensor Circuit Range/Performance
P0722 Output Speed Sensor Circuit No Signal
P0723 Output Speed Sensor Circuit Intermittent
P0724 Torque Converter/Brake Switch "B" Circuit High
P0725 Engine Speed Input Circuit Malfunction
P0726 Engine Speed Input Circuit Range/Performance
P0727 Engine Speed Input Circuit No Signal
P0728 Engine Speed Input Circuit Intermittent
P0730 Incorrect Gear Ratio
P0731 Gear no. 1 Incorrect Ratio
P0732 Gear no. 2 Incorrect Ratio
P0733 Gear no. 3 Incorrect Ratio
P0734 Gear no. 4 Incorrect Ratio
P0735 Gear no. 5 Incorrect Ratio
P0736 Reverse Incorrect Ratio
P0740 Torque Converter Clutch Circuit Malfunction
P0741 Torque Converter Clutch Circuit Performance or Stuck Off
P0742 Torque Converter Clutch Circuit Stuck On
P0743 Torque Converter Clutch Circuit Electrical
P0744 Torque Converter Clutch Circuit Intermittent
P0745 Pressure Control Solenoid Malfunction
P0746 Pressure Control Solenoid Performance or Stuck Off
P0747 Pressure Control Solenoid Stuck On
P0748 Pressure Control Solenoid Electrical
P0749 Pressure Control Solenoid Intermittent
P0750 Shift Solenoid "A" Malfunction
P0751 Shift Solenoid "A" Performance or Stuck Off
P0752 Shift Solenoid "A" Stuck On
P0753 Shift Solenoid "A" Electrical
P0754 Shift Solenoid "A" Intermittent
P0755 Shift Solenoid "B" Malfunction
P0756 Shift Solenoid "B" Performance or Stuck Off

P0757 Shift Solenoid "B" Stuck On
P0758 Shift Solenoid "B" Electrical
P0759 Shift Solenoid "B" Intermittent
P0760 Shift Solenoid "C" Malfunction
P0761 Shift Solenoid "C" Performance Or Stuck Off
P0762 Shift Solenoid "C" Stuck On
P0763 Shift Solenoid "C" Electrical
P0764 Shift Solenoid "C" Intermittent
P0765 Shift Solenoid "D" Malfunction
P0766 Shift Solenoid "D" Performance Or Stuck Off
P0767 Shift Solenoid "D" Stuck On
P0768 Shift Solenoid "D" Electrical
P0769 Shift Solenoid "D" Intermittent
P0770 Shift Solenoid "E" Malfunction
P0771 Shift Solenoid "E" Performance Or Stuck Off
P0772 Shift Solenoid "E" Stuck On
P0773 Shift Solenoid "E" Electrical
P0774 Shift Solenoid "E" Intermittent
P0780 Shift Malfunction
P0781 1–2 Shift Malfunction
P0782 2–3 Shift Malfunction
P0783 3–4 Shift Malfunction
P0784 4–5 Shift Malfunction
P0785 Shift/Timing Solenoid Malfunction
P0786 Shift/Timing Solenoid Range/Performance
P0787 Shift/Timing Solenoid Low
P0788 Shift/Timing Solenoid High
P0789 Shift/Timing Solenoid Intermittent
P0790 Normal/Performance Switch Circuit Malfunction
P0801 Reverse Inhibit Control Circuit Malfunction
P0803 1–4 Upshift (Skip Shift) Solenoid Control Circuit Malfunction
P0804 1–4 Upshift (Skip Shift) Lamp Control Circuit Malfunction
P1120 Secondary Throttle Position Sensor Circuit Fault
P1125 Tandem Throttle Position Sensor Circuit Fault
P1210 Traction Control System Signal Fault
P1220 Fuel Pump Control Module Fault
P1320 Ignition Control Signal Fault
P1336 Crankshaft Position Sensor Circuit Fault
P1400 EGR/EVAP Control Solenoid Circuit Fault
P1401 EGR Temperature Sensor Circuit Fault
P1443 EVAP Canister Control Vacuum Switch Circuit Fault
P1445 EVAP Purge Volume Control Valve Circuit Fault
P1605 TCM A/T Diagnosis Communication Line Fault
P1705 Throttle Position Sensor (Switch) Circuit Fault
P1760 Overrun Clutch Solenoid Valve Circuit Fault
P1900 Cooling Fan Control Circuit Fault

Isuzu

➡**This section also provides coverage for the Honda Passport since it shares a platform with the Isuzu Rodeo.**

READING CODES

Reading the control module memory is one of the first steps in OBD II system diagnostics. This step should be initially performed to determine the general nature of the fault. Subsequent readings will determine if the fault has been cleared.

Reading codes can be performed by any of the methods below:

• Read the control module memory with the Generic Scan Tool (GST)

• Read the control module memory with the vehicle manufacturer's specific tester

To read the fault codes, connect the scan tool or tester according to the manufacturer's instructions. Follow the manufacturer's specified procedure for reading the codes.

CLEARING CODES

Control module reset procedures are a very important part of OBD II System diagnostics. This step should be done at the end of any fault code repair and at the end of any driveability repair.

Clearing codes can be performed by any of the methods below:

• Clear the control module memory with the Generic Scan Tool (GST)

• Clear the control module memory with the vehicle manufacturer's specific tester

• Turn the ignition off and remove the negative battery cable for at least 1 minute.

Removing the negative battery cable may cause other systems in the vehicle to lose their memory. Prior to removing the cable, ensure you have the proper reset codes for radios and alarms.

➡**The MIL will may also be de-activated for some codes if the vehicle completes three consecutive trips without a fault detected with vehicle conditions similar to those present during the fault.**

DIAGNOSTIC TROUBLE CODES

P0000 No Failures
P0100 Mass or Volume Air Flow Circuit Malfunction
P0101 Mass or Volume Air Flow Circuit Range/Performance Problem
P0102 Mass or Volume Air Flow Circuit Low Input
P0103 Mass or Volume Air Flow Circuit High Input
P0104 Mass or Volume Air Flow Circuit Intermittent
P0105 Manifold Absolute Pressure/Barometric Pressure Circuit Malfunction
P0106 Manifold Absolute Pressure/Barometric Pressure Circuit Range/Performance Problem
P0107 Manifold Absolute Pressure/Barometric Pressure Circuit Low Input
P0108 Manifold Absolute Pressure/Barometric Pressure Circuit High Input
P0109 Manifold Absolute Pressure/Barometric Pressure Circuit Intermittent
P0110 Intake Air Temperature Circuit Malfunction
P0111 Intake Air Temperature Circuit Range/Performance Problem
P0112 Intake Air Temperature Circuit Low Input
P0113 Intake Air Temperature Circuit High Input
P0114 Intake Air Temperature Circuit Intermittent
P0115 Engine Coolant Temperature Circuit Malfunction
P0116 Engine Coolant Temperature Circuit Range/Performance Problem
P0117 Engine Coolant Temperature Circuit Low Input
P0118 Engine Coolant Temperature Circuit High Input
P0119 Engine Coolant Temperature Circuit Intermittent
P0120 Throttle/Pedal Position Sensor/Switch "A" Circuit Malfunction

P0121 Throttle/Pedal Position Sensor/Switch "A" Circuit Range/Performance Problem
P0122 Throttle/Pedal Position Sensor/Switch "A" Circuit Low Input
P0123 Throttle/Pedal Position Sensor/Switch "A" Circuit High Input
P0124 Throttle/Pedal Position Sensor/Switch "A" Circuit Intermittent
P0125 Insufficient Coolant Temperature For Closed Loop Fuel Control
P0126 Insufficient Coolant Temperature For Stable Operation
P0130 O_2 Circuit Malfunction (Bank no. 1 Sensor no. 1)
P0131 O_2 Sensor Circuit Low Voltage (Bank no. 1 Sensor no. 1)
P0132 O_2 Sensor Circuit High Voltage (Bank no. 1 Sensor no. 1)
P0133 O_2 Sensor Circuit Slow Response (Bank no. 1 Sensor no. 1)
P0134 O_2 Sensor Circuit No Activity Detected (Bank no. 1 Sensor no. 1)
P0135 O_2 Sensor Heater Circuit Malfunction (Bank no. 1 Sensor no. 1)
P0136 O_2 Sensor Circuit Malfunction (Bank no. 1 Sensor no. 2)
P0137 O_2 Sensor Circuit Low Voltage (Bank no. 1 Sensor no. 2)
P0138 O_2 Sensor Circuit High Voltage (Bank no. 1 Sensor no. 2)
P0139 O_2 Sensor Circuit Slow Response (Bank no. 1 Sensor no. 2)
P0140 O_2 Sensor Circuit No Activity Detected (Bank no. 1 Sensor no. 2)
P0141 O_2 Sensor Heater Circuit Malfunction (Bank no. 1 Sensor no. 2)
P0142 O_2 Sensor Circuit Malfunction (Bank no. 1 Sensor no. 3)
P0143 O_2 Sensor Circuit Low Voltage (Bank no. 1 Sensor no. 3)
P0144 O_2 Sensor Circuit High Voltage (Bank no. 1 Sensor no. 3)
P0145 O_2 Sensor Circuit Slow Response (Bank no. 1 Sensor no. 3)
P0146 O_2 Sensor Circuit No Activity Detected (Bank no. 1 Sensor no. 3)
P0147 O_2 Sensor Heater Circuit Malfunction (Bank no. 1 Sensor no. 3)
P0150 O_2 Sensor Circuit Malfunction (Bank no. 2 Sensor no. 1)
P0151 O_2 Sensor Circuit Low Voltage (Bank no. 2 Sensor no. 1)
P0152 O_2 Sensor Circuit High Voltage (Bank no. 2 Sensor no. 1)
P0153 O_2 Sensor Circuit Slow Response (Bank no. 2 Sensor no. 1)
P0154 O_2 Sensor Circuit No Activity Detected (Bank no. 2 Sensor no. 1)
P0155 O_2 Sensor Heater Circuit Malfunction (Bank no. 2 Sensor no. 1)
P0156 O_2 Sensor Circuit Malfunction (Bank no. 2 Sensor no. 2)
P0157 O_2 Sensor Circuit Low Voltage (Bank no. 2 Sensor no. 2)
P0158 O_2 Sensor Circuit High Voltage (Bank no. 2 Sensor no. 2)
P0159 O_2 Sensor Circuit Slow Response (Bank no. 2 Sensor no. 2)
P0160 O_2 Sensor Circuit No Activity Detected (Bank no. 2 Sensor no. 2)
P0161 O_2 Sensor Heater Circuit Malfunction (Bank no. 2 Sensor no. 2)
P0162 O_2 Sensor Circuit Malfunction (Bank no. 2 Sensor no. 3)
P0163 O_2 Sensor Circuit Low Voltage (Bank no. 2 Sensor no. 3)
P0164 O_2 Sensor Circuit High Voltage (Bank no. 2 Sensor no. 3)
P0165 O_2 Sensor Circuit Slow Response (Bank no. 2 Sensor no. 3)

P0166 O_2 Sensor Circuit No Activity Detected (Bank no. 2 Sensor no. 3)

P0167 O_2 Sensor Heater Circuit Malfunction (Bank no. 2 Sensor no. 3)

P0170 Fuel Trim Malfunction (Bank no. 1)

P0171 System Too Lean (Bank no. 1)

P0172 System Too Rich (Bank no. 1)

P0173 Fuel Trim Malfunction (Bank no. 2)

P0174 System Too Lean (Bank no. 2)

P0175 System Too Rich (Bank no. 2)

P0176 Fuel Composition Sensor Circuit Malfunction

P0177 Fuel Composition Sensor Circuit Range/Performance

P0178 Fuel Composition Sensor Circuit Low Input

P0179 Fuel Composition Sensor Circuit High Input

P0180 Fuel Temperature Sensor "A" Circuit Malfunction

P0181 Fuel Temperature Sensor "A" Circuit Range/Performance

P0182 Fuel Temperature Sensor "A" Circuit Low Input

P0183 Fuel Temperature Sensor "A" Circuit High Input

P0184 Fuel Temperature Sensor "A" Circuit Intermittent

P0185 Fuel Temperature Sensor "B" Circuit Malfunction

P0186 Fuel Temperature Sensor "B" Circuit Range/Performance

P0187 Fuel Temperature Sensor "B" Circuit Low Input

P0188 Fuel Temperature Sensor "B" Circuit High Input

P0189 Fuel Temperature Sensor "B" Circuit Intermittent

P0190 Fuel Rail Pressure Sensor Circuit Malfunction

P0191 Fuel Rail Pressure Sensor Circuit Range/Performance

P0192 Fuel Rail Pressure Sensor Circuit Low Input

P0193 Fuel Rail Pressure Sensor Circuit High Input

P0194 Fuel Rail Pressure Sensor Circuit Intermittent

P0195 Engine Oil Temperature Sensor Malfunction

P0196 Engine Oil Temperature Sensor Range/Performance

P0197 Engine Oil Temperature Sensor Low

P0198 Engine Oil Temperature Sensor High

P0199 Engine Oil Temperature Sensor Intermittent

P0200 Injector Circuit Malfunction

P0201 Injector Circuit Malfunction—Cylinder no. 1

P0202 Injector Circuit Malfunction—Cylinder no. 2

P0203 Injector Circuit Malfunction—Cylinder no. 3

P0204 Injector Circuit Malfunction—Cylinder no. 4

P0205 Injector Circuit Malfunction—Cylinder no. 5

P0206 Injector Circuit Malfunction—Cylinder no. 6

P0207 Injector Circuit Malfunction—Cylinder no. 7

P0208 Injector Circuit Malfunction—Cylinder no. 8

P0209 Injector Circuit Malfunction—Cylinder no. 9

P0210 Injector Circuit Malfunction—Cylinder no. 10

P0211 Injector Circuit Malfunction—Cylinder no. 11

P0212 Injector Circuit Malfunction—Cylinder no. 12

P0213 Cold Start Injector no. 1 Malfunction

P0214 Cold Start Injector no. 2 Malfunction

P0215 Engine Shutoff Solenoid Malfunction

P0216 Injection Timing Control Circuit Malfunction

P0217 Engine Over Temperature Condition

P0218 Transmission Over Temperature Condition

P0219 Engine Over Speed Condition

P0220 Throttle/Pedal Position Sensor/Switch "B" Circuit Malfunction

P0221 Throttle/Pedal Position Sensor/Switch "B" Circuit Range/Performance Problem

P0222 Throttle/Pedal Position Sensor/Switch "B" Circuit Low Input

P0223 Throttle/Pedal Position Sensor/Switch "B" Circuit High Input

P0224 Throttle/Pedal Position Sensor/Switch "B" Circuit Intermittent

P0225 Throttle/Pedal Position Sensor/Switch "C" Circuit Malfunction

P0226 Throttle/Pedal Position Sensor/Switch "C" Circuit Range/Performance Problem

P0227 Throttle/Pedal Position Sensor/Switch "C" Circuit Low Input

P0228 Throttle/Pedal Position Sensor/Switch "C" Circuit High Input

P0229 Throttle/Pedal Position Sensor/Switch "C" Circuit Intermittent

P0230 Fuel Pump Primary Circuit Malfunction

P0231 Fuel Pump Secondary Circuit Low

P0232 Fuel Pump Secondary Circuit High

P0233 Fuel Pump Secondary Circuit Intermittent

P0234 Engine Over Boost Condition

P0261 Cylinder no. 1 Injector Circuit Low

P0262 Cylinder no. 1 Injector Circuit High

P0263 Cylinder no. 1 Contribution/Balance Fault

P0264 Cylinder no. 2 Injector Circuit Low

P0265 Cylinder no. 2 Injector Circuit High

P0266 Cylinder no. 2 Contribution/Balance Fault

P0267 Cylinder no. 3 Injector Circuit Low

P0268 Cylinder no. 3 Injector Circuit High

P0269 Cylinder no. 3 Contribution/Balance Fault

P0270 Cylinder no. 4 Injector Circuit Low

P0271 Cylinder no. 4 Injector Circuit High

P0272 Cylinder no. 4 Contribution/Balance Fault

P0273 Cylinder no. 5 Injector Circuit Low

P0274 Cylinder no. 5 Injector Circuit High

P0275 Cylinder no. 5 Contribution/Balance Fault

P0276 Cylinder no. 6 Injector Circuit Low

P0277 Cylinder no. 6 Injector Circuit High

P0278 Cylinder no. 6 Contribution/Balance Fault

P0279 Cylinder no. 7 Injector Circuit Low

P0280 Cylinder no. 7 Injector Circuit High

P0281 Cylinder no. 7 Contribution/Balance Fault

P0282 Cylinder no. 8 Injector Circuit Low

P0283 Cylinder no. 8 Injector Circuit High

P0284 Cylinder no. 8 Contribution/Balance Fault

P0285 Cylinder no. 9 Injector Circuit Low

P0286 Cylinder no. 9 Injector Circuit High

P0287 Cylinder no. 9 Contribution/Balance Fault

P0288 Cylinder no. 10 Injector Circuit Low

P0289 Cylinder no. 10 Injector Circuit High

P0290 Cylinder no. 10 Contribution/Balance Fault

P0291 Cylinder no. 11 Injector Circuit Low

P0292 Cylinder no. 11 Injector Circuit High

P0293 Cylinder no. 11 Contribution/Balance Fault

P0294 Cylinder no. 12 Injector Circuit Low

P0295 Cylinder no. 12 Injector Circuit High

P0296 Cylinder no. 12 Contribution/Balance Fault

P0300 Random/Multiple Cylinder Misfire Detected

P0301 Cylinder no. 1—Misfire Detected

P0302 Cylinder no. 2—Misfire Detected

P0303 Cylinder no. 3—Misfire Detected

P0304 Cylinder no. 4—Misfire Detected

P0305 Cylinder no. 5—Misfire Detected

P0306 Cylinder no. 6—Misfire Detected

P0307 Cylinder no. 7—Misfire Detected

P0308 Cylinder no. 8—Misfire Detected

P0309 Cylinder no. 9—Misfire Detected
P0310 Cylinder no. 10—Misfire Detected
P0311 Cylinder no. 11—Misfire Detected
P0312 Cylinder no. 12—Misfire Detected
P0320 Ignition/Distributor Engine Speed Input Circuit Malfunction
P0321 Ignition/Distributor Engine Speed Input Circuit Range/Performance
P0322 Ignition/Distributor Engine Speed Input Circuit No Signal
P0323 Ignition/Distributor Engine Speed Input Circuit Intermittent
P0325 Knock Sensor no. 1—Circuit Malfunction (Bank no. 1 or Single Sensor)
P0326 Knock Sensor no. 1—Circuit Range/Performance (Bank no. 1 or Single Sensor)
P0327 Knock Sensor no. 1—Circuit Low Input (Bank no. 1 or Single Sensor)
P0328 Knock Sensor no. 1—Circuit High Input (Bank no. 1 or Single Sensor)
P0329 Knock Sensor no. 1—Circuit Input Intermittent (Bank no. 1 or Single Sensor)
P0330 Knock Sensor no. 2—Circuit Malfunction (Bank no. 2)
P0331 Knock Sensor no. 2—Circuit Range/Performance (Bank no. 2)
P0332 Knock Sensor no. 2—Circuit Low Input (Bank no. 2)
P0333 Knock Sensor no. 2—Circuit High Input (Bank no. 2)
P0334 Knock Sensor no. 2—Circuit Input Intermittent (Bank no. 2)
P0335 Crankshaft Position Sensor "A" Circuit Malfunction
P0336 Crankshaft Position Sensor "A" Circuit Range/Performance
P0337 Crankshaft Position Sensor "A" Circuit Low Input
P0338 Crankshaft Position Sensor "A" Circuit High Input
P0339 Crankshaft Position Sensor "A" Circuit Intermittent
P0340 Camshaft Position Sensor Circuit Malfunction
P0341 Camshaft Position Sensor Circuit Range/Performance
P0342 Camshaft Position Sensor Circuit Low Input
P0343 Camshaft Position Sensor Circuit High Input
P0344 Camshaft Position Sensor Circuit Intermittent
P0350 Ignition Coil Primary/Secondary Circuit Malfunction
P0351 Ignition Coil "A" Primary/Secondary Circuit Malfunction
P0352 Ignition Coil "B" Primary/Secondary Circuit Malfunction
P0353 Ignition Coil "C" Primary/Secondary Circuit Malfunction
P0354 Ignition Coil "D" Primary/Secondary Circuit Malfunction
P0355 Ignition Coil "E" Primary/Secondary Circuit Malfunction
P0356 Ignition Coil "F" Primary/Secondary Circuit Malfunction
P0357 Ignition Coil "G" Primary/Secondary Circuit Malfunction
P0358 Ignition Coil "H" Primary/Secondary Circuit Malfunction
P0359 Ignition Coil "I" Primary/Secondary Circuit Malfunction
P0360 Ignition Coil "J" Primary/Secondary Circuit Malfunction
P0361 Ignition Coil "K" Primary/Secondary Circuit Malfunction
P0362 Ignition Coil "L" Primary/Secondary Circuit Malfunction
P0370 Timing Reference High Resolution Signal "A" Malfunction
P0371 Timing Reference High Resolution Signal "A" Too Many Pulses
P0372 Timing Reference High Resolution Signal "A" Too Few Pulses
P0373 Timing Reference High Resolution Signal "A" Intermittent/Erratic Pulses
P0374 Timing Reference High Resolution Signal "A" No Pulses

P0375 Timing Reference High Resolution Signal "B" Malfunction
P0376 Timing Reference High Resolution Signal "B" Too Many Pulses
P0377 Timing Reference High Resolution Signal "B" Too Few Pulses
P0378 Timing Reference High Resolution Signal "B" Intermittent/Erratic Pulses
P0379 Timing Reference High Resolution Signal "B" No Pulses
P0380 Glow Plug/Heater Circuit "A" Malfunction
P0381 Glow Plug/Heater Indicator Circuit Malfunction
P0382 Glow Plug/Heater Circuit "B" Malfunction
P0385 Crankshaft Position Sensor "B" Circuit Malfunction
P0386 Crankshaft Position Sensor "B" Circuit Range/Performance
P0387 Crankshaft Position Sensor "B" Circuit Low Input
P0388 Crankshaft Position Sensor "B" Circuit High Input
P0389 Crankshaft Position Sensor "B" Circuit Intermittent
P0400 Exhaust Gas Recirculation Flow Malfunction
P0401 Exhaust Gas Recirculation Flow Insufficient Detected
P0402 Exhaust Gas Recirculation Flow Excessive Detected
P0403 Exhaust Gas Recirculation Circuit Malfunction
P0404 Exhaust Gas Recirculation Circuit Range/Performance
P0405 Exhaust Gas Recirculation Sensor "A" Circuit Low
P0406 Exhaust Gas Recirculation Sensor "A" Circuit High
P0407 Exhaust Gas Recirculation Sensor "B" Circuit Low
P0408 Exhaust Gas Recirculation Sensor "B" Circuit High
P0410 Secondary Air Injection System Malfunction
P0411 Secondary Air Injection System Incorrect Flow Detected
P0412 Secondary Air Injection System Switching Valve "A" Circuit Malfunction
P0413 Secondary Air Injection System Switching Valve "A" Circuit Open
P0414 Secondary Air Injection System Switching Valve "A" Circuit Shorted
P0415 Secondary Air Injection System Switching Valve "B" Circuit Malfunction
P0416 Secondary Air Injection System Switching Valve "B" Circuit Open
P0417 Secondary Air Injection System Switching Valve "B" Circuit Shorted
P0418 Secondary Air Injection System Relay "A" Circuit Malfunction
P0419 Secondary Air Injection System Relay "B" Circuit Malfunction
P0420 Catalyst System Efficiency Below Threshold (Bank no. 1)
P0421 Warm Up Catalyst Efficiency Below Threshold (Bank no. 1)
P0422 Main Catalyst Efficiency Below Threshold (Bank no. 1)
P0423 Heated Catalyst Efficiency Below Threshold (Bank no. 1)
P0424 Heated Catalyst Temperature Below Threshold (Bank no. 1)
P0430 Catalyst System Efficiency Below Threshold (Bank no. 2)
P0431 Warm Up Catalyst Efficiency Below Threshold (Bank no. 2)
P0432 Main Catalyst Efficiency Below Threshold (Bank no. 2)
P0433 Heated Catalyst Efficiency Below Threshold (Bank no. 2)
P0434 Heated Catalyst Temperature Below Threshold (Bank no. 2)
P0440 Evaporative Emission Control System Malfunction
P0441 Evaporative Emission Control System Incorrect Purge Flow
P0442 Evaporative Emission Control System Leak Detected (Small Leak)
P0443 Evaporative Emission Control System Purge Control Valve Circuit Malfunction

P0444 Evaporative Emission Control System Purge Control Valve Circuit Open

P0445 Evaporative Emission Control System Purge Control Valve Circuit Shorted

P0446 Evaporative Emission Control System Vent Control Circuit Malfunction

P0447 Evaporative Emission Control System Vent Control Circuit Open

P0448 Evaporative Emission Control System Vent Control Circuit Shorted

P0449 Evaporative Emission Control System Vent Valve/Solenoid Circuit Malfunction

P0450 Evaporative Emission Control System Pressure Sensor Malfunction

P0451 Evaporative Emission Control System Pressure Sensor Range/Performance

P0452 Evaporative Emission Control System Pressure Sensor Low Input

P0453 Evaporative Emission Control System Pressure Sensor High Input

P0454 Evaporative Emission Control System Pressure Sensor Intermittent

P0455 Evaporative Emission Control System Leak Detected (Gross Leak)

P0460 Fuel Level Sensor Circuit Malfunction

P0461 Fuel Level Sensor Circuit Range/Performance

P0462 Fuel Level Sensor Circuit Low Input

P0463 Fuel Level Sensor Circuit High Input

P0464 Fuel Level Sensor Circuit Intermittent

P0465 Purge Flow Sensor Circuit Malfunction

P0466 Purge Flow Sensor Circuit Range/Performance

P0467 Purge Flow Sensor Circuit Low Input

P0468 Purge Flow Sensor Circuit High Input

P0469 Purge Flow Sensor Circuit Intermittent

P0470 Exhaust Pressure Sensor Malfunction

P0471 Exhaust Pressure Sensor Range/Performance

P0472 Exhaust Pressure Sensor Low

P0473 Exhaust Pressure Sensor High

P0474 Exhaust Pressure Sensor Intermittent

P0475 Exhaust Pressure Control Valve Malfunction

P0476 Exhaust Pressure Control Valve Range/Performance

P0477 Exhaust Pressure Control Valve Low

P0478 Exhaust Pressure Control Valve High

P0479 Exhaust Pressure Control Valve Intermittent

P0480 Cooling Fan no. 1 Control Circuit Malfunction

P0481 Cooling Fan no. 2 Control Circuit Malfunction

P0482 Cooling Fan no. 3 Control Circuit Malfunction

P0483 Cooling Fan Rationality Check Malfunction

P0484 Cooling Fan Circuit Over Current

P0485 Cooling Fan Power/Ground Circuit Malfunction

P0500 Vehicle Speed Sensor Malfunction

P0501 Vehicle Speed Sensor Range/Performance

P0502 Vehicle Speed Sensor Circuit Low Input

P0503 Vehicle Speed Sensor Intermittent/Erratic/High

P0505 Idle Control System Malfunction

P0506 Idle Control System RPM Lower Than Expected

P0507 Idle Control System RPM Higher Than Expected

P0510 Closed Throttle Position Switch Malfunction

P0520 Engine Oil Pressure Sensor/Switch Circuit Malfunction

P0521 Engine Oil Pressure Sensor/Switch Range/Performance

P0522 Engine Oil Pressure Sensor/Switch Low Voltage

P0523 Engine Oil Pressure Sensor/Switch High Voltage

P0530 A/C Refrigerant Pressure Sensor Circuit Malfunction

P0531 A/C Refrigerant Pressure Sensor Circuit Range/Performance

P0532 A/C Refrigerant Pressure Sensor Circuit Low Input

P0533 A/C Refrigerant Pressure Sensor Circuit High Input

P0534 A/C Refrigerant Charge Loss

P0550 Power Steering Pressure Sensor Circuit Malfunction

P0551 Power Steering Pressure Sensor Circuit Range/Performance

P0552 Power Steering Pressure Sensor Circuit Low Input

P0553 Power Steering Pressure Sensor Circuit High Input

P0554 Power Steering Pressure Sensor Circuit Intermittent

P0560 System Voltage Malfunction

P0561 System Voltage Unstable

P0562 System Voltage Low

P0563 System Voltage High

P0565 Cruise Control On Signal Malfunction

P0566 Cruise Control Off Signal Malfunction

P0567 Cruise Control Resume Signal Malfunction

P0568 Cruise Control Set Signal Malfunction

P0569 Cruise Control Coast Signal Malfunction

P0570 Cruise Control Accel Signal Malfunction

P0571 Cruise Control/Brake Switch "A" Circuit Malfunction

P0572 Cruise Control/Brake Switch "A" Circuit Low

P0573 Cruise Control/Brake Switch "A" Circuit High

P0574 Through P0580 Reserved for Cruise Codes

P0600 Serial Communication Link Malfunction

P0601 Internal Control Module Memory Check Sum Error

P0602 Control Module Programming Error

P0603 Internal Control Module Keep Alive Memory (KAM) Error

P0604 Internal Control Module Random Access Memory (RAM) Error

P0605 Internal Control Module Read Only Memory (ROM) Error

P0606 PCM Processor Fault

P0608 Control Module VSS Output "A" Malfunction

P0609 Control Module VSS Output "B" Malfunction

P0620 Generator Control Circuit Malfunction

P0621 Generator Lamp "L" Control Circuit Malfunction

P0622 Generator Field "F" Control Circuit Malfunction

P0650 Malfunction Indicator Lamp (MIL) Control Circuit Malfunction

P0654 Engine RPM Output Circuit Malfunction

P0655 Engine Hot Lamp Output Control Circuit Malfunction

P0656 Fuel Level Output Circuit Malfunction

P0700 Transmission Control System Malfunction

P0701 Transmission Control System Range/Performance

P0702 Transmission Control System Electrical

P0703 Torque Converter/Brake Switch "B" Circuit Malfunction

P0704 Clutch Switch Input Circuit Malfunction

P0705 Transmission Range Sensor Circuit Malfunction (PRNDL Input)

P0706 Transmission Range Sensor Circuit Range/Performance

P0707 Transmission Range Sensor Circuit Low Input

P0708 Transmission Range Sensor Circuit High Input

P0709 Transmission Range Sensor Circuit Intermittent

P0710 Transmission Fluid Temperature Sensor Circuit Malfunction

P0711 Transmission Fluid Temperature Sensor Circuit Range/Performance

P0712 Transmission Fluid Temperature Sensor Circuit Low Input

P0713 Transmission Fluid Temperature Sensor Circuit High Input

P0714 Transmission Fluid Temperature Sensor Circuit Intermittent
P0715 Input/Turbine Speed Sensor Circuit Malfunction
P0716 Input/Turbine Speed Sensor Circuit Range/Performance
P0717 Input/Turbine Speed Sensor Circuit No Signal
P0718 Input/Turbine Speed Sensor Circuit Intermittent
P0719 Torque Converter/Brake Switch "B" Circuit Low
P0720 Output Speed Sensor Circuit Malfunction
P0721 Output Speed Sensor Circuit Range/Performance
P0722 Output Speed Sensor Circuit No Signal
P0723 Output Speed Sensor Circuit Intermittent
P0724 Torque Converter/Brake Switch "B" Circuit High
P0725 Engine Speed Input Circuit Malfunction
P0726 Engine Speed Input Circuit Range/Performance
P0727 Engine Speed Input Circuit No Signal
P0728 Engine Speed Input Circuit Intermittent
P0730 Incorrect Gear Ratio
P0731 Gear no. 1 Incorrect Ratio
P0732 Gear no. 2 Incorrect Ratio
P0733 Gear no. 3 Incorrect Ratio
P0734 Gear no. 4 Incorrect Ratio
P0735 Gear no. 5 Incorrect Ratio
P0736 Reverse Incorrect Ratio
P0740 Torque Converter Clutch Circuit Malfunction
P0741 Torque Converter Clutch Circuit Performance or Stuck Off
P0742 Torque Converter Clutch Circuit Stuck On
P0743 Torque Converter Clutch Circuit Electrical
P0744 Torque Converter Clutch Circuit Intermittent
P0745 Pressure Control Solenoid Malfunction
P0746 Pressure Control Solenoid Performance or Stuck Off
P0747 Pressure Control Solenoid Stuck On
P0748 Pressure Control Solenoid Electrical
P0749 Pressure Control Solenoid Intermittent
P0750 Shift Solenoid "A" Malfunction
P0751 Shift Solenoid "A" Performance or Stuck Off
P0752 Shift Solenoid "A" Stuck On
P0753 Shift Solenoid "A" Electrical
P0754 Shift Solenoid "A" Intermittent
P0755 Shift Solenoid "B" Malfunction
P0756 Shift Solenoid "B" Performance or Stuck Off
P0757 Shift Solenoid "B" Stuck On
P0758 Shift Solenoid "B" Electrical
P0759 Shift Solenoid "B" Intermittent
P0760 Shift Solenoid "C" Malfunction
P0761 Shift Solenoid "C" Performance Or Stuck Off
P0762 Shift Solenoid "C" Stuck On
P0763 Shift Solenoid "C" Electrical
P0764 Shift Solenoid "C" Intermittent
P0765 Shift Solenoid "D" Malfunction
P0766 Shift Solenoid "D" Performance Or Stuck Off
P0767 Shift Solenoid "D" Stuck On
P0768 Shift Solenoid "D" Electrical
P0769 Shift Solenoid "D" Intermittent
P0770 Shift Solenoid "E" Malfunction
P0771 Shift Solenoid "E" Performance Or Stuck Off
P0772 Shift Solenoid "E" Stuck On
P0773 Shift Solenoid "E" Electrical
P0774 Shift Solenoid "E" Intermittent
P0780 Shift Malfunction
P0781 1–2 Shift Malfunction
P0782 2–3 Shift Malfunction

P0783 3–4 Shift Malfunction
P0784 4–5 Shift Malfunction
P0785 Shift/Timing Solenoid Malfunction
P0786 Shift/Timing Solenoid Range/Performance
P0787 Shift/Timing Solenoid Low
P0788 Shift/Timing Solenoid High
P0789 Shift/Timing Solenoid Intermittent
P0790 Normal/Performance Switch Circuit Malfunction
P0801 Reverse Inhibit Control Circuit Malfunction
P0803 1–4 Upshift (Skip Shift) Solenoid Control Circuit Malfunction
P0804 1–4 Upshift (Skip Shift) Lamp Control Circuit Malfunction
P1106 Map Sensor Circuit Intermittent High Voltage
P1107 MAP Sensor Circuit Intermittent Low Voltage
P1111 IAT Sensor Circuit Intermittent High Voltage
P1112 IAT Sensor Circuit Intermittent Low Voltage
P1114 ECT Sensor Circuit Intermittent Low Voltage
P1115 ECT Sensor Circuit Intermittent High Voltage
P1121 TP Sensor Circuit Intermittent High Voltage
P1122 TP Sensor Circuit Intermittent Low Voltage
P1133 HO$_2$S-11 Insufficient Switching (Bank 1 Sensor 1)
P1134 HO$_2$S-11 Transition Time Ratio (Bank 1 Sensor 1)
P1153 HO$_2$S-21 Insufficient Switching (Bank 2 Sensor l)
P1154 HO$_2$S-21 Transition Time Ratio (Bank 2 Sensor 1)
P1171 Fuel System Lean During Acceleration
P1391 G-Acceleration Sensor Intermittent Low Voltage
P1390 G-Acceleration (Low G) Sensor Performance
P1392 Rough Road G-Sensor Circuit Low Voltage
P1393 Rough Road G-Sensor Circuit High Voltage
P1394 G-Acceleration Sensor Intermittent High Voltage
P1406 EGR Valve Pintle Position Sensor Circuit Fault
P1441 EVAP System Flow During Non-Purge
P1442 EVAP System Flow During Non-Purge
P1508 Idle Speed Control System-Low
P1509 Idle Speed Control System-High
P1618 Serial Peripheral Interface Communication Error
P1640 Output Driver Module 'A' Fault
P1790 PCM ROM (Transmission Side) Check Sum Error
P1792 PCM EEPROM (Transmission Side) Check Sum Error
P1835 Kick Down Switch Always On
P1850 Brake Band Apply Solenoid Electrical Fault
P1860 TCC PWM Solenoid Electrical Fault
P1870 Transmission Component Slipping

Kia

READING CODES

Reading the control module memory is on of the first steps in OBD II system diagnostics.

This step should be initially performed to determine the general nature of the fault. Subsequent readings will determine if the fault has been cleared.

Reading codes can be performed by any of the methods below:
- Read the control module memory with the Generic Scan Tool (GST)
- Read the control module memory with the vehicle manufacturer's specific tester

To read the fault codes, connect the scan tool or tester according to the manufacturer's instructions. Follow the manufacturer's specified procedure for reading the codes.

CLEARING CODES

Control module reset procedures are a very important part of OBD II System diagnostics. This step should be done at the end of any fault code repair and at the end of any driveability repair.

Clearing codes can be performed by any of the methods below:

• Clear the control module memory with the Generic Scan Tool (GST)

• Clear the control module memory with the vehicle manufacturer's specific tester

• Turn the ignition **OFF** and remove the negative battery cable for at least 1 minute.

Removing the negative battery cable may cause other systems in the vehicle to lose their memory. Prior to removing the cable, ensure you have the proper reset codes for radios and alarms.

➡**The MIL will may also be de-activated for some codes if the vehicle completes three consecutive trips without a fault detected with vehicle conditions similar to those present during the fault.**

DIAGNOSTIC TROUBLE CODES

P0000 No Failures
P0100 Mass or Volume Air Flow Circuit Malfunction
P0101 Mass or Volume Air Flow Circuit Range/Performance Problem
P0102 Mass or Volume Air Flow Circuit Low Input
P0103 Mass or Volume Air Flow Circuit High Input
P0104 Mass or Volume Air Flow Circuit Intermittent
P0105 Manifold Absolute Pressure/Barometric Pressure Circuit Malfunction
P0106 Manifold Absolute Pressure/Barometric Pressure Circuit Range/Performance Problem
P0107 Manifold Absolute Pressure/Barometric Pressure Circuit Low Input
P0108 Manifold Absolute Pressure/Barometric Pressure Circuit High Input
P0109 Manifold Absolute Pressure/Barometric Pressure Circuit Intermittent
P0110 Intake Air Temperature Circuit Malfunction
P0111 Intake Air Temperature Circuit Range/Performance Problem
P0112 Intake Air Temperature Circuit Low Input
P0113 Intake Air Temperature Circuit High Input
P0114 Intake Air Temperature Circuit Intermittent
P0115 Engine Coolant Temperature Circuit Malfunction
P0116 Engine Coolant Temperature Circuit Range/Performance Problem
P0117 Engine Coolant Temperature Circuit Low Input
P0118 Engine Coolant Temperature Circuit High Input
P0119 Engine Coolant Temperature Circuit Intermittent
P0120 Throttle/Pedal Position Sensor/Switch "A" Circuit Malfunction
P0121 Throttle/Pedal Position Sensor/Switch "A" Circuit Range/Performance Problem
P0122 Throttle/Pedal Position Sensor/Switch "A" Circuit Low Input
P0123 Throttle/Pedal Position Sensor/Switch "A" Circuit High Input
P0124 Throttle/Pedal Position Sensor/Switch "A" Circuit Intermittent

P0125 Insufficient Coolant Temperature For Closed Loop Fuel Control
P0126 Insufficient Coolant Temperature For Stable Operation
P0130 O_2 Circuit Malfunction (Bank no. 1 Sensor no. 1)
P0131 O_2 Sensor Circuit Low Voltage (Bank no. 1 Sensor no. 1)
P0132 O_2 Sensor Circuit High Voltage (Bank no. 1 Sensor no. 1)
P0133 O_2 Sensor Circuit Slow Response (Bank no. 1 Sensor no. 1)
P0134 O_2 Sensor Circuit No Activity Detected (Bank no. 1 Sensor no. 1)
P0135 O_2 Sensor Heater Circuit Malfunction (Bank no. 1 Sensor no. 1)
P0136 O_2 Sensor Circuit Malfunction (Bank no. 1 Sensor no. 2)
P0137 O_2 Sensor Circuit Low Voltage (Bank no. 1 Sensor no. 2)
P0138 O_2 Sensor Circuit High Voltage (Bank no. 1 Sensor no. 2)
P0139 O_2 Sensor Circuit Slow Response (Bank no. 1 Sensor no. 2)
P0140 O_2 Sensor Circuit No Activity Detected (Bank no. 1 Sensor no. 2)
P0141 O_2 Sensor Heater Circuit Malfunction (Bank no. 1 Sensor no. 2)
P0142 O_2 Sensor Circuit Malfunction (Bank no. 1 Sensor no. 3)
P0143 O_2 Sensor Circuit Low Voltage (Bank no. 1 Sensor no. 3)
P0144 O_2 Sensor Circuit High Voltage (Bank no. 1 Sensor no. 3)
P0145 O_2 Sensor Circuit Slow Response (Bank no. 1 Sensor no. 3)
P0146 O_2 Sensor Circuit No Activity Detected (Bank no. 1 Sensor no. 3)
P0147 O_2 Sensor Heater Circuit Malfunction (Bank no. 1 Sensor no. 3)
P0150 O_2 Sensor Circuit Malfunction (Bank no. 2 Sensor no. 1)
P0151 O_2 Sensor Circuit Low Voltage (Bank no. 2 Sensor no. 1)
P0152 O_2 Sensor Circuit High Voltage (Bank no. 2 Sensor no. 1)
P0153 O_2 Sensor Circuit Slow Response (Bank no. 2 Sensor no. 1)
P0154 O_2 Sensor Circuit No Activity Detected (Bank no. 2 Sensor no. 1)
P0155 O_2 Sensor Heater Circuit Malfunction (Bank no. 2 Sensor no. 1)
P0156 O_2 Sensor Circuit Malfunction (Bank no. 2 Sensor no. 2)
P0157 O_2 Sensor Circuit Low Voltage (Bank no. 2 Sensor no. 2)
P0158 O_2 Sensor Circuit High Voltage (Bank no. 2 Sensor no. 2)
P0159 O_2 Sensor Circuit Slow Response (Bank no. 2 Sensor no. 2)
P0160 O_2 Sensor Circuit No Activity Detected (Bank no. 2 Sensor no. 2)
P0161 O_2 Sensor Heater Circuit Malfunction (Bank no. 2 Sensor no. 2)
P0162 O_2 Sensor Circuit Malfunction (Bank no. 2 Sensor no. 3)
P0163 O_2 Sensor Circuit Low Voltage (Bank no. 2 Sensor no. 3)
P0164 O_2 Sensor Circuit High Voltage (Bank no. 2 Sensor no. 3)
P0165 O_2 Sensor Circuit Slow Response (Bank no. 2 Sensor no. 3)
P0166 O_2 Sensor Circuit No Activity Detected (Bank no. 2 Sensor no. 3)
P0167 O_2 Sensor Heater Circuit Malfunction (Bank no. 2 Sensor no. 3)
P0170 Fuel Trim Malfunction (Bank no. 1)
P0171 System Too Lean (Bank no. 1)
P0172 System Too Rich (Bank no. 1)
P0173 Fuel Trim Malfunction (Bank no. 2)
P0174 System Too Lean (Bank no. 2)

P0175 System Too Rich (Bank no. 2)
P0176 Fuel Composition Sensor Circuit Malfunction
P0177 Fuel Composition Sensor Circuit Range/Performance
P0178 Fuel Composition Sensor Circuit Low Input
P0179 Fuel Composition Sensor Circuit High Input
P0180 Fuel Temperature Sensor "A" Circuit Malfunction
P0181 Fuel Temperature Sensor "A" Circuit Range/Performance
P0182 Fuel Temperature Sensor "A" Circuit Low Input
P0183 Fuel Temperature Sensor "A" Circuit High Input
P0184 Fuel Temperature Sensor "A" Circuit Intermittent
P0185 Fuel Temperature Sensor "B" Circuit Malfunction
P0186 Fuel Temperature Sensor "B" Circuit Range/Performance
P0187 Fuel Temperature Sensor "B" Circuit Low Input
P0188 Fuel Temperature Sensor "B" Circuit High Input
P0189 Fuel Temperature Sensor "B" Circuit Intermittent
P0190 Fuel Rail Pressure Sensor Circuit Malfunction
P0191 Fuel Rail Pressure Sensor Circuit Range/Performance
P0192 Fuel Rail Pressure Sensor Circuit Low Input
P0193 Fuel Rail Pressure Sensor Circuit High Input
P0194 Fuel Rail Pressure Sensor Circuit Intermittent
P0195 Engine Oil Temperature Sensor Malfunction
P0196 Engine Oil Temperature Sensor Range/Performance
P0197 Engine Oil Temperature Sensor Low
P0198 Engine Oil Temperature Sensor High
P0199 Engine Oil Temperature Sensor Intermittent
P0200 Injector Circuit Malfunction
P0201 Injector Circuit Malfunction—Cylinder no. 1
P0202 Injector Circuit Malfunction—Cylinder no. 2
P0203 Injector Circuit Malfunction—Cylinder no. 3
P0204 Injector Circuit Malfunction—Cylinder no. 4
P0205 Injector Circuit Malfunction—Cylinder no. 5
P0206 Injector Circuit Malfunction—Cylinder no. 6
P0207 Injector Circuit Malfunction—Cylinder no. 7
P0208 Injector Circuit Malfunction—Cylinder no. 8
P0209 Injector Circuit Malfunction—Cylinder no. 9
P0210 Injector Circuit Malfunction—Cylinder no. 10
P0211 Injector Circuit Malfunction—Cylinder no. 11
P0212 Injector Circuit Malfunction—Cylinder no. 12
P0213 Cold Start Injector no. 1 Malfunction
P0214 Cold Start Injector no. 2 Malfunction
P0215 Engine Shutoff Solenoid Malfunction
P0216 Injection Timing Control Circuit Malfunction
P0217 Engine Over Temperature Condition
P0218 Transmission Over Temperature Condition
P0219 Engine Over Speed Condition
P0220 Throttle/Pedal Position Sensor/Switch "B" Circuit Malfunction
P0221 Throttle/Pedal Position Sensor/Switch "B" Circuit Range/Performance Problem
P0222 Throttle/Pedal Position Sensor/Switch "B" Circuit Low Input
P0223 Throttle/Pedal Position Sensor/Switch "B" Circuit High Input
P0224 Throttle/Pedal Position Sensor/Switch "B" Circuit Intermittent
P0225 Throttle/Pedal Position Sensor/Switch "C" Circuit Malfunction
P0226 Throttle/Pedal Position Sensor/Switch "C" Circuit Range/Performance Problem
P0227 Throttle/Pedal Position Sensor/Switch "C" Circuit Low Input
P0228 Throttle/Pedal Position Sensor/Switch "C" Circuit High Input
P0229 Throttle/Pedal Position Sensor/Switch "C" Circuit Intermittent
P0230 Fuel Pump Primary Circuit Malfunction
P0231 Fuel Pump Secondary Circuit Low
P0232 Fuel Pump Secondary Circuit High
P0233 Fuel Pump Secondary Circuit Intermittent
P0234 Engine Over Boost Condition
P0261 Cylinder no. 1 Injector Circuit Low
P0262 Cylinder no. 1 Injector Circuit High
P0263 Cylinder no. 1 Contribution/Balance Fault
P0264 Cylinder no. 2 Injector Circuit Low
P0265 Cylinder no. 2 Injector Circuit High
P0266 Cylinder no. 2 Contribution/Balance Fault
P0267 Cylinder no. 3 Injector Circuit Low
P0268 Cylinder no. 3 Injector Circuit High
P0269 Cylinder no. 3 Contribution/Balance Fault
P0270 Cylinder no. 4 Injector Circuit Low
P0271 Cylinder no. 4 Injector Circuit High
P0272 Cylinder no. 4 Contribution/Balance Fault
P0273 Cylinder no. 5 Injector Circuit Low
P0274 Cylinder no. 5 Injector Circuit High
P0275 Cylinder no. 5 Contribution/Balance Fault
P0276 Cylinder no. 6 Injector Circuit Low
P0277 Cylinder no. 6 Injector Circuit High
P0278 Cylinder no. 6 Contribution/Balance Fault
P0279 Cylinder no. 7 Injector Circuit Low
P0280 Cylinder no. 7 Injector Circuit High
P0281 Cylinder no. 7 Contribution/Balance Fault
P0282 Cylinder no. 8 Injector Circuit Low
P0283 Cylinder no. 8 Injector Circuit High
P0284 Cylinder no. 8 Contribution/Balance Fault
P0285 Cylinder no. 9 Injector Circuit Low
P0286 Cylinder no. 9 Injector Circuit High
P0287 Cylinder no. 9 Contribution/Balance Fault
P0288 Cylinder no. 10 Injector Circuit Low
P0289 Cylinder no. 10 Injector Circuit High
P0290 Cylinder no. 10 Contribution/Balance Fault
P0291 Cylinder no. 11 Injector Circuit Low
P0292 Cylinder no. 11 Injector Circuit High
P0293 Cylinder no. 11 Contribution/Balance Fault
P0294 Cylinder no. 12 Injector Circuit Low
P0295 Cylinder no. 12 Injector Circuit High
P0296 Cylinder no. 12 Contribution/Balance Fault
P0300 Random/Multiple Cylinder Misfire Detected
P0301 Cylinder no. 1—Misfire Detected
P0302 Cylinder no. 2—Misfire Detected
P0303 Cylinder no. 3—Misfire Detected
P0304 Cylinder no. 4—Misfire Detected
P0305 Cylinder no. 5—Misfire Detected
P0306 Cylinder no. 6—Misfire Detected
P0307 Cylinder no. 7—Misfire Detected
P0308 Cylinder no. 8—Misfire Detected
P0309 Cylinder no. 9—Misfire Detected
P0310 Cylinder no. 10—Misfire Detected
P0311 Cylinder no. 11—Misfire Detected
P0312 Cylinder no. 12—Misfire Detected
P0320 Ignition/Distributor Engine Speed Input Circuit Malfunction
P0321 Ignition/Distributor Engine Speed Input Circuit Range/Performance

P0322 Ignition/Distributor Engine Speed Input Circuit No Signal
P0323 Ignition/Distributor Engine Speed Input Circuit Intermittent
P0325 Knock Sensor no. 1—Circuit Malfunction (Bank no. 1 or Single Sensor)
P0326 Knock Sensor no. 1—Circuit Range/Performance (Bank no. 1 or Single Sensor)
P0327 Knock Sensor no. 1—Circuit Low Input (Bank no. 1 or Single Sensor)
P0328 Knock Sensor no. 1—Circuit High Input (Bank no. 1 or Single Sensor)
P0329 Knock Sensor no. 1—Circuit Input Intermittent (Bank no. 1 or Single Sensor)
P0330 Knock Sensor no. 2—Circuit Malfunction (Bank no. 2)
P0331 Knock Sensor no. 2—Circuit Range/Performance (Bank no. 2)
P0332 Knock Sensor no. 2—Circuit Low Input (Bank no. 2)
P0333 Knock Sensor no. 2—Circuit High Input (Bank no. 2)
P0334 Knock Sensor no. 2—Circuit Input Intermittent (Bank no. 2)
P0335 Crankshaft Position Sensor "A" Circuit Malfunction
P0336 Crankshaft Position Sensor "A" Circuit Range/Performance
P0337 Crankshaft Position Sensor "A" Circuit Low Input
P0338 Crankshaft Position Sensor "A" Circuit High Input
P0339 Crankshaft Position Sensor "A" Circuit Intermittent
P0340 Camshaft Position Sensor Circuit Malfunction
P0341 Camshaft Position Sensor Circuit Range/Performance
P0342 Camshaft Position Sensor Circuit Low Input
P0343 Camshaft Position Sensor Circuit High Input
P0344 Camshaft Position Sensor Circuit Intermittent
P0350 Ignition Coil Primary/Secondary Circuit Malfunction
P0351 Ignition Coil "A" Primary/Secondary Circuit Malfunction
P0352 Ignition Coil "B" Primary/Secondary Circuit Malfunction
P0353 Ignition Coil "C" Primary/Secondary Circuit Malfunction
P0354 Ignition Coil "D" Primary/Secondary Circuit Malfunction
P0355 Ignition Coil "E" Primary/Secondary Circuit Malfunction
P0356 Ignition Coil "F" Primary/Secondary Circuit Malfunction
P0357 Ignition Coil "G" Primary/Secondary Circuit Malfunction
P0358 Ignition Coil "H" Primary/Secondary Circuit Malfunction
P0359 Ignition Coil "I" Primary/Secondary Circuit Malfunction
P0360 Ignition Coil "J" Primary/Secondary Circuit Malfunction
P0361 Ignition Coil "K" Primary/Secondary Circuit Malfunction
P0362 Ignition Coil "L" Primary/Secondary Circuit Malfunction
P0370 Timing Reference High Resolution Signal "A" Malfunction
P0371 Timing Reference High Resolution Signal "A" Too Many Pulses
P0372 Timing Reference High Resolution Signal "A" Too Few Pulses
P0373 Timing Reference High Resolution Signal "A" Intermittent/Erratic Pulses
P0374 Timing Reference High Resolution Signal "A" No Pulses
P0375 Timing Reference High Resolution Signal "B" Malfunction
P0376 Timing Reference High Resolution Signal "B" Too Many Pulses
P0377 Timing Reference High Resolution Signal "B" Too Few Pulses
P0378 Timing Reference High Resolution Signal "B" Intermittent/Erratic Pulses
P0379 Timing Reference High Resolution Signal "B" No Pulses

P0380 Glow Plug/Heater Circuit "A" Malfunction
P0381 Glow Plug/Heater Indicator Circuit Malfunction
P0382 Glow Plug/Heater Circuit "B" Malfunction
P0385 Crankshaft Position Sensor "B" Circuit Malfunction
P0386 Crankshaft Position Sensor "B" Circuit Range/Performance
P0387 Crankshaft Position Sensor "B" Circuit Low Input
P0388 Crankshaft Position Sensor "B" Circuit High Input
P0389 Crankshaft Position Sensor "B" Circuit Intermittent
P0400 Exhaust Gas Recirculation Flow Malfunction
P0401 Exhaust Gas Recirculation Flow Insufficient Detected
P0402 Exhaust Gas Recirculation Flow Excessive Detected
P0403 Exhaust Gas Recirculation Circuit Malfunction
P0404 Exhaust Gas Recirculation Circuit Range/Performance
P0405 Exhaust Gas Recirculation Sensor "A" Circuit Low
P0406 Exhaust Gas Recirculation Sensor "A" Circuit High
P0407 Exhaust Gas Recirculation Sensor "B" Circuit Low
P0408 Exhaust Gas Recirculation Sensor "B" Circuit High
P0410 Secondary Air Injection System Malfunction
P0411 Secondary Air Injection System Incorrect Flow Detected
P0412 Secondary Air Injection System Switching Valve "A" Circuit Malfunction
P0413 Secondary Air Injection System Switching Valve "A" Circuit Open
P0414 Secondary Air Injection System Switching Valve "A" Circuit Shorted
P0415 Secondary Air Injection System Switching Valve "B" Circuit Malfunction
P0416 Secondary Air Injection System Switching Valve "B" Circuit Open
P0417 Secondary Air Injection System Switching Valve "B" Circuit Shorted
P0418 Secondary Air Injection System Relay "A" Circuit Malfunction
P0419 Secondary Air Injection System Relay "B" Circuit Malfunction
P0420 Catalyst System Efficiency Below Threshold (Bank no. 1)
P0421 Warm Up Catalyst Efficiency Below Threshold (Bank no. 1)
P0422 Main Catalyst Efficiency Below Threshold (Bank no. 1)
P0423 Heated Catalyst Efficiency Below Threshold (Bank no. 1)
P0424 Heated Catalyst Temperature Below Threshold (Bank no. 1)
P0430 Catalyst System Efficiency Below Threshold (Bank no. 2)
P0431 Warm Up Catalyst Efficiency Below Threshold (Bank no. 2)
P0432 Main Catalyst Efficiency Below Threshold (Bank no. 2)
P0433 Heated Catalyst Efficiency Below Threshold (Bank no. 2)
P0434 Heated Catalyst Temperature Below Threshold (Bank no. 2)
P0440 Evaporative Emission Control System Malfunction
P0441 Evaporative Emission Control System Incorrect Purge Flow
P0442 Evaporative Emission Control System Leak Detected (Small Leak)
P0443 Evaporative Emission Control System Purge Control Valve Circuit Malfunction
P0444 Evaporative Emission Control System Purge Control Valve Circuit Open
P0445 Evaporative Emission Control System Purge Control Valve Circuit Shorted
P0446 Evaporative Emission Control System Vent Control Circuit Malfunction
P0447 Evaporative Emission Control System Vent Control Circuit Open

P0448 Evaporative Emission Control System Vent Control Circuit Shorted

P0449 Evaporative Emission Control System Vent Valve/Solenoid Circuit Malfunction

P0450 Evaporative Emission Control System Pressure Sensor Malfunction

P0451 Evaporative Emission Control System Pressure Sensor Range/Performance

P0452 Evaporative Emission Control System Pressure Sensor Low Input

P0453 Evaporative Emission Control System Pressure Sensor High Input

P0454 Evaporative Emission Control System Pressure Sensor Intermittent

P0455 Evaporative Emission Control System Leak Detected (Gross Leak)

P0460 Fuel Level Sensor Circuit Malfunction

P0461 Fuel Level Sensor Circuit Range/Performance

P0462 Fuel Level Sensor Circuit Low Input

P0463 Fuel Level Sensor Circuit High Input

P0464 Fuel Level Sensor Circuit Intermittent

P0465 Purge Flow Sensor Circuit Malfunction

P0466 Purge Flow Sensor Circuit Range/Performance

P0467 Purge Flow Sensor Circuit Low Input

P0468 Purge Flow Sensor Circuit High Input

P0469 Purge Flow Sensor Circuit Intermittent

P0470 Exhaust Pressure Sensor Malfunction

P0471 Exhaust Pressure Sensor Range/Performance

P0472 Exhaust Pressure Sensor Low

P0473 Exhaust Pressure Sensor High

P0474 Exhaust Pressure Sensor Intermittent

P0475 Exhaust Pressure Control Valve Malfunction

P0476 Exhaust Pressure Control Valve Range/Performance

P0477 Exhaust Pressure Control Valve Low

P0478 Exhaust Pressure Control Valve High

P0479 Exhaust Pressure Control Valve Intermittent

P0480 Cooling Fan no. 1 Control Circuit Malfunction

P0481 Cooling Fan no. 2 Control Circuit Malfunction

P0482 Cooling Fan no. 3 Control Circuit Malfunction

P0483 Cooling Fan Rationality Check Malfunction

P0484 Cooling Fan Circuit Over Current

P0485 Cooling Fan Power/Ground Circuit Malfunction

P0500 Vehicle Speed Sensor Malfunction

P0501 Vehicle Speed Sensor Range/Performance

P0502 Vehicle Speed Sensor Circuit Low Input

P0503 Vehicle Speed Sensor Intermittent/Erratic/High

P0505 Idle Control System Malfunction

P0506 Idle Control System RPM Lower Than Expected

P0507 Idle Control System RPM Higher Than Expected

P0510 Closed Throttle Position Switch Malfunction

P0520 Engine Oil Pressure Sensor/Switch Circuit Malfunction

P0521 Engine Oil Pressure Sensor/Switch Range/Performance

P0522 Engine Oil Pressure Sensor/Switch Low Voltage

P0523 Engine Oil Pressure Sensor/Switch High Voltage

P0530 A/C Refrigerant Pressure Sensor Circuit Malfunction

P0531 A/C Refrigerant Pressure Sensor Circuit Range/Performance

P0532 A/C Refrigerant Pressure Sensor Circuit Low Input

P0533 A/C Refrigerant Pressure Sensor Circuit High Input

P0534 A/C Refrigerant Charge Loss

P0550 Power Steering Pressure Sensor Circuit Malfunction

P0551 Power Steering Pressure Sensor Circuit Range/Performance

P0552 Power Steering Pressure Sensor Circuit Low Input

P0553 Power Steering Pressure Sensor Circuit High Input

P0554 Power Steering Pressure Sensor Circuit Intermittent

P0560 System Voltage Malfunction

P0561 System Voltage Unstable

P0562 System Voltage Low

P0563 System Voltage High

P0565 Cruise Control On Signal Malfunction

P0566 Cruise Control Off Signal Malfunction

P0567 Cruise Control Resume Signal Malfunction

P0568 Cruise Control Set Signal Malfunction

P0569 Cruise Control Coast Signal Malfunction

P0570 Cruise Control Accel Signal Malfunction

P0571 Cruise Control/Brake Switch "A" Circuit Malfunction

P0572 Cruise Control/Brake Switch "A" Circuit Low

P0573 Cruise Control/Brake Switch "A" Circuit High

P0574 Through P0580 Reserved for Cruise Codes

P0600 Serial Communication Link Malfunction

P0601 Internal Control Module Memory Check Sum Error

P0602 Control Module Programming Error

P0603 Internal Control Module Keep Alive Memory (KAM) Error

P0604 Internal Control Module Random Access Memory (RAM) Error

P0605 Internal Control Module Read Only Memory (ROM) Error

P0606 PCM Processor Fault

P0608 Control Module VSS Output "A" Malfunction

P0609 Control Module VSS Output "B" Malfunction

P0620 Generator Control Circuit Malfunction

P0621 Generator Lamp "L" Control Circuit Malfunction

P0622 Generator Field "F" Control Circuit Malfunction

P0650 Malfunction Indicator Lamp (MIL) Control Circuit Malfunction

P0654 Engine RPM Output Circuit Malfunction

P0655 Engine Hot Lamp Output Control Circuit Malfunction

P0656 Fuel Level Output Circuit Malfunction

P0700 Transmission Control System Malfunction

P0701 Transmission Control System Range/Performance

P0702 Transmission Control System Electrical

P0703 Torque Converter/Brake Switch "B" Circuit Malfunction

P0704 Clutch Switch Input Circuit Malfunction

P0705 Transmission Range Sensor Circuit Malfunction (PRNDL Input)

P0706 Transmission Range Sensor Circuit Range/Performance

P0707 Transmission Range Sensor Circuit Low Input

P0708 Transmission Range Sensor Circuit High Input

P0709 Transmission Range Sensor Circuit Intermittent

P0710 Transmission Fluid Temperature Sensor Circuit Malfunction

P0711 Transmission Fluid Temperature Sensor Circuit Range/Performance

P0712 Transmission Fluid Temperature Sensor Circuit Low Input

P0713 Transmission Fluid Temperature Sensor Circuit High Input

P0714 Transmission Fluid Temperature Sensor Circuit Intermittent

P0715 Input/Turbine Speed Sensor Circuit Malfunction

P0716 Input/Turbine Speed Sensor Circuit Range/Performance

P0717 Input/Turbine Speed Sensor Circuit No Signal

P0718 Input/Turbine Speed Sensor Circuit Intermittent

P0719 Torque Converter/Brake Switch "B" Circuit Low
P0720 Output Speed Sensor Circuit Malfunction
P0721 Output Speed Sensor Circuit Range/Performance
P0722 Output Speed Sensor Circuit No Signal
P0723 Output Speed Sensor Circuit Intermittent
P0724 Torque Converter/Brake Switch "B" Circuit High
P0725 Engine Speed Input Circuit Malfunction
P0726 Engine Speed Input Circuit Range/Performance
P0727 Engine Speed Input Circuit No Signal
P0728 Engine Speed Input Circuit Intermittent
P0730 Incorrect Gear Ratio
P0731 Gear no. 1 Incorrect Ratio
P0732 Gear no. 2 Incorrect Ratio
P0733 Gear no. 3 Incorrect Ratio
P0734 Gear no. 4 Incorrect Ratio
P0735 Gear no. 5 Incorrect Ratio
P0736 Reverse Incorrect Ratio
P0740 Torque Converter Clutch Circuit Malfunction
P0741 Torque Converter Clutch Circuit Performance or Stuck Off
P0742 Torque Converter Clutch Circuit Stuck On
P0743 Torque Converter Clutch Circuit Electrical
P0744 Torque Converter Clutch Circuit Intermittent
P0745 Pressure Control Solenoid Malfunction
P0746 Pressure Control Solenoid Performance or Stuck Off
P0747 Pressure Control Solenoid Stuck On
P0748 Pressure Control Solenoid Electrical
P0749 Pressure Control Solenoid Intermittent
P0750 Shift Solenoid "A" Malfunction
P0751 Shift Solenoid "A" Performance or Stuck Off
P0752 Shift Solenoid "A" Stuck On
P0753 Shift Solenoid "A" Electrical
P0754 Shift Solenoid "A" Intermittent
P0755 Shift Solenoid "B" Malfunction
P0756 Shift Solenoid "B" Performance or Stuck Off
P0757 Shift Solenoid "B" Stuck On
P0758 Shift Solenoid "B" Electrical
P0759 Shift Solenoid "B" Intermittent
P0760 Shift Solenoid "C" Malfunction
P0761 Shift Solenoid "C" Performance Or Stuck Off
P0762 Shift Solenoid "C" Stuck On
P0763 Shift Solenoid "C" Electrical
P0764 Shift Solenoid "C" Intermittent
P0765 Shift Solenoid "D" Malfunction
P0766 Shift Solenoid "D" Performance Or Stuck Off
P0767 Shift Solenoid "D" Stuck On
P0768 Shift Solenoid "D" Electrical
P0769 Shift Solenoid "D" Intermittent
P0770 Shift Solenoid "E" Malfunction
P0771 Shift Solenoid "E" Performance Or Stuck Off
P0772 Shift Solenoid "E" Stuck On
P0773 Shift Solenoid "E" Electrical
P0774 Shift Solenoid "E" Intermittent
P0780 Shift Malfunction
P0781 1–2 Shift Malfunction
P0782 2–3 Shift Malfunction
P0783 3–4 Shift Malfunction
P0784 4–5 Shift Malfunction
P0785 Shift/Timing Solenoid Malfunction
P0786 Shift/Timing Solenoid Range/Performance
P0787 Shift/Timing Solenoid Low
P0788 Shift/Timing Solenoid High
P0789 Shift/Timing Solenoid Intermittent

P0790 Normal/Performance Switch Circuit Malfunction
P0801 Reverse Inhibit Control Circuit Malfunction
P0803 1–4 Upshift (Skip Shift) Solenoid Control Circuit Malfunction
P0804 1–4 Upshift (Skip Shift) Lamp Control Circuit Malfunction
P1102 HO$_2$S-11 Heater Circuit High Voltage
P1105 HO$_2$S-12 Heater Circuit High Voltage
P1115 HO$_2$S-11 Heater Circuit Low Voltage
P1117 HO$_2$S-12 Heater Circuit Low Voltage
P1123 Long Term Fuel Trim Adaptive Air System Low
P1124 Long Term Fuel Trim Adaptive Air System High
P1127 Long Term Fuel Trim Multiplicative Air System Low
P1128 Long Term Fuel Trim Multiplicative Air System High
P1140 Load Calculation Cross Check
P1170 HO$_2$S-11 Circuit Voltage Stuck At Mid-Range
P1195 EGR Boost Or Pressure Sensor Circuit Fault
P1196 Ignition Switch Start Circuit Fault
P1213 Fuel Injector 1, 2, 3 Or 4 Circuit High Voltage
P1214 Fuel Injector 1, 2, 3 Or 4 Circuit High Voltage
P1215 Fuel Injector 1, 2, 3 Or 4 Circuit High Voltage
P1216 Fuel Injector 1, 2, 3 Or 4 Circuit High Voltage
P1225 Fuel Injector 1, 2, 5 Or 4 Circuit Low Voltage
P1226 Fuel Injector 1, 2, 5 Or 4 Circuit Low Voltage
P1227 Fuel Injector 1, 2, 5 Or 4 Circuit Low Voltage
P1228 Fuel Injector 1, 2, 5 Or 4 Circuit Low Voltage
P1250 Pressure Regulator Control Solenoid Circuit Fault
P1345 No SGC (CMP) Signal To PCM
P1386 Knock Sensor Control Zero Test
P1401 EGR Control Solenoid Circuit Signal Low
P1402 EGR Control Solenoid Circuit Signal High
P1402 EGR Valve Position Sensor Circuit Fault
P1410 EVAP Purge Control Solenoid Circuit High Voltage
P1412 EGR Differential Pressure Sensor Signal Low
P1413 EGR Differential Pressure Sensor Signal High
P1425 EVAP Purge Control Solenoid Circuit Low Voltage
P1449 Canister Drain Cut Valve Solenoid Circuit Fault
P1455 Fuel Tank Sending Unit Circuit Fault
P1458 Air Conditioning Compressor Clutch Signal Fault
P1485 EGR Vent Control Solenoid Circuit Fault
P1486 EGR Vacuum Control Solenoid Circuit Fault
P1487 EGR Boost Sensor Solenoid Circuit Fault
P1510 Idle Air Control Valve Closing Coil High Voltage
P1513 Idle Air Control Valve Closing Coil Low Voltage
P1515 A/T To M/T Codification
P1523 VICS Solenoid Valve Circuit Fault
P1552 Idle Air Control Valve Opening Coil Low Voltage
P1553 Idle Air Control Valve Opening Coil High Voltage
P1606 Chassis Accelerator Sensor Signal Circuit Fault
P1608 PCM Internal Fault
P1611 MIL Request Circuit Low Voltage
P1614 MIL Request Circuit High Voltage
P1616 Chassis Accelerator Sensor Signal Low Voltage
P1617 Chassis Accelerator Sensor Signal High Voltage
P1624 TCM to PCM MIL Request Circuit Fault
P1655 Unused Power Stage "B"
P1660 Unused Power Stage 'A'
P1660 Unused Power Stage 'B'
P1665 Power Stage Group 'A'
P1743 Torque Converter Clutch Solenoid Circuit Fault
P1794 Battery Or Circuit Fault
P1797 Clutch Pedal Switch (MT) Or PIN Switch Circuit Fault

Land Rover

READING CODES

Reading the control module memory is one of the first steps in OBD II system diagnostics.

This step should be initially performed to determine the general nature of the fault. Subsequent readings will determine if the fault has been cleared.

Reading codes can be performed by any of the methods below:

• Read the control module memory with the Generic Scan Tool (GST)

• Read the control module memory with the vehicle manufacturer's specific tester

To read the fault codes, connect the scan tool or tester according to the manufacturer's instructions. Follow the manufacturer's specified procedure for reading the codes.

CLEARING CODES

Control module reset procedures are a very important part of OBD II System diagnostics. This step should be done at the end of any fault code repair and at the end of any driveability repair.

Clearing codes can be performed by any of the methods below:

• Clear the control module memory with the Generic Scan Tool (GST)

• Clear the control module memory with the vehicle manufacturer's specific tester

• Turn the ignition off and remove the negative battery cable for at least 1 minute.

Removing the negative battery cable may cause other systems in the vehicle to lose their memory. Prior to removing the cable, ensure you have the proper reset codes for radios and alarms.

➡**The MIL will may also be de-activated for some codes if the vehicle completes three consecutive trips without a fault detected with vehicle conditions similar to those present during the fault.**

DIAGNOSTIC TROUBLE CODES

P0100 Mass or Volume Air Flow Circuit Malfunction
P0101 Mass or Volume Air Flow Circuit Range/Performance Problem
P0102 Mass or Volume Air Flow Circuit Low Input
P0103 Mass or Volume Air Flow Circuit High Input
P0104 Mass or Volume Air Flow Circuit Intermittent
P0105 Manifold Absolute Pressure/Barometric Pressure Circuit Malfunction
P0106 Manifold Absolute Pressure/Barometric Pressure Circuit Range/Performance Problem
P0107 Manifold Absolute Pressure/Barometric Pressure Circuit Low Input
P0108 Manifold Absolute Pressure/Barometric Pressure Circuit High Input
P0109 Manifold Absolute Pressure/Barometric Pressure Circuit Intermittent
P0110 Intake Air Temperature Circuit Malfunction
P0111 Intake Air Temperature Circuit Range/Performance Problem
P0112 Intake Air Temperature Circuit Low Input

P0113 Intake Air Temperature Circuit High Input
P0114 Intake Air Temperature Circuit Intermittent
P0115 Engine Coolant Temperature Circuit Malfunction
P0116 Engine Coolant Temperature Circuit Range/Performance Problem
P0117 Engine Coolant Temperature Circuit Low Input
P0118 Engine Coolant Temperature Circuit High Input
P0119 Engine Coolant Temperature Circuit Intermittent
P0120 Throttle/Pedal Position Sensor/Switch "A" Circuit Malfunction
P0121 Throttle/Pedal Position Sensor/Switch "A" Circuit Range/Performance Problem
P0122 Throttle/Pedal Position Sensor/Switch "A" Circuit Low Input
P0123 Throttle/Pedal Position Sensor/Switch "A" Circuit High Input
P0124 Throttle/Pedal Position Sensor/Switch "A" Circuit Intermittent
P0125 Insufficient Coolant Temperature For Closed Loop Fuel Control
P0126 Insufficient Coolant Temperature For Stable Operation
P0130 O_2 Circuit Malfunction (Bank no. 1 Sensor no. 1)
P0131 O_2 Sensor Circuit Low Voltage (Bank no. 1 Sensor no. 1)
P0132 O_2 Sensor Circuit High Voltage (Bank no. 1 Sensor no. 1)
P0133 O_2 Sensor Circuit Slow Response (Bank no. 1 Sensor no. 1)
P0134 O_2 Sensor Circuit No Activity Detected (Bank no. 1 Sensor no. 1)
P0135 O_2 Sensor Heater Circuit Malfunction (Bank no. 1 Sensor no. 1)
P0136 O_2 Sensor Circuit Malfunction (Bank no. 1 Sensor no. 2)
P0137 O_2 Sensor Circuit Low Voltage (Bank no. 1 Sensor no. 2)
P0138 O_2 Sensor Circuit High Voltage (Bank no. 1 Sensor no. 2)
P0139 O_2 Sensor Circuit Slow Response (Bank no. 1 Sensor no. 2)
P0140 O_2 Sensor Circuit No Activity Detected (Bank no. 1 Sensor no. 2)
P0141 O_2 Sensor Heater Circuit Malfunction (Bank no. 1 Sensor no. 2)
P0142 O_2 Sensor Circuit Malfunction (Bank no. 1 Sensor no. 3)
P0143 O_2 Sensor Circuit Low Voltage (Bank no. 1 Sensor no. 3)
P0144 O_2 Sensor Circuit High Voltage (Bank no. 1 Sensor no. 3)
P0145 O_2 Sensor Circuit Slow Response (Bank no. 1 Sensor no. 3)
P0146 O_2 Sensor Circuit No Activity Detected (Bank no. 1 Sensor no. 3)
P0147 O_2 Sensor Heater Circuit Malfunction (Bank no. 1 Sensor no. 3)
P0150 O_2 Sensor Circuit Malfunction (Bank no. 2 Sensor no. 1)
P0151 O_2 Sensor Circuit Low Voltage (Bank no. 2 Sensor no. 1)
P0152 O_2 Sensor Circuit High Voltage (Bank no. 2 Sensor no. 1)
P0153 O_2 Sensor Circuit Slow Response (Bank no. 2 Sensor no. 1)
P0154 O_2 Sensor Circuit No Activity Detected (Bank no. 2 Sensor no. 1)
P0155 O_2 Sensor Heater Circuit Malfunction (Bank no. 2 Sensor no. 1)
P0156 O_2 Sensor Circuit Malfunction (Bank no. 2 Sensor no. 2)
P0157 O_2 Sensor Circuit Low Voltage (Bank no. 2 Sensor no. 2)
P0158 O_2 Sensor Circuit High Voltage (Bank no. 2 Sensor no. 2)
P0159 O_2 Sensor Circuit Slow Response (Bank no. 2 Sensor no. 2)

P0160 O_2 Sensor Circuit No Activity Detected (Bank no. 2 Sensor no. 2)

P0161 O_2 Sensor Heater Circuit Malfunction (Bank no. 2 Sensor no. 2)

P0162 O_2 Sensor Circuit Malfunction (Bank no. 2 Sensor no. 3)

P0163 O_2 Sensor Circuit Low Voltage (Bank no. 2 Sensor no. 3)

P0164 O_2 Sensor Circuit High Voltage (Bank no. 2 Sensor no. 3)

P0165 O_2 Sensor Circuit Slow Response (Bank no. 2 Sensor no. 3)

P0166 O_2 Sensor Circuit No Activity Detected (Bank no. 2 Sensor no. 3)

P0167 O_2 Sensor Heater Circuit Malfunction (Bank no. 2 Sensor no. 3)

P0170 Fuel Trim Malfunction (Bank no. 1)

P0171 System Too Lean (Bank no. 1)

P0172 System Too Rich (Bank no. 1)

P0173 Fuel Trim Malfunction (Bank no. 2)

P0174 System Too Lean (Bank no. 2)

P0175 System Too Rich (Bank no. 2)

P0176 Fuel Composition Sensor Circuit Malfunction

P0177 Fuel Composition Sensor Circuit Range/Performance

P0178 Fuel Composition Sensor Circuit Low Input

P0179 Fuel Composition Sensor Circuit High Input

P0180 Fuel Temperature Sensor "A" Circuit Malfunction

P0181 Fuel Temperature Sensor "A" Circuit Range/Performance

P0182 Fuel Temperature Sensor "A" Circuit Low Input

P0183 Fuel Temperature Sensor "A" Circuit High Input

P0184 Fuel Temperature Sensor "A" Circuit Intermittent

P0185 Fuel Temperature Sensor "B" Circuit Malfunction

P0186 Fuel Temperature Sensor "B" Circuit Range/Performance

P0187 Fuel Temperature Sensor "B" Circuit Low Input

P0188 Fuel Temperature Sensor "B" Circuit High Input

P0189 Fuel Temperature Sensor "B" Circuit Intermittent

P0190 Fuel Rail Pressure Sensor Circuit Malfunction

P0191 Fuel Rail Pressure Sensor Circuit Range/Performance

P0192 Fuel Rail Pressure Sensor Circuit Low Input

P0193 Fuel Rail Pressure Sensor Circuit High Input

P0194 Fuel Rail Pressure Sensor Circuit Intermittent

P0195 Engine Oil Temperature Sensor Malfunction

P0196 Engine Oil Temperature Sensor Range/Performance

P0197 Engine Oil Temperature Sensor Low

P0198 Engine Oil Temperature Sensor High

P0199 Engine Oil Temperature Sensor Intermittent

P0200 Injector Circuit Malfunction

P0201 Injector Circuit Malfunction—Cylinder no. 1

P0202 Injector Circuit Malfunction—Cylinder no. 2

P0203 Injector Circuit Malfunction—Cylinder no. 3

P0204 Injector Circuit Malfunction—Cylinder no. 4

P0205 Injector Circuit Malfunction—Cylinder no. 5

P0206 Injector Circuit Malfunction—Cylinder no. 6

P0207 Injector Circuit Malfunction—Cylinder no. 7

P0208 Injector Circuit Malfunction—Cylinder no. 8

P0209 Injector Circuit Malfunction—Cylinder no. 9

P0210 Injector Circuit Malfunction—Cylinder no. 10

P0211 Injector Circuit Malfunction—Cylinder no. 11

P0212 Injector Circuit Malfunction—Cylinder no. 12

P0213 Cold Start Injector no. 1 Malfunction

P0214 Cold Start Injector no. 2 Malfunction

P0215 Engine Shutoff Solenoid Malfunction

P0216 Injection Timing Control Circuit Malfunction

P0217 Engine Over Temperature Condition

P0218 Transmission Over Temperature Condition

P0219 Engine Over Speed Condition

P0220 Throttle/Pedal Position Sensor/Switch "B" Circuit Malfunction

P0221 Throttle/Pedal Position Sensor/Switch "B" Circuit Range/Performance Problem

P0222 Throttle/Pedal Position Sensor/Switch "B" Circuit Low Input

P0223 Throttle/Pedal Position Sensor/Switch "B" Circuit High Input

P0224 Throttle/Pedal Position Sensor/Switch "B" Circuit Intermittent

P0225 Throttle/Pedal Position Sensor/Switch "C" Circuit Malfunction

P0226 Throttle/Pedal Position Sensor/Switch "C" Circuit Range/Performance Problem

P0227 Throttle/Pedal Position Sensor/Switch "C" Circuit Low Input

P0228 Throttle/Pedal Position Sensor/Switch "C" Circuit High Input

P0229 Throttle/Pedal Position Sensor/Switch "C" Circuit Intermittent

P0230 Fuel Pump Primary Circuit Malfunction

P0231 Fuel Pump Secondary Circuit Low

P0232 Fuel Pump Secondary Circuit High

P0233 Fuel Pump Secondary Circuit Intermittent

P0234 Engine Over Boost Condition

P0261 Cylinder no. 1 Injector Circuit Low

P0262 Cylinder no. 1 Injector Circuit High

P0263 Cylinder no. 1 Contribution/Balance Fault

P0264 Cylinder no. 2 Injector Circuit Low

P0265 Cylinder no. 2 Injector Circuit High

P0266 Cylinder no. 2 Contribution/Balance Fault

P0267 Cylinder no. 3 Injector Circuit Low

P0268 Cylinder no. 3 Injector Circuit High

P0269 Cylinder no. 3 Contribution/Balance Fault

P0270 Cylinder no. 4 Injector Circuit Low

P0271 Cylinder no. 4 Injector Circuit High

P0272 Cylinder no. 4 Contribution/Balance Fault

P0273 Cylinder no. 5 Injector Circuit Low

P0274 Cylinder no. 5 Injector Circuit High

P0275 Cylinder no. 5 Contribution/Balance Fault

P0276 Cylinder no. 6 Injector Circuit Low

P0277 Cylinder no. 6 Injector Circuit High

P0278 Cylinder no. 6 Contribution/Balance Fault

P0279 Cylinder no. 7 Injector Circuit Low

P0280 Cylinder no. 7 Injector Circuit High

P0281 Cylinder no. 7 Contribution/Balance Fault

P0282 Cylinder no. 8 Injector Circuit Low

P0283 Cylinder no. 8 Injector Circuit High

P0284 Cylinder no. 8 Contribution/Balance Fault

P0285 Cylinder no. 9 Injector Circuit Low

P0286 Cylinder no. 9 Injector Circuit High

P0287 Cylinder no. 9 Contribution/Balance Fault

P0288 Cylinder no. 10 Injector Circuit Low

P0289 Cylinder no. 10 Injector Circuit High

P0290 Cylinder no. 10 Contribution/Balance Fault

P0291 Cylinder no. 11 Injector Circuit Low

P0292 Cylinder no. 11 Injector Circuit High

P0293 Cylinder no. 11 Contribution/Balance Fault

P0294 Cylinder no. 12 Injector Circuit Low

P0295 Cylinder no. 12 Injector Circuit High

P0296 Cylinder no. 12 Contribution/Balance Fault

P0300 Random/Multiple Cylinder Misfire Detected
P0301 Cylinder no. 1—Misfire Detected
P0302 Cylinder no. 2—Misfire Detected
P0303 Cylinder no. 3—Misfire Detected
P0304 Cylinder no. 4—Misfire Detected
P0305 Cylinder no. 5—Misfire Detected
P0306 Cylinder no. 6—Misfire Detected
P0307 Cylinder no. 7—Misfire Detected
P0308 Cylinder no. 8—Misfire Detected
P0309 Cylinder no. 9—Misfire Detected
P0310 Cylinder no. 10—Misfire Detected
P0311 Cylinder no. 11—Misfire Detected
P0312 Cylinder no. 12—Misfire Detected
P0320 Ignition/Distributor Engine Speed Input Circuit Malfunction
P0321 Ignition/Distributor Engine Speed Input Circuit Range/Performance
P0322 Ignition/Distributor Engine Speed Input Circuit No Signal
P0323 Ignition/Distributor Engine Speed Input Circuit Intermittent
P0325 Knock Sensor no. 1—Circuit Malfunction (Bank no. 1 or Single Sensor)
P0326 Knock Sensor no. 1—Circuit Range/Performance (Bank no. 1 or Single Sensor)
P0327 Knock Sensor no. 1—Circuit Low Input (Bank no. 1 or Single Sensor)
P0328 Knock Sensor no. 1—Circuit High Input (Bank no. 1 or Single Sensor)
P0329 Knock Sensor no. 1—Circuit Input Intermittent (Bank no. 1 or Single Sensor)
P0330 Knock Sensor no. 2—Circuit Malfunction (Bank no. 2)
P0331 Knock Sensor no. 2—Circuit Range/Performance (Bank no. 2)
P0332 Knock Sensor no. 2—Circuit Low Input (Bank no. 2)
P0333 Knock Sensor no. 2—Circuit High Input (Bank no. 2)
P0334 Knock Sensor no. 2—Circuit Input Intermittent (Bank no. 2)
P0335 Crankshaft Position Sensor "A" Circuit Malfunction
P0336 Crankshaft Position Sensor "A" Circuit Range/Performance
P0337 Crankshaft Position Sensor "A" Circuit Low Input
P0338 Crankshaft Position Sensor "A" Circuit High Input
P0339 Crankshaft Position Sensor "A" Circuit Intermittent
P0340 Camshaft Position Sensor Circuit Malfunction
P0341 Camshaft Position Sensor Circuit Range/Performance
P0342 Camshaft Position Sensor Circuit Low Input
P0343 Camshaft Position Sensor Circuit High Input
P0344 Camshaft Position Sensor Circuit Intermittent
P0350 Ignition Coil Primary/Secondary Circuit Malfunction
P0351 Ignition Coil "A" Primary/Secondary Circuit Malfunction
P0352 Ignition Coil "B" Primary/Secondary Circuit Malfunction
P0353 Ignition Coil "C" Primary/Secondary Circuit Malfunction
P0354 Ignition Coil "D" Primary/Secondary Circuit Malfunction
P0355 Ignition Coil "E" Primary/Secondary Circuit Malfunction
P0356 Ignition Coil "F" Primary/Secondary Circuit Malfunction
P0357 Ignition Coil "G" Primary/Secondary Circuit Malfunction
P0358 Ignition Coil "H" Primary/Secondary Circuit Malfunction
P0359 Ignition Coil "I" Primary/Secondary Circuit Malfunction
P0360 Ignition Coil "J" Primary/Secondary Circuit Malfunction
P0361 Ignition Coil "K" Primary/Secondary Circuit Malfunction
P0362 Ignition Coil "L" Primary/Secondary Circuit Malfunction

P0370 Timing Reference High Resolution Signal "A" Malfunction
P0371 Timing Reference High Resolution Signal "A" Too Many Pulses
P0372 Timing Reference High Resolution Signal "A" Too Few Pulses
P0373 Timing Reference High Resolution Signal "A" Intermittent/Erratic Pulses
P0374 Timing Reference High Resolution Signal "A" No Pulses
P0375 Timing Reference High Resolution Signal "B" Malfunction
P0376 Timing Reference High Resolution Signal "B" Too Many Pulses
P0377 Timing Reference High Resolution Signal "B" Too Few Pulses
P0378 Timing Reference High Resolution Signal "B" Intermittent/Erratic Pulses
P0379 Timing Reference High Resolution Signal "B" No Pulses
P0380 Glow Plug/Heater Circuit "A" Malfunction
P0381 Glow Plug/Heater Indicator Circuit Malfunction
P0382 Glow Plug/Heater Circuit "B" Malfunction
P0385 Crankshaft Position Sensor "B" Circuit Malfunction
P0386 Crankshaft Position Sensor "B" Circuit Range/Performance
P0387 Crankshaft Position Sensor "B" Circuit Low Input
P0388 Crankshaft Position Sensor "B" Circuit High Input
P0389 Crankshaft Position Sensor "B" Circuit Intermittent
P0400 Exhaust Gas Recirculation Flow Malfunction
P0401 Exhaust Gas Recirculation Flow Insufficient Detected
P0402 Exhaust Gas Recirculation Flow Excessive Detected
P0403 Exhaust Gas Recirculation Circuit Malfunction
P0404 Exhaust Gas Recirculation Circuit Range/Performance
P0405 Exhaust Gas Recirculation Sensor "A" Circuit Low
P0406 Exhaust Gas Recirculation Sensor "A" Circuit High
P0407 Exhaust Gas Recirculation Sensor "B" Circuit Low
P0408 Exhaust Gas Recirculation Sensor "B" Circuit High
P0410 Secondary Air Injection System Malfunction
P0411 Secondary Air Injection System Incorrect Flow Detected
P0412 Secondary Air Injection System Switching Valve "A" Circuit Malfunction
P0413 Secondary Air Injection System Switching Valve "A" Circuit Open
P0414 Secondary Air Injection System Switching Valve "A" Circuit Shorted
P0415 Secondary Air Injection System Switching Valve "B" Circuit Malfunction
P0416 Secondary Air Injection System Switching Valve "B" Circuit Open
P0417 Secondary Air Injection System Switching Valve "B" Circuit Shorted
P0418 Secondary Air Injection System Relay "A" Circuit Malfunction
P0419 Secondary Air Injection System Relay "B" Circuit Malfunction
P0420 Catalyst System Efficiency Below Threshold (Bank no. 1)
P0421 Warm Up Catalyst Efficiency Below Threshold (Bank no. 1)
P0422 Main Catalyst Efficiency Below Threshold (Bank no. 1)
P0423 Heated Catalyst Efficiency Below Threshold (Bank no. 1)
P0424 Heated Catalyst Temperature Below Threshold (Bank no. 1)
P0430 Catalyst System Efficiency Below Threshold (Bank no. 2)
P0431 Warm Up Catalyst Efficiency Below Threshold (Bank no. 2)
P0432 Main Catalyst Efficiency Below Threshold (Bank no. 2)

P0433 Heated Catalyst Efficiency Below Threshold (Bank no. 2)
P0434 Heated Catalyst Temperature Below Threshold (Bank no. 2)
P0440 Evaporative Emission Control System Malfunction
P0441 Evaporative Emission Control System Incorrect Purge Flow
P0442 Evaporative Emission Control System Leak Detected (Small Leak)
P0443 Evaporative Emission Control System Purge Control Valve Circuit Malfunction
P0444 Evaporative Emission Control System Purge Control Valve Circuit Open
P0445 Evaporative Emission Control System Purge Control Valve Circuit Shorted
P0446 Evaporative Emission Control System Vent Control Circuit Malfunction
P0447 Evaporative Emission Control System Vent Control Circuit Open
P0448 Evaporative Emission Control System Vent Control Circuit Shorted
P0449 Evaporative Emission Control System Vent Valve/Solenoid Circuit Malfunction
P0450 Evaporative Emission Control System Pressure Sensor Malfunction
P0451 Evaporative Emission Control System Pressure Sensor Range/Performance
P0452 Evaporative Emission Control System Pressure Sensor Low Input
P0453 Evaporative Emission Control System Pressure Sensor High Input
P0454 Evaporative Emission Control System Pressure Sensor Intermittent
P0455 Evaporative Emission Control System Leak Detected (Gross Leak)
P0460 Fuel Level Sensor Circuit Malfunction
P0461 Fuel Level Sensor Circuit Range/Performance
P0462 Fuel Level Sensor Circuit Low Input
P0463 Fuel Level Sensor Circuit High Input
P0464 Fuel Level Sensor Circuit Intermittent
P0465 Purge Flow Sensor Circuit Malfunction
P0466 Purge Flow Sensor Circuit Range/Performance
P0467 Purge Flow Sensor Circuit Low Input
P0468 Purge Flow Sensor Circuit High Input
P0469 Purge Flow Sensor Circuit Intermittent
P0470 Exhaust Pressure Sensor Malfunction
P0471 Exhaust Pressure Sensor Range/Performance
P0472 Exhaust Pressure Sensor Low
P0473 Exhaust Pressure Sensor High
P0474 Exhaust Pressure Sensor Intermittent
P0475 Exhaust Pressure Control Valve Malfunction
P0476 Exhaust Pressure Control Valve Range/Performance
P0477 Exhaust Pressure Control Valve Low
P0478 Exhaust Pressure Control Valve High
P0479 Exhaust Pressure Control Valve Intermittent
P0480 Cooling Fan no. 1 Control Circuit Malfunction
P0481 Cooling Fan no. 2 Control Circuit Malfunction
P0482 Cooling Fan no. 3 Control Circuit Malfunction
P0483 Cooling Fan Rationality Check Malfunction
P0484 Cooling Fan Circuit Over Current
P0485 Cooling Fan Power/Ground Circuit Malfunction
P0500 Vehicle Speed Sensor Malfunction
P0501 Vehicle Speed Sensor Range/Performance
P0502 Vehicle Speed Sensor Circuit Low Input

P0503 Vehicle Speed Sensor Intermittent/Erratic/High
P0505 Idle Control System Malfunction
P0506 Idle Control System RPM Lower Than Expected
P0507 Idle Control System RPM Higher Than Expected
P0510 Closed Throttle Position Switch Malfunction
P0520 Engine Oil Pressure Sensor/Switch Circuit Malfunction
P0521 Engine Oil Pressure Sensor/Switch Range/Performance
P0522 Engine Oil Pressure Sensor/Switch Low Voltage
P0523 Engine Oil Pressure Sensor/Switch High Voltage
P0530 A/C Refrigerant Pressure Sensor Circuit Malfunction
P0531 A/C Refrigerant Pressure Sensor Circuit Range/Performance
P0532 A/C Refrigerant Pressure Sensor Circuit Low Input
P0533 A/C Refrigerant Pressure Sensor Circuit High Input
P0534 A/C Refrigerant Charge Loss
P0550 Power Steering Pressure Sensor Circuit Malfunction
P0551 Power Steering Pressure Sensor Circuit Range/Performance
P0552 Power Steering Pressure Sensor Circuit Low Input
P0553 Power Steering Pressure Sensor Circuit High Input
P0554 Power Steering Pressure Sensor Circuit Intermittent
P0560 System Voltage Malfunction
P0561 System Voltage Unstable
P0562 System Voltage Low
P0563 System Voltage High
P0565 Cruise Control On Signal Malfunction
P0566 Cruise Control Off Signal Malfunction
P0567 Cruise Control Resume Signal Malfunction
P0568 Cruise Control Set Signal Malfunction
P0569 Cruise Control Coast Signal Malfunction
P0570 Cruise Control Accel Signal Malfunction
P057 1 Cruise Control/Brake Switch "A" Circuit Malfunction
P0572 Cruise Control/Brake Switch "A" Circuit Low
P0573 Cruise Control/Brake Switch "A" Circuit High
P0574 Through P0580 Reserved for Cruise Codes
P0600 Serial Communication Link Malfunction
P0601 Internal Control Module Memory Check Sum Error
P0602 Control Module Programming Error
P0603 Internal Control Module Keep Alive Memory (KAM) Error
P0604 Internal Control Module Random Access Memory (RAM) Error
P0605 Internal Control Module Read Only Memory (ROM) Error
P0606 PCM Processor Fault
P0608 Control Module VSS Output "A" Malfunction
P0609 Control Module VSS Output "B" Malfunction
P0620 Generator Control Circuit Malfunction
P0621 Generator Lamp "L" Control Circuit Malfunction
P0622 Generator Field "F" Control Circuit Malfunction
P0650 Malfunction Indicator Lamp (MIL) Control Circuit Malfunction
P0654 Engine RPM Output Circuit Malfunction
P0655 Engine Hot Lamp Output Control Circuit Malfunction
P0656 Fuel Level Output Circuit Malfunction
P0700 Transmission Control System Malfunction
P0701 Transmission Control System Range/Performance
P0702 Transmission Control System Electrical
P0703 Torque Converter/Brake Switch "B" Circuit Malfunction
P0704 Clutch Switch Input Circuit Malfunction
P0705 Transmission Range Sensor Circuit Malfunction (PRNDL Input)
P0706 Transmission Range Sensor Circuit Range/Performance
P0707 Transmission Range Sensor Circuit Low Input

P0708 Transmission Range Sensor Circuit High Input

P0709 Transmission Range Sensor Circuit Intermittent

P0710 Transmission Fluid Temperature Sensor Circuit Malfunction

P0711 Transmission Fluid Temperature Sensor Circuit Range/Performance

P0712 Transmission Fluid Temperature Sensor Circuit Low Input

P0713 Transmission Fluid Temperature Sensor Circuit High Input

P0714 Transmission Fluid Temperature Sensor Circuit Intermittent

P0715 Input/Turbine Speed Sensor Circuit Malfunction

P0716 Input/Turbine Speed Sensor Circuit Range/Performance

P0717 Input/Turbine Speed Sensor Circuit No Signal

P0718 Input/Turbine Speed Sensor Circuit Intermittent

P0719 Torque Converter/Brake Switch "B" Circuit Low

P0720 Output Speed Sensor Circuit Malfunction

P0721 Output Speed Sensor Circuit Range/Performance

P0722 Output Speed Sensor Circuit No Signal

P0723 Output Speed Sensor Circuit Intermittent

P0724 Torque Converter/Brake Switch "B" Circuit High

P0725 Engine Speed Input Circuit Malfunction

P0726 Engine Speed Input Circuit Range/Performance

P0727 Engine Speed Input Circuit No Signal

P0728 Engine Speed Input Circuit Intermittent

P0730 Incorrect Gear Ratio

P0731 Gear no. 1 Incorrect Ratio

P0732 Gear no. 2 Incorrect Ratio

P0733 Gear no. 3 Incorrect Ratio

P0734 Gear no. 4 Incorrect Ratio

P0735 Gear no. 5 Incorrect Ratio

P0736 Reverse Incorrect Ratio

P0740 Torque Converter Clutch Circuit Malfunction

P0741 Torque Converter Clutch Circuit Performance or Stuck Off

P0742 Torque Converter Clutch Circuit Stuck On

P0743 Torque Converter Clutch Circuit Electrical

P0744 Torque Converter Clutch Circuit Intermittent

P0745 Pressure Control Solenoid Malfunction

P0746 Pressure Control Solenoid Performance or Stuck Off

P0747 Pressure Control Solenoid Stuck On

P0748 Pressure Control Solenoid Electrical

P0749 Pressure Control Solenoid Intermittent

P0750 Shift Solenoid "A" Malfunction

P0751 Shift Solenoid "A" Performance or Stuck Off

P0752 Shift Solenoid "A" Stuck On

P0753 Shift Solenoid "A" Electrical

P0754 Shift Solenoid "A" Intermittent

P0755 Shift Solenoid "B" Malfunction

P0756 Shift Solenoid "B" Performance or Stuck Off

P0757 Shift Solenoid "B" Stuck On

P0758 Shift Solenoid "B" Electrical

P0759 Shift Solenoid "B" Intermittent

P0760 Shift Solenoid "C" Malfunction

P0761 Shift Solenoid "C" Performance Or Stuck Off

P0762 Shift Solenoid "C" Stuck On

P0763 Shift Solenoid "C" Electrical

P0764 Shift Solenoid "C" Intermittent

P0765 Shift Solenoid "D" Malfunction

P0766 Shift Solenoid "D" Performance Or Stuck Off

P0767 Shift Solenoid "D" Stuck On

P0768 Shift Solenoid "D" Electrical

P0769 Shift Solenoid "D" Intermittent

P0770 Shift Solenoid "E" Malfunction

P0771 Shift Solenoid "E" Performance Or Stuck Off

P0772 Shift Solenoid "E" Stuck On

P0773 Shift Solenoid "E" Electrical

P0774 Shift Solenoid "E" Intermittent

P0780 Shift Malfunction

P0781 1–2 Shift Malfunction

P0782 2–3 Shift Malfunction

P0783 3–4 Shift Malfunction

P0784 4–5 Shift Malfunction

P0785 Shift/Timing Solenoid Malfunction

P0786 Shift/Timing Solenoid Range/Performance

P0787 Shift/Timing Solenoid Low

P0788 Shift/Timing Solenoid High

P0789 Shift/Timing Solenoid Intermittent

P0790 Normal/Performance Switch Circuit Malfunction

P0801 Reverse Inhibit Control Circuit Malfunction

P0803 1–4 Upshift (Skip Shift) Solenoid Control Circuit Malfunction

P0804 1–4 Upshift (Skip Shift) Lamp Control Circuit Malfunction

Lexus

READING CODES

Reading the control module memory is on of the first steps in OBD II system diagnostics.

This step should be initially performed to determine the general nature of the fault. Subsequent readings will determine if the fault has been cleared.

Reading codes can be performed by any of the methods below:

• Read the control module memory with the Generic Scan Tool (GST)

• Read the control module memory with the vehicle manufacturer's specific tester

To read the fault codes, connect the scan tool or tester according to the manufacturer's instructions. Follow the manufacturer's specified procedure for reading the codes.

CLEARING CODES

Control module reset procedures are a very important part of OBD II System diagnostics. This step should be done at the end of any fault code repair and at the end of any driveability repair.

Clearing codes can be performed by any of the methods below:

• Clear the control module memory with the Generic Scan Tool (GST)

• Clear the control module memory with the vehicle manufacturer's specific tester

• Turn the ignition **OFF** and remove the negative battery cable for at least 1 minute.

Removing the negative battery cable may cause other systems in the vehicle to lose their memory. Prior to removing the cable, ensure you have the proper reset codes for radios and alarms.

➡**The MIL will may also be de-activated for some codes if the vehicle completes three consecutive trips without a fault detected with vehicle conditions similar to those present during the fault.**

DIAGNOSTIC TROUBLE CODES

P0000 No Failures
P0100 Mass or Volume Air Flow Circuit Malfunction
P0101 Mass or Volume Air Flow Circuit Range/Performance Problem
P0102 Mass or Volume Air Flow Circuit Low Input
P0103 Mass or Volume Air Flow Circuit High Input
P0104 Mass or Volume Air Flow Circuit Intermittent
P0105 Manifold Absolute Pressure/Barometric Pressure Circuit Malfunction
P0106 Manifold Absolute Pressure/Barometric Pressure Circuit Range/Performance Problem
P0107 Manifold Absolute Pressure/Barometric Pressure Circuit Low Input
P0108 Manifold Absolute Pressure/Barometric Pressure Circuit High Input
P0109 Manifold Absolute Pressure/Barometric Pressure Circuit Intermittent
P0110 Intake Air Temperature Circuit Malfunction
P0111 Intake Air Temperature Circuit Range/Performance Problem
P0112 Intake Air Temperature Circuit Low Input
P0113 Intake Air Temperature Circuit High Input
P0114 Intake Air Temperature Circuit Intermittent
P0115 Engine Coolant Temperature Circuit Malfunction
P0116 Engine Coolant Temperature Circuit Range/Performance Problem
P0117 Engine Coolant Temperature Circuit Low Input
P0118 Engine Coolant Temperature Circuit High Input
P0119 Engine Coolant Temperature Circuit Intermittent
P0120 Throttle/Pedal Position Sensor/Switch "A" Circuit Malfunction
P0121 Throttle/Pedal Position Sensor/Switch "A" Circuit Range/Performance Problem
P0122 Throttle/Pedal Position Sensor/Switch "A" Circuit Low Input
P0123 Throttle/Pedal Position Sensor/Switch "A" Circuit High Input
P0124 Throttle/Pedal Position Sensor/Switch "A" Circuit Intermittent
P0125 Insufficient Coolant Temperature For Closed Loop Fuel Control
P0126 Insufficient Coolant Temperature For Stable Operation
P0130 O_2 Circuit Malfunction (Bank no. 1 Sensor no. 1)
P0131 O_2 Sensor Circuit Low Voltage (Bank no. 1 Sensor no. 1)
P0132 O_2 Sensor Circuit High Voltage (Bank no. 1 Sensor no. 1)
P0133 O_2 Sensor Circuit Slow Response (Bank no. 1 Sensor no. 1)
P0134 O_2 Sensor Circuit No Activity Detected (Bank no. 1 Sensor no. 1)
P0135 O_2 Sensor Heater Circuit Malfunction (Bank no. 1 Sensor no. 1)
P0136 O_2 Sensor Circuit Malfunction (Bank no. 1 Sensor no. 2)
P0137 O_2 Sensor Circuit Low Voltage (Bank no. 1 Sensor no. 2)
P0138 O_2 Sensor Circuit High Voltage (Bank no. 1 Sensor no. 2)
P0139 O_2 Sensor Circuit Slow Response (Bank no. 1 Sensor no. 2)
P0140 O_2 Sensor Circuit No Activity Detected (Bank no. 1 Sensor no. 2)
P0141 O_2 Sensor Heater Circuit Malfunction (Bank no. 1 Sensor no. 2)

P0142 O_2 Sensor Circuit Malfunction (Bank no. 1 Sensor no. 3)
P0143 O_2 Sensor Circuit Low Voltage (Bank no. 1 Sensor no. 3)
P0144 O_2 Sensor Circuit High Voltage (Bank no. 1 Sensor no. 3)
P0145 O_2 Sensor Circuit Slow Response (Bank no. 1 Sensor no. 3)
P0146 O_2 Sensor Circuit No Activity Detected (Bank no. 1 Sensor no. 3)
P0147 O_2 Sensor Heater Circuit Malfunction (Bank no. 1 Sensor no. 3)
P0150 O_2 Sensor Circuit Malfunction (Bank no. 2 Sensor no. 1)
P0151 O_2 Sensor Circuit Low Voltage (Bank no. 2 Sensor no. 1)
P0152 O_2 Sensor Circuit High Voltage (Bank no. 2 Sensor no. 1)
P0153 O_2 Sensor Circuit Slow Response (Bank no. 2 Sensor no. 1)
P0154 O_2 Sensor Circuit No Activity Detected (Bank no. 2 Sensor no. 1)
P0155 O_2 Sensor Heater Circuit Malfunction (Bank no. 2 Sensor no. 1)
P0156 O_2 Sensor Circuit Malfunction (Bank no. 2 Sensor no. 2)
P0157 O_2 Sensor Circuit Low Voltage (Bank no. 2 Sensor no. 2)
P0158 O_2 Sensor Circuit High Voltage (Bank no. 2 Sensor no. 2)
P0159 O_2 Sensor Circuit Slow Response (Bank no. 2 Sensor no. 2)
P0160 O_2 Sensor Circuit No Activity Detected (Bank no. 2 Sensor no. 2)
P0161 O_2 Sensor Heater Circuit Malfunction (Bank no. 2 Sensor no. 2)
P0162 O_2 Sensor Circuit Malfunction (Bank no. 2 Sensor no. 3)
P0163 O_2 Sensor Circuit Low Voltage (Bank no. 2 Sensor no. 3)
P0164 O_2 Sensor Circuit High Voltage (Bank no. 2 Sensor no. 3)
P0165 O_2 Sensor Circuit Slow Response (Bank no. 2 Sensor no. 3)
P0166 O_2 Sensor Circuit No Activity Detected (Bank no. 2 Sensor no. 3)
P0167 O_2 Sensor Heater Circuit Malfunction (Bank no. 2 Sensor no. 3)
P0170 Fuel Trim Malfunction (Bank no. 1)
P0171 System Too Lean (Bank no. 1)
P0172 System Too Rich (Bank no. 1)
P0173 Fuel Trim Malfunction (Bank no. 2)
P0174 System Too Lean (Bank no. 2)
P0175 System Too Rich (Bank no. 2)
P0176 Fuel Composition Sensor Circuit Malfunction
P0177 Fuel Composition Sensor Circuit Range/Performance
P0178 Fuel Composition Sensor Circuit Low Input
P0179 Fuel Composition Sensor Circuit High Input
P0180 Fuel Temperature Sensor "A" Circuit Malfunction
P0181 Fuel Temperature Sensor "A" Circuit Range/Performance
P0182 Fuel Temperature Sensor "A" Circuit Low Input
P0183 Fuel Temperature Sensor "A" Circuit High Input
P0184 Fuel Temperature Sensor "A" Circuit Intermittent
P0185 Fuel Temperature Sensor "B" Circuit Malfunction
P0186 Fuel Temperature Sensor "B" Circuit Range/Performance
P0187 Fuel Temperature Sensor "B" Circuit Low Input
P0188 Fuel Temperature Sensor "B" Circuit High Input
P0189 Fuel Temperature Sensor "B" Circuit Intermittent
P0190 Fuel Rail Pressure Sensor Circuit Malfunction
P0191 Fuel Rail Pressure Sensor Circuit Range/Performance
P0192 Fuel Rail Pressure Sensor Circuit Low Input
P0193 Fuel Rail Pressure Sensor Circuit High Input
P0194 Fuel Rail Pressure Sensor Circuit Intermittent
P0195 Engine Oil Temperature Sensor Malfunction

P0196 Engine Oil Temperature Sensor Range/Performance
P0197 Engine Oil Temperature Sensor Low
P0198 Engine Oil Temperature Sensor High
P0199 Engine Oil Temperature Sensor Intermittent
P0200 Injector Circuit Malfunction
P0201 Injector Circuit Malfunction—Cylinder no. 1
P0202 Injector Circuit Malfunction—Cylinder no. 2
P0203 Injector Circuit Malfunction—Cylinder no. 3
P0204 Injector Circuit Malfunction—Cylinder no. 4
P0205 Injector Circuit Malfunction—Cylinder no. 5
P0206 Injector Circuit Malfunction—Cylinder no. 6
P0207 Injector Circuit Malfunction—Cylinder no. 7
P0208 Injector Circuit Malfunction—Cylinder no. 8
P0209 Injector Circuit Malfunction—Cylinder no. 9
P0210 Injector Circuit Malfunction—Cylinder no. 10
P0211 Injector Circuit Malfunction—Cylinder no. 11
P0212 Injector Circuit Malfunction—Cylinder no. 12
P0213 Cold Start Injector no. 1 Malfunction
P0214 Cold Start Injector no. 2 Malfunction
P0215 Engine Shutoff Solenoid Malfunction
P0216 Injection Timing Control Circuit Malfunction
P0217 Engine Over Temperature Condition
P0218 Transmission Over Temperature Condition
P0219 Engine Over Speed Condition
P0220 Throttle/Pedal Position Sensor/Switch "B" Circuit Malfunction
P0221 Throttle/Pedal Position Sensor/Switch "B" Circuit Range/Performance Problem
P0222 Throttle/Pedal Position Sensor/Switch "B" Circuit Low Input
P0223 Throttle/Pedal Position Sensor/Switch "B" Circuit High Input
P0224 Throttle/Pedal Position Sensor/Switch "B" Circuit Intermittent
P0225 Throttle/Pedal Position Sensor/Switch "C" Circuit Malfunction
P0226 Throttle/Pedal Position Sensor/Switch "C" Circuit Range/Performance Problem
P0227 Throttle/Pedal Position Sensor/Switch "C" Circuit Low Input
P0228 Throttle/Pedal Position Sensor/Switch "C" Circuit High Input
P0229 Throttle/Pedal Position Sensor/Switch "C" Circuit Intermittent
P0230 Fuel Pump Primary Circuit Malfunction
P0231 Fuel Pump Secondary Circuit Low
P0232 Fuel Pump Secondary Circuit High
P0233 Fuel Pump Secondary Circuit Intermittent
P0234 Engine Over Boost Condition
P0261 Cylinder no. 1 Injector Circuit Low
P0262 Cylinder no. 1 Injector Circuit High
P0263 Cylinder no. 1 Contribution/Balance Fault
P0264 Cylinder no. 2 Injector Circuit Low
P0265 Cylinder no. 2 Injector Circuit High
P0266 Cylinder no. 2 Contribution/Balance Fault
P0267 Cylinder no. 3 Injector Circuit Low
P0268 Cylinder no. 3 Injector Circuit High
P0269 Cylinder no. 3 Contribution/Balance Fault
P0270 Cylinder no. 4 Injector Circuit Low
P0271 Cylinder no. 4 Injector Circuit High
P0272 Cylinder no. 4 Contribution/Balance Fault
P0273 Cylinder no. 5 Injector Circuit Low

P0274 Cylinder no. 5 Injector Circuit High
P0275 Cylinder no. 5 Contribution/Balance Fault
P0276 Cylinder no. 6 Injector Circuit Low
P0277 Cylinder no. 6 Injector Circuit High
P0278 Cylinder no. 6 Contribution/Balance Fault
P0279 Cylinder no. 7 Injector Circuit Low
P0280 Cylinder no. 7 Injector Circuit High
P0281 Cylinder no. 7 Contribution/Balance Fault
P0282 Cylinder no. 8 Injector Circuit Low
P0283 Cylinder no. 8 Injector Circuit High
P0284 Cylinder no. 8 Contribution/Balance Fault
P0285 Cylinder no. 9 Injector Circuit Low
P0286 Cylinder no. 9 Injector Circuit High
P0287 Cylinder no. 9 Contribution/Balance Fault
P0288 Cylinder no. 10 Injector Circuit Low
P0289 Cylinder no. 10 Injector Circuit High
P0290 Cylinder no. 10 Contribution/Balance Fault
P0291 Cylinder no. 11 Injector Circuit Low
P0292 Cylinder no. 11 Injector Circuit High
P0293 Cylinder no. 11 Contribution/Balance Fault
P0294 Cylinder no. 12 Injector Circuit Low
P0295 Cylinder no. 12 Injector Circuit High
P0296 Cylinder no. 12 Contribution/Balance Fault
P0300 Random/Multiple Cylinder Misfire Detected
P0301 Cylinder no. 1—Misfire Detected
P0302 Cylinder no. 2—Misfire Detected
P0303 Cylinder no. 3—Misfire Detected
P0304 Cylinder no. 4—Misfire Detected
P0305 Cylinder no. 5—Misfire Detected
P0306 Cylinder no. 6—Misfire Detected
P0307 Cylinder no. 7—Misfire Detected
P0308 Cylinder no. 8—Misfire Detected
P0309 Cylinder no. 9—Misfire Detected
P0310 Cylinder no. 10—Misfire Detected
P0311 Cylinder no. 11—Misfire Detected
P0312 Cylinder no. 12—Misfire Detected
P0320 Ignition/Distributor Engine Speed Input Circuit Malfunction
P0321 Ignition/Distributor Engine Speed Input Circuit Range/Performance
P0322 Ignition/Distributor Engine Speed Input Circuit No Signal
P0323 Ignition/Distributor Engine Speed Input Circuit Intermittent
P0325 Knock Sensor no. 1—Circuit Malfunction (Bank no. 1 or Single Sensor)
P0326 Knock Sensor no. 1—Circuit Range/Performance (Bank no. 1 or Single Sensor)
P0327 Knock Sensor no. 1—Circuit Low Input (Bank no. 1 or Single Sensor)
P0328 Knock Sensor no. 1—Circuit High Input (Bank no. 1 or Single Sensor)
P0329 Knock Sensor no. 1—Circuit Input Intermittent (Bank no. 1 or Single Sensor)
P0330 Knock Sensor no. 2—Circuit Malfunction (Bank no. 2)
P0331 Knock Sensor no. 2—Circuit Range/Performance (Bank no. 2)
P0332 Knock Sensor no. 2—Circuit Low Input (Bank no. 2)
P0333 Knock Sensor no. 2—Circuit High Input (Bank no. 2)
P0334 Knock Sensor no. 2—Circuit Input Intermittent (Bank no. 2)
P0335 Crankshaft Position Sensor "A" Circuit Malfunction

P0336 Crankshaft Position Sensor "A" Circuit Range/Performance

P0337 Crankshaft Position Sensor "A" Circuit Low Input

P0338 Crankshaft Position Sensor "A" Circuit High Input

P0339 Crankshaft Position Sensor "A" Circuit Intermittent

P0340 Camshaft Position Sensor Circuit Malfunction

P0341 Camshaft Position Sensor Circuit Range/Performance

P0342 Camshaft Position Sensor Circuit Low Input

P0343 Camshaft Position Sensor Circuit High Input

P0344 Camshaft Position Sensor Circuit Intermittent

P0350 Ignition Coil Primary/Secondary Circuit Malfunction

P0351 Ignition Coil "A" Primary/Secondary Circuit Malfunction

P0352 Ignition Coil "B" Primary/Secondary Circuit Malfunction

P0353 Ignition Coil "C" Primary/Secondary Circuit Malfunction

P0354 Ignition Coil "D" Primary/Secondary Circuit Malfunction

P0355 Ignition Coil "E" Primary/Secondary Circuit Malfunction

P0356 Ignition Coil "F" Primary/Secondary Circuit Malfunction

P0357 Ignition Coil "G" Primary/Secondary Circuit Malfunction

P0358 Ignition Coil "H" Primary/Secondary Circuit Malfunction

P0359 Ignition Coil "I" Primary/Secondary Circuit Malfunction

P0360 Ignition Coil "J" Primary/Secondary Circuit Malfunction

P0361 Ignition Coil "K" Primary/Secondary Circuit Malfunction

P0362 Ignition Coil "L" Primary/Secondary Circuit Malfunction

P0370 Timing Reference High Resolution Signal "A" Malfunction

P0371 Timing Reference High Resolution Signal "A" Too Many Pulses

P0372 Timing Reference High Resolution Signal "A" Too Few Pulses

P0373 Timing Reference High Resolution Signal "A" Intermittent/Erratic Pulses

P0374 Timing Reference High Resolution Signal "A" No Pulses

P0375 Timing Reference High Resolution Signal "B" Malfunction

P0376 Timing Reference High Resolution Signal "B" Too Many Pulses

P0377 Timing Reference High Resolution Signal "B" Too Few Pulses

P0378 Timing Reference High Resolution Signal "B" Intermittent/Erratic Pulses

P0379 Timing Reference High Resolution Signal "B" No Pulses

P0380 Glow Plug/Heater Circuit "A" Malfunction

P0381 Glow Plug/Heater Indicator Circuit Malfunction

P0382 Glow Plug/Heater Circuit "B" Malfunction

P0385 Crankshaft Position Sensor "B" Circuit Malfunction

P0386 Crankshaft Position Sensor "B" Circuit Range/Performance

P0387 Crankshaft Position Sensor "B" Circuit Low Input

P0388 Crankshaft Position Sensor "B" Circuit High Input

P0389 Crankshaft Position Sensor "B" Circuit Intermittent

P0400 Exhaust Gas Recirculation Flow Malfunction

P0401 Exhaust Gas Recirculation Flow Insufficient Detected

P0402 Exhaust Gas Recirculation Flow Excessive Detected

P0403 Exhaust Gas Recirculation Circuit Malfunction

P0404 Exhaust Gas Recirculation Circuit Range/Performance

P0405 Exhaust Gas Recirculation Sensor "A" Circuit Low

P0406 Exhaust Gas Recirculation Sensor "A" Circuit High

P0407 Exhaust Gas Recirculation Sensor "B" Circuit Low

P0408 Exhaust Gas Recirculation Sensor "B" Circuit High

P0410 Secondary Air Injection System Malfunction

P0411 Secondary Air Injection System Incorrect Flow Detected

P0412 Secondary Air Injection System Switching Valve "A" Circuit Malfunction

P0413 Secondary Air Injection System Switching Valve "A" Circuit Open

P0414 Secondary Air Injection System Switching Valve "A" Circuit Shorted

P0415 Secondary Air Injection System Switching Valve "B" Circuit Malfunction

P0416 Secondary Air Injection System Switching Valve "B" Circuit Open

P0417 Secondary Air Injection System Switching Valve "B" Circuit Shorted

P0418 Secondary Air Injection System Relay "A" Circuit Malfunction

P0419 Secondary Air Injection System Relay "B" Circuit Malfunction

P0420 Catalyst System Efficiency Below Threshold (Bank no. 1)

P0421 Warm Up Catalyst Efficiency Below Threshold (Bank no. 1)

P0422 Main Catalyst Efficiency Below Threshold (Bank no. 1)

P0423 Heated Catalyst Efficiency Below Threshold (Bank no. 1)

P0424 Heated Catalyst Temperature Below Threshold (Bank no. 1)

P0430 Catalyst System Efficiency Below Threshold (Bank no. 2)

P0431 Warm Up Catalyst Efficiency Below Threshold (Bank no. 2)

P0432 Main Catalyst Efficiency Below Threshold (Bank no. 2)

P0433 Heated Catalyst Efficiency Below Threshold (Bank no. 2)

P0434 Heated Catalyst Temperature Below Threshold (Bank no. 2)

P0440 Evaporative Emission Control System Malfunction

P0441 Evaporative Emission Control System Incorrect Purge Flow

P0442 Evaporative Emission Control System Leak Detected (Small Leak)

P0443 Evaporative Emission Control System Purge Control Valve Circuit Malfunction

P0444 Evaporative Emission Control System Purge Control Valve Circuit Open

P0445 Evaporative Emission Control System Purge Control Valve Circuit Shorted

P0446 Evaporative Emission Control System Vent Control Circuit Malfunction

P0447 Evaporative Emission Control System Vent Control Circuit Open

P0448 Evaporative Emission Control System Vent Control Circuit Shorted

P0449 Evaporative Emission Control System Vent Valve/Solenoid Circuit Malfunction

P0450 Evaporative Emission Control System Pressure Sensor Malfunction

P0451 Evaporative Emission Control System Pressure Sensor Range/Performance

P0452 Evaporative Emission Control System Pressure Sensor Low Input

P0453 Evaporative Emission Control System Pressure Sensor High Input

P0454 Evaporative Emission Control System Pressure Sensor Intermittent

P0455 Evaporative Emission Control System Leak Detected (Gross Leak)

P0460 Fuel Level Sensor Circuit Malfunction

P0461 Fuel Level Sensor Circuit Range/Performance

P0462 Fuel Level Sensor Circuit Low Input
P0463 Fuel Level Sensor Circuit High Input
P0464 Fuel Level Sensor Circuit Intermittent
P0465 Purge Flow Sensor Circuit Malfunction
P0466 Purge Flow Sensor Circuit Range/Performance
P0467 Purge Flow Sensor Circuit Low Input
P0468 Purge Flow Sensor Circuit High Input
P0469 Purge Flow Sensor Circuit Intermittent
P0470 Exhaust Pressure Sensor Malfunction
P0471 Exhaust Pressure Sensor Range/Performance
P0472 Exhaust Pressure Sensor Low
P0473 Exhaust Pressure Sensor High
P0474 Exhaust Pressure Sensor Intermittent
P0475 Exhaust Pressure Control Valve Malfunction
P0476 Exhaust Pressure Control Valve Range/Performance
P0477 Exhaust Pressure Control Valve Low
P0478 Exhaust Pressure Control Valve High
P0479 Exhaust Pressure Control Valve Intermittent
P0480 Cooling Fan no. 1 Control Circuit Malfunction
P0481 Cooling Fan no. 2 Control Circuit Malfunction
P0482 Cooling Fan no. 3 Control Circuit Malfunction
P0483 Cooling Fan Rationality Check Malfunction
P0484 Cooling Fan Circuit Over Current
P0485 Cooling Fan Power/Ground Circuit Malfunction
P0500 Vehicle Speed Sensor Malfunction
P0501 Vehicle Speed Sensor Range/Performance
P0502 Vehicle Speed Sensor Circuit Low Input
P0503 Vehicle Speed Sensor Intermittent/Erratic/High
P0505 Idle Control System Malfunction
P0506 Idle Control System RPM Lower Than Expected
P0507 Idle Control System RPM Higher Than Expected
P0510 Closed Throttle Position Switch Malfunction
P0520 Engine Oil Pressure Sensor/Switch Circuit Malfunction
P0521 Engine Oil Pressure Sensor/Switch Range/Performance
P0522 Engine Oil Pressure Sensor/Switch Low Voltage
P0523 Engine Oil Pressure Sensor/Switch High Voltage
P0530 A/C Refrigerant Pressure Sensor Circuit Malfunction
P0531 A/C Refrigerant Pressure Sensor Circuit Range/Performance
P0532 A/C Refrigerant Pressure Sensor Circuit Low Input
P0533 A/C Refrigerant Pressure Sensor Circuit High Input
P0534 A/C Refrigerant Charge Loss
P0550 Power Steering Pressure Sensor Circuit Malfunction
P0551 Power Steering Pressure Sensor Circuit Range/Performance
P0552 Power Steering Pressure Sensor Circuit Low Input
P0553 Power Steering Pressure Sensor Circuit High Input
P0554 Power Steering Pressure Sensor Circuit Intermittent
P0560 System Voltage Malfunction
P0561 System Voltage Unstable
P0562 System Voltage Low
P0563 System Voltage High
P0565 Cruise Control On Signal Malfunction
P0566 Cruise Control Off Signal Malfunction
P0567 Cruise Control Resume Signal Malfunction
P0568 Cruise Control Set Signal Malfunction
P0569 Cruise Control Coast Signal Malfunction
P0570 Cruise Control Accel Signal Malfunction
P0571 Cruise Control/Brake Switch "A" Circuit Malfunction
P0572 Cruise Control/Brake Switch "A" Circuit Low
P0573 Cruise Control/Brake Switch "A" Circuit High
P0574 **Through P0580** Reserved for Cruise Codes

P0600 Serial Communication Link Malfunction
P0601 Internal Control Module Memory Check Sum Error
P0602 Control Module Programming Error
P0603 Internal Control Module Keep Alive Memory (KAM) Error
P0604 Internal Control Module Random Access Memory (RAM) Error
P0605 Internal Control Module Read Only Memory (ROM) Error
P0606 PCM Processor Fault
P0608 Control Module VSS Output "A" Malfunction
P0609 Control Module VSS Output "B" Malfunction
P0620 Generator Control Circuit Malfunction
P0621 Generator Lamp "L" Control Circuit Malfunction
P0622 Generator Field "F" Control Circuit Malfunction
P0650 Malfunction Indicator Lamp (MIL) Control Circuit Malfunction
P0654 Engine RPM Output Circuit Malfunction
P0655 Engine Hot Lamp Output Control Circuit Malfunction
P0656 Fuel Level Output Circuit Malfunction
P0700 Transmission Control System Malfunction
P0701 Transmission Control System Range/Performance
P0702 Transmission Control System Electrical
P0703 Torque Converter/Brake Switch "B" Circuit Malfunction
P0704 Clutch Switch Input Circuit Malfunction
P0705 Transmission Range Sensor Circuit Malfunction (PRNDL Input)
P0706 Transmission Range Sensor Circuit Range/Performance
P0707 Transmission Range Sensor Circuit Low Input
P0708 Transmission Range Sensor Circuit High Input
P0709 Transmission Range Sensor Circuit Intermittent
P0710 Transmission Fluid Temperature Sensor Circuit Malfunction
P0711 Transmission Fluid Temperature Sensor Circuit Range/Performance
P0712 Transmission Fluid Temperature Sensor Circuit Low Input
P0713 Transmission Fluid Temperature Sensor Circuit High Input
P0714 Transmission Fluid Temperature Sensor Circuit Intermittent
P0715 Input/Turbine Speed Sensor Circuit Malfunction
P0716 Input/Turbine Speed Sensor Circuit Range/Performance
P0717 Input/Turbine Speed Sensor Circuit No Signal
P0718 Input/Turbine Speed Sensor Circuit Intermittent
P0719 Torque Converter/Brake Switch "B" Circuit Low
P0720 Output Speed Sensor Circuit Malfunction
P0721 Output Speed Sensor Circuit Range/Performance
P0722 Output Speed Sensor Circuit No Signal
P0723 Output Speed Sensor Circuit Intermittent
P0724 Torque Converter/Brake Switch "B" Circuit High
P0725 Engine Speed Input Circuit Malfunction
P0726 Engine Speed Input Circuit Range/Performance
P0727 Engine Speed Input Circuit No Signal
P0728 Engine Speed Input Circuit Intermittent
P0730 Incorrect Gear Ratio
P0731 Gear no. 1 Incorrect Ratio
P0732 Gear no. 2 Incorrect Ratio
P0733 Gear no. 3 Incorrect Ratio
P0734 Gear no. 4 Incorrect Ratio
P0735 Gear no. 5 Incorrect Ratio
P0736 Reverse Incorrect Ratio
P0740 Torque Converter Clutch Circuit Malfunction

P0741 Torque Converter Clutch Circuit Performance or Stuck Off

P0742 Torque Converter Clutch Circuit Stuck On

P0743 Torque Converter Clutch Circuit Electrical

P0744 Torque Converter Clutch Circuit Intermittent

P0745 Pressure Control Solenoid Malfunction

P0746 Pressure Control Solenoid Performance or Stuck Off

P0747 Pressure Control Solenoid Stuck On

P0748 Pressure Control Solenoid Electrical

P0749 Pressure Control Solenoid Intermittent

P0750 Shift Solenoid "A" Malfunction

P0751 Shift Solenoid "A" Performance or Stuck Off

P0752 Shift Solenoid "A" Stuck On

P0753 Shift Solenoid "A" Electrical

P0754 Shift Solenoid "A" Intermittent

P0755 Shift Solenoid "B" Malfunction

P0756 Shift Solenoid "B" Performance or Stuck Off

P0757 Shift Solenoid "B" Stuck On

P0758 Shift Solenoid "B" Electrical

P0759 Shift Solenoid "B" Intermittent

P0760 Shift Solenoid "C" Malfunction

P0761 Shift Solenoid "C" Performance Or Stuck Off

P0762 Shift Solenoid "C" Stuck On

P0763 Shift Solenoid "C" Electrical

P0764 Shift Solenoid "C" Intermittent

P0765 Shift Solenoid "D" Malfunction

P0766 Shift Solenoid "D" Performance Or Stuck Off

P0767 Shift Solenoid "D" Stuck On

P0768 Shift Solenoid "D" Electrical

P0769 Shift Solenoid "D" Intermittent

P0770 Shift Solenoid "E" Malfunction

P0771 Shift Solenoid "E" Performance Or Stuck Off

P0772 Shift Solenoid "E" Stuck On

P0773 Shift Solenoid "E" Electrical

P0774 Shift Solenoid "E" Intermittent

P0780 Shift Malfunction

P0781 1–2 Shift Malfunction

P0782 2–3 Shift Malfunction

P0783 3–4 Shift Malfunction

P0784 4–5 Shift Malfunction

P0785 Shift/Timing Solenoid Malfunction

P0786 Shift/Timing Solenoid Range/Performance

P0787 Shift/Timing Solenoid Low

P0788 Shift/Timing Solenoid High

P0789 Shift/Timing Solenoid Intermittent

P0790 Normal/Performance Switch Circuit Malfunction

P0801 Reverse Inhibit Control Circuit Malfunction

P0803 1–4 Upshift (Skip Shift) Solenoid Control Circuit Malfunction

P0804 1–4 Upshift (Skip Shift) Lamp Control Circuit Malfunction

P1100 Barometric Pressure Sensor Circuit Fault

P1200 Fuel Pump Relay Circuit Fault

P1300 Igniter Circuit Fault (Bank 1)

P1305 Igniter Circuit Fault (Bank 2)

P1335 Crankshaft Position Sensor Circuit Fault

P1400 Sub-Throttle Position Sensor Circuit Fault

P1401 Sub-Throttle Position Sensor Performance

P1500 Starter Signal Circuit Fault

P1510 Air Volume Too Low With Supercharger On

P1600 PCM Battery Back-up Circuit Fault

P1605 Knock Control CPU Fault

P1700 Vehicle Speed Sensor Circuit Fault

P1705 Direct Clutch Speed Sensor Circuit Fault

P1765 Linear Shift Solenoid Circuit Fault

P1780 Park Neutral Position Switch Fault

Mazda

READING CODES

Reading the control module memory is on of the first steps in OBD II system diagnostics.

This step should be initially performed to determine the general nature of the fault. Subsequent readings will determine if the fault has been cleared.

Reading codes can be performed by any of the methods below:
- Read the control module memory with the Generic Scan Tool (GST)
- Read the control module memory with the vehicle manufacturer's specific tester

To read the fault codes, connect the scan tool or tester according to the manufacturer's instructions. Follow the manufacturer's specified procedure for reading the codes.

CLEARING CODES

Control module reset procedures are a very important part of OBD II System diagnostics. This step should be done at the end of any fault code repair and at the end of any driveability repair.

Clearing codes can be performed by any of the methods below:
- Clear the control module memory with the Generic Scan Tool (GST)
- Clear the control module memory with the vehicle manufacturer's specific tester
- Turn the ignition **OFF** and remove the negative battery cable for at least 1 minute.

Removing the negative battery cable may cause other systems in the vehicle to lose their memory. Prior to removing the cable, ensure you have the proper reset codes for radios and alarms.

➡**The MIL will may also be de-activated for some codes if the vehicle completes three consecutive trips without a fault detected with vehicle conditions similar to those present during the fault.**

DIAGNOSTIC TROUBLE CODES

P0000 No Failures

P0100 Mass or Volume Air Flow Circuit Malfunction

P0101 Mass or Volume Air Flow Circuit Range/Performance Problem

P0102 Mass or Volume Air Flow Circuit Low Input

P0103 Mass or Volume Air Flow Circuit High Input

P0104 Mass or Volume Air Flow Circuit Intermittent

P0105 Manifold Absolute Pressure/Barometric Pressure Circuit Malfunction

P0106 Manifold Absolute Pressure/Barometric Pressure Circuit Range/Performance Problem

P0107 Manifold Absolute Pressure/Barometric Pressure Circuit Low Input

P0108 Manifold Absolute Pressure/Barometric Pressure Circuit High Input

P0109 Manifold Absolute Pressure/Barometric Pressure Circuit Intermittent

P0110 Intake Air Temperature Circuit Malfunction

P0111 Intake Air Temperature Circuit Range/Performance Problem

P0112 Intake Air Temperature Circuit Low Input

P0113 Intake Air Temperature Circuit High Input

P0114 Intake Air Temperature Circuit Intermittent

P0115 Engine Coolant Temperature Circuit Malfunction

P0116 Engine Coolant Temperature Circuit Range/Performance Problem

P0117 Engine Coolant Temperature Circuit Low Input

P0118 Engine Coolant Temperature Circuit High Input

P0119 Engine Coolant Temperature Circuit Intermittent

P0120 Throttle/Pedal Position Sensor/Switch "A" Circuit Malfunction

P0121 Throttle/Pedal Position Sensor/Switch "A" Circuit Range/Performance Problem

P0122 Throttle/Pedal Position Sensor/Switch "A" Circuit Low Input

P0123 Throttle/Pedal Position Sensor/Switch "A" Circuit High Input

P0124 Throttle/Pedal Position Sensor/Switch "A" Circuit Intermittent

P0125 Insufficient Coolant Temperature For Closed Loop Fuel Control

P0126 Insufficient Coolant Temperature For Stable Operation

P0130 O_2 Circuit Malfunction (Bank no. 1 Sensor no. 1)

P0131 O_2 Sensor Circuit Low Voltage (Bank no. 1 Sensor no. 1)

P0132 O_2 Sensor Circuit High Voltage (Bank no. 1 Sensor no. 1)

P0133 O_2 Sensor Circuit Slow Response (Bank no. 1 Sensor no. 1)

P0134 O_2 Sensor Circuit No Activity Detected (Bank no. 1 Sensor no. 1)

P0135 O_2 Sensor Heater Circuit Malfunction (Bank no. 1 Sensor no. 1)

P0136 O_2 Sensor Circuit Malfunction (Bank no. 1 Sensor no. 2)

P0137 O_2 Sensor Circuit Low Voltage (Bank no. 1 Sensor no. 2)

P0138 O_2 Sensor Circuit High Voltage (Bank no. 1 Sensor no. 2)

P0139 O_2 Sensor Circuit Slow Response (Bank no. 1 Sensor no. 2)

P0140 O_2 Sensor Circuit No Activity Detected (Bank no. 1 Sensor no. 2)

P0141 O_2 Sensor Heater Circuit Malfunction (Bank no. 1 Sensor no. 2)

P0142 O_2 Sensor Circuit Malfunction (Bank no. 1 Sensor no. 3)

P0143 O_2 Sensor Circuit Low Voltage (Bank no. 1 Sensor no. 3)

P0144 O_2 Sensor Circuit High Voltage (Bank no. 1 Sensor no. 3)

P0145 O_2 Sensor Circuit Slow Response (Bank no. 1 Sensor no. 3)

P0146 O_2 Sensor Circuit No Activity Detected (Bank no. 1 Sensor no. 3)

P0147 O_2 Sensor Heater Circuit Malfunction (Bank no. 1 Sensor no. 3)

P0150 O_2 Sensor Circuit Malfunction (Bank no. 2 Sensor no. 1)

P0151 O_2 Sensor Circuit Low Voltage (Bank no. 2 Sensor no. 1)

P0152 O_2 Sensor Circuit High Voltage (Bank no. 2 Sensor no. 1)

P0153 O_2 Sensor Circuit Slow Response (Bank no. 2 Sensor no. 1)

P0154 O_2 Sensor Circuit No Activity Detected (Bank no. 2 Sensor no. 1)

P0155 O_2 Sensor Heater Circuit Malfunction (Bank no. 2 Sensor no. 1)

P0156 O_2 Sensor Circuit Malfunction (Bank no. 2 Sensor no. 2)

P0157 O_2 Sensor Circuit Low Voltage (Bank no. 2 Sensor no. 2)

P0158 O_2 Sensor Circuit High Voltage (Bank no. 2 Sensor no. 2)

P0159 O_2 Sensor Circuit Slow Response (Bank no. 2 Sensor no. 2)

P0160 O_2 Sensor Circuit No Activity Detected (Bank no. 2 Sensor no. 2)

P0161 O_2 Sensor Heater Circuit Malfunction (Bank no. 2 Sensor no. 2)

P0162 O_2 Sensor Circuit Malfunction (Bank no. 2 Sensor no. 3)

P0163 O_2 Sensor Circuit Low Voltage (Bank no. 2 Sensor no. 3)

P0164 O_2 Sensor Circuit High Voltage (Bank no. 2 Sensor no. 3)

P0165 O_2 Sensor Circuit Slow Response (Bank no. 2 Sensor no. 3)

P0166 O_2 Sensor Circuit No Activity Detected (Bank no. 2 Sensor no. 3)

P0167 O_2 Sensor Heater Circuit Malfunction (Bank no. 2 Sensor no. 3)

P0170 Fuel Trim Malfunction (Bank no. 1)

P0171 System Too Lean (Bank no. 1)

P0172 System Too Rich (Bank no. 1)

P0173 Fuel Trim Malfunction (Bank no. 2)

P0174 System Too Lean (Bank no. 2)

P0175 System Too Rich (Bank no. 2)

P0176 Fuel Composition Sensor Circuit Malfunction

P0177 Fuel Composition Sensor Circuit Range/Performance

P0178 Fuel Composition Sensor Circuit Low Input

P0179 Fuel Composition Sensor Circuit High Input

P0180 Fuel Temperature Sensor "A" Circuit Malfunction

P0181 Fuel Temperature Sensor "A" Circuit Range/Performance

P0182 Fuel Temperature Sensor "A" Circuit Low Input

P0183 Fuel Temperature Sensor "A" Circuit High Input

P0184 Fuel Temperature Sensor "A" Circuit Intermittent

P0185 Fuel Temperature Sensor "B" Circuit Malfunction

P0186 Fuel Temperature Sensor "B" Circuit Range/Performance

P0187 Fuel Temperature Sensor "B" Circuit Low Input

P0188 Fuel Temperature Sensor "B" Circuit High Input

P0189 Fuel Temperature Sensor "B" Circuit Intermittent

P0190 Fuel Rail Pressure Sensor Circuit Malfunction

P0191 Fuel Rail Pressure Sensor Circuit Range/Performance

P0192 Fuel Rail Pressure Sensor Circuit Low Input

P0193 Fuel Rail Pressure Sensor Circuit High Input

P0194 Fuel Rail Pressure Sensor Circuit Intermittent

P0195 Engine Oil Temperature Sensor Malfunction

P0196 Engine Oil Temperature Sensor Range/Performance

P0197 Engine Oil Temperature Sensor Low

P0198 Engine Oil Temperature Sensor High

P0199 Engine Oil Temperature Sensor Intermittent

P0200 Injector Circuit Malfunction

P0201 Injector Circuit Malfunction—Cylinder no. 1

P0202 Injector Circuit Malfunction—Cylinder no. 2

P0203 Injector Circuit Malfunction—Cylinder no. 3

P0204 Injector Circuit Malfunction—Cylinder no. 4

P0205 Injector Circuit Malfunction—Cylinder no. 5
P0206 Injector Circuit Malfunction—Cylinder no. 6
P0207 Injector Circuit Malfunction—Cylinder no. 7
P0208 Injector Circuit Malfunction—Cylinder no. 8
P0209 Injector Circuit Malfunction—Cylinder no. 9
P0210 Injector Circuit Malfunction—Cylinder no. 10
P0211 Injector Circuit Malfunction—Cylinder no. 11
P0212 Injector Circuit Malfunction—Cylinder no. 12
P0213 Cold Start Injector no. 1 Malfunction
P0214 Cold Start Injector no. 2 Malfunction
P0215 Engine Shutoff Solenoid Malfunction
P0216 Injection Timing Control Circuit Malfunction
P0217 Engine Over Temperature Condition
P0218 Transmission Over Temperature Condition
P0219 Engine Over Speed Condition
P0220 Throttle/Pedal Position Sensor/Switch "B" Circuit Malfunction
P0221 Throttle/Pedal Position Sensor/Switch "B" Circuit Range/Performance Problem
P0222 Throttle/Pedal Position Sensor/Switch "B" Circuit Low Input
P0223 Throttle/Pedal Position Sensor/Switch "B" Circuit High Input
P0224 Throttle/Pedal Position Sensor/Switch "B" Circuit Intermittent
P0225 Throttle/Pedal Position Sensor/Switch "C" Circuit Malfunction
P0226 Throttle/Pedal Position Sensor/Switch "C" Circuit Range/Performance Problem
P0227 Throttle/Pedal Position Sensor/Switch "C" Circuit Low Input
P0228 Throttle/Pedal Position Sensor/Switch "C" Circuit High Input
P0229 Throttle/Pedal Position Sensor/Switch "C" Circuit Intermittent
P0230 Fuel Pump Primary Circuit Malfunction
P0231 Fuel Pump Secondary Circuit Low
P0232 Fuel Pump Secondary Circuit High
P0233 Fuel Pump Secondary Circuit Intermittent
P0234 Engine Over Boost Condition
P0261 Cylinder no. 1 Injector Circuit Low
P0262 Cylinder no. 1 Injector Circuit High
P0263 Cylinder no. 1 Contribution/Balance Fault
P0264 Cylinder no. 2 Injector Circuit Low
P0265 Cylinder no. 2 Injector Circuit High
P0266 Cylinder no. 2 Contribution/Balance Fault
P0267 Cylinder no. 3 Injector Circuit Low
P0268 Cylinder no. 3 Injector Circuit High
P0269 Cylinder no. 3 Contribution/Balance Fault
P0270 Cylinder no. 4 Injector Circuit Low
P0271 Cylinder no. 4 Injector Circuit High
P0272 Cylinder no. 4 Contribution/Balance Fault
P0273 Cylinder no. 5 Injector Circuit Low
P0274 Cylinder no. 5 Injector Circuit High
P0275 Cylinder no. 5 Contribution/Balance Fault
P0276 Cylinder no. 6 Injector Circuit Low
P0277 Cylinder no. 6 Injector Circuit High
P0278 Cylinder no. 6 Contribution/Balance Fault
P0279 Cylinder no. 7 Injector Circuit Low
P0280 Cylinder no. 7 Injector Circuit High
P0281 Cylinder no. 7 Contrution/Balance Fault
P0282 Cylinder no. 8 Injector Circuit Low

P0283 Cylinder no. 8 Injector Circuit High
P0284 Cylinder no. 8 Contribution/Balance Fault
P0285 Cylinder no. 9 Injector Circuit Low
P0286 Cylinder no. 9 Injector Circuit High
P0287 Cylinder no. 9 Contribution/Balance Fault
P0288 Cylinder no. 10 Injector Circuit Low
P0289 Cylinder no. 10 Injector Circuit High
P0290 Cylinder no. 10 Contribution/Balance Fault
P0291 Cylinder no. 11 Injector Circuit Low
P0292 Cylinder no. 11 Injector Circuit High
P0293 Cylinder no. 11 Contribution/Balance Fault
P0294 Cylinder no. 12 Injector Circuit Low
P0295 Cylinder no. 12 Injector Circuit High
P0296 Cylinder no. 12 Contribution/Balance Fault
P0300 Random/Multiple Cylinder Misfire Detected
P0301 Cylinder no. 1—Misfire Detected
P0302 Cylinder no. 2—Misfire Detected
P0303 Cylinder no. 3—Misfire Detected
P0304 Cylinder no. 4—Misfire Detected
P0305 Cylinder no. 5—Misfire Detected
P0306 Cylinder no. 6—Misfire Detected
P0307 Cylinder no. 7—Misfire Detected
P0308 Cylinder no. 8—Misfire Detected
P0309 Cylinder no. 9—Misfire Detected
P0310 Cylinder no. 10—Misfire Detected
P0311 Cylinder no. 11—Misfire Detected
P0312 Cylinder no. 12—Misfire Detected
P0320 Ignition/Distributor Engine Speed Input Circuit Malfunction
P0321 Ignition/Distributor Engine Speed Input Circuit Range/Performance
P0322 Ignition/Distributor Engine Speed Input Circuit No Signal
P0323 Ignition/Distributor Engine Speed Input Circuit Intermittent
P0325 Knock Sensor no. 1—Circuit Malfunction (Bank no. 1 or Single Sensor)
P0326 Knock Sensor no. 1—Circuit Range/Performance (Bank no. 1 or Single Sensor)
P0327 Knock Sensor no. 1—Circuit Low Input (Bank no. 1 or Single Sensor)
P0328 Knock Sensor no. 1—Circuit High Input (Bank no. 1 or Single Sensor)
P0329 Knock Sensor no. 1—Circuit Input Intermittent (Bank no. 1 or Single Sensor)
P0330 Knock Sensor no. 2—Circuit Malfunction (Bank no. 2)
P0331 Knock Sensor no. 2—Circuit Range/Performance (Bank no. 2)
P0332 Knock Sensor no. 2—Circuit Low Input (Bank no. 2)
P0333 Knock Sensor no. 2—Circuit High Input (Bank no. 2)
P0334 Knock Sensor no. 2—Circuit Input Intermittent (Bank no. 2)
P0335 Crankshaft Position Sensor "A" Circuit Malfunction
P0336 Crankshaft Position Sensor "A" Circuit Range/Performance
P0337 Crankshaft Position Sensor "A" Circuit Low Input
P0338 Crankshaft Position Sensor "A" Circuit High Input
P0339 Crankshaft Position Sensor "A" Circuit Intermittent
P0340 Camshaft Position Sensor Circuit Malfunction
P0341 Camshaft Position Sensor Circuit Range/Performance
P0342 Camshaft Position Sensor Circuit Low Input
P0343 Camshaft Position Sensor Circuit High Input
P0344 Camshaft Position Sensor Circuit Intermittent

P0350 Ignition Coil Primary/Secondary Circuit Malfunction
P0351 Ignition Coil "A" Primary/Secondary Circuit Malfunction
P0352 Ignition Coil "B" Primary/Secondary Circuit Malfunction
P0353 Ignition Coil "C" Primary/Secondary Circuit Malfunction
P0354 Ignition Coil "D" Primary/Secondary Circuit Malfunction
P0355 Ignition Coil "E" Primary/Secondary Circuit Malfunction
P0356 Ignition Coil "F" Primary/Secondary Circuit Malfunction
P0357 Ignition Coil "G" Primary/Secondary Circuit Malfunction
P0358 Ignition Coil "H" Primary/Secondary Circuit Malfunction
P0359 Ignition Coil "I" Primary/Secondary Circuit Malfunction
P0360 Ignition Coil "J" Primary/Secondary Circuit Malfunction
P0361 Ignition Coil "K" Primary/Secondary Circuit Malfunction
P0362 Ignition Coil "L" Primary/Secondary Circuit Malfunction
P0370 Timing Reference High Resolution Signal "A" Malfunction
P0371 Timing Reference High Resolution Signal "A" Too Many Pulses
P0372 Timing Reference High Resolution Signal "A" Too Few Pulses
P0373 Timing Reference High Resolution Signal "A" Intermittent/Erratic Pulses
P0374 Timing Reference High Resolution Signal "A" No Pulses
P0375 Timing Reference High Resolution Signal "B" Malfunction
P0376 Timing Reference High Resolution Signal "B" Too Many Pulses
P0377 Timing Reference High Resolution Signal "B" Too Few Pulses
P0378 Timing Reference High Resolution Signal "B" Intermittent/Erratic Pulses
P0379 Timing Reference High Resolution Signal "B" No Pulses
P0380 Glow Plug/Heater Circuit "A" Malfunction
P0381 Glow Plug/Heater Indicator Circuit Malfunction
P0382 Glow Plug/Heater Circuit "B" Malfunction
P0385 Crankshaft Position Sensor "B" Circuit Malfunction
P0386 Crankshaft Position Sensor "B" Circuit Range/Performance
P0387 Crankshaft Position Sensor "B" Circuit Low Input
P0388 Crankshaft Position Sensor "B" Circuit High Input
P0389 Crankshaft Position Sensor "B" Circuit Intermittent
P0400 Exhaust Gas Recirculation Flow Malfunction
P0401 Exhaust Gas Recirculation Flow Insufficient Detected
P0402 Exhaust Gas Recirculation Flow Excessive Detected
P0403 Exhaust Gas Recirculation Circuit Malfunction
P0404 Exhaust Gas Recirculation Circuit Range/Performance
P0405 Exhaust Gas Recirculation Sensor "A" Circuit Low
P0406 Exhaust Gas Recirculation Sensor "A" Circuit High
P0407 Exhaust Gas Recirculation Sensor "B" Circuit Low
P0408 Exhaust Gas Recirculation Sensor "B" Circuit High
P0410 Secondary Air Injection System Malfunction
P0411 Secondary Air Injection System Incorrect Flow Detected
P0412 Secondary Air Injection System Switching Valve "A" Circuit Malfunction
P0413 Secondary Air Injection System Switching Valve "A" Circuit Open
P0414 Secondary Air Injection System Switching Valve "A" Circuit Shorted
P0415 Secondary Air Injection System Switching Valve "B" Circuit Malfunction
P0416 Secondary Air Injection System Switching Valve "B" Circuit Open

P0417 Secondary Air Injection System Switching Valve "B" Circuit Shorted
P0418 Secondary Air Injection System Relay "A" Circuit Malfunction
P0419 Secondary Air Injection System Relay "B" Circuit Malfunction
P0420 Catalyst System Efficiency Below Threshold (Bank no. 1)
P0421 Warm Up Catalyst Efficiency Below Threshold (Bank no. 1)
P0422 Main Catalyst Efficiency Below Threshold (Bank no. 1)
P0423 Heated Catalyst Efficiency Below Threshold (Bank no. 1)
P0424 Heated Catalyst Temperature Below Threshold (Bank no. 1)
P0430 Catalyst System Efficiency Below Threshold (Bank no. 2)
P0431 Warm Up Catalyst Efficiency Below Threshold (Bank no. 2)
P0432 Main Catalyst Efficiency Below Threshold (Bank no. 2)
P0433 Heated Catalyst Efficiency Below Threshold (Bank no. 2)
P0434 Heated Catalyst Temperature Below Threshold (Bank no. 2)
P0440 Evaporative Emission Control System Malfunction
P0441 Evaporative Emission Control System Incorrect Purge Flow
P0442 Evaporative Emission Control System Leak Detected (Small Leak)
P0443 Evaporative Emission Control System Purge Control Valve Circuit Malfunction
P0444 Evaporative Emission Control System Purge Control Valve Circuit Open
P0445 Evaporative Emission Control System Purge Control Valve Circuit Shorted
P0446 Evaporative Emission Control System Vent Control Circuit Malfunction
P0447 Evaporative Emission Control System Vent Control Circuit Open
P0448 Evaporative Emission Control System Vent Control Circuit Shorted
P0449 Evaporative Emission Control System Vent Valve/Solenoid Circuit Malfunction
P0450 Evaporative Emission Control System Pressure Sensor Malfunction
P0451 Evaporative Emission Control System Pressure Sensor Range/Performance
P0452 Evaporative Emission Control System Pressure Sensor Low Input
P0453 Evaporative Emission Control System Pressure Sensor High Input
P0454 Evaporative Emission Control System Pressure Sensor Intermittent
P0455 Evaporative Emission Control System Leak Detected (Gross Leak)
P0460 Fuel Level Sensor Circuit Malfunction
P0461 Fuel Level Sensor Circuit Range/Performance
P0462 Fuel Level Sensor Circuit Low Input
P0463 Fuel Level Sensor Circuit High Input
P0464 Fuel Level Sensor Circuit Intermittent
P0465 Purge Flow Sensor Circuit Malfunction
P0466 Purge Flow Sensor Circuit Range/Performance
P0467 Purge Flow Sensor Circuit Low Input
P0468 Purge Flow Sensor Circuit High Input
P0469 Purge Flow Sensor Circuit Intermittent
P0470 Exhaust Pressure Sensor Malfunction
P0471 Exhaust Pressure Sensor Range/Performance
P0472 Exhaust Pressure Sensor Low
P0473 Exhaust Pressure Sensor High
P0474 Exhaust Pressure Sensor Intermittent

P0475 Exhaust Pressure Control Valve Malfunction
P0476 Exhaust Pressure Control Valve Range/Performance
P0477 Exhaust Pressure Control Valve Low
P0478 Exhaust Pressure Control Valve High
P0479 Exhaust Pressure Control Valve Intermittent
P0480 Cooling Fan no. 1 Control Circuit Malfunction
P0481 Cooling Fan no. 2 Control Circuit Malfunction
P0482 Cooling Fan no. 3 Control Circuit Malfunction
P0483 Cooling Fan Rationality Check Malfunction
P0484 Cooling Fan Circuit Over Current
P0485 Cooling Fan Power/Ground Circuit Malfunction
P0500 Vehicle Speed Sensor Malfunction
P0501 Vehicle Speed Sensor Range/Performance
P0502 Vehicle Speed Sensor Circuit Low Input
P0503 Vehicle Speed Sensor Intermittent/Erratic/High
P0505 Idle Control System Malfunction
P0506 Idle Control System RPM Lower Than Expected
P0507 Idle Control System RPM Higher Than Expected
P0510 Closed Throttle Position Switch Malfunction
P0520 Engine Oil Pressure Sensor/Switch Circuit Malfunction
P0521 Engine Oil Pressure Sensor/Switch Range/Performance
P0522 Engine Oil Pressure Sensor/Switch Low Voltage
P0523 Engine Oil Pressure Sensor/Switch High Voltage
P0530 A/C Refrigerant Pressure Sensor Circuit Malfunction
P0531 A/C Refrigerant Pressure Sensor Circuit Range/Performance
P0532 A/C Refrigerant Pressure Sensor Circuit Low Input
P0533 A/C Refrigerant Pressure Sensor Circuit High Input
P0534 A/C Refrigerant Charge Loss
P0550 Power Steering Pressure Sensor Circuit Malfunction
P0551 Power Steering Pressure Sensor Circuit Range/Performance
P0552 Power Steering Pressure Sensor Circuit Low Input
P0553 Power Steering Pressure Sensor Circuit High Input
P0554 Power Steering Pressure Sensor Circuit Intermittent
P0560 System Voltage Malfunction
P0561 System Voltage Unstable
P0562 System Voltage Low
P0563 System Voltage High
P0565 Cruise Control On Signal Malfunction
P0566 Cruise Control Off Signal Malfunction
P0567 Cruise Control Resume Signal Malfunction
P0568 Cruise Control Set Signal Malfunction
P0569 Cruise Control Coast Signal Malfunction
P0570 Cruise Control Accel Signal Malfunction
P0571 Cruise Control/Brake Switch "A" Circuit Malfunction
P0572 Cruise Control/Brake Switch "A" Circuit Low
P0573 Cruise Control/Brake Switch "A" Circuit High
P0574 **Through** **P0580** Reserved for Cruise Codes
P0600 Serial Communication Link Malfunction
P0601 Internal Control Module Memory Check Sum Error
P0602 Control Module Programming Error
P0603 Internal Control Module Keep Alive Memory (KAM) Error
P0604 Internal Control Module Random Access Memory (RAM) Error
P0605 Internal Control Module Read Only Memory (ROM) Error
P0606 PCM Processor Fault
P0608 Control Module VSS Output "A" Malfunction
P0609 Control Module VSS Output "B" Malfunction
P0620 Generator Control Circuit Malfunction
P0621 Generator Lamp "L" Control Circuit Malfunction
P0622 Generator Field "F" Control Circuit Malfunction

P0650 Malfunction Indicator Lamp (MIL) Control Circuit Malfunction
P0654 Engine RPM Output Circuit Malfunction
P0655 Engine Hot Lamp Output Control Circuit Malfunction
P0656 Fuel Level Output Circuit Malfunction
P0700 Transmission Control System Malfunction
P0701 Transmission Control System Range/Performance
P0702 Transmission Control System Electrical
P0703 Torque Converter/Brake Switch "B" Circuit Malfunction
P0704 Clutch Switch Input Circuit Malfunction
P0705 Transmission Range Sensor Circuit Malfunction (PRNDL Input)
P0706 Transmission Range Sensor Circuit Range/Performance
P0707 Transmission Range Sensor Circuit Low Input
P0708 Transmission Range Sensor Circuit High Input
P0709 Transmission Range Sensor Circuit Intermittent
P0710 Transmission Fluid Temperature Sensor Circuit Malfunction
P0711 Transmission Fluid Temperature Sensor Circuit Range/Performance
P0712 Transmission Fluid Temperature Sensor Circuit Low Input
P0713 Transmission Fluid Temperature Sensor Circuit High Input
P0714 Transmission Fluid Temperature Sensor Circuit Intermittent
P0715 Input/Turbine Speed Sensor Circuit Malfunction
P0716 Input/Turbine Speed Sensor Circuit Range/Performance
P0717 Input/Turbine Speed Sensor Circuit No Signal
P0718 Input/Turbine Speed Sensor Circuit Intermittent
P0719 Torque Converter/Brake Switch "B" Circuit Low
P0720 Output Speed Sensor Circuit Malfunction
P0721 Output Speed Sensor Circuit Range/Performance
P0722 Output Speed Sensor Circuit No Signal
P0723 Output Speed Sensor Circuit Intermittent
P0724 Torque Converter/Brake Switch "B" Circuit High
P0725 Engine Speed Input Circuit Malfunction
P0726 Engine Speed Input Circuit Range/Performance
P0727 Engine Speed Input Circuit No Signal
P0728 Engine Speed Input Circuit Intermittent
P0730 Incorrect Gear Ratio
P0731 Gear no. 1 Incorrect Ratio
P0732 Gear no. 2 Incorrect Ratio
P0733 Gear no. 3 Incorrect Ratio
P0734 Gear no. 4 Incorrect Ratio
P0735 Gear no. 5 Incorrect Ratio
P0736 Reverse Incorrect Ratio
P0740 Torque Converter Clutch Circuit Malfunction
P0741 Torque Converter Clutch Circuit Performance or Stuck Off
P0742 Torque Converter Clutch Circuit Stuck On
P0743 Torque Converter Clutch Circuit Electrical
P0744 Torque Converter Clutch Circuit Intermittent
P0745 Pressure Control Solenoid Malfunction
P0746 Pressure Control Solenoid Performance or Stuck Off
P0747 Pressure Control Solenoid Stuck On
P0748 Pressure Control Solenoid Electrical
P0749 Pressure Control Solenoid Intermittent
P0750 Shift Solenoid "A" Malfunction
P0751 Shift Solenoid "A" Performance or Stuck Off
P0752 Shift Solenoid "A" Stuck On

P0753 Shift Solenoid "A" Electrical
P0754 Shift Solenoid "A" Intermittent
P0755 Shift Solenoid "B" Malfunction
P0756 Shift Solenoid "B" Performance or Stuck Off
P0757 Shift Solenoid "B" Stuck On
P0758 Shift Solenoid "B" Electrical
P0759 Shift Solenoid "B" Intermittent
P0760 Shift Solenoid "C" Malfunction
P0761 Shift Solenoid "C" Performance Or Stuck Off
P0762 Shift Solenoid "C" Stuck On
P0763 Shift Solenoid "C" Electrical
P0764 Shift Solenoid "C" Intermittent
P0765 Shift Solenoid "D" Malfunction
P0766 Shift Solenoid "D" Performance Or Stuck Off
P0767 Shift Solenoid "D" Stuck On
P0768 Shift Solenoid "D" Electrical
P0769 Shift Solenoid "D" Intermittent
P0770 Shift Solenoid "E" Malfunction
P0771 Shift Solenoid "E" Performance Or Stuck Off
P0772 Shift Solenoid "E" Stuck On
P0773 Shift Solenoid "E" Electrical
P0774 Shift Solenoid "E" Intermittent
P0780 Shift Malfunction
P0781 1–2 Shift Malfunction
P0782 2–3 Shift Malfunction
P0783 3–4 Shift Malfunction
P0784 4–5 Shift Malfunction
P0785 Shift/Timing Solenoid Malfunction
P0786 Shift/Timing Solenoid Range/Performance
P0787 Shift/Timing Solenoid Low
P0788 Shift/Timing Solenoid High
P0789 Shift/Timing Solenoid Intermittent
P0790 Normal/Performance Switch Circuit Malfunction
P0801 Reverse Inhibit Control Circuit Malfunction
P0803 1–4 Upshift (Skip Shift) Solenoid Control Circuit Malfunction
P0804 1–4 Upshift (Skip Shift) Lamp Control Circuit Malfunction
P1000 OBD II Monitor Testing Not Complete More Driving Required
P1001 Key On Engine Running (KOER) Self-Test Not Able To Complete, KOER Aborted
P1100 Mass Air Flow (MAF) Sensor Intermittent
P1101 Mass Air Flow (MAF) Sensor Out Of Self-Test Range
P1110 Intake Air Temperature (IAT) Sensor Signal Circuit Fault
P1112 Intake Air Temperature (IAT) Sensor Intermittent
P1113 Intake Air Temperature (IAT) Sensor Intermittent
P1116 Engine Coolant Temperature (ECT) Sensor Out Of Self-Test Range
P1117 Engine Coolant Temperature (ECT) Sensor Intermittent
P1120 Throttle Position (TP) Sensor Out Of Range (Low)
P1121 Throttle Position (TP) Sensor Inconsistent With MAF Sensor
P1124 Throttle Position (TP) Sensor Out Of Self-Test Range
P1125 Throttle Position (TP) Sensor Circuit Intermittent
P1127 Exhaust Not Warm Enough, Downstream Heated Oxygen Sensors (HO2S) Not Tested
P1128 Upstream Heated Oxygen Sensors (HO2S) Swapped From Bank To Bank
P1129 Downstream Heated Oxygen Sensors (HO2S) Swapped From Bank To Bank

P1130 Lack Of Upstream Heated Oxygen Sensor (HO2S 11) Switch, Adaptive Fuel At Limit (Bank #1)
P1131 Lack Of Upstream Heated Oxygen Sensor (HO2S 11) Switch, Sensor Indicates Lean (Bank #1)
P1132 Lack Of Upstream Heated Oxygen Sensor (HO2S 11) Switch, Sensor Indicates Rich (Bank#1)
P1137 Lack Of Downstream Heated Oxygen Sensor (HO2S 12) Switch, Sensor Indicates Lean (Bank#1)
P1138 Lack Of Downstream Heated Oxygen Sensor (HO2S 12) Switch, Sensor Indicates Rich (Bank#1)
P1150 Lack Of Upstream Heated Oxygen Sensor (HO2S 21) Switch, Adaptive Fuel At Limit (Bank #2)
P1151 Lack Of Upstream Heated Oxygen Sensor (HO2S 21) Switch, Sensor Indicates Lean (Bank#2)
P1152 Lack Of Upstream Heated Oxygen Sensor (HO2S 21) Switch, Sensor Indicates Rich (Bank #2)
P1170 (HO2S 11) Signal Remained Unchanged For More Than 20 Seconds After Closed Loop
P1173 Feedback A/F Mixture Control (HO2S 21) Signal Remained Unchanged For More Than 20 Seconds After Closed Loop
P1195 Barometric (BARO) Pressure Sensor Circuit Malfunction (Signal Is From EGR Boost Sensor)
P1196 Starter Switch Circuit Malfunction
P1235 Fuel Pump Control Out Of Range (MIL DTC)
P1236 Fuel Pump Control Out Of Range (No MIL)
P1250 Fuel Pressure Regulator Control (FPRC) Solenoid Malfunction
P1252 Fuel Pressure Regulator Control (FPRC) Solenoid Malfunction
P1260 THEFT Detected—Engine Disabled
P1270 Engine RPM Or Vehicle Speed Limiter Reached
P1345 No Camshaft Position Sensor Signal
P1351 Ignition Diagnostic Monitor (IDM) Circuit Input Malfunction
P1351 Indicates Ignition System Malfunction
P1352 Indicates Ignition System Malfunction
P1353 Indicates Ignition System Malfunction
P1354 Indicates Ignition System Malfunction
P1358 Ignition Diagnostic Monitor (IDM) Signal Out Of Self-Test Range
P1359 Spark Output Circuit Malfunction
P1360 Ignition Coil "A" Secondary Circuit Fault
P1361 Ignition Coil "A" Secondary Circuit Fault
P1362 Ignition Coil "A" Secondary Circuit Fault
P1364 Spark Output Circuit Malfunction
P1365 Ignition Coil Secondary Circuit Fault
P1390 Octane Adjust (OCT ADJ) Out Of Self-Test Range
P1400 Differential Pressure Feedback EGR (DPFE) Sensor Circuit Low Voltage Detected
P1401 Differential Pressure Feedback EGR (DPFE) Sensor Circuit High Voltage Detected/EGR Temperature Sensor
P1402 EGR Valve Position Sensor Open Or Short
P1405 Differential Pressure Feedback EGR (DPFE) Sensor Upstream Hose Off Or Plugged
P1406 Differential Pressure Feedback EGR (DPFE) Sensor Downstream Hose Off Or Plugged
P1407 Exhaust Gas Recirculation (EGR) No Flow Detected (Valve Stuck Closed Or Inoperative)
P1408 Exhaust Gas Recirculation (EGR) Flow Out Of Self-Test Range
P1409 Electronic Vacuum Regulator (EVR) Control Circuit Malfunction

P1443 Evaporative Emission Control System—Vacuum System, Purge Control Solenoid Or Purge Control Valve Malfunction

P1444 Purge Flow Sensor (PFS) Circuit Low Input

P1445 Purge Flow Sensor (PFS) Circuit High Input

P1449 Evaporative Emission Control System Unable To Hold Vacuum

P1455 Evaporative Emission Control System Control Leak Detected (Gross Leak)

P1460 Wide Open Throttle Air Conditioning Cut-Off Circuit Malfunction

P1464 Air Conditioning (A/C) Demand Out Of Self-Test Range/A/C On During KOER Or CCT Test

P1474 Low Fan Control Primary Circuit Malfunction

P1485 EGR Control Solenoid Open Or Short

P1486 EGR Vent Solenoid Open Or Short

P1487 EGR Boost Check Solenoid Open Or Short

P1500 Vehicle Speed Sensor (VSS) Circuit Intermittent

P1501 Vehicle Speed Sensor (VSS) Out Of Self-Test Range/Vehicle Moved During Test

P1502 Invalid Self Test—Auxiliary Powertrain Control Module (APCM) Functioning

P1504 Idle Air Control (IAC) Circuit Malfunction

P1505 Idle Air Control (IAC) System At Adaptive Clip

P1506 Idle Air Control (IAC) Overspeed Error

P1507 Idle Air Control (IAC) Underspeed Error

P1508 Bypass Air Solenoid "1" Circuit Fault

P1509 Bypass Air Solenoid "2" Circuit Fault

P1521 Variable Resonance Induction System (VRIS) Solenoid #1 Open Or Short

P1522 Variable Resonance Induction System (VRIS) Solenoid #2 Open Or Short

P1523 High Speed Inlet Air (HSIA) Solenoid Open Or Short

P1524 Charge Air Cooler Bypass Solenoid Circuit Fault

P1525 ABV Vacuum Solenoid Circuit Fault

P1526 ABV Vent Solenoid Circuit Fault

P1529 Atmospheric balance Air Control Valve Fault

P1540 ABV System Fault

P1601 Serial Communication Error

P1602 Serial Communication Error

P1605 Powertrain Control Module (PCM)—Keep Alive Memory (KAM) Test Error

P1608 PCM Internal Circuit Malfunction

P1609 PCM Internal Circuit Malfunction

P1627 Serial Communication Error

P1628 Serial Communication Error

P1650 Power Steering Pressure (PSP) Switch Out Of Self-Test Range

P1651 Power Steering Pressure (PSP) Switch Input Malfunction

P1701 Reverse Engagement Error

P1703 Brake On/Off (BOO) Switch Out Of Self-Test Range

P1705 Transmission Range (TR) Sensor Out Of Self-Test Range

P1706 High Vehicle Speed In Park

P1709 Park Or Neutral Position (PNP) Or Clutch Pedal Position (CPP) Switch Out Of Self-Test Range

P1711 Transmission Fluid Temperature (TFT) Sensor Out Of Self-Test Range

P1720 Vehicle Speed Sensor (VSS) Circuit Malfunction

P1729 4x4 Low Switch Error

P1741 Torque Converter Clutch (TCC) Control Error

P1742 Torque Converter Clutch (TCC) Solenoid Failed On (Turns On MIL)

P1743 Torque Converter Clutch (TCC) Solenoid Failed On (Turns On TCIL)

P1746 Electronic Pressure Control (EPC) Solenoid Open Circuit (Low Input)

P1747 Electronic Pressure Control (EPC) Solenoid Short Circuit (High Input)

P1749 Electronic Pressure Control (EPC) Solenoid Failed Low

P1751 Shift Solenoid#1 (SS1) Performance

P1754 Coast Clutch Solenoid (CCS) Circuit Malfunction

P1756 Shift Solenoid#2 (SS2) Performance

P1761 Shift Solenoid #(SS2) Performance

P1780 Transmission Control Switch (TCS) Circuit Out Of Self-Test Range

P1781 4x4 Low Switch, Out Of Self-Test Range

P1783 Transmission Over Temperature Condition

P1794 PCM Battery Direct Power Circuit Fault

P1797 P/N Switch Open or Short Circuit Fault

Mercedes-Benz

READING CODES

Reading the control module memory is one of the first steps in OBD II system diagnostics. This step should be initially performed to determine the general nature of the fault. Subsequent readings will determine if the fault has been cleared.

Reading codes can be performed by any of the methods below:

• Read the control module memory with the Generic Scan Tool (GST)

• Read the control module memory with the vehicle manufacturer's specific tester

To read the fault codes, connect the scan tool or tester according to the manufacturer's instructions. Follow the manufacturer's specified procedure for reading the codes.

CLEARING CODES

Control module reset procedures are a very important part of OBD II System diagnostics. This step should be done at the end of any fault code repair and at the end of any driveability repair.

Clearing codes can be performed by any of the methods below:

• Clear the control module memory with the Generic Scan Tool (GST)

• Clear the control module memory with the vehicle manufacturer's specific tester

• Turn the ignition off and remove the negative battery cable for at least 1 minute.

Removing the negative battery cable may cause other systems in the vehicle to lose their memory. Prior to removing the cable, ensure you have the proper reset codes for radios and alarms.

➡**The MIL will may also be de-activated for some codes if the vehicle completes three consecutive trips without a fault detected with vehicle conditions similar to those present during the fault.**

DIAGNOSTIC TROUBLE CODES

P0100 Mass or Volume Air Flow Circuit Malfunction

P0101 Mass or Volume Air Flow Circuit Range/Performance Problem

P0102 Mass or Volume Air Flow Circuit Low Input

P0103 Mass or Volume Air Flow Circuit High Input
P0104 Mass or Volume Air Flow Circuit Intermittent
P0105 Manifold Absolute Pressure/Barometric Pressure Circuit Malfunction
P0106 Manifold Absolute Pressure/Barometric Pressure Circuit Range/Performance Problem
P0107 Manifold Absolute Pressure/Barometric Pressure Circuit Low Input
P0108 Manifold Absolute Pressure/Barometric Pressure Circuit High Input
P0109 Manifold Absolute Pressure/Barometric Pressure Circuit Intermittent
P0110 Intake Air Temperature Circuit Malfunction
P0111 Intake Air Temperature Circuit Range/Performance Problem
P0112 Intake Air Temperature Circuit Low Input
P0113 Intake Air Temperature Circuit High Input
P0114 Intake Air Temperature Circuit Intermittent
P0115 Engine Coolant Temperature Circuit Malfunction
P0116 Engine Coolant Temperature Circuit Range/Performance Problem
P0117 Engine Coolant Temperature Circuit Low Input
P0118 Engine Coolant Temperature Circuit High Input
P0119 Engine Coolant Temperature Circuit Intermittent
P0120 Throttle/Pedal Position Sensor/Switch "A" Circuit Malfunction
P0121 Throttle/Pedal Position Sensor/Switch "A" Circuit Range/Performance Problem
P0122 Throttle/Pedal Position Sensor/Switch "A" Circuit Low Input
P0123 Throttle/Pedal Position Sensor/Switch "A" Circuit High Input
P0124 Throttle/Pedal Position Sensor/Switch "A" Circuit Intermittent
P0125 Insufficient Coolant Temperature For Closed Loop Fuel Control
P0126 Insufficient Coolant Temperature For Stable Operation
P0130 O_2 Circuit Malfunction (Bank no. 1 Sensor no. 1)
P0131 O_2 Sensor Circuit Low Voltage (Bank no. 1 Sensor no. 1)
P0132 O_2 Sensor Circuit High Voltage (Bank no. 1 Sensor no. 1)
P0133 O_2 Sensor Circuit Slow Response (Bank no. 1 Sensor no. 1)
P0134 O_2 Sensor Circuit No Activity Detected (Bank no. 1 Sensor no. 1)
P0135 O_2 Sensor Heater Circuit Malfunction (Bank no. 1 Sensor no. 1)
P0136 O_2 Sensor Circuit Malfunction (Bank no. 1 Sensor no. 2)
P0137 O_2 Sensor Circuit Low Voltage (Bank no. 1 Sensor no. 2)
P0138 O_2 Sensor Circuit High Voltage (Bank no. 1 Sensor no. 2)
P0139 O_2 Sensor Circuit Slow Response (Bank no. 1 Sensor no. 2)
P0140 O_2 Sensor Circuit No Activity Detected (Bank no. 1 Sensor no. 2)
P0141 O_2 Sensor Heater Circuit Malfunction (Bank no. 1 Sensor no. 2)
P0142 O_2 Sensor Circuit Malfunction (Bank no. 1 Sensor no. 3)
P0143 O_2 Sensor Circuit Low Voltage (Bank no. 1 Sensor no. 3)
P0144 O_2 Sensor Circuit High Voltage (Bank no. 1 Sensor no. 3)
P0145 O_2 Sensor Circuit Slow Response (Bank no. 1 Sensor no. 3)
P0146 O_2 Sensor Circuit No Activity Detected (Bank no. 1 Sensor no. 3)
P0147 O_2 Sensor Heater Circuit Malfunction (Bank no. 1 Sensor no. 3)
P0150 O_2 Sensor Circuit Malfunction (Bank no. 2 Sensor no. 1)
P0151 O_2 Sensor Circuit Low Voltage (Bank no. 2 Sensor no. 1)
P0152 O_2 Sensor Circuit High Voltage (Bank no. 2 Sensor no. 1)
P0153 O_2 Sensor Circuit Slow Response (Bank no. 2 Sensor no. 1)
P0154 O_2 Sensor Circuit No Activity Detected (Bank no. 2 Sensor no. 1)
P0155 O_2 Sensor Heater Circuit Malfunction (Bank no. 2 Sensor no. 1)
P0156 O_2 Sensor Circuit Malfunction (Bank no. 2 Sensor no. 2)
P0157 O_2 Sensor Circuit Low Voltage (Bank no. 2 Sensor no. 2)
P0158 O_2 Sensor Circuit High Voltage (Bank no. 2 Sensor no. 2)
P0159 O_2 Sensor Circuit Slow Response (Bank no. 2 Sensor no. 2)
P0160 O_2 Sensor Circuit No Activity Detected (Bank no. 2 Sensor no. 2)
P0161 O_2 Sensor Heater Circuit Malfunction (Bank no. 2 Sensor no. 2)
P0162 O_2 Sensor Circuit Malfunction (Bank no. 2 Sensor no. 3)
P0163 O_2 Sensor Circuit Low Voltage (Bank no. 2 Sensor no. 3)
P0164 O_2 Sensor Circuit High Voltage (Bank no. 2 Sensor no. 3)
P0165 O_2 Sensor Circuit Slow Response (Bank no. 2 Sensor no. 3)
P0166 O_2 Sensor Circuit No Activity Detected (Bank no. 2 Sensor no. 3)
P0167 O_2 Sensor Heater Circuit Malfunction (Bank no. 2 Sensor no. 3)
P0170 Fuel Trim Malfunction (Bank no. 1)
P0171 System Too Lean (Bank no. 1)
P0172 System Too Rich (Bank no. 1)
P0173 Fuel Trim Malfunction (Bank no. 2)
P0174 System Too Lean (Bank no. 2)
P0175 System Too Rich (Bank no. 2)
P0176 Fuel Composition Sensor Circuit Malfunction
P0177 Fuel Composition Sensor Circuit Range/Performance
P0178 Fuel Composition Sensor Circuit Low Input
P0179 Fuel Composition Sensor Circuit High Input
P0180 Fuel Temperature Sensor "A" Circuit Malfunction
P0181 Fuel Temperature Sensor "A" Circuit Range/Performance
P0182 Fuel Temperature Sensor "A" Circuit Low Input
P0183 Fuel Temperature Sensor "A" Circuit High Input
P0184 Fuel Temperature Sensor "A" Circuit Intermittent
P0185 Fuel Temperature Sensor "B" Circuit Malfunction
P0186 Fuel Temperature Sensor "B" Circuit Range/Performance
P0187 Fuel Temperature Sensor "B" Circuit Low Input
P0188 Fuel Temperature Sensor "B" Circuit High Input
P0189 Fuel Temperature Sensor "B" Circuit Intermittent
P0190 Fuel Rail Pressure Sensor Circuit Malfunction
P0191 Fuel Rail Pressure Sensor Circuit Range/Performance
P0192 Fuel Rail Pressure Sensor Circuit Low Input
P0193 Fuel Rail Pressure Sensor Circuit High Input
P0194 Fuel Rail Pressure Sensor Circuit Intermittent
P0195 Engine Oil Temperature Sensor Malfunction
P0196 Engine Oil Temperature Sensor Range/Performance
P0197 Engine Oil Temperature Sensor Low
P0198 Engine Oil Temperature Sensor High
P0199 Engine Oil Temperature Sensor Intermittent
P0200 Injector Circuit Malfunction
P0201 Injector Circuit Malfunction—Cylinder no. 1
P0202 Injector Circuit Malfunction—Cylinder no. 2

P0203 Injector Circuit Malfunction—Cylinder no. 3
P0204 Injector Circuit Malfunction—Cylinder no. 4
P0205 Injector Circuit Malfunction—Cylinder no. 5
P0206 Injector Circuit Malfunction—Cylinder no. 6
P0207 Injector Circuit Malfunction—Cylinder no. 7
P0208 Injector Circuit Malfunction—Cylinder no. 8
P0209 Injector Circuit Malfunction—Cylinder no. 9
P0210 Injector Circuit Malfunction—Cylinder no. 10
P0211 Injector Circuit Malfunction—Cylinder no. 11
P0212 Injector Circuit Malfunction—Cylinder no. 12
P0213 Cold Start Injector no. 1 Malfunction
P0214 Cold Start Injector no. 2 Malfunction
P0215 Engine Shutoff Solenoid Malfunction
P0216 Injection Timing Control Circuit Malfunction
P0217 Engine Over Temperature Condition
P0218 Transmission Over Temperature Condition
P0219 Engine Over Speed Condition
P0220 Throttle/Pedal Position Sensor/Switch "B" Circuit Malfunction
P0221 Throttle/Pedal Position Sensor/Switch "B" Circuit Range/Performance Problem
P0222 Throttle/Pedal Position Sensor/Switch "B" Circuit Low Input
P0223 Throttle/Pedal Position Sensor/Switch "B" Circuit High Input
P0224 Throttle/Pedal Position Sensor/Switch "B" Circuit Intermittent
P0225 Throttle/Pedal Position Sensor/Switch "C" Circuit Malfunction
P0226 Throttle/Pedal Position Sensor/Switch "C" Circuit Range/Performance Problem
P0227 Throttle/Pedal Position Sensor/Switch "C" Circuit Low Input
P0228 Throttle/Pedal Position Sensor/Switch "C" Circuit High Input
P0229 Throttle/Pedal Position Sensor/Switch "C" Circuit Intermittent
P0230 Fuel Pump Primary Circuit Malfunction
P0231 Fuel Pump Secondary Circuit Low
P0232 Fuel Pump Secondary Circuit High
P0233 Fuel Pump Secondary Circuit Intermittent
P0234 Engine Over Boost Condition
P0261 Cylinder no. 1 Injector Circuit Low
P0262 Cylinder no. 1 Injector Circuit High
P0263 Cylinder no. 1 Contribution/Balance Fault
P0264 Cylinder no. 2 Injector Circuit Low
P0265 Cylinder no. 2 Injector Circuit High
P0266 Cylinder no. 2 Contribution/Balance Fault
P0267 Cylinder no. 3 Injector Circuit Low
P0268 Cylinder no. 3 Injector Circuit High
P0269 Cylinder no. 3 Contribution/Balance Fault
P0270 Cylinder no. 4 Injector Circuit Low
P0271 Cylinder no. 4 Injector Circuit High
P0272 Cylinder no. 4 Contribution/Balance Fault
P0273 Cylinder no. 5 Injector Circuit Low
P0274 Cylinder no. 5 Injector Circuit High
P0275 Cylinder no. 5 Contribution/Balance Fault
P0276 Cylinder no. 6 Injector Circuit Low
P0277 Cylinder no. 6 Injector Circuit High
P0278 Cylinder no. 6 Contribution/Balance Fault
P0279 Cylinder no. 7 Injector Circuit Low
P0280 Cylinder no. 7 Injector Circuit High

P0281 Cylinder no. 7 Contribution/Balance Fault
P0282 Cylinder no. 8 Injector Circuit Low
P0283 Cylinder no. 8 Injector Circuit High
P0284 Cylinder no. 8 Contribution/Balance Fault
P0285 Cylinder no. 9 Injector Circuit Low
P0286 Cylinder no. 9 Injector Circuit High
P0287 Cylinder no. 9 Contribution/Balance Fault
P0288 Cylinder no. 10 Injector Circuit Low
P0289 Cylinder no. 10 Injector Circuit High
P0290 Cylinder no. 10 Contribution/Balance Fault
P0291 Cylinder no. 11 Injector Circuit Low
P0292 Cylinder no. 11 Injector Circuit High
P0293 Cylinder no. 11 Contribution/Balance Fault
P0294 Cylinder no. 12 Injector Circuit Low
P0295 Cylinder no. 12 Injector Circuit High
P0296 Cylinder no. 12 Contribution/Balance Fault
P0300 Random/Multiple Cylinder Misfire Detected
P0301 Cylinder no. 1—Misfire Detected
P0302 Cylinder no. 2—Misfire Detected
P0303 Cylinder no. 3—Misfire Detected
P0304 Cylinder no. 4—Misfire Detected
P0305 Cylinder no. 5—Misfire Detected
P0306 Cylinder no. 6—Misfire Detected
P0307 Cylinder no. 7—Misfire Detected
P0308 Cylinder no. 8—Misfire Detected
P0309 Cylinder no. 9—Misfire Detected
P0310 Cylinder no. 10—Misfire Detected
P0311 Cylinder no. 11—Misfire Detected
P0312 Cylinder no. 12—Misfire Detected
P0320 Ignition/Distributor Engine Speed Input Circuit Malfunction
P0321 Ignition/Distributor Engine Speed Input Circuit Range/Performance
P0322 Ignition/Distributor Engine Speed Input Circuit No Signal
P0323 Ignition/Distributor Engine Speed Input Circuit Intermittent
P0325 Knock Sensor no. 1—Circuit Malfunction (Bank no. 1 or Single Sensor)
P0326 Knock Sensor no. 1—Circuit Range/Performance (Bank no. 1 or Single Sensor)
P0327 Knock Sensor no. 1—Circuit Low Input (Bank no. 1 or Single Sensor)
P0328 Knock Sensor no. 1—Circuit High Input (Bank no. 1 or Single Sensor)
P0329 Knock Sensor no. 1—Circuit Input Intermittent (Bank no. 1 or Single Sensor)
P0330 Knock Sensor no. 2—Circuit Malfunction (Bank no. 2)
P0331 Knock Sensor no. 2—Circuit Range/Performance (Bank no. 2)
P0332 Knock Sensor no. 2—Circuit Low Input (Bank no. 2)
P0333 Knock Sensor no. 2—Circuit High Input (Bank no. 2)
P0334 Knock Sensor no. 2—Circuit Input Intermittent (Bank no. 2)
P0335 Crankshaft Position Sensor "A" Circuit Malfunction
P0336 Crankshaft Position Sensor "A" Circuit Range/Performance
P0337 Crankshaft Position Sensor "A" Circuit Low Input
P0338 Crankshaft Position Sensor "A" Circuit High Input
P0339 Crankshaft Position Sensor "A" Circuit Intermittent
P0340 Camshaft Position Sensor Circuit Malfunction
P0341 Camshaft Position Sensor Circuit Range/Performance
P0342 Camshaft Position Sensor Circuit Low Input

P0343 Camshaft Position Sensor Circuit High Input
P0344 Camshaft Position Sensor Circuit Intermittent
P0350 Ignition Coil Primary/Secondary Circuit Malfunction
P0351 Ignition Coil "A" Primary/Secondary Circuit Malfunction
P0352 Ignition Coil "B" Primary/Secondary Circuit Malfunction
P0353 Ignition Coil "C" Primary/Secondary Circuit Malfunction
P0354 Ignition Coil "D" Primary/Secondary Circuit Malfunction
P0355 Ignition Coil "E" Primary/Secondary Circuit Malfunction
P0356 Ignition Coil "F" Primary/Secondary Circuit Malfunction
P0357 Ignition Coil "G" Primary/Secondary Circuit Malfunction
P0358 Ignition Coil "H" Primary/Secondary Circuit Malfunction
P0359 Ignition Coil "I" Primary/Secondary Circuit Malfunction
P0360 Ignition Coil "J" Primary/Secondary Circuit Malfunction
P0361 Ignition Coil "K" Primary/Secondary Circuit Malfunction
P0362 Ignition Coil "L" Primary/Secondary Circuit Malfunction
P0370 Timing Reference High Resolution Signal "A" Malfunction
P0371 Timing Reference High Resolution Signal "A" Too Many Pulses
P0372 Timing Reference High Resolution Signal "A" Too Few Pulses
P0373 Timing Reference High Resolution Signal "A" Intermittent/Erratic Pulses
P0374 Timing Reference High Resolution Signal "A" No Pulses
P0375 Timing Reference High Resolution Signal "B" Malfunction
P0376 Timing Reference High Resolution Signal "B" Too Many Pulses
P0377 Timing Reference High Resolution Signal "B" Too Few Pulses
P0378 Timing Reference High Resolution Signal "B" Intermittent/Erratic Pulses
P0379 Timing Reference High Resolution Signal "B" No Pulses
P0380 Glow Plug/Heater Circuit "A" Malfunction
P0381 Glow Plug/Heater Indicator Circuit Malfunction
P0382 Glow Plug/Heater Circuit "B" Malfunction
P0385 Crankshaft Position Sensor "B" Circuit Malfunction
P0386 Crankshaft Position Sensor "B" Circuit Range/Performance
P0387 Crankshaft Position Sensor "B" Circuit Low Input
P0388 Crankshaft Position Sensor "B" Circuit High Input
P0389 Crankshaft Position Sensor "B" Circuit Intermittent
P0400 Exhaust Gas Recirculation Flow Malfunction
P0401 Exhaust Gas Recirculation Flow Insufficient Detected
P0402 Exhaust Gas Recirculation Flow Excessive Detected
P0403 Exhaust Gas Recirculation Circuit Malfunction
P0404 Exhaust Gas Recirculation Circuit Range/Performance
P0405 Exhaust Gas Recirculation Sensor "A" Circuit Low
P0406 Exhaust Gas Recirculation Sensor "A" Circuit High
P0407 Exhaust Gas Recirculation Sensor "B" Circuit Low
P0408 Exhaust Gas Recirculation Sensor "B" Circuit High
P0410 Secondary Air Injection System Malfunction
P0411 Secondary Air Injection System Incorrect Flow Detected
P0412 Secondary Air Injection System Switching Valve "A" Circuit Malfunction
P0413 Secondary Air Injection System Switching Valve "A" Circuit Open
P0414 Secondary Air Injection System Switching Valve "A" Circuit Shorted
P0415 Secondary Air Injection System Switching Valve "B" Circuit Malfunction

P0416 Secondary Air Injection System Switching Valve "B" Circuit Open
P0417 Secondary Air Injection System Switching Valve "B" Circuit Shorted
P0418 Secondary Air Injection System Relay "A" Circuit Malfunction
P0419 Secondary Air Injection System Relay "B" Circuit Malfunction
P0420 Catalyst System Efficiency Below Threshold (Bank no. 1)
P0421 Warm Up Catalyst Efficiency Below Threshold (Bank no. 1)
P0422 Main Catalyst Efficiency Below Threshold (Bank no. 1)
P0423 Heated Catalyst Efficiency Below Threshold (Bank no. 1)
P0424 Heated Catalyst Temperature Below Threshold (Bank no. 1)
P0430 Catalyst System Efficiency Below Threshold (Bank no. 2)
P0431 Warm Up Catalyst Efficiency Below Threshold (Bank no. 2)
P0432 Main Catalyst Efficiency Below Threshold (Bank no. 2)
P0433 Heated Catalyst Efficiency Below Threshold (Bank no. 2)
P0434 Heated Catalyst Temperature Below Threshold (Bank no. 2)
P0440 Evaporative Emission Control System Malfunction
P0441 Evaporative Emission Control System Incorrect Purge Flow
P0442 Evaporative Emission Control System Leak Detected (Small Leak)
P0443 Evaporative Emission Control System Purge Control Valve Circuit Malfunction
P0444 Evaporative Emission Control System Purge Control Valve Circuit Open
P0445 Evaporative Emission Control System Purge Control Valve Circuit Shorted
P0446 Evaporative Emission Control System Vent Control Circuit Malfunction
P0447 Evaporative Emission Control System Vent Control Circuit Open
P0448 Evaporative Emission Control System Vent Control Circuit Shorted
P0449 Evaporative Emission Control System Vent Valve/Solenoid Circuit Malfunction
P0450 Evaporative Emission Control System Pressure Sensor Malfunction
P0451 Evaporative Emission Control System Pressure Sensor Range/Performance
P0452 Evaporative Emission Control System Pressure Sensor Low Input
P0453 Evaporative Emission Control System Pressure Sensor High Input
P0454 Evaporative Emission Control System Pressure Sensor Intermittent
P0455 Evaporative Emission Control System Leak Detected (Gross Leak)
P0460 Fuel Level Sensor Circuit Malfunction
P0461 Fuel Level Sensor Circuit Range/Performance
P0462 Fuel Level Sensor Circuit Low Input
P0463 Fuel Level Sensor Circuit High Input
P0464 Fuel Level Sensor Circuit Intermittent
P0465 Purge Flow Sensor Circuit Malfunction
P0466 Purge Flow Sensor Circuit Range/Performance
P0467 Purge Flow Sensor Circuit Low Input
P0468 Purge Flow Sensor Circuit High Input
P0469 Purge Flow Sensor Circuit Intermittent
P0470 Exhaust Pressure Sensor Malfunction
P0471 Exhaust Pressure Sensor Range/Performance
P0472 Exhaust Pressure Sensor Low

P0473 Exhaust Pressure Sensor High
P0474 Exhaust Pressure Sensor Intermittent
P0475 Exhaust Pressure Control Valve Malfunction
P0476 Exhaust Pressure Control Valve Range/Performance
P0477 Exhaust Pressure Control Valve Low
P0478 Exhaust Pressure Control Valve High
P0479 Exhaust Pressure Control Valve Intermittent
P0480 Cooling Fan no. 1 Control Circuit Malfunction
P0481 Cooling Fan no. 2 Control Circuit Malfunction
P0482 Cooling Fan no. 3 Control Circuit Malfunction
P0483 Cooling Fan Rationality Check Malfunction
P0484 Cooling Fan Circuit Over Current
P0485 Cooling Fan Power/Ground Circuit Malfunction
P0500 Vehicle Speed Sensor Malfunction
P0501 Vehicle Speed Sensor Range/Performance
P0502 Vehicle Speed Sensor Circuit Low Input
P0503 Vehicle Speed Sensor Intermittent/Erratic/High
P0505 Idle Control System Malfunction
P0506 Idle Control System RPM Lower Than Expected
P0507 Idle Control System RPM Higher Than Expected
P0510 Closed Throttle Position Switch Malfunction
P0520 Engine Oil Pressure Sensor/Switch Circuit Malfunction
P0521 Engine Oil Pressure Sensor/Switch Range/Performance
P0522 Engine Oil Pressure Sensor/Switch Low Voltage
P0523 Engine Oil Pressure Sensor/Switch High Voltage
P0530 A/C Refrigerant Pressure Sensor Circuit Malfunction
P0531 A/C Refrigerant Pressure Sensor Circuit Range/Performance
P0532 A/C Refrigerant Pressure Sensor Circuit Low Input
P0533 A/C Refrigerant Pressure Sensor Circuit High Input
P0534 A/C Refrigerant Charge Loss
P0550 Power Steering Pressure Sensor Circuit Malfunction
P0551 Power Steering Pressure Sensor Circuit Range/Performance
P0552 Power Steering Pressure Sensor Circuit Low Input
P0553 Power Steering Pressure Sensor Circuit High Input
P0554 Power Steering Pressure Sensor Circuit Intermittent
P0560 System Voltage Malfunction
P0561 System Voltage Unstable
P0562 System Voltage Low
P0563 System Voltage High
P0565 Cruise Control On Signal Malfunction
P0566 Cruise Control Off Signal Malfunction
P0567 Cruise Control Resume Signal Malfunction
P0568 Cruise Control Set Signal Malfunction
P0569 Cruise Control Coast Signal Malfunction
P0570 Cruise Control Accel Signal Malfunction
P0571 Cruise Control/Brake Switch "A" Circuit Malfunction
P0572 Cruise Control/Brake Switch "A" Circuit Low
P0573 Cruise Control/Brake Switch "A" Circuit High
P0574 **Through P0580** Reserved for Cruise Codes
P0600 Serial Communication Link Malfunction
P0601 Internal Control Module Memory Check Sum Error
P0602 Control Module Programming Error
P0603 Internal Control Module Keep Alive Memory (KAM) Error
P0604 Internal Control Module Random Access Memory (RAM) Error
P0605 Internal Control Module Read Only Memory (ROM) Error
P0606 PCM Processor Fault
P0608 Control Module VSS Output "A" Malfunction
P0609 Control Module VSS Output "B" Malfunction
P0620 Generator Control Circuit Malfunction

P0621 Generator Lamp "L" Control Circuit Malfunction
P0622 Generator Field "F" Control Circuit Malfunction
P0650 Malfunction Indicator Lamp (MIL) Control Circuit Malfunction
P0654 Engine RPM Output Circuit Malfunction
P0655 Engine Hot Lamp Output Control Circuit Malfunction
P0656 Fuel Level Output Circuit Malfunction
P0700 Transmission Control System Malfunction
P0701 Transmission Control System Range/Performance
P0702 Transmission Control System Electrical
P0703 Torque Converter/Brake Switch "B" Circuit Malfunction
P0704 Clutch Switch Input Circuit Malfunction
P0705 Transmission Range Sensor Circuit Malfunction (PRNDL Input)
P0706 Transmission Range Sensor Circuit Range/Performance
P0707 Transmission Range Sensor Circuit Low Input
P0708 Transmission Range Sensor Circuit High Input
P0709 Transmission Range Sensor Circuit Intermittent
P0710 Transmission Fluid Temperature Sensor Circuit Malfunction
P0711 Transmission Fluid Temperature Sensor Circuit Range/Performance
P0712 Transmission Fluid Temperature Sensor Circuit Low Input
P0713 Transmission Fluid Temperature Sensor Circuit High Input
P0714 Transmission Fluid Temperature Sensor Circuit Intermittent
P0715 Input/Turbine Speed Sensor Circuit Malfunction
P0716 Input/Turbine Speed Sensor Circuit Range/Performance
P0717 Input/Turbine Speed Sensor Circuit No Signal
P0718 Input/Turbine Speed Sensor Circuit Intermittent
P0719 Torque Converter/Brake Switch "B" Circuit Low
P0720 Output Speed Sensor Circuit Malfunction
P0721 Output Speed Sensor Circuit Range/Performance
P0722 Output Speed Sensor Circuit No Signal
P0723 Output Speed Sensor Circuit Intermittent
P0724 Torque Converter/Brake Switch "B" Circuit High
P0725 Engine Speed Input Circuit Malfunction
P0726 Engine Speed Input Circuit Range/Performance
P0727 Engine Speed Input Circuit No Signal
P0728 Engine Speed Input Circuit Intermittent
P0730 Incorrect Gear Ratio
P0731 Gear no. 1 Incorrect Ratio
P0732 Gear no. 2 Incorrect Ratio
P0733 Gear no. 3 Incorrect Ratio
P0734 Gear no. 4 Incorrect Ratio
P0735 Gear no. 5 Incorrect Ratio
P0736 Reverse Incorrect Ratio
P0740 Torque Converter Clutch Circuit Malfunction
P0741 Torque Converter Clutch Circuit Performance or Stuck Off
P0742 Torque Converter Clutch Circuit Stuck On
P0743 Torque Converter Clutch Circuit Electrical
P0744 Torque Converter Clutch Circuit Intermittent
P0745 Pressure Control Solenoid Malfunction
P0746 Pressure Control Solenoid Performance or Stuck Off
P0747 Pressure Control Solenoid Stuck On
P0748 Pressure Control Solenoid Electrical
P0749 Pressure Control Solenoid Intermittent
P0750 Shift Solenoid "A" Malfunction
P0751 Shift Solenoid "A" Performance or Stuck Off
P0752 Shift Solenoid "A" Stuck On

P0753 Shift Solenoid "A" Electrical
P0754 Shift Solenoid "A" Intermittent
P0755 Shift Solenoid "B" Malfunction
P0756 Shift Solenoid "B" Performance or Stuck Off
P0757 Shift Solenoid "B" Stuck On
P0758 Shift Solenoid "B" Electrical
P0759 Shift Solenoid "B" Intermittent
P0760 Shift Solenoid "C" Malfunction
P0761 Shift Solenoid "C" Performance Or Stuck Off
P0762 Shift Solenoid "C" Stuck On
P0763 Shift Solenoid "C" Electrical
P0764 Shift Solenoid "C" Intermittent
P0765 Shift Solenoid "D" Malfunction
P0766 Shift Solenoid "D" Performance Or Stuck Off
P0767 Shift Solenoid "D" Stuck On
P0768 Shift Solenoid "D" Electrical
P0769 Shift Solenoid "D" Intermittent
P0770 Shift Solenoid "E" Malfunction
P0771 Shift Solenoid "E" Performance Or Stuck Off
P0772 Shift Solenoid "E" Stuck On
P0773 Shift Solenoid "E" Electrical
P0774 Shift Solenoid "E" Intermittent
P0780 Shift Malfunction
P0781 1–2 Shift Malfunction
P0782 2–3 Shift Malfunction
P0783 3–4 Shift Malfunction
P0784 4–5 Shift Malfunction
P0785 Shift/Timing Solenoid Malfunction
P0786 Shift/Timing Solenoid Range/Performance
P0787 Shift/Timing Solenoid Low
P0788 Shift/Timing Solenoid High
P0789 Shift/Timing Solenoid Intermittent
P0790 Normal/Performance Switch Circuit Malfunction
P0801 Reverse Inhibit Control Circuit Malfunction
P0803 1–4 Upshift (Skip Shift) Solenoid Control Circuit Malfunction
P0804 1–4 Upshift (Skip Shift) Lamp Control Circuit Malfunction

Mitsubishi

READING CODES

With a Scan Tool

Reading the control module memory is on of the first steps in OBD II system diagnostics. This step should be initially performed to determine the general nature of the fault. Subsequent readings will determine if the fault has been cleared.

Reading codes can be performed by any of the methods below:
• Read the control module memory with the Generic Scan Tool (GST)
• Read the control module memory with the vehicle manufacturer's specific tester

To read the fault codes, connect the scan tool or tester according to the manufacturer's instructions. Follow the manufacturer's specified procedure for reading the codes.

Without a Scan Tool

The Diagnostic Link Connector (DLC), under the left-hand side of the instrument panel, must be located to retrieve any DTC's.

In 1996, all Mitsubishi switched from an arbitrary code listing and format, to the federally regulated On Board Diagnostics 2nd Generation (OBD II) code system. Normally, OBD II equipped vehicles do not have the option of allowing the person servicing the vehicle to flash the codes out with a voltmeter; usually a scan tool is necessary to retrieve OBD II codes. Mitsubishi, however, does provide this option.

The Federal government decided that it was time to create a standard for vehicle diagnostic systems codes for ease of servicing and to insure that certain of the vehicle's systems were being monitored for emissions purposes. Since OBD II codes are standardized (they all contain one letter and four numbers), they are easy to decipher.

The OBD II system in the Mitsubishi models is designed so that it will flash the DTC's out on a voltmeter (even though a scan tool is better). However, the first two characters of the code are not used. This is because the transaxle is a part of the powertrain, so all transaxle related codes will begin with a P. Also, since there are no overlapping numbers between SAE and Mitsubishi codes, the second digit is also not necessary.

The system flashes the codes out in a series of flashes in three groups, each group corresponding to one of the three last digits of the OBD II code. Therefore, Code P0753 would be flashed out in seven flashes, followed by five flashes, then by three flashes. Each group of flashes is separated by a brief pause. All of the flashes are of the same duration, with the only exception being zero. Zero is represented by a long flash. Therefore, seven flashes, one long flash, two flashes would indicate a P0702 code (shorted TP sensor circuit).

To retrieve the codes, perform the following:

1. Perform the preliminary inspection, located earlier in this section. This is very important, since a loose or disconnected wire,

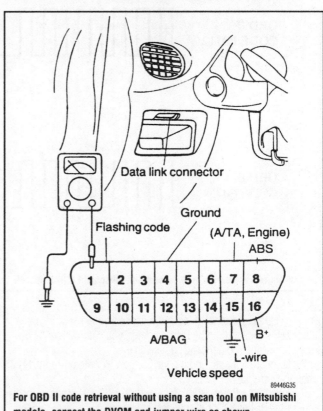

For OBD II code retrieval without using a scan tool on Mitsubishi models, connect the DVOM and jumper wire as shown

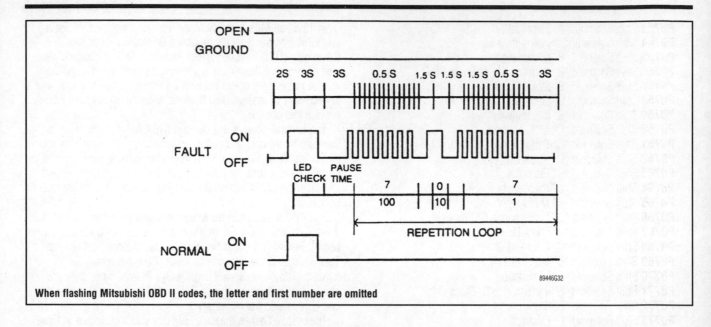

When flashing Mitsubishi OBD II codes, the letter and first number are omitted

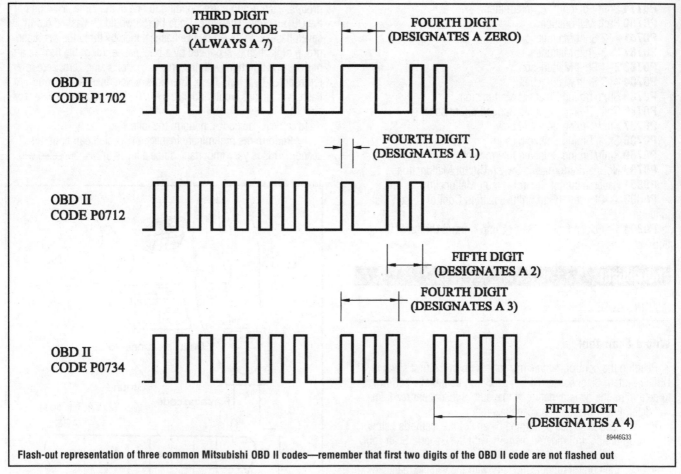

Flash-out representation of three common Mitsubishi OBD II codes—remember that first two digits of the OBD II code are not flashed out

or corroded connector terminals can cause a whole slew of unrelated DTC's to be stored by the computer; you will waste a lot of time performing a diagnostic "goose chase."

2. Grab some paper and a pencil or pen to write down the DTC's when they are flashed out.

3. Locate the Diagnostic Link Connector (DLC), which is located under the left-hand side of the instrument panel.

4. Start the engine and drive the vehicle until the transaxle goes into the failsafe mode.

5. Park the vehicle, but do not turn the ignition **OFF**. Allow it to idle.

6. Attach a voltmeter (analog or digital) to the test terminals on the Diagnostic Link Connector (DLC). The negative lead should be attached to terminal 4 and the positive lead to terminal 1.

7. Observe the voltmeter and count the flashes (or arm sweeps if using an analog voltmeter); note the applicable codes.

8. After all of the DTC(s) have been retrieved, fix the applicable problems, clear the codes, drive the vehicle, and perform the retrieval procedure again to ensure that all of the codes are gone.

CLEARING CODES

With a Scan Tool

Control module reset procedures are a very important part of OBD II System diagnostics.

This step should be done at the end of any fault code repair and at the end of any driveability repair.

Clearing codes can be performed by any of the methods below:

- Clear the control module memory with the Generic Scan Tool (GST)
- Clear the control module memory with the vehicle manufacturer's specific tester

➡**The MIL will may also be de-activated for some codes if the vehicle completes three consecutive trips without a fault detected with vehicle conditions similar to those present during the fault.**

Without a Scan Tool

➡**It is not recommended that the negative battery cable be disconnected to clear DTC's. Doing so will clear all settings from the vehicle's computer system, resulting in lost radio presets, seat memories, anti-theft codes, driveability parameters, etc., and there are better ways of spending your afternoon than resetting all of these.**

1. Turn the ignition switch **OFF**.
2. Disconnect the wiring harness connector from the control system computer.
3. Wait at least one minute, then reconnect the computer. The codes should now be cleared.

If there are still codes present, either the codes were not properly cleared (Are the codes identical to those flashed out previously?) or the underlying problem is still there (Are only some of the codes the same as previously?)

DIAGNOSTIC TROUBLE CODES

P0000 No Failures
P0100 Mass or Volume Air Flow Circuit Malfunction
P0101 Mass or Volume Air Flow Circuit Range/Performance Problem
P0102 Mass or Volume Air Flow Circuit Low Input
P0103 Mass or Volume Air Flow Circuit High Input
P0104 Mass or Volume Air Flow Circuit Intermittent
P0105 Manifold Absolute Pressure/Barometric Pressure Circuit Malfunction
P0106 Manifold Absolute Pressure/Barometric Pressure Circuit Range/Performance Problem
P0107 Manifold Absolute Pressure/Barometric Pressure Circuit Low Input
P0108 Manifold Absolute Pressure/Barometric Pressure Circuit High Input
P0109 Manifold Absolute Pressure/Barometric Pressure Circuit Intermittent

P0110 Intake Air Temperature Circuit Malfunction
P0111 Intake Air Temperature Circuit Range/Performance Problem
P0112 Intake Air Temperature Circuit Low Input
P0113 Intake Air Temperature Circuit High Input
P0114 Intake Air Temperature Circuit Intermittent
P0115 Engine Coolant Temperature Circuit Malfunction
P0116 Engine Coolant Temperature Circuit Range/Performance Problem
P0117 Engine Coolant Temperature Circuit Low Input
P0118 Engine Coolant Temperature Circuit High Input
P0119 Engine Coolant Temperature Circuit Intermittent
P0120 Throttle/Pedal Position Sensor/Switch "A" Circuit Malfunction
P0121 Throttle/Pedal Position Sensor/Switch "A" Circuit Range/Performance Problem
P0122 Throttle/Pedal Position Sensor/Switch "A" Circuit Low Input
P0123 Throttle/Pedal Position Sensor/Switch "A" Circuit High Input
P0124 Throttle/Pedal Position Sensor/Switch "A" Circuit Intermittent
P0125 Insufficient Coolant Temperature For Closed Loop Fuel Control
P0126 Insufficient Coolant Temperature For Stable Operation
P0130 O_2 Circuit Malfunction (Bank no. 1 Sensor no. 1)
P0131 O_2 Sensor Circuit Low Voltage (Bank no. 1 Sensor no. 1)
P0132 O_2 Sensor Circuit High Voltage (Bank no. 1 Sensor no. 1)
P0133 O_2 Sensor Circuit Slow Response (Bank no. 1 Sensor no. 1)
P0134 O_2 Sensor Circuit No Activity Detected (Bank no. 1 Sensor no. 1)
P0135 O_2 Sensor Heater Circuit Malfunction (Bank no. 1 Sensor no. 1)
P0136 O_2 Sensor Circuit Malfunction (Bank no. 1 Sensor no. 2)
P0137 O_2 Sensor Circuit Low Voltage (Bank no. 1 Sensor no. 2)
P0138 O_2 Sensor Circuit High Voltage (Bank no. 1 Sensor no. 2)
P0139 O_2 Sensor Circuit Slow Response (Bank no. 1 Sensor no. 2)
P0140 O_2 Sensor Circuit No Activity Detected (Bank no. 1 Sensor no. 2)
P0141 O_2 Sensor Heater Circuit Malfunction (Bank no. 1 Sensor no. 2)
P0142 O_2 Sensor Circuit Malfunction (Bank no. 1 Sensor no. 3)
P0143 O_2 Sensor Circuit Low Voltage (Bank no. 1 Sensor no. 3)
P0144 O_2 Sensor Circuit High Voltage (Bank no. 1 Sensor no. 3)
P0145 O_2 Sensor Circuit Slow Response (Bank no. 1 Sensor no. 3)
P0146 O_2 Sensor Circuit No Activity Detected (Bank no. 1 Sensor no. 3)
P0147 O_2 Sensor Heater Circuit Malfunction (Bank no. 1 Sensor no. 3)
P0150 O_2 Sensor Circuit Malfunction (Bank no. 2 Sensor no. 1)
P0151 O_2 Sensor Circuit Low Voltage (Bank no. 2 Sensor no. 1)
P0152 O_2 Sensor Circuit High Voltage (Bank no. 2 Sensor no. 1)

P0153 O_2 Sensor Circuit Slow Response (Bank no. 2 Sensor no. 1)

P0154 O_2 Sensor Circuit No Activity Detected (Bank no. 2 Sensor no. 1)

P0155 O_2 Sensor Heater Circuit Malfunction (Bank no. 2 Sensor no. 1)

P0156 O_2 Sensor Circuit Malfunction (Bank no. 2 Sensor no. 2)

P0157 O_2 Sensor Circuit Low Voltage (Bank no. 2 Sensor no. 2)

P0158 O_2 Sensor Circuit High Voltage (Bank no. 2 Sensor no. 2)

P0159 O_2 Sensor Circuit Slow Response (Bank no. 2 Sensor no. 2)

P0160 O_2 Sensor Circuit No Activity Detected (Bank no. 2 Sensor no. 2)

P0161 O_2 Sensor Heater Circuit Malfunction (Bank no. 2 Sensor no. 2)

P0162 O_2 Sensor Circuit Malfunction (Bank no. 2 Sensor no. 3)

P0163 O_2 Sensor Circuit Low Voltage (Bank no. 2 Sensor no. 3)

P0164 O_2 Sensor Circuit High Voltage (Bank no. 2 Sensor no. 3)

P0165 O_2 Sensor Circuit Slow Response (Bank no. 2 Sensor no. 3)

P0166 O_2 Sensor Circuit No Activity Detected (Bank no. 2 Sensor no. 3)

P0167 O_2 Sensor Heater Circuit Malfunction (Bank no. 2 Sensor no. 3)

P0170 Fuel Trim Malfunction (Bank no. 1)

P0171 System Too Lean (Bank no. 1)

P0172 System Too Rich (Bank no. 1)

P0173 Fuel Trim Malfunction (Bank no. 2)

P0174 System Too Lean (Bank no. 2)

P0175 System Too Rich (Bank no. 2)

P0176 Fuel Composition Sensor Circuit Malfunction

P0177 Fuel Composition Sensor Circuit Range/Performance

P0178 Fuel Composition Sensor Circuit Low Input

P0179 Fuel Composition Sensor Circuit High Input

P0180 Fuel Temperature Sensor "A" Circuit Malfunction

P0181 Fuel Temperature Sensor "A" Circuit Range/Performance

P0182 Fuel Temperature Sensor "A" Circuit Low Input

P0183 Fuel Temperature Sensor "A" Circuit High Input

P0184 Fuel Temperature Sensor "A" Circuit Intermittent

P0185 Fuel Temperature Sensor "B" Circuit Malfunction

P0186 Fuel Temperature Sensor "B" Circuit Range/Performance

P0187 Fuel Temperature Sensor "B" Circuit Low Input

P0188 Fuel Temperature Sensor "B" Circuit High Input

P0189 Fuel Temperature Sensor "B" Circuit Intermittent

P0190 Fuel Rail Pressure Sensor Circuit Malfunction

P0191 Fuel Rail Pressure Sensor Circuit Range/Performance

P0192 Fuel Rail Pressure Sensor Circuit Low Input

P0193 Fuel Rail Pressure Sensor Circuit High Input

P0194 Fuel Rail Pressure Sensor Circuit Intermittent

P0195 Engine Oil Temperature Sensor Malfunction

P0196 Engine Oil Temperature Sensor Range/Performance

P0197 Engine Oil Temperature Sensor Low

P0198 Engine Oil Temperature Sensor High

P0199 Engine Oil Temperature Sensor Intermittent

P0200 Injector Circuit Malfunction

P0201 Injector Circuit Malfunction—Cylinder no. 1

P0202 Injector Circuit Malfunction—Cylinder no. 2

P0203 Injector Circuit Malfunction—Cylinder no. 3

P0204 Injector Circuit Malfunction—Cylinder no. 4

P0205 Injector Circuit Malfunction—Cylinder no. 5

P0206 Injector Circuit Malfunction—Cylinder no. 6

P0207 Injector Circuit Malfunction—Cylinder no. 7

P0208 Injector Circuit Malfunction—Cylinder no. 8

P0209 Injector Circuit Malfunction—Cylinder no. 9

P0210 Injector Circuit Malfunction—Cylinder no. 10

P0211 Injector Circuit Malfunction—Cylinder no. 11

P0212 Injector Circuit Malfunction—Cylinder no. 12

P0213 Cold Start Injector no. 1 Malfunction

P0214 Cold Start Injector no. 2 Malfunction

P0215 Engine Shutoff Solenoid Malfunction

P0216 Injection Timing Control Circuit Malfunction

P0217 Engine Over Temperature Condition

P0218 Transmission Over Temperature Condition

P0219 Engine Over Speed Condition

P0220 Throttle/Pedal Position Sensor/Switch "B" Circuit Malfunction

P0221 Throttle/Pedal Position Sensor/Switch "B" Circuit Range/Performance Problem

P0222 Throttle/Pedal Position Sensor/Switch "B" Circuit Low Input

P0223 Throttle/Pedal Position Sensor/Switch "B" Circuit High Input

P0224 Throttle/Pedal Position Sensor/Switch "B" Circuit Intermittent

P0225 Throttle/Pedal Position Sensor/Switch "C" Circuit Malfunction

P0226 Throttle/Pedal Position Sensor/Switch "C" Circuit Range/Performance Problem

P0227 Throttle/Pedal Position Sensor/Switch "C" Circuit Low Input

P0228 Throttle/Pedal Position Sensor/Switch "C" Circuit High Input

P0229 Throttle/Pedal Position Sensor/Switch "C" Circuit Intermittent

P0230 Fuel Pump Primary Circuit Malfunction

P0231 Fuel Pump Secondary Circuit Low

P0232 Fuel Pump Secondary Circuit High

P0233 Fuel Pump Secondary Circuit Intermittent

P0234 Engine Over Boost Condition

P0261 Cylinder no. 1 Injector Circuit Low

P0262 Cylinder no. 1 Injector Circuit High

P0263 Cylinder no. 1 Contribution/Balance Fault

P0264 Cylinder no. 2 Injector Circuit Low

P0265 Cylinder no. 2 Injector Circuit High

P0266 Cylinder no. 2 Contribution/Balance Fault

P0267 Cylinder no. 3 Injector Circuit Low

P0268 Cylinder no. 3 Injector Circuit High

P0269 Cylinder no. 3 Contribution/Balance Fault

P0270 Cylinder no. 4 Injector Circuit Low

P0271 Cylinder no. 4 Injector Circuit High

P0272 Cylinder no. 4 Contribution/Balance Fault

P0273 Cylinder no. 5 Injector Circuit Low

P0274 Cylinder no. 5 Injector Circuit High

P0275 Cylinder no. 5 Contribution/Balance Fault

P0276 Cylinder no. 6 Injector Circuit Low

P0277 Cylinder no. 6 Injector Circuit High

P0278 Cylinder no. 6 Contribution/Balance Fault

P0279 Cylinder no. 7 Injector Circuit Low
P0280 Cylinder no. 7 Injector Circuit High
P0281 Cylinder no. 7 Contribution/Balance Fault
P0282 Cylinder no. 8 Injector Circuit Low
P0283 Cylinder no. 8 Injector Circuit High
P0284 Cylinder no. 8 Contribution/Balance Fault
P0285 Cylinder no. 9 Injector Circuit Low
P0286 Cylinder no. 9 Injector Circuit High
P0287 Cylinder no. 9 Contribution/Balance Fault
P0288 Cylinder no. 10 Injector Circuit Low
P0289 Cylinder no. 10 Injector Circuit High
P0290 Cylinder no. 10 Contribution/Balance Fault
P0291 Cylinder no. 11 Injector Circuit Low
P0292 Cylinder no. 11 Injector Circuit High
P0293 Cylinder no. 11 Contribution/Balance Fault
P0294 Cylinder no. 12 Injector Circuit Low
P0295 Cylinder no. 12 Injector Circuit High
P0296 Cylinder no. 12 Contribution/Balance Fault
P0300 Random/Multiple Cylinder Misfire Detected
P0301 Cylinder no. 1—Misfire Detected
P0302 Cylinder no. 2—Misfire Detected
P0303 Cylinder no. 3—Misfire Detected
P0304 Cylinder no. 4—Misfire Detected
P0305 Cylinder no. 5—Misfire Detected
P0306 Cylinder no. 6—Misfire Detected
P0307 Cylinder no. 7—Misfire Detected
P0308 Cylinder no. 8—Misfire Detected
P0309 Cylinder no. 9—Misfire Detected
P0310 Cylinder no. 10—Misfire Detected
P0311 Cylinder no. 11—Misfire Detected
P0312 Cylinder no. 12—Misfire Detected
P0320 Ignition/Distributor Engine Speed Input Circuit Malfunction
P0321 Ignition/Distributor Engine Speed Input Circuit Range/Performance
P0322 Ignition/Distributor Engine Speed Input Circuit No Signal
P0323 Ignition/Distributor Engine Speed Input Circuit Intermittent
P0325 Knock Sensor no. 1—Circuit Malfunction (Bank no. 1 or Single Sensor)
P0326 Knock Sensor no. 1—Circuit Range/Performance (Bank no. 1 or Single Sensor)
P0327 Knock Sensor no. 1—Circuit Low Input (Bank no. 1 or Single Sensor)
P0328 Knock Sensor no. 1—Circuit High Input (Bank no. 1 or Single Sensor)
P0329 Knock Sensor no. 1—Circuit Input Intermittent (Bank no. 1 or Single Sensor)
P0330 Knock Sensor no. 2—Circuit Malfunction (Bank no. 2)
P0331 Knock Sensor no. 2—Circuit Range/Performance (Bank no. 2)
P0332 Knock Sensor no. 2—Circuit Low Input (Bank no. 2)
P0333 Knock Sensor no. 2—Circuit High Input (Bank no. 2)
P0334 Knock Sensor no. 2—Circuit Input Intermittent (Bank no. 2)
P0335 Crankshaft Position Sensor "A" Circuit Malfunction
P0336 Crankshaft Position Sensor "A" Circuit Range/Performance
P0337 Crankshaft Position Sensor "A" Circuit Low Input
P0338 Crankshaft Position Sensor "A" Circuit High Input

P0339 Crankshaft Position Sensor "A" Circuit Intermittent
P0340 Camshaft Position Sensor Circuit Malfunction
P0341 Camshaft Position Sensor Circuit Range/Performance
P0342 Camshaft Position Sensor Circuit Low Input
P0343 Camshaft Position Sensor Circuit High Input
P0344 Camshaft Position Sensor Circuit Intermittent
P0350 Ignition Coil Primary/Secondary Circuit Malfunction
P0351 Ignition Coil "A" Primary/Secondary Circuit Malfunction
P0352 Ignition Coil "B" Primary/Secondary Circuit Malfunction
P0353 Ignition Coil "C" Primary/Secondary Circuit Malfunction
P0354 Ignition Coil "D" Primary/Secondary Circuit Malfunction
P0355 Ignition Coil "E" Primary/Secondary Circuit Malfunction
P0356 Ignition Coil "F" Primary/Secondary Circuit Malfunction
P0357 Ignition Coil "G" Primary/Secondary Circuit Malfunction
P0358 Ignition Coil "H" Primary/Secondary Circuit Malfunction
P0359 Ignition Coil "I" Primary/Secondary Circuit Malfunction
P0360 Ignition Coil "J" Primary/Secondary Circuit Malfunction
P0361 Ignition Coil "K" Primary/Secondary Circuit Malfunction
P0362 Ignition Coil "L" Primary/Secondary Circuit Malfunction
P0370 Timing Reference High Resolution Signal "A" Malfunction
P0371 Timing Reference High Resolution Signal "A" Too Many Pulses
P0372 Timing Reference High Resolution Signal "A" Too Few Pulses
P0373 Timing Reference High Resolution Signal "A" Intermittent/Erratic Pulses
P0374 Timing Reference High Resolution Signal "A" No Pulses
P0375 Timing Reference High Resolution Signal "B" Malfunction
P0376 Timing Reference High Resolution Signal "B" Too Many Pulses
P0377 Timing Reference High Resolution Signal "B" Too Few Pulses
P0378 Timing Reference High Resolution Signal "B" Intermittent/Erratic Pulses
P0379 Timing Reference High Resolution Signal "B" No Pulses
P0380 Glow Plug/Heater Circuit "A" Malfunction
P0381 Glow Plug/Heater Indicator Circuit Malfunction
P0382 Glow Plug/Heater Circuit "B" Malfunction
P0385 Crankshaft Position Sensor "B" Circuit Malfunction
P0386 Crankshaft Position Sensor "B" Circuit Range/Performance
P0387 Crankshaft Position Sensor "B" Circuit Low Input
P0388 Crankshaft Position Sensor "B" Circuit High Input
P0389 Crankshaft Position Sensor "B" Circuit Intermittent
P0400 Exhaust Gas Recirculation Flow Malfunction
P0401 Exhaust Gas Recirculation Flow Insufficient Detected

P0402 Exhaust Gas Recirculation Flow Excessive Detected
P0403 Exhaust Gas Recirculation Circuit Malfunction
P0404 Exhaust Gas Recirculation Circuit Range/Performance
P0405 Exhaust Gas Recirculation Sensor "A" Circuit Low
P0406 Exhaust Gas Recirculation Sensor "A" Circuit High
P0407 Exhaust Gas Recirculation Sensor "B" Circuit Low
P0408 Exhaust Gas Recirculation Sensor "B" Circuit High
P0410 Secondary Air Injection System Malfunction
P0411 Secondary Air Injection System Incorrect Flow Detected
P0412 Secondary Air Injection System Switching Valve "A" Circuit Malfunction
P0413 Secondary Air Injection System Switching Valve "A" Circuit Open
P0414 Secondary Air Injection System Switching Valve "A" Circuit Shorted
P0415 Secondary Air Injection System Switching Valve "B" Circuit Malfunction
P0416 Secondary Air Injection System Switching Valve "B" Circuit Open
P0417 Secondary Air Injection System Switching Valve "B" Circuit Shorted
P0418 Secondary Air Injection System Relay "A" Circuit Malfunction
P0419 Secondary Air Injection System Relay "B" Circuit Malfunction
P0420 Catalyst System Efficiency Below Threshold (Bank no. 1)
P0421 Warm Up Catalyst Efficiency Below Threshold (Bank no. 1)
P0422 Main Catalyst Efficiency Below Threshold (Bank no. 1)
P0423 Heated Catalyst Efficiency Below Threshold (Bank no. 1)
P0424 Heated Catalyst Temperature Below Threshold (Bank no. 1)
P0430 Catalyst System Efficiency Below Threshold (Bank no. 2)
P0431 Warm Up Catalyst Efficiency Below Threshold (Bank no. 2)
P0432 Main Catalyst Efficiency Below Threshold (Bank no. 2)
P0433 Heated Catalyst Efficiency Below Threshold (Bank no. 2)
P0434 Heated Catalyst Temperature Below Threshold (Bank no. 2)
P0440 Evaporative Emission Control System Malfunction
P0441 Evaporative Emission Control System Incorrect Purge Flow
P0442 Evaporative Emission Control System Leak Detected (Small Leak)
P0443 Evaporative Emission Control System Purge Control Valve Circuit Malfunction
P0444 Evaporative Emission Control System Purge Control Valve Circuit Open
P0445 Evaporative Emission Control System Purge Control Valve Circuit Shorted
P0446 Evaporative Emission Control System Vent Control Circuit Malfunction
P0447 Evaporative Emission Control System Vent Control Circuit Open
P0448 Evaporative Emission Control System Vent Control Circuit Shorted
P0449 Evaporative Emission Control System Vent Valve/Solenoid Circuit Malfunction

P0450 Evaporative Emission Control System Pressure Sensor Malfunction
P0451 Evaporative Emission Control System Pressure Sensor Range/Performance
P0452 Evaporative Emission Control System Pressure Sensor Low Input
P0453 Evaporative Emission Control System Pressure Sensor High Input
P0454 Evaporative Emission Control System Pressure Sensor Intermittent
P0455 Evaporative Emission Control System Leak Detected (Gross Leak)
P0460 Fuel Level Sensor Circuit Malfunction
P0461 Fuel Level Sensor Circuit Range/Performance
P0462 Fuel Level Sensor Circuit Low Input
P0463 Fuel Level Sensor Circuit High Input
P0464 Fuel Level Sensor Circuit Intermittent
P0465 Purge Flow Sensor Circuit Malfunction
P0466 Purge Flow Sensor Circuit Range/Performance
P0467 Purge Flow Sensor Circuit Low Input
P0468 Purge Flow Sensor Circuit High Input
P0469 Purge Flow Sensor Circuit Intermittent
P0470 Exhaust Pressure Sensor Malfunction
P0471 Exhaust Pressure Sensor Range/Performance
P0472 Exhaust Pressure Sensor Low
P0473 Exhaust Pressure Sensor High
P0474 Exhaust Pressure Sensor Intermittent
P0475 Exhaust Pressure Control Valve Malfunction
P0476 Exhaust Pressure Control Valve Range/Performance
P0477 Exhaust Pressure Control Valve Low
P0478 Exhaust Pressure Control Valve High
P0479 Exhaust Pressure Control Valve Intermittent
P0480 Cooling Fan no. 1 Control Circuit Malfunction
P0481 Cooling Fan no. 2 Control Circuit Malfunction
P0482 Cooling Fan no. 3 Control Circuit Malfunction
P0483 Cooling Fan Rationality Check Malfunction
P0484 Cooling Fan Circuit Over Current
P0485 Cooling Fan Power/Ground Circuit Malfunction
P0500 Vehicle Speed Sensor Malfunction
P0501 Vehicle Speed Sensor Range/Performance
P0502 Vehicle Speed Sensor Circuit Low Input
P0503 Vehicle Speed Sensor Intermittent/Erratic/High
P0505 Idle Control System Malfunction
P0506 Idle Control System RPM Lower Than Expected
P0507 Idle Control System RPM Higher Than Expected
P0510 Closed Throttle Position Switch Malfunction
P0520 Engine Oil Pressure Sensor/Switch Circuit Malfunction
P0521 Engine Oil Pressure Sensor/Switch Range/Performance
P0522 Engine Oil Pressure Sensor/Switch Low Voltage
P0523 Engine Oil Pressure Sensor/Switch High Voltage
P0530 A/C Refrigerant Pressure Sensor Circuit Malfunction
P0531 A/C Refrigerant Pressure Sensor Circuit Range/Performance
P0532 A/C Refrigerant Pressure Sensor Circuit Low Input
P0533 A/C Refrigerant Pressure Sensor Circuit High Input
P0534 A/C Refrigerant Charge Loss
P0550 Power Steering Pressure Sensor Circuit Malfunction
P0551 Power Steering Pressure Sensor Circuit Range/Performance
P0552 Power Steering Pressure Sensor Circuit Low Input

P0553 Power Steering Pressure Sensor Circuit High Input
P0554 Power Steering Pressure Sensor Circuit Intermittent
P0560 System Voltage Malfunction
P0561 System Voltage Unstable
P0562 System Voltage Low
P0563 System Voltage High
P0565 Cruise Control On Signal Malfunction
P0566 Cruise Control Off Signal Malfunction
P0567 Cruise Control Resume Signal Malfunction
P0568 Cruise Control Set Signal Malfunction
P0569 Cruise Control Coast Signal Malfunction
P0570 Cruise Control Accel Signal Malfunction
P0571 Cruise Control/Brake Switch "A" Circuit Malfunction
P0572 Cruise Control/Brake Switch "A" Circuit Low
P0573 Cruise Control/Brake Switch "A" Circuit High
P0574 Through P0580 Reserved for Cruise Codes
P0600 Serial Communication Link Malfunction
P0601 Internal Control Module Memory Check Sum Error
P0602 Control Module Programming Error
P0603 Internal Control Module Keep Alive Memory (KAM) Error
P0604 Internal Control Module Random Access Memory (RAM) Error
P0605 Internal Control Module Read Only Memory (ROM) Error
P0606 PCM Processor Fault
P0608 Control Module VSS Output "A" Malfunction
P0609 Control Module VSS Output "B" Malfunction
P0620 Generator Control Circuit Malfunction
P0621 Generator Lamp "L" Control Circuit Malfunction
P0622 Generator Field "F" Control Circuit Malfunction
P0650 Malfunction Indicator Lamp (MIL) Control Circuit Malfunction
P0654 Engine RPM Output Circuit Malfunction
P0655 Engine Hot Lamp Output Control Circuit Malfunction
P0656 Fuel Level Output Circuit Malfunction
P0700 Transmission Control System Malfunction
P0701 Transmission Control System Range/Performance
P0702 Transmission Control System Electrical
P0703 Torque Converter/Brake Switch "B" Circuit Malfunction
P0704 Clutch Switch Input Circuit Malfunction
P0705 Transmission Range Sensor Circuit Malfunction (PRNDL Input)
P0706 Transmission Range Sensor Circuit Range/Performance
P0707 Transmission Range Sensor Circuit Low Input
P0708 Transmission Range Sensor Circuit High Input
P0709 Transmission Range Sensor Circuit Intermittent
P0710 Transmission Fluid Temperature Sensor Circuit Malfunction
P0711 Transmission Fluid Temperature Sensor Circuit Range/Performance
P0712 Transmission Fluid Temperature Sensor Circuit Low Input
P0713 Transmission Fluid Temperature Sensor Circuit High Input
P0714 Transmission Fluid Temperature Sensor Circuit Intermittent
P0715 Input/Turbine Speed Sensor Circuit Malfunction
P0716 Input/Turbine Speed Sensor Circuit Range/Performance

P0717 Input/Turbine Speed Sensor Circuit No Signal
P0718 Input/Turbine Speed Sensor Circuit Intermittent
P0719 Torque Converter/Brake Switch "B" Circuit Low
P0720 Output Speed Sensor Circuit Malfunction
P0721 Output Speed Sensor Circuit Range/Performance
P0722 Output Speed Sensor Circuit No Signal
P0723 Output Speed Sensor Circuit Intermittent
P0724 Torque Converter/Brake Switch "B" Circuit High
P0725 Engine Speed Input Circuit Malfunction
P0726 Engine Speed Input Circuit Range/Performance
P0727 Engine Speed Input Circuit No Signal
P0728 Engine Speed Input Circuit Intermittent
P0730 Incorrect Gear Ratio
P0731 Gear no. 1 Incorrect Ratio
P0732 Gear no. 2 Incorrect Ratio
P0733 Gear no. 3 Incorrect Ratio
P0734 Gear no. 4 Incorrect Ratio
P0735 Gear no. 5 Incorrect Ratio
P0736 Reverse Incorrect Ratio
P0740 Torque Converter Clutch Circuit Malfunction
P0741 Torque Converter Clutch Circuit Performance or Stuck Off
P0742 Torque Converter Clutch Circuit Stuck On
P0743 Torque Converter Clutch Circuit Electrical
P0744 Torque Converter Clutch Circuit Intermittent
P0745 Pressure Control Solenoid Malfunction
P0746 Pressure Control Solenoid Performance or Stuck Off
P0747 Pressure Control Solenoid Stuck On
P0748 Pressure Control Solenoid Electrical
P0749 Pressure Control Solenoid Intermittent
P0750 Shift Solenoid "A" Malfunction
P0751 Shift Solenoid "A" Performance or Stuck Off
P0752 Shift Solenoid "A" Stuck On
P0753 Shift Solenoid "A" Electrical
P0754 Shift Solenoid "A" Intermittent
P0755 Shift Solenoid "B" Malfunction
P0756 Shift Solenoid "B" Performance or Stuck Off
P0757 Shift Solenoid "B" Stuck On
P0758 Shift Solenoid "B" Electrical
P0759 Shift Solenoid "B" Intermittent
P0760 Shift Solenoid "C" Malfunction
P0761 Shift Solenoid "C" Performance Or Stuck Off
P0762 Shift Solenoid "C" Stuck On
P0763 Shift Solenoid "C" Electrical
P0764 Shift Solenoid "C" Intermittent
P0765 Shift Solenoid "D" Malfunction
P0766 Shift Solenoid "D" Performance Or Stuck Off
P0767 Shift Solenoid "D" Stuck On
P0768 Shift Solenoid "D" Electrical
P0769 Shift Solenoid "D" Intermittent
P0770 Shift Solenoid "E" Malfunction
P0771 Shift Solenoid "E" Performance Or Stuck Off
P0772 Shift Solenoid "E" Stuck On
P0773 Shift Solenoid "E" Electrical
P0774 Shift Solenoid "E" Intermittent
P0780 Shift Malfunction
P0781 1–2 Shift Malfunction
P0782 2–3 Shift Malfunction
P0783 3–4 Shift Malfunction
P0784 4–5 Shift Malfunction
P0785 Shift/Timing Solenoid Malfunction
P0786 Shift/Timing Solenoid Range/Performance

P0787 Shift/Timing Solenoid Low
P0788 Shift/Timing Solenoid High
P0789 Shift/Timing Solenoid Intermittent
P0790 Normal/Performance Switch Circuit Malfunction
P0801 Reverse Inhibit Control Circuit Malfunction
P0803 1–4 Upshift (Skip Shift) Solenoid Control Circuit Malfunction
P0804 1–4 Upshift (Skip Shift) Lamp Control Circuit Malfunction
P1100 Induction Control Motor Position Sensor Fault
P1101 Traction Control Vacuum Solenoid Circuit Fault
P1102 Traction Control Ventilation Solenoid Circuit Fault
P1103 Turbocharger Waste Gate Actuator Circuit Fault
P1104 Turbocharger Waste Gate Solenoid Circuit Fault
P1105 Fuel Pressure Solenoid Circuit Fault
P1294 Target Idle Speed Not Reached
P1295 No 5-Volt Supply To TP Sensor
P1296 No 5-Volt Supply To MAP Sensor
P1297 No Change In MAP From Start To Run
P1300 Ignition Timing Adjustment Circuit

P1390 Timing Belt Skipped One Tooth Or More
P1391 Intermittent Loss Of CMP Or CKP Sensor Signals
P1400 Manifold Differential Pressure Sensor Fault
P1443 EVAP Purge Control Solenoid "2" Circuit Fault
P1486 EVAP Leak Monitor Pinched Hose Detected
P1487 High Speed Radiator Fan Control Relay Circuit Fault
P1490 Low Speed Fan Control Relay Fault
P1492 Battery Temperature Sensor High Voltage
P1494 EVAP Ventilation Switch Or Mechanical Fault
P1495 EVAP Ventilation Solenoid Circuit Fault
P1496 5-Volt Supply Output Too Low
P1500 Generator FR Terminal Circuit Fault
P1600 PCM-TCM Serial Communication Link Circuit Fault
P1696 PCM Failure- EEPROM Write Denied
P1715 No CCD Messages From TCM
P1750 TCM Pulse Generator Circuit Fault
P1791 Pressure Control, Shift Control, TCC Solenoid Fault
P1899 PCM ECT Level Signal to TCM Circuit Fault
P1989 High Speed Condenser Fan Control Relay Fault

Code	Output pattern (for voltmeter)	Cause	Remedy
P1702	A5AT005E	Shorted throttle position sensor circuit	o Check the throttle position sensor connector o check the throttle position sensor itself o Check the closed throttle position switch o Check the throttle position sensor wiring harness o Check the wiring between ECM and throttle position sensor
P1701	A5AT005F	Open throttle position sensor circuit	
P1704	A5AT005H	Throttle position sensor malfunction Improperly adjusted throttle position sensor	
P0712	A5AT005I	Open fluid temperature sensor circuit	o Fluid temperature sensor connector inspection o Fluid temperature sensor inspection o Fluid temperature sensor wiring harness inspection
P0713	A5AT005J	Shorted fluid temperature sensor circuit	
P1709	A5AT005K	Open kickdown servo switch circuit Shorted kickdown servo switch circuit	o Check the kickdown servo switch connector o Check the kickdown servo switch o Check the kickdown servo switch wiring harness

89446G27

Mitsubishi flash out DTC's, 1 of 4—Type 4 (OBD II) Codes

Code	Output pattern (for voltmeter)	Cause	Remedy
P0727	A5AT005L	Open ignition pulse pickup cable circuit	o Check the ignition pulse signal line o Check the wiring between ECM and ignisiton system
P1714	A5AT005M	Short-circuited or improperly adjusted closed throttle position switch	o Check the closed throttle position switch connector o Check the closed throttle position switch itself o Adjust the closed throttle position switch o Check the closed throttle position switch wiring harness
P0717	A5AT005N	Open-circuited pulse generator A	o Check the pulse generator A and pulse generator B o Check the vehicle speed reed switch (for chattering) o Check the pulse generator A and B wiring harness
P0722	A5AT005O	Open-circuited pulse generator B	
P0707	A5AT005Z	No input signal	o Check the transaxle range switch o Check the transaxle range wiring harness o Check the manual control cable
P0708	A5ATA05A	More than two input signals	
P0752	A5AT005P	Open shift control solenoid vave A circuit	o Check the solenoid valve connector o Check the shift control solenoid valve A o Check the shift control solenoid valve A wiring harness
P0753	A5AT005Q	Shorted shift control solenoid valve A circuit	
P0757	A5AT005R	Open shift control solenoid valve B circuit	o Check the shift control solenoid valve connector o Check the shift control solenoid valve B wiring harness o Check the shift control solenoid valve B
P0758	A5AT005S	Short shift control solenoid valve B circuit	
P0747	A5AT005T	Open pressure control solenoid valve circuit	o Check the pressure control solenoid valve o Check the pressure control solenoid valve wiring harness
P0748	A5AT005U	Shorted pressure control solenoid valve circuit	

89446G28

Mitsubishi flash out DTC's, 2 of 4—Type 4 (OBD II) Codes

Code	Output pattern (for voltmeter)	Cause	Remedy
P0743	A5AT005V	Open circuit in damper clutch control solenoid valve	o Inspection of solenoid valve connector
P0742	A5AT005W	Short circuit in damper clutch control solenoid valve	o Individual inspection of damper clutch control solenoid valve
P0740	A5AT005X	Defect in the damper clutch system	o Check the damper clutch control solenoid valve wiring harness
P1744	A5AT005Y		o Chck the TCM o Inspection of damper clutch hydraulic system
P0731	A5ATA05B	Shifting to first gear does not match the engine speed	o Check the pulse generator A and pulse generator B connector
P0732	A5ATA05C	Shifting to second gear does not match the engine speed	o Check the pulse generator A and puls generator B o Check the one way clutch or rear clutch o Check the pulse generator wiring harness o Kickdown brake slippage
P0733	A5ATA05D	Shifting to third gear does not match the engine speed	o Check the rear clutch or control system o Check the pulse generator A and pulse generator B connector o Check the pulse generator A and pulse generator B o Check the pulse generator wiring harness o Check the rear clutch slippage or control system o Check the front clutch slippage or control system
P0734	A5ATA05E	Shifting to fourth gear does not match the engine speed	o Check the pulse generator A and B connector o Check the pulse generator A and B o Kickdown brake slippage o Check the end clutch or control system o Check the pulse generator wiring harness
-		Normal	

89446G29

Mitsubishi flash out DTC's, 3 of 4—Type 4 (OBD II) Codes

FAIL-SAFE ITEM

Code No.	Output code — Output pattern (for voltmeter)	Description	Fail-safe	Note (relation to diagnostic trouble code)
P0717	A5AT005N	Open-circuited pulse generator A	Locked in third (D) or second (2,L)	When code No.0717 is generated fourth time
P0722	A5AT005O	Open-circuited pulse generator B	Locked in third (D) or second (2,L)	When code No.0722 is generated fourth time
P0752	A5AT005P	Open-circuited or shorted shift control solenoid valve A	Lock in third	When code No.0752 or 0753 is generated fourth time
P0753	A5AT005Q			
P0757	A5AT005R	Open-circuited or shorted shift control solenoid valve B	Lock in third gear	When code No.0757 or 0758 is generated fourth time
P0758	A5AT005S			
P0747	A5AT005T	Open-circuited or shorted pressure control solenoid valve	Locked in third (D) or second (2,L)	When code No.0747 or 0748 is generated fourth time
P0748	A5AT005U			
P0731	A5ATA05B	Gear shifting does not match the engine speed	Locked in third (D) or second (2,L)	When either code No.0731, 0732, 0733 or 0734 is generated fourth time
P0732	A5ATA05C			
P0733	A5ATA05D			
P0734	A5ATA05E			

89446G30

Mitsubishi flash out DTC's, 4 of 4—Type 4 (OBD II) Codes

Nissan

➡**This section also provides coverage for the Mercury Villager since it shares a platform with the Nissan Quest**

READING CODES

With Scan Tool

Reading the control module memory is on of the first steps in OBD II system diagnostics. This step should be initially performed to determine the general nature of the fault. Subsequent readings will determine if the fault has been cleared.

Reading codes can be performed by any of the methods below:
• Read the control module memory with the Generic Scan Tool (GST)
• Read the control module memory with the vehicle manufacturer's specific tester

To read the fault codes, connect the scan tool or tester according to the manufacturer's instructions. Follow the manufacturer's specified procedure for reading the codes.

Without Scan Tool

The ECM is capable of outputting data in four different modes, depending on the position of the mode switch and the ignition key. Modes are switched by turning the mode screw on the side of the ECM, near the red LED. Additional modes are accessed by turning the ignition key on or off.

The ECM is located forward of the center console, behind an access panel on the Altima, and in the passenger's side kick panel on the 240SX.

With the ECM set in Mode 1 and the ignition in the**ON** position, a malfunction indicator lamp bulb check may be performed. When the engine is started, the ECM will illuminate the indicator lamps as a warning of a fault in the system.

Mode 2 is set by turning the mode selector screw fully clockwise, waiting 2 seconds, then turning the screw fully counterclockwise. With the ignition in the**ON** position, self-diagnostic results will be output as a series of lamp flashes. When the engine is started, the oxygen sensor monitor function is enabled and the red LED on the ECM is used to determine proper oxygen sensor function.

1. Remove the access cover and locate the mode adjusting screw and LED on the ECM.

2. Turn the ignition switch**ON** , but do not start the engine. Both the LED and the malfunction indicator lamp on the instrument panel should be illuminated. This is a bulb check.

3. Start the engine.

➡**Switching modes is not possible while the engine is running.**

4. If the LED or malfunction indicator lamp illuminates, there is a fault in the system.

5. Turn the mode selector screw fully clockwise. Wait 2 seconds, then turn the screw fully counterclockwise.

6. The diagnostic trouble codes will now be read from the ECM memory. They will appear as flashes of the malfunction indicator lamp, or the ECM's LED.

7. After all codes have been read, turn the mode selector screw fully clockwise to erase the codes.

➡**Turn the mode adjusting screw to the fully counterclockwise position whenever the vehicle is in use.**

8. Turn the ignition **OFF**.

➡**When the ignition switch is turnedOFF during diagnosis, power to the ECM will drop after approximately 5 seconds. The diagnosis will automatically return to Mode 1 at this time.**

CLEARING CODES

With Scan Tool

Control module reset procedures are a very important part of OBD II System diagnostics. This step should be done at the end of any fault code repair and at the end of any driveability repair.

Clearing codes can be performed by any of the methods below:
• Clear the control module memory with the Generic Scan Tool (GST)
• Clear the control module memory with the vehicle manufacturer's specific tester

➡**The MIL will may also be de-activated for some codes if the vehicle completes three consecutive trips without a fault detected with vehicle conditions similar to those present during the fault.**

Without Scan Tool

The easiest way to clear trouble codes without a scan tool is to turn the mode selector screw fully clockwise after all codes have been read.

➡**Turn the mode adjusting screw to the fully counterclockwise position whenever the vehicle is in use.**

Codes may also be erased by turning the ignition **OFF** and remove the negative battery cable for at least 1 minute. However, removing the negative battery cable may cause other systems in the vehicle to lose their memory. Prior to removing the cable, ensure you have the proper reset codes for radios and alarms.

DIAGNOSTIC TROUBLE CODES

P0000 No Failures
P0100 Mass or Volume Air Flow Circuit Malfunction
P0101 Mass or Volume Air Flow Circuit Range/Performance Problem
P0102 Mass or Volume Air Flow Circuit Low Input
P0103 Mass or Volume Air Flow Circuit High Input
P0104 Mass or Volume Air Flow Circuit Intermittent
P0105 Manifold Absolute Pressure/Barometric Pressure Circuit Malfunction
P0106 Manifold Absolute Pressure/Barometric Pressure Circuit Range/Performance Problem
P0107 Manifold Absolute Pressure/Barometric Pressure Circuit Low Input
P0108 Manifold Absolute Pressure/Barometric Pressure Circuit High Input
P0109 Manifold Absolute Pressure/Barometric Pressure Circuit Intermittent
P0110 Intake Air Temperature Circuit Malfunction
P0111 Intake Air Temperature Circuit Range/Performance Problem
P0112 Intake Air Temperature Circuit Low Input
P0113 Intake Air Temperature Circuit High Input
P0114 Intake Air Temperature Circuit Intermittent

P0115 Engine Coolant Temperature Circuit Malfunction
P0116 Engine Coolant Temperature Circuit Range/Performance Problem
P0117 Engine Coolant Temperature Circuit Low Input
P0118 Engine Coolant Temperature Circuit High Input
P0119 Engine Coolant Temperature Circuit Intermittent
P0120 Throttle/Pedal Position Sensor/Switch "A" Circuit Malfunction
P0121 Throttle/Pedal Position Sensor/Switch "A" Circuit Range/Performance Problem
P0122 Throttle/Pedal Position Sensor/Switch "A" Circuit Low Input
P0123 Throttle/Pedal Position Sensor/Switch "A" Circuit High Input
P0124 Throttle/Pedal Position Sensor/Switch "A" Circuit Intermittent
P0125 Insufficient Coolant Temperature For Closed Loop Fuel Control
P0126 Insufficient Coolant Temperature For Stable Operation
P0130 O_2 Circuit Malfunction (Bank no. 1 Sensor no. 1)
P0131 O_2 Sensor Circuit Low Voltage (Bank no. 1 Sensor no. 1)
P0132 O_2 Sensor Circuit High Voltage (Bank no. 1 Sensor no. 1)
P0133 O_2 Sensor Circuit Slow Response (Bank no. 1 Sensor no. 1)
P0134 O_2 Sensor Circuit No Activity Detected (Bank no. 1 Sensor no. 1)
P0135 O_2 Sensor Heater Circuit Malfunction (Bank no. 1 Sensor no. 1)
P0136 O_2 Sensor Circuit Malfunction (Bank no. 1 Sensor no. 2)
P0137 O_2 Sensor Circuit Low Voltage (Bank no. 1 Sensor no. 2)
P0138 O_2 Sensor Circuit High Voltage (Bank no. 1 Sensor no. 2)
P0139 O_2 Sensor Circuit Slow Response (Bank no. 1 Sensor no. 2)
P0140 O_2 Sensor Circuit No Activity Detected (Bank no. 1 Sensor no. 2)
P0141 O_2 Sensor Heater Circuit Malfunction (Bank no. 1 Sensor no. 2)
P0142 O_2 Sensor Circuit Malfunction (Bank no. 1 Sensor no. 3)
P0143 O_2 Sensor Circuit Low Voltage (Bank no. 1 Sensor no. 3)
P0144 O_2 Sensor Circuit High Voltage (Bank no. 1 Sensor no. 3)
P0145 O_2 Sensor Circuit Slow Response (Bank no. 1 Sensor no. 3)
P0146 O_2 Sensor Circuit No Activity Detected (Bank no. 1 Sensor no. 3)
P0147 O_2 Sensor Heater Circuit Malfunction (Bank no. 1 Sensor no. 3)
P0150 O_2 Sensor Circuit Malfunction (Bank no. 2 Sensor no. 1)
P0151 O_2 Sensor Circuit Low Voltage (Bank no. 2 Sensor no. 1)
P0152 O_2 Sensor Circuit High Voltage (Bank no. 2 Sensor no. 1)
P0153 O_2 Sensor Circuit Slow Response (Bank no. 2 Sensor no. 1)
P0154 O_2 Sensor Circuit No Activity Detected (Bank no. 2 Sensor no. 1)
P0155 O_2 Sensor Heater Circuit Malfunction (Bank no. 2 Sensor no. 1)
P0156 O_2 Sensor Circuit Malfunction (Bank no. 2 Sensor no. 2)
P0157 O_2 Sensor Circuit Low Voltage (Bank no. 2 Sensor no. 2)
P0158 O_2 Sensor Circuit High Voltage (Bank no. 2 Sensor no. 2)
P0159 O_2 Sensor Circuit Slow Response (Bank no. 2 Sensor no. 2)
P0160 O_2 Sensor Circuit No Activity Detected (Bank no. 2 Sensor no. 2)

P0161 O_2 Sensor Heater Circuit Malfunction (Bank no. 2 Sensor no. 2)
P0162 O_2 Sensor Circuit Malfunction (Bank no. 2 Sensor no. 3)
P0163 O_2 Sensor Circuit Low Voltage (Bank no. 2 Sensor no. 3)
P0164 O_2 Sensor Circuit High Voltage (Bank no. 2 Sensor no. 3)
P0165 O_2 Sensor Circuit Slow Response (Bank no. 2 Sensor no. 3)
P0166 O_2 Sensor Circuit No Activity Detected (Bank no. 2 Sensor no. 3)
P0167 O_2 Sensor Heater Circuit Malfunction (Bank no. 2 Sensor no. 3)
P0170 Fuel Trim Malfunction (Bank no. 1)
P0171 System Too Lean (Bank no. 1)
P0172 System Too Rich (Bank no. 1)
P0173 Fuel Trim Malfunction (Bank no. 2)
P0174 System Too Lean (Bank no. 2)
P0175 System Too Rich (Bank no. 2)
P0176 Fuel Composition Sensor Circuit Malfunction
P0177 Fuel Composition Sensor Circuit Range/Performance
P0178 Fuel Composition Sensor Circuit Low Input
P0179 Fuel Composition Sensor Circuit High Input
P0180 Fuel Temperature Sensor "A" Circuit Malfunction
P0181 Fuel Temperature Sensor "A" Circuit Range/Performance
P0182 Fuel Temperature Sensor "A" Circuit Low Input
P0183 Fuel Temperature Sensor "A" Circuit High Input
P0184 Fuel Temperature Sensor "A" Circuit Intermittent
P0185 Fuel Temperature Sensor "B" Circuit Malfunction
P0186 Fuel Temperature Sensor "B" Circuit Range/Performance
P0187 Fuel Temperature Sensor "B" Circuit Low Input
P0188 Fuel Temperature Sensor "B" Circuit High Input
P0189 Fuel Temperature Sensor "B" Circuit Intermittent
P0190 Fuel Rail Pressure Sensor Circuit Malfunction
P0191 Fuel Rail Pressure Sensor Circuit Range/Performance
P0192 Fuel Rail Pressure Sensor Circuit Low Input
P0193 Fuel Rail Pressure Sensor Circuit High Input
P0194 Fuel Rail Pressure Sensor Circuit Intermittent
P0195 Engine Oil Temperature Sensor Malfunction
P0196 Engine Oil Temperature Sensor Range/Performance
P0197 Engine Oil Temperature Sensor Low
P0198 Engine Oil Temperature Sensor High
P0199 Engine Oil Temperature Sensor Intermittent
P0200 Injector Circuit Malfunction
P0201 Injector Circuit Malfunction—Cylinder no. 1
P0202 Injector Circuit Malfunction—Cylinder no. 2
P0203 Injector Circuit Malfunction—Cylinder no. 3
P0204 Injector Circuit Malfunction—Cylinder no. 4
P0205 Injector Circuit Malfunction—Cylinder no. 5
P0206 Injector Circuit Malfunction—Cylinder no. 6
P0207 Injector Circuit Malfunction—Cylinder no. 7
P0208 Injector Circuit Malfunction—Cylinder no. 8
P0209 Injector Circuit Malfunction—Cylinder no. 9
P0210 Injector Circuit Malfunction—Cylinder no. 10
P0211 Injector Circuit Malfunction—Cylinder no. 11
P0212 Injector Circuit Malfunction—Cylinder no. 12
P0213 Cold Start Injector no. 1 Malfunction
P0214 Cold Start Injector no. 2 Malfunction
P0215 Engine Shutoff Solenoid Malfunction
P0216 Injection Timing Control Circuit Malfunction
P0217 Engine Over Temperature Condition
P0218 Transmission Over Temperature Condition
P0219 Engine Over Speed Condition

P0220 Throttle/Pedal Position Sensor/Switch "B" Circuit Malfunction

P0221 Throttle/Pedal Position Sensor/Switch "B" Circuit Range/Performance Problem

P0222 Throttle/Pedal Position Sensor/Switch "B" Circuit Low Input

P0223 Throttle/Pedal Position Sensor/Switch "B" Circuit High Input

P0224 Throttle/Pedal Position Sensor/Switch "B" Circuit Intermittent

P0225 Throttle/Pedal Position Sensor/Switch "C" Circuit Malfunction

P0226 Throttle/Pedal Position Sensor/Switch "C" Circuit Range/Performance Problem

P0227 Throttle/Pedal Position Sensor/Switch "C" Circuit Low Input

P0228 Throttle/Pedal Position Sensor/Switch "C" Circuit High Input

P0229 Throttle/Pedal Position Sensor/Switch "C" Circuit Intermittent

P0230 Fuel Pump Primary Circuit Malfunction
P0231 Fuel Pump Secondary Circuit Low
P0232 Fuel Pump Secondary Circuit High
P0233 Fuel Pump Secondary Circuit Intermittent
P0234 Engine Over Boost Condition
P0261 Cylinder no. 1 Injector Circuit Low
P0262 Cylinder no. 1 Injector Circuit High
P0263 Cylinder no. 1 Contribution/Balance Fault
P0264 Cylinder no. 2 Injector Circuit Low
P0265 Cylinder no. 2 Injector Circuit High
P0266 Cylinder no. 2 Contribution/Balance Fault
P0267 Cylinder no. 3 Injector Circuit Low
P0268 Cylinder no. 3 Injector Circuit High
P0269 Cylinder no. 3 Contribution/Balance Fault
P0270 Cylinder no. 4 Injector Circuit Low
P0271 Cylinder no. 4 Injector Circuit High
P0272 Cylinder no. 4 Contribution/Balance Fault
P0273 Cylinder no. 5 Injector Circuit Low
P0274 Cylinder no. 5 Injector Circuit High
P0275 Cylinder no. 5 Contribution/Balance Fault
P0276 Cylinder no. 6 Injector Circuit Low
P0277 Cylinder no. 6 Injector Circuit High
P0278 Cylinder no. 6 Contribution/Balance Fault
P0279 Cylinder no. 7 Injector Circuit Low
P0280 Cylinder no. 7 Injector Circuit High
P0281 Cylinder no. 7 Contribution/Balance Fault
P0282 Cylinder no. 8 Injector Circuit Low
P0283 Cylinder no. 8 Injector Circuit High
P0284 Cylinder no. 8 Contribution/Balance Fault
P0285 Cylinder no. 9 Injector Circuit Low
P0286 Cylinder no. 9 Injector Circuit High
P0287 Cylinder no. 9 Contribution/Balance Fault
P0288 Cylinder no. 10 Injector Circuit Low
P0289 Cylinder no. 10 Injector Circuit High
P0290 Cylinder no. 10 Contribution/Balance Fault
P0291 Cylinder no. 11 Injector Circuit Low
P0292 Cylinder no. 11 Injector Circuit High
P0293 Cylinder no. 11 Contribution/Balance Fault
P0294 Cylinder no. 12 Injector Circuit Low
P0295 Cylinder no. 12 Injector Circuit High
P0296 Cylinder no. 12 Contribution/Balance Fault
P0300 Random/Multiple Cylinder Misfire Detected

P0301 Cylinder no. 1—Misfire Detected
P0302 Cylinder no. 2—Misfire Detected
P0303 Cylinder no. 3—Misfire Detected
P0304 Cylinder no. 4—Misfire Detected
P0305 Cylinder no. 5—Misfire Detected
P0306 Cylinder no. 6—Misfire Detected
P0307 Cylinder no. 7—Misfire Detected
P0308 Cylinder no. 8—Misfire Detected
P0309 Cylinder no. 9—Misfire Detected
P0310 Cylinder no. 10—Misfire Detected
P0311 Cylinder no. 11—Misfire Detected
P0312 Cylinder no. 12—Misfire Detected
P0320 Ignition/Distributor Engine Speed Input Circuit Malfunction

P0321 Ignition/Distributor Engine Speed Input Circuit Range/Performance

P0322 Ignition/Distributor Engine Speed Input Circuit No Signal

P0323 Ignition/Distributor Engine Speed Input Circuit Intermittent

P0325 Knock Sensor no. 1—Circuit Malfunction (Bank no. 1 or Single Sensor)

P0326 Knock Sensor no. 1—Circuit Range/Performance (Bank no. 1 or Single Sensor)

P0327 Knock Sensor no. 1—Circuit Low Input (Bank no. 1 or Single Sensor)

P0328 Knock Sensor no. 1—Circuit High Input (Bank no. 1 or Single Sensor)

P0329 Knock Sensor no. 1—Circuit Input Intermittent (Bank no. 1 or Single Sensor)

P0330 Knock Sensor no. 2—Circuit Malfunction (Bank no. 2)

P0331 Knock Sensor no. 2—Circuit Range/Performance (Bank no. 2)

P0332 Knock Sensor no. 2—Circuit Low Input (Bank no. 2)
P0333 Knock Sensor no. 2—Circuit High Input (Bank no. 2)
P0334 Knock Sensor no. 2—Circuit Input Intermittent (Bank no. 2)

P0335 Crankshaft Position Sensor "A" Circuit Malfunction
P0336 Crankshaft Position Sensor "A" Circuit Range/Performance

P0337 Crankshaft Position Sensor "A" Circuit Low Input
P0338 Crankshaft Position Sensor "A" Circuit High Input
P0339 Crankshaft Position Sensor "A" Circuit Intermittent
P0340 Camshaft Position Sensor Circuit Malfunction
P0341 Camshaft Position Sensor Circuit Range/Performance
P0342 Camshaft Position Sensor Circuit Low Input
P0343 Camshaft Position Sensor Circuit High Input
P0344 Camshaft Position Sensor Circuit Intermittent
P0350 Ignition Coil Primary/Secondary Circuit Malfunction
P0351 Ignition Coil "A" Primary/Secondary Circuit Malfunction
P0352 Ignition Coil "B" Primary/Secondary Circuit Malfunction
P0353 Ignition Coil "C" Primary/Secondary Circuit Malfunction
P0354 Ignition Coil "D" Primary/Secondary Circuit Malfunction
P0355 Ignition Coil "E" Primary/Secondary Circuit Malfunction
P0356 Ignition Coil "F" Primary/Secondary Circuit Malfunction
P0357 Ignition Coil "G" Primary/Secondary Circuit Malfunction
P0358 Ignition Coil "H" Primary/Secondary Circuit Malfunction
P0359 Ignition Coil "I" Primary/Secondary Circuit Malfunction
P0360 Ignition Coil "J" Primary/Secondary Circuit Malfunction
P0361 Ignition Coil "K" Primary/Secondary Circuit Malfunction
P0362 Ignition Coil "L" Primary/Secondary Circuit Malfunction
P0370 Timing Reference High Resolution Signal "A" Malfunction

P0371 Timing Reference High Resolution Signal "A" Too Many Pulses

P0372 Timing Reference High Resolution Signal "A" Too Few Pulses

P0373 Timing Reference High Resolution Signal "A" Intermittent/Erratic Pulses

P0374 Timing Reference High Resolution Signal "A" No Pulses

P0375 Timing Reference High Resolution Signal "B" Malfunction

P0376 Timing Reference High Resolution Signal "B" Too Many Pulses

P0377 Timing Reference High Resolution Signal "B" Too Few Pulses

P0378 Timing Reference High Resolution Signal "B" Intermittent/Erratic Pulses

P0379 Timing Reference High Resolution Signal "B" No Pulses

P0380 Glow Plug/Heater Circuit "A" Malfunction

P0381 Glow Plug/Heater Indicator Circuit Malfunction

P0382 Glow Plug/Heater Circuit "B" Malfunction

P0385 Crankshaft Position Sensor "B" Circuit Malfunction

P0386 Crankshaft Position Sensor "B" Circuit Range/Performance

P0387 Crankshaft Position Sensor "B" Circuit Low Input

P0388 Crankshaft Position Sensor "B" Circuit High Input

P0389 Crankshaft Position Sensor "B" Circuit Intermittent

P0400 Exhaust Gas Recirculation Flow Malfunction

P0401 Exhaust Gas Recirculation Flow Insufficient Detected

P0402 Exhaust Gas Recirculation Flow Excessive Detected

P0403 Exhaust Gas Recirculation Circuit Malfunction

P0404 Exhaust Gas Recirculation Circuit Range/Performance

P0405 Exhaust Gas Recirculation Sensor "A" Circuit Low

P0406 Exhaust Gas Recirculation Sensor "A" Circuit High

P0407 Exhaust Gas Recirculation Sensor "B" Circuit Low

P0408 Exhaust Gas Recirculation Sensor "B" Circuit High

P0410 Secondary Air Injection System Malfunction

P0411 Secondary Air Injection System Incorrect Flow Detected

P0412 Secondary Air Injection System Switching Valve "A" Circuit Malfunction

P0413 Secondary Air Injection System Switching Valve "A" Circuit Open

P0414 Secondary Air Injection System Switching Valve "A" Circuit Shorted

P0415 Secondary Air Injection System Switching Valve "B" Circuit Malfunction

P0416 Secondary Air Injection System Switching Valve "B" Circuit Open

P0417 Secondary Air Injection System Switching Valve "B" Circuit Shorted

P0418 Secondary Air Injection System Relay "A" Circuit Malfunction

P0419 Secondary Air Injection System Relay "B" Circuit Malfunction

P0420 Catalyst System Efficiency Below Threshold (Bank no. 1)

P0421 Warm Up Catalyst Efficiency Below Threshold (Bank no. 1)

P0422 Main Catalyst Efficiency Below Threshold (Bank no. 1)

P0423 Heated Catalyst Efficiency Below Threshold (Bank no. 1)

P0424 Heated Catalyst Temperature Below Threshold (Bank no. 1)

P0430 Catalyst System Efficiency Below Threshold (Bank no. 2)

P0431 Warm Up Catalyst Efficiency Below Threshold (Bank no. 2)

P0432 Main Catalyst Efficiency Below Threshold (Bank no. 2)

P0433 Heated Catalyst Efficiency Below Threshold (Bank no. 2)

P0434 Heated Catalyst Temperature Below Threshold (Bank no. 2)

P0440 Evaporative Emission Control System Malfunction

P0441 Evaporative Emission Control System Incorrect Purge Flow

P0442 Evaporative Emission Control System Leak Detected (Small Leak)

P0443 Evaporative Emission Control System Purge Control Valve Circuit Malfunction

P0444 Evaporative Emission Control System Purge Control Valve Circuit Open

P0445 Evaporative Emission Control System Purge Control Valve Circuit Shorted

P0446 Evaporative Emission Control System Vent Control Circuit Malfunction

P0447 Evaporative Emission Control System Vent Control Circuit Open

P0448 Evaporative Emission Control System Vent Control Circuit Shorted

P0449 Evaporative Emission Control System Vent Valve/Solenoid Circuit Malfunction

P0450 Evaporative Emission Control System Pressure Sensor Malfunction

P0451 Evaporative Emission Control System Pressure Sensor Range/Performance

P0452 Evaporative Emission Control System Pressure Sensor Low Input

P0453 Evaporative Emission Control System Pressure Sensor High Input

P0454 Evaporative Emission Control System Pressure Sensor Intermittent

P0455 Evaporative Emission Control System Leak Detected (Gross Leak)

P0460 Fuel Level Sensor Circuit Malfunction

P0461 Fuel Level Sensor Circuit Range/Performance

P0462 Fuel Level Sensor Circuit Low Input

P0463 Fuel Level Sensor Circuit High Input

P0464 Fuel Level Sensor Circuit Intermittent

P0465 Purge Flow Sensor Circuit Malfunction

P0466 Purge Flow Sensor Circuit Range/Performance

P0467 Purge Flow Sensor Circuit Low Input

P0468 Purge Flow Sensor Circuit High Input

P0469 Purge Flow Sensor Circuit Intermittent

P0470 Exhaust Pressure Sensor Malfunction

P0471 Exhaust Pressure Sensor Range/Performance

P0472 Exhaust Pressure Sensor Low

P0473 Exhaust Pressure Sensor High

P0474 Exhaust Pressure Sensor Intermittent

P0475 Exhaust Pressure Control Valve Malfunction

P0476 Exhaust Pressure Control Valve Range/Performance

P0477 Exhaust Pressure Control Valve Low

P0478 Exhaust Pressure Control Valve High

P0479 Exhaust Pressure Control Valve Intermittent

P0480 Cooling Fan no. 1 Control Circuit Malfunction

P0481 Cooling Fan no. 2 Control Circuit Malfunction

P0482 Cooling Fan no. 3 Control Circuit Malfunction

P0483 Cooling Fan Rationality Check Malfunction

P0484 Cooling Fan Circuit Over Current

P0485 Cooling Fan Power/Ground Circuit Malfunction

P0500 Vehicle Speed Sensor Malfunction

P0501 Vehicle Speed Sensor Range/Performance

P0502 Vehicle Speed Sensor Circuit Low Input

P0503 Vehicle Speed Sensor Intermittent/Erratic/High
P0505 Idle Control System Malfunction
P0506 Idle Control System RPM Lower Than Expected
P0507 Idle Control System RPM Higher Than Expected
P0510 Closed Throttle Position Switch Malfunction
P0520 Engine Oil Pressure Sensor/Switch Circuit Malfunction
P0521 Engine Oil Pressure Sensor/Switch Range/Performance
P0522 Engine Oil Pressure Sensor/Switch Low Voltage
P0523 Engine Oil Pressure Sensor/Switch High Voltage
P0530 A/C Refrigerant Pressure Sensor Circuit Malfunction
P0531 A/C Refrigerant Pressure Sensor Circuit Range/Performance
P0532 A/C Refrigerant Pressure Sensor Circuit Low Input
P0533 A/C Refrigerant Pressure Sensor Circuit High Input
P0534 A/C Refrigerant Charge Loss
P0550 Power Steering Pressure Sensor Circuit Malfunction
P0551 Power Steering Pressure Sensor Circuit Range/Performance
P0552 Power Steering Pressure Sensor Circuit Low Input
P0553 Power Steering Pressure Sensor Circuit High Input
P0554 Power Steering Pressure Sensor Circuit Intermittent
P0560 System Voltage Malfunction
P0561 System Voltage Unstable
P0562 System Voltage Low
P0563 System Voltage High
P0565 Cruise Control On Signal Malfunction
P0566 Cruise Control Off Signal Malfunction
P0567 Cruise Control Resume Signal Malfunction
P0568 Cruise Control Set Signal Malfunction
P0569 Cruise Control Coast Signal Malfunction
P0570 Cruise Control Accel Signal Malfunction
P0571 Cruise Control/Brake Switch "A" Circuit Malfunction
P0572 Cruise Control/Brake Switch "A" Circuit Low
P0573 Cruise Control/Brake Switch "A" Circuit High
P0574 Through P0580 Reserved for Cruise Codes
P0600 Serial Communication Link Malfunction
P0601 Internal Control Module Memory Check Sum Error
P0602 Control Module Programming Error
P0603 Internal Control Module Keep Alive Memory (KAM) Error
P0604 Internal Control Module Random Access Memory (RAM) Error
P0605 Internal Control Module Read Only Memory (ROM) Error
P0606 PCM Processor Fault
P0608 Control Module VSS Output "A" Malfunction
P0609 Control Module VSS Output "B" Malfunction
P0620 Generator Control Circuit Malfunction
P0621 Generator Lamp "L" Control Circuit Malfunction
P0622 Generator Field "F" Control Circuit Malfunction
P0650 Malfunction Indicator Lamp (MIL) Control Circuit Malfunction
P0654 Engine RPM Output Circuit Malfunction
P0655 Engine Hot Lamp Output Control Circuit Malfunction
P0656 Fuel Level Output Circuit Malfunction
P0700 Transmission Control System Malfunction
P0701 Transmission Control System Range/Performance
P0702 Transmission Control System Electrical
P0703 Torque Converter/Brake Switch "B" Circuit Malfunction
P0704 Clutch Switch Input Circuit Malfunction
P0705 Transmission Range Sensor Circuit Malfunction (PRNDL Input)
P0706 Transmission Range Sensor Circuit Range/Performance
P0707 Transmission Range Sensor Circuit Low Input

P0708 Transmission Range Sensor Circuit High Input
P0709 Transmission Range Sensor Circuit Intermittent
P0710 Transmission Fluid Temperature Sensor Circuit Malfunction
P0711 Transmission Fluid Temperature Sensor Circuit Range/Performance
P0712 Transmission Fluid Temperature Sensor Circuit Low Input
P0713 Transmission Fluid Temperature Sensor Circuit High Input
P0714 Transmission Fluid Temperature Sensor Circuit Intermittent
P0715 Input/Turbine Speed Sensor Circuit Malfunction
P0716 Input/Turbine Speed Sensor Circuit Range/Performance
P0717 Input/Turbine Speed Sensor Circuit No Signal
P0718 Input/Turbine Speed Sensor Circuit Intermittent
P0719 Torque Converter/Brake Switch "B" Circuit Low
P0720 Output Speed Sensor Circuit Malfunction
P0721 Output Speed Sensor Circuit Range/Performance
P0722 Output Speed Sensor Circuit No Signal
P0723 Output Speed Sensor Circuit Intermittent
P0724 Torque Converter/Brake Switch "B" Circuit High
P0725 Engine Speed Input Circuit Malfunction
P0726 Engine Speed Input Circuit Range/Performance
P0727 Engine Speed Input Circuit No Signal
P0728 Engine Speed Input Circuit Intermittent
P0730 Incorrect Gear Ratio
P0731 Gear no. 1 Incorrect Ratio
P0732 Gear no. 2 Incorrect Ratio
P0733 Gear no. 3 Incorrect Ratio
P0734 Gear no. 4 Incorrect Ratio
P0735 Gear no. 5 Incorrect Ratio
P0736 Reverse Incorrect Ratio
P0740 Torque Converter Clutch Circuit Malfunction
P0741 Torque Converter Clutch Circuit Performance or Stuck Off
P0742 Torque Converter Clutch Circuit Stuck On
P0743 Torque Converter Clutch Circuit Electrical
P0744 Torque Converter Clutch Circuit Intermittent
P0745 Pressure Control Solenoid Malfunction
P0746 Pressure Control Solenoid Performance or Stuck Off
P0747 Pressure Control Solenoid Stuck On
P0748 Pressure Control Solenoid Electrical
P0749 Pressure Control Solenoid Intermittent
P0750 Shift Solenoid "A" Malfunction
P0751 Shift Solenoid "A" Performance or Stuck Off
P0752 Shift Solenoid "A" Stuck On
P0753 Shift Solenoid "A" Electrical
P0754 Shift Solenoid "A" Intermittent
P0755 Shift Solenoid "B" Malfunction
P0756 Shift Solenoid "B" Performance or Stuck Off
P0757 Shift Solenoid "B" Stuck On
P0758 Shift Solenoid "B" Electrical
P0759 Shift Solenoid "B" Intermittent
P0760 Shift Solenoid "C" Malfunction
P0761 Shift Solenoid "C" Performance Or Stuck Off
P0762 Shift Solenoid "C" Stuck On
P0763 Shift Solenoid "C" Electrical
P0764 Shift Solenoid "C" Intermittent
P0765 Shift Solenoid "D" Malfunction
P0766 Shift Solenoid "D" Performance Or Stuck Off
P0767 Shift Solenoid "D" Stuck On

P0768 Shift Solenoid "D" Electrical
P0769 Shift Solenoid "D" Intermittent
P0770 Shift Solenoid "E" Malfunction
P0771 Shift Solenoid "E" Performance Or Stuck Off
P0772 Shift Solenoid "E" Stuck On
P0773 Shift Solenoid "E" Electrical
P0774 Shift Solenoid "E" Intermittent
P0780 Shift Malfunction
P0781 1–2 Shift Malfunction
P0782 2–3 Shift Malfunction
P0783 3–4 Shift Malfunction
P0784 4–5 Shift Malfunction
P0785 Shift/Timing Solenoid Malfunction
P0786 Shift/Timing Solenoid Range/Performance
P0787 Shift/Timing Solenoid Low
P0788 Shift/Timing Solenoid High
P0789 Shift/Timing Solenoid Intermittent
P0790 Normal/Performance Switch Circuit Malfunction
P0801 Reverse Inhibit Control Circuit Malfunction
P0803 1–4 Upshift (Skip Shift) Solenoid Control Circuit Malfunction
P0804 1–4 Upshift (Skip Shift) Lamp Control Circuit Malfunction
P1120 Secondary Throttle Position Sensor Circuit Fault
P1125 Tandem Throttle Position Sensor Circuit Fault
P1210 Traction Control System Signal Fault
P1220 Fuel Pump Control Module Fault
P1320 Ignition Control Signal Fault
P1336 Crankshaft Position Sensor Circuit Fault
P1400 EGR/EVAP Control Solenoid Circuit Fault
P1401 EGR Temperature Sensor Circuit Fault
P1443 EVAP Canister Control Vacuum Switch Circuit Fault
P1445 EVAP Purge Volume Control Valve Circuit Fault
P1605 TCM AT Diagnosis Communication Line Fault
P1705 Throttle Position Sensor (Switch) Circuit Fault
P1760 Overrun Clutch Solenoid Valve Circuit Fault
P1900 Cooling Fan Control Circuit Fault

TROUBLE CODE EQUIVALENT LIST

0505 No Self Diagnostic Failure Indicated
0102 Mass or Volume Air Flow Circuit Malfunction
0401 Intake Air Temperature Circuit Malfunction
0103 Engine Coolant Temperature Circuit Malfunction
0403 Throttle/Pedal Position Sensor/Switch "A" Circuit Malfunction
0908 Insufficient Coolant Temperature For Closed Loop Fuel Control
0303 O_2 Circuit Malfunction
0307 Closed Loop Control
0901 O_2 Sensor Heater Circuit Malfunction (Bank no. 1 Sensor no. 1)
0707 O_2 Sensor Circuit Malfunction (Bank no. 1 Sensor no. 2)
0902 O_2 Sensor Heater Circuit Malfunction (Bank no. 1 Sensor no. 2)
0115 System Too Lean (Bank no. 1)
0114 System Too Rich (Bank no. 1)
0701 Random/Multiple Cylinder Misfire Detected
0608 Cylinder no. 1—Misfire Detected
0607 Cylinder no. 2—Misfire Detected
0606 Cylinder no. 3—Misfire Detected
0605 Cylinder no. 4—Misfire Detected

0304 Knock Sensor no. 1—Circuit Malfunction (Bank no. 1 or Single Sensor)
0802 Crankshaft Position Sensor "A" Circuit Malfunction
0101 Camshaft Position Sensor Circuit Malfunction
0302 Exhaust Gas Recirculation Flow Malfunction
0306 Exhaust Gas Recirculation Flow Excessive Detected
0702 Catalyst System Efficiency Below Threshold (Bank no. 1)
0104 Vehicle Speed Sensor Malfunction
0205 Idle Control System Malfunction
0301 Internal Control Module Read Only Memory (ROM) Error
1003 Transmission Range Sensor Circuit Malfunction (PRNDL Input)
1101 Inhibitor Switch Circuit
1208 Transmission Fluid Temperature Sensor Circuit Malfunction
1102 Output Speed Sensor Circuit Malfunction
1207 Engine Speed Input Circuit Malfunction
1103 Gear no. 1 Incorrect Ratio
1104 Gear no. 2 Incorrect Ratio
1105 Gear no. 3 Incorrect Ratio
1106 Gear no. 4 Incorrect Ratio
1204 Torque Converter Clutch Circuit Malfunction
1205 Pressure Control Solenoid Malfunction
1108 Shift Solenoid "A" Malfunction
1201 Shift Solenoid "B" Malfunction
0201 Ignition Control Signal Fault
0905 Crankshaft Position Sensor Circuit Fault
1005 EGR/EVAP Control Solenoid Circuit Fault
0305 EGR Temperature Sensor Circuit Fault
0804 TCM AT Diagnosis Communication Line Fault
1206 Throttle Position Sensor (Switch) Circuit Fault
1203 Overrun Clutch Solenoid Valve Circuit Fault
1308 Cooling Fan Control Circuit Fault

Porsche

READING CODES

Reading the control module memory is on of the first steps in OBD II system diagnostics. This step should be initially performed to determine the general nature of the fault. Subsequent readings will determine if the fault has been cleared.

Reading codes can be performed by any of the methods below:

• Read the control module memory with the Generic Scan Tool (GST)

• Read the control module memory with the vehicle manufacturer's specific tester

To read the fault codes, connect the scan tool or tester according to the manufacturer's instructions. Follow the manufacturer's specified procedure for reading the codes.

CLEARING CODES

Control module reset procedures are a very important part of OBD II System diagnostics. This step should be done at the end of any fault code repair and at the end of any driveability repair.

Clearing codes can be performed by any of the methods below:

• Clear the control module memory with the Generic Scan Tool (GST)

• Clear the control module memory with the vehicle manufacturer's specific tester

• Turn the ignition**OFF** and remove the negative battery cable for at least 1 minute.

Removing the negative battery cable may cause other systems in the vehicle to lose their memory. Prior to removing the cable, ensure you have the proper reset codes for radios and alarms.

➡**The MIL will may also be de-activated for some codes if the vehicle completes three consecutive trips without a fault detected with vehicle conditions similar to those present during the fault.**

DIAGNOSTIC TROUBLE CODES

P0000 No Failures
P0100 Mass or Volume Air Flow Circuit Malfunction
P0101 Mass or Volume Air Flow Circuit Range/Performance Problem
P0102 Mass or Volume Air Flow Circuit Low Input
P0103 Mass or Volume Air Flow Circuit High Input
P0104 Mass or Volume Air Flow Circuit Intermittent
P0105 Manifold Absolute Pressure/Barometric Pressure Circuit Malfunction
P0106 Manifold Absolute Pressure/Barometric Pressure Circuit Range/Performance Problem
P0107 Manifold Absolute Pressure/Barometric Pressure Circuit Low Input
P0108 Manifold Absolute Pressure/Barometric Pressure Circuit High Input
P0109 Manifold Absolute Pressure/Barometric Pressure Circuit Intermittent
P0110 Intake Air Temperature Circuit Malfunction
P0111 Intake Air Temperature Circuit Range/Performance Problem
P0112 Intake Air Temperature Circuit Low Input
P0113 Intake Air Temperature Circuit High Input
P0114 Intake Air Temperature Circuit Intermittent
P0115 Engine Coolant Temperature Circuit Malfunction
P0116 Engine Coolant Temperature Circuit Range/Performance Problem
P0117 Engine Coolant Temperature Circuit Low Input
P0118 Engine Coolant Temperature Circuit High Input
P0119 Engine Coolant Temperature Circuit Intermittent
P0120 Throttle/Pedal Position Sensor/Switch "A" Circuit Malfunction
P0121 Throttle/Pedal Position Sensor/Switch "A" Circuit Range/Performance Problem
P0122 Throttle/Pedal Position Sensor/Switch "A" Circuit Low Input
P0123 Throttle/Pedal Position Sensor/Switch "A" Circuit High Input
P0124 Throttle/Pedal Position Sensor/Switch "A" Circuit Intermittent
P0125 Insufficient Coolant Temperature For Closed Loop Fuel Control
P0126 Insufficient Coolant Temperature For Stable Operation
P0130 O_2 Circuit Malfunction (Bank no. 1 Sensor no. 1)
P0131 O_2 Sensor Circuit Low Voltage (Bank no. 1 Sensor no. 1)
P0132 O_2 Sensor Circuit High Voltage (Bank no. 1 Sensor no. 1)
P0133 O_2 Sensor Circuit Slow Response (Bank no. 1 Sensor no. 1)
P0134 O_2 Sensor Circuit No Activity Detected (Bank no. 1 Sensor no. 1)

P0135 O_2 Sensor Heater Circuit Malfunction (Bank no. 1 Sensor no. 1)
P0136 O_2 Sensor Circuit Malfunction (Bank no. 1 Sensor no. 2)
P0137 O_2 Sensor Circuit Low Voltage (Bank no. 1 Sensor no. 2)
P0138 O_2 Sensor Circuit High Voltage (Bank no. 1 Sensor no. 2)
P0139 O_2 Sensor Circuit Slow Response (Bank no. 1 Sensor no. 2)
P0140 O_2 Sensor Circuit No Activity Detected (Bank no. 1 Sensor no. 2)
P0141 O_2 Sensor Heater Circuit Malfunction (Bank no. 1 Sensor no. 2)
P0142 O_2 Sensor Circuit Malfunction (Bank no. 1 Sensor no. 3)
P0143 O_2 Sensor Circuit Low Voltage (Bank no. 1 Sensor no. 3)
P0144 O_2 Sensor Circuit High Voltage (Bank no. 1 Sensor no. 3)
P0145 O_2 Sensor Circuit Slow Response (Bank no. 1 Sensor no. 3)
P0146 O_2 Sensor Circuit No Activity Detected (Bank no. 1 Sensor no. 3)
P0147 O_2 Sensor Heater Circuit Malfunction (Bank no. 1 Sensor no. 3)
P0150 O_2 Sensor Circuit Malfunction (Bank no. 2 Sensor no. 1)
P0151 O_2 Sensor Circuit Low Voltage (Bank no. 2 Sensor no. 1)
P0152 O_2 Sensor Circuit High Voltage (Bank no. 2 Sensor no. 1)
P0153 O_2 Sensor Circuit Slow Response (Bank no. 2 Sensor no. 1)
P0154 O_2 Sensor Circuit No Activity Detected (Bank no. 2 Sensor no. 1)
P0155 O_2 Sensor Heater Circuit Malfunction (Bank no. 2 Sensor no. 1)
P0156 O_2 Sensor Circuit Malfunction (Bank no. 2 Sensor no. 2)
P0157 O_2 Sensor Circuit Low Voltage (Bank no. 2 Sensor no. 2)
P0158 O_2 Sensor Circuit High Voltage (Bank no. 2 Sensor no. 2)
P0159 O_2 Sensor Circuit Slow Response (Bank no. 2 Sensor no. 2)
P0160 O_2 Sensor Circuit No Activity Detected (Bank no. 2 Sensor no. 2)
P0161 O_2 Sensor Heater Circuit Malfunction (Bank no. 2 Sensor no. 2)
P0162 O_2 Sensor Circuit Malfunction (Bank no. 2 Sensor no. 3)
P0163 O_2 Sensor Circuit Low Voltage (Bank no. 2 Sensor no. 3)
P0164 O_2 Sensor Circuit High Voltage (Bank no. 2 Sensor no. 3)
P0165 O_2 Sensor Circuit Slow Response (Bank no. 2 Sensor no. 3)
P0166 O_2 Sensor Circuit No Activity Detected (Bank no. 2 Sensor no. 3)
P0167 O_2 Sensor Heater Circuit Malfunction (Bank no. 2 Sensor no. 3)
P0170 Fuel Trim Malfunction (Bank no. 1)
P0171 System Too Lean (Bank no. 1)
P0172 System Too Rich (Bank no. 1)
P0173 Fuel Trim Malfunction (Bank no. 2)
P0174 System Too Lean (Bank no. 2)
P0175 System Too Rich (Bank no. 2)
P0176 Fuel Composition Sensor Circuit Malfunction
P0177 Fuel Composition Sensor Circuit Range/Performance
P0178 Fuel Composition Sensor Circuit Low Input
P0179 Fuel Composition Sensor Circuit High Input
P0180 Fuel Temperature Sensor "A" Circuit Malfunction
P0181 Fuel Temperature Sensor "A" Circuit Range/Performance
P0182 Fuel Temperature Sensor "A" Circuit Low Input
P0183 Fuel Temperature Sensor "A" Circuit High Input
P0184 Fuel Temperature Sensor "A" Circuit Intermittent

P0185 Fuel Temperature Sensor "B" Circuit Malfunction
P0186 Fuel Temperature Sensor "B" Circuit Range/Performance
P0187 Fuel Temperature Sensor "B" Circuit Low Input
P0188 Fuel Temperature Sensor "B" Circuit High Input
P0189 Fuel Temperature Sensor "B" Circuit Intermittent
P0190 Fuel Rail Pressure Sensor Circuit Malfunction
P0191 Fuel Rail Pressure Sensor Circuit Range/Performance
P0192 Fuel Rail Pressure Sensor Circuit Low Input
P0193 Fuel Rail Pressure Sensor Circuit High Input
P0194 Fuel Rail Pressure Sensor Circuit Intermittent
P0195 Engine Oil Temperature Sensor Malfunction
P0196 Engine Oil Temperature Sensor Range/Performance
P0197 Engine Oil Temperature Sensor Low
P0198 Engine Oil Temperature Sensor High
P0199 Engine Oil Temperature Sensor Intermittent
P0200 Injector Circuit Malfunction
P0201 Injector Circuit Malfunction—Cylinder no. 1
P0202 Injector Circuit Malfunction—Cylinder no. 2
P0203 Injector Circuit Malfunction—Cylinder no. 3
P0204 Injector Circuit Malfunction—Cylinder no. 4
P0205 Injector Circuit Malfunction—Cylinder no. 5
P0206 Injector Circuit Malfunction—Cylinder no. 6
P0207 Injector Circuit Malfunction—Cylinder no. 7
P0208 Injector Circuit Malfunction—Cylinder no. 8
P0209 Injector Circuit Malfunction—Cylinder no. 9
P0210 Injector Circuit Malfunction—Cylinder no. 10
P0211 Injector Circuit Malfunction—Cylinder no. 11
P0212 Injector Circuit Malfunction—Cylinder no. 12
P0213 Cold Start Injector no. 1 Malfunction
P0214 Cold Start Injector no. 2 Malfunction
P0215 Engine Shutoff Solenoid Malfunction
P0216 Injection Timing Control Circuit Malfunction
P0217 Engine Over Temperature Condition
P0218 Transmission Over Temperature Condition
P0219 Engine Over Speed Condition
P0220 Throttle/Pedal Position Sensor/Switch "B" Circuit Malfunction
P0221 Throttle/Pedal Position Sensor/Switch "B" Circuit Range/Performance Problem
P0222 Throttle/Pedal Position Sensor/Switch "B" Circuit Low Input
P0223 Throttle/Pedal Position Sensor/Switch "B" Circuit High Input
P0224 Throttle/Pedal Position Sensor/Switch "B" Circuit Intermittent
P0225 Throttle/Pedal Position Sensor/Switch "C" Circuit Malfunction
P0226 Throttle/Pedal Position Sensor/Switch "C" Circuit Range/Performance Problem
P0227 Throttle/Pedal Position Sensor/Switch "C" Circuit Low Input
P0228 Throttle/Pedal Position Sensor/Switch "C" Circuit High Input
P0229 Throttle/Pedal Position Sensor/Switch "C" Circuit Intermittent
P0230 Fuel Pump Primary Circuit Malfunction
P0231 Fuel Pump Secondary Circuit Low
P0232 Fuel Pump Secondary Circuit High
P0233 Fuel Pump Secondary Circuit Intermittent
P0234 Engine Over Boost Condition
P0261 Cylinder no. 1 Injector Circuit Low
P0262 Cylinder no. 1 Injector Circuit High

P0263 Cylinder no. 1 Contribution/Balance Fault
P0264 Cylinder no. 2 Injector Circuit Low
P0265 Cylinder no. 2 Injector Circuit High
P0266 Cylinder no. 2 Contribution/Balance Fault
P0267 Cylinder no. 3 Injector Circuit Low
P0268 Cylinder no. 3 Injector Circuit High
P0269 Cylinder no. 3 Contribution/Balance Fault
P0270 Cylinder no. 4 Injector Circuit Low
P0271 Cylinder no. 4 Injector Circuit High
P0272 Cylinder no. 4 Contribution/Balance Fault
P0273 Cylinder no. 5 Injector Circuit Low
P0274 Cylinder no. 5 Injector Circuit High
P0275 Cylinder no. 5 Contribution/Balance Fault
P0276 Cylinder no. 6 Injector Circuit Low
P0277 Cylinder no. 6 Injector Circuit High
P0278 Cylinder no. 6 Contribution/Balance Fault
P0279 Cylinder no. 7 Injector Circuit Low
P0280 Cylinder no. 7 Injector Circuit High
P0281 Cylinder no. 7 Contribution/Balance Fault
P0282 Cylinder no. 8 Injector Circuit Low
P0283 Cylinder no. 8 Injector Circuit High
P0284 Cylinder no. 8 Contribution/Balance Fault
P0285 Cylinder no. 9 Injector Circuit Low
P0286 Cylinder no. 9 Injector Circuit High
P0287 Cylinder no. 9 Contribution/Balance Fault
P0288 Cylinder no. 10 Injector Circuit Low
P0289 Cylinder no. 10 Injector Circuit High
P0290 Cylinder no. 10 Contribution/Balance Fault
P0291 Cylinder no. 11 Injector Circuit Low
P0292 Cylinder no. 11 Injector Circuit High
P0293 Cylinder no. 11 Contribution/Balance Fault
P0294 Cylinder no. 12 Injector Circuit Low
P0295 Cylinder no. 12 Injector Circuit High
P0296 Cylinder no. 12 Contribution/Balance Fault
P0300 Random/Multiple Cylinder Misfire Detected
P0301 Cylinder no. 1—Misfire Detected
P0302 Cylinder no. 2—Misfire Detected
P0303 Cylinder no. 3—Misfire Detected
P0304 Cylinder no. 4—Misfire Detected
P0305 Cylinder no. 5—Misfire Detected
P0306 Cylinder no. 6—Misfire Detected
P0307 Cylinder no. 7—Misfire Detected
P0308 Cylinder no. 8—Misfire Detected
P0309 Cylinder no. 9—Misfire Detected
P0310 Cylinder no. 10—Misfire Detected
P0311 Cylinder no. 11—Misfire Detected
P0312 Cylinder no. 12—Misfire Detected
P0320 Ignition/Distributor Engine Speed Input Circuit Malfunction
P0321 Ignition/Distributor Engine Speed Input Circuit Range/Performance
P0322 Ignition/Distributor Engine Speed Input Circuit No Signal
P0323 Ignition/Distributor Engine Speed Input Circuit Intermittent
P0325 Knock Sensor no. 1—Circuit Malfunction (Bank no. 1 or Single Sensor)
P0326 Knock Sensor no. 1—Circuit Range/Performance (Bank no. 1 or Single Sensor)
P0327 Knock Sensor no. 1—Circuit Low Input (Bank no. 1 or Single Sensor)
P0328 Knock Sensor no. 1—Circuit High Input (Bank no. 1 or Single Sensor)

P0329 Knock Sensor no. 1—Circuit Input Intermittent (Bank no. 1 or Single Sensor)

P0330 Knock Sensor no. 2—Circuit Malfunction (Bank no. 2)

P0331 Knock Sensor no. 2—Circuit Range/Performance (Bank no. 2)

P0332 Knock Sensor no. 2—Circuit Low Input (Bank no. 2)

P0333 Knock Sensor no. 2—Circuit High Input (Bank no. 2)

P0334 Knock Sensor no. 2—Circuit Input Intermittent (Bank no. 2)

P0335 Crankshaft Position Sensor "A" Circuit Malfunction

P0336 Crankshaft Position Sensor "A" Circuit Range/Performance

P0337 Crankshaft Position Sensor "A" Circuit Low Input

P0338 Crankshaft Position Sensor "A" Circuit High Input

P0339 Crankshaft Position Sensor "A" Circuit Intermittent

P0340 Camshaft Position Sensor Circuit Malfunction

P0341 Camshaft Position Sensor Circuit Range/Performance

P0342 Camshaft Position Sensor Circuit Low Input

P0343 Camshaft Position Sensor Circuit High Input

P0344 Camshaft Position Sensor Circuit Intermittent

P0350 Ignition Coil Primary/Secondary Circuit Malfunction

P0351 Ignition Coil "A" Primary/Secondary Circuit Malfunction

P0352 Ignition Coil "B" Primary/Secondary Circuit Malfunction

P0353 Ignition Coil "C" Primary/Secondary Circuit Malfunction

P0354 Ignition Coil "D" Primary/Secondary Circuit Malfunction

P0355 Ignition Coil "E" Primary/Secondary Circuit Malfunction

P0356 Ignition Coil "F" Primary/Secondary Circuit Malfunction

P0357 Ignition Coil "G" Primary/Secondary Circuit Malfunction

P0358 Ignition Coil "H" Primary/Secondary Circuit Malfunction

P0359 Ignition Coil "I" Primary/Secondary Circuit Malfunction

P0360 Ignition Coil "J" Primary/Secondary Circuit Malfunction

P0361 Ignition Coil "K" Primary/Secondary Circuit Malfunction

P0362 Ignition Coil "L" Primary/Secondary Circuit Malfunction

P0370 Timing Reference High Resolution Signal "A" Malfunction

P0371 Timing Reference High Resolution Signal "A" Too Many Pulses

P0372 Timing Reference High Resolution Signal "A" Too Few Pulses

P0373 Timing Reference High Resolution Signal "A" Intermittent/Erratic Pulses

P0374 Timing Reference High Resolution Signal "A" No Pulses

P0375 Timing Reference High Resolution Signal "B" Malfunction

P0376 Timing Reference High Resolution Signal "B" Too Many Pulses

P0377 Timing Reference High Resolution Signal "B" Too Few Pulses

P0378 Timing Reference High Resolution Signal "B" Intermittent/Erratic Pulses

P0379 Timing Reference High Resolution Signal "B" No Pulses

P0380 Glow Plug/Heater Circuit "A" Malfunction

P0381 Glow Plug/Heater Indicator Circuit Malfunction

P0382 Glow Plug/Heater Circuit "B" Malfunction

P0385 Crankshaft Position Sensor "B" Circuit Malfunction

P0386 Crankshaft Position Sensor "B" Circuit Range/Performance

P0387 Crankshaft Position Sensor "B" Circuit Low Input

P0388 Crankshaft Position Sensor "B" Circuit High Input

P0389 Crankshaft Position Sensor "B" Circuit Intermittent

P0400 Exhaust Gas Recirculation Flow Malfunction

P0401 Exhaust Gas Recirculation Flow Insufficient Detected

P0402 Exhaust Gas Recirculation Flow Excessive Detected

P0403 Exhaust Gas Recirculation Circuit Malfunction

P0404 Exhaust Gas Recirculation Circuit Range/Performance

P0405 Exhaust Gas Recirculation Sensor "A" Circuit Low

P0406 Exhaust Gas Recirculation Sensor "A" Circuit High

P0407 Exhaust Gas Recirculation Sensor "B" Circuit Low

P0408 Exhaust Gas Recirculation Sensor "B" Circuit High

P0410 Secondary Air Injection System Malfunction

P0411 Secondary Air Injection System Incorrect Flow Detected

P0412 Secondary Air Injection System Switching Valve "A" Circuit Malfunction

P0413 Secondary Air Injection System Switching Valve "A" Circuit Open

P0414 Secondary Air Injection System Switching Valve "A" Circuit Shorted

P0415 Secondary Air Injection System Switching Valve "B" Circuit Malfunction

P0416 Secondary Air Injection System Switching Valve "B" Circuit Open

P0417 Secondary Air Injection System Switching Valve "B" Circuit Shorted

P0418 Secondary Air Injection System Relay "A" Circuit Malfunction

P0419 Secondary Air Injection System Relay "B" Circuit Malfunction

P0420 Catalyst System Efficiency Below Threshold (Bank no. 1)

P0421 Warm Up Catalyst Efficiency Below Threshold (Bank no. 1)

P0422 Main Catalyst Efficiency Below Threshold (Bank no. 1)

P0423 Heated Catalyst Efficiency Below Threshold (Bank no. 1)

P0424 Heated Catalyst Temperature Below Threshold (Bank no. 1)

P0430 Catalyst System Efficiency Below Threshold (Bank no. 2)

P0431 Warm Up Catalyst Efficiency Below Threshold (Bank no. 2)

P0432 Main Catalyst Efficiency Below Threshold (Bank no. 2)

P0433 Heated Catalyst Efficiency Below Threshold (Bank no. 2)

P0434 Heated Catalyst Temperature Below Threshold (Bank no. 2)

P0440 Evaporative Emission Control System Malfunction

P0441 Evaporative Emission Control System Incorrect Purge Flow

P0442 Evaporative Emission Control System Leak Detected (Small Leak)

P0443 Evaporative Emission Control System Purge Control Valve Circuit Malfunction

P0444 Evaporative Emission Control System Purge Control Valve Circuit Open

P0445 Evaporative Emission Control System Purge Control Valve Circuit Shorted

P0446 Evaporative Emission Control System Vent Control Circuit Malfunction

P0447 Evaporative Emission Control System Vent Control Circuit Open

P0448 Evaporative Emission Control System Vent Control Circuit Shorted

P0449 Evaporative Emission Control System Vent Valve/Solenoid Circuit Malfunction

P0450 Evaporative Emission Control System Pressure Sensor Malfunction

P0451 Evaporative Emission Control System Pressure Sensor Range/Performance

P0452 Evaporative Emission Control System Pressure Sensor Low Input

P0453 Evaporative Emission Control System Pressure Sensor High Input

P0454 Evaporative Emission Control System Pressure Sensor Intermittent

P0455 Evaporative Emission Control System Leak Detected (Gross Leak)

P0460 Fuel Level Sensor Circuit Malfunction

P0461 Fuel Level Sensor Circuit Range/Performance

P0462 Fuel Level Sensor Circuit Low Input

P0463 Fuel Level Sensor Circuit High Input

P0464 Fuel Level Sensor Circuit Intermittent

P0465 Purge Flow Sensor Circuit Malfunction

P0466 Purge Flow Sensor Circuit Range/Performance

P0467 Purge Flow Sensor Circuit Low Input

P0468 Purge Flow Sensor Circuit High Input

P0469 Purge Flow Sensor Circuit Intermittent

P0470 Exhaust Pressure Sensor Malfunction

P0471 Exhaust Pressure Sensor Range/Performance

P0472 Exhaust Pressure Sensor Low

P0473 Exhaust Pressure Sensor High

P0474 Exhaust Pressure Sensor Intermittent

P0475 Exhaust Pressure Control Valve Malfunction

P0476 Exhaust Pressure Control Valve Range/Performance

P0477 Exhaust Pressure Control Valve Low

P0478 Exhaust Pressure Control Valve High

P0479 Exhaust Pressure Control Valve Intermittent

P0480 Cooling Fan no. 1 Control Circuit Malfunction

P0481 Cooling Fan no. 2 Control Circuit Malfunction

P0482 Cooling Fan no. 3 Control Circuit Malfunction

P0483 Cooling Fan Rationality Check Malfunction

P0484 Cooling Fan Circuit Over Current

P0485 Cooling Fan Power/Ground Circuit Malfunction

P0500 Vehicle Speed Sensor Malfunction

P0501 Vehicle Speed Sensor Range/Performance

P0502 Vehicle Speed Sensor Circuit Low Input

P0503 Vehicle Speed Sensor Intermittent/Erratic/High

P0505 Idle Control System Malfunction

P0506 Idle Control System RPM Lower Than Expected

P0507 Idle Control System RPM Higher Than Expected

P0510 Closed Throttle Position Switch Malfunction

P0520 Engine Oil Pressure Sensor/Switch Circuit Malfunction

P0521 Engine Oil Pressure Sensor/Switch Range/Performance

P0522 Engine Oil Pressure Sensor/Switch Low Voltage

P0523 Engine Oil Pressure Sensor/Switch High Voltage

P0530 A/C Refrigerant Pressure Sensor Circuit Malfunction

P0531 A/C Refrigerant Pressure Sensor Circuit Range/Performance

P0532 A/C Refrigerant Pressure Sensor Circuit Low Input

P0533 A/C Refrigerant Pressure Sensor Circuit High Input

P0534 A/C Refrigerant Charge Loss

P0550 Power Steering Pressure Sensor Circuit Malfunction

P0551 Power Steering Pressure Sensor Circuit Range/Performance

P0552 Power Steering Pressure Sensor Circuit Low Input

P0553 Power Steering Pressure Sensor Circuit High Input

P0554 Power Steering Pressure Sensor Circuit Intermittent

P0560 System Voltage Malfunction

P0561 System Voltage Unstable

P0562 System Voltage Low

P0563 System Voltage High

P0565 Cruise Control On Signal Malfunction

P0566 Cruise Control Off Signal Malfunction

P0567 Cruise Control Resume Signal Malfunction

P0568 Cruise Control Set Signal Malfunction

P0569 Cruise Control Coast Signal Malfunction

P0570 Cruise Control Accel Signal Malfunction

P0571 Cruise Control/Brake Switch "A" Circuit Malfunction

P0572 Cruise Control/Brake Switch "A" Circuit Low

P0573 Cruise Control/Brake Switch "A" Circuit High

P0574 Through P0580 Reserved for Cruise Codes

P0600 Serial Communication Link Malfunction

P0601 Internal Control Module Memory Check Sum Error

P0602 Control Module Programming Error

P0603 Internal Control Module Keep Alive Memory (KAM) Error

P0604 Internal Control Module Random Access Memory (RAM) Error

P0605 Internal Control Module Read Only Memory (ROM) Error

P0606 PCM Processor Fault

P0608 Control Module VSS Output "A" Malfunction

P0609 Control Module VSS Output "B" Malfunction

P0620 Generator Control Circuit Malfunction

P0621 Generator Lamp "L" Control Circuit Malfunction

P0622 Generator Field "F" Control Circuit Malfunction

P0650 Malfunction Indicator Lamp (MIL) Control Circuit Malfunction

P0654 Engine RPM Output Circuit Malfunction

P0655 Engine Hot Lamp Output Control Circuit Malfunction

P0656 Fuel Level Output Circuit Malfunction

P0700 Transmission Control System Malfunction

P0701 Transmission Control System Range/Performance

P0702 Transmission Control System Electrical

P0703 Torque Converter/Brake Switch "B" Circuit Malfunction

P0704 Clutch Switch Input Circuit Malfunction

P0705 Transmission Range Sensor Circuit Malfunction (PRNDL Input)

P0706 Transmission Range Sensor Circuit Range/Performance

P0707 Transmission Range Sensor Circuit Low Input

P0708 Transmission Range Sensor Circuit High Input

P0709 Transmission Range Sensor Circuit Intermittent

P0710 Transmission Fluid Temperature Sensor Circuit Malfunction

P0711 Transmission Fluid Temperature Sensor Circuit Range/Performance

P0712 Transmission Fluid Temperature Sensor Circuit Low Input

P0713 Transmission Fluid Temperature Sensor Circuit High Input

P0714 Transmission Fluid Temperature Sensor Circuit Intermittent

P0715 Input/Turbine Speed Sensor Circuit Malfunction

P0716 Input/Turbine Speed Sensor Circuit Range/Performance

P0717 Input/Turbine Speed Sensor Circuit No Signal

P0718 Input/Turbine Speed Sensor Circuit Intermittent

P0719 Torque Converter/Brake Switch "B" Circuit Low

P0720 Output Speed Sensor Circuit Malfunction

P0721 Output Speed Sensor Circuit Range/Performance

P0722 Output Speed Sensor Circuit No Signal

P0723 Output Speed Sensor Circuit Intermittent

P0724 Torque Converter/Brake Switch "B" Circuit High

P0725 Engine Speed Input Circuit Malfunction

P0726 Engine Speed Input Circuit Range/Performance

P0727 Engine Speed Input Circuit No Signal

P0728 Engine Speed Input Circuit Intermittent

P0730 Incorrect Gear Ratio

P0731 Gear no. 1 Incorrect Ratio

P0732 Gear no. 2 Incorrect Ratio

P0733 Gear no. 3 Incorrect Ratio
P0734 Gear no. 4 Incorrect Ratio
P0735 Gear no. 5 Incorrect Ratio
P0736 Reverse Incorrect Ratio
P0740 Torque Converter Clutch Circuit Malfunction
P0741 Torque Converter Clutch Circuit Performance or Stuck Off
P0742 Torque Converter Clutch Circuit Stuck On
P0743 Torque Converter Clutch Circuit Electrical
P0744 Torque Converter Clutch Circuit Intermittent
P0745 Pressure Control Solenoid Malfunction
P0746 Pressure Control Solenoid Performance or Stuck Off
P0747 Pressure Control Solenoid Stuck On
P0748 Pressure Control Solenoid Electrical
P0749 Pressure Control Solenoid Intermittent
P0750 Shift Solenoid "A" Malfunction
P0751 Shift Solenoid "A" Performance or Stuck Off
P0752 Shift Solenoid "A" Stuck On
P0753 Shift Solenoid "A" Electrical
P0754 Shift Solenoid "A" Intermittent
P0755 Shift Solenoid "B" Malfunction
P0756 Shift Solenoid "B" Performance or Stuck Off
P0757 Shift Solenoid "B" Stuck On
P0758 Shift Solenoid "B" Electrical
P0759 Shift Solenoid "B" Intermittent
P0760 Shift Solenoid "C" Malfunction
P0761 Shift Solenoid "C" Performance Or Stuck Off
P0762 Shift Solenoid "C" Stuck On
P0763 Shift Solenoid "C" Electrical
P0764 Shift Solenoid "C" Intermittent
P0765 Shift Solenoid "D" Malfunction
P0766 Shift Solenoid "D" Performance Or Stuck Off
P0767 Shift Solenoid "D" Stuck On
P0768 Shift Solenoid "D" Electrical
P0769 Shift Solenoid "D" Intermittent
P0770 Shift Solenoid "E" Malfunction
P0771 Shift Solenoid "E" Performance Or Stuck Off
P0772 Shift Solenoid "E" Stuck On
P0773 Shift Solenoid "E" Electrical
P0774 Shift Solenoid "E" Intermittent
P0780 Shift Malfunction
P0781 1–2 Shift Malfunction
P0782 2–3 Shift Malfunction
P0783 3–4 Shift Malfunction
P0784 4–5 Shift Malfunction
P0785 Shift/Timing Solenoid Malfunction
P0786 Shift/Timing Solenoid Range/Performance
P0787 Shift/Timing Solenoid Low
P0788 Shift/Timing Solenoid High
P0789 Shift/Timing Solenoid Intermittent
P0790 Normal/Performance Switch Circuit Malfunction
P0801 Reverse Inhibit Control Circuit Malfunction
P0803 1–4 Upshift (Skip Shift) Solenoid Control Circuit Malfunction
P0804 1–4 Upshift (Skip Shift) Lamp Control Circuit Malfunction
P1102 Oxygen Sensor Heating
P1105 Oxygen Sensor Heating
P1107 Oxygen Sensor Heating
P1110 Oxygen Sensor Heating
P1115 Oxygen Sensor Heating
P1117 Oxygen Sensor Heating
P1119 Oxygen Sensor Heating

P1121 Oxygen Sensor Heating
P1123 Oxygen Sensing Heating
P1124 Oxygen Sensing
P1125 Oxygen Sensing
P1126 Oxygen Sensing
P1127 Oxygen Sensing
P1128 Oxygen Sensing
P1129 Oxygen Sensing
P1130 Oxygen Sensing
P1136 Oxygen Sensing
P1137 Oxygen Sensing
P1138 Oxygen Sensing
P1139 Oxygen Sensing
P1140 Load Signal
P1157 Engine Compartment Temperature
P1158 Engine Compartment Temperature
P1213 Fuel Injector, Cylinder 1
P1214 Fuel Injector, Cylinder 2
P1215 Fuel Injector, Cylinder 3
P1216 Fuel Injector, Cylinder 4
P1217 Fuel Injector, Cylinder 5
P1218 Fuel Injector, Cylinder 6
P1225 Fuel Injector, Cylinder 1
P1226 Fuel Injector, Cylinder 2
P1227 Fuel Injector, Cylinder 3
P1228 Fuel Injector, Cylinder 4
P1229 Fuel Injector, Cylinder 5
P1230 Fuel Injector, Cylinder 6
P1237 Fuel Injector, Cylinder 1
P1238 Fuel Injector, Cylinder 2
P1239 Fuel Injector, Cylinder 3
P1240 Fuel injector, Cylinder 4
P1241 Fuel Injector, Cylinder 5
P1242 Fuel Injector, Cylinder 6
P1265 Airbag Signal
P1275 Oxygen Sensor Aging Ahead of Three Way Catalytic Converter
P1276 Oxygen Sensor Aging Ahead of Three Way Catalytic Converter
P1313 Misfire Cylinder 1, Emission Related
P1314 Misfire Cylinder 2, Emission Related
P1315 Misfire Cylinder 3, Emission Related
P1316 Misfire, Cylinder 4, Emission Related
P1317 Misfire, Cylinder 5, Emission Related
P1318 Misfire, Cylinder 6, Emission Related
P1319 Misfire Emission Related
P1324 Timing Chain out of Position, Bank 2
P1340 Timing Chain out of Position, Bank 1
P1384 Knock Sensor 1
P1385 Knock Sensor 2
P1386 Knock Sensor Test Pulse
P1386 Knock Control Test Pulse
P1397 Camshaft Position Sensor 2
P1411 Secondary Air Injection System
P1455 A/C Compressor Control
P1456 A/C Compressor Control
P1457 A/C Compressor Control
P1458 A/C Compressor Signal
P1501 Fuel Pump Relay End-Stage
P1502 Fuel Pump Relay End-Stage
P1510 Idle Air Control Valve
P1513 Idle Air Control Valve

P1514 Idle Control Valve
P1515 Intake Manifold Resonance Flap
P1516 Intake Manifold Resonance Flap
P1524 Camshaft Adjustment, Bank 2
P1530 Camshaft Adjustment, Bank 1
P1531 Camshaft Adjustment, Bank 1
P1539 Camshaft Adjustment, Bank 2
P1541 Fuel Pump Relay End-Stage
P1551 Idle Air Control Valve
P1552 Idle Air Control Valve
P1553 Idle Air Control Valve
P1555 Charge Pressure Characteristics
P1556 Charge Deviations
P1557 Charge Deviations
P1570 Immobilizer
P1571 Immobilizer
P1585 Misfire With Empty Fuel Tank
P1593 Intake Manifold Length Tuning 2
P1594 Intake Manifold Length Tuning 2
P1595 Intake Manifold Length Tuning 2
P1600 Voltage Supply
P1601 Voltage Supply
P1602 Voltage Supply
P1610 MIL Activated Externally
P1611 MIL Activated Externally
P1614 MIL Activated Externally
P1640 Engine Control Module
P1656 Coolant Shutoff Valve
P1671 Engine Compartment Purge Fan End-Stage
P1673 Fan End-Stage
P1689 Engine Control Module
P1691 Malfunction Indicator Lamp
P1692 Malfunction Indicator Lamp
P1693 Malfunction Indicator Lamp
P1704 Kickdown Switch
P1710 Speed Signal, Right Front
P1715 Speed Signal, Left Front
P1744 Manual Program Switch
P1746 Control Unit Defective (Relay)
P1748 Control unit defective (relay sticks)
P1749 Version coding
P1750 Voltage supply, solenoid valve pressure regulators 1
P1761 Shiftlock P/N
P1762 Shiftlock P/N
P1764 Instrument cluster triggering
P1765 Throttle-valve information error
P1770 Load signal from ECM
P1782 Engine Engagement
P1813 Pressure regulator 1
P1818 Pressure regulator 2
P1823 Pressure regulator 3
P1828 Pressure regulator 4

Saab

READING CODES

Reading the control module memory is on of the first steps in OBD II system diagnostics. This step should be initially performed to determine the general nature of the fault. Subsequent readings will determine if the fault has been cleared.

Reading codes can be performed by any of the methods below:
• Read the control module memory with the Generic Scan Tool (GST)
• Read the control module memory with the vehicle manufacturer's specific tester

To read the fault codes, connect the scan tool or tester according to the manufacturer's instructions. Follow the manufacturer's specified procedure for reading the codes.

CLEARING CODES

Control module reset procedures are a very important part of OBD II System diagnostics. This step should be done at the end of any fault code repair and at the end of any driveability repair.

Clearing codes can be performed by any of the methods below:
• Clear the control module memory with the Generic Scan Tool (GST)
• Clear the control module memory with the vehicle manufacturer's specific tester
• Turn the ignition **OFF** and remove the negative battery cable for at least 1 minute.

Removing the negative battery cable may cause other systems in the vehicle to lose their memory. Prior to removing the cable, ensure you have the proper reset codes for radios and alarms.

➡**The MIL will may also be de-activated for some codes if the vehicle completes three consecutive trips without a fault detected with vehicle conditions similar to those present during the fault.**

DIAGNOSTIC TROUBLE CODES

P0000 No Failures
P0100 Mass or Volume Air Flow Circuit Malfunction
P0101 Mass or Volume Air Flow Circuit Range/Performance Problem
P0102 Mass or Volume Air Flow Circuit Low Input
P0103 Mass or Volume Air Flow Circuit High Input
P0104 Mass or Volume Air Flow Circuit Intermittent
P0105 Manifold Absolute Pressure/Barometric Pressure Circuit Malfunction
P0106 Manifold Absolute Pressure/Barometric Pressure Circuit Range/Performance Problem
P0107 Manifold Absolute Pressure/Barometric Pressure Circuit Low Input
P0108 Manifold Absolute Pressure/Barometric Pressure Circuit High Input
P0109 Manifold Absolute Pressure/Barometric Pressure Circuit Intermittent
P0110 Intake Air Temperature Circuit Malfunction
P0111 Intake Air Temperature Circuit Range/Performance Problem
P0112 Intake Air Temperature Circuit Low Input
P0113 Intake Air Temperature Circuit High Input
P0114 Intake Air Temperature Circuit Intermittent
P0115 Engine Coolant Temperature Circuit Malfunction
P0116 Engine Coolant Temperature Circuit Range/Performance Problem
P0117 Engine Coolant Temperature Circuit Low Input
P0118 Engine Coolant Temperature Circuit High Input
P0119 Engine Coolant Temperature Circuit Intermittent
P0120 Throttle/Pedal Position Sensor/Switch "A" Circuit Malfunction

P0121 Throttle/Pedal Position Sensor/Switch "A" Circuit Range/Performance Problem

P0122 Throttle/Pedal Position Sensor/Switch "A" Circuit Low Input

P0123 Throttle/Pedal Position Sensor/Switch "A" Circuit High Input

P0124 Throttle/Pedal Position Sensor/Switch "A" Circuit Intermittent

P0125 Insufficient Coolant Temperature For Closed Loop Fuel Control

P0126 Insufficient Coolant Temperature For Stable Operation

P0130 O_2 Circuit Malfunction (Bank no. 1 Sensor no. 1)

P0131 O_2 Sensor Circuit Low Voltage (Bank no. 1 Sensor no. 1)

P0132 O_2 Sensor Circuit High Voltage (Bank no. 1 Sensor no. 1)

P0133 O_2 Sensor Circuit Slow Response (Bank no. 1 Sensor no. 1)

P0134 O_2 Sensor Circuit No Activity Detected (Bank no. 1 Sensor no. 1)

P0135 O_2 Sensor Heater Circuit Malfunction (Bank no. 1 Sensor no. 1)

P0136 O_2 Sensor Circuit Malfunction (Bank no. 1 Sensor no. 2)

P0137 O_2 Sensor Circuit Low Voltage (Bank no. 1 Sensor no. 2)

P0138 O_2 Sensor Circuit High Voltage (Bank no. 1 Sensor no. 2)

P0139 O_2 Sensor Circuit Slow Response (Bank no. 1 Sensor no. 2)

P0140 O_2 Sensor Circuit No Activity Detected (Bank no. 1 Sensor no. 2)

P0141 O_2 Sensor Heater Circuit Malfunction (Bank no. 1 Sensor no. 2)

P0142 O_2 Sensor Circuit Malfunction (Bank no. 1 Sensor no. 3)

P0143 O_2 Sensor Circuit Low Voltage (Bank no. 1 Sensor no. 3)

P0144 O_2 Sensor Circuit High Voltage (Bank no. 1 Sensor no. 3)

P0145 O_2 Sensor Circuit Slow Response (Bank no. 1 Sensor no. 3)

P0146 O_2 Sensor Circuit No Activity Detected (Bank no. 1 Sensor no. 3)

P0147 O_2 Sensor Heater Circuit Malfunction (Bank no. 1 Sensor no. 3)

P0150 O_2 Sensor Circuit Malfunction (Bank no. 2 Sensor no. 1)

P0151 O_2 Sensor Circuit Low Voltage (Bank no. 2 Sensor no. 1)

P0152 O_2 Sensor Circuit High Voltage (Bank no. 2 Sensor no. 1)

P0153 O_2 Sensor Circuit Slow Response (Bank no. 2 Sensor no. 1)

P0154 O_2 Sensor Circuit No Activity Detected (Bank no. 2 Sensor no. 1)

P0155 O_2 Sensor Heater Circuit Malfunction (Bank no. 2 Sensor no. 1)

P0156 O_2 Sensor Circuit Malfunction (Bank no. 2 Sensor no. 2)

P0157 O_2 Sensor Circuit Low Voltage (Bank no. 2 Sensor no. 2)

P0158 O_2 Sensor Circuit High Voltage (Bank no. 2 Sensor no. 2)

P0159 O_2 Sensor Circuit Slow Response (Bank no. 2 Sensor no. 2)

P0160 O_2 Sensor Circuit No Activity Detected (Bank no. 2 Sensor no. 2)

P0161 O_2 Sensor Heater Circuit Malfunction (Bank no. 2 Sensor no. 2)

P0162 O_2 Sensor Circuit Malfunction (Bank no. 2 Sensor no. 3)

P0163 O_2 Sensor Circuit Low Voltage (Bank no. 2 Sensor no. 3)

P0164 O_2 Sensor Circuit High Voltage (Bank no. 2 Sensor no. 3)

P0165 O_2 Sensor Circuit Slow Response (Bank no. 2 Sensor no. 3)

P0166 O_2 Sensor Circuit No Activity Detected (Bank no. 2 Sensor no. 3)

P0167 O_2 Sensor Heater Circuit Malfunction (Bank no. 2 Sensor no. 3)

P0170 Fuel Trim Malfunction (Bank no. 1)

P0171 System Too Lean (Bank no. 1)

P0172 System Too Rich (Bank no. 1)

P0173 Fuel Trim Malfunction (Bank no. 2)

P0174 System Too Lean (Bank no. 2)

P0175 System Too Rich (Bank no. 2)

P0176 Fuel Composition Sensor Circuit Malfunction

P0177 Fuel Composition Sensor Circuit Range/Performance

P0178 Fuel Composition Sensor Circuit Low Input

P0179 Fuel Composition Sensor Circuit High Input

P0180 Fuel Temperature Sensor "A" Circuit Malfunction

P0181 Fuel Temperature Sensor "A" Circuit Range/Performance

P0182 Fuel Temperature Sensor "A" Circuit Low Input

P0183 Fuel Temperature Sensor "A" Circuit High Input

P0184 Fuel Temperature Sensor "A" Circuit Intermittent

P0185 Fuel Temperature Sensor "B" Circuit Malfunction

P0186 Fuel Temperature Sensor "B" Circuit Range/Performance

P0187 Fuel Temperature Sensor "B" Circuit Low Input

P0188 Fuel Temperature Sensor "B" Circuit High Input

P0189 Fuel Temperature Sensor "B" Circuit Intermittent

P0190 Fuel Rail Pressure Sensor Circuit Malfunction

P0191 Fuel Rail Pressure Sensor Circuit Range/Performance

P0192 Fuel Rail Pressure Sensor Circuit Low Input

P0193 Fuel Rail Pressure Sensor Circuit High Input

P0194 Fuel Rail Pressure Sensor Circuit Intermittent

P0195 Engine Oil Temperature Sensor Malfunction

P0196 Engine Oil Temperature Sensor Range/Performance

P0197 Engine Oil Temperature Sensor Low

P0198 Engine Oil Temperature Sensor High

P0199 Engine Oil Temperature Sensor Intermittent

P0200 Injector Circuit Malfunction

P0201 Injector Circuit Malfunction—Cylinder no. 1

P0202 Injector Circuit Malfunction—Cylinder no. 2

P0203 Injector Circuit Malfunction—Cylinder no. 3

P0204 Injector Circuit Malfunction—Cylinder no. 4

P0205 Injector Circuit Malfunction—Cylinder no. 5

P0206 Injector Circuit Malfunction—Cylinder no. 6

P0207 Injector Circuit Malfunction—Cylinder no. 7

P0208 Injector Circuit Malfunction—Cylinder no. 8

P0209 Injector Circuit Malfunction—Cylinder no. 9

P0210 Injector Circuit Malfunction—Cylinder no. 10

P0211 Injector Circuit Malfunction—Cylinder no. 11

P0212 Injector Circuit Malfunction—Cylinder no. 12

P0213 Cold Start Injector no. 1 Malfunction

P0214 Cold Start Injector no. 2 Malfunction

P0215 Engine Shutoff Solenoid Malfunction

P0216 Injection Timing Control Circuit Malfunction

P0217 Engine Over Temperature Condition

P0218 Transmission Over Temperature Condition

P0219 Engine Over Speed Condition

P0220 Throttle/Pedal Position Sensor/Switch "B" Circuit Malfunction

P0221 Throttle/Pedal Position Sensor/Switch "B" Circuit Range/Performance Problem

P0222 Throttle/Pedal Position Sensor/Switch "B" Circuit Low Input

P0223 Throttle/Pedal Position Sensor/Switch "B" Circuit High Input

P0224 Throttle/Pedal Position Sensor/Switch "B" Circuit Intermittent

P0225 Throttle/Pedal Position Sensor/Switch "C" Circuit Malfunction

P0226 Throttle/Pedal Position Sensor/Switch "C" Circuit Range/Performance Problem

P0227 Throttle/Pedal Position Sensor/Switch "C" Circuit Low Input

P0228 Throttle/Pedal Position Sensor/Switch "C" Circuit High Input

P0229 Throttle/Pedal Position Sensor/Switch "C" Circuit Intermittent

P0230 Fuel Pump Primary Circuit Malfunction

P0231 Fuel Pump Secondary Circuit Low

P0232 Fuel Pump Secondary Circuit High

P0233 Fuel Pump Secondary Circuit Intermittent

P0234 Engine Over Boost Condition

P0261 Cylinder no. 1 Injector Circuit Low

P0262 Cylinder no. 1 Injector Circuit High

P0263 Cylinder no. 1 Contribution/Balance Fault

P0264 Cylinder no. 2 Injector Circuit Low

P0265 Cylinder no. 2 Injector Circuit High

P0266 Cylinder no. 2 Contribution/Balance Fault

P0267 Cylinder no. 3 Injector Circuit Low

P0268 Cylinder no. 3 Injector Circuit High

P0269 Cylinder no. 3 Contribution/Balance Fault

P0270 Cylinder no. 4 Injector Circuit Low

P0271 Cylinder no. 4 Injector Circuit High

P0272 Cylinder no. 4 Contribution/Balance Fault

P0273 Cylinder no. 5 Injector Circuit Low

P0274 Cylinder no. 5 Injector Circuit High

P0275 Cylinder no. 5 Contribution/Balance Fault

P0276 Cylinder no. 6 Injector Circuit Low

P0277 Cylinder no. 6 Injector Circuit High

P0278 Cylinder no. 6 Contribution/Balance Fault

P0279 Cylinder no. 7 Injector Circuit Low

P0280 Cylinder no. 7 Injector Circuit High

P0281 Cylinder no. 7 Contribution/Balance Fault

P0282 Cylinder no. 8 Injector Circuit Low

P0283 Cylinder no. 8 Injector Circuit High

P0284 Cylinder no. 8 Contribution/Balance Fault

P0285 Cylinder no. 9 Injector Circuit Low

P0286 Cylinder no. 9 Injector Circuit High

P0287 Cylinder no. 9 Contribution/Balance Fault

P0288 Cylinder no. 10 Injector Circuit Low

P0289 Cylinder no. 10 Injector Circuit High

P0290 Cylinder no. 10 Contribution/Balance Fault

P0291 Cylinder no. 11 Injector Circuit Low

P0292 Cylinder no. 11 Injector Circuit High

P0293 Cylinder no. 11 Contribution/Balance Fault

P0294 Cylinder no. 12 Injector Circuit Low

P0295 Cylinder no. 12 Injector Circuit High

P0296 Cylinder no. 12 Contribution/Balance Fault

P0300 Random/Multiple Cylinder Misfire Detected

P0301 Cylinder no. 1—Misfire Detected

P0302 Cylinder no. 2—Misfire Detected

P0303 Cylinder no. 3—Misfire Detected

P0304 Cylinder no. 4—Misfire Detected

P0305 Cylinder no. 5—Misfire Detected

P0306 Cylinder no. 6—Misfire Detected

P0307 Cylinder no. 7—Misfire Detected

P0308 Cylinder no. 8—Misfire Detected

P0309 Cylinder no. 9—Misfire Detected

P0310 Cylinder no. 10—Misfire Detected

P0311 Cylinder no. 11—Misfire Detected

P0312 Cylinder no. 12—Misfire Detected

P0320 Ignition/Distributor Engine Speed Input Circuit Malfunction

P0321 Ignition/Distributor Engine Speed Input Circuit Range/Performance

P0322 Ignition/Distributor Engine Speed Input Circuit No Signal

P0323 Ignition/Distributor Engine Speed Input Circuit Intermittent

P0325 Knock Sensor no. 1—Circuit Malfunction (Bank no. 1 or Single Sensor)

P0326 Knock Sensor no. 1—Circuit Range/Performance (Bank no. 1 or Single Sensor)

P0327 Knock Sensor no. 1—Circuit Low Input (Bank no. 1 or Single Sensor)

P0328 Knock Sensor no. 1—Circuit High Input (Bank no. 1 or Single Sensor)

P0329 Knock Sensor no. 1—Circuit Input Intermittent (Bank no. 1 or Single Sensor)

P0330 Knock Sensor no. 2—Circuit Malfunction (Bank no. 2)

P0331 Knock Sensor no. 2—Circuit Range/Performance (Bank no. 2)

P0332 Knock Sensor no. 2—Circuit Low Input (Bank no. 2)

P0333 Knock Sensor no. 2—Circuit High Input (Bank no. 2)

P0334 Knock Sensor no. 2—Circuit Input Intermittent (Bank no. 2)

P0335 Crankshaft Position Sensor "A" Circuit Malfunction

P0336 Crankshaft Position Sensor "A" Circuit Range/Performance

P0337 Crankshaft Position Sensor "A" Circuit Low Input

P0338 Crankshaft Position Sensor "A" Circuit High Input

P0339 Crankshaft Position Sensor "A" Circuit Intermittent

P0340 Camshaft Position Sensor Circuit Malfunction

P0341 Camshaft Position Sensor Circuit Range/Performance

P0342 Camshaft Position Sensor Circuit Low Input

P0343 Camshaft Position Sensor Circuit High Input

P0344 Camshaft Position Sensor Circuit Intermittent

P0350 Ignition Coil Primary/Secondary Circuit Malfunction

P0351 Ignition Coil "A" Primary/Secondary Circuit Malfunction

P0352 Ignition Coil "B" Primary/Secondary Circuit Malfunction

P0353 Ignition Coil "C" Primary/Secondary Circuit Malfunction

P0354 Ignition Coil "D" Primary/Secondary Circuit Malfunction

P0355 Ignition Coil "E" Primary/Secondary Circuit Malfunction

P0356 Ignition Coil "F" Primary/Secondary Circuit Malfunction

P0357 Ignition Coil "G" Primary/Secondary Circuit Malfunction

P0358 Ignition Coil "H" Primary/Secondary Circuit Malfunction

P0359 Ignition Coil "I" Primary/Secondary Circuit Malfunction

P0360 Ignition Coil "J" Primary/Secondary Circuit Malfunction

P0361 Ignition Coil "K" Primary/Secondary Circuit Malfunction

P0362 Ignition Coil "L" Primary/Secondary Circuit Malfunction

P0370 Timing Reference High Resolution Signal "A" Malfunction

P0371 Timing Reference High Resolution Signal "A" Too Many Pulses

P0372 Timing Reference High Resolution Signal "A" Too Few Pulses

P0373 Timing Reference High Resolution Signal "A" Intermittent/Erratic Pulses

P0374 Timing Reference High Resolution Signal "A" No Pulses

P0375 Timing Reference High Resolution Signal "B" Malfunction

P0376 Timing Reference High Resolution Signal "B" Too Many Pulses

P0377 Timing Reference High Resolution Signal "B" Too Few Pulses

P0378 Timing Reference High Resolution Signal "B" Intermittent/Erratic Pulses

P0379 Timing Reference High Resolution Signal "B" No Pulses

P0380 Glow Plug/Heater Circuit "A" Malfunction

P0381 Glow Plug/Heater Indicator Circuit Malfunction

P0382 Glow Plug/Heater Circuit "B" Malfunction

P0385 Crankshaft Position Sensor "B" Circuit Malfunction

P0386 Crankshaft Position Sensor "B" Circuit Range/Performance

P0387 Crankshaft Position Sensor "B" Circuit Low Input

P0388 Crankshaft Position Sensor "B" Circuit High Input

P0389 Crankshaft Position Sensor "B" Circuit Intermittent

P0400 Exhaust Gas Recirculation Flow Malfunction

P0401 Exhaust Gas Recirculation Flow Insufficient Detected

P0402 Exhaust Gas Recirculation Flow Excessive Detected

P0403 Exhaust Gas Recirculation Circuit Malfunction

P0404 Exhaust Gas Recirculation Circuit Range/Performance

P0405 Exhaust Gas Recirculation Sensor "A" Circuit Low

P0406 Exhaust Gas Recirculation Sensor "A" Circuit High

P0407 Exhaust Gas Recirculation Sensor "B" Circuit Low

P0408 Exhaust Gas Recirculation Sensor "B" Circuit High

P0410 Secondary Air Injection System Malfunction

P0411 Secondary Air Injection System Incorrect Flow Detected

P0412 Secondary Air Injection System Switching Valve "A" Circuit Malfunction

P0413 Secondary Air Injection System Switching Valve "A" Circuit Open

P0414 Secondary Air Injection System Switching Valve "A" Circuit Shorted

P0415 Secondary Air Injection System Switching Valve "B" Circuit Malfunction

P0416 Secondary Air Injection System Switching Valve "B" Circuit Open

P0417 Secondary Air Injection System Switching Valve "B" Circuit Shorted

P0418 Secondary Air Injection System Relay "A" Circuit Malfunction

P0419 Secondary Air Injection System Relay "B" Circuit Malfunction

P0420 Catalyst System Efficiency Below Threshold (Bank no. 1)

P0421 Warm Up Catalyst Efficiency Below Threshold (Bank no. 1)

P0422 Main Catalyst Efficiency Below Threshold (Bank no. 1)

P0423 Heated Catalyst Efficiency Below Threshold (Bank no. 1)

P0424 Heated Catalyst Temperature Below Threshold (Bank no. 1)

P0430 Catalyst System Efficiency Below Threshold (Bank no. 2)

P0431 Warm Up Catalyst Efficiency Below Threshold (Bank no. 2)

P0432 Main Catalyst Efficiency Below Threshold (Bank no. 2)

P0433 Heated Catalyst Efficiency Below Threshold (Bank no. 2)

P0434 Heated Catalyst Temperature Below Threshold (Bank no. 2)

P0440 Evaporative Emission Control System Malfunction

P0441 Evaporative Emission Control System Incorrect Purge Flow

P0442 Evaporative Emission Control System Leak Detected (Small Leak)

P0443 Evaporative Emission Control System Purge Control Valve Circuit Malfunction

P0444 Evaporative Emission Control System Purge Control Valve Circuit Open

P0445 Evaporative Emission Control System Purge Control Valve Circuit Shorted

P0446 Evaporative Emission Control System Vent Control Circuit Malfunction

P0447 Evaporative Emission Control System Vent Control Circuit Open

P0448 Evaporative Emission Control System Vent Control Circuit Shorted

P0449 Evaporative Emission Control System Vent Valve/Solenoid Circuit Malfunction

P0450 Evaporative Emission Control System Pressure Sensor Malfunction

P0451 Evaporative Emission Control System Pressure Sensor Range/Performance

P0452 Evaporative Emission Control System Pressure Sensor Low Input

P0453 Evaporative Emission Control System Pressure Sensor High Input

P0454 Evaporative Emission Control System Pressure Sensor Intermittent

P0455 Evaporative Emission Control System Leak Detected (Gross Leak)

P0460 Fuel Level Sensor Circuit Malfunction

P0461 Fuel Level Sensor Circuit Range/Performance

P0462 Fuel Level Sensor Circuit Low Input

P0463 Fuel Level Sensor Circuit High Input

P0464 Fuel Level Sensor Circuit Intermittent

P0465 Purge Flow Sensor Circuit Malfunction

P0466 Purge Flow Sensor Circuit Range/Performance

P0467 Purge Flow Sensor Circuit Low Input

P0468 Purge Flow Sensor Circuit High Input

P0469 Purge Flow Sensor Circuit Intermittent

P0470 Exhaust Pressure Sensor Malfunction

P0471 Exhaust Pressure Sensor Range/Performance

P0472 Exhaust Pressure Sensor Low

P0473 Exhaust Pressure Sensor High

P0474 Exhaust Pressure Sensor Intermittent

P0475 Exhaust Pressure Control Valve Malfunction

P0476 Exhaust Pressure Control Valve Range/Performance

P0477 Exhaust Pressure Control Valve Low

P0478 Exhaust Pressure Control Valve High

P0479 Exhaust Pressure Control Valve Intermittent

P0480 Cooling Fan no. 1 Control Circuit Malfunction

P0481 Cooling Fan no. 2 Control Circuit Malfunction

P0482 Cooling Fan no. 3 Control Circuit Malfunction

P0483 Cooling Fan Rationality Check Malfunction

P0484 Cooling Fan Circuit Over Current

P0485 Cooling Fan Power/Ground Circuit Malfunction

P0500 Vehicle Speed Sensor Malfunction

P0501 Vehicle Speed Sensor Range/Performance

P0502 Vehicle Speed Sensor Circuit Low Input

P0503 Vehicle Speed Sensor Intermittent/Erratic/High

P0505 Idle Control System Malfunction

P0506 Idle Control System RPM Lower Than Expected

P0507 Idle Control System RPM Higher Than Expected

P0510 Closed Throttle Position Switch Malfunction

P0520 Engine Oil Pressure Sensor/Switch Circuit Malfunction

P0521 Engine Oil Pressure Sensor/Switch Range/Performance

P0522 Engine Oil Pressure Sensor/Switch Low Voltage

P0523 Engine Oil Pressure Sensor/Switch High Voltage

P0530 A/C Refrigerant Pressure Sensor Circuit Malfunction
P0531 A/C Refrigerant Pressure Sensor Circuit Range/Performance
P0532 A/C Refrigerant Pressure Sensor Circuit Low Input
P0533 A/C Refrigerant Pressure Sensor Circuit High Input
P0534 A/C Refrigerant Charge Loss
P0550 Power Steering Pressure Sensor Circuit Malfunction
P0551 Power Steering Pressure Sensor Circuit Range/Performance
P0552 Power Steering Pressure Sensor Circuit Low Input
P0553 Power Steering Pressure Sensor Circuit High Input
P0554 Power Steering Pressure Sensor Circuit Intermittent
P0560 System Voltage Malfunction
P0561 System Voltage Unstable
P0562 System Voltage Low
P0563 System Voltage High
P0565 Cruise Control On Signal Malfunction
P0566 Cruise Control Off Signal Malfunction
P0567 Cruise Control Resume Signal Malfunction
P0568 Cruise Control Set Signal Malfunction
P0569 Cruise Control Coast Signal Malfunction
P0570 Cruise Control Accel Signal Malfunction
P0571 Cruise Control/Brake Switch "A" Circuit Malfunction
P0572 Cruise Control/Brake Switch "A" Circuit Low
P0573 Cruise Control/Brake Switch "A" Circuit High
P0574 Through P0580 Reserved for Cruise Codes
P0600 Serial Communication Link Malfunction
P0601 Internal Control Module Memory Check Sum Error
P0602 Control Module Programming Error
P0603 Internal Control Module Keep Alive Memory (KAM) Error
P0604 Internal Control Module Random Access Memory (RAM) Error
P0605 Internal Control Module Read Only Memory (ROM) Error
P0606 PCM Processor Fault
P0608 Control Module VSS Output "A" Malfunction
P0609 Control Module VSS Output "B" Malfunction
P0620 Generator Control Circuit Malfunction
P0621 Generator Lamp "L" Control Circuit Malfunction
P0622 Generator Field "F" Control Circuit Malfunction
P0650 Malfunction Indicator Lamp (MIL) Control Circuit Malfunction
P0654 Engine RPM Output Circuit Malfunction
P0655 Engine Hot Lamp Output Control Circuit Malfunction
P0656 Fuel Level Output Circuit Malfunction
P0700 Transmission Control System Malfunction
P0701 Transmission Control System Range/Performance
P0702 Transmission Control System Electrical
P0703 Torque Converter/Brake Switch "B" Circuit Malfunction
P0704 Clutch Switch Input Circuit Malfunction
P0705 Transmission Range Sensor Circuit Malfunction (PRNDL Input)
P0706 Transmission Range Sensor Circuit Range/Performance
P0707 Transmission Range Sensor Circuit Low Input
P0708 Transmission Range Sensor Circuit High Input
P0709 Transmission Range Sensor Circuit Intermittent
P0710 Transmission Fluid Temperature Sensor Circuit Malfunction
P0711 Transmission Fluid Temperature Sensor Circuit Range/Performance
P0712 Transmission Fluid Temperature Sensor Circuit Low Input

P0713 Transmission Fluid Temperature Sensor Circuit High Input
P0714 Transmission Fluid Temperature Sensor Circuit Intermittent
P0715 Input/Turbine Speed Sensor Circuit Malfunction
P0716 Input/Turbine Speed Sensor Circuit Range/Performance
P0717 Input/Turbine Speed Sensor Circuit No Signal
P0718 Input/Turbine Speed Sensor Circuit Intermittent
P0719 Torque Converter/Brake Switch "B" Circuit Low
P0720 Output Speed Sensor Circuit Malfunction
P0721 Output Speed Sensor Circuit Range/Performance
P0722 Output Speed Sensor Circuit No Signal
P0723 Output Speed Sensor Circuit Intermittent
P0724 Torque Converter/Brake Switch "B" Circuit High
P0725 Engine Speed Input Circuit Malfunction
P0726 Engine Speed Input Circuit Range/Performance
P0727 Engine Speed Input Circuit No Signal
P0728 Engine Speed Input Circuit Intermittent
P0730 Incorrect Gear Ratio
P0731 Gear no. 1 Incorrect Ratio
P0732 Gear no. 2 Incorrect Ratio
P0733 Gear no. 3 Incorrect Ratio
P0734 Gear no. 4 Incorrect Ratio
P0735 Gear no. 5 Incorrect Ratio
P0736 Reverse Incorrect Ratio
P0740 Torque Converter Clutch Circuit Malfunction
P0741 Torque Converter Clutch Circuit Performance or Stuck Off
P0742 Torque Converter Clutch Circuit Stuck On
P0743 Torque Converter Clutch Circuit Electrical
P0744 Torque Converter Clutch Circuit Intermittent
P0745 Pressure Control Solenoid Malfunction
P0746 Pressure Control Solenoid Performance or Stuck Off
P0747 Pressure Control Solenoid Stuck On
P0748 Pressure Control Solenoid Electrical
P0749 Pressure Control Solenoid Intermittent
P0750 Shift Solenoid "A" Malfunction
P0751 Shift Solenoid "A" Performance or Stuck Off
P0752 Shift Solenoid "A" Stuck On
P0753 Shift Solenoid "A" Electrical
P0754 Shift Solenoid "A" Intermittent
P0755 Shift Solenoid "B" Malfunction
P0756 Shift Solenoid "B" Performance or Stuck Off
P0757 Shift Solenoid "B" Stuck On
P0758 Shift Solenoid "B" Electrical
P0759 Shift Solenoid "B" Intermittent
P0760 Shift Solenoid "C" Malfunction
P0761 Shift Solenoid "C" Performance Or Stuck Off
P0762 Shift Solenoid "C" Stuck On
P0763 Shift Solenoid "C" Electrical
P0764 Shift Solenoid "C" Intermittent
P0765 Shift Solenoid "D" Malfunction
P0766 Shift Solenoid "D" Performance Or Stuck Off
P0767 Shift Solenoid "D" Stuck On
P0768 Shift Solenoid "D" Electrical
P0769 Shift Solenoid "D" Intermittent
P0770 Shift Solenoid "E" Malfunction
P0771 Shift Solenoid "E" Performance Or Stuck Off
P0772 Shift Solenoid "E" Stuck On
P0773 Shift Solenoid "E" Electrical
P0774 Shift Solenoid "E" Intermittent
P0780 Shift Malfunction

P0781 1–2 Shift Malfunction
P0782 2–3 Shift Malfunction
P0783 3–4 Shift Malfunction
P0784 4–5 Shift Malfunction
P0785 Shift/Timing Solenoid Malfunction
P0786 Shift/Timing Solenoid Range/Performance
P0787 Shift/Timing Solenoid Low
P0788 Shift/Timing Solenoid High
P0789 Shift/Timing Solenoid Intermittent
P0790 Normal/Performance Switch Circuit Malfunction
P0801 Reverse Inhibit Control Circuit Malfunction
P0803 1–4 Upshift (Skip Shift) Solenoid Control Circuit Malfunction
P0804 1–4 Upshift (Skip Shift) Lamp Control Circuit Malfunction
P1102 Front heated oxygen sensor, bank 1, control module input. Current in preheating circuit much too high.
P1105 Rear heated oxygen sensor, bank 1, control module input. Current in preheating circuit much too high.
P1115 Front heated oxygen sensor, bank 1, control module input. Current in preheating circuit much too low.
P1117 Rear heated oxygen sensor, bank 1, control module input. Current in preheating circuit much too low.
P1123 Additive adaptation, bank 1. Min. value.
P1124 Additive adaptation, bank 1. Max value.
P1125 Additive adaptation, bank 2. Min. value.
P1126 Additive adaptation, bank 2. Max value.
P1127 Multiplicative adaptation, bank 1. Min. value.
P1128 Multiplicative adaptation, bank 1. Max value.
P1129 Multiplicative adaptation, bank 2. Min. value.
P1130 Multiplicative adaptation, bank 2. Max value.
P1170 Closed loop. Malfunction.
P1171 Closed loop. Lean mixture.
P1172 Closed loop. Rich mixture.
P1213 Injector, cylinder 1, control module output. Shorting to battery positive (B+).
P1214 Injector, cylinder 2, control module output. Shorting to battery positive (B+).
P1215 Injector, cylinder 3, control module output. Shorting to battery positive (B+).
P1216 Injector, cylinder 4, control module output. Shorting to battery positive (B+).
P1217 Injector, cylinder 5, control module output. Shorting to battery positive (B+).
P1218 Injector, cylinder 6, control module output. Shorting to battery positive (B+).
P1225 Injector, cylinder 1, control module output. Open circuit or shorting to ground.
P1226 Injector, cylinder 2, control module output. Open circuit or shorting to ground.
P1227 Injector, cylinder 3, control module output. Open circuit or shorting to ground.
P1228 Injector, cylinder 4, control module output. Open circuit or shorting to ground.
P1229 Injector, cylinder 5, control module output. Open circuit or shorting to ground.
P1230 Injector, cylinder 6, control module output. Open circuit or shorting to ground.
P1386 Control module, electronic circuitry for processing knock sensor signals Internal fault.
P1396 Crankshaft position sensor, control module input. Malfunctioning. slotted ring has too many ribs.

P1410 EVAP canister purge valve, control module output. Shorting to battery positive (B+).
P1416 Tank level. Low level in conjunction with misfiring or fault in fuel system.
P1425 EVAP canister purge module output. Shorting to ground.
P1426 EVAP canister purge module output. Open circuit.
P1500 Battery voltage outside limits.
P1501 Fuel pump relay, control module output. Shorting to ground.
P1502 Fuel pump relay, control module output. Shorting to battery positive (B+)
P1510 Idle air control valve, open control module output. Shorting to battery positive (B+).
P1513 Idle air control valve, open control module output. Shorting to ground.
P1514 Idle air control valve, open function, control module output. Open circuit.
P1541 Fuel pump relay, control module out put.
P1549 Boost pressure control. Malfunction.
P1551 Idle air control valve, close function, control module output. Open circuit.
P1552 Idle air control valve, close function, control module output. Shorting to ground.
P1553 Idle air control valve, close function, control module output. Shorting to battery positive (B+).
P1576 Brake light switch. Shorting to battery positive (B+).
P1577 Brake light switch. Open circuit.
P1585 Fuel less than 10 liters.
P1611 CHECK ENGINE request, input signal to control module. Shorting to ground.
P1616 Rough road sensor. Control module input low, shorting to ground.
P1617 Rough road sensor. Control module input high; open circuit or shorting to battery positive (B+).
P1624 The automatic transmission has a stored emission-related fault.
P1664 Shift up, output signal from control module. Malfunction.
P1665 Intermittent fault which cannot identify any other specific diagnostic trouble
    code.
P1669 TCS active, input signal to control module.
P1670 Intermittent fault which cannot identify any other specific diagnostic trouble
    code.
P1675 Intermittent fault which cannot identify any other specific diagnostic trouble code.
P1680 Relay, secondary air injection. Control module output, open circuit or short circuit.
P1691 CHECK ENGINE, output signal from On control module. Open circuit.
P1692 CHECK ENGINE, output signal from On control module. Shorting to ground.
P1693 CHECK ENGINE, output signal from On control module. Open circuit or shorting to ground or battery positive (B+).

Subaru

READING CODES

Reading the control module memory is on of the first steps in OBD II system diagnostics.

This step should be initially performed to determine the general nature of the fault. Subsequent readings will determine if the fault has been cleared.

Reading codes can be performed by any of the methods below:
- Read the control module memory with the Generic Scan Tool (GST)
- Read the control module memory with the vehicle manufacturer's specific tester

To read the fault codes, connect the scan tool or tester according to the manufacturer's instructions. Follow the manufacturer's specified procedure for reading the codes.

CLEARING CODES

Control module reset procedures are a very important part of OBD II System diagnostics. This step should be done at the end of any fault code repair and at the end of any driveability repair.

Clearing codes can be performed by any of the methods below:
- Clear the control module memory with the Generic Scan Tool (GST)
- Clear the control module memory with the vehicle manufacturer's specific tester
- Turn the ignition**OFF** and disconnect the negative battery cable for at least 1 minute.

Removing the negative battery cable may cause other systems in the vehicle to lose their memory. Prior to removing the cable, ensure you have the proper reset codes for radios and alarms.

➡**The MIL will may also be de-activated for some codes if the vehicle completes three consecutive trips without a fault detected with vehicle conditions similar to those present during the fault.**

DIAGNOSTIC TROUBLE CODES

P0000 No Failures
P0100 Mass or Volume Air Flow Circuit Malfunction
P0101 Mass or Volume Air Flow Circuit Range/Performance Problem
P0102 Mass or Volume Air Flow Circuit Low Input
P0103 Mass or Volume Air Flow Circuit High Input
P0104 Mass or Volume Air Flow Circuit Intermittent
P0105 Manifold Absolute Pressure/Barometric Pressure Circuit Malfunction
P0106 Manifold Absolute Pressure/Barometric Pressure Circuit Range/Performance Problem
P0107 Manifold Absolute Pressure/Barometric Pressure Circuit Low Input
P0108 Manifold Absolute Pressure/Barometric Pressure Circuit High Input
P0109 Manifold Absolute Pressure/Barometric Pressure Circuit Intermittent
P0110 Intake Air Temperature Circuit Malfunction
P0111 Intake Air Temperature Circuit Range/Performance Problem
P0112 Intake Air Temperature Circuit Low Input
P0113 Intake Air Temperature Circuit High Input
P0114 Intake Air Temperature Circuit Intermittent
P0115 Engine Coolant Temperature Circuit Malfunction
P0116 Engine Coolant Temperature Circuit Range/Performance Problem
P0117 Engine Coolant Temperature Circuit Low Input
P0118 Engine Coolant Temperature Circuit High Input

P0119 Engine Coolant Temperature Circuit Intermittent
P0120 Throttle/Pedal Position Sensor/Switch "A" Circuit Malfunction
P0121 Throttle/Pedal Position Sensor/Switch "A" Circuit Range/Performance Problem
P0122 Throttle/Pedal Position Sensor/Switch "A" Circuit Low Input
P0123 Throttle/Pedal Position Sensor/Switch "A" Circuit High Input
P0124 Throttle/Pedal Position Sensor/Switch "A" Circuit Intermittent
P0125 Insufficient Coolant Temperature For Closed Loop Fuel Control
P0126 Insufficient Coolant Temperature For Stable Operation
P0130 O_2 Circuit Malfunction (Bank no. 1 Sensor no. 1)
P0131 O_2 Sensor Circuit Low Voltage (Bank no. 1 Sensor no. 1)
P0132 O_2 Sensor Circuit High Voltage (Bank no. 1 Sensor no. 1)
P0133 O_2 Sensor Circuit Slow Response (Bank no. 1 Sensor no. 1)
P0134 O_2 Sensor Circuit No Activity Detected (Bank no. 1 Sensor no. 1)
P0135 O_2 Sensor Heater Circuit Malfunction (Bank no. 1 Sensor no. 1)
P0136 O_2 Sensor Circuit Malfunction (Bank no. 1 Sensor no. 2)
P0137 O_2 Sensor Circuit Low Voltage (Bank no. 1 Sensor no. 2)
P0138 O_2 Sensor Circuit High Voltage (Bank no. 1 Sensor no. 2)
P0139 O_2 Sensor Circuit Slow Response (Bank no. 1 Sensor no. 2)
P0140 O_2 Sensor Circuit No Activity Detected (Bank no. 1 Sensor no. 2)
P0141 O_2 Sensor Heater Circuit Malfunction (Bank no. 1 Sensor no. 2)
P0142 O_2 Sensor Circuit Malfunction (Bank no. 1 Sensor no. 3)
P0143 O_2 Sensor Circuit Low Voltage (Bank no. 1 Sensor no. 3)
P0144 O_2 Sensor Circuit High Voltage (Bank no. 1 Sensor no. 3)
P0145 O_2 Sensor Circuit Slow Response (Bank no. 1 Sensor no. 3)
P0146 O_2 Sensor Circuit No Activity Detected (Bank no. 1 Sensor no. 3)
P0147 O_2 Sensor Heater Circuit Malfunction (Bank no. 1 Sensor no. 3)
P0150 O_2 Sensor Circuit Malfunction (Bank no. 2 Sensor no. 1)
P0151 O_2 Sensor Circuit Low Voltage (Bank no. 2 Sensor no. 1)
P0152 O_2 Sensor Circuit High Voltage (Bank no. 2 Sensor no. 1)
P0153 O_2 Sensor Circuit Slow Response (Bank no. 2 Sensor no. 1)
P0154 O_2 Sensor Circuit No Activity Detected (Bank no. 2 Sensor no. 1)
P0155 O_2 Sensor Heater Circuit Malfunction (Bank no. 2 Sensor no. 1)
P0156 O_2 Sensor Circuit Malfunction (Bank no. 2 Sensor no. 2)
P0157 O_2 Sensor Circuit Low Voltage (Bank no. 2 Sensor no. 2)
P0158 O_2 Sensor Circuit High Voltage (Bank no. 2 Sensor no. 2)
P0159 O_2 Sensor Circuit Slow Response (Bank no. 2 Sensor no. 2)
P0160 O_2 Sensor Circuit No Activity Detected (Bank no. 2 Sensor no. 2)
P0161 O_2 Sensor Heater Circuit Malfunction (Bank no. 2 Sensor no. 2)
P0162 O_2 Sensor Circuit Malfunction (Bank no. 2 Sensor no. 3)
P0163 O_2 Sensor Circuit Low Voltage (Bank no. 2 Sensor no. 3)
P0164 O_2 Sensor Circuit High Voltage (Bank no. 2 Sensor no. 3)

P0165 O_2 Sensor Circuit Slow Response (Bank no. 2 Sensor no. 3)

P0166 O_2 Sensor Circuit No Activity Detected (Bank no. 2 Sensor no. 3)

P0167 O_2 Sensor Heater Circuit Malfunction (Bank no. 2 Sensor no. 3)

P0170 Fuel Trim Malfunction (Bank no. 1)

P0171 System Too Lean (Bank no. 1)

P0172 System Too Rich (Bank no. 1)

P0173 Fuel Trim Malfunction (Bank no. 2)

P0174 System Too Lean (Bank no. 2)

P0175 System Too Rich (Bank no. 2)

P0176 Fuel Composition Sensor Circuit Malfunction

P0177 Fuel Composition Sensor Circuit Range/Performance

P0178 Fuel Composition Sensor Circuit Low Input

P0179 Fuel Composition Sensor Circuit High Input

P0180 Fuel Temperature Sensor "A" Circuit Malfunction

P0181 Fuel Temperature Sensor "A" Circuit Range/Performance

P0182 Fuel Temperature Sensor "A" Circuit Low Input

P0183 Fuel Temperature Sensor "A" Circuit High Input

P0184 Fuel Temperature Sensor "A" Circuit Intermittent

P0185 Fuel Temperature Sensor "B" Circuit Malfunction

P0186 Fuel Temperature Sensor "B" Circuit Range/Performance

P0187 Fuel Temperature Sensor "B" Circuit Low Input

P0188 Fuel Temperature Sensor "B" Circuit High Input

P0189 Fuel Temperature Sensor "B" Circuit Intermittent

P0190 Fuel Rail Pressure Sensor Circuit Malfunction

P0191 Fuel Rail Pressure Sensor Circuit Range/Performance

P0192 Fuel Rail Pressure Sensor Circuit Low Input

P0193 Fuel Rail Pressure Sensor Circuit High Input

P0194 Fuel Rail Pressure Sensor Circuit Intermittent

P0195 Engine Oil Temperature Sensor Malfunction

P0196 Engine Oil Temperature Sensor Range/Performance

P0197 Engine Oil Temperature Sensor Low

P0198 Engine Oil Temperature Sensor High

P0199 Engine Oil Temperature Sensor Intermittent

P0200 Injector Circuit Malfunction

P0201 Injector Circuit Malfunction—Cylinder no. 1

P0202 Injector Circuit Malfunction—Cylinder no. 2

P0203 Injector Circuit Malfunction—Cylinder no. 3

P0204 Injector Circuit Malfunction—Cylinder no. 4

P0205 Injector Circuit Malfunction—Cylinder no. 5

P0206 Injector Circuit Malfunction—Cylinder no. 6

P0207 Injector Circuit Malfunction—Cylinder no. 7

P0208 Injector Circuit Malfunction—Cylinder no. 8

P0209 Injector Circuit Malfunction—Cylinder no. 9

P0210 Injector Circuit Malfunction—Cylinder no. 10

P0211 Injector Circuit Malfunction—Cylinder no. 11

P0212 Injector Circuit Malfunction—Cylinder no. 12

P0213 Cold Start Injector no. 1 Malfunction

P0214 Cold Start Injector no. 2 Malfunction

P0215 Engine Shutoff Solenoid Malfunction

P0216 Injection Timing Control Circuit Malfunction

P0217 Engine Over Temperature Condition

P0218 Transmission Over Temperature Condition

P0219 Engine Over Speed Condition

P0220 Throttle/Pedal Position Sensor/Switch "B" Circuit Malfunction

P0221 Throttle/Pedal Position Sensor/Switch "B" Circuit Range/Performance Problem

P0222 Throttle/Pedal Position Sensor/Switch "B" Circuit Low Input

P0223 Throttle/Pedal Position Sensor/Switch "B" Circuit High Input

P0224 Throttle/Pedal Position Sensor/Switch "B" Circuit Intermittent

P0225 Throttle/Pedal Position Sensor/Switch "C" Circuit Malfunction

P0226 Throttle/Pedal Position Sensor/Switch "C" Circuit Range/Performance Problem

P0227 Throttle/Pedal Position Sensor/Switch "C" Circuit Low Input

P0228 Throttle/Pedal Position Sensor/Switch "C" Circuit High Input

P0229 Throttle/Pedal Position Sensor/Switch "C" Circuit Intermittent

P0230 Fuel Pump Primary Circuit Malfunction

P0231 Fuel Pump Secondary Circuit Low

P0232 Fuel Pump Secondary Circuit High

P0233 Fuel Pump Secondary Circuit Intermittent

P0234 Engine Over Boost Condition

P0261 Cylinder no. 1 Injector Circuit Low

P0262 Cylinder no. 1 Injector Circuit High

P0263 Cylinder no. 1 Contribution/Balance Fault

P0264 Cylinder no. 2 Injector Circuit Low

P0265 Cylinder no. 2 Injector Circuit High

P0266 Cylinder no. 2 Contribution/Balance Fault

P0267 Cylinder no. 3 Injector Circuit Low

P0268 Cylinder no. 3 Injector Circuit High

P0269 Cylinder no. 3 Contribution/Balance Fault

P0270 Cylinder no. 4 Injector Circuit Low

P0271 Cylinder no. 4 Injector Circuit High

P0272 Cylinder no. 4 Contribution/Balance Fault

P0273 Cylinder no. 5 Injector Circuit Low

P0274 Cylinder no. 5 Injector Circuit High

P0275 Cylinder no. 5 Contribution/Balance Fault

P0276 Cylinder no. 6 Injector Circuit Low

P0277 Cylinder no. 6 Injector Circuit High

P0278 Cylinder no. 6 Contribution/Balance Fault

P0279 Cylinder no. 7 Injector Circuit Low

P0280 Cylinder no. 7 Injector Circuit High

P0281 Cylinder no. 7 Contribution/Balance Fault

P0282 Cylinder no. 8 Injector Circuit Low

P0283 Cylinder no. 8 Injector Circuit High

P0284 Cylinder no. 8 Contribution/Balance Fault

P0285 Cylinder no. 9 Injector Circuit Low

P0286 Cylinder no. 9 Injector Circuit High

P0287 Cylinder no. 9 Contribution/Balance Fault

P0288 Cylinder no. 10 Injector Circuit Low

P0289 Cylinder no. 10 Injector Circuit High

P0290 Cylinder no. 10 Contribution/Balance Fault

P0291 Cylinder no. 11 Injector Circuit Low

P0292 Cylinder no. 11 Injector Circuit High

P0293 Cylinder no. 11 Contribution/Balance Fault

P0294 Cylinder no. 12 Injector Circuit Low

P0295 Cylinder no. 12 Injector Circuit High

P0296 Cylinder no. 12 Contribution/Balance Fault

P0300 Random/Multiple Cylinder Misfire Detected

P0301 Cylinder no. 1—Misfire Detected

P0302 Cylinder no. 2—Misfire Detected

P0303 Cylinder no. 3—Misfire Detected

P0304 Cylinder no. 4—Misfire Detected

P0305 Cylinder no. 5—Misfire Detected

P0306 Cylinder no. 6—Misfire Detected

P0307 Cylinder no. 7—Misfire Detected
P0308 Cylinder no. 8—Misfire Detected
P0309 Cylinder no. 9—Misfire Detected
P0310 Cylinder no. 10—Misfire Detected
P0311 Cylinder no. 11—Misfire Detected
P0312 Cylinder no. 12—Misfire Detected
P0320 Ignition/Distributor Engine Speed Input Circuit Malfunction
P0321 Ignition/Distributor Engine Speed Input Circuit Range/Performance
P0322 Ignition/Distributor Engine Speed Input Circuit No Signal
P0323 Ignition/Distributor Engine Speed Input Circuit Intermittent
P0325 Knock Sensor no. 1—Circuit Malfunction (Bank no. 1 or Single Sensor)
P0326 Knock Sensor no. 1—Circuit Range/Performance (Bank no. 1 or Single Sensor)
P0327 Knock Sensor no. 1—Circuit Low Input (Bank no. 1 or Single Sensor)
P0328 Knock Sensor no. 1—Circuit High Input (Bank no. 1 or Single Sensor)
P0329 Knock Sensor no. 1—Circuit Input Intermittent (Bank no. 1 or Single Sensor)
P0330 Knock Sensor no. 2—Circuit Malfunction (Bank no. 2)
P0331 Knock Sensor no. 2—Circuit Range/Performance (Bank no. 2)
P0332 Knock Sensor no. 2—Circuit Low Input (Bank no. 2)
P0333 Knock Sensor no. 2—Circuit High Input (Bank no. 2)
P0334 Knock Sensor no. 2—Circuit Input Intermittent (Bank no. 2)
P0335 Crankshaft Position Sensor "A" Circuit Malfunction
P0336 Crankshaft Position Sensor "A" Circuit Range/Performance
P0337 Crankshaft Position Sensor "A" Circuit Low Input
P0338 Crankshaft Position Sensor "A" Circuit High Input
P0339 Crankshaft Position Sensor "A" Circuit Intermittent
P0340 Camshaft Position Sensor Circuit Malfunction
P0341 Camshaft Position Sensor Circuit Range/Performance
P0342 Camshaft Position Sensor Circuit Low Input
P0343 Camshaft Position Sensor Circuit High Input
P0344 Camshaft Position Sensor Circuit Intermittent
P0350 Ignition Coil Primary/Secondary Circuit Malfunction
P0351 Ignition Coil "A" Primary/Secondary Circuit Malfunction
P0352 Ignition Coil "B" Primary/Secondary Circuit Malfunction
P0353 Ignition Coil "C" Primary/Secondary Circuit Malfunction
P0354 Ignition Coil "D" Primary/Secondary Circuit Malfunction
P0355 Ignition Coil "E" Primary/Secondary Circuit Malfunction
P0356 Ignition Coil "F" Primary/Secondary Circuit Malfunction
P0357 Ignition Coil "G" Primary/Secondary Circuit Malfunction
P0358 Ignition Coil "H" Primary/Secondary Circuit Malfunction
P0359 Ignition Coil "I" Primary/Secondary Circuit Malfunction
P0360 Ignition Coil "J" Primary/Secondary Circuit Malfunction
P0361 Ignition Coil "K" Primary/Secondary Circuit Malfunction
P0362 Ignition Coil "L" Primary/Secondary Circuit Malfunction
P0370 Timing Reference High Resolution Signal "A" Malfunction
P0371 Timing Reference High Resolution Signal "A" Too Many Pulses
P0372 Timing Reference High Resolution Signal "A" Too Few Pulses
P0373 Timing Reference High Resolution Signal "A" Intermittent/Erratic Pulses

P0374 Timing Reference High Resolution Signal "A" No Pulses
P0375 Timing Reference High Resolution Signal "B" Malfunction
P0376 Timing Reference High Resolution Signal "B" Too Many Pulses
P0377 Timing Reference High Resolution Signal "B" Too Few Pulses
P0378 Timing Reference High Resolution Signal "B" Intermittent/Erratic Pulses
P0379 Timing Reference High Resolution Signal "B" No Pulses
P0380 Glow Plug/Heater Circuit "A" Malfunction
P0381 Glow Plug/Heater Indicator Circuit Malfunction
P0382 Glow Plug/Heater Circuit "B" Malfunction
P0385 Crankshaft Position Sensor "B" Circuit Malfunction
P0386 Crankshaft Position Sensor "B" Circuit Range/Performance
P0387 Crankshaft Position Sensor "B" Circuit Low Input
P0388 Crankshaft Position Sensor "B" Circuit High Input
P0389 Crankshaft Position Sensor "B" Circuit Intermittent
P0400 Exhaust Gas Recirculation Flow Malfunction
P0401 Exhaust Gas Recirculation Flow Insufficient Detected
P0402 Exhaust Gas Recirculation Flow Excessive Detected
P0403 Exhaust Gas Recirculation Circuit Malfunction
P0404 Exhaust Gas Recirculation Circuit Range/Performance
P0405 Exhaust Gas Recirculation Sensor "A" Circuit Low
P0406 Exhaust Gas Recirculation Sensor "A" Circuit High
P0407 Exhaust Gas Recirculation Sensor "B" Circuit Low
P0408 Exhaust Gas Recirculation Sensor "B" Circuit High
P0410 Secondary Air Injection System Malfunction
P0411 Secondary Air Injection System Incorrect Flow Detected
P0412 Secondary Air Injection System Switching Valve "A" Circuit Malfunction
P0413 Secondary Air Injection System Switching Valve "A" Circuit Open
P0414 Secondary Air Injection System Switching Valve "A" Circuit Shorted
P0415 Secondary Air Injection System Switching Valve "B" Circuit Malfunction
P0416 Secondary Air Injection System Switching Valve "B" Circuit Open
P0417 Secondary Air Injection System Switching Valve "B" Circuit Shorted
P0418 Secondary Air Injection System Relay "A" Circuit Malfunction
P0419 Secondary Air Injection System Relay "B" Circuit Malfunction
P0420 Catalyst System Efficiency Below Threshold (Bank no. 1)
P0421 Warm Up Catalyst Efficiency Below Threshold (Bank no. 1)
P0422 Main Catalyst Efficiency Below Threshold (Bank no. 1)
P0423 Heated Catalyst Efficiency Below Threshold (Bank no. 1)
P0424 Heated Catalyst Temperature Below Threshold (Bank no. 1)
P0430 Catalyst System Efficiency Below Threshold (Bank no. 2)
P0431 Warm Up Catalyst Efficiency Below Threshold (Bank no. 2)
P0432 Main Catalyst Efficiency Below Threshold (Bank no. 2)
P0433 Heated Catalyst Efficiency Below Threshold (Bank no. 2)
P0434 Heated Catalyst Temperature Below Threshold (Bank no. 2)
P0440 Evaporative Emission Control System Malfunction
P0441 Evaporative Emission Control System Incorrect Purge Flow
P0442 Evaporative Emission Control System Leak Detected (Small Leak)

P0443 Evaporative Emission Control System Purge Control Valve Circuit Malfunction

P0444 Evaporative Emission Control System Purge Control Valve Circuit Open

P0445 Evaporative Emission Control System Purge Control Valve Circuit Shorted

P0446 Evaporative Emission Control System Vent Control Circuit Malfunction

P0447 Evaporative Emission Control System Vent Control Circuit Open

P0448 Evaporative Emission Control System Vent Control Circuit Shorted

P0449 Evaporative Emission Control System Vent Valve/Solenoid Circuit Malfunction

P0450 Evaporative Emission Control System Pressure Sensor Malfunction

P0451 Evaporative Emission Control System Pressure Sensor Range/Performance

P0452 Evaporative Emission Control System Pressure Sensor Low Input

P0453 Evaporative Emission Control System Pressure Sensor High Input

P0454 Evaporative Emission Control System Pressure Sensor Intermittent

P0455 Evaporative Emission Control System Leak Detected (Gross Leak)

P0460 Fuel Level Sensor Circuit Malfunction

P0461 Fuel Level Sensor Circuit Range/Performance

P0462 Fuel Level Sensor Circuit Low Input

P0463 Fuel Level Sensor Circuit High Input

P0464 Fuel Level Sensor Circuit Intermittent

P0465 Purge Flow Sensor Circuit Malfunction

P0466 Purge Flow Sensor Circuit Range/Performance

P0467 Purge Flow Sensor Circuit Low Input

P0468 Purge Flow Sensor Circuit High Input

P0469 Purge Flow Sensor Circuit Intermittent

P0470 Exhaust Pressure Sensor Malfunction

P0471 Exhaust Pressure Sensor Range/Performance

P0472 Exhaust Pressure Sensor Low

P0473 Exhaust Pressure Sensor High

P0474 Exhaust Pressure Sensor Intermittent

P0475 Exhaust Pressure Control Valve Malfunction

P0476 Exhaust Pressure Control Valve Range/Performance

P0477 Exhaust Pressure Control Valve Low

P0478 Exhaust Pressure Control Valve High

P0479 Exhaust Pressure Control Valve Intermittent

P0480 Cooling Fan no. 1 Control Circuit Malfunction

P0481 Cooling Fan no. 2 Control Circuit Malfunction

P0482 Cooling Fan no. 3 Control Circuit Malfunction

P0483 Cooling Fan Rationality Check Malfunction

P0484 Cooling Fan Circuit Over Current

P0485 Cooling Fan Power/Ground Circuit Malfunction

P0500 Vehicle Speed Sensor Malfunction

P0501 Vehicle Speed Sensor Range/Performance

P0502 Vehicle Speed Sensor Circuit Low Input

P0503 Vehicle Speed Sensor Intermittent/Erratic/High

P0505 Idle Control System Malfunction

P0506 Idle Control System RPM Lower Than Expected

P0507 Idle Control System RPM Higher Than Expected

P0510 Closed Throttle Position Switch Malfunction

P0520 Engine Oil Pressure Sensor/Switch Circuit Malfunction

P0521 Engine Oil Pressure Sensor/Switch Range/Performance

P0522 Engine Oil Pressure Sensor/Switch Low Voltage

P0523 Engine Oil Pressure Sensor/Switch High Voltage

P0530 A/C Refrigerant Pressure Sensor Circuit Malfunction

P0531 A/C Refrigerant Pressure Sensor Circuit Range/Performance

P0532 A/C Refrigerant Pressure Sensor Circuit Low Input

P0533 A/C Refrigerant Pressure Sensor Circuit High Input

P0534 A/C Refrigerant Charge Loss

P0550 Power Steering Pressure Sensor Circuit Malfunction

P0551 Power Steering Pressure Sensor Circuit Range/Performance

P0552 Power Steering Pressure Sensor Circuit Low Input

P0553 Power Steering Pressure Sensor Circuit High Input

P0554 Power Steering Pressure Sensor Circuit Intermittent

P0560 System Voltage Malfunction

P0561 System Voltage Unstable

P0562 System Voltage Low

P0563 System Voltage High

P0565 Cruise Control On Signal Malfunction

P0566 Cruise Control Off Signal Malfunction

P0567 Cruise Control Resume Signal Malfunction

P0568 Cruise Control Set Signal Malfunction

P0569 Cruise Control Coast Signal Malfunction

P0570 Cruise Control Accel Signal Malfunction

P0571 Cruise Control/Brake Switch "A" Circuit Malfunction

P0572 Cruise Control/Brake Switch "A" Circuit Low

P0573 Cruise Control/Brake Switch "A" Circuit High

P0574 Through P0580 Reserved for Cruise Codes

P0600 Serial Communication Link Malfunction

P0601 Internal Control Module Memory Check Sum Error

P0602 Control Module Programming Error

P0603 Internal Control Module Keep Alive Memory (KAM) Error

P0604 Internal Control Module Random Access Memory (RAM) Error

P0605 Internal Control Module Read Only Memory (ROM) Error

P0606 PCM Processor Fault

P0608 Control Module VSS Output "A" Malfunction

P0609 Control Module VSS Output "B" Malfunction

P0620 Generator Control Circuit Malfunction

P0621 Generator Lamp "L" Control Circuit Malfunction

P0622 Generator Field "F" Control Circuit Malfunction

P0650 Malfunction Indicator Lamp (MIL) Control Circuit Malfunction

P0654 Engine RPM Output Circuit Malfunction

P0655 Engine Hot Lamp Output Control Circuit Malfunction

P0656 Fuel Level Output Circuit Malfunction

P0700 Transmission Control System Malfunction

P0701 Transmission Control System Range/Performance

P0702 Transmission Control System Electrical

P0703 Torque Converter/Brake Switch "B" Circuit Malfunction

P0704 Clutch Switch Input Circuit Malfunction

P0705 Transmission Range Sensor Circuit Malfunction (PRNDL Input)

P0706 Transmission Range Sensor Circuit Range/Performance

P0707 Transmission Range Sensor Circuit Low Input

P0708 Transmission Range Sensor Circuit High Input

P0709 Transmission Range Sensor Circuit Intermittent

P0710 Transmission Fluid Temperature Sensor Circuit Malfunction

P0711 Transmission Fluid Temperature Sensor Circuit Range/Performance

P0712 Transmission Fluid Temperature Sensor Circuit Low Input
P0713 Transmission Fluid Temperature Sensor Circuit High Input
P0714 Transmission Fluid Temperature Sensor Circuit Intermittent
P0715 Input/Turbine Speed Sensor Circuit Malfunction
P0716 Input/Turbine Speed Sensor Circuit Range/Performance
P0717 Input/Turbine Speed Sensor Circuit No Signal
P0718 Input/Turbine Speed Sensor Circuit Intermittent
P0719 Torque Converter/Brake Switch "B" Circuit Low
P0720 Output Speed Sensor Circuit Malfunction
P0721 Output Speed Sensor Circuit Range/Performance
P0722 Output Speed Sensor Circuit No Signal
P0723 Output Speed Sensor Circuit Intermittent
P0724 Torque Converter/Brake Switch "B" Circuit High
P0725 Engine Speed Input Circuit Malfunction
P0726 Engine Speed Input Circuit Range/Performance
P0727 Engine Speed Input Circuit No Signal
P0728 Engine Speed Input Circuit Intermittent
P0730 Incorrect Gear Ratio
P0731 Gear no. 1 Incorrect Ratio
P0732 Gear no. 2 Incorrect Ratio
P0733 Gear no. 3 Incorrect Ratio
P0734 Gear no. 4 Incorrect Ratio
P0735 Gear no. 5 Incorrect Ratio
P0736 Reverse Incorrect Ratio
P0740 Torque Converter Clutch Circuit Malfunction
P0741 Torque Converter Clutch Circuit Performance or Stuck Off
P0742 Torque Converter Clutch Circuit Stuck On
P0743 Torque Converter Clutch Circuit Electrical
P0744 Torque Converter Clutch Circuit Intermittent
P0745 Pressure Control Solenoid Malfunction
P0746 Pressure Control Solenoid Performance or Stuck Off
P0747 Pressure Control Solenoid Stuck On
P0748 Pressure Control Solenoid Electrical
P0749 Pressure Control Solenoid Intermittent
P0750 Shift Solenoid "A" Malfunction
P0751 Shift Solenoid "A" Performance or Stuck Off
P0752 Shift Solenoid "A" Stuck On
P0753 Shift Solenoid "A" Electrical
P0754 Shift Solenoid "A" Intermittent
P0755 Shift Solenoid "B" Malfunction
P0756 Shift Solenoid "B" Performance or Stuck Off
P0757 Shift Solenoid "B" Stuck On
P0758 Shift Solenoid "B" Electrical
P0759 Shift Solenoid "B" Intermittent
P0760 Shift Solenoid "C" Malfunction
P0761 Shift Solenoid "C" Performance Or Stuck Off
P0762 Shift Solenoid "C" Stuck On
P0763 Shift Solenoid "C" Electrical
P0764 Shift Solenoid "C" Intermittent
P0765 Shift Solenoid "D" Malfunction
P0766 Shift Solenoid "D" Performance Or Stuck Off
P0767 Shift Solenoid "D" Stuck On
P0768 Shift Solenoid "D" Electrical
P0769 Shift Solenoid "D" Intermittent
P0770 Shift Solenoid "E" Malfunction
P0771 Shift Solenoid "E" Performance Or Stuck Off
P0772 Shift Solenoid "E" Stuck On
P0773 Shift Solenoid "E" Electrical

P0774 Shift Solenoid "E" Intermittent
P0780 Shift Malfunction
P0781 1–2 Shift Malfunction
P0782 2–3 Shift Malfunction
P0783 3–4 Shift Malfunction
P0784 4–5 Shift Malfunction
P0785 Shift/Timing Solenoid Malfunction
P0786 Shift/Timing Solenoid Range/Performance
P0787 Shift/Timing Solenoid Low
P0788 Shift/Timing Solenoid High
P0789 Shift/Timing Solenoid Intermittent
P0790 Normal/Performance Switch Circuit Malfunction
P0801 Reverse Inhibit Control Circuit Malfunction
P0803 1–4 Upshift (Skip Shift) Solenoid Control Circuit Malfunction
P0804 1–4 Upshift (Skip Shift) Lamp Control Circuit Malfunction
P1100 Starter Switch Circuit Fault
P1101 Neutral Position Switch Circuit Fault (MT)

Suzuki

READING CODES

Reading the control module memory is on of the first steps in OBD II system diagnostics. This step should be initially performed to determine the general nature of the fault. Subsequent readings will determine if the fault has been cleared.

Reading codes can be performed by any of the methods below:
• Read the control module memory with the Generic Scan Tool (GST)
• Read the control module memory with the vehicle manufacturer's specific tester

To read the fault codes, connect the scan tool or tester according to the manufacturer's instructions. Follow the manufacturer's specified procedure for reading the codes.

CLEARING CODES

Control module reset procedures are a very important part of OBD II System diagnostics. This step should be done at the end of any fault code repair and at the end of any driveability repair.

Clearing codes can be performed by any of the methods below:
• Clear the control module memory with the Generic Scan Tool (GST)
• Clear the control module memory with the vehicle manufacturer's specific tester
• Turn the ignition**OFF** and remove the negative battery cable for at least 1 minute.

Removing the negative battery cable may cause other systems in the vehicle to lose their memory. Prior to removing the cable, ensure you have the proper reset codes for radios and alarms.

➡**The MIL will may also be de-activated for some codes if the vehicle completes three consecutive trips without a fault detected with vehicle conditions similar to those present during the fault.**

DIAGNOSTIC TROUBLE CODES

P0000 No Failures
P0100 Mass or Volume Air Flow Circuit Malfunction

P0101 Mass or Volume Air Flow Circuit Range/Performance Problem

P0102 Mass or Volume Air Flow Circuit Low Input

P0103 Mass or Volume Air Flow Circuit High Input

P0104 Mass or Volume Air Flow Circuit Intermittent

P0105 Manifold Absolute Pressure/Barometric Pressure Circuit Malfunction

P0106 Manifold Absolute Pressure/Barometric Pressure Circuit Range/Performance Problem

P0107 Manifold Absolute Pressure/Barometric Pressure Circuit Low Input

P0108 Manifold Absolute Pressure/Barometric Pressure Circuit High Input

P0109 Manifold Absolute Pressure/Barometric Pressure Circuit Intermittent

P0110 Intake Air Temperature Circuit Malfunction

P0111 Intake Air Temperature Circuit Range/Performance Problem

P0112 Intake Air Temperature Circuit Low Input

P0113 Intake Air Temperature Circuit High Input

P0114 Intake Air Temperature Circuit Intermittent

P0115 Engine Coolant Temperature Circuit Malfunction

P0116 Engine Coolant Temperature Circuit Range/Performance Problem

P0117 Engine Coolant Temperature Circuit Low Input

P0118 Engine Coolant Temperature Circuit High Input

P0119 Engine Coolant Temperature Circuit Intermittent

P0120 Throttle/Pedal Position Sensor/Switch "A" Circuit Malfunction

P0121 Throttle/Pedal Position Sensor/Switch "A" Circuit Range/Performance Problem

P0122 Throttle/Pedal Position Sensor/Switch "A" Circuit Low Input

P0123 Throttle/Pedal Position Sensor/Switch "A" Circuit High Input

P0124 Throttle/Pedal Position Sensor/Switch "A" Circuit Intermittent

P0125 Insufficient Coolant Temperature For Closed Loop Fuel Control

P0126 Insufficient Coolant Temperature For Stable Operation

P0130 O_2 Circuit Malfunction (Bank no. 1 Sensor no. 1)

P0131 O_2 Sensor Circuit Low Voltage (Bank no. 1 Sensor no. 1)

P0132 O_2 Sensor Circuit High Voltage (Bank no. 1 Sensor no. 1)

P0133 O_2 Sensor Circuit Slow Response (Bank no. 1 Sensor no. 1)

P0134 O_2 Sensor Circuit No Activity Detected (Bank no. 1 Sensor no. 1)

P0135 O_2 Sensor Heater Circuit Malfunction (Bank no. 1 Sensor no. 1)

P0136 O_2 Sensor Circuit Malfunction (Bank no. 1 Sensor no. 2)

P0137 O_2 Sensor Circuit Low Voltage (Bank no. 1 Sensor no. 2)

P0138 O_2 Sensor Circuit High Voltage (Bank no. 1 Sensor no. 2)

P0139 O_2 Sensor Circuit Slow Response (Bank no. 1 Sensor no. 2)

P0140 O_2 Sensor Circuit No Activity Detected (Bank no. 1 Sensor no. 2)

P0141 O_2 Sensor Heater Circuit Malfunction (Bank no. 1 Sensor no. 2)

P0142 O_2 Sensor Circuit Malfunction (Bank no. 1 Sensor no. 3)

P0143 O_2 Sensor Circuit Low Voltage (Bank no. 1 Sensor no. 3)

P0144 O_2 Sensor Circuit High Voltage (Bank no. 1 Sensor no. 3)

P0145 O_2 Sensor Circuit Slow Response (Bank no. 1 Sensor no. 3)

P0146 O_2 Sensor Circuit No Activity Detected (Bank no. 1 Sensor no. 3)

P0147 O_2 Sensor Heater Circuit Malfunction (Bank no. 1 Sensor no. 3)

P0150 O_2 Sensor Circuit Malfunction (Bank no. 2 Sensor no. 1)

P0151 O_2 Sensor Circuit Low Voltage (Bank no. 2 Sensor no. 1)

P0152 O_2 Sensor Circuit High Voltage (Bank no. 2 Sensor no. 1)

P0153 O_2 Sensor Circuit Slow Response (Bank no. 2 Sensor no. 1)

P0154 O_2 Sensor Circuit No Activity Detected (Bank no. 2 Sensor no. 1)

P0155 O_2 Sensor Heater Circuit Malfunction (Bank no. 2 Sensor no. 1)

P0156 O_2 Sensor Circuit Malfunction (Bank no. 2 Sensor no. 2)

P0157 O_2 Sensor Circuit Low Voltage (Bank no. 2 Sensor no. 2)

P0158 O_2 Sensor Circuit High Voltage (Bank no. 2 Sensor no. 2)

P0159 O_2 Sensor Circuit Slow Response (Bank no. 2 Sensor no. 2)

P0160 O_2 Sensor Circuit No Activity Detected (Bank no. 2 Sensor no. 2)

P0161 O_2 Sensor Heater Circuit Malfunction (Bank no. 2 Sensor no. 2)

P0162 O_2 Sensor Circuit Malfunction (Bank no. 2 Sensor no. 3)

P0163 O_2 Sensor Circuit Low Voltage (Bank no. 2 Sensor no. 3)

P0164 O_2 Sensor Circuit High Voltage (Bank no. 2 Sensor no. 3)

P0165 O_2 Sensor Circuit Slow Response (Bank no. 2 Sensor no. 3)

P0166 O_2 Sensor Circuit No Activity Detected (Bank no. 2 Sensor no. 3)

P0167 O_2 Sensor Heater Circuit Malfunction (Bank no. 2 Sensor no. 3)

P0170 Fuel Trim Malfunction (Bank no. 1)

P0171 System Too Lean (Bank no. 1)

P0172 System Too Rich (Bank no. 1)

P0173 Fuel Trim Malfunction (Bank no. 2)

P0174 System Too Lean (Bank no. 2)

P0175 System Too Rich (Bank no. 2)

P0176 Fuel Composition Sensor Circuit Malfunction

P0177 Fuel Composition Sensor Circuit Range/Performance

P0178 Fuel Composition Sensor Circuit Low Input

P0179 Fuel Composition Sensor Circuit High Input

P0180 Fuel Temperature Sensor "A" Circuit Malfunction

P0181 Fuel Temperature Sensor "A" Circuit Range/Performance

P0182 Fuel Temperature Sensor "A" Circuit Low Input

P0183 Fuel Temperature Sensor "A" Circuit High Input

P0184 Fuel Temperature Sensor "A" Circuit Intermittent

P0185 Fuel Temperature Sensor "B" Circuit Malfunction

P0186 Fuel Temperature Sensor "B" Circuit Range/Performance

P0187 Fuel Temperature Sensor "B" Circuit Low Input

P0188 Fuel Temperature Sensor "B" Circuit High Input

P0189 Fuel Temperature Sensor "B" Circuit Intermittent

P0190 Fuel Rail Pressure Sensor Circuit Malfunction

P0191 Fuel Rail Pressure Sensor Circuit Range/Performance

P0192 Fuel Rail Pressure Sensor Circuit Low Input

P0193 Fuel Rail Pressure Sensor Circuit High Input

P0194 Fuel Rail Pressure Sensor Circuit Intermittent

P0195 Engine Oil Temperature Sensor Malfunction

P0196 Engine Oil Temperature Sensor Range/Performance

P0197 Engine Oil Temperature Sensor Low

P0198 Engine Oil Temperature Sensor High

P0199 Engine Oil Temperature Sensor Intermittent
P0200 Injector Circuit Malfunction
P0201 Injector Circuit Malfunction—Cylinder no. 1
P0202 Injector Circuit Malfunction—Cylinder no. 2
P0203 Injector Circuit Malfunction—Cylinder no. 3
P0204 Injector Circuit Malfunction—Cylinder no. 4
P0205 Injector Circuit Malfunction—Cylinder no. 5
P0206 Injector Circuit Malfunction—Cylinder no. 6
P0207 Injector Circuit Malfunction—Cylinder no. 7
P0208 Injector Circuit Malfunction—Cylinder no. 8
P0209 Injector Circuit Malfunction—Cylinder no. 9
P0210 Injector Circuit Malfunction—Cylinder no. 10
P0211 Injector Circuit Malfunction—Cylinder no. 11
P0212 Injector Circuit Malfunction—Cylinder no. 12
P0213 Cold Start Injector no. 1 Malfunction
P0214 Cold Start Injector no. 2 Malfunction
P0215 Engine Shutoff Solenoid Malfunction
P0216 Injection Timing Control Circuit Malfunction
P0217 Engine Over Temperature Condition
P0218 Transmission Over Temperature Condition
P0219 Engine Over Speed Condition
P0220 Throttle/Pedal Position Sensor/Switch "B" Circuit Malfunction
P0221 Throttle/Pedal Position Sensor/Switch "B" Circuit Range/Performance Problem
P0222 Throttle/Pedal Position Sensor/Switch "B" Circuit Low Input
P0223 Throttle/Pedal Position Sensor/Switch "B" Circuit High Input
P0224 Throttle/Pedal Position Sensor/Switch "B" Circuit Intermittent
P0225 Throttle/Pedal Position Sensor/Switch "C" Circuit Malfunction
P0226 Throttle/Pedal Position Sensor/Switch "C" Circuit Range/Performance Problem
P0227 Throttle/Pedal Position Sensor/Switch "C" Circuit Low Input
P0228 Throttle/Pedal Position Sensor/Switch "C" Circuit High Input
P0229 Throttle/Pedal Position Sensor/Switch "C" Circuit Intermittent
P0230 Fuel Pump Primary Circuit Malfunction
P0231 Fuel Pump Secondary Circuit Low
P0232 Fuel Pump Secondary Circuit High
P0233 Fuel Pump Secondary Circuit Intermittent
P0234 Engine Over Boost Condition
P0261 Cylinder no. 1 Injector Circuit Low
P0262 Cylinder no. 1 Injector Circuit High
P0263 Cylinder no. 1 Contribution/Balance Fault
P0264 Cylinder no. 2 Injector Circuit Low
P0265 Cylinder no. 2 Injector Circuit High
P0266 Cylinder no. 2 Contribution/Balance Fault
P0267 Cylinder no. 3 Injector Circuit Low
P0268 Cylinder no. 3 Injector Circuit High
P0269 Cylinder no. 3 Contribution/Balance Fault
P0270 Cylinder no. 4 Injector Circuit Low
P0271 Cylinder no. 4 Injector Circuit High
P0272 Cylinder no. 4 Contribution/Balance Fault
P0273 Cylinder no. 5 Injector Circuit Low
P0274 Cylinder no. 5 Injector Circuit High
P0275 Cylinder no. 5 Contribution/Balance Fault
P0276 Cylinder no. 6 Injector Circuit Low

P0277 Cylinder no. 6 Injector Circuit High
P0278 Cylinder no. 6 Contribution/Balance Fault
P0279 Cylinder no. 7 Injector Circuit Low
P0280 Cylinder no. 7 Injector Circuit High
P0281 Cylinder no. 7 Contribution/Balance Fault
P0282 Cylinder no. 8 Injector Circuit Low
P0283 Cylinder no. 8 Injector Circuit High
P0284 Cylinder no. 8 Contribution/Balance Fault
P0285 Cylinder no. 9 Injector Circuit Low
P0286 Cylinder no. 9 Injector Circuit High
P0287 Cylinder no. 9 Contribution/Balance Fault
P0288 Cylinder no. 10 Injector Circuit Low
P0289 Cylinder no. 10 Injector Circuit High
P0290 Cylinder no. 10 Contribution/Balance Fault
P0291 Cylinder no. 11 Injector Circuit Low
P0292 Cylinder no. 11 Injector Circuit High
P0293 Cylinder no. 11 Contribution/Balance Fault
P0294 Cylinder no. 12 Injector Circuit Low
P0295 Cylinder no. 12 Injector Circuit High
P0296 Cylinder no. 12 Contribution/Balance Fault
P0300 Random/Multiple Cylinder Misfire Detected
P0301 Cylinder no. 1—Misfire Detected
P0302 Cylinder no. 2—Misfire Detected
P0303 Cylinder no. 3—Misfire Detected
P0304 Cylinder no. 4—Misfire Detected
P0305 Cylinder no. 5—Misfire Detected
P0306 Cylinder no. 6—Misfire Detected
P0307 Cylinder no. 7—Misfire Detected
P0308 Cylinder no. 8—Misfire Detected
P0309 Cylinder no. 9—Misfire Detected
P0310 Cylinder no. 10—Misfire Detected
P0311 Cylinder no. 11—Misfire Detected
P0312 Cylinder no. 12—Misfire Detected
P0320 Ignition/Distributor Engine Speed Input Circuit Malfunction
P0321 Ignition/Distributor Engine Speed Input Circuit Range/Performance
P0322 Ignition/Distributor Engine Speed Input Circuit No Signal
P0323 Ignition/Distributor Engine Speed Input Circuit Intermittent
P0325 Knock Sensor no. 1—Circuit Malfunction (Bank no. 1 or Single Sensor)
P0326 Knock Sensor no. 1—Circuit Range/Performance (Bank no. 1 or Single Sensor)
P0327 Knock Sensor no. 1—Circuit Low Input (Bank no. 1 or Single Sensor)
P0328 Knock Sensor no. 1—Circuit High Input (Bank no. 1 or Single Sensor)
P0329 Knock Sensor no. 1—Circuit Input Intermittent (Bank no. 1 or Single Sensor)
P0330 Knock Sensor no. 2—Circuit Malfunction (Bank no. 2)
P0331 Knock Sensor no. 2—Circuit Range/Performance (Bank no. 2)
P0332 Knock Sensor no. 2—Circuit Low Input (Bank no. 2)
P0333 Knock Sensor no. 2—Circuit High Input (Bank no. 2)
P0334 Knock Sensor no. 2—Circuit Input Intermittent (Bank no. 2)
P0335 Crankshaft Position Sensor "A" Circuit Malfunction
P0336 Crankshaft Position Sensor "A" Circuit Range/Performance
P0337 Crankshaft Position Sensor "A" Circuit Low Input
P0338 Crankshaft Position Sensor "A" Circuit High Input

P0339 Crankshaft Position Sensor "A" Circuit Intermittent
P0340 Camshaft Position Sensor Circuit Malfunction
P0341 Camshaft Position Sensor Circuit Range/Performance
P0342 Camshaft Position Sensor Circuit Low Input
P0343 Camshaft Position Sensor Circuit High Input
P0344 Camshaft Position Sensor Circuit Intermittent
P0350 Ignition Coil Primary/Secondary Circuit Malfunction
P0351 Ignition Coil "A" Primary/Secondary Circuit Malfunction
P0352 Ignition Coil "B" Primary/Secondary Circuit Malfunction
P0353 Ignition Coil "C" Primary/Secondary Circuit Malfunction
P0354 Ignition Coil "D" Primary/Secondary Circuit Malfunction
P0355 Ignition Coil "E" Primary/Secondary Circuit Malfunction
P0356 Ignition Coil "F" Primary/Secondary Circuit Malfunction
P0357 Ignition Coil "G" Primary/Secondary Circuit Malfunction
P0358 Ignition Coil "H" Primary/Secondary Circuit Malfunction
P0359 Ignition Coil "I" Primary/Secondary Circuit Malfunction
P0360 Ignition Coil "J" Primary/Secondary Circuit Malfunction
P0361 Ignition Coil "K" Primary/Secondary Circuit Malfunction
P0362 Ignition Coil "L" Primary/Secondary Circuit Malfunction
P0370 Timing Reference High Resolution Signal "A" Malfunction
P0371 Timing Reference High Resolution Signal "A" Too Many Pulses
P0372 Timing Reference High Resolution Signal "A" Too Few Pulses
P0373 Timing Reference High Resolution Signal "A" Intermittent/Erratic Pulses
P0374 Timing Reference High Resolution Signal "A" No Pulses
P0375 Timing Reference High Resolution Signal "B" Malfunction
P0376 Timing Reference High Resolution Signal "B" Too Many Pulses
P0377 Timing Reference High Resolution Signal "B" Too Few Pulses
P0378 Timing Reference High Resolution Signal "B" Intermittent/Erratic Pulses
P0379 Timing Reference High Resolution Signal "B" No Pulses
P0380 Glow Plug/Heater Circuit "A" Malfunction
P0381 Glow Plug/Heater Indicator Circuit Malfunction
P0382 Glow Plug/Heater Circuit "B" Malfunction
P0385 Crankshaft Position Sensor "B" Circuit Malfunction
P0386 Crankshaft Position Sensor "B" Circuit Range/Performance
P0387 Crankshaft Position Sensor "B" Circuit Low Input
P0388 Crankshaft Position Sensor "B" Circuit High Input
P0389 Crankshaft Position Sensor "B" Circuit Intermittent
P0400 Exhaust Gas Recirculation Flow Malfunction
P0401 Exhaust Gas Recirculation Flow Insufficient Detected
P0402 Exhaust Gas Recirculation Flow Excessive Detected
P0403 Exhaust Gas Recirculation Circuit Malfunction
P0404 Exhaust Gas Recirculation Circuit Range/Performance
P0405 Exhaust Gas Recirculation Sensor "A" Circuit Low
P0406 Exhaust Gas Recirculation Sensor "A" Circuit High
P0407 Exhaust Gas Recirculation Sensor "B" Circuit Low
P0408 Exhaust Gas Recirculation Sensor "B" Circuit High
P0410 Secondary Air Injection System Malfunction

P0411 Secondary Air Injection System Incorrect Flow Detected
P0412 Secondary Air Injection System Switching Valve "A" Circuit Malfunction
P0413 Secondary Air Injection System Switching Valve "A" Circuit Open
P0414 Secondary Air Injection System Switching Valve "A" Circuit Shorted
P0415 Secondary Air Injection System Switching Valve "B" Circuit Malfunction
P0416 Secondary Air Injection System Switching Valve "B" Circuit Open
P0417 Secondary Air Injection System Switching Valve "B" Circuit Shorted
P0418 Secondary Air Injection System Relay "A" Circuit Malfunction
P0419 Secondary Air Injection System Relay "B" Circuit Malfunction
P0420 Catalyst System Efficiency Below Threshold (Bank no. 1)
P0421 Warm Up Catalyst Efficiency Below Threshold (Bank no. 1)
P0422 Main Catalyst Efficiency Below Threshold (Bank no. 1)
P0423 Heated Catalyst Efficiency Below Threshold (Bank no. 1)
P0424 Heated Catalyst Temperature Below Threshold (Bank no. 1)
P0430 Catalyst System Efficiency Below Threshold (Bank no. 2)
P0431 Warm Up Catalyst Efficiency Below Threshold (Bank no. 2)
P0432 Main Catalyst Efficiency Below Threshold (Bank no. 2)
P0433 Heated Catalyst Efficiency Below Threshold (Bank no. 2)
P0434 Heated Catalyst Temperature Below Threshold (Bank no. 2)
P0440 Evaporative Emission Control System Malfunction
P0441 Evaporative Emission Control System Incorrect Purge Flow
P0442 Evaporative Emission Control System Leak Detected (Small Leak)
P0443 Evaporative Emission Control System Purge Control Valve Circuit Malfunction
P0444 Evaporative Emission Control System Purge Control Valve Circuit Open
P0445 Evaporative Emission Control System Purge Control Valve Circuit Shorted
P0446 Evaporative Emission Control System Vent Control Circuit Malfunction
P0447 Evaporative Emission Control System Vent Control Circuit Open
P0448 Evaporative Emission Control System Vent Control Circuit Shorted
P0449 Evaporative Emission Control System Vent Valve/Solenoid Circuit Malfunction
P0450 Evaporative Emission Control System Pressure Sensor Malfunction
P0451 Evaporative Emission Control System Pressure Sensor Range/Performance
P0452 Evaporative Emission Control System Pressure Sensor Low Input
P0453 Evaporative Emission Control System Pressure Sensor High Input
P0454 Evaporative Emission Control System Pressure Sensor Intermittent
P0455 Evaporative Emission Control System Leak Detected (Gross Leak)
P0460 Fuel Level Sensor Circuit Malfunction
P0461 Fuel Level Sensor Circuit Range/Performance
P0462 Fuel Level Sensor Circuit Low Input
P0463 Fuel Level Sensor Circuit High Input

P0464 Fuel Level Sensor Circuit Intermittent
P0465 Purge Flow Sensor Circuit Malfunction
P0466 Purge Flow Sensor Circuit Range/Performance
P0467 Purge Flow Sensor Circuit Low Input
P0468 Purge Flow Sensor Circuit High Input
P0469 Purge Flow Sensor Circuit Intermittent
P0470 Exhaust Pressure Sensor Malfunction
P0471 Exhaust Pressure Sensor Range/Performance
P0472 Exhaust Pressure Sensor Low
P0473 Exhaust Pressure Sensor High
P0474 Exhaust Pressure Sensor Intermittent
P0475 Exhaust Pressure Control Valve Malfunction
P0476 Exhaust Pressure Control Valve Range/Performance
P0477 Exhaust Pressure Control Valve Low
P0478 Exhaust Pressure Control Valve High
P0479 Exhaust Pressure Control Valve Intermittent
P0480 Cooling Fan no. 1 Control Circuit Malfunction
P0481 Cooling Fan no. 2 Control Circuit Malfunction
P0482 Cooling Fan no. 3 Control Circuit Malfunction
P0483 Cooling Fan Rationality Check Malfunction
P0484 Cooling Fan Circuit Over Current
P0485 Cooling Fan Power/Ground Circuit Malfunction
P0500 Vehicle Speed Sensor Malfunction
P0501 Vehicle Speed Sensor Range/Performance
P0502 Vehicle Speed Sensor Circuit Low Input
P0503 Vehicle Speed Sensor Intermittent/Erratic/High
P0505 Idle Control System Malfunction
P0506 Idle Control System RPM Lower Than Expected
P0507 Idle Control System RPM Higher Than Expected
P0510 Closed Throttle Position Switch Malfunction
P0520 Engine Oil Pressure Sensor/Switch Circuit Malfunction
P0521 Engine Oil Pressure Sensor/Switch Range/Performance
P0522 Engine Oil Pressure Sensor/Switch Low Voltage
P0523 Engine Oil Pressure Sensor/Switch High Voltage
P0530 A/C Refrigerant Pressure Sensor Circuit Malfunction
P0531 A/C Refrigerant Pressure Sensor Circuit Range/Performance
P0532 A/C Refrigerant Pressure Sensor Circuit Low Input
P0533 A/C Refrigerant Pressure Sensor Circuit High Input
P0534 A/C Refrigerant Charge Loss
P0550 Power Steering Pressure Sensor Circuit Malfunction
P0551 Power Steering Pressure Sensor Circuit Range/Performance
P0552 Power Steering Pressure Sensor Circuit Low Input
P0553 Power Steering Pressure Sensor Circuit High Input
P0554 Power Steering Pressure Sensor Circuit Intermittent
P0560 System Voltage Malfunction
P0561 System Voltage Unstable
P0562 System Voltage Low
P0563 System Voltage High
P0565 Cruise Control On Signal Malfunction
P0566 Cruise Control Off Signal Malfunction
P0567 Cruise Control Resume Signal Malfunction
P0568 Cruise Control Set Signal Malfunction
P0569 Cruise Control Coast Signal Malfunction
P0570 Cruise Control Accel Signal Malfunction
P0571 Cruise Control/Brake Switch "A" Circuit Malfunction
P0572 Cruise Control/Brake Switch "A" Circuit Low
P0573 Cruise Control/Brake Switch "A" Circuit High
P0574 **Through** P0580 Reserved for Cruise Codes
P0600 Serial Communication Link Malfunction
P0601 Internal Control Module Memory Check Sum Error

P0602 Control Module Programming Error
P0603 Internal Control Module Keep Alive Memory (KAM) Error
P0604 Internal Control Module Random Access Memory (RAM) Error
P0605 Internal Control Module Read Only Memory (ROM) Error
P0606 PCM Processor Fault
P0608 Control Module VSS Output "A" Malfunction
P0609 Control Module VSS Output "B" Malfunction
P0620 Generator Control Circuit Malfunction
P0621 Generator Lamp "L" Control Circuit Malfunction
P0622 Generator Field "F" Control Circuit Malfunction
P0650 Malfunction Indicator Lamp (MIL) Control Circuit Malfunction
P0654 Engine RPM Output Circuit Malfunction
P0655 Engine Hot Lamp Output Control Circuit Malfunction
P0656 Fuel Level Output Circuit Malfunction
P0700 Transmission Control System Malfunction
P0701 Transmission Control System Range/Performance
P0702 Transmission Control System Electrical
P0703 Torque Converter/Brake Switch "B" Circuit Malfunction
P0704 Clutch Switch Input Circuit Malfunction
P0705 Transmission Range Sensor Circuit Malfunction (PRNDL Input)
P0706 Transmission Range Sensor Circuit Range/Performance
P0707 Transmission Range Sensor Circuit Low Input
P0708 Transmission Range Sensor Circuit High Input
P0709 Transmission Range Sensor Circuit Intermittent
P0710 Transmission Fluid Temperature Sensor Circuit Malfunction
P0711 Transmission Fluid Temperature Sensor Circuit Range/Performance
P0712 Transmission Fluid Temperature Sensor Circuit Low Input
P0713 Transmission Fluid Temperature Sensor Circuit High Input
P0714 Transmission Fluid Temperature Sensor Circuit Intermittent
P0715 Input/Turbine Speed Sensor Circuit Malfunction
P0716 Input/Turbine Speed Sensor Circuit Range/Performance
P0717 Input/Turbine Speed Sensor Circuit No Signal
P0718 Input/Turbine Speed Sensor Circuit Intermittent
P0719 Torque Converter/Brake Switch "B" Circuit Low
P0720 Output Speed Sensor Circuit Malfunction
P0721 Output Speed Sensor Circuit Range/Performance
P0722 Output Speed Sensor Circuit No Signal
P0723 Output Speed Sensor Circuit Intermittent
P0724 Torque Converter/Brake Switch "B" Circuit High
P0725 Engine Speed Input Circuit Malfunction
P0726 Engine Speed Input Circuit Range/Performance
P0727 Engine Speed Input Circuit No Signal
P0728 Engine Speed Input Circuit Intermittent
P0730 Incorrect Gear Ratio
P0731 Gear no. 1 Incorrect Ratio
P0732 Gear no. 2 Incorrect Ratio
P0733 Gear no. 3 Incorrect Ratio
P0734 Gear no. 4 Incorrect Ratio
P0735 Gear no. 5 Incorrect Ratio
P0736 Reverse Incorrect Ratio
P0740 Torque Converter Clutch Circuit Malfunction
P0741 Torque Converter Clutch Circuit Performance or Stuck Off
P0742 Torque Converter Clutch Circuit Stuck On

P0743 Torque Converter Clutch Circuit Electrical
P0744 Torque Converter Clutch Circuit Intermittent
P0745 Pressure Control Solenoid Malfunction
P0746 Pressure Control Solenoid Performance or Stuck Off
P0747 Pressure Control Solenoid Stuck On
P0748 Pressure Control Solenoid Electrical
P0749 Pressure Control Solenoid Intermittent
P0750 Shift Solenoid "A" Malfunction
P0751 Shift Solenoid "A" Performance or Stuck Off
P0752 Shift Solenoid "A" Stuck On
P0753 Shift Solenoid "A" Electrical
P0754 Shift Solenoid "A" Intermittent
P0755 Shift Solenoid "B" Malfunction
P0756 Shift Solenoid "B" Performance or Stuck Off
P0757 Shift Solenoid "B" Stuck On
P0758 Shift Solenoid "B" Electrical
P0759 Shift Solenoid "B" Intermittent
P0760 Shift Solenoid "C" Malfunction
P0761 Shift Solenoid "C" Performance Or Stuck Off
P0762 Shift Solenoid "C" Stuck On
P0763 Shift Solenoid "C" Electrical
P0764 Shift Solenoid "C" Intermittent
P0765 Shift Solenoid "D" Malfunction
P0766 Shift Solenoid "D" Performance Or Stuck Off
P0767 Shift Solenoid "D" Stuck On
P0768 Shift Solenoid "D" Electrical
P0769 Shift Solenoid "D" Intermittent
P0770 Shift Solenoid "E" Malfunction
P0771 Shift Solenoid "E" Performance Or Stuck Off
P0772 Shift Solenoid "E" Stuck On
P0773 Shift Solenoid "E" Electrical
P0774 Shift Solenoid "E" Intermittent
P0780 Shift Malfunction
P0781 1–2 Shift Malfunction
P0782 2–3 Shift Malfunction
P0783 3–4 Shift Malfunction
P0784 4–5 Shift Malfunction
P0785 Shift/Timing Solenoid Malfunction
P0786 Shift/Timing Solenoid Range/Performance
P0787 Shift/Timing Solenoid Low
P0788 Shift/Timing Solenoid High
P0789 Shift/Timing Solenoid Intermittent
P0790 Normal/Performance Switch Circuit Malfunction
P0801 Reverse Inhibit Control Circuit Malfunction
P0803 1–4 Upshift (Skip Shift) Solenoid Control Circuit Malfunction
P0804 1–4 Upshift (Skip Shift) Lamp Control Circuit Malfunction
P1250 EFI Heater Circuit Fault
P1408 Manifold Differential Pressure Sensor Circuit Fault
P1410 Fuel Tank Pressure Control Solenoid Circuit Fault
P1450 Barometric Pressure Sensor Circuit Fault
P1451 Barometric Pressure Sensor Performance
P1460 Cooling Fan Control System Fault
P1500 Starter Signal Circuit Fault
P1510 Back-up Power Supply Fault
P1530 Ignition Timing Adjustment Switch Circuit
P1600 PCM Battery Circuit Fault
P1700 TCM Throttle Position Sensor Circuit Fault
P1705 TCM ECT Circuit Fault
P1715 PNP Switch Circuit Fault
P1717 AT Drive Range Signal Circuit Fault

Toyota

READING CODES

Reading the control module memory is on of the first steps in OBD II system diagnostics. This step should be initially performed to determine the general nature of the fault. Subsequent readings will determine if the fault has been cleared.

Reading codes can be performed by any of the methods below:

• Read the control module memory with the Generic Scan Tool (GST)

• Read the control module memory with the vehicle manufacturer's specific tester

To read the fault codes, connect the scan tool or tester according to the manufacturer's instructions. Follow the manufacturer's specified procedure for reading the codes.

CLEARING CODES

Control module reset procedures are a very important part of OBD II System diagnostics. This step should be done at the end of any fault code repair and at the end of any driveability repair.

Clearing codes can be performed by any of the methods below:

• Clear the control module memory with the Generic Scan Tool (GST)

• Clear the control module memory with the vehicle manufacturer's specific tester

• Turn the ignition**OFF** and remove the negative battery cable for at least 1 minute.

Removing the negative battery cable may cause other systems in the vehicle to lose their memory. Prior to removing the cable, ensure you have the proper reset codes for radios and alarms.

➡**The MIL will may also be de-activated for some codes if the vehicle completes three consecutive trips without a fault detected with vehicle conditions similar to those present during the fault.**

DIAGNOSTIC TROUBLE CODES

P0000 No Failures
P0100 Mass or Volume Air Flow Circuit Malfunction
P0101 Mass or Volume Air Flow Circuit Range/Performance Problem
P0102 Mass or Volume Air Flow Circuit Low Input
P0103 Mass or Volume Air Flow Circuit High Input
P0104 Mass or Volume Air Flow Circuit Intermittent
P0105 Manifold Absolute Pressure/Barometric Pressure Circuit Malfunction
P0106 Manifold Absolute Pressure/Barometric Pressure Circuit Range/Performance Problem
P0107 Manifold Absolute Pressure/Barometric Pressure Circuit Low Input
P0108 Manifold Absolute Pressure/Barometric Pressure Circuit High Input
P0109 Manifold Absolute Pressure/Barometric Pressure Circuit Intermittent
P0110 Intake Air Temperature Circuit Malfunction
P0111 Intake Air Temperature Circuit Range/Performance Problem
P0112 Intake Air Temperature Circuit Low Input
P0113 Intake Air Temperature Circuit High Input
P0114 Intake Air Temperature Circuit Intermittent
P0115 Engine Coolant Temperature Circuit Malfunction

P0116 Engine Coolant Temperature Circuit Range/Performance Problem

P0117 Engine Coolant Temperature Circuit Low Input

P0118 Engine Coolant Temperature Circuit High Input

P0119 Engine Coolant Temperature Circuit Intermittent

P0120 Throttle/Pedal Position Sensor/Switch "A" Circuit Malfunction

P0121 Throttle/Pedal Position Sensor/Switch "A" Circuit Range/Performance Problem

P0122 Throttle/Pedal Position Sensor/Switch "A" Circuit Low Input

P0123 Throttle/Pedal Position Sensor/Switch "A" Circuit High Input

P0124 Throttle/Pedal Position Sensor/Switch "A" Circuit Intermittent

P0125 Insufficient Coolant Temperature For Closed Loop Fuel Control

P0126 Insufficient Coolant Temperature For Stable Operation

P0130 O_2 Circuit Malfunction (Bank no. 1 Sensor no. 1)

P0131 O_2 Sensor Circuit Low Voltage (Bank no. 1 Sensor no. 1)

P0132 O_2 Sensor Circuit High Voltage (Bank no. 1 Sensor no. 1)

P0133 O_2 Sensor Circuit Slow Response (Bank no. 1 Sensor no. 1)

P0134 O_2 Sensor Circuit No Activity Detected (Bank no. 1 Sensor no. 1)

P0135 O_2 Sensor Heater Circuit Malfunction (Bank no. 1 Sensor no. 1)

P0136 O_2 Sensor Circuit Malfunction (Bank no. 1 Sensor no. 2)

P0137 O_2 Sensor Circuit Low Voltage (Bank no. 1 Sensor no. 2)

P0138 O_2 Sensor Circuit High Voltage (Bank no. 1 Sensor no. 2)

P0139 O_2 Sensor Circuit Slow Response (Bank no. 1 Sensor no. 2)

P0140 O_2 Sensor Circuit No Activity Detected (Bank no. 1 Sensor no. 2)

P0141 O_2 Sensor Heater Circuit Malfunction (Bank no. 1 Sensor no. 2)

P0142 O_2 Sensor Circuit Malfunction (Bank no. 1 Sensor no. 3)

P0143 O_2 Sensor Circuit Low Voltage (Bank no. 1 Sensor no. 3)

P0144 O_2 Sensor Circuit High Voltage (Bank no. 1 Sensor no. 3)

P0145 O_2 Sensor Circuit Slow Response (Bank no. 1 Sensor no. 3)

P0146 O_2 Sensor Circuit No Activity Detected (Bank no. 1 Sensor no. 3)

P0147 O_2 Sensor Heater Circuit Malfunction (Bank no. 1 Sensor no. 3)

P0150 O_2 Sensor Circuit Malfunction (Bank no. 2 Sensor no. 1)

P0151 O_2 Sensor Circuit Low Voltage (Bank no. 2 Sensor no. 1)

P0152 O_2 Sensor Circuit High Voltage (Bank no. 2 Sensor no. 1)

P0153 O_2 Sensor Circuit Slow Response (Bank no. 2 Sensor no. 1)

P0154 O_2 Sensor Circuit No Activity Detected (Bank no. 2 Sensor no. 1)

P0155 O_2 Sensor Heater Circuit Malfunction (Bank no. 2 Sensor no. 1)

P0156 O_2 Sensor Circuit Malfunction (Bank no. 2 Sensor no. 2)

P0157 O_2 Sensor Circuit Low Voltage (Bank no. 2 Sensor no. 2)

P0158 O_2 Sensor Circuit High Voltage (Bank no. 2 Sensor no. 2)

P0159 O_2 Sensor Circuit Slow Response (Bank no. 2 Sensor no. 2)

P0160 O_2 Sensor Circuit No Activity Detected (Bank no. 2 Sensor no. 2)

P0161 O_2 Sensor Heater Circuit Malfunction (Bank no. 2 Sensor no. 2)

P0162 O_2 Sensor Circuit Malfunction (Bank no. 2 Sensor no. 3)

P0163 O_2 Sensor Circuit Low Voltage (Bank no. 2 Sensor no. 3)

P0164 O_2 Sensor Circuit High Voltage (Bank no. 2 Sensor no. 3)

P0165 O_2 Sensor Circuit Slow Response (Bank no. 2 Sensor no. 3)

P0166 O_2 Sensor Circuit No Activity Detected (Bank no. 2 Sensor no. 3)

P0167 O_2 Sensor Heater Circuit Malfunction (Bank no. 2 Sensor no. 3)

P0170 Fuel Trim Malfunction (Bank no. 1)

P0171 System Too Lean (Bank no. 1)

P0172 System Too Rich (Bank no. 1)

P0173 Fuel Trim Malfunction (Bank no. 2)

P0174 System Too Lean (Bank no. 2)

P0175 System Too Rich (Bank no. 2)

P0176 Fuel Composition Sensor Circuit Malfunction

P0177 Fuel Composition Sensor Circuit Range/Performance

P0178 Fuel Composition Sensor Circuit Low Input

P0179 Fuel Composition Sensor Circuit High Input

P0180 Fuel Temperature Sensor "A" Circuit Malfunction

P0181 Fuel Temperature Sensor "A" Circuit Range/Performance

P0182 Fuel Temperature Sensor "A" Circuit Low Input

P0183 Fuel Temperature Sensor "A" Circuit High Input

P0184 Fuel Temperature Sensor "A" Circuit Intermittent

P0185 Fuel Temperature Sensor "B" Circuit Malfunction

P0186 Fuel Temperature Sensor "B" Circuit Range/Performance

P0187 Fuel Temperature Sensor "B" Circuit Low Input

P0188 Fuel Temperature Sensor "B" Circuit High Input

P0189 Fuel Temperature Sensor "B" Circuit Intermittent

P0190 Fuel Rail Pressure Sensor Circuit Malfunction

P0191 Fuel Rail Pressure Sensor Circuit Range/Performance

P0192 Fuel Rail Pressure Sensor Circuit Low Input

P0193 Fuel Rail Pressure Sensor Circuit High Input

P0194 Fuel Rail Pressure Sensor Circuit Intermittent

P0195 Engine Oil Temperature Sensor Malfunction

P0196 Engine Oil Temperature Sensor Range/Performance

P0197 Engine Oil Temperature Sensor Low

P0198 Engine Oil Temperature Sensor High

P0199 Engine Oil Temperature Sensor Intermittent

P0200 Injector Circuit Malfunction

P0201 Injector Circuit Malfunction—Cylinder no. 1

P0202 Injector Circuit Malfunction—Cylinder no. 2

P0203 Injector Circuit Malfunction—Cylinder no. 3

P0204 Injector Circuit Malfunction—Cylinder no. 4

P0205 Injector Circuit Malfunction—Cylinder no. 5

P0206 Injector Circuit Malfunction—Cylinder no. 6

P0207 Injector Circuit Malfunction—Cylinder no. 7

P0208 Injector Circuit Malfunction—Cylinder no. 8

P0209 Injector Circuit Malfunction—Cylinder no. 9

P0210 Injector Circuit Malfunction—Cylinder no. 10

P0211 Injector Circuit Malfunction—Cylinder no. 11

P0212 Injector Circuit Malfunction—Cylinder no. 12

P0213 Cold Start Injector no. 1 Malfunction

P0214 Cold Start Injector no. 2 Malfunction

P0215 Engine Shutoff Solenoid Malfunction

P0216 Injection Timing Control Circuit Malfunction

P0217 Engine Over Temperature Condition

P0218 Transmission Over Temperature Condition

P0219 Engine Over Speed Condition

P0220 Throttle/Pedal Position Sensor/Switch "B" Circuit Malfunction

P0221 Throttle/Pedal Position Sensor/Switch "B" Circuit Range/Performance Problem

P0222 Throttle/Pedal Position Sensor/Switch "B" Circuit Low Input

P0223 Throttle/Pedal Position Sensor/Switch "B" Circuit High Input

P0224 Throttle/Pedal Position Sensor/Switch "B" Circuit Intermittent

P0225 Throttle/Pedal Position Sensor/Switch "C" Circuit Malfunction

P0226 Throttle/Pedal Position Sensor/Switch "C" Circuit Range/Performance Problem

P0227 Throttle/Pedal Position Sensor/Switch "C" Circuit Low Input

P0228 Throttle/Pedal Position Sensor/Switch "C" Circuit High Input

P0229 Throttle/Pedal Position Sensor/Switch "C" Circuit Intermittent

P0230 Fuel Pump Primary Circuit Malfunction

P0231 Fuel Pump Secondary Circuit Low

P0232 Fuel Pump Secondary Circuit High

P0233 Fuel Pump Secondary Circuit Intermittent

P0234 Engine Over Boost Condition

P0261 Cylinder no. 1 Injector Circuit Low

P0262 Cylinder no. 1 Injector Circuit High

P0263 Cylinder no. 1 Contribution/Balance Fault

P0264 Cylinder no. 2 Injector Circuit Low

P0265 Cylinder no. 2 Injector Circuit High

P0266 Cylinder no. 2 Contribution/Balance Fault

P0267 Cylinder no. 3 Injector Circuit Low

P0268 Cylinder no. 3 Injector Circuit High

P0269 Cylinder no. 3 Contribution/Balance Fault

P0270 Cylinder no. 4 Injector Circuit Low

P0271 Cylinder no. 4 Injector Circuit High

P0272 Cylinder no. 4 Contribution/Balance Fault

P0273 Cylinder no. 5 Injector Circuit Low

P0274 Cylinder no. 5 Injector Circuit High

P0275 Cylinder no. 5 Contribution/Balance Fault

P0276 Cylinder no. 6 Injector Circuit Low

P0277 Cylinder no. 6 Injector Circuit High

P0278 Cylinder no. 6 Contribution/Balance Fault

P0279 Cylinder no. 7 Injector Circuit Low

P0280 Cylinder no. 7 Injector Circuit High

P0281 Cylinder no. 7 Contribution/Balance Fault

P0282 Cylinder no. 8 Injector Circuit Low

P0283 Cylinder no. 8 Injector Circuit High

P0284 Cylinder no. 8 Contribution/Balance Fault

P0285 Cylinder no. 9 Injector Circuit Low

P0286 Cylinder no. 9 Injector Circuit High

P0287 Cylinder no. 9 Contribution/Balance Fault

P0288 Cylinder no. 10 Injector Circuit Low

P0289 Cylinder no. 10 Injector Circuit High

P0290 Cylinder no. 10 Contribution/Balance Fault

P0291 Cylinder no. 11 Injector Circuit Low

P0292 Cylinder no. 11 Injector Circuit High

P0293 Cylinder no. 11 Contribution/Balance Fault

P0294 Cylinder no. 12 Injector Circuit Low

P0295 Cylinder no. 12 Injector Circuit High

P0296 Cylinder no. 12 Contribution/Balance Fault

P0300 Random/Multiple Cylinder Misfire Detected

P0301 Cylinder no. 1—Misfire Detected

P0302 Cylinder no. 2—Misfire Detected

P0303 Cylinder no. 3—Misfire Detected

P0304 Cylinder no. 4—Misfire Detected

P0305 Cylinder no. 5—Misfire Detected

P0306 Cylinder no. 6—Misfire Detected

P0307 Cylinder no. 7—Misfire Detected

P0308 Cylinder no. 8—Misfire Detected

P0309 Cylinder no. 9—Misfire Detected

P0310 Cylinder no. 10—Misfire Detected

P0311 Cylinder no. 11—Misfire Detected

P0312 Cylinder no. 12—Misfire Detected

P0320 Ignition/Distributor Engine Speed Input Circuit Malfunction

P0321 Ignition/Distributor Engine Speed Input Circuit Range/Performance

P0322 Ignition/Distributor Engine Speed Input Circuit No Signal

P0323 Ignition/Distributor Engine Speed Input Circuit Intermittent

P0325 Knock Sensor no. 1—Circuit Malfunction (Bank no. 1 or Single Sensor)

P0326 Knock Sensor no. 1—Circuit Range/Performance (Bank no. 1 or Single Sensor)

P0327 Knock Sensor no. 1—Circuit Low Input (Bank no. 1 or Single Sensor)

P0328 Knock Sensor no. 1—Circuit High Input (Bank no. 1 or Single Sensor)

P0329 Knock Sensor no. 1—Circuit Input Intermittent (Bank no. 1 or Single Sensor)

P0330 Knock Sensor no. 2—Circuit Malfunction (Bank no. 2)

P0331 Knock Sensor no. 2—Circuit Range/Performance (Bank no. 2)

P0332 Knock Sensor no. 2—Circuit Low Input (Bank no. 2)

P0333 Knock Sensor no. 2—Circuit High Input (Bank no. 2)

P0334 Knock Sensor no. 2—Circuit Input Intermittent (Bank no. 2)

P0335 Crankshaft Position Sensor "A" Circuit Malfunction

P0336 Crankshaft Position Sensor "A" Circuit Range/Performance

P0337 Crankshaft Position Sensor "A" Circuit Low Input

P0338 Crankshaft Position Sensor "A" Circuit High Input

P0339 Crankshaft Position Sensor "A" Circuit Intermittent

P0340 Camshaft Position Sensor Circuit Malfunction

P0341 Camshaft Position Sensor Circuit Range/Performance

P0342 Camshaft Position Sensor Circuit Low Input

P0343 Camshaft Position Sensor Circuit High Input

P0344 Camshaft Position Sensor Circuit Intermittent

P0350 Ignition Coil Primary/Secondary Circuit Malfunction

P0351 Ignition Coil "A" Primary/Secondary Circuit Malfunction

P0352 Ignition Coil "B" Primary/Secondary Circuit Malfunction

P0353 Ignition Coil "C" Primary/Secondary Circuit Malfunction

P0354 Ignition Coil "D" Primary/Secondary Circuit Malfunction

P0355 Ignition Coil "E" Primary/Secondary Circuit Malfunction

P0356 Ignition Coil "F" Primary/Secondary Circuit Malfunction

P0357 Ignition Coil "G" Primary/Secondary Circuit Malfunction

P0358 Ignition Coil "H" Primary/Secondary Circuit Malfunction

P0359 Ignition Coil "I" Primary/Secondary Circuit Malfunction

P0360 Ignition Coil "J" Primary/Secondary Circuit Malfunction

P0361 Ignition Coil "K" Primary/Secondary Circuit Malfunction

P0362 Ignition Coil "L" Primary/Secondary Circuit Malfunction

P0370 Timing Reference High Resolution Signal "A" Malfunction

P0371 Timing Reference High Resolution Signal "A" Too Many Pulses

P0372 Timing Reference High Resolution Signal "A" Too Few Pulses

P0373 Timing Reference High Resolution Signal "A" Intermittent/Erratic Pulses

P0374 Timing Reference High Resolution Signal "A" No Pulses

P0375 Timing Reference High Resolution Signal "B" Malfunction

P0376 Timing Reference High Resolution Signal "B" Too Many Pulses

P0377 Timing Reference High Resolution Signal "B" Too Few Pulses

P0378 Timing Reference High Resolution Signal "B" Intermittent/Erratic Pulses

P0379 Timing Reference High Resolution Signal "B" No Pulses

P0380 Glow Plug/Heater Circuit "A" Malfunction

P0381 Glow Plug/Heater Indicator Circuit Malfunction

P0382 Glow Plug/Heater Circuit "B" Malfunction

P0385 Crankshaft Position Sensor "B" Circuit Malfunction

P0386 Crankshaft Position Sensor "B" Circuit Range/Performance

P0387 Crankshaft Position Sensor "B" Circuit Low Input

P0388 Crankshaft Position Sensor "B" Circuit High Input

P0389 Crankshaft Position Sensor "B" Circuit Intermittent

P0400 Exhaust Gas Recirculation Flow Malfunction

P0401 Exhaust Gas Recirculation Flow Insufficient Detected

P0402 Exhaust Gas Recirculation Flow Excessive Detected

P0403 Exhaust Gas Recirculation Circuit Malfunction

P0404 Exhaust Gas Recirculation Circuit Range/Performance

P0405 Exhaust Gas Recirculation Sensor "A" Circuit Low

P0406 Exhaust Gas Recirculation Sensor "A" Circuit High

P0407 Exhaust Gas Recirculation Sensor "B" Circuit Low

P0408 Exhaust Gas Recirculation Sensor "B" Circuit High

P0410 Secondary Air Injection System Malfunction

P0411 Secondary Air Injection System Incorrect Flow Detected

P0412 Secondary Air Injection System Switching Valve "A" Circuit Malfunction

P0413 Secondary Air Injection System Switching Valve "A" Circuit Open

P0414 Secondary Air Injection System Switching Valve "A" Circuit Shorted

P0415 Secondary Air Injection System Switching Valve "B" Circuit Malfunction

P0416 Secondary Air Injection System Switching Valve "B" Circuit Open

P0417 Secondary Air Injection System Switching Valve "B" Circuit Shorted

P0418 Secondary Air Injection System Relay "A" Circuit Malfunction

P0419 Secondary Air Injection System Relay "B" Circuit Malfunction

P0420 Catalyst System Efficiency Below Threshold (Bank no. 1)

P0421 Warm Up Catalyst Efficiency Below Threshold (Bank no. 1)

P0422 Main Catalyst Efficiency Below Threshold (Bank no. 1)

P0423 Heated Catalyst Efficiency Below Threshold (Bank no. 1)

P0424 Heated Catalyst Temperature Below Threshold (Bank no. 1)

P0430 Catalyst System Efficiency Below Threshold (Bank no. 2)

P0431 Warm Up Catalyst Efficiency Below Threshold (Bank no. 2)

P0432 Main Catalyst Efficiency Below Threshold (Bank no. 2)

P0433 Heated Catalyst Efficiency Below Threshold (Bank no. 2)

P0434 Heated Catalyst Temperature Below Threshold (Bank no. 2)

P0440 Evaporative Emission Control System Malfunction

P0441 Evaporative Emission Control System Incorrect Purge Flow

P0442 Evaporative Emission Control System Leak Detected (Small Leak)

P0443 Evaporative Emission Control System Purge Control Valve Circuit Malfunction

P0444 Evaporative Emission Control System Purge Control Valve Circuit Open

P0445 Evaporative Emission Control System Purge Control Valve Circuit Shorted

P0446 Evaporative Emission Control System Vent Control Circuit Malfunction

P0447 Evaporative Emission Control System Vent Control Circuit Open

P0448 Evaporative Emission Control System Vent Control Circuit Shorted

P0449 Evaporative Emission Control System Vent Valve/Solenoid Circuit Malfunction

P0450 Evaporative Emission Control System Pressure Sensor Malfunction

P0451 Evaporative Emission Control System Pressure Sensor Range/Performance

P0452 Evaporative Emission Control System Pressure Sensor Low Input

P0453 Evaporative Emission Control System Pressure Sensor High Input

P0454 Evaporative Emission Control System Pressure Sensor Intermittent

P0455 Evaporative Emission Control System Leak Detected (Gross Leak)

P0460 Fuel Level Sensor Circuit Malfunction

P0461 Fuel Level Sensor Circuit Range/Performance

P0462 Fuel Level Sensor Circuit Low Input

P0463 Fuel Level Sensor Circuit High Input

P0464 Fuel Level Sensor Circuit Intermittent

P0465 Purge Flow Sensor Circuit Malfunction

P0466 Purge Flow Sensor Circuit Range/Performance

P0467 Purge Flow Sensor Circuit Low Input

P0468 Purge Flow Sensor Circuit High Input

P0469 Purge Flow Sensor Circuit Intermittent

P0470 Exhaust Pressure Sensor Malfunction

P0471 Exhaust Pressure Sensor Range/Performance

P0472 Exhaust Pressure Sensor Low

P0473 Exhaust Pressure Sensor High

P0474 Exhaust Pressure Sensor Intermittent

P0475 Exhaust Pressure Control Valve Malfunction

P0476 Exhaust Pressure Control Valve Range/Performance

P0477 Exhaust Pressure Control Valve Low

P0478 Exhaust Pressure Control Valve High

P0479 Exhaust Pressure Control Valve Intermittent

P0480 Cooling Fan no. 1 Control Circuit Malfunction

P0481 Cooling Fan no. 2 Control Circuit Malfunction

P0482 Cooling Fan no. 3 Control Circuit Malfunction

P0483 Cooling Fan Rationality Check Malfunction

P0484 Cooling Fan Circuit Over Current

P0485 Cooling Fan Power/Ground Circuit Malfunction

P0500 Vehicle Speed Sensor Malfunction

P0501 Vehicle Speed Sensor Range/Performance

P0502 Vehicle Speed Sensor Circuit Low Input

P0503 Vehicle Speed Sensor Intermittent/Erratic/High

P0505 Idle Control System Malfunction

P0506 Idle Control System RPM Lower Than Expected

P0507 Idle Control System RPM Higher Than Expected

P0510 Closed Throttle Position Switch Malfunction

P0520 Engine Oil Pressure Sensor/Switch Circuit Malfunction

P0521 Engine Oil Pressure Sensor/Switch Range/Performance

P0522 Engine Oil Pressure Sensor/Switch Low Voltage

P0523 Engine Oil Pressure Sensor/Switch High Voltage

P0530 A/C Refrigerant Pressure Sensor Circuit Malfunction

P0531 A/C Refrigerant Pressure Sensor Circuit Range/Performance

P0532 A/C Refrigerant Pressure Sensor Circuit Low Input

P0533 A/C Refrigerant Pressure Sensor Circuit High Input
P0534 A/C Refrigerant Charge Loss
P0550 Power Steering Pressure Sensor Circuit Malfunction
P0551 Power Steering Pressure Sensor Circuit Range/Performance
P0552 Power Steering Pressure Sensor Circuit Low Input
P0553 Power Steering Pressure Sensor Circuit High Input
P0554 Power Steering Pressure Sensor Circuit Intermittent
P0560 System Voltage Malfunction
P0561 System Voltage Unstable
P0562 System Voltage Low
P0563 System Voltage High
P0565 Cruise Control On Signal Malfunction
P0566 Cruise Control Off Signal Malfunction
P0567 Cruise Control Resume Signal Malfunction
P0568 Cruise Control Set Signal Malfunction
P0569 Cruise Control Coast Signal Malfunction
P0570 Cruise Control Accel Signal Malfunction
P0571 Cruise Control/Brake Switch "A" Circuit Malfunction
P0572 Cruise Control/Brake Switch "A" Circuit Low
P0573 Cruise Control/Brake Switch "A" Circuit High
P0574 **Through P0580** Reserved for Cruise Codes
P0600 Serial Communication Link Malfunction
P0601 Internal Control Module Memory Check Sum Error
P0602 Control Module Programming Error
P0603 Internal Control Module Keep Alive Memory (KAM) Error
P0604 Internal Control Module Random Access Memory (RAM) Error
P0605 Internal Control Module Read Only Memory (ROM) Error
P0606 PCM Processor Fault
P0608 Control Module VSS Output "A" Malfunction
P0609 Control Module VSS Output "B" Malfunction
P0620 Generator Control Circuit Malfunction
P0621 Generator Lamp "L" Control Circuit Malfunction
P0622 Generator Field "F" Control Circuit Malfunction
P0650 Malfunction Indicator Lamp (MIL) Control Circuit Malfunction
P0654 Engine RPM Output Circuit Malfunction
P0655 Engine Hot Lamp Output Control Circuit Malfunction
P0656 Fuel Level Output Circuit Malfunction
P0700 Transmission Control System Malfunction
P0701 Transmission Control System Range/Performance
P0702 Transmission Control System Electrical
P0703 Torque Converter/Brake Switch "B" Circuit Malfunction
P0704 Clutch Switch Input Circuit Malfunction
P0705 Transmission Range Sensor Circuit Malfunction (PRNDL Input)
P0706 Transmission Range Sensor Circuit Range/Performance
P0707 Transmission Range Sensor Circuit Low Input
P0708 Transmission Range Sensor Circuit High Input
P0709 Transmission Range Sensor Circuit Intermittent
P0710 Transmission Fluid Temperature Sensor Circuit Malfunction
P0711 Transmission Fluid Temperature Sensor Circuit Range/Performance
P0712 Transmission Fluid Temperature Sensor Circuit Low Input
P0713 Transmission Fluid Temperature Sensor Circuit High Input
P0714 Transmission Fluid Temperature Sensor Circuit Intermittent
P0715 Input/Turbine Speed Sensor Circuit Malfunction
P0716 Input/Turbine Speed Sensor Circuit Range/Performance
P0717 Input/Turbine Speed Sensor Circuit No Signal
P0718 Input/Turbine Speed Sensor Circuit Intermittent

P0719 Torque Converter/Brake Switch "B" Circuit Low
P0720 Output Speed Sensor Circuit Malfunction
P0721 Output Speed Sensor Circuit Range/Performance
P0722 Output Speed Sensor Circuit No Signal
P0723 Output Speed Sensor Circuit Intermittent
P0724 Torque Converter/Brake Switch "B" Circuit High
P0725 Engine Speed Input Circuit Malfunction
P0726 Engine Speed Input Circuit Range/Performance
P0727 Engine Speed Input Circuit No Signal
P0728 Engine Speed Input Circuit Intermittent
P0730 Incorrect Gear Ratio
P0731 Gear no. 1 Incorrect Ratio
P0732 Gear no. 2 Incorrect Ratio
P0733 Gear no. 3 Incorrect Ratio
P0734 Gear no. 4 Incorrect Ratio
P0735 Gear no. 5 Incorrect Ratio
P0736 Reverse Incorrect Ratio
P0740 Torque Converter Clutch Circuit Malfunction
P0741 Torque Converter Clutch Circuit Performance or Stuck Off
P0742 Torque Converter Clutch Circuit Stuck On
P0743 Torque Converter Clutch Circuit Electrical
P0744 Torque Converter Clutch Circuit Intermittent
P0745 Pressure Control Solenoid Malfunction
P0746 Pressure Control Solenoid Performance or Stuck Off
P0747 Pressure Control Solenoid Stuck On
P0748 Pressure Control Solenoid Electrical
P0749 Pressure Control Solenoid Intermittent
P0750 Shift Solenoid "A" Malfunction
P0751 Shift Solenoid "A" Performance or Stuck Off
P0752 Shift Solenoid "A" Stuck On
P0753 Shift Solenoid "A" Electrical
P0754 Shift Solenoid "A" Intermittent
P0755 Shift Solenoid "B" Malfunction
P0756 Shift Solenoid "B" Performance or Stuck Off
P0757 Shift Solenoid "B" Stuck On
P0758 Shift Solenoid "B" Electrical
P0759 Shift Solenoid "B" Intermittent
P0760 Shift Solenoid "C" Malfunction
P0761 Shift Solenoid "C" Performance Or Stuck Off
P0762 Shift Solenoid "C" Stuck On
P0763 Shift Solenoid "C" Electrical
P0764 Shift Solenoid "C" Intermittent
P0765 Shift Solenoid "D" Malfunction
P0766 Shift Solenoid "D" Performance Or Stuck Off
P0767 Shift Solenoid "D" Stuck On
P0768 Shift Solenoid "D" Electrical
P0769 Shift Solenoid "D" Intermittent
P0770 Shift Solenoid "E" Malfunction
P0771 Shift Solenoid "E" Performance Or Stuck Off
P0772 Shift Solenoid "E" Stuck On
P0773 Shift Solenoid "E" Electrical
P0774 Shift Solenoid "E" Intermittent
P0780 Shift Malfunction
P0781 1–2 Shift Malfunction
P0782 2–3 Shift Malfunction
P0783 3–4 Shift Malfunction
P0784 4–5 Shift Malfunction
P0785 Shift/Timing Solenoid Malfunction
P0786 Shift/Timing Solenoid Range/Performance
P0787 Shift/Timing Solenoid Low
P0788 Shift/Timing Solenoid High
P0789 Shift/Timing Solenoid Intermittent

P0790 Normal/Performance Switch Circuit Malfunction
P0801 Reverse Inhibit Control Circuit Malfunction
P0803 1–4 Upshift (Skip Shift) Solenoid Control Circuit Malfunction
P0804 1–4 Upshift (Skip Shift) Lamp Control Circuit Malfunction
P1100 Barometric Pressure Sensor Circuit Fault
P1200 Fuel Pump Relay Circuit Fault
P1300 Igniter Circuit Fault (Bank 1)
P1305 Igniter Circuit Fault (Bank 2)
P1335 Crankshaft Position Sensor Circuit Fault
P1400 Sub-Throttle Position Sensor Circuit Fault
P1401 Sub-Throttle Position Sensor Performance
P1500 Starter Signal Circuit Fault
P1510 Air Volume Too Low With Supercharger On
P1600 PCM Battery Back-up Circuit Fault
P1605 Knock Control CPU Fault
P1700 Vehicle Speed Sensor Circuit Fault
P1705 Direct Clutch Speed Sensor Circuit Fault
P1765 Linear Shift Solenoid Circuit Fault
P1780 Park Neutral Position Switch Fault

Volkswagen

READING CODES

Reading the control module memory is one of the first steps in OBD II system diagnostics. This step should be initially performed to determine the general nature of the fault. Subsequent readings will determine if the fault has been cleared.

Reading codes can be performed by any of the methods below:
• Read the control module memory with the Generic Scan Tool (GST)
• Read the control module memory with the VAG 1551 scan tool/tester—if the VAG 1551 is being used, refer to the trouble code equivalent list for the proper code identification

To read the fault codes, connect the scan tool or tester according to the manufacturer's instructions. Follow the manufacturer's specified procedure for reading the codes.

CLEARING CODES

➡**This procedure is applicable to generic and manufacturer specific codes.**

Control module reset procedures are a very important part of OBD II System diagnostics. This step should be done at the end of any fault code repair and at the end of any driveability repair.

Clearing codes can be performed by any of the methods below:
• Clear the control module memory with the Generic Scan Tool (GST)
• Clear the control module memory with the vehicle manufacturer's specific tester
• Turn the ignition **OFF** and remove the negative battery cable for at least 1 minute.

Removing the negative battery cable may cause other systems in the vehicle to lose their memory. Prior to removing the cable, ensure you have the proper reset codes for radios and alarms.

➡**The MIL will may also be de-activated for some codes if the vehicle completes three consecutive trips without a fault detected with vehicle conditions similar to those present during the fault.**

DIAGNOSTIC TROUBLE CODES

P0000 No Failures
P0100 Mass or Volume Air Flow Circuit Malfunction
P0101 Mass or Volume Air Flow Circuit Range/Performance Problem
P0102 Mass or Volume Air Flow Circuit Low Input
P0103 Mass or Volume Air Flow Circuit High Input
P0104 Mass or Volume Air Flow Circuit Intermittent
P0105 Manifold Absolute Pressure/Barometric Pressure Circuit Malfunction
P0106 Manifold Absolute Pressure/Barometric Pressure Circuit Range/Performance Problem
P0107 Manifold Absolute Pressure/Barometric Pressure Circuit Low Input
P0108 Manifold Absolute Pressure/Barometric Pressure Circuit High Input
P0109 Manifold Absolute Pressure/Barometric Pressure Circuit Intermittent
P0110 Intake Air Temperature Circuit Malfunction
P0111 Intake Air Temperature Circuit Range/Performance Problem
P0112 Intake Air Temperature Circuit Low Input
P0113 Intake Air Temperature Circuit High Input
P0114 Intake Air Temperature Circuit Intermittent
P0115 Engine Coolant Temperature Circuit Malfunction
P0116 Engine Coolant Temperature Circuit Range/Performance Problem
P0117 Engine Coolant Temperature Circuit Low Input
P0118 Engine Coolant Temperature Circuit High Input
P0119 Engine Coolant Temperature Circuit Intermittent
P0120 Throttle/Pedal Position Sensor/Switch "A" Circuit Malfunction
P0121 Throttle/Pedal Position Sensor/Switch "A" Circuit Range/Performance Problem
P0122 Throttle/Pedal Position Sensor/Switch "A" Circuit Low Input
P0123 Throttle/Pedal Position Sensor/Switch "A" Circuit High Input
P0124 Throttle/Pedal Position Sensor/Switch "A" Circuit Intermittent
P0125 Insufficient Coolant Temperature For Closed Loop Fuel Control
P0126 Insufficient Coolant Temperature For Stable Operation
P0130 O_2 Circuit Malfunction (Bank no. 1 Sensor no. 1)
P0131 O_2 Sensor Circuit Low Voltage (Bank no. 1 Sensor no. 1)
P0132 O_2 Sensor Circuit High Voltage (Bank no. 1 Sensor no. 1)
P0133 O_2 Sensor Circuit Slow Response (Bank no. 1 Sensor no. 1)
P0134 O_2 Sensor Circuit No Activity Detected (Bank no. 1 Sensor no. 1)
P0135 O_2 Sensor Heater Circuit Malfunction (Bank no. 1 Sensor no. 1)
P0136 O_2 Sensor Circuit Malfunction (Bank no. 1 Sensor no. 2)
P0137 O_2 Sensor Circuit Low Voltage (Bank no. 1 Sensor no. 2)
P0138 O_2 Sensor Circuit High Voltage (Bank no. 1 Sensor no. 2)
P0139 O_2 Sensor Circuit Slow Response (Bank no. 1 Sensor no. 2)
P0140 O_2 Sensor Circuit No Activity Detected (Bank no. 1 Sensor no. 2)
P0141 O_2 Sensor Heater Circuit Malfunction (Bank no. 1 Sensor no. 2)

P0142 O_2 Sensor Circuit Malfunction (Bank no. 1 Sensor no. 3)
P0143 O_2 Sensor Circuit Low Voltage (Bank no. 1 Sensor no. 3)
P0144 O_2 Sensor Circuit High Voltage (Bank no. 1 Sensor no. 3)
P0145 O_2 Sensor Circuit Slow Response (Bank no. 1 Sensor no. 3)
P0146 O_2 Sensor Circuit No Activity Detected (Bank no. 1 Sensor no. 3)
P0147 O_2 Sensor Heater Circuit Malfunction (Bank no. 1 Sensor no. 3)
P0150 O_2 Sensor Circuit Malfunction (Bank no. 2 Sensor no. 1)
P0151 O_2 Sensor Circuit Low Voltage (Bank no. 2 Sensor no. 1)
P0152 O_2 Sensor Circuit High Voltage (Bank no. 2 Sensor no. 1)
P0153 O_2 Sensor Circuit Slow Response (Bank no. 2 Sensor no. 1)
P0154 O_2 Sensor Circuit No Activity Detected (Bank no. 2 Sensor no. 1)
P0155 O_2 Sensor Heater Circuit Malfunction (Bank no. 2 Sensor no. 1)
P0156 O_2 Sensor Circuit Malfunction (Bank no. 2 Sensor no. 2)
P0157 O_2 Sensor Circuit Low Voltage (Bank no. 2 Sensor no. 2)
P0158 O_2 Sensor Circuit High Voltage (Bank no. 2 Sensor no. 2)
P0159 O_2 Sensor Circuit Slow Response (Bank no. 2 Sensor no. 2)
P0160 O_2 Sensor Circuit No Activity Detected (Bank no. 2 Sensor no. 2)
P0161 O_2 Sensor Heater Circuit Malfunction (Bank no. 2 Sensor no. 2)
P0162 O_2 Sensor Circuit Malfunction (Bank no. 2 Sensor no. 3)
P0163 O_2 Sensor Circuit Low Voltage (Bank no. 2 Sensor no. 3)
P0164 O_2 Sensor Circuit High Voltage (Bank no. 2 Sensor no. 3)
P0165 O_2 Sensor Circuit Slow Response (Bank no. 2 Sensor no. 3)
P0166 O_2 Sensor Circuit No Activity Detected (Bank no. 2 Sensor no. 3)
P0167 O_2 Sensor Heater Circuit Malfunction (Bank no. 2 Sensor no. 3)
P0170 Fuel Trim Malfunction (Bank no. 1)
P0171 System Too Lean (Bank no. 1)
P0172 System Too Rich (Bank no. 1)
P0173 Fuel Trim Malfunction (Bank no. 2)
P0174 System Too Lean (Bank no. 2)
P0175 System Too Rich (Bank no. 2)
P0176 Fuel Composition Sensor Circuit Malfunction
P0177 Fuel Composition Sensor Circuit Range/Performance
P0178 Fuel Composition Sensor Circuit Low Input
P0179 Fuel Composition Sensor Circuit High Input
P0180 Fuel Temperature Sensor "A" Circuit Malfunction
P0181 Fuel Temperature Sensor "A" Circuit Range/Performance
P0182 Fuel Temperature Sensor "A" Circuit Low Input
P0183 Fuel Temperature Sensor "A" Circuit High Input
P0184 Fuel Temperature Sensor "A" Circuit Intermittent
P0185 Fuel Temperature Sensor "B" Circuit Malfunction
P0186 Fuel Temperature Sensor "B" Circuit Range/Performance
P0187 Fuel Temperature Sensor "B" Circuit Low Input
P0188 Fuel Temperature Sensor "B" Circuit High Input
P0189 Fuel Temperature Sensor "B" Circuit Intermittent
P0190 Fuel Rail Pressure Sensor Circuit Malfunction
P0191 Fuel Rail Pressure Sensor Circuit Range/Performance
P0192 Fuel Rail Pressure Sensor Circuit Low Input
P0193 Fuel Rail Pressure Sensor Circuit High Input
P0194 Fuel Rail Pressure Sensor Circuit Intermittent
P0195 Engine Oil Temperature Sensor Malfunction

P0196 Engine Oil Temperature Sensor Range/Performance
P0197 Engine Oil Temperature Sensor Low
P0198 Engine Oil Temperature Sensor High
P0199 Engine Oil Temperature Sensor Intermittent
P0200 Injector Circuit Malfunction
P0201 Injector Circuit Malfunction—Cylinder no. 1
P0202 Injector Circuit Malfunction—Cylinder no. 2
P0203 Injector Circuit Malfunction—Cylinder no. 3
P0204 Injector Circuit Malfunction—Cylinder no. 4
P0205 Injector Circuit Malfunction—Cylinder no. 5
P0206 Injector Circuit Malfunction—Cylinder no. 6
P0207 Injector Circuit Malfunction—Cylinder no. 7
P0208 Injector Circuit Malfunction—Cylinder no. 8
P0209 Injector Circuit Malfunction—Cylinder no. 9
P0210 Injector Circuit Malfunction—Cylinder no. 10
P0211 Injector Circuit Malfunction—Cylinder no. 11
P0212 Injector Circuit Malfunction—Cylinder no. 12
P0213 Cold Start Injector no. 1 Malfunction
P0214 Cold Start Injector no. 2 Malfunction
P0215 Engine Shutoff Solenoid Malfunction
P0216 Injection Timing Control Circuit Malfunction
P0217 Engine Over Temperature Condition
P0218 Transmission Over Temperature Condition
P0219 Engine Over Speed Condition
P0220 Throttle/Pedal Position Sensor/Switch "B" Circuit Malfunction
P0221 Throttle/Pedal Position Sensor/Switch "B" Circuit Range/Performance Problem
P0222 Throttle/Pedal Position Sensor/Switch "B" Circuit Low Input
P0223 Throttle/Pedal Position Sensor/Switch "B" Circuit High Input
P0224 Throttle/Pedal Position Sensor/Switch "B" Circuit Intermittent
P0225 Throttle/Pedal Position Sensor/Switch "C" Circuit Malfunction
P0226 Throttle/Pedal Position Sensor/Switch "C" Circuit Range/Performance Problem
P0227 Throttle/Pedal Position Sensor/Switch "C" Circuit Low Input
P0228 Throttle/Pedal Position Sensor/Switch "C" Circuit High Input
P0229 Throttle/Pedal Position Sensor/Switch "C" Circuit Intermittent
P0230 Fuel Pump Primary Circuit Malfunction
P0231 Fuel Pump Secondary Circuit Low
P0232 Fuel Pump Secondary Circuit High
P0233 Fuel Pump Secondary Circuit Intermittent
P0234 Engine Over Boost Condition
P0261 Cylinder no. 1 Injector Circuit Low
P0262 Cylinder no. 1 Injector Circuit High
P0263 Cylinder no. 1 Contribution/Balance Fault
P0264 Cylinder no. 2 Injector Circuit Low
P0265 Cylinder no. 2 Injector Circuit High
P0266 Cylinder no. 2 Contribution/Balance Fault
P0267 Cylinder no. 3 Injector Circuit Low
P0268 Cylinder no. 3 Injector Circuit High
P0269 Cylinder no. 3 Contribution/Balance Fault
P0270 Cylinder no. 4 Injector Circuit Low
P0271 Cylinder no. 4 Injector Circuit High
P0272 Cylinder no. 4 Contribution/Balance Fault
P0273 Cylinder no. 5 Injector Circuit Low

P0274 Cylinder no. 5 Injector Circuit High
P0275 Cylinder no. 5 Contribution/Balance Fault
P0276 Cylinder no. 6 Injector Circuit Low
P0277 Cylinder no. 6 Injector Circuit High
P0278 Cylinder no. 6 Contribution/Balance Fault
P0279 Cylinder no. 7 Injector Circuit Low
P0280 Cylinder no. 7 Injector Circuit High
P0281 Cylinder no. 7 Contribution/Balance Fault
P0282 Cylinder no. 8 Injector Circuit Low
P0283 Cylinder no. 8 Injector Circuit High
P0284 Cylinder no. 8 Contribution/Balance Fault
P0285 Cylinder no. 9 Injector Circuit Low
P0286 Cylinder no. 9 Injector Circuit High
P0287 Cylinder no. 9 Contribution/Balance Fault
P0288 Cylinder no. 10 Injector Circuit Low
P0289 Cylinder no. 10 Injector Circuit High
P0290 Cylinder no. 10 Contribution/Balance Fault
P0291 Cylinder no. 11 Injector Circuit Low
P0292 Cylinder no. 11 Injector Circuit High
P0293 Cylinder no. 11 Contribution/Balance Fault
P0294 Cylinder no. 12 Injector Circuit Low
P0295 Cylinder no. 12 Injector Circuit High
P0296 Cylinder no. 12 Contribution/Balance Fault
P0300 Random/Multiple Cylinder Misfire Detected
P0301 Cylinder no. 1—Misfire Detected
P0302 Cylinder no. 2—Misfire Detected
P0303 Cylinder no. 3—Misfire Detected
P0304 Cylinder no. 4—Misfire Detected
P0305 Cylinder no. 5—Misfire Detected
P0306 Cylinder no. 6—Misfire Detected
P0307 Cylinder no. 7—Misfire Detected
P0308 Cylinder no. 8—Misfire Detected
P0309 Cylinder no. 9—Misfire Detected
P0310 Cylinder no. 10—Misfire Detected
P0311 Cylinder no. 11—Misfire Detected
P0312 Cylinder no. 12—Misfire Detected
P0320 Ignition/Distributor Engine Speed Input Circuit Malfunction
P0321 Ignition/Distributor Engine Speed Input Circuit Range/Performance
P0322 Ignition/Distributor Engine Speed Input Circuit No Signal
P0323 Ignition/Distributor Engine Speed Input Circuit Intermittent
P0325 Knock Sensor no. 1—Circuit Malfunction (Bank no. 1 or Single Sensor)
P0326 Knock Sensor no. 1—Circuit Range/Performance (Bank no. 1 or Single Sensor)
P0327 Knock Sensor no. 1—Circuit Low Input (Bank no. 1 or Single Sensor)
P0328 Knock Sensor no. 1—Circuit High Input (Bank no. 1 or Single Sensor)
P0329 Knock Sensor no. 1—Circuit Input Intermittent (Bank no. 1 or Single Sensor)
P0330 Knock Sensor no. 2—Circuit Malfunction (Bank no. 2)
P0331 Knock Sensor no. 2—Circuit Range/Performance (Bank no. 2)
P0332 Knock Sensor no. 2—Circuit Low Input (Bank no. 2)
P0333 Knock Sensor no. 2—Circuit High Input (Bank no. 2)
P0334 Knock Sensor no. 2—Circuit Input Intermittent (Bank no. 2)
P0335 Crankshaft Position Sensor "A" Circuit Malfunction
P0336 Crankshaft Position Sensor "A" Circuit Range/Performance

P0337 Crankshaft Position Sensor "A" Circuit Low Input
P0338 Crankshaft Position Sensor "A" Circuit High Input
P0339 Crankshaft Position Sensor "A" Circuit Intermittent
P0340 Camshaft Position Sensor Circuit Malfunction
P0341 Camshaft Position Sensor Circuit Range/Performance
P0342 Camshaft Position Sensor Circuit Low Input
P0343 Camshaft Position Sensor Circuit High Input
P0344 Camshaft Position Sensor Circuit Intermittent
P0350 Ignition Coil Primary/Secondary Circuit Malfunction
P0351 Ignition Coil "A" Primary/Secondary Circuit Malfunction
P0352 Ignition Coil "B" Primary/Secondary Circuit Malfunction
P0353 Ignition Coil "C" Primary/Secondary Circuit Malfunction
P0354 Ignition Coil "D" Primary/Secondary Circuit Malfunction
P0355 Ignition Coil "E" Primary/Secondary Circuit Malfunction
P0356 Ignition Coil "F" Primary/Secondary Circuit Malfunction
P0357 Ignition Coil "G" Primary/Secondary Circuit Malfunction
P0358 Ignition Coil "H" Primary/Secondary Circuit Malfunction
P0359 Ignition Coil "I" Primary/Secondary Circuit Malfunction
P0360 Ignition Coil "J" Primary/Secondary Circuit Malfunction
P0361 Ignition Coil "K" Primary/Secondary Circuit Malfunction
P0362 Ignition Coil "L" Primary/Secondary Circuit Malfunction
P0370 Timing Reference High Resolution Signal "A" Malfunction
P0371 Timing Reference High Resolution Signal "A" Too Many Pulses
P0372 Timing Reference High Resolution Signal "A" Too Few Pulses
P0373 Timing Reference High Resolution Signal "A" Intermittent/Erratic Pulses
P0374 Timing Reference High Resolution Signal "A" No Pulses
P0375 Timing Reference High Resolution Signal "B" Malfunction
P0376 Timing Reference High Resolution Signal "B" Too Many Pulses
P0377 Timing Reference High Resolution Signal "B" Too Few Pulses
P0378 Timing Reference High Resolution Signal "B" Intermittent/Erratic Pulses
P0379 Timing Reference High Resolution Signal "B" No Pulses
P0380 Glow Plug/Heater Circuit "A" Malfunction
P0381 Glow Plug/Heater Indicator Circuit Malfunction
P0382 Glow Plug/Heater Circuit "B" Malfunction
P0385 Crankshaft Position Sensor "B" Circuit Malfunction
P0386 Crankshaft Position Sensor "B" Circuit Range/Performance
P0387 Crankshaft Position Sensor "B" Circuit Low Input
P0388 Crankshaft Position Sensor "B" Circuit High Input
P0389 Crankshaft Position Sensor "B" Circuit Intermittent
P0400 Exhaust Gas Recirculation Flow Malfunction
P0401 Exhaust Gas Recirculation Flow Insufficient Detected
P0402 Exhaust Gas Recirculation Flow Excessive Detected
P0403 Exhaust Gas Recirculation Circuit Malfunction
P0404 Exhaust Gas Recirculation Circuit Range/Performance
P0405 Exhaust Gas Recirculation Sensor "A" Circuit Low
P0406 Exhaust Gas Recirculation Sensor "A" Circuit High
P0407 Exhaust Gas Recirculation Sensor "B" Circuit Low
P0408 Exhaust Gas Recirculation Sensor "B" Circuit High
P0410 Secondary Air Injection System Malfunction
P0411 Secondary Air Injection System Incorrect Flow Detected
P0412 Secondary Air Injection System Switching Valve "A" Circuit Malfunction
P0413 Secondary Air Injection System Switching Valve "A" Circuit Open
P0414 Secondary Air Injection System Switching Valve "A" Circuit Shorted

P0415 Secondary Air Injection System Switching Valve "B" Circuit Malfunction

P0416 Secondary Air Injection System Switching Valve "B" Circuit Open

P0417 Secondary Air Injection System Switching Valve "B" Circuit Shorted

P0418 Secondary Air Injection System Relay "A" Circuit Malfunction

P0419 Secondary Air Injection System Relay "B" Circuit Malfunction

P0420 Catalyst System Efficiency Below Threshold (Bank no. 1)

P0421 Warm Up Catalyst Efficiency Below Threshold (Bank no. 1)

P0422 Main Catalyst Efficiency Below Threshold (Bank no. 1)

P0423 Heated Catalyst Efficiency Below Threshold (Bank no. 1)

P0424 Heated Catalyst Temperature Below Threshold (Bank no. 1)

P0430 Catalyst System Efficiency Below Threshold (Bank no. 2)

P0431 Warm Up Catalyst Efficiency Below Threshold (Bank no. 2)

P0432 Main Catalyst Efficiency Below Threshold (Bank no. 2)

P0433 Heated Catalyst Efficiency Below Threshold (Bank no. 2)

P0434 Heated Catalyst Temperature Below Threshold (Bank no. 2)

P0440 Evaporative Emission Control System Malfunction

P0441 Evaporative Emission Control System Incorrect Purge Flow

P0442 Evaporative Emission Control System Leak Detected (Small Leak)

P0443 Evaporative Emission Control System Purge Control Valve Circuit Malfunction

P0444 Evaporative Emission Control System Purge Control Valve Circuit Open

P0445 Evaporative Emission Control System Purge Control Valve Circuit Shorted

P0446 Evaporative Emission Control System Vent Control Circuit Malfunction

P0447 Evaporative Emission Control System Vent Control Circuit Open

P0448 Evaporative Emission Control System Vent Control Circuit Shorted

P0449 Evaporative Emission Control System Vent Valve/Solenoid Circuit Malfunction

P0450 Evaporative Emission Control System Pressure Sensor Malfunction

P0451 Evaporative Emission Control System Pressure Sensor Range/Performance

P0452 Evaporative Emission Control System Pressure Sensor Low Input

P0453 Evaporative Emission Control System Pressure Sensor High Input

P0454 Evaporative Emission Control System Pressure Sensor Intermittent

P0455 Evaporative Emission Control System Leak Detected (Gross Leak)

P0460 Fuel Level Sensor Circuit Malfunction

P0461 Fuel Level Sensor Circuit Range/Performance

P0462 Fuel Level Sensor Circuit Low Input

P0463 Fuel Level Sensor Circuit High Input

P0464 Fuel Level Sensor Circuit Intermittent

P0465 Purge Flow Sensor Circuit Malfunction

P0466 Purge Flow Sensor Circuit Range/Performance

P0467 Purge Flow Sensor Circuit Low Input

P0468 Purge Flow Sensor Circuit High Input

P0469 Purge Flow Sensor Circuit Intermittent

P0470 Exhaust Pressure Sensor Malfunction

P0471 Exhaust Pressure Sensor Range/Performance

P0472 Exhaust Pressure Sensor Low

P0473 Exhaust Pressure Sensor High

P0474 Exhaust Pressure Sensor Intermittent

P0475 Exhaust Pressure Control Valve Malfunction

P0476 Exhaust Pressure Control Valve Range/Performance

P0477 Exhaust Pressure Control Valve Low

P0478 Exhaust Pressure Control Valve High

P0479 Exhaust Pressure Control Valve Intermittent

P0480 Cooling Fan no. 1 Control Circuit Malfunction

P0481 Cooling Fan no. 2 Control Circuit Malfunction

P0482 Cooling Fan no. 3 Control Circuit Malfunction

P0483 Cooling Fan Rationality Check Malfunction

P0484 Cooling Fan Circuit Over Current

P0485 Cooling Fan Power/Ground Circuit Malfunction

P0500 Vehicle Speed Sensor Malfunction

P0501 Vehicle Speed Sensor Range/Performance

P0502 Vehicle Speed Sensor Circuit Low Input

P0503 Vehicle Speed Sensor Intermittent/Erratic/High

P0505 Idle Control System Malfunction

P0506 Idle Control System RPM Lower Than Expected

P0507 Idle Control System RPM Higher Than Expected

P0510 Closed Throttle Position Switch Malfunction

P0520 Engine Oil Pressure Sensor/Switch Circuit Malfunction

P0521 Engine Oil Pressure Sensor/Switch Range/Performance

P0522 Engine Oil Pressure Sensor/Switch Low Voltage

P0523 Engine Oil Pressure Sensor/Switch High Voltage

P0530 A/C Refrigerant Pressure Sensor Circuit Malfunction

P0531 A/C Refrigerant Pressure Sensor Circuit Range/Performance

P0532 A/C Refrigerant Pressure Sensor Circuit Low Input

P0533 A/C Refrigerant Pressure Sensor Circuit High Input

P0534 A/C Refrigerant Charge Loss

P0550 Power Steering Pressure Sensor Circuit Malfunction

P0551 Power Steering Pressure Sensor Circuit Range/Performance

P0552 Power Steering Pressure Sensor Circuit Low Input

P0553 Power Steering Pressure Sensor Circuit High Input

P0554 Power Steering Pressure Sensor Circuit Intermittent

P0560 System Voltage Malfunction

P0561 System Voltage Unstable

P0562 System Voltage Low

P0563 System Voltage High

P0565 Cruise Control On Signal Malfunction

P0566 Cruise Control Off Signal Malfunction

P0567 Cruise Control Resume Signal Malfunction

P0568 Cruise Control Set Signal Malfunction

P0569 Cruise Control Coast Signal Malfunction

P0570 Cruise Control Accel Signal Malfunction

P0571 Cruise Control/Brake Switch "A" Circuit Malfunction

P0572 Cruise Control/Brake Switch "A" Circuit Low

P0573 Cruise Control/Brake Switch "A" Circuit High

P0574 Through P0580 Reserved for Cruise Codes

P0600 Serial Communication Link Malfunction

P0601 Internal Control Module Memory Check Sum Error

P0602 Control Module Programming Error

P0603 Internal Control Module Keep Alive Memory (KAM) Error

P0604 Internal Control Module Random Access Memory (RAM) Error

P0605 Internal Control Module Read Only Memory (ROM) Error

P0606 PCM Processor Fault

P0608 Control Module VSS Output "A" Malfunction

P0609 Control Module VSS Output "B" Malfunction

P0620 Generator Control Circuit Malfunction

P0621 Generator Lamp "L" Control Circuit Malfunction

P0622 Generator Field "F" Control Circuit Malfunction

P0650 Malfunction Indicator Lamp (MIL) Control Circuit Malfunction

P0654 Engine RPM Output Circuit Malfunction

P0655 Engine Hot Lamp Output Control Circuit Malfunction

P0656 Fuel Level Output Circuit Malfunction

P0700 Transmission Control System Malfunction

P0701 Transmission Control System Range/Performance

P0702 Transmission Control System Electrical

P0703 Torque Converter/Brake Switch "B" Circuit Malfunction

P0704 Clutch Switch Input Circuit Malfunction

P0705 Transmission Range Sensor Circuit Malfunction (PRNDL Input)

P0706 Transmission Range Sensor Circuit Range/Performance

P0707 Transmission Range Sensor Circuit Low Input

P0708 Transmission Range Sensor Circuit High Input

P0709 Transmission Range Sensor Circuit Intermittent

P0710 Transmission Fluid Temperature Sensor Circuit Malfunction

P0711 Transmission Fluid Temperature Sensor Circuit Range/Performance

P0712 Transmission Fluid Temperature Sensor Circuit Low Input

P0713 Transmission Fluid Temperature Sensor Circuit High Input

P0714 Transmission Fluid Temperature Sensor Circuit Intermittent

P0715 Input/Turbine Speed Sensor Circuit Malfunction

P0716 Input/Turbine Speed Sensor Circuit Range/Performance

P0717 Input/Turbine Speed Sensor Circuit No Signal

P0718 Input/Turbine Speed Sensor Circuit Intermittent

P0719 Torque Converter/Brake Switch "B" Circuit Low

P0720 Output Speed Sensor Circuit Malfunction

P0721 Output Speed Sensor Circuit Range/Performance

P0722 Output Speed Sensor Circuit No Signal

P0723 Output Speed Sensor Circuit Intermittent

P0724 Torque Converter/Brake Switch "B" Circuit High

P0725 Engine Speed Input Circuit Malfunction

P0726 Engine Speed Input Circuit Range/Performance

P0727 Engine Speed Input Circuit No Signal

P0728 Engine Speed Input Circuit Intermittent

P0730 Incorrect Gear Ratio

P0731 Gear no. 1 Incorrect Ratio

P0732 Gear no. 2 Incorrect Ratio

P0733 Gear no. 3 Incorrect Ratio

P0734 Gear no. 4 Incorrect Ratio

P0735 Gear no. 5 Incorrect Ratio

P0736 Reverse Incorrect Ratio

P0740 Torque Converter Clutch Circuit Malfunction

P0741 Torque Converter Clutch Circuit Performance or Stuck Off

P0742 Torque Converter Clutch Circuit Stuck On

P0743 Torque Converter Clutch Circuit Electrical

P0744 Torque Converter Clutch Circuit Intermittent

P0745 Pressure Control Solenoid Malfunction

P0746 Pressure Control Solenoid Performance or Stuck Off

P0747 Pressure Control Solenoid Stuck On

P0748 Pressure Control Solenoid Electrical

P0749 Pressure Control Solenoid Intermittent

P0750 Shift Solenoid "A" Malfunction

P0751 Shift Solenoid "A" Performance or Stuck Off

P0752 Shift Solenoid "A" Stuck On

P0753 Shift Solenoid "A" Electrical

P0754 Shift Solenoid "A" Intermittent

P0755 Shift Solenoid "B" Malfunction

P0756 Shift Solenoid "B" Performance or Stuck Off

P0757 Shift Solenoid "B" Stuck On

P0758 Shift Solenoid "B" Electrical

P0759 Shift Solenoid "B" Intermittent

P0760 Shift Solenoid "C" Malfunction

P0761 Shift Solenoid "C" Performance Or Stuck Off

P0762 Shift Solenoid "C" Stuck On

P0763 Shift Solenoid "C" Electrical

P0764 Shift Solenoid "C" Intermittent

P0765 Shift Solenoid "D" Malfunction

P0766 Shift Solenoid "D" Performance Or Stuck Off

P0767 Shift Solenoid "D" Stuck On

P0768 Shift Solenoid "D" Electrical

P0769 Shift Solenoid "D" Intermittent

P0770 Shift Solenoid "E" Malfunction

P0771 Shift Solenoid "E" Performance Or Stuck Off

P0772 Shift Solenoid "E" Stuck On

P0773 Shift Solenoid "E" Electrical

P0774 Shift Solenoid "E" Intermittent

P0780 Shift Malfunction

P0781 1–2 Shift Malfunction

P0782 2–3 Shift Malfunction

P0783 3–4 Shift Malfunction

P0784 4–5 Shift Malfunction

P0785 Shift/Timing Solenoid Malfunction

P0786 Shift/Timing Solenoid Range/Performance

P0787 Shift/Timing Solenoid Low

P0788 Shift/Timing Solenoid High

P0789 Shift/Timing Solenoid Intermittent

P0790 Normal/Performance Switch Circuit Malfunction

P0801 Reverse Inhibit Control Circuit Malfunction

P0803 1–4 Upshift (Skip Shift) Solenoid Control Circuit Malfunction

P0804 1–4 Upshift (Skip Shift) Lamp Control Circuit Malfunction

P1102 Oxygen Sensor Heating Circuit, Bank 1-Sensor 1 Short to B+

P1105 Oxygen Sensor Heating Circuit, Bank 1-Sensor 2 Short to B+

P1107 Oxygen Sensor Heating Circuit, Bank 2-Sensor I Short to B+

P1110 Oxygen Sensor Heating Circuit, Bank 2-Sensor 2 Short to B+

P1127 Long Term Fuel Trim Multiplicative, Bank 1 System Too Rich

P1128 Long Term Fuel Trim Multiplicative, Bank 1 System Too Lean

P1129 Long Term Fuel Trim Multiplicative, Bank2 System too Rich

P1130 Long Term Fuel Trim Multiplicative, Bank2 System too Lean

P1136 Long Term Fuel Trim Additive, Bank 1 System Too Lean

P1137 Long Term Fuel Trim Additive, Bank 1 System Too Rich

P1138 Long Term Fuel Trim Additive Fuel, Bank I System too Lean

P1139 Long Term Fuel Trim Additive Fuel, Bank 1 System too Rich

P1141 Load Calculation Cross Check Range/Performance

P1176 Oxygen Correction Behind Catalyst, B1 Limit Attained

P1177 Oxygen Correction Behind Catalyst. 82 Limit Attained

P1196 Oxygen Sensor Heater Circuit, Bank 1-Sensor 1 Electrical Malfunction

P1197 Oxygen Sensor Heater Circuit, Bank 2-Sensor I Electrical Malfunction

P1198 Oxygen Sensor Heater Circuit, Bankl-Sensor2 Electrical Malfunction

P1198 Oxygen Sensor Heater Circuit, Bank 1-Sensor 2 Electrical Malfunction

P1199 Oxygen Sensor Heater Circuit, Bank 2-Sensor 2 Electrical Malfunction
P1201 Cylinder 1, Fuel Injection Circuit Electrical Malfunction
P1202 Cylinder 2, Fuel Injection Circuit Electrical Malfunction
P1203 Cylinder 3, Fuel Injection Circuit Electrical Malfunction
P1204 Cylinder 4, Fuel Injection Circuit Electrical Malfunction
P1205 Cylinder 5, Fuel Injection Circuit Electrical Malfunction
P1206 Cylinder 6, Fuel Injection Circuit Electrical Malfunction
P1207 Cylinder 7, Fuel Injection Circuit Electrical Malfunction
P1208 Cylinder 8, Fuel Injection Circuit Electrical Malfunction
P1213 Cylinder I-Fuel Injection Circuit Short to B+
P1214 Cylinder 2-Fuel Injection Circuit Short to B+
P1215 Cylinder 3 Fuel Injection Circuit Short to B+
P1216 Cylinder 4 Fuel Injection Circuit Short to B+
P1217 Cylinder 5 Fuel Injection Circuit Short to B+
P1218 Cylinder 6 Fuel Injection Circuit Short to B+
P1219 Cylinder 7, Fuel Injection Circuit Short to B+
P1219 Cylinder 8, Fuel Injection Circuit Short to B+
P1225 Cylinder I Fuel Injection Circuit Short to Ground
P1226 Cylinder 2 Fuel Injection Circuit Short to Ground
P1227 Cylinder 3 Fuel Injection Circuit Short to Ground
P1228 Cylinder 4 Fuel Injection Circuit Short to Ground
P1229 Cylinder 5 Fuel Injection Circuit Short to Ground
P1230 Cylinder 6 Fuel Injection Circuit Short to Ground
P1237 Cylinder I Fuel Injection Circuit Open Circuit
P1238 Cylinder 2 Fuel Injection Circuit Open Circuit
P1239 Cylinder 3 Fuel Injection Circuit Open Circuit
P1240 Cylinder 4 Fuel Injection Circuit Open Circuit
P1241 Cylinder 5 Fuel Injection Circuit Open Circuit
P1242 Cylinder 6 Fuel Injection Circuit Open Circuit
P1250 Fuel Level Too Low
P1280 Fuel Injection Air Control Valve Circuit Flow too Low
P1283 Fuel Injection Air Control Valve Circuit Electrical Malfunction
P1325 Cylinder I Knock Control Limit Attained
P1326 Cylinder 2 Knock Control Limit Attained
P1327 Cylinder 3 Knock Control Limit Attained
P1328 Cylinder 4 Knock Control Limit Attained
P1329 Cylinder 5 Knock Control Limit Attained
P1330 Cylinder 6 Knock Control Limit Attained
P1331 Cylinder 7, Knock Control Limit Attained
P1332 Cylinder 8, Knock Control Limit Attained
P1337 Camshaft Position Sensor, Bank 1 Short to Ground
P1338 Camshaft Position Sensor, Bank 1 Open Circuit/Short to B+
P1340 Boost Pressure Control Valve Short to B+
P1386 Internal Control Module Knock Control Circuit Error
P1391 Camshaft Position Sensor, Bank 2 Short to Ground
P1392 Camshaft Position Sensor, Bank 2 Open Circuit/Short to B+
P1410 Tank Ventilation Valve Circuit Short to B+
P1420 Secondary Air Injection Module Short To B+
P1421 Secondary Air Injection Module Short To Ground
P1421 Secondary Air Injection Valve Circuit Short to Ground
P1422 Secondary Air Injection System Control Valve Circuit Short To B+
P1422 Secondary Air Injection Valve Circuit Short to B+
P1425 Tank Vent Valve Short To Ground
P1425 Tank Vent Valve Short to Ground
P1426 Tank Vent Valve Open
P1426 Tank Vent Valve Open

P1432 Secondary Air Injection Valve Open
P1433 Secondary Air Injection System Pump Relay Circuit Open
P1434 Secondary Air Injection System Pump Relay Circuit Short to B+
P1435 Secondary Air Injection System Pump Relay Circuit Short to Ground
P1436 Secondary Air Injection System Pump Relay Circuit Electrical Malfunction
P1450 Secondary Air Injection System Circuit Short To B+
P1451 Secondary Air Injection System Circuit Short To Ground
P1452 Secondary Air Injection System Open Circuit
P1471 EVAP Emission Control LDP Circuit Short to B+
P1472 EVAP Emission Control LDP Circuit Short to Ground
P1473 EVAP Emission Control LDP Circuit Open Circuit
P1475 EVAP Emission Control LDP Circuit Malfunction/Signal Circuit Open
P1476 EVAP Emission Control LDP Circuit Malfunction/Insufficient Vacuum
P1477 EVAP Emission Control LDP Circuit Malfunction
P1500 Fuel Pump Relay Circuit Electrical Malfunction
P1501 Fuel Pump Relay Circuit Short to Ground
P1502 Fuel Pump Relay Circuit Short to B+
P1505 Closed Throttle Position Switch Does Not Close/Open Circuit
P1506 Closed Throttle Position Switch Does Not Open/Short to Ground
P1507 Idle System Learned Value Lower Limit Attained
P1508 Idle System Learned Value Upper Limit Attained
P1512 Intake Manifold Changeover Valve Circuit, Short to B+
P1515 Intake Manifold Changeover Valve Circuit, Short to Ground
P1516 Intake Manifold Changeover Valve Circuit, Open
P1519 Intake Camshaft Control, Bank I Malfunction
P1522 Intake Camshaft Control, Bank 2 Malfunction
P1543 Throttle Actuation Potentiometer Signal Too Low
P1544 Throttle Actuation Potentiometer Signal Too High
P1545 Throttle Position Control Malfunction
P1547 Boost Pressure Control Valve Short to Ground
P1548 Boost Pressure Control Valve Open
P1555 Charge Pressure Upper Limit Exceeded
P1556 Charge Pressure Negative Deviation
P1557 Charge Pressure Positive Deviation
P1558 Throttle Actuator Electrical Malfunction
P1559 Idle Speed Control Throttle Position Adaptation Malfunction
P1560 Maximum Engine Speed Exceeded
P1564 Idle Speed Control, Throttle Position Low Voltage During Adaptation
P1580 Throttle Actuator (B1) Malfunction
P1582 Idle Adaptation At Limit
P1602 Power Supply (B+) Terminal 30 Low Voltage
P1606 Rough Road Spec Engine Torque ABS-ECU Electrical Malfunction
P1611 MIL Call-up Circuit/Transmission Control Module Short to Ground
P1612 Electronic Control Module Incorrect Coding
P1613 MIL Call-up Circuit Open/Short to B+
P1624 MIL Request Signal Active
P1625 CAN-Bus Implausible Message from Transmission Control
P1626 CAN-Bus Missing Message from Transmission Control
P1640 Internal Control Module (EEPROM) Error
P1640 Internal Control Module (EEPROM) Error

P1681 Control Unit Programming not Finished
P1690 Malfunction Indicator Light Malfunction
P1693 Malfunction Indicator Light Short to B+
P1778 Solenoid EV7 Electrical Malfunction
P1780 Engine Intervention Readable

TROUBLE CODE EQUIVALENT LIST

The following codes are displayed by Volkswagen's VAG 1551 scan tool/tester.

16486 Mass or Volume Air Flow Circuit Low Input
16487 Mass or Volume Air Flow Circuit High Input
16491 Manifold Absolute Pressure or Barometric Pressure Low Input
16492 Manifold Absolute Pressure or Barometric Pressure High Input
16496 Intake Air Temperature Circuit Low Input
16497 Intake Air Temperature Circuit High Input
16500 Engine Coolant Temperature Circuit Range/Performance
16501 Engine Coolant Temperature Circuit Low Input
16502 Oxygen Engine Coolant Temperature Circuit High Input
16504 Throttle Position Sensor A Circuit Malfunction
16505 Throttle/Pedal Position Sensor A Circuit Range/Performance
16506 Throttle/Pedal Position Sensor A Circuit Low Input
16507 Throttle/Pedal Position Sensor A Circuit High Input
16509 Insufficient Coolant Temperature For Closed Loop Fuel Control
16514 Oxygen Sensor Circuit, Bank 1-Sensor 1 Malfunction
16515 Oxygen Sensor Circuit, Bank 1-Sensor 1 Low Voltage
16516 Oxygen Sensor Circuit, Bank 1-Sensor 1 High Voltage
16517 Oxygen Sensor Circuit, Bank 1-Sensor 1 Slow Response
16518 Oxygen Sensor Circuit, Bank 1-Sensor 1 No Activity Detected
16519 Oxygen Sensor Heater Circuit, Bank 1-Sensor 1 Malfunction
16520 Oxygen Sensor Circuit, Bank 1-Sensor 2 Malfunction
16521 Oxygen Sensor Circuit, Bank 1-Sensor 2 Low Voltage
16522 Oxygen Sensor Circuit, Bank 1-Sensor 2 High Voltage
16524 Oxygen Sensor Circuit, Bank 1-Sensor 2 No Activity Detected
16525 Oxygen Sensor Heater Circuit, Bank 1-Sensor 2 Malfunction
16534 Oxygen Sensor Circuit, Bank 2-Sensor 1 Malfunction
16535 Oxygen Sensor Circuit, Bank 2-Sensor 1 Low Voltage
16536 Oxygen Sensor Circuit, Bank 2-Sensor 1 High Voltage
16537 Oxygen Sensor Circuit, Bank 2-Sensor 1 Slow Response
16538 Oxygen Sensor Circuit, Bank 2-Sensor 1 No Activity Detected
16540 Oxygen Sensor Circuit, Bank 2-Sensor 2 Malfunction
16541 Oxygen Sensor Circuit, Bank 2-Sensor 2 Low Voltage
16542 Oxygen Sensor Circuit, Bank 2-Sensor 2 High Voltage
16544 Oxygen Sensor Circuit, Bank 2-Sensor 2 No Activity Detected
16555 Oxygen System Too Lean, Bank 1
16556 Oxygen System Too Rich, Bank 1
16684 Random Multiple Misfire Detected
16685 Cylinder 1 Misfire Detected
16686 Cylinder 2 Misfire Detected
16687 Cylinder 3 Misfire Detected
16688 Cylinder 4 Misfire Detected
16689 Cylinder 5 Misfire Detected
16690 Cylinder 6 Misfire Detected
16705 Ignition Distributor Engine Speed Input Circuit Range/Performance

16706 Ignition /Distributor Engine Speed Input Circuit No Signal
16711 Knock sensor 1 Circuit Low Input
16712 Knock Sensor Circuit, High Input
16716 Knock sensor 2 Circuit Low Input
16716 Knock Sensor Circuit, Low Input
16717 Knock Sensor 2 Circuit, High Input
16725 Camshaft Position Sensor Circuit Range/Performance
16795 Secondary Air Injection System Incorrect Flow Detected
16806 Main Catalyst Efficiency Below Threshold (Bank 1)
16806 Main Catalyst, Bank I Efficiency Below Threshold
16824 Evaporative Emission Control System Malfunction
16825 EVAP Emission Contr. Sys. Incorrect Purge Flow
16826 EVAP Emission Contr. Sys. (Small Leak) Leak Detected
16839 EVAP Emission Contr. Sys. (Gross Leak) Leak Detected
16885 Vehicle Speed Sensor Range/Performance
16890 Idle Control System RPM Lower Than Expected
16891 Idle Control System RPM Higher Than Expected
16894 Closed Throttle Position Switch Malfunction
16944 System Voltage Malfunction
16946 System Voltage Low Voltage
16947 System Voltage High Voltage
16985 Internal Contr. Module Memory Check Sum Error
16988 Internal Contr. Module Random Access Memory (RAM) Error
16989 Internal Control Module Read Only Memory (ROM) Error
17091 Transmission Range Sensor Circuit Low Input
17092 Transmission Range Sensor Circuit High Input
17099 Input Turbine Speed Sensor Circuit Malfunction
17106 Output Speed Sensor Circuit No Signal
17109 Engine Speed Input Circuit Malfunction
17132 Pressure Control Solenoid Electrical
17137 Shift Solenoid A Electrical
17142 Shift Solenoid B Electrical
17147 Shift Solenoid C Electrical
17152 Shift Solenoid D Electrical
17157 Shift Solenoid E Electrical
16684 Random/Multiple Cylinder Misfire Detected
16685 Cylinder I Misfire Detected
16686 Cylinder 2 Misfire Detected
17510 Oxygen Sensor Heating Circuit, Bank 1-Sensor 1 Short to B+
17513 Oxygen Sensor Heating Circuit, Bank 1-Sensor 2 Short to B+
17515 Oxygen Sensor Heating Circuit, Bank 2-Sensor I Short to B+
17518 Oxygen Sensor Heating Circuit, Bank 2-Sensor 2 Short to B+
17535 Long Term Fuel Trim Multiplicative, Bank 1 System Too Rich
17536 Long Term Fuel Trim Multiplicative, Bank 1 System Too Lean
17537 Long Term Fuel Trim Multiplicative, Bank2 System too Rich
17538 Long Term Fuel Trim Multiplicative, Bank2 System too Lean
17544 Long Term Fuel Trim Additive, Bank 1 System Too Lean
17545 Long Term Fuel Trim Additive, Bank 1 System Too Rich
17546 Long Term Fuel Trim Additive Fuel, Bank I System too Lean
17547 Long Term Fuel Trim Additive Fuel, Bank 1 System too Rich
17549 Load Calculation Cross Check Range/Performance
17584 Oxygen Correction Behind Catalyst, B1 Limit Attained
17585 Oxygen Correction Behind Catalyst. 82 Limit Attained

17604 Oxygen Sensor Heater Circuit, Bank 1-Sensor 1 Electrical Malfunction

17605 Oxygen Sensor Heater Circuit, Bank 2-Sensor I Electrical Malfunction

17606 Oxygen Sensor Heater Circuit, Bankl-Sensor2 Electrical Malfunction

17606 Oxygen Sensor Heater Circuit, Bank 1-Sensor 2 Electrical Malfunction

17607 Oxygen Sensor Heater Circuit, Bank 2-Sensor 2 Electrical Malfunction

17609 Cylinder 1, Fuel Injection Circuit Electrical Malfunction

17610 Cylinder 2, Fuel Injection Circuit Electrical Malfunction

17611 Cylinder 3, Fuel Injection Circuit Electrical Malfunction

17612 Cylinder 4, Fuel Injection Circuit Electrical Malfunction

17613 Cylinder 5, Fuel Injection Circuit Electrical Malfunction

17614 Cylinder 6, Fuel Injection Circuit Electrical Malfunction

17615 Cylinder 7, Fuel Injection Circuit Electrical Malfunction

17616 Cylinder 8, Fuel Injection Circuit Electrical Malfunction

17621 Cylinder I-Fuel Injection Circuit Short to B+

17622 Cylinder 2-Fuel Injection Circuit Short to B+

17623 Cylinder 3 Fuel Injection Circuit Short to B+

17624 Cylinder 4 Fuel Injection Circuit Short to B+

17625 Cylinder 5 Fuel Injection Circuit Short to B+

17626 Cylinder 6 Fuel Injection Circuit Short to B+

17627 Cylinder 7, Fuel Injection Circuit Short to B+

17628 Cylinder 8, Fuel Injection Circuit Short to B+

17633 Cylinder I Fuel Injection Circuit Short to Ground

17634 Cylinder 2 Fuel Injection Circuit Short to Ground

17635 Cylinder 3 Fuel Injection Circuit Short to Ground

17636 Cylinder 4 Fuel Injection Circuit Short to Ground

17637 Cylinder 5 Fuel Injection Circuit Short to Ground

17638 Cylinder 6 Fuel Injection Circuit Short to Ground

17645 Cylinder I Fuel Injection Circuit Open Circuit

17646 Cylinder 2 Fuel Injection Circuit Open Circuit

17647 Cylinder 3 Fuel Injection Circuit Open Circuit

17648 Cylinder 4 Fuel Injection Circuit Open Circuit

17649 Cylinder 5 Fuel Injection Circuit Open Circuit

17650 Cylinder 6 Fuel Injection Circuit Open Circuit

17658 Fuel Level Too Low

17688 Fuel Injection Air Control Valve Circuit Flow too Low

17691 Fuel Injection Air Control Valve Circuit Electrical Malfunction

17733 Cylinder I Knock Control Limit Attained

17734 Cylinder 2 Knock Control Limit Attained

17735 Cylinder 3 Knock Control Limit Attained

17736 Cylinder 4 Knock Control Limit Attained

17737 Cylinder 5 Knock Control Limit Attained

17738 Cylinder 6 Knock Control Limit Attained

17739 Cylinder 7 Knock Control Limit Attained

17740 Cylinder 8 Knock Control Limit Attained

17745 Camshaft Position Sensor, Bank 1 Short to Ground

17746 Camshaft Position Sensor, Bank 1 Open Circuit/Short to B+

17954 Boost Pressure Control Valve Short to B+

17794 Internal Control Module Knock Control Circuit Error

17799 Camshaft Position Sensor, Bank 2 Short to Ground

17800 Camshaft Position Sensor, Bank 2 Open Circuit/Short to B+

17818 Tank Ventilation Valve Circuit Short to B+

17818 Tank Ventilation Valve Circuit Short to B+

17828 Secondary Air Injection Module Short To B+

17829 Secondary Air Injection Module Short To Ground

17829 Secondary Air Injection Valve Circuit Short to Ground

17830 Secondary Air Injection System Control Valve Circuit Short To B+

17830 Secondary Air Injection Valve Circuit Short to B+

17833 Tank Vent Valve Short To Ground

17833 Tank Vent Valve Short to Ground

17834 Tank Vent Valve Open

17834 Tank Vent Valve Open

17840 Secondary Air Injection Valve Open

17841 Secondary Air Injection System Pump Relay Circuit Open

17842 Secondary Air Injection System Pump Relay Circuit Short to B+

17843 Secondary Air Injection System Pump Relay Circuit Short to Ground

17844 Secondary Air Injection System Pump Relay Circuit Electrical Malfunction

17858 Secondary Air Injection System Circuit Short To B+

17859 Secondary Air Injection System Circuit Short To Ground

17860 Secondary Air Injection System Open Circuit

17879 EVAP Emission Control LDP Circuit Short to B+

17880 EVAP Emission Control LDP Circuit Short to Ground

17881 EVAP Emission Control LDP Circuit Open Circuit

17883 EVAP Emission Control LDP Circuit Malfunction/Signal Circuit Open

17884 EVAP Emission Control LDP Circuit Malfunction/Insufficient Vacuum

17885 EVAP Emission Control LDP Circuit Malfunction

17908 Fuel Pump Relay Circuit Electrical Malfunction

17909 Fuel Pump Relay Circuit Short to Ground

17910 Fuel Pump Relay Circuit Short to B+

17913 Closed Throttle Position Switch Does Not Close/Open Circuit

17914 Closed Throttle Position Switch Does Not Open/Short to Ground

17915 Idle System Learned Value Lower Limit Attained

17916 Idle System Learned Value Upper Limit Attained

17920 Intake Manifold Changeover Valve Circuit, Short to B+

17923 Intake Manifold Changeover Valve Circuit, Short to Ground

17924 Intake Manifold Changeover Valve Circuit, Open

17927 Intake Camshaft Control, Bank 2 Malfunction

17951 Throttle Actuation Potentiometer Signal Too Low

17952 Throttle Actuation Potentiometer Signal Too High

17953 Throttle Position Control Malfunction

17955 Boost Pressure Control Valve Short to Ground

17956 Boost Pressure Control Valve Open

17963 Charge Pressure Upper Limit Exceeded

17964 Charge Pressure Negative Deviation

17965 Charge Pressure Positive Deviation

17966 Throttle Actuator Electrical Malfunction

17967 Idle Speed Control Throttle Position Adaptation Malfunction

17968 Maximum Engine Speed Exceeded

17972 Idle Speed Control, Throttle Position Low Voltage During Adaptation

17988 Throttle Actuator (B1) Malfunction

17990 Idle Adaptation At Limit

18010 Power Supply (B+) Terminal 30 Low Voltage

18014 Rough Road Spec Engine Torque ABS-ECU Electrical Malfunction

18019 MIL Call-up Circuit/Transmission Control Module Short to Ground

18020 Electronic Control Module Incorrect Coding
18021 MIL Call-up Circuit Open/Short to B+
18032 MIL Request Signal Active
18033 CAN-Bus Implausible Message from Transmission Control
18034 CAN-Bus Missing Message from Transmission Control
18048 Internal Control Module (EEPROM) Error
18048 Internal Control Module (EEPROM) Error
18089 Control Unit Programming not Finished
18098 Malfunction Indicator Light Malfunction
18101 Malfunction Indicator Light Short to B+
18186 Solenoid EV7 Electrical Malfunction
18188 Engine Intervention Readable

Volvo

READING CODES

➡**This procedure is applicable to generic and manufacturer specific codes.**

Reading the control module memory is on of the first steps in OBD II system diagnostics. This step should be initially performed to determine the general nature of the fault. Subsequent readings will determine if the fault has been cleared.

Reading codes can be performed by any of the methods below:
• Read the control module memory with the Generic Scan Tool (GST)
• Read the control module memory with the vehicle manufacturer's specific tester

To read the fault codes, connect the scan tool or tester according to the manufacturer's instructions. Follow the manufacturer's specified procedure for reading the codes.

CLEARING CODES

➡**This procedure is applicable to generic and manufacturer specific codes.**

Control module reset procedures are a very important part of OBD II System diagnostics. This step should be done at the end of any fault code repair and at the end of any driveability repair.

Clearing codes can be performed by any of the methods below:
• Clear the control module memory with the Generic Scan Tool (GST)
• Clear the control module memory with the vehicle manufacturer's specific tester
• Turn the ignition **OFF** and remove the negative battery cable for at least 1 minute.

Removing the negative battery cable may cause other systems in the vehicle to lose their memory. Prior to removing the cable, ensure you have the proper reset codes for radios and alarms.

➡**The MIL will may also be de-activated for some codes if the vehicle completes three consecutive trips without a fault detected with vehicle conditions similar to those present during the fault.**

DIAGNOSTIC TROUBLE CODES

P0000 No Failures
P0100 Mass or Volume Air Flow Circuit Malfunction
P0101 Mass or Volume Air Flow Circuit Range/Performance Problem

P0102 Mass or Volume Air Flow Circuit Low Input
P0103 Mass or Volume Air Flow Circuit High Input
P0104 Mass or Volume Air Flow Circuit Intermittent
P0105 Manifold Absolute Pressure/Barometric Pressure Circuit Malfunction
P0106 Manifold Absolute Pressure/Barometric Pressure Circuit Range/Performance Problem
P0107 Manifold Absolute Pressure/Barometric Pressure Circuit Low Input
P0108 Manifold Absolute Pressure/Barometric Pressure Circuit High Input
P0109 Manifold Absolute Pressure/Barometric Pressure Circuit Intermittent
P0110 Intake Air Temperature Circuit Malfunction
P0111 Intake Air Temperature Circuit Range/Performance Problem
P0112 Intake Air Temperature Circuit Low Input
P0113 Intake Air Temperature Circuit High Input
P0114 Intake Air Temperature Circuit Intermittent
P0115 Engine Coolant Temperature Circuit Malfunction
P0116 Engine Coolant Temperature Circuit Range/Performance Problem
P0117 Engine Coolant Temperature Circuit Low Input
P0118 Engine Coolant Temperature Circuit High Input
P0119 Engine Coolant Temperature Circuit Intermittent
P0120 Throttle/Pedal Position Sensor/Switch "A" Circuit Malfunction
P0121 Throttle/Pedal Position Sensor/Switch "A" Circuit Range/Performance Problem
P0122 Throttle/Pedal Position Sensor/Switch "A" Circuit Low Input
P0123 Throttle/Pedal Position Sensor/Switch "A" Circuit High Input
P0124 Throttle/Pedal Position Sensor/Switch "A" Circuit Intermittent
P0125 Insufficient Coolant Temperature For Closed Loop Fuel Control
P0126 Insufficient Coolant Temperature For Stable Operation
P0130 O_2 Circuit Malfunction (Bank no. 1 Sensor no. 1)
P0131 O_2 Sensor Circuit Low Voltage (Bank no. 1 Sensor no. 1)
P0132 O_2 Sensor Circuit High Voltage (Bank no. 1 Sensor no. 1)
P0133 O_2 Sensor Circuit Slow Response (Bank no. 1 Sensor no. 1)
P0134 O_2 Sensor Circuit No Activity Detected (Bank no. 1 Sensor no. 1)
P0135 O_2 Sensor Heater Circuit Malfunction (Bank no. 1 Sensor no. 1)
P0136 O_2 Sensor Circuit Malfunction (Bank no. 1 Sensor no. 2)
P0137 O_2 Sensor Circuit Low Voltage (Bank no. 1 Sensor no. 2)
P0138 O_2 Sensor Circuit High Voltage (Bank no. 1 Sensor no. 2)
P0139 O_2 Sensor Circuit Slow Response (Bank no. 1 Sensor no. 2)
P0140 O_2 Sensor Circuit No Activity Detected (Bank no. 1 Sensor no. 2)
P0141 O_2 Sensor Heater Circuit Malfunction (Bank no. 1 Sensor no. 2)
P0142 O_2 Sensor Circuit Malfunction (Bank no. 1 Sensor no. 3)
P0143 O_2 Sensor Circuit Low Voltage (Bank no. 1 Sensor no. 3)
P0144 O_2 Sensor Circuit High Voltage (Bank no. 1 Sensor no. 3)
P0145 O_2 Sensor Circuit Slow Response (Bank no. 1 Sensor no. 3)
P0146 O_2 Sensor Circuit No Activity Detected (Bank no. 1 Sensor no. 3)
P0147 O_2 Sensor Heater Circuit Malfunction (Bank no. 1 Sensor no. 3)

P0150 O_2 Sensor Circuit Malfunction (Bank no. 2 Sensor no. 1)
P0151 O_2 Sensor Circuit Low Voltage (Bank no. 2 Sensor no. 1)
P0152 O_2 Sensor Circuit High Voltage (Bank no. 2 Sensor no. 1)
P0153 O_2 Sensor Circuit Slow Response (Bank no. 2 Sensor no. 1)
P0154 O_2 Sensor Circuit No Activity Detected (Bank no. 2 Sensor no. 1)
P0155 O_2 Sensor Heater Circuit Malfunction (Bank no. 2 Sensor no. 1)
P0156 O_2 Sensor Circuit Malfunction (Bank no. 2 Sensor no. 2)
P0157 O_2 Sensor Circuit Low Voltage (Bank no. 2 Sensor no. 2)
P0158 O_2 Sensor Circuit High Voltage (Bank no. 2 Sensor no. 2)
P0159 O_2 Sensor Circuit Slow Response (Bank no. 2 Sensor no. 2)
P0160 O_2 Sensor Circuit No Activity Detected (Bank no. 2 Sensor no. 2)
P0161 O_2 Sensor Heater Circuit Malfunction (Bank no. 2 Sensor no. 2)
P0162 O_2 Sensor Circuit Malfunction (Bank no. 2 Sensor no. 3)
P0163 O_2 Sensor Circuit Low Voltage (Bank no. 2 Sensor no. 3)
P0164 O_2 Sensor Circuit High Voltage (Bank no. 2 Sensor no. 3)
P0165 O_2 Sensor Circuit Slow Response (Bank no. 2 Sensor no. 3)
P0166 O_2 Sensor Circuit No Activity Detected (Bank no. 2 Sensor no. 3)
P0167 O_2 Sensor Heater Circuit Malfunction (Bank no. 2 Sensor no. 3)
P0170 Fuel Trim Malfunction (Bank no. 1)
P0171 System Too Lean (Bank no. 1)
P0172 System Too Rich (Bank no. 1)
P0173 Fuel Trim Malfunction (Bank no. 2)
P0174 System Too Lean (Bank no. 2)
P0175 System Too Rich (Bank no. 2)
P0176 Fuel Composition Sensor Circuit Malfunction
P0177 Fuel Composition Sensor Circuit Range/Performance
P0178 Fuel Composition Sensor Circuit Low Input
P0179 Fuel Composition Sensor Circuit High Input
P0180 Fuel Temperature Sensor "A" Circuit Malfunction
P0181 Fuel Temperature Sensor "A" Circuit Range/Performance
P0182 Fuel Temperature Sensor "A" Circuit Low Input
P0183 Fuel Temperature Sensor "A" Circuit High Input
P0184 Fuel Temperature Sensor "A" Circuit Intermittent
P0185 Fuel Temperature Sensor "B" Circuit Malfunction
P0186 Fuel Temperature Sensor "B" Circuit Range/Performance
P0187 Fuel Temperature Sensor "B" Circuit Low Input
P0188 Fuel Temperature Sensor "B" Circuit High Input
P0189 Fuel Temperature Sensor "B" Circuit Intermittent
P0190 Fuel Rail Pressure Sensor Circuit Malfunction
P0191 Fuel Rail Pressure Sensor Circuit Range/Performance
P0192 Fuel Rail Pressure Sensor Circuit Low Input
P0193 Fuel Rail Pressure Sensor Circuit High Input
P0194 Fuel Rail Pressure Sensor Circuit Intermittent
P0195 Engine Oil Temperature Sensor Malfunction
P0196 Engine Oil Temperature Sensor Range/Performance
P0197 Engine Oil Temperature Sensor Low
P0198 Engine Oil Temperature Sensor High
P0199 Engine Oil Temperature Sensor Intermittent
P0200 Injector Circuit Malfunction
P0201 Injector Circuit Malfunction—Cylinder no. 1
P0202 Injector Circuit Malfunction—Cylinder no. 2
P0203 Injector Circuit Malfunction—Cylinder no. 3
P0204 Injector Circuit Malfunction—Cylinder no. 4

P0205 Injector Circuit Malfunction—Cylinder no. 5
P0206 Injector Circuit Malfunction—Cylinder no. 6
P0207 Injector Circuit Malfunction—Cylinder no. 7
P0208 Injector Circuit Malfunction—Cylinder no. 8
P0209 Injector Circuit Malfunction—Cylinder no. 9
P0210 Injector Circuit Malfunction—Cylinder no. 10
P0211 Injector Circuit Malfunction—Cylinder no. 11
P0212 Injector Circuit Malfunction—Cylinder no. 12
P0213 Cold Start Injector no. 1 Malfunction
P0214 Cold Start Injector no. 2 Malfunction
P0215 Engine Shutoff Solenoid Malfunction
P0216 Injection Timing Control Circuit Malfunction
P0217 Engine Over Temperature Condition
P0218 Transmission Over Temperature Condition
P0219 Engine Over Speed Condition
P0220 Throttle/Pedal Position Sensor/Switch "B" Circuit Malfunction
P0221 Throttle/Pedal Position Sensor/Switch "B" Circuit Range/Performance Problem
P0222 Throttle/Pedal Position Sensor/Switch "B" Circuit Low Input
P0223 Throttle/Pedal Position Sensor/Switch "B" Circuit High Input
P0224 Throttle/Pedal Position Sensor/Switch "B" Circuit Intermittent
P0225 Throttle/Pedal Position Sensor/Switch "C" Circuit Malfunction
P0226 Throttle/Pedal Position Sensor/Switch "C" Circuit Range/Performance Problem
P0227 Throttle/Pedal Position Sensor/Switch "C" Circuit Low Input
P0228 Throttle/Pedal Position Sensor/Switch "C" Circuit High Input
P0229 Throttle/Pedal Position Sensor/Switch "C" Circuit Intermittent
P0230 Fuel Pump Primary Circuit Malfunction
P0231 Fuel Pump Secondary Circuit Low
P0232 Fuel Pump Secondary Circuit High
P0233 Fuel Pump Secondary Circuit Intermittent
P0234 Engine Over Boost Condition
P0261 Cylinder no. 1 Injector Circuit Low
P0262 Cylinder no. 1 Injector Circuit High
P0263 Cylinder no. 1 Contribution/Balance Fault
P0264 Cylinder no. 2 Injector Circuit Low
P0265 Cylinder no. 2 Injector Circuit High
P0266 Cylinder no. 2 Contribution/Balance Fault
P0267 Cylinder no. 3 Injector Circuit Low
P0268 Cylinder no. 3 Injector Circuit High
P0269 Cylinder no. 3 Contribution/Balance Fault
P0270 Cylinder no. 4 Injector Circuit Low
P0271 Cylinder no. 4 Injector Circuit High
P0272 Cylinder no. 4 Contribution/Balance Fault
P0273 Cylinder no. 5 Injector Circuit Low
P0274 Cylinder no. 5 Injector Circuit High
P0275 Cylinder no. 5 Contribution/Balance Fault
P0276 Cylinder no. 6 Injector Circuit Low
P0277 Cylinder no. 6 Injector Circuit High
P0278 Cylinder no. 6 Contribution/Balance Fault
P0279 Cylinder no. 7 Injector Circuit Low
P0280 Cylinder no. 7 Injector Circuit High
P0281 Cylinder no. 7 Contribution/Balance Fault
P0282 Cylinder no. 8 Injector Circuit Low
P0283 Cylinder no. 8 Injector Circuit High

P0284 Cylinder no. 8 Contribution/Balance Fault
P0285 Cylinder no. 9 Injector Circuit Low
P0286 Cylinder no. 9 Injector Circuit High
P0287 Cylinder no. 9 Contribution/Balance Fault
P0288 Cylinder no. 10 Injector Circuit Low
P0289 Cylinder no. 10 Injector Circuit High
P0290 Cylinder no. 10 Contribution/Balance Fault
P0291 Cylinder no. 11 Injector Circuit Low
P0292 Cylinder no. 11 Injector Circuit High
P0293 Cylinder no. 11 Contribution/Balance Fault
P0294 Cylinder no. 12 Injector Circuit Low
P0295 Cylinder no. 12 Injector Circuit High
P0296 Cylinder no. 12 Contribution/Balance Fault
P0300 Random/Multiple Cylinder Misfire Detected
P0301 Cylinder no. 1—Misfire Detected
P0302 Cylinder no. 2—Misfire Detected
P0303 Cylinder no. 3—Misfire Detected
P0304 Cylinder no. 4—Misfire Detected
P0305 Cylinder no. 5—Misfire Detected
P0306 Cylinder no. 6—Misfire Detected
P0307 Cylinder no. 7—Misfire Detected
P0308 Cylinder no. 8—Misfire Detected
P0309 Cylinder no. 9—Misfire Detected
P0310 Cylinder no. 10—Misfire Detected
P0311 Cylinder no. 11—Misfire Detected
P0312 Cylinder no. 12—Misfire Detected
P0320 Ignition/Distributor Engine Speed Input Circuit Malfunction
P0321 Ignition/Distributor Engine Speed Input Circuit Range/Performance
P0322 Ignition/Distributor Engine Speed Input Circuit No Signal
P0323 Ignition/Distributor Engine Speed Input Circuit Intermittent
P0325 Knock Sensor no. 1—Circuit Malfunction (Bank no. 1 or Single Sensor)
P0326 Knock Sensor no. 1—Circuit Range/Performance (Bank no. 1 or Single Sensor)
P0327 Knock Sensor no. 1—Circuit Low Input (Bank no. 1 or Single Sensor)
P0328 Knock Sensor no. 1—Circuit High Input (Bank no. 1 or Single Sensor)
P0329 Knock Sensor no. 1—Circuit Input Intermittent (Bank no. 1 or Single Sensor)
P0330 Knock Sensor no. 2—Circuit Malfunction (Bank no. 2)
P0331 Knock Sensor no. 2—Circuit Range/Performance (Bank no. 2)
P0332 Knock Sensor no. 2—Circuit Low Input (Bank no. 2)
P0333 Knock Sensor no. 2—Circuit High Input (Bank no. 2)
P0334 Knock Sensor no. 2—Circuit Input Intermittent (Bank no. 2)
P0335 Crankshaft Position Sensor "A" Circuit Malfunction
P0336 Crankshaft Position Sensor "A" Circuit Range/Performance
P0337 Crankshaft Position Sensor "A" Circuit Low Input
P0338 Crankshaft Position Sensor "A" Circuit High Input
P0339 Crankshaft Position Sensor "A" Circuit Intermittent
P0340 Camshaft Position Sensor Circuit Malfunction
P0341 Camshaft Position Sensor Circuit Range/Performance
P0342 Camshaft Position Sensor Circuit Low Input
P0343 Camshaft Position Sensor Circuit High Input
P0344 Camshaft Position Sensor Circuit Intermittent
P0350 Ignition Coil Primary/Secondary Circuit Malfunction
P0351 Ignition Coil "A" Primary/Secondary Circuit Malfunction
P0352 Ignition Coil "B" Primary/Secondary Circuit Malfunction
P0353 Ignition Coil "C" Primary/Secondary Circuit Malfunction

P0354 Ignition Coil "D" Primary/Secondary Circuit Malfunction
P0355 Ignition Coil "E" Primary/Secondary Circuit Malfunction
P0356 Ignition Coil "F" Primary/Secondary Circuit Malfunction
P0357 Ignition Coil "G" Primary/Secondary Circuit Malfunction
P0358 Ignition Coil "H" Primary/Secondary Circuit Malfunction
P0359 Ignition Coil "I" Primary/Secondary Circuit Malfunction
P0360 Ignition Coil "J" Primary/Secondary Circuit Malfunction
P0361 Ignition Coil "K" Primary/Secondary Circuit Malfunction
P0362 Ignition Coil "L" Primary/Secondary Circuit Malfunction
P0370 Timing Reference High Resolution Signal "A" Malfunction
P0371 Timing Reference High Resolution Signal "A" Too Many Pulses
P0372 Timing Reference High Resolution Signal "A" Too Few Pulses
P0373 Timing Reference High Resolution Signal "A" Intermittent/Erratic Pulses
P0374 Timing Reference High Resolution Signal "A" No Pulses
P0375 Timing Reference High Resolution Signal "B" Malfunction
P0376 Timing Reference High Resolution Signal "B" Too Many Pulses
P0377 Timing Reference High Resolution Signal "B" Too Few Pulses
P0378 Timing Reference High Resolution Signal "B" Intermittent/Erratic Pulses
P0379 Timing Reference High Resolution Signal "B" No Pulses
P0380 Glow Plug/Heater Circuit "A" Malfunction
P0381 Glow Plug/Heater Indicator Circuit Malfunction
P0382 Glow Plug/Heater Circuit "B" Malfunction
P0385 Crankshaft Position Sensor "B" Circuit Malfunction
P0386 Crankshaft Position Sensor "B" Circuit Range/Performance
P0387 Crankshaft Position Sensor "B" Circuit Low Input
P0388 Crankshaft Position Sensor "B" Circuit High Input
P0389 Crankshaft Position Sensor "B" Circuit Intermittent
P0400 Exhaust Gas Recirculation Flow Malfunction
P0401 Exhaust Gas Recirculation Flow Insufficient Detected
P0402 Exhaust Gas Recirculation Flow Excessive Detected
P0403 Exhaust Gas Recirculation Circuit Malfunction
P0404 Exhaust Gas Recirculation Circuit Range/Performance
P0405 Exhaust Gas Recirculation Sensor "A" Circuit Low
P0406 Exhaust Gas Recirculation Sensor "A" Circuit High
P0407 Exhaust Gas Recirculation Sensor "B" Circuit Low
P0408 Exhaust Gas Recirculation Sensor "B" Circuit High
P0410 Secondary Air Injection System Malfunction
P0411 Secondary Air Injection System Incorrect Flow Detected
P0412 Secondary Air Injection System Switching Valve "A" Circuit Malfunction
P0413 Secondary Air Injection System Switching Valve "A" Circuit Open
P0414 Secondary Air Injection System Switching Valve "A" Circuit Shorted
P0415 Secondary Air Injection System Switching Valve "B" Circuit Malfunction
P0416 Secondary Air Injection System Switching Valve "B" Circuit Open
P0417 Secondary Air Injection System Switching Valve "B" Circuit Shorted
P0418 Secondary Air Injection System Relay "A" Circuit Malfunction
P0419 Secondary Air Injection System Relay "B" Circuit Malfunction
P0420 Catalyst System Efficiency Below Threshold (Bank no. 1)

P0421 Warm Up Catalyst Efficiency Below Threshold (Bank no. 1)
P0422 Main Catalyst Efficiency Below Threshold (Bank no. 1)
P0423 Heated Catalyst Efficiency Below Threshold (Bank no. 1)
P0424 Heated Catalyst Temperature Below Threshold (Bank no. 1)
P0430 Catalyst System Efficiency Below Threshold (Bank no. 2)
P0431 Warm Up Catalyst Efficiency Below Threshold (Bank no. 2)
P0432 Main Catalyst Efficiency Below Threshold (Bank no. 2)
P0433 Heated Catalyst Efficiency Below Threshold (Bank no. 2)
P0434 Heated Catalyst Temperature Below Threshold (Bank no. 2)
P0440 Evaporative Emission Control System Malfunction
P0441 Evaporative Emission Control System Incorrect Purge Flow
P0442 Evaporative Emission Control System Leak Detected (Small Leak)
P0443 Evaporative Emission Control System Purge Control Valve Circuit Malfunction
P0444 Evaporative Emission Control System Purge Control Valve Circuit Open
P0445 Evaporative Emission Control System Purge Control Valve Circuit Shorted
P0446 Evaporative Emission Control System Vent Control Circuit Malfunction
P0447 Evaporative Emission Control System Vent Control Circuit Open
P0448 Evaporative Emission Control System Vent Control Circuit Shorted
P0449 Evaporative Emission Control System Vent Valve/Solenoid Circuit Malfunction
P0450 Evaporative Emission Control System Pressure Sensor Malfunction
P0451 Evaporative Emission Control System Pressure Sensor Range/Performance
P0452 Evaporative Emission Control System Pressure Sensor Low Input
P0453 Evaporative Emission Control System Pressure Sensor High Input
P0454 Evaporative Emission Control System Pressure Sensor Intermittent
P0455 Evaporative Emission Control System Leak Detected (Gross Leak)
P0460 Fuel Level Sensor Circuit Malfunction
P0461 Fuel Level Sensor Circuit Range/Performance
P0462 Fuel Level Sensor Circuit Low Input
P0463 Fuel Level Sensor Circuit High Input
P0464 Fuel Level Sensor Circuit Intermittent
P0465 Purge Flow Sensor Circuit Malfunction
P0466 Purge Flow Sensor Circuit Range/Performance
P0467 Purge Flow Sensor Circuit Low Input
P0468 Purge Flow Sensor Circuit High Input
P0469 Purge Flow Sensor Circuit Intermittent
P0470 Exhaust Pressure Sensor Malfunction
P0471 Exhaust Pressure Sensor Range/Performance
P0472 Exhaust Pressure Sensor Low
P0473 Exhaust Pressure Sensor High
P0474 Exhaust Pressure Sensor Intermittent
P0475 Exhaust Pressure Control Valve Malfunction
P0476 Exhaust Pressure Control Valve Range/Performance
P0477 Exhaust Pressure Control Valve Low
P0478 Exhaust Pressure Control Valve High
P0479 Exhaust Pressure Control Valve Intermittent
P0480 Cooling Fan no. 1 Control Circuit Malfunction
P0481 Cooling Fan no. 2 Control Circuit Malfunction
P0482 Cooling Fan no. 3 Control Circuit Malfunction

P0483 Cooling Fan Rationality Check Malfunction
P0484 Cooling Fan Circuit Over Current
P0485 Cooling Fan Power/Ground Circuit Malfunction
P0500 Vehicle Speed Sensor Malfunction
P0501 Vehicle Speed Sensor Range/Performance
P0502 Vehicle Speed Sensor Circuit Low Input
P0503 Vehicle Speed Sensor Intermittent/Erratic/High
P0505 Idle Control System Malfunction
P0506 Idle Control System RPM Lower Than Expected
P0507 Idle Control System RPM Higher Than Expected
P0510 Closed Throttle Position Switch Malfunction
P0520 Engine Oil Pressure Sensor/Switch Circuit Malfunction
P0521 Engine Oil Pressure Sensor/Switch Range/Performance
P0522 Engine Oil Pressure Sensor/Switch Low Voltage
P0523 Engine Oil Pressure Sensor/Switch High Voltage
P0530 A/C Refrigerant Pressure Sensor Circuit Malfunction
P0531 A/C Refrigerant Pressure Sensor Circuit Range/Performance
P0532 A/C Refrigerant Pressure Sensor Circuit Low Input
P0533 A/C Refrigerant Pressure Sensor Circuit High Input
P0534 A/C Refrigerant Charge Loss
P0550 Power Steering Pressure Sensor Circuit Malfunction
P0551 Power Steering Pressure Sensor Circuit Range/Performance
P0552 Power Steering Pressure Sensor Circuit Low Input
P0553 Power Steering Pressure Sensor Circuit High Input
P0554 Power Steering Pressure Sensor Circuit Intermittent
P0560 System Voltage Malfunction
P0561 System Voltage Unstable
P0562 System Voltage Low
P0563 System Voltage High
P0565 Cruise Control On Signal Malfunction
P0566 Cruise Control Off Signal Malfunction
P0567 Cruise Control Resume Signal Malfunction
P0568 Cruise Control Set Signal Malfunction
P0569 Cruise Control Coast Signal Malfunction
P0570 Cruise Control Accel Signal Malfunction
P0571 Cruise Control/Brake Switch "A" Circuit Malfunction
P0572 Cruise Control/Brake Switch "A" Circuit Low
P0573 Cruise Control/Brake Switch "A" Circuit High
P0574 Through P0580 Reserved for Cruise Codes
P0600 Serial Communication Link Malfunction
P0601 Internal Control Module Memory Check Sum Error
P0602 Control Module Programming Error
P0603 Internal Control Module Keep Alive Memory (KAM) Error
P0604 Internal Control Module Random Access Memory (RAM) Error
P0605 Internal Control Module Read Only Memory (ROM) Error
P0606 PCM Processor Fault
P0608 Control Module VSS Output "A" Malfunction
P0609 Control Module VSS Output "B" Malfunction
P0620 Generator Control Circuit Malfunction
P0621 Generator Lamp "L" Control Circuit Malfunction
P0622 Generator Field "F" Control Circuit Malfunction
P0650 Malfunction Indicator Lamp (MIL) Control Circuit Malfunction
P0654 Engine RPM Output Circuit Malfunction
P0655 Engine Hot Lamp Output Control Circuit Malfunction
P0656 Fuel Level Output Circuit Malfunction
P0700 Transmission Control System Malfunction
P0701 Transmission Control System Range/Performance
P0702 Transmission Control System Electrical
P0703 Torque Converter/Brake Switch "B" Circuit Malfunction

P0704 Clutch Switch Input Circuit Malfunction
P0705 Transmission Range Sensor Circuit Malfunction (PRNDL Input)
P0706 Transmission Range Sensor Circuit Range/Performance
P0707 Transmission Range Sensor Circuit Low Input
P0708 Transmission Range Sensor Circuit High Input
P0709 Transmission Range Sensor Circuit Intermittent
P0710 Transmission Fluid Temperature Sensor Circuit Malfunction
P0711 Transmission Fluid Temperature Sensor Circuit Range/Performance
P0712 Transmission Fluid Temperature Sensor Circuit Low Input
P0713 Transmission Fluid Temperature Sensor Circuit High Input
P0714 Transmission Fluid Temperature Sensor Circuit Intermittent
P0715 Input/Turbine Speed Sensor Circuit Malfunction
P0716 Input/Turbine Speed Sensor Circuit Range/Performance
P0717 Input/Turbine Speed Sensor Circuit No Signal
P0718 Input/Turbine Speed Sensor Circuit Intermittent
P0719 Torque Converter/Brake Switch "B" Circuit Low
P0720 Output Speed Sensor Circuit Malfunction
P0721 Output Speed Sensor Circuit Range/Performance
P0722 Output Speed Sensor Circuit No Signal
P0723 Output Speed Sensor Circuit Intermittent
P0724 Torque Converter/Brake Switch "B" Circuit High
P0725 Engine Speed Input Circuit Malfunction
P0726 Engine Speed Input Circuit Range/Performance
P0727 Engine Speed Input Circuit No Signal
P0728 Engine Speed Input Circuit Intermittent
P0730 Incorrect Gear Ratio
P0731 Gear no. 1 Incorrect Ratio
P0732 Gear no. 2 Incorrect Ratio
P0733 Gear no. 3 Incorrect Ratio
P0734 Gear no. 4 Incorrect Ratio
P0735 Gear no. 5 Incorrect Ratio
P0736 Reverse Incorrect Ratio
P0740 Torque Converter Clutch Circuit Malfunction
P0741 Torque Converter Clutch Circuit Performance or Stuck Off
P0742 Torque Converter Clutch Circuit Stuck On
P0743 Torque Converter Clutch Circuit Electrical
P0744 Torque Converter Clutch Circuit Intermittent
P0745 Pressure Control Solenoid Malfunction
P0746 Pressure Control Solenoid Performance or Stuck Off
P0747 Pressure Control Solenoid Stuck On
P0748 Pressure Control Solenoid Electrical
P0749 Pressure Control Solenoid Intermittent
P0750 Shift Solenoid "A" Malfunction
P0751 Shift Solenoid "A" Performance or Stuck Off
P0752 Shift Solenoid "A" Stuck On
P0753 Shift Solenoid "A" Electrical
P0754 Shift Solenoid "A" Intermittent
P0755 Shift Solenoid "B" Malfunction
P0756 Shift Solenoid "B" Performance or Stuck Off
P0757 Shift Solenoid "B" Stuck On
P0758 Shift Solenoid "B" Electrical
P0759 Shift Solenoid "B" Intermittent
P0760 Shift Solenoid "C" Malfunction
P0761 Shift Solenoid "C" Performance Or Stuck Off
P0762 Shift Solenoid "C" Stuck On

P0763 Shift Solenoid "C" Electrical
P0764 Shift Solenoid "C" Intermittent
P0765 Shift Solenoid "D" Malfunction
P0766 Shift Solenoid "D" Performance Or Stuck Off
P0767 Shift Solenoid "D" Stuck On
P0768 Shift Solenoid "D" Electrical
P0769 Shift Solenoid "D" Intermittent
P0770 Shift Solenoid "E" Malfunction
P0771 Shift Solenoid "E" Performance Or Stuck Off
P0772 Shift Solenoid "E" Stuck On
P0773 Shift Solenoid "E" Electrical
P0774 Shift Solenoid "E" Intermittent
P0780 Shift Malfunction
P0781 1–2 Shift Malfunction
P0782 2–3 Shift Malfunction
P0783 3–4 Shift Malfunction
P0784 4–5 Shift Malfunction
P0785 Shift/Timing Solenoid Malfunction
P0786 Shift/Timing Solenoid Range/Performance
P0787 Shift/Timing Solenoid Low
P0788 Shift/Timing Solenoid High
P0789 Shift/Timing Solenoid Intermittent
P0790 Normal/Performance Switch Circuit Malfunction
P0801 Reverse Inhibit Control Circuit Malfunction
P0803 1–4 Upshift (Skip Shift) Solenoid Control Circuit Malfunction
P0804 1–4 Upshift (Skip Shift) Lamp Control Circuit Malfunction
P1307 Accelerometer signal
P1308 Accelerometer signal
P1326 Fault in engine control module (ECM) knock control circuit.
P1327 Fault in engine control module (ECM) knock control circuit.
P1328 Fault in engine control module (ECM) knock control circuit.
P1329 Fault in engine control module (ECM) knock control circuit.
P1401 Fault in engine control module (ECM) engine coolant temperature (ECT) sensor circuit NTC switching
P1403 Fault in engine control module (ECM) control module box temperature sensor
P1404 Fault in engine control module (ECM) control module box temperature sensor
P1405 Temperature warning greater than 230 degrees F
P1406 Temperature warning greater than 212 degrees F
P1505 Idle air control (IAC) valve opening signal
P1506 Idle air control (IAC) valve opening signal
P1507 Idle air control (IAC) valve closing signal
P1508 Idle air control (IAC) valve closing signal
P1604 Ignition discharge module (IDM) group D
P1605 Ignition discharge module (IDM) group E
P1617 Cable fault between AW 50–42 transmission control module (TCM) and Motronic 4.4 engine control module (ECM) (lamp lights)
PI618 Cable fault between AW 50-Q2 transmission control module (TCM) and Motronic 4.4 engine control module (ECM) (lamp lights)
P1619 Engine cooling fan (FC) low-speed, signal
P1620 Engine cooling fan (FC) low-speed, signal
P1621 Diagnostic trouble code (DTC) in automatic transmission control module(TCM)

MAINTENANCE LIGHT (MIL) RESETTING

Acura

1995 LEGEND, 1995 VIGOR, & 1995–96 2.5TL & 3.2TL MODELS

The Maintenance Reminder Indicator informs you when it is time for scheduled maintenance. When it is near 7,500 miles (12,000 km) since the last maintenance, the indicator will turn yellow. If you exceed 7,500 miles (12,000 km), the indicator will turn red. The indicator can be reset by inserting the ignition key or other similar object into the slot below the indicator. This will extinguish the indicator for the next 7500 miles (12,000 km).

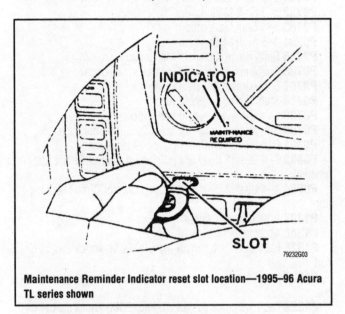

Maintenance Reminder Indicator reset slot location—1995–96 Acura TL series shown

1995–97 INTEGRA & 1996–97 3.5RL MODELS

The Maintenance Reminder Indicator reminds you that it is time for scheduled maintenance. For the first 6,000 miles (9,600 km) after the Maintenance Required Indicator is reset, it will come on for

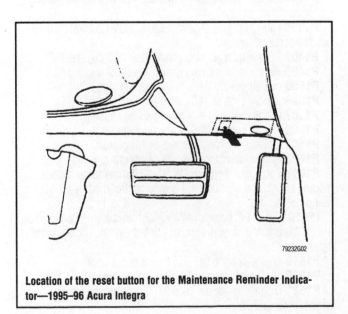

Location of the reset button for the Maintenance Reminder Indicator—1995–96 Acura Integra

2 seconds when you turn the ignition **ON**. Between 6,000 miles (9,600 km) and 7,500 miles (12,000 km) this indicator will light for two seconds when you first turn the ignition **ON**, then flash for ten seconds. If you exceed 7,500 miles (12,000 km) without having the scheduled maintenance performed, this indicator will remain on as a constant reminder. Reset the indicator by pressing the reset button. This button is located on the bottom of the dashboard to the right of the steering column.

Audi

1995–96 100, A4, A6, S4 & S6 MODELS

Some models are equipped with a Service Reminder Indicator located in the trip recorder. Whenever a routine scheduled service is due, the particular type of service flashes for a few seconds in place of the trip recorder. The Service Reminder Indicator will alert the driver approximately 620 miles (1000 km) or 10 days prior to the actual service time is required. The display will continue to flash for about 60 seconds after the engine has been started. If desired, it can be switched to the trip recorder before this time has elapsed by pressing the reset button next to the speedometer. The following service displays will be shown when required. When the required service has be performed, the indicator must be reset by the dealer. A special VAG 1551/1 scan tool is required to perform the reset procedure.

OEL—Oil change service required
In 1—Inspection service required
In 2—Additional service work required

BMW

1995–96 MODELS

The on-board computer is used to evaluate mileage, average engine speed, engine and coolant temperatures, as well as other computer input factors that determine maintenance intervals. There are 5 green, 1 yellow and 1 red LED's used to remind the driver of oil changes and other maintenance services. The green LED's will be illuminated when the ignition is in the **ON** position and the engine is not running. There will not be as many green LED's illuminated when maintenance time gets closer. A yellow LED that is illuminated when the engine is running, will indicate maintenance is now due. The red LED will be illuminated when the service interval has been exceeded by approximately 1000 miles (1600 km). There is a service interval reset tool manufactured by the Assenmacher Tool Company, tool number 62–1–100 and with the aid of an additional adapter, the tool can be used on the 1995–96 models.

Chrysler Corporation

The Emission Maintenance Reminder (EMR) light is now referred to as a Service Reminder Indicator (SRI) lamp. It is located in the dash and is labeled MAINT REQD. It is used on 5.9L V8 HDC engine and 8.0L V10 gas powered engine vehicles only. The SRI lamp will illuminate at the 60,000 and 82,000 mileage (96,000 and 131,000 km) marks and will remain ON until it is reset. Perform the required maintenance before resetting the lamp. Failure to adhere to part replacement or service required may be a violation of federal

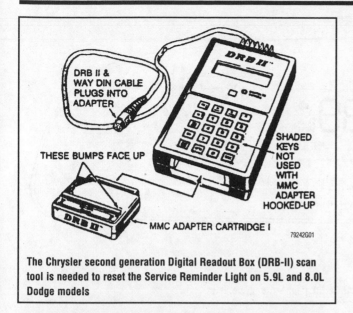

The Chrysler second generation Digital Readout Box (DRB-II) scan tool is needed to reset the Service Reminder Light on 5.9L and 8.0L Dodge models

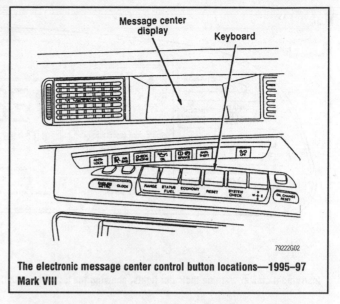

The electronic message center control button locations—1995–97 Mark VIII

law. Resetting the SRI lamp requires the use of the second generation Digital Readout Box (DRB-II) scan tool or equivalent. Consult the scan tool's instruction guide for this procedure.

Ford Motor Company

1995 CONTINENTAL

The 1995 Continental is equipped with a SERVICE INTERVAL REMINDER light. During the SYSTEM CHECK sequence, the SERVICE symbol activates and displays the miles (kilometers) left until the next normal service is due. After the necessary service is complete, reset the reminder as follows:
1. Press the **SYSTEM CHECK** button.
2. Press the **RESET** button.
3. Press the **SYSTEM CHECK** and the **RESET** button at the same time. The display should now show 7,200 miles (11,580 km). The mileage until the next service will count down from this point.

1995–97 MARK VIII

The 1995–97 Mark VIII continues to use the CHANGE OIL SOON or OIL CHANGE REQUIRED light. When the oil life left is between five percent and zero percent, CHANGE OIL SOON will be displayed on the message center. When oil life reaches zero percent, the OIL CHANGE REQUIRED message will be displayed. The message center indicator will indicate the percentage of oil life left during the System Check (during start up). This percentage is based on the driver's driving history and the amount of time since the last oil change. In order to ensure oil life left indications, the driver should only perform the OIL CHANGE RESET procedure after every oil change. Reset the system by pressing the **OIL CHANGE RESET** switch and hold for 5 seconds. After a successful reset the message center will display oil life indicators. The CHANGE OIL SOON or OIL CHANGE REQUIRED message will disappear after the 5 second interval.

1996–97 EXPLORER

The 1996–97 Explorer is equipped with the CHANGE OIL SOON or OIL CHANGE REQUIRED light. When oil life left is between five

percent and zero percent, CHANGE OIL SOON will be displayed on the message center. When oil life reaches zero percent, the OIL CHANGE REQUIRED message will be displayed. The message center indicator will indicate the percent of oil life left during the System Check. This percentage is based on the driver's driving history and the time since the last oil change. In order to ensure oil life left indications, the driver should only perform the OIL CHANGE RESET procedure after every oil change. Reset the system by pressing the OIL CHANGE RESET switch and holding it for 5 seconds. After a successful reset the Message Center will display oil life indicators. The CHANGE OIL SOON or OIL CHANGE REQUIRED message will disappear after the 5 second interval.

General Motors Corporation

B BODY VEHICLES

The General Motors B body class designation includes the Caprice, the Impala, the Roadmaster and the Fleetwood.

After changing the engine oil, reset the engine oil life indicator whether the CHANGE OIL warning/indicator lamp illuminated or not.
1. Reset the engine oil life monitor as follows:
   a. Turn the ignition switch to the **ON** position, but don't start the engine.
   b. Depress the accelerator pedal to the Wide Open Throttle (WOT) position and release it three times within five seconds.
   c. If the CHANGE OIL warning/indicator lamp blinks twice, then goes out, the system has been reset. If the CHANGE OIL warning/indicator lamp does not reset, turn the ignition switch to the **OFF** position and repeat the procedure.

C & H AND G BODY VEHICLES

The General Motors C & H body class designations include the LeSabre, the Park Ave., the Eighty Eight, the Ninety Eight, the Bonneville, the Regency and the LSS. The G body designation refers to the Riviera and the Aurora.

These vehicles may be equipped with an ENGINE OIL LIFE INDEX (EOLI) in the display located on the Drivers Information Center (DIC). The Powertrain Control Module (PCM) determines

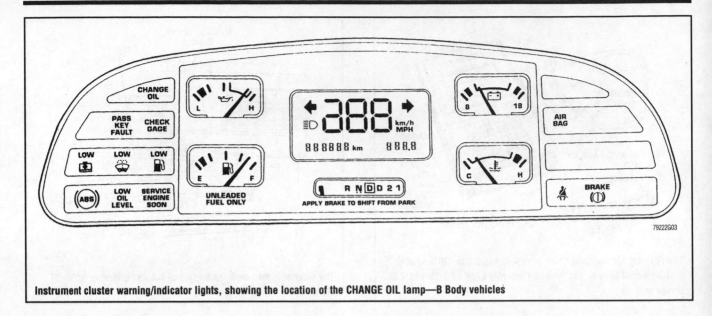

Instrument cluster warning/indicator lights, showing the location of the CHANGE OIL lamp—B Body vehicles

approximately when the engine oil should be changed by calculating information based on vehicle speed, coolant temperature and engine RPM. Once the PCM determines it is time to change the engine oil, it will illuminate the CHANGE OIL SOON light on the DIC. This indicates that the remaining oil life is below 10 percent, (not to be confused with oil level). On a new vehicle, or one that has just been reset, the oil life is 100 percent. This percentage will slowly decrease based on inputs the PCM receives. When oil life reaches the 0 percent mark, the PCM will illuminate the CHANGE OIL NOW light on the DIC. At this time both messages will be displayed accompanied by a slow 5 second audible chime. The reset button must be pressed to acknowledge each of these messages. If a steady CHANGE OIL SOON light persists, Diagnostic Trouble Code (DTC) 61 has been set, indicating a possible shorted switch. Remaining oil life percentage can be displayed by pressing the **OIL** button on the DIC and advancing through the messages until the oil life index is displayed. The oil life index will not detect abnormal conditions, such as excessively dusty conditions or engine malfunctions that could otherwise affect engine oil life.

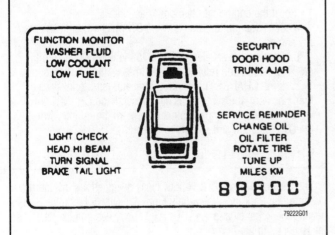

Driver Information Center (DIC) display—most General Motors vehicles

Reset the EOLI as follows:

1. Acknowledge all diagnostic messages in the Drivers Information Center by pressing the **RESET** button.

2. Press the **SEL** button on the left to select OIL. Press the **SEL** button on the right, if necessary, to display oil life.

3. Press and hold the **RESET** button for about 5 seconds. Once the oil life index has been reset, a RESET message will be displayed, then oil life will change to 100 percent. Be careful not reset the oil life accidentally at any time other than when the oil has just been changed. It can not be reset accurately until the next oil change.

E & K BODY VEHICLES

Engine Oil

The General Motors E & K body class designations include the DeVille, the Eldorado, the Seville, and the DeVille Concours.

These vehicles are equipped with an Engine Oil Life Index (EOLI) feature as part of the Driver Information Center (DIC) display. Engine oil life is displayed through engine data as the EOLI and as a CHANGE ENGINE OIL message. The EOLI is displayed following a number between 0 (zero) and 100. This is the percentage of oil life remaining, based on driving conditions, engine oil temperature, and mileage driven since the last time the oil life indicator was reset. When the oil life index reaches 10 percent or less a CHANGE OIL SOON message will appear as a reminder to schedule an oil change. When the oil life index reaches 0, the CHANGE ENGINE OIL message will appear indicating that the oil should be changed within the next 200 miles (320 km). After the oil has been changed, display the EOLI message by pressing the **INFORMATION** button several times. Press and hold the **RESET** button until the display shows 100. This will reset the oil life index. The CHANGE ENGINE OIL message will remain off until the next oil change is needed. The percentage of oil life remaining may be checked at any time by pressing the **INFORMATION** button several times until the EOLI appears.

Transaxle Fluid

➡**This procedure is only applicable for 1995–96 models. Also, on 1996 models, a scan tool is necessary to reset the CHANGE TRANS FLUID indicator.**

All Cadillacs with 4.6L Northstar engines and 4T80-E transaxles are equipped with a transaxle fluid change indicator. A CHANGE TRANS FLUID message will display on the Information Center when the Powertrain Control Module (PCM) monitors actual operating conditions and displays the CHANGE TRANS FLUID message due either to calculations based on those conditions, or at the 100,000 mile (160,000 km) mark. Change the fluid by removing the lower pan and the sidecover drain plug. Change both the transmission fluid and the filter every 15,000 miles (25,000 km), if the vehicle is mainly driven under one or more of these conditions:

- In heavy city traffic where outside temperatures regularly reach 90° F (32° C) or higher.
- On high or mountain terrain.
- Frequent trailer pulling.
- Used for delivery service.

If the vehicle is not used under any of the preceding conditions, change both the fluid and filter (or service the screen) every 100,000 miles (160,000 km).

1. Reset the light on 1995 models as follows:

a. When the CHANGE TRANS FLUID message appears, change the fluid in both the pan and sidecover.

b. Turn the key **ON** with the engine stopped. Press and hold the **OFF** and **REAR DEFOG** buttons on the climate control simultaneously until the TRANS FLUID RESET message appears in the Information Center (between 5–20 seconds) The system is now reset.

2. Reset the indicator on 1996 models as follows:

a. When the CHANGE TRANS FLUID message appears, change the fluid in both the pan and side cover.

b. To reset the transaxle oil life indicator it requires the use of a scan tool. Set the scan tool to the MISC TEST, then perform the OIL LIFE RESET followed by the TRANS OIL ST. The system is now reset.

Y BODY VEHICLES

The General Motors Y body class designation refers to the Corvette.

1995–96 Vehicles

1. Turn the key to the **ON** position, but do not start the engine.
2. Press the **ENG MET** button on the trip monitor and release. Press the button again within 5 seconds.
3. Within 5 seconds of pressing the button the second time, press and hold the **GAUGES** button on the trip monitor. The CHANGE OIL light will flash.
4. Hold the **GAUGES** button until the CHANGE OIL light stops flashing and extinguishes. The monitor is reset at this point. Repeat this procedure if the light does not go out.

1997–99 Vehicles

The 1997–99 Corvette is equipped with an Oil Life Monitor. The Oil Life Monitor is used to inform the driver when the next service is required. A scan tool can be used to monitor and display the amount of oil life left (shown as a percentage).

➡**Repair any Throttle Position (TP) or Accelerator Pedal Position (APP) sensor diagnostic trouble codes before resetting the oil life monitor.**

1. Turn the ignition **ON**, but do not start the engine.

2. Depress the accelerator pedal to the Wide Open Throttle (WOT) position and release it three times within five seconds.
3. If the Oil Life Monitor is not reset, perform this procedure again.
4. Change the oil and the filter.

Geo

1995 TRACKER

The CHECK ENGINE light is utilized to indicate periodic service intervals or to indicate there is a problem in the engine management system. On federal Tracker vehicles, the CHECK ENGINE light will come ON while the engine is running at the 50,000 mile (80,000 km), 80,000 mile (128,000 km), and 100,000 mile (160,000 km) marks. This alerts the driver that it is time for a scheduled service. When servicing is completed the light must be reset by turning the cancel switch, located behind lower trim panel near the steering column, off.

On 1995 California Tracker vehicles, the CHECK ENGINE light will come ON only when a fault code is sensed in the engine management system. When the problem has been corrected, clear the codes and the light should go out. A cancel switch is not used on these vehicles.

On all vehicles equipped with a CHECK ENGINE light, the light will illuminate when a fault is sensed in the engine management system. The fault will be memorized by the ECU for recall later. Once the fault has been recalled and the cause of the fault repaired, the memory should be erased. Disconnect the negative battery cable for 60 seconds or more. Reconnect the cable and start the engine. Allow the engine to reach operating temperature and ground the diagnostic switch. Check for a Code 12 and disconnect the diagnostic switch. The codes will be reset at this point.

Honda

EXCEPT PASSPORT

Vehicles equipped with a Maintenance Reminder Indicator will illuminate when it is time for scheduled maintenance. When it is near

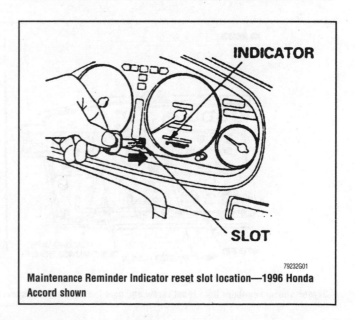

Maintenance Reminder Indicator reset slot location—1996 Honda Accord shown

7,500 miles (12,000 km) since the last maintenance, the indicator will turn yellow. If you exceed 7,500 miles (12,000 km), the indicator will turn red. The indicator can be reset by inserting the ignition key, or other similar object, into the slot below the indicator. This will extinguish the indicator for the next 7,500 miles (12,000 km).

PASSPORT

The 1995 Passport DX model is equipped with an Oxygen Sensor Life Indicator light. It is no longer used in 1996–99 models. The oxygen sensor must be replaced after 90,000 miles (144,000 km) of vehicle operation. When the odometer reading reaches 90,000 miles (144,000 km), the oxygen sensor life indicator light (02) will illuminate to remind the driver to change the oxygen sensor. After replacing the oxygen sensor, the oxygen sensor life indicator light must be reset to remind the driver to replace the oxygen sensor after the next 90,000 miles (144,000 km).

➡**The reset screw is located in the back of the instrument cluster.**

1. Perform the reset procedure as follows:
   a. Remove the instrument panel cluster assembly.
   b. Remove the masking tape from hole **B**.
   c. Remove the screw from hole **A** and install it in hole **B**.
   d. Apply new masking tape over hole **A**.

➡**The above procedure assumes that the oxygen sensor is being replaced for the first time (after 90,000 miles/144,000 km). For subsequent reset procedures (at the next 90,000 miles/144,000 km), hole positions will alternate accordingly from the procedure presented here.**

Isuzu

1995 AMIGO, PICK-UP, TROOPER & RODEO

The 1995 Amigo, Pick-Up, Trooper and Rodeo are equipped with an oxygen sensor maintenance light, located in the dash panel, that will illuminate every 90,000 miles (144,000 km), indicating that the oxygen sensor must be replaced. This light was used up through

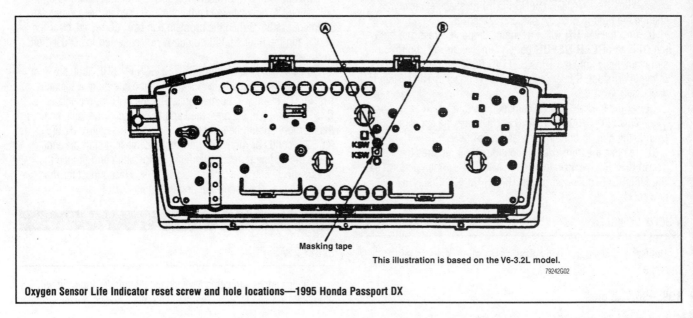

Masking tape

This illustration is based on the V6-3.2L model.

79242G02

Oxygen Sensor Life Indicator reset screw and hole locations—1995 Honda Passport DX

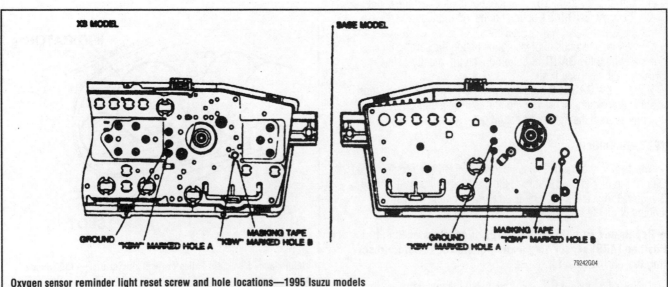

XB MODEL

BASE MODEL

GROUND "KSW" MARKED HOLE A "KSW" MARKED HOLE B MASKING TAPE

GROUND "KSW" MARKED HOLE A "KSW" MARKED HOLE B MASKING TAPE

79242G04

Oxygen sensor reminder light reset screw and hole locations—1995 Isuzu models

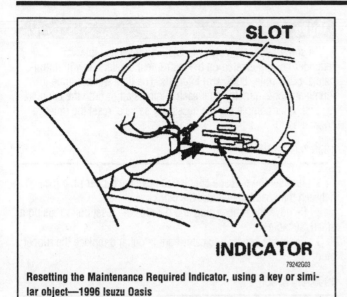

Resetting the Maintenance Required Indicator, using a key or similar object—1996 Isuzu Oasis

1995, but was dropped for 1996 models. After the sensor has been replaced and the emission system checked, reset the maintenance light as follows:

1. To reset the reminder light, proceed as follows:

   a. Remove the instrument cluster assembly.

   b. Working on the backside of the instrument cluster, remove the masking tape over hole **B** in the instrument cluster.

   c. Remove the screw from hole **A** of the instrument cluster and place that screw in hole **B** of the instrument cluster.

   d. After switching the screw from hole **A** to hole **B**, be sure to place a piece of masking tape over hole **A** of the instrument cluster.

➡**At the next 90,000 mile (144,000 km) interval the screw hole positions will be the opposite of the previous replacement.**

OASIS

The 1996–97 Oasis is equipped with a maintenance reminder indicator located on the instrument panel. It lets the driver know it is time for a scheduled maintenance. When it is near 7,500 miles (12,000 km) since the last maintenance, the indicator will turn yellow. If you exceed 7,500 miles (12,000 km), the indicator will turn red. When the required maintenance has been performed the indicator can be reset by inserting the ignition key or other similar object into the slot below the indicator. This will extinguish the indicator for the next 7500 miles (12,000 km).

The 1998–99 Oasis is equipped with a maintenance reminder indicator located on the instrument panel. It lets the driver know it is time for a scheduled maintenance. The maintenance reminder light will blink for ten seconds when the ignition is first turned **ON** between 6,000 to 7,500 miles (9500 to 12,000 km) since the last maintenance service, after 7,500 miles (12,000 km) the light will stay ON for ten seconds. When the required maintenance has been

performed the indicator can be reset by pushing in and holding the select/reset switch for more than ten seconds with the ignition switch **ON**.

Land Rover

1995–96 MODELS

Vehicles equipped with Maintenance Indicator Light (MIL), will illuminate indicating it's time for service or there is a problem in the engine management system. After service is completed, reset the maintenance light.

The service reminder light can only be only reset using a special scanner tool. The scanner tool is connected to the Diagnostic Link Connector (DLC). Consult the scanner tool manufacturer's instructions for resetting the service light.

Mitsubishi

1995–96 MODELS

The 1995–96 Mitsubishi models use a maintenance warning light in some of the light vehicles. This warning light now illuminates as MAINT. REQD., an abbreviation for maintenance required. The warning light will illuminate after 50,000 miles (80,000 km), 80,000 miles (128,000 km), 100,000 miles (160,000 km) and 120,000 miles (192,000 km). The maintenance required light is used as a reminder to inspect the emission control system and to perform the required emission service related to the mileage interval. After completing the necessary emission service, the maintenance light can be reset as follows:

The reset procedure and reset switch locations have remained the same as previous vehicles. On some models, the reset switch is located on the back of the instrument cluster near the speedometer cable junction. On other models, the reset switch is located on the lower right-hand corner of the instrument cluster, behind the face panel. After the switch is located, slide the switch knob to the other side to reset the maintenance warning light.

➡**After 120,000 miles (192,000 km), the warning light bulb should be removed to prevent it from continuing to illuminate.**

Saab

1995–96 900 SERIES

The 900 series uses an information center called the Saab Information Display (SID). The TIME FOR SERVICE light will illuminate in the SID when it is time for scheduled service. Reset the light after service is completed as follows: press and hold the CLEAR button for at least 8 seconds until TIME FOR SERVICE appears on the display and an audible signal is heard. After 21 starts of the engine the TIME FOR SERVICE reminder message will be canceled.

Suzuki

1995 SIDEKICK & SAMURAI

The CHECK ENGINE light is utilized to indicate periodic service intervals or to indicate there is a problem in the engine management system. On Samurai and federal Sidekick vehicles only, the CHECK ENGINE light will come ON while the engine is running at the 50,000 mile (80,000 km), 80,000 mile (128,000 km), and 100,000 mile (160,000 km) marks. This alerts the driver that it is time for a scheduled service. When servicing is completed the light must be reset by turning the cancel switch, located behind lower trim panel near the steering column, off.

On 1995 California Sidekick vehicles, the CHECK ENGINE light will come ON only when a fault code is sensed in the engine management system. When the problem has been corrected, clear the codes and the light should go out. A cancel switch is not used on these vehicles.

On all vehicles equipped with a CHECK ENGINE light, the light will illuminate when a fault is sensed in the engine management system. The fault will be memorized by the ECU for recall later. Once the fault has been recalled and the cause of the fault repaired, the memory should be erased. Disconnect the negative battery cable for 60 seconds or more. Reconnect the cable and start the engine. Allow the engine to reach operating temperature and ground the diagnostic switch. Check for a Code 12 and disconnect the diagnostic switch. The codes will be reset at this point.

Volvo

On some 1995–96 Volvo models, an oil service interval reminder light is located on the instrument cluster and will illuminate at 5000 mile (8000 km) intervals. The light will continue to illuminate for 2 minutes after each engine start or until the counter is reset. After completing the necessary service, reset the mileage counter as follows:

1995 740, 940 & 960

1. To reset the mileage counter, remove the rubber plug located between the speedometer and the clock.
2. Remove the rubber plug and depress the reset button, using a small, suitable rod.
3. Verify the service indicator light is out and replace the rubber plug.

1996 850 & 960

The service reminder light can only be reset using a special scanner tool. The scanner tool is connected to the Diagnostic Link Connector (DLC). Consult the manufacturer's instructions for resetting the service light.

OXYGEN (O₂) SENSORS

2

GENERAL INFORMATION

An Oxygen (O₂) sensor is an input device used by the engine control computer to monitor the amount of oxygen in the exhaust gas stream. This information is used by the computer, along with other inputs, to fine-tune the air/fuel mixture so that the engine can run with the greatest efficiency in all conditions. The O₂ sensor sends this information to the computer in the form of a 100–900 millivolt (mV) reference signal, which is actually created by the O₂ sensor itself through chemical interactions between the sensor tip material (zirconium dioxide in almost all cases) and the oxygen levels in the exhaust gas stream and ambient atmosphere gas. At operating temperatures, approximately 1100°F (600°C), the element becomes a semiconductor. Essentially, through the differing levels of oxygen in the exhaust gas stream and in the surrounding atmosphere, the sensor creates a voltage signal which is directly and consistently related to the concentration of oxygen in the exhaust stream. Typically, a higher than normal amount of oxygen in the exhaust stream indicates that not all of the available oxygen was used in the combustion process, because there was not enough fuel (lean condition) present. Inversely, a lower than normal concentration of oxygen in the exhaust stream indicates that a large amount was used in the combustion process, because a larger than necessary amount of fuel was present (rich condition). Thus, the engine control computer can correct the amount of fuel introduced into the combustion chambers.

Since the control computer uses the O₂ sensor output voltage as an indication of the oxygen concentration, and the oxygen concentration directly affects O₂ sensor output, the signal voltage from the sensor to the computer fluctuates constantly. This fluctuation is caused by the nature of the interaction between the computer and the O₂ sensor, which follows a general pattern: detect, compare, compensate, detect, compare, compensate, etc. This means that when the computer detects a lean signal from the O₂ sensor, it compares the reading with known parameters stored within its memory. It calculates that there is too much oxygen present in the exhaust gases, so it compensates by adding more fuel to the air/fuel mixture. This, in turn, causes the O₂ sensor to send a rich signal to the computer, which then compares this new signal, and adjusts the air/fuel mixture again. This pattern constantly repeats itself: detect rich, compare, compensate lean, detect lean, compare, compensate rich, etc. Since the O₂ sensor fluctuates between rich and lean, and because the lean limit for sensor output is 100 mV and the rich limit is 900 mV, the proper voltage signal from a normally functioning O₂ sensor consistently fluctuates between 100–300 and 700–900 mV.

➡**The sensor voltage may never quite reach 100 or 900 mV, but it should fluctuate from at least below 300 mV to above 700 mV, and the mid-point of the fluctuations should be centered around 500 mV.**

To improve O₂ sensor efficiency, newer O₂ sensors were designed with a built-in heating element, and were called Heated O₂ (HO₂) sensors. This heating element was incorporated into the sensor so that the sensor would reach optimal operating temperature quicker, meaning that the O₂ sensor output signal could be used by the engine control computer sooner. Because the sensor reaches

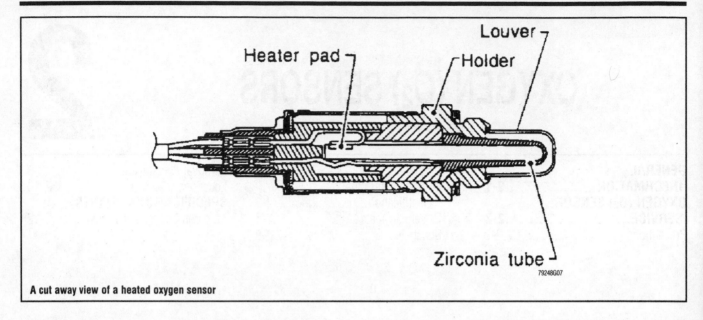

A cut away view of a heated oxygen sensor

optimal temperature quicker, modern vehicles enjoy improved driveability and fuel economy even before the engine reaches normal operating temperature.

Although a few manufacturers changed earlier, in 1995 all vehicles were required to implement a new set of engine control parameters, referred to as On-Board Diagnostics second generation (OBD-II). This updated system (based on the former OBD-I), called for additional O_2 sensors to be used after the catalytic converter, so that catalytic converter efficiency could be measured by the vehicle's engine control computer. The O_2 sensors mounted in the exhaust system after the catalytic converters are not used to affect air/fuel mixture; they are used solely to monitor catalytic converter efficiency.

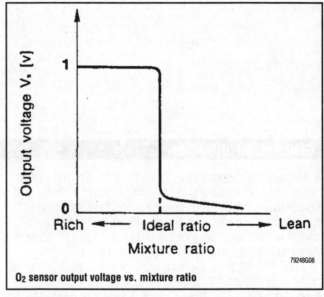

O_2 sensor output voltage vs. mixture ratio

OXYGEN (O₂) SENSOR SERVICE

Precautions

When testing or servicing an O_2 sensor you will need to start and warm the engine to operating temperature in order to either perform the necessary testing procedures or to easily remove the sensor from its fitting. This will create a situation in which you will be working around a **HOT** exhaust system. The following is a list of precautions to consider during this service:

• Do not pierce any wires when testing an O_2 sensor, as this can lead to wiring harness damage. Backprobe the connector, when necessary.

• While testing the sensor, be sure to keep out of the way of moving engine components, such as the cooling fan. Refrain from wearing loose clothing which may become tangled in moving engine components.

• Safety glasses must be worn at all times when working on, or near, the exhaust system. Older exhaust systems may be covered with loose rust particles which can shower you when disturbed. These particles are more than a nuisance and can injure your eye.

• Be cautious when working on and around the hot exhaust system. Painful burns will result if skin is exposed to the exhaust system pipes or manifolds.

• The O_2 sensor may be difficult to remove when the engine temperature is below 120°F (48°C). Excessive force may damage the threads in the exhaust manifold or pipe, therefore always start the engine and allow it to reach normal operating temperature prior to removal.

• Since O_2 sensors are usually designed with a permanently-attached wiring pigtail (this allows the wiring harness and sensor connectors to be positioned away from the hot exhaust system), it may be necessary to use a socket or wrench that is designed specifically for this purpose. Before purchasing such a socket, be sure that you can't save some money by using a box end wrench for sensor removal.

Testing

The best, and most accurate method to test the operation of an O₂ sensor is with the use of either an oscilloscope or a Diagnostic Scan Tool (DST), following their specific instructions for testing. It is possible, however, to test whether the O₂ sensor is functioning properly within general parameters using a Digital Volt-Ohmmeter (DVOM), also referred to as a Digital Multi-Meter (DMM). Newer DMM's are often designed to perform many advanced diagnostic functions, and some are even constructed to be used as an oscilloscope. Two in-vehicle testing procedures, and one bench test procedure, will be provided for the common zirconium dioxide oxygen sensor. The first in-vehicle test makes use of a standard DVOM with a 10 megohm impedance, whereas the second in-vehicle test presented necessitates the usage of an advanced DMM with MIN/MAX/Average functions. Both of these in-vehicle test procedures are likely to set Diagnostic Trouble Codes (DTC's) in the engine control computer. Therefore, after testing, be sure to clear all DTC's before retesting the sensor, if necessary.

These are some of the common DTC's which may be set during testing:

- Open in the O₂ sensor circuit
- Constant low voltage in the O₂ sensor circuit
- Constant high voltage in the O₂ sensor circuit
- Other fuel system problems could set a O₂ sensor code

➡ **Because an improperly functioning fuel delivery and/or control system can adversely affect the O₂ sensor voltage output signal, testing only the O₂ sensor is an inaccurate method for diagnosing an engine driveability problem.**

If after testing the sensor, the sensor is thought to be defective because of high or low readings, be sure to check that the fuel delivery and engine management system is working properly before condemning the O₂ sensor. Otherwise, the new O₂ sensor may continue to register the same high or low readings.

Often, by testing the O₂ sensor, another problem in the engine control management system can be diagnosed. If the sensor appears to be defective while installed in the vehicle, perform the bench test. If the sensor functions properly during the bench test, chances are that there may be a larger problem in the vehicle's fuel delivery and/or control system.

Many things can cause an O₂ sensor to fail, including old age, antifreeze contamination, physical damage, prolonged exposure to overly-rich exhaust gases, and exposure to silicone sealant fumes. Be sure to remedy any such condition prior to installing a new sensor, otherwise the new sensor may be damaged as well.

➡ **Perform a visual inspection of the sensor. Black sooty deposits may indicate a rich air/fuel mixture, brown deposits may indicate an oil consumption problem, and white gritty deposits may indicate an internal coolant leak. All of these conditions can destroy a new sensor if not corrected before installation.**

O₂ SENSOR TERMINAL IDENTIFICATION

The easiest method for determining sensor terminal identification is to use a wiring diagram for the vehicle and engine in question. However, if a wiring diagram is not available there is a method for determining terminal identification. Throughout the testing procedures, the following terms will be used for clarity:

- Vehicle harness connector—this refers to the connector on the wires which are attached to the vehicle; NOT the connector at the end of the sensor pigtail.
- Sensor pigtail connector—this refers to the connector attached to the sensor itself.
- O₂ circuit—this refers to the circuit in a Heated O₂ (HO₂) sensor which corresponds to the oxygen-sensing function of the sensor; NOT the heating element circuit.
- Heating circuit—this refers to the circuit in a HO₂ sensor which is designed to warm the HO₂ sensor quickly to improve driveability.
- Sensor Output (SOUT) terminal—this is the terminal which corresponds to the O₂ circuit output. This is the terminal which will register the millivolt signals created by the sensor based upon the amount of oxygen in the exhaust gas stream.
- Sensor Ground (SGND) terminal—when a sensor is so equipped, this refers to the O₂ circuit ground terminal. Many O₂ sensors are not equipped with a ground wire, rather they utilize the exhaust system for the ground circuit.
- Heating Power (HPWR) terminal—this terminal corresponds to the circuit which provides the O₂ sensor heating circuit with power when the ignition key is turned to the ON or RUN positions.
- Heating Ground (HGND) terminal—this is the terminal connected to the heating circuit ground wire.

One Wire Sensor

One wire sensors are by far the easiest to determine sensor terminal identification, but this is self-evident. On one wire O₂ sensors, the single wire terminal is the SOUT and the exhaust system is used to provide the sensor ground pathway. Proceed to the test procedures.

Two Wire Sensor

On two wire sensors, one of the connector terminals is the SOUT and the other is the SGND. To determine which one is which, perform the following:

1. Locate the O₂ sensor and its pigtail connector. It may be necessary to raise and safely support the vehicle to gain access to the connector.

2. Start the engine and allow it to warm up to normal operating temperature, then turn the engine **OFF**.

3. Using a DVOM set to read 100–900 mV (millivolts) DC, backprobe the positive DVOM lead to one of the unidentified terminals and attach the negative lead to a good engine ground.

❊❊ CAUTION

While the engine is running, keep clear of all moving and hot components. Do not wear loose clothing. Otherwise severe personal injury or death may occur.

4. Have an assistant restart the engine and allow it to idle.
5. Check the DVOM for voltage.
6. If no voltage is evident, check your DVOM leads to ensure that they are properly connected to the terminal and engine ground. If still no voltage is evident at the first terminal, move the positive meter lead to backprobe the second terminal.
7. If voltage is now present, the positive meter lead is attached to the SOUT terminal. The remaining terminal is the SGND terminal. If still no voltage is evident, either the O₂ sensor is defective or the

meter leads are not making adequate contact with the engine ground and terminal contacts; clean the contacts and retest. If still no voltage is evident, the sensor is defective.

8. Have your assistant turn the engine **OFF**.
9. Label the sensor pigtail SOUT and SGND terminals.
10. Proceed to the test procedures.

Three Wire Sensor

➡**Three wire sensors are HO₂ sensors.**

On three wire sensors, one of the connector terminals is the SOUT, one of the terminals is the HPWR and the other is the HGND. The SGND is achieved through the exhaust system, as with the one wire O₂ sensor. To identify the three terminals, perform the following:

1. Locate the O₂ sensor and its pigtail connector. It may be necessary to raise and safely support the vehicle to gain access to the connector.
2. Disengage the sensor pigtail connector from the vehicle harness connector.
3. Using a DVOM set to read 12 volts, attach the DVOM ground lead to a good engine ground.
4. Have an assistant turn the ignition switch **ON** without actually starting the engine.
5. Probe all three terminals in the vehicle harness connector. One of the terminals should exhibit 12 volts of power with the ignition key **ON**; this is the HPWR terminal.
   a. If the HPWR terminal was identified, note which of the sensor harness connector terminals is the HPWR, then match the vehicle harness connector to the sensor pigtail connector. Label the corresponding sensor pigtail connector terminal with HPWR.
   b. If none of the terminals showed 12 volts of power, locate and test the heater relay or fuse. Then, perform Steps 3–5 again.
6. Start the engine and allow it to warm up to normal operating temperature, then turn the engine **OFF**.
7. Have your assistant turn the ignition **OFF**.
8. Using the DVOM set to measure resistance (ohms), attach one of the leads to the HPWR terminal of the sensor pigtail connector. Use the other lead to probe the two remaining terminals of the sensor pigtail connector, one at a time. The DVOM should show continuity with only one of the remaining unidentified terminals; this is the HGND terminal. The remaining terminal is the SOUT.
   a. If continuity was found with only one of the two unidentified terminals, label the HGND and SOUT terminals on the sensor pigtail connector.
   b. If no continuity was evident, or if continuity was evident from both unidentified terminals, the O₂ sensor is defective.
9. All three wire terminals should now be labeled on the sensor pigtail connector. Proceed with the test procedures.

Four Wire Sensor

➡**Four wire sensors are HO₂ sensors.**

On four wire sensors, one of the connector terminals is the SOUT, one of the terminals is the SGND, one of the terminals is the HPWR and the other is the HGND. To identify the four terminals, perform the following:

1. Locate the O₂ sensor and its pigtail connector. It may be necessary to raise and safely support the vehicle to gain access to the connector.
2. Disengage the sensor pigtail connector from the vehicle harness connector.

3. Using a DVOM set to read 12 volts, attach the DVOM ground lead to a good engine ground.
4. Have an assistant turn the ignition switch **ON** without actually starting the engine.
5. Probe all four terminals in the vehicle harness connector. One of the terminals should exhibit 12 volts of power with the ignition key **ON**; this is the HPWR terminal.
   a. If the HPWR terminal was identified, note which of the sensor harness connector terminals is the HPWR, then match the vehicle harness connector to the sensor pigtail connector. Label the corresponding sensor pigtail connector terminal with HPWR.
   b. If none of the terminals showed 12 volts of power, locate and test the heater relay or fuse. Then, perform Steps 3–5 again.
6. Have your assistant turn the ignition **OFF**.
7. Using the DVOM set to measure resistance (ohms), attach one of the leads to the HPWR terminal of the sensor pigtail connector. Use the other lead to probe the three remaining terminals of the sensor pigtail connector, one at a time. The DVOM should show continuity with only one of the remaining unidentified terminals; this is the HGND terminal.
   a. If continuity was found with only one of the two unidentified terminals, label the HGND terminal on the sensor pigtail connector.
   b. If no continuity was evident, or if continuity was evident from all unidentified terminals, the O₂ sensor is defective.
   c. If continuity was found at two of the other terminals, the sensor is probably defective. However, the sensor may not necessarily be defective, because it may have been designed with the two ground wires joined inside the sensor in case one of the ground wires is damaged; the other circuit could still function properly. Though, this is highly unlikely. A wiring diagram is necessary in this particular case to know whether the sensor was so designed.
8. Reattach the sensor pigtail connector to the vehicle harness connector.
9. Start the engine and allow it to warm up to normal operating temperature, then turn the engine **OFF**.
10. Using a DVOM set to read 100–900 mV (millivolts) DC, backprobe the negative DVOM lead to one of the unidentified terminals and the positive lead to the other unidentified terminal.

※※ CAUTION

While the engine is running, keep clear of all moving and hot components. Do not wear loose clothing. Otherwise severe personal injury or death may occur.

11. Have an assistant restart the engine and allow it to idle.
12. Check the DVOM for voltage.
    a. If no voltage is evident, check your DVOM leads to ensure that they are properly connected to the terminals. If still no voltage is evident at either of the terminals, either the terminals were accidentally marked incorrectly or the sensor is defective.
    b. If voltage is present, but the polarity is reversed (the DVOM will show a negative voltage amount), turn the engine **OFF** and swap the two DVOM leads on the terminals. Start the engine and ensure that the voltage now shows the proper polarity.
    c. If voltage is evident and is the proper polarity, the positive DVOM lead is attached to the SOUT and the negative lead to the SGND terminals.
13. Have your assistant turn the engine **OFF**.
14. Label the sensor pigtail SOUT and SGND terminals.

IN-VEHICLE TESTS

✳✳ WARNING

Never apply voltage to the O₂ circuit of the sensor, otherwise it may be damaged. Also, never connect an ohmmeter (or a DVOM set on the ohm function) to both of the O₂ circuit terminals (SOUT and SGND) of the sensor pigtail connector; it may damage the sensor.

Test 1 makes use of a standard DVOM with a 10 megohm impedance, whereas Test 2 necessitates the usage of an advanced Digital Multi-Meter (DMM) with MIN/MAX/Average functions or a sliding bar graph function. Both of these in-vehicle test procedures are likely to set Diagnostic Trouble Codes (DTC's) in the engine control computer. Therefore, after testing, be sure to clear all DTC's before retesting the sensor, if necessary. The third in-vehicle test is designed for the use of a scan tool or oscilloscope. The fourth test (Heating Circuit Test) is designed to check the function of the heating circuit in a HO₂ sensor.

➡ If the O₂ sensor being tested is designed to use the exhaust system for the SGND, excessive corrosion between the exhaust and the O₂ sensor may affect sensor functioning.

The in-vehicle tests may be performed for O₂ sensors located in the exhaust system after the catalytic converter. However, the O₂ sensors located behind the catalytic converter will not fluctuate like the sensors mounted before the converter, because the converter, when functioning properly, emits a steady amount of oxygen. If the O₂ sensor mounted after the catalytic converter exhibits a fluctuating signal (like other O₂ sensors), the catalytic converter is most likely defective.

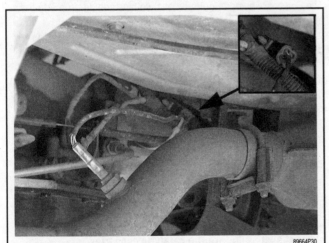

To test the O₂ sensor, locate it and its connector (inset), which should be positioned away from the exhaust system to prevent heat damage

Test 1—Digital Volt-Ohmmeter (DVOM)

This test will not only verify proper sensor functioning, but is also designed to ensure the engine control computer and associated wiring is functioning properly as well.

1. Start the engine and allow it to warm up to normal operating temperature.

➡ If you are using the opening of the thermostat to gauge normal operating temperature, be forewarned: a defective thermostat can open too early and prevent the engine from reaching normal operating temperature. This can cause a slightly rich condition in the exhaust, which can throw the O₂ sensor readings off slightly.

2. Turn the ignition switch OFF, then locate the O₂ sensor pigtail connector.

3. Perform a visual inspection of the connector to ensure it is properly engaged and all terminals are straight, tight and free from corrosion or damage.

4. Disengage the sensor pigtail connector from the vehicle harness connector.

5. On sensors equipped with a SGND terminal (sensors which do not use the exhaust system for the sensor ground pathway), connect a jumper wire to the SGND terminal and to a good, clean engine ground (preferably the negative terminal of the battery).

6. Using a DVOM set to read DC voltage, attach the positive lead to the SOUT terminal of the sensor pigtail connector, and the DVOM negative lead to a good engine ground.

✳✳ CAUTION

While the engine is running, keep clear of all moving and hot components. Do not wear loose clothing. Otherwise severe personal injury or death may occur.

7. Have an assistant start the engine and hold it at approximately 2,000 rpm. Wait at least 1 minute before commencing with the test to allow the O₂ sensor to sufficiently warm up.

➡ Some carbureted Asian models may not switch into closed loop operation until engine speed is above 2,500 rpm.

8. Using a jumper wire, connect the SOUT terminal of the **vehicle harness connector** to a good engine ground. This will fool the engine control computer into thinking it is receiving a lean signal from the O₂ sensor, and, therefore, the computer will enrichen the air/fuel ratio. With the SOUT terminal so grounded, the DVOM should register at least 800 mV, as the control computer adds additional fuel to the air/fuel ratio.

9. While observing the DVOM, disconnect the vehicle harness connector SOUT jumper wire from the engine ground. Use the jumper wire to apply slightly less than 1 volt to the SOUT terminal of the vehicle harness connector. One method to do this is by grasping and squeezing the end of the jumper between your forefinger and thumb of one hand while touching the positive terminal of the battery post with your other hand. This allows your body to act as a resistor for the battery positive voltage, and fools the engine control computer into thinking it is receiving a rich signal. Or, use a mostly-drained AA battery by connecting the positive terminal of the AA battery to the jumper wire and the negative terminal of the battery to a good engine ground. (Another jumper wire may be necessary to do this.) The computer should lean the air/fuel mixture out. This lean mixture should register as 150 mV or less on the DVOM.

10. If the DVOM did not register millivoltages as indicated, the problem may be either the sensor, the engine control computer or the associated wiring. Perform the following to determine which is the defective component:

   a. Remove the vehicle harness connector SOUT jumper wire.

   b. While observing the DVOM, artificially enrich the air/fuel charge using propane. The DVOM reading should register higher than normal millivoltages. (Normal voltage for an ideal air/fuel mixture is approximately 450–550 mV DC). Then, lean the air/fuel intake charger by either disconnecting one of the fuel injector wiring harness connectors (to prevent the injector from delivering fuel) or by detaching one or two vacuum lines (to add additional non-metered air into the engine). The DVOM should now register lower than normal millivoltages. If the DVOM functioned as indicated, the problem lies elsewhere in the fuel delivery and control system. If the DVOM readings were still unresponsive, the O$_2$ sensor is defective; replace the sensor and retest.

➡**Poor wire connections and/or ground circuits may shift a normal O$_2$ sensor's millivoltage readings up into the rich range or down into the lean range. It is a good idea to check the wire condition and continuity before replacing a component which will not fix the problem. A voltage drop test between the sensor case and ground which reveals 14–16 mV, or more, indicates a probable bad ground.**

11. Turn the engine **OFF**, remove the DVOM and all associated jumper wires. Reattach the vehicle harness connector to the sensor pigtail connector. If applicable, reattach the fuel injector wiring connector and/or the vacuum line(s).

12. Clear any DTC's present in the engine control computer memory, as necessary.

Test 2—Digital Multi-Meter (DMM)

This test method is a more straight forward O$_2$ sensor test, and does not test the engine control computer's response to the O$_2$ sensor signal. The use of a DMM with the MIN/MAX/Average function or sliding bar graph/wave function is necessary for this test. Don't forget that the O$_2$ sensor mounted after the catalytic converter (if equipped) will not fluctuate like the other O$_2$ sensor(s) will.

1. Start the engine and allow it to warm up to normal operating temperature.

➡**If you are using the opening of the thermostat to gauge normal operating temperature, be forewarned: a defective thermostat can open too early and prevent the engine from reaching normal operating temperature. This can cause a slightly rich condition in the exhaust, which can throw the O$_2$ sensor readings off slightly.**

2. Turn the ignition switch **OFF**, then locate the O$_2$ sensor pigtail connector.

3. Perform a visual inspection of the connector to ensure it is properly engaged and all terminals are straight, tight and free from corrosion or damage.

4. Backprobe the O$_2$ sensor connector terminals. Attach the DMM positive test lead to the SOUT terminal of the sensor pigtail connector and the negative lead to either the SGND terminal of the sensor pigtail connector (if equipped—refer to the terminal identification procedures earlier in this section for clarification) or to a good, clean engine ground.

5. Activate the MIN/MAX/Average or sliding bar graph/wave function on the DMM.

✳✳ CAUTION

While the engine is running, keep clear of all moving and hot components. Do not wear loose clothing. Otherwise severe personal injury or death may occur.

6. Have an assistant start the engine and wait a few minutes before commencing with the test to allow the O$_2$ sensor to sufficiently warm up.

7. Read the minimum, maximum and average readings exhibited by the O$_2$ sensor, or observe the bar graph/wave form. The average reading for a properly functioning O$_2$ sensor is be approximately 450–550 mV DC. The minimum and maximum readings should vary more than 300–600 mV. A typical O$_2$ sensor can fluctuate from as low as 100 mV to as high as 900 mV; if the sensor range of fluctuation is not large enough, the sensor is defective. Also, if the fluctuation range is biased up or down in the scale. For example, if the fluctuation range is 400 mV to 900 mV the sensor is defective, because the readings are pushed up into the rich range (as long as the fuel delivery system is functioning properly). The same goes for a fluctuation range pushed down into the lean range. The mid-point of the fluctuation range should be around 400–500 mV. Finally, if the O$_2$ sensor voltage fluctuates too slowly (usually the voltage wave should oscillate past the mid-way point of 500 mV several times per second) the sensor is defective. (Technician's refer to this state as "lazy.")

➡**Poor wire connections and/or ground circuits may shift a normal O$_2$ sensor's millivoltage readings up into the rich range or down into the lean range. It is a good idea to check the wire condition and continuity before replacing a component which will not fix the problem. A voltage drop test between the sensor case and ground which reveals 14–16 mV, or more, indicates a probable bad ground.**

8. Using the propane method, enrichen the air/fuel mixture and observe the DMM readings. The average O$_2$ sensor output signal voltage should rise into the rich range.

9. Lean the air/fuel mixture by either disconnecting a fuel injector wiring harness connector or by disconnecting a vacuum line. The O$_2$ sensor average output signal voltage should drop into the lean range.

10. If the O$_2$ sensor did not react as indicated, the sensor is defective and should be replaced.

11. Turn the engine **OFF**, remove the DMM and all associated jumper wires. Reattach the vehicle harness connector to the sensor pigtail connector. If applicable, reattach the fuel injector wiring connector and/or the vacuum line(s).

12. Clear any DTC's present in the engine control computer memory, as necessary.

Test 3—Oscilloscope

This test is designed for the use of an oscilloscope to test the functioning of an O$_2$ sensor.

➡**This test is only applicable for O$_2$ sensors mounted in the exhaust system before the catalytic converter.**

1. Start the engine and allow it to reach normal operating temperature.

2. Turn the engine **OFF**, and locate the O$_2$ sensor connector. Backprobe the scope lead to the O$_2$ sensor connector SOUT terminal. Refer to the manufacturer's instructions for more information on attaching the scope to the vehicle.

3. Turn the scope ON.

4. Set the oscilloscope amplitude to 200 mV per division, and the time to 1 second per division. Use the 1:1 setting of the probe, and be sure to connect the scope's ground lead to a good, clean

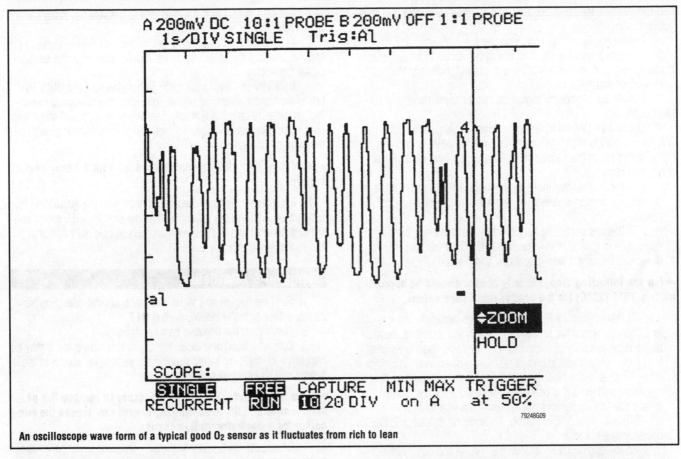

An oscilloscope wave form of a typical good O₂ sensor as it fluctuates from rich to lean

engine ground. Set the signal function to automatic or internal triggering.

5. Start the engine and run it at 2,000 rpm.

6. The oscilloscope should display a wave form, representative of the O₂ sensor switching between lean (100–300 mV) and rich (700–900 mV). The sensor should switch between rich and lean, or lean and rich (crossing the mid-point of 500 mV) several times per second. Also, the range of each wave should reach at least above 700 mV and below 300 mV. However, an occasional low peak is acceptable.

7. Force the air/fuel mixture rich by introducing propane into the engine, then observe the oscilloscope readings. The fluctuating range of the O₂ sensor should climb into the rich range.

8. Lean the air/fuel mixture out by either detaching a vacuum line or by disengaging one of the fuel injector's wiring connectors. Watch the scope readings; the O₂ sensor wave form should drop toward the lean range.

9. If the O₂ sensor's wave form does not fluctuate adequately, is not centered around 500 mV during normal engine operation, does not climb toward the rich range when propane is added to the engine, or does not drop toward the lean range when a vacuum hose or fuel injector connector is detached, the sensor is defective.

10. Reattach the fuel injector connector or vacuum hose.

11. Disconnect the oscilloscope from the vehicle.

Heating Circuit Test

The heating circuit in an O₂ sensor is designed only to heat the sensor quicker than a non-heated sensor. This provides an advantage of increased engine driveability and fuel economy while the engine temperature is still below normal operating temperature,

because the fuel management system can enter closed loop operation (more efficient than open loop operation) sooner.

Therefore, if the heating element goes bad, the O₂ sensor may still function properly once the sensor warms up to its normal temperature. This will take longer than normal and may cause mild driveability-related problems while the engine has not reached normal operating temperature.

If the heating element is found to be defective, replace the O₂ sensor without wasting your time testing the O₂ circuit; if necessary,

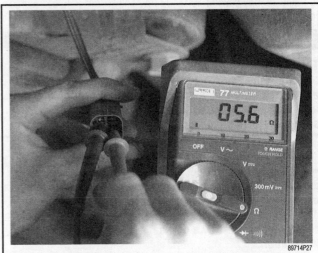

The heating circuit of the O₂ sensor can be tested with a DMM set to measure resistance

you can perform the O_2 circuit test with the new O_2 sensor and save yourself some time.

1. Locate the O_2 sensor pigtail connector.

2. Perform a visual inspection of the connector to ensure it is properly engaged and all terminals are straight, tight and free from corrosion or damage.

3. Disengage the sensor pigtail connector from the vehicle harness connector.

4. Using a DVOM set to read resistance (ohms), attach one DVOM test lead to the HPWR terminal, and the other lead to the HGND terminal, of the sensor pigtail connector, then observe the resistance readings.

    a. If there is no continuity between the HPWR and HGND terminals, the sensor is defective. Replace it with a new one and retest.

    b. If there is continuity between the two terminals, but the resistance is greater than approximately 20 ohms, the sensor is defective. Replace it with a new one and retest.

➡**For the following step, the HO₂ sensor should be approximately 75°F (23°C) for the proper resistance values.**

    c. If there is continuity between the two terminals and it is less than 20 ohms, the sensor is probably not defective. Because of the large diversity of engine control systems used in vehicles today, O_2 sensor heating circuit resistance specifications change often. Generally, the amount of resistance an O_2 sensor heating circuit should exhibit is between 2–9 ohms. However, some manufacturer's O_2 sensors may show resistance as high as 15–20 ohms. As a rule of thumb, 20 ohms of resistance is the upper limit allowable.

5. Turn the engine **OFF**, remove the DVOM and all associated jumper wires. Reattach the vehicle harness connector to the sensor pigtail connector.

6. Clear any DTC's present in the engine control computer memory, as necessary.

BENCH TEST

➡**Utilize one of the in-vehicle tests before performing this test.**

This test is designed to test an O_2 sensor which does not seem to fluctuate fully beyond 400–700 mV. The sensor is to be secured in a table-mounted vise.

✺ CAUTION

This test can be very dangerous. Take the necessary precautions when working with a propane torch. Ensure that all combustible substances are removed from the work area and have a fire extinguisher ready at all times. Be sure to wear the appropriate protective clothing as well.

1. Remove the O_2 sensor.

➡**Perform a visual inspection of the sensor. Black sooty deposits may indicate a rich air/fuel mixture, brown deposits may indicate an oil consumption problem, and white gritty deposits may indicate an internal coolant leak. All of these conditions can destroy a new sensor if not corrected before installation.**

2. Position the sensor in a vise so that the vise holds the sensor by the hex portion of its case.

3. Attach one lead of a DVOM set to read DC millivoltages to the sensor case and the other lead to the SOUT terminal of the sensor pigtail connector.

4. Carefully use a propane torch to heat the tip (and ONLY the tip) of the sensor. Once the sensor reaches close to normal operating temperature range, alternately heat the sensor up and allow it to cool down; the sensor output voltage signal should change with the temperature change.

➡**This may also clean a sensor covered with a heavy coat of carbon.**

5. If the sensor voltage does not change with the fluctuation in temperature, replace the sensor with a new one. Install the new sensor and perform one of the in-vehicle tests to rule out additional fuel management system faults.

Removal & Installation

1. Start the engine and allow it to reach normal operating temperature, then turn the ignition switch **OFF**.

2. Disconnect the negative battery cable.

3. Open the hood and locate the O_2 sensor connector. It may be necessary to raise and safely support the vehicle for access to the sensor and its connector.

➡**On a few models, it may be necessary to remove the passenger seat and lift the carpeting in order to access the connector for a downstream O₂ sensor.**

4. Disengage the O_2 sensor pigtail connector from the vehicle harness connector.

➡**There are generally two methods used to mount an O₂ sensor in the exhaust system: either the O₂ sensor is threaded directly into the exhaust component (screw-in type), or the O₂ sensor is retained by a flange and two nuts or bolts (flange type).**

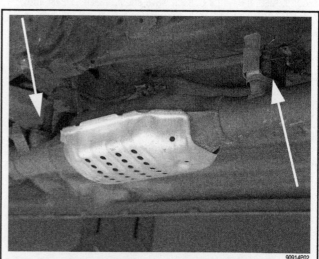

Since sensor locations vary between vehicles, the first step in removal is to locate the O₂ sensors (arrows) . . .

. . . and the sensor connector (2), which is usually near the O₂ sensor (1), but removed enough from the heat of the exhaust system

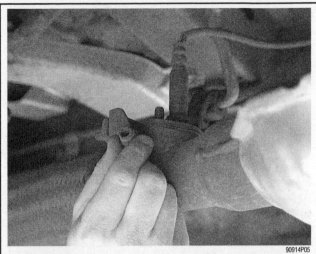

. . . which happen to be nuts in this particular case—some models may use bolts rather than nuts

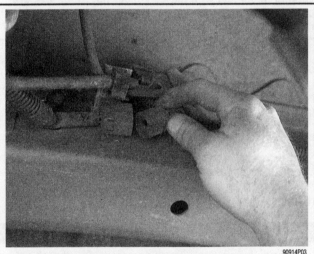

Disengage the sensor pigtail connector half from the vehicle harness connector half

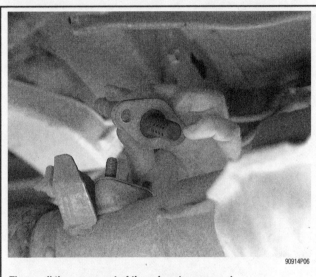

Then, pull the sensor out of the exhaust component

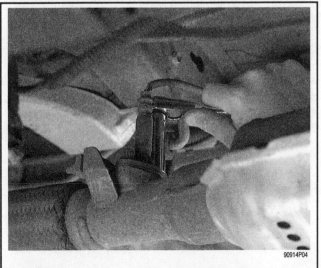

For flange type sensors, loosen the hold-down fasteners . . .

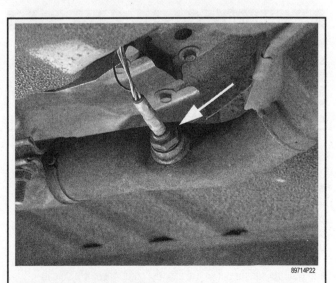

For screw-in type sensors (arrow) . . .

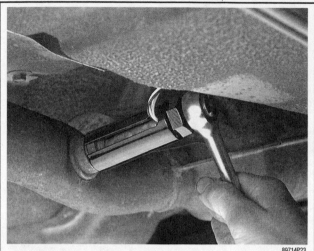

. . . either use a box end wrench to loosen the sensor or a socket designed expressly for this purpose . . .

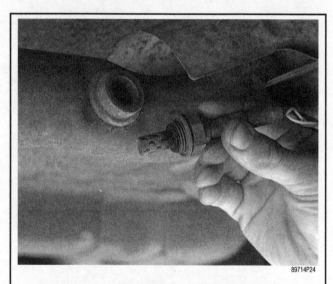

. . . then remove the sensor from the exhaust component

❋❋ WARNING

To prevent damaging a screw-in type O₂ sensor, if excessive force is needed to remove the sensor lubricate it with penetrating oil prior to removal. Also, be sure to protect the tip of the sensor; O₂ sensor tips are very sensitive and may be easily damaged if allowed to strike or come in contact with other objects.

5. Remove the sensor, as follows:
• For screw-in type sensors—Since O₂ sensors are usually designed with a permanently-attached wiring pigtail (this allows the wiring harness and sensor connectors to be positioned away from the hot exhaust system), it may be necessary to use a socket or wrench that is designed specifically for this purpose. Before purchasing such a socket, be sure that you can't save some money by using a box end wrench for sensor removal.
• For flange type sensors—Loosen the hold-down nuts or bolts and pull the sensor out of the exhaust component. Be sure to

remove and discard the old sensor gasket, if equipped. You will need a new gasket for installation.

6. Perform a visual inspection of the sensor. Black sooty deposits may indicate a rich air/fuel mixture, brown deposits may indicate an oil consumption problem, and white gritty deposits may indicate an internal coolant leak. All of these conditions can destroy a new sensor if not corrected before installation.

To install:

7. Install the sensor, as follows:

➡**A special anti-seize compound is used on most screw-in type O₂ sensor threads, and is designed to ease O₂ sensor removal. New sensors usually have the compound already applied to the threads. However, if installing the old O₂ sensor or the new sensor did not come with compound, apply a thin coating of electrically-conductive anti-seize compound to the sensor threads.**

❋❋ WARNING

Be sure to prevent any of the anti-seize compound from coming in contact with the O₂ sensor tip. Also, take precautions to protect the sensor tip from physical damage during installation.

• For screw-in type sensors—Install the sensor in the mounting boss, then tighten it securely.
• For flange type sensors—Position a new sensor gasket on the exhaust component and insert the sensor. Tighten the hold-down fasteners securely and evenly.

8. Reattach the sensor pigtail connector to the vehicle harness connector.

9. Lower the vehicle.

10. Connect the negative battery cable.

11. Start the engine and ensure no Diagnostic Trouble Codes (DTC's) are set.

Locations

Generally, there are only five different locations in the exhaust system where O₂ sensors are positioned. The five locations have been given numbers and will be used in the accompanying charts to identify the positions of O₂ sensors in most vehicles.

Due to mid-year production changes or factory inconsistencies, all models may not be covered. If a vehicle you are servicing is not covered in the charts, inspect the exhaust system (while cold!) in the five general locations to find the applicable O₂ sensors.

➡**On models equipped with dual exhaust systems, there may be up to 4 or 5 O₂ sensors in the exhaust system. Be sure to locate all of them before commencing with any testing or service.**

The five locations are as follows:
• Location No.1—exhaust manifold or down pipe.
• Location No.2—both exhaust manifolds or down pipes of a V-type engine.
• Location No.3—exhaust collector.
• Location No.4—outlet of the catalytic converter.
• Location No.5—both the inlet and outlet of catalytic converter. This location is used to monitor the efficiency of the catalytic converter.

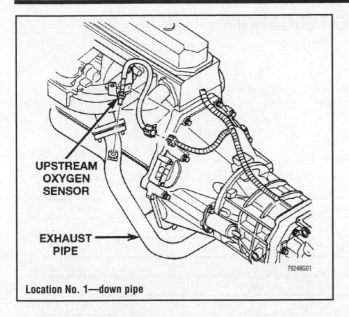

Location No. 1—down pipe

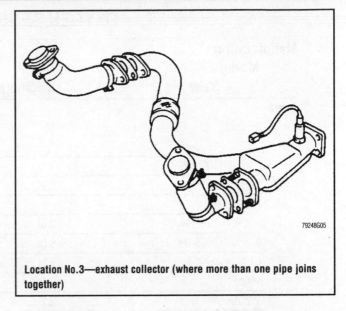

Location No.3—exhaust collector (where more than one pipe joins together)

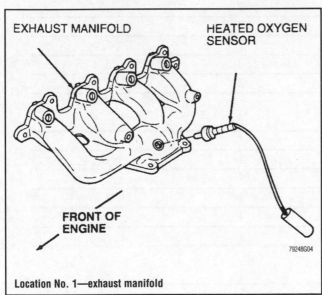

Location No. 1—exhaust manifold

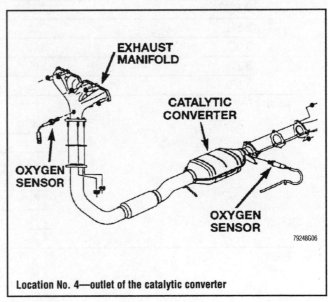

Location No. 4—outlet of the catalytic converter

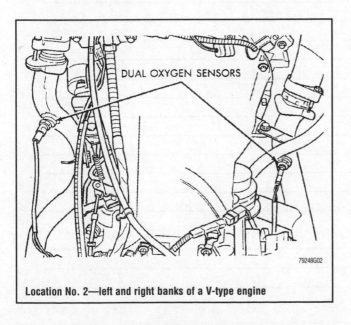

Location No. 2—left and right banks of a V-type engine

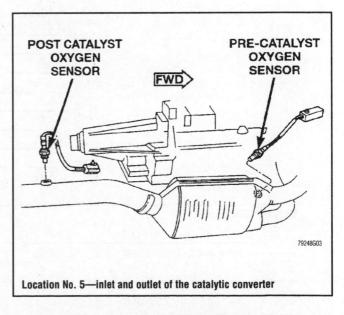

Location No. 5—inlet and outlet of the catalytic converter

OXYGEN SENSOR LOCATIONS

Manufacturer Model Year	Engine(s)	No. of Sensors	Location(s)
Acura			
Integra			
1990-95	1.7L/1.8L	1	1
1996-99	1.8L	2	5
Legend	3.2L	2	2
NSX			
1990-94	3.0L	2	2
1995-99	3.0L	4	2, 4
SLX	3.2L/3.5L	4	2, 5
Vigor	2.5L	1	1
2.2CL	2.2L	2	5
2.3CL	2.3L	2	5
2.5TL	2.5L	2	3, 4
3.0CL	3.0L	2	3, 4
3.2TL	3.2L	3	2, 4
3.5RL	3.5L	3	2, 4
Audi			
	1.8L	2	1, 4
	2.0L	1	1
	2.2L	1	1
	2.3L	1	1
	2.8L	2	2, 4
	3.7L	4	5
	4.2L	4	5
BMW			
	1.8L	1	1
	1.9L	2	2
	2.3L	1	3
	2.5L	2	5
	2.8L	2	5
	3.0L	4	5
	3.2L	4	5
	3.5L	1	1
	4.0L	4	5
	4.4L	4	5
	5.0L	2	2
	5.4L	4	5
	5.6L	4	5
Chrysler Corporation			
Acclaim, Lebaron, and Spirit	All	1	1
Avenger and Sebring Coupe	2.0L	2	1, 4
	2.5L	3	2, 4

OXYGEN SENSOR LOCATIONS

Manufacturer Model Year	Engine(s)	No. of Sensors	Location(s)
Chrysler Corporation (cont.)			
Chrysler Imports			
1990-95	1.5L	1	1
	1.8L	1	1
	2.4L	1	1
1996	1.5L	2	1, 4
	1.8L	2	1, 4
	2.4L	2	3, 4
Chrysler TC	2.2L/3.0L	1	1
Daytona	All	1	1
Dynasty and New Yorker	All	1	1
Fifth Ave and Imperial	3.3L/3.8L	1	1
JA/JX Vehicles ①	All	2	1, 4
Laser and Talon			
1990-94	All	1	1
1995-99	All	2	1, 4
LH Vehicles ②			
1994-95	3.3L	1	1
	3.5L	1	1
1996-99	All	4	2,4
Monaco and Premier	All	1	1
Minivans			
1990-95	All	1	1
1996-99	All	2	1, 4
Plymouth Neon	All	2	1, 4
Stealth	non-turbo	1	1
	turbo	2	2
Trucks and Vans			
1990-95	All	1	1
1996-99	2.5L	2	1, 4
	3.9L	2	1, 4
	5.2L	2	1, 4
	5.9L	2	1, 4
	5.9L HDC	4	2, 5
	8.0L HDC	4	2, 5
Ford Motor Company			
Aspire			
1995	1.3L	1	1
1996-97	1.3L	2	1, 4
Capri	1.6L	1	1
Contour, Mystique and 1999 Cougar	2.0L	2	1, 4
	2.5L	3	1, 4

91272C01

OXYGEN SENSOR LOCATIONS

Manufacturer Model Year	Engine(s)	No. of Sensors	Location(s)
Ford Motor Company (cont.)			
Cougar (1995-97)	3.8L	2	1, 4
	4.6L	4	5
Escort, Tracer and ZX2			
1990	1.9L	1	1
1991-94	1.8L	1	1
	1.9L	1	1
1995-97	1.8L	2	1, 4
	1.9L	2	1, 4
1998-99	2.0L	2	1, 4
Festiva	1.3L	1	1
Full-size Cars			
1990-94	3.8L	1	1
	5.0L/5.8L	1	1
1995-99	4.6L	4	5
Full-size Trucks and Vans			
1990-94	All	1	1
1995-96	4.9L	1	1
	5.0L	3	2, 4
	5.8L	3	2, 4
Full-size Trucks and Vans (cont.)			
1997-99	4.2L	2	3, 4
	4.6L	4	2, 4
	5.4L	4	2, 4
	6.8L	4	2, 4
Light Trucks and Vans			
1990-95	All	1	1
1996-99	2.3L	2	1, 4
	2.5L	2	1, 4
	3.0L	3	2, 4
	4.0L (E)	4	2, 4
	4.0L (X)	4	2, 4
	5.0L	4	2, 4
Minivans			
1990-94	All	1	1
1995-99	3.0L	4	2, 4
	3.8L	4	2, 4
Mustang			
1990-94	2.3L/3.8L	1	1
	5.0L	2	2
1995	3.8L	2	1, 4
1996-99	All	4	2, 4

OXYGEN SENSOR LOCATIONS

Manufacturer Model Year	Engine(s)	No. of Sensors	Location(s)
Ford Motor Company (cont.)			
Probe			
1990-95	2.0L	1	1
	2.5L	2	2
1996-97	2.0L	2	1, 4
	2.5L	4	2, 4
Taurus and Sable			
1990-94	2.5L	1	1
	except 2.5L	2	2
1995	All	2	1, 4
1996-99	All	4	2, 4
Tempo and Topaz	2.3L	1	1
	3.0L	2	2
Villager			
1995	3.0L	1	1
1996-99	3.0L	2	1, 4
General Motors Corporation			
A Body			
1990-95	2.2L	1	1
	3.1L	1	1
1996	2.2L	2	1, 4
	3.1L	2	1, 4
B Body			
1990-95	4.3L	2	2
	5.7L	2	2
1996	4.3L	4	2, 4
	5.7L	4	2,4
C & H Bodies			
1990-95	3.8L	1	1
1996-99	3.8L	2	1, 4
D Body	All	1	1
E & K Bodies			
1990-95	4.6L	2	2
	4.9L	2	2
1996-99	4.6L	4	5
F Body			
1990-95	3.4L	2	2
	5.7L	2	2
1996-99	3.8L	4	5
	5.7L	4	5

91272C03

OXYGEN SENSOR LOCATIONS

Manufacturer Model Year	Engine(s)	No. of Sensors	Location(s)
General Motors Corporation (cont.)			
Full-size Trucks and Vans			
1995-99	4.3L	4	2, 5
	5.0L	4	2, 5
	5.7L	4	2, 5
	7.4L	4	2, 5
G Body			
1995	3.8L	1	1
	4.0L	1	1
1996-99	3.8L	2	1, 4
	4.0L	2	1, 4
GEO/Chevrolet			
Except Tracker			
1990-95	All	1	1
1996-99	All	2	1, 4
Tracker			
1990-95	1.3L	1	1
1996-99	1.6L	2	1, 4
	1.8L	2	1, 4
J Body			
1990-95	2.2L	1	1
	2.3L	1	1
1996-99	2.2L	2	5
	2.4L	2	5
L Body			
1990-95	2.2L	1	1
	3.1L	1	1
1996	2.2L	2	5
	3.1L	2	5
Light Trucks and Vans			
1990-94	All	1	1
1995	2.2L	2	1, 4
	4.3L	4	2, 5
1996-99	2.2L	2	1, 4
	4.3L	4	2, 5
L/N Body	2.4L	2	1, 4
	3.1L	2	1, 4
Minivans			
1990-95	3.1L	1	1
	3.8L	1	1
1996-99	3.4L	2	1, 4

OXYGEN SENSOR LOCATIONS

Manufacturer Model Year	Engine(s)	No. of Sensors	Location(s)
General Motors Corporation (cont.)			
N Body			
1990-95	2.3L	1	1
	3.1L	1	1
1996-99	2.4L	2	1, 4
	3.1L	2	1, 4
Pontiac Lemans	1.6L/2.0L	1	1
V Body	3.0L	4	5
W Body			
1990-95	3.1L	1	1
	3.4L	1	1
1996-99	3.1L	2	5
	3.4L	2	5
	3.8L	2	5
Y Body			
1990-94	VIN J	2	2
	except VIN J	1	1
1995	5.7L	2	2
1996-99	5.7L	4	5
Honda			
Accord			
1990-95	2.2L	1	1
1996-99	2.2L	2	1, 4
	2.7L	2	5
Civc			
1990-95	All	1	1
1996-99	All	2	1, 4
Del Sol			
1990-95	All	1	1
1996-97	All	2	1, 4
Prelude			
1990-95	All	1	1
1996-99	All	2	1, 4
Trucks and Sport Utility Vehicles			
	2.3L	1	1
	2.6L	1	1
	3.2L	1	3
1996-99	2.0L	2	3, 4
	2.2L	2	3, 4
	2.6L	2	4, 3
	3.2L	4	2, 5

91272C05

OXYGEN SENSOR LOCATIONS

Manufacturer Model Year	Engine(s)	No. of Sensors	Location(s)
Hyundai			
1990-94	All	1	1
1995-99	1.5L	2	1, 4
	1.8L	1	1
	2.0L	1 (2 CA)	1 (4 CA)
	3.0L	1 (2 CA)	1 (4 CA)
Infiniti			
G20			
1990-95	2.0L	1	1
1996	2.0L	2	1, 4
I30			
1996	3.0L	2	1, 4
1997-99	3.0L	3	2, 4
J30			
1990-95	3.0L	2	2
1996-97	3.0L	4	2, 4
M30	3.0L	1	1
Q45			
1990-94	4.5L	1	1
1995-99	4.5L	4	2, 4
Trucks and Sport Utility Vehicles	2.4L	2	1, 4
	3.0L	4	2, 4
	3.3L	4	2, 4
	4.3L	4	2, 4
Isuzu			
Full Size Trucks/Sport Utility Vehicles			
1990-95	2.2L	1	1
	2.3L	1	1
	2.6L	1	1
	3.2L	1	3
1996-99	2.0L	2	3, 4
	2.2L	2	3, 4
	2.6L	2	4, 3
	3.2L	4	2, 5
Light Trucks and Vans			
1990-94	All	1	1
1995	2.2L	2	1, 4
	4.3L	4	2, 5
1996-99	2.2L	2	1, 4
	4.3L	4	2, 5

91272C06

OXYGEN SENSOR LOCATIONS

Manufacturer Model Year	Engine(s)	No. of Sensors	Location(s)
Jaguar			
1990-94	All	1	1
1995-99	4.0L	2	5
	6.0L	4	2, 4
Jeep			
1990-95	2.5L	1	1
	4.0L	1	1
	5.2L	1	1
1996-99	2.5L	2	1, 4
	4.0L	2	1, 4
	5.2L	3	1, 5
	5.9L	3	1, 5
Kia			
Sephia			
1994-95	1.8L	1	1
1996-99	1.8L	2	1, 4
Sportage			
1994-95	2.0L	1	1
1996-99	2.0L	2	1, 4
Land Rover			
1990-95	3.9L	2	2
	4.0L	2	2
1996-99	4.0L	4	2, 4
	4.6L	4	2, 4
Lexus			
ES 250	2.5L	1, 2CA	1, (4 CA)
ES 300			
1990-94	3.0L	2, 3CA	2, (4CA)
1995-99	3.0L	3	2, 4
GS/SC 300			
1990-94	3.0L	2, 3CA	2, (4CA)
1995-99	3.0L	3	2, 4
GS/LS/SC 400			
1990-94	4.0L	2, 3CA	2, (4CA)
1995-99	4.0L	4	2, 4
LX 450	4.5L	2	1,4
Mazda			
929	3.0L	2	2
MX3	1.6L	2, 4 CA	2, (4 CA)
MX6/626			
1990-95	2.0L	1	1
	2.5L	2	2

91272C07

OXYGEN SENSOR LOCATIONS

Manufacturer Model Year	Engine(s)	No. of Sensors	Location(s)
Mazda (cont.)			
MX6/626 (cont.)			
1996-97	2.0L	2	1, 4
	2.5L	3	2, 4
RX-7	1.3L	1	1
Miata			
1990-94	All	1	1
1995-99	1.8L	2	1, 4
Millenia	2.0L	4	2, 4
	2.3L	4	2, 4
Mini-vans			
1995-99	3.0L	2	1, 4
Protegé and 323			
1990-94	All	1	1
1995-99	1.5L	2	1, 4
	1.8L	2	1, 4
Trucks and Vans			
1990-95	All	1	1
1996-99	2.3L	2	1, 4
	2.5L	2	1, 4
	3.0L	3	2, 4
	4.0L (E)	4	2, 4
	4.0L (X)	4	2, 4
	5.0L	4	2, 4
Mercedes-Benz			
	2.2L (111)	2	5
	2.3L (111)	2	5
	2.8L (104)	2	5
	3.2L (104)	2	5
	3.2L (112)	4	5
	4.2L (119)	4	5
	4.3L (113)	4	5
	5.0L (119)	4	5
Mitsubishi			
Diamante			
1990-94	3.0L/3.5L	1	1
1995-99	3.0L/3.5L	2	3, 4
Eclipse			
1990-94	All	1	1
1995-99	All	2	1, 4
Galant			
1990-94	2.4L	1	1
1995-99	2.4L	2	3, 4

91272C08

OXYGEN SENSOR LOCATIONS

Manufacturer Model Year	Engine(s)	No. of Sensors	Location(s)
Mitsubishi (cont.)			
Mirage			
1990-94	All	1	1
1995-99	All	2	3, 4
Precis	1.5L	1	1
Trucks and Sport Utility Vehicles			
1990-94	2.4L	1	1
	3.0L	1	1
	3.5L	1	1
1995-99	2.4L	2	1, 4
	3.0L	3	2, 4
	3.5L	3	2, 4
3000GT			
1990-94	3.0L	1	1
1995-99	3.0L	3	2, 4
Nissan			
Altima			
1990-94	2.4L	1	1
1995-99	2.4L	2	1, 4
Axxess	All	1	1
Maxima			
1990-94	3.0L	1	1
1995-96	3.0L	2	1, 4
1997-99	3.0L	3	2, 4
Quest			
1995	3.0L	1	1
1996-99	3.0L	2	1, 4
Sentra, NX and 200SX			
1990-94	All	1	1
1995-99	1.6L	2	1, 4
	2.0L	2	1, 4
Stanza	All	1	1
Trucks and Sport Utility Vehicles			
1990-94	2.4L	1	1
	3.0L	1	1
	3.3L	1	1
	4.3L	1	1
1995-99	2.4L	2	1, 4
	3.0L	4	2, 4
	3.3L	4	2, 4
	4.3L	4	2, 4

OXYGEN SENSOR LOCATIONS

Manufacturer Model Year	Engine(s)	No. of Sensors	Location(s)
Nissan (cont.)			
240SX			
1990-94	2.4L	1	1
1995-99	2.4L	2	1, 4
300ZX			
1990-94	3.0L	2	2
1995-99	3.0L	4	2, 4
Porsche			
911	3.6L	4	5
928	5.4L	1	3
968	3.0L	2	5
Boxster	2.5L	4	5
Saab			
900 and 9000			
1990-94	4 Cyl.	1	1
	6 Cyl.	1	1
1995-97	4 Cyl.	2	1, 4
	6 Cyl.	3	2, 4
9-3 and 9-5			
1998-99	4 Cyl.	2	1, 4
	6 Cyl.	3	2, 4
Saturn			
Coupe, Sedan and Wagons			
1991-95	1.9L	1	1
1996-99	1.9L	2	1, 4
Subaru			
Except Forester			
1990-94	1.8L	1	1
	2.2L	1	1
	3.3L	2	2
1995	1.8L	1	3
	2.2L	2	3, 4
	3.3L	3	2, 4
1996-99	1.8L	2	3, 4
	2.2L	2	3, 4
	2.5L	2	3, 4
	3.3L	3	2, 4
Forester	2.5L	3	3, 4
Suzuki			
Cars			
1990-94	All	1	1
1995-99	All	2	1, 4

91272C10

OXYGEN SENSOR LOCATIONS

Manufacturer Model Year	Engine(s)	No. of Sensors	Location(s)
Suzuki (cont.)			
Sport Utillity Vehicles			
1990-95	1.3L	1	1
1996-99	1.6L	2	1, 4
	1.8L	2	1, 4
Toyota			
Avalon	All	3	2, 4
Camry			
1990-94	All	1	1
1995-99	2.2L	2	1, 4
	3.0L	3	2, 4
Celica			
1990-94	All	1	1
1995-99	All	2	1, 4
Corolla			
1990-95	All	1 (2 CA)	1 (4 CA)
1996-99	All	2	1, 4
Cressida	All	1, 2 CA	1,(4 CA)
MR2	2.0L	1, 2 CA	1,(4 CA)
	2.2L	1, 2 CA	1,(4 CA)
Paseo			
1990-95	1.5L	1, 2 CA	1, (4 CA)
1996-98	1.5L	2	1, 4
Supra			
1990-94	3.0L	1 (2 CA)	1 (4 CA)
1995-99	3.0L	3	2, 4
Tercel			
1990-94	1.5L	1, 2 CA	1, (4 CA)
1995-99	1.5L	2	1, 4
Trucks and Sport Utility Vehicles			
1990-94	2.4L	1 (2 CA)	1 (4 CA)
	2.7L	1 (2 CA)	1 (4 CA)
	3.0L	1 (2 CA)	1 (4 CA)
	3.4L	1 (2 CA)	1 (4 CA)
	4.5L	1 (2 CA)	1 (4 CA)
1995	2.4L	1 (2 CA)	1 (4 CA)
	2.7L	2	3, 4
	3.0L	2	3, 4
	3.4L	2	3, 4
	4.5L	2	1,4

OXYGEN SENSOR LOCATIONS

Manufacturer Model Year	Engine(s)	No. of Sensors	Location(s)
Toyota (cont.)			
Trucks and Sport Utility Vehicles (cont.)			
1996-99	2.0L	2	1, 4
	2.4L	2	3, 4
	2.7L	2	3, 4
	3.0L	2	3, 4
	3.4L	2	3, 4
	4.5L	2	1, 4
	4.7L	2	3, 4
Volkswagen			
Cabrio			
1994	All	1	1
1995-99	2.0L	2	1, 4
Cabriolet	All	1	1
Golf			
1990-94	All	1	1
1995-99	2.0L	2	1, 4
GTI			
1990-94	All	1	1
1995-99	2.0L	2	1, 4
	2.8L	2	1, 4
Jetta			
1990-94	All	1	1
1995-99	1.8L	2	1, 4
	2.0L	2	1, 4
	2.8L	2	1, 4
New Beetle	2.0L	2	1, 4
Passat			
1990-94	All	1	1
1995-99	1.8L	2	1, 4
	2.8L	2	1, 4
Volvo			
200 Series	All	1	1
700 Series	All	1	1
800 Series			
1993-94	All	1	1
1995-99	All	2	5
900 Series			
1991-94	All	1	1
1995-99	All	2	5
C70/S70/V70	All	2	5
S90/V90	All	2	5

① Chrysler JA class designation refers to the Chrysler Cirrus, Plymouth Breeze, and Dodge Stratus.
   Chrysler JX class designation refers to the Chrysler Sebring Convertible.
② Chrysler LH class designation refers to the Chrysler Concorde, LHS, New Yorker, Dodge Intrepid, and Eagle Vision.

91272C12

CONVERSION FACTORS

LENGTH–DISTANCE

Inches (in.)	x 25.4	= Millimeters (mm)	x .0394	= Inches
Feet (ft.)	x .305	= Meters (m)	x 3.281	= Feet
Miles	x 1.609	= Kilometers (km)	x .0621	= Miles

VOLUME

Cubic Inches (in3)	x 16.387	= Cubic Centimeters	x .061	= in3
IMP Pints (IMP pt.)	x .568	= Liters (L)	x 1.76	= IMP pt.
IMP Quarts (IMP qt.)	x 1.137	= Liters (L)	x .88	= IMP qt.
IMP Gallons (IMP gal.)	x 4.546	= Liters (L)	x .22	= IMP gal.
IMP Quarts (IMP qt.)	x 1.201	= US Quarts (US qt.)	x .833	= IMP qt.
IMP Gallons (IMP gal.)	x 1.201	= US Gallons (US gal.)	x .833	= IMP gal.
Fl. Ounces	x 29.573	= Milliliters	x .034	= Ounces
US Pints (US pt.)	x .473	= Liters (L)	x 2.113	= Pints
US Quarts (US qt.)	x .946	= Liters (L)	x 1.057	= Quarts
US Gallons (US gal.)	x 3.785	= Liters (L)	x .264	= Gallons

MASS–WEIGHT

Ounces (oz.)	x 28.35	= Grams (g)	x .035	= Ounces
Pounds (lb.)	x .454	= Kilograms (kg)	x 2.205	= Pounds

PRESSURE

Pounds Per Sq. In. (psi)	x 6.895	= Kilopascals (kPa)	x .145	= psi
Inches of Mercury (Hg)	x .4912	= psi	x 2.036	= Hg
Inches of Mercury (Hg)	x 3.377	= Kilopascals (kPa)	x .2961	= Hg
Inches of Water (H$_2$O)	x .07355	= Inches of Mercury	x 13.783	= H$_2$O
Inches of Water (H$_2$O)	x .03613	= psi	x 27.684	= H$_2$O
Inches of Water (H$_2$O)	x .248	= Kilopascals (kPa)	x 4.026	= H$_2$O

TORQUE

Pounds–Force Inches (in–lb)	x .113	= Newton Meters (N·m)	x 8.85	= in–lb
Pounds–Force Feet (ft–lb)	x 1.356	= Newton Meters (N·m)	x .738	= ft–lb

VELOCITY

Miles Per Hour (MPH)	x 1.609	= Kilometers Per Hour (KPH)	x .621	= MPH

POWER

Horsepower (Hp)	x .745	= Kilowatts	x 1.34	= Horsepower

FUEL CONSUMPTION*

Miles Per Gallon IMP (MPG)	x .354	= Kilometers Per Liter (Km/L)
Kilometers Per Liter (Km/L)	x 2.352	= IMP MPG
Miles Per Gallon US (MPG)	x .425	= Kilometers Per Liter (Km/L)
Kilometers Per Liter (Km/L)	x 2.352	= US MPG

*It is common to covert from miles per gallon (mpg) to liters/100 kilometers (1/100 km), where mpg (IMP) x 1/100 km = 282 and mpg (US) x 1/100 km = 235.

TEMPERATURE

Degree Fahrenheit (°F)	= (°C x 1.8) + 32
Degree Celsius (°C)	= (°F – 32) x .56

TCCS1044

Standard Torque Specifications and Fastener Markings

In the absence of specific torques, the following chart can be used as a guide to the maximum safe torque of a particular size/grade of fastener.

- There is no torque difference for fine or coarse threads.
- Torque values are based on clean, dry threads. Reduce the value by 10% if threads are oiled prior to assembly.
- The torque required for aluminum components or fasteners is considerably less.

U.S. Bolts

SAE Grade Number	1 or 2			5			6 or 7		
Number of lines always 2 less than the grade number.									
Bolt Size (Inches)—(Thread)	Maximum Torque			Maximum Torque			Maximum Torque		
	Ft./Lbs.	Kgm	Nm	Ft./Lbs.	Kgm	Nm	Ft./Lbs.	Kgm	Nm
¼ — 20	5	0.7	6.8	8	1.1	10.8	10	1.4	13.5
— 28	6	0.8	8.1	10	1.4	13.6			
5/16 — 18	11	1.5	14.9	17	2.3	23.0	19	2.6	25.8
— 24	13	1.8	17.6	19	2.6	25.7			
⅜ — 16	18	2.5	24.4	31	4.3	42.0	34	4.7	46.0
— 24	20	2.75	27.1	35	4.8	47.5			
7/16 — 14	28	3.8	37.0	49	6.8	66.4	55	7.6	74.5
— 20	30	4.2	40.7	55	7.6	74.5			
½ — 13	39	5.4	52.8	75	10.4	101.7	85	11.75	115.2
— 20	41	5.7	55.6	85	11.7	115.2			
9/16 — 12	51	7.0	69.2	110	15.2	149.1	120	16.6	162.7
— 18	55	7.6	74.5	120	16.6	162.7			
⅝ — 11	83	11.5	112.5	150	20.7	203.3	167	23.0	226.5
— 18	95	13.1	128.8	170	23.5	230.5			
¾ — 10	105	14.5	142.3	270	37.3	366.0	280	38.7	379.6
— 16	115	15.9	155.9	295	40.8	400.0			
⅞ — 9	160	22.1	216.9	395	54.6	535.5	440	60.9	596.5
— 14	175	24.2	237.2	435	60.1	589.7			
1 — 8	236	32.5	318.6	590	81.6	799.9	660	91.3	894.8
— 14	250	34.6	338.9	660	91.3	849.8			

Metric Bolts

Relative Strength Marking	4.6, 4.8			8.8		
Bolt Markings						
Bolt Size Thread Size x Pitch (mm)	Maximum Torque			Maximum Torque		
	Ft./Lbs.	Kgm	Nm	Ft./Lbs.	Kgm	Nm
6 x 1.0	2–3	.2–.4	3–4	3–6	4–.8	5–8
8 x 1.25	6–8	.8–1	8–12	9–14	1.2–1.9	13–19
10 x 1.25	12–17	1.5–2.3	16–23	20–29	2.7–4.0	27–39
12 x 1.25	21–32	2.9–4.4	29–43	35–53	4.8–7.3	47–72
14 x 1.5	35–52	4.8–7.1	48–70	57–85	7.8–11.7	77–110
16 x 1.5	51–77	7.0–10.6	67–100	90–120	12.4–16.5	130–160
18 x 1.5	74–110	10.2–15.1	100–150	130–170	17.9–23.4	180–230
20 x 1.5	110–140	15.1–19.3	150–190	190–240	26.2–46.9	160–320
22 x 1.5	150–190	22.0–26.2	200–260	250–320	34.5–44.1	340–430
24 x 1.5	190–240	26.2–46.9	260–320	310–410	42.7–56.5	420–550

TCCS1098

ENGLISH/METRIC CONVERSION: LENGTH

To convert inches (in.) to millimeters (mm), multiply the number of inches by 25.4
To convert millimeters (mm) to inches (in.), multiply the number of millimeters by 0.04

Inches		Millimeters	Inches		Millimeters	Inches		Millimeters
Fraction	Decimal	Decimal	Fraction	Decimal	Decimal	Fraction	Decimal	Decimal
1/64	0.016	0.397	11/32	0.344	8.731	11/16	0.688	17.463
1/32	0.031	0.794	23/64	0.359	9.128	45/64	0.703	17.859
3/64	0.047	1.191	3/8	0.375	9.525	23/32	0.719	18.256
1/16	0.063	1.588	25/64	0.391	9.922	47/64	0.734	18.653
5/64	0.078	1.984	13/32	0.406	10.319	3/4	0.750	19.050
3/32	0.094	2.381	27/64	0.422	10.716	49/64	0.766	19.447
7/64	0.109	2.778	7/16	0.438	11.113	25/32	0.781	19.844
1/8	0.125	3.175	29/64	0.453	11.509	51/64	0.797	20.241
9/64	0.141	3.572	15/32	0.469	11.906	13/16	0.813	20.638
5/32	0.156	3.969	31/64	0.484	12.303	53/64	0.828	21.034
11/64	0.172	4.366	1/2	0.500	12.700	27/32	0.844	21.431
3/16	0.188	4.763	33/64	0.516	13.097	55/64	0.859	21.828
13/64	0.203	5.159	17/32	0.531	13.494	7/8	0.875	22.225
7/32	0.219	5.556	35/64	0.547	13.891	57/64	0.891	22.622
15/64	0.234	5.953	9/16	0.563	14.288	29/32	0.906	23.019
1/4	0.250	6.350	37/64	0.578	14.684	59/64	0.922	23.416
17/64	0.266	6.747	19/32	0.594	15.081	15/16	0.938	23.813
9/32	0.281	7.144	39/64	0.609	15.478	61/64	0.953	24.209
19/64	0.297	7.541	5/8	0.625	15.875	31/32	0.969	24.606
5/16	0.313	7.938	41/64	0.641	16.272	63/64	0.984	25.003
21/64	0.328	8.334	21/32	0.656	16.669	1/1	1.000	25.400
			43/64	0.672	17.066			

Inches	Millimeters	Inches	Millimeters	Inches	Millimeters	Inches	Millimeters
0.0001	0.00254	0.005	0.1270	0.09	2.286	4	101.6
0.0002	0.00508	0.006	0.1524	0.1	2.54	5	127.0
0.0003	0.00762	0.007	0.1778	0.2	5.08	6	152.4
0.0004	0.01016	0.008	0.2032	0.3	7.62	7	177.8
0.0005	0.01270	0.009	0.2286	0.4	10.16	8	203.2
0.0006	0.01524	0.01	0.254	0.5	12.70	9	228.6
0.0007	0.01778	0.02	0.508	0.6	15.24	10	254.0
0.0008	0.02032	0.03	0.762	0.7	17.78	11	279.4
0.0009	0.02286	0.04	1.016	0.8	20.32	12	304.8
0.001	0.0254	0.05	1.270	0.9	22.86	13	330.2
0.002	0.0508	0.06	1.524	1	25.4	14	355.6
0.003	0.0762	0.07	1.778	2	50.8	15	381.0
0.004	0.1016	0.08	2.032	3	76.2	16	406.4

9300BA03